ASP.NET Core 3
框架揭秘（上册）

蒋金楠 ◎著

电子工业出版社
Publishing House of Electronics Industry
北京·BEIJING

内 容 简 介

本书主要阐述 ASP.NET Core 最核心的部分——请求处理管道。通过阅读本书,读者可以深刻、系统地了解 ASP.NET Core 应用在启动过程中管道的构建方式,以及请求在管道中的处理流程。本书还详细讲述了 .NET Core 跨平台的本质,以及多个常用的基础框架(如依赖注入、文件信息、配置选项和诊断日志等)。本书还对大部分原生的中间件提供了系统性介绍,采用"编程体验"、"总体设计"、"具体实现"和"灵活运用"的流程,使读者可以循序渐进地学习 ASP.NET Core 的每个功能模块。

本书可供所有 .NET 从业人员阅读与参考。

未经许可,不得以任何方式复制或抄袭本书之部分或全部内容。
版权所有,侵权必究。

图书在版编目(CIP)数据

ASP.NET Core 3 框架揭秘. 上册 / 蒋金楠著. —北京:电子工业出版社,2020.5
ISBN 978-7-121-38462-2

Ⅰ. ①A… Ⅱ. ①蒋… Ⅲ. ①网页制作工具-程序设计 Ⅳ. ①TP393.092.2

中国版本图书馆 CIP 数据核字(2020)第 027867 号

责任编辑:张春雨　　　特约编辑:田学清
印　　刷:三河市良远印务有限公司
装　　订:三河市良远印务有限公司
出版发行:电子工业出版社
　　　　　北京市海淀区万寿路 173 信箱　　邮编:100036
开　　本:787×980　1/16　　印张:57.25　　字数:1390 千字
版　　次:2020 年 5 月第 1 版
印　　次:2020 年 5 月第 3 次印刷
定　　价:199.00 元(上下册)

凡所购买电子工业出版社图书有缺损问题,请向购买书店调换。若书店售缺,请与本社发行部联系,联系及邮购电话:(010)88254888,88258888。
质量投诉请发邮件至 zlts@phei.com.cn,盗版侵权举报请发邮件至 dbqq@phei.com.cn。
本书咨询联系方式:010-51260888-819,faq@phei.com.cn。

前言

写作源起

计算机图书市场存在一系列介绍 ASP.NET Web Forms、ASP.NET MVC、ASP.NET Web API 的图书，但是找不到一本专门介绍 ASP.NET 自身框架的图书，作为一名拥有 17 年工作经验的 .NET 开发者，笔者对此感到十分困惑。上述这些 Web 开发框架都是建立在 ASP.NET 底层框架之上的，底层 ASP.NET 框架才是根基所在。过去笔者接触过很多资深的 ASP.NET 开发人员，发现他们对 ASP.NET 框架大都没有进行深入了解。

2014 年，出版《ASP.NET MVC 5 框架揭秘》之后，笔者原本打算写"ASP.NET 框架揭秘"。但在新书准备过程中，微软推出了 ASP.NET Core（当时被称为 ASP.NET 5，还没有 .NET Core 的概念）。所以，笔者将研究重点转移到 ASP.NET Core。

本书耗时 5 年左右，笔者投入了大量心血。2015 年年初，笔者开始了本书的写作，微软在 2016 年 6 月正式发布 .NET Core 1.0 时，本书的绝大部分内容就已经完成。随后，微软不断推出新的版本，本书的内容也在不断快速"迭代"中。本书正文部分共计 800 多页，但笔者在写作过程中删除的部分不少于这个数字。

有人认为自己每天只是做一些简单的编程工作，根本没有必要去了解底层原理和设计方面的内容。其实，不论我们从事何种层次的工作，最根本的目的只有一个——解决问题。解决方案分两种：一种是"扬汤止沸"，另一种是"釜底抽薪"。看到锅里不断沸腾的水，大多数人会选择不断地往锅里浇冷水，笔者希望这本书能够使读者看到锅底熊熊燃烧的薪火。

本书内容

ASP.NET Core 是一个全新的 Web 开发平台，为我们构建了一个可复用和可定制的请求处理管道，微软在它上面构建了 MVC、SignalR、GRPC、Orleans 这样广泛使用的 Web 框架，我们也可以利用它构建自己的 Web 框架（笔者曾经通过 ASP.NET Core 构建了一款 GraphQL 框架）。本书只关注最本质的东西，即 ASP.NET Core 请求处理管道，并不会涉及上述这些 Web 框架。本书的内容主要划分为如下 4 个部分。

跨平台的开发体验和实现原理

.NET Core 与传统 .NET Framework 最大的区别是跨平台，作为开篇入门材料，第 1 章通过几个简单的 Hello World 程序，让读者可以体验如何在 Windows、macOS、Linux 平台上开发 .NET Core 应用，以及通过 Docker 容器部署 ASP.NET Core 应用的乐趣。第 2 章将告诉读者 .NET Core 的跨平台究竟是如何实现的。

基础框架

ASP.NET Core 框架依赖于一些基础框架，其中最重要的是注入框架。由于依赖注入框架不但是构建 ASP.NET Core 请求处理管道的基石，而且依赖注入也是 ASP.NET Core 应用的基本编程模式，所以本书的第 3 章和第 4 章对依赖注入原理及依赖注入框架的设计与编程方式进行了详细介绍。

ASP.NET Core 应用具有很多读取文件内容的场景，所以它构建了一个抽象的文件系统，第 5 章会对这个文件系统的设计模型和两种实现方式（物理文件系统和程序集内嵌文件系统）进行详细介绍。

.NET Core 针对"配置"的支持是传统 .NET 开发人员所不能想象的，所以采用两章的篇幅对这一主题进行讲解：第 6 章旨在介绍支持多种数据源的配置系统；不论是开发 ASP.NET Core 应用还是组件，都可以采用 Options 模式来读取配置选项，第 7 章会着重讲述这种强类型的配置选项编程方式。

.NET Core 在错误诊断方面为我们提供了多种选择，第 8 章介绍了 5 种常用的记录诊断日志的方式。.NET Core 还提供了一个支持多种输出渠道的日志系统，该日志系统在第 9 章进行了详细的介绍。

管道详解

.NET Core 的服务承载系统用来承载那些需要长时间运行的服务，ASP.NET Core 作为最重要的服务类型被承载于该系统中，第 10 章会对该服务承载系统进行系统介绍。由于请求处理管道是本书的核心所在，所以采用 3 章的篇幅进行介绍：第 11 章主要从编程模型的角度来认识管道；

第 12 章提供了一个极简版的模拟框架来展示 ASP.NET Core 框架的总体设计；第 13 章以这个模拟框架为基础，采用渐进的方式补充一些遗漏的细节，进而将 ASP.NET Core 框架真实的管道展现在读者眼前。

中间件

ASP.NET Core 框架的请求处理管道由服务器和中间件组成，管道利用服务器来监听和接收请求，并完成最终对请求的响应，应用针对请求的处理则体现在有序排列的中间件上。微软为我们提供了一系列原生的中间件，对这些中间件的介绍全部在下册。

这部分涉及用来处理文件请求（第 14 章）、路由（第 15 章）、异常（第 16 章）的中间件，也包括用来响应缓存（第 17 章）和会话（第 18 章）的中间件，还包括用来实现认证（第 19 章）、授权（第 20 章）、跨域资源共享（第 21 章）等与安全相关的中间件。

这部分还介绍了针对本地化（第 22 章）和健康检查（第 23 章）的中间件。除此之外，这部分还介绍了用来实现主机名过滤、HTTP 重写、设置基础路径等功能的中间件，这些零散的中间件全部在第 24 章进行介绍。

写作特点

本书是揭秘系列的第 6 本书。在过去的十来年里，笔者得到了很多热心读者的反馈，这些反馈对书中的内容基本上都持正面评价，但对写作技巧和表达方式的评价则不尽相同。每个作者都有属于自己的写作风格，每个读者的学习思维方式也不尽相同，两者很难出现百分之百的契合，但笔者还是决定在本书上做出改变。

本书内容采用了不一样的组织方式，笔者认为这样的方式更符合系统地学习一门全新技术的"流程"。对于每个模块，笔者采用"体验先行"的原则，提供一些简单的实例演示，使读者对当前模块的基本功能特性和编程模式具有大致的了解。同时，在编程体验中抽取一些核心对象，并利用它们构建当前模块的抽象模型，使读者只要读懂了这个模型也就了解了当前模块的总体设计。接下来我们从抽象转向具体，进一步深入介绍抽象模型的实现原理。为了使读者能够在真实项目中灵活自如地运用当前模块，笔者介绍了一些面向应用的扩展和最佳实践。总体来说，本书采用"编程体验"、"总体设计"、"具体实现"和"灵活运用"的流程，使读者能循序渐进地学习 ASP.NET Core 的每个功能模块。

本书综合运用 3 种不同的"语言"（文字语言、图表语言和编程语言）来讲述每个技术主题。一图胜千言，笔者在每章都精心设计了很多图表，这些具象的图表能够帮助读者理解技术模块的总体设计、执行流程和交互方式。除了利用编程语言描述应用编程接口（API），本书还提供了近200个实例，这些实例具有不同的作用，有的是为了演示某个实用的编程技巧或者最佳实践，有的是为了强调一些容易忽视但很重要的技术细节，有的是为了探测和证明所述的论点。

本书在很多地方会展示一些类型的代码，但是这些代码和真正的源代码是有差异的，两者的差异缘于以下几个原因：第一，源代码在版本更替中一直在发生改变；第二，由于篇幅的限制，笔者刻意删除了一些细枝末节的代码，如针对参数的验证、诊断日志的输出和异常处理等；第三，很多源代码其实都具有优化的空间。综上所述，本书提供的代码片段旨在揭示设计原理和实现逻辑，不是为了向读者展示源代码。

目标读者

虽然本书关注的是 ASP.NET Core 自身框架提供的请求处理管道，而不是具体某个应用编程框架（如 MVC、SignalR、GRPC 等），但是本书适合所有 .NET 技术从业人员阅读。

笔者认为任何好的设计都应该是简单的，唯有简单的设计才能应对后续版本更替中出现的复杂问题。从这个意义上讲，ASP.NET 框架就是好的设计。因为自正式推出的那一刻起，ASP.NET 框架的总体设计基本上没有发生改变。ASP.NET Core 的设计同样是好的设计，其简单的管道式设计在未来的版本更替中也不会发生太大的改变，既然是好的设计，它就应该是简单的。

正如上面所说，本书采用渐进式的写作方式，那些完全没有接触过 ASP.NET Core 的开发人员也可以通过本书深入、系统地掌握这门技术。由于本书提供的大部分内容都是独一无二的，即使是资深的 .NET 开发设计人员，也能在书中找到很多不甚了解的盲点。

关于作者

蒋金楠，同程艺龙技术专家。知名 IT 博主（多年来一直排名博客园第一位），拥有个人微信公众号"大内老A"；2007—2018 年连续 12 次被评为微软 MVP（最有价值专家），也是少数跨多领域（Solutions Architect、Connected System、Microsoft Integration 和 ASP.NET/IIS 等）的 MVP 之一；畅销 IT 图书作者，先后出版了《WCF 全面解析》、《ASP.NET MVC 4 框架揭秘》、《ASP.NET MVC 5 框架揭秘》和《ASP.NET Web API 2 框架揭秘》等著作。

致谢

本书得以顺利出版离不开博文视点张春雨团队的辛勤努力，他们的专业水准和责任心为本书提供了质量保证。此外，徐妍妍在本书写作过程中做了大量的校对工作，在此表示衷心感谢。

本书支持

由于本书是随着 ASP.NET Core 一起成长起来的，并且随着 ASP.NET Core 的版本更替进行了多次"迭代"，所以书中某些内容最初是根据旧版本编写的，新版本对应的内容发生改变后相应内容可能没有及时更新。对于 ASP.NET Core 的每次版本升级，笔者基本上会尽可能将书中的内容做相应的更改，但其中难免有所疏漏。由于笔者的能力和时间有限，书中难免存在不足之处，恳请广大读者批评指正。

笔者博客：http://www.cnblogs.com/artech

笔者微博：http://www.weibo.com/artech

笔者电子邮箱：jinnan@outlook.com

笔者微信公众号：大内老 A

读者服务

- 获取本书配套素材、代码、视频、习题、模板、教程、课件资源
- 获取更多技术专家分享视频与学习资源
- 加入读者交流群，与更多读者互动、与本书作者互动

扫码回复：38462

目 录

第 1 章 全新的开发体验 1
  1.1 Windows 平台 1
    1.1.1 构建开发环境 1
    1.1.2 利用命令行创建 .NET Core 应用 2
    1.1.3 ASP.NET Core 应用 6
    1.1.4 ASP.NET Core MVC 应用 13
  1.2 macOS 用户 17
    1.2.1 构建开发环境 17
    1.2.2 利用命令行创建 .NET Core 应用 17
    1.2.3 ASP.NET Core MVC 应用 18
  1.3 Linux 19
    1.3.1 启用 Linux 子系统 20
    1.3.2 构建开发环境 22
    1.3.3 利用命令行创建 ASP.NET Core 应用 22
  1.4 Docker 24

第 2 章 跨平台的奥秘 27
  2.1 历史的枷锁 27
    2.1.1 Windows 下的 .NET 27
    2.1.2 非 Windows 下的 .NET 31
  2.2 复用之伤 34
    2.2.1 源代码复用 34
    2.2.2 程序集复用 36

2.3 全新的布局 44
2.3.1 跨平台的 .NET Core 44
2.3.2 统一的 BCL 51
2.3.3 展望未来 56

第 3 章 依赖注入（上篇）...58
3.1 控制反转 58
3.1.1 流程控制的反转 58
3.1.2 好莱坞法则 61
3.1.3 流程定制 62
3.2 IoC 模式 62
3.2.1 模板方法 63
3.2.2 工厂方法 64
3.2.3 抽象工厂 66
3.3 依赖注入 68
3.3.1 由容器提供对象 68
3.3.2 3 种依赖注入方式 69
3.3.3 Service Locator 模式 72
3.4 一个简易版的依赖注入容器 74
3.4.1 编程体验 74
3.4.2 设计与实现 79
3.4.3 扩展方法 85

第 4 章 依赖注入（下篇）...89
4.1 利用容器提供服务 89
4.1.1 服务的注册与消费 89
4.1.2 生命周期 93
4.1.3 针对服务注册的验证 96
4.2 服务注册 99
4.2.1 ServiceDescriptor 99
4.2.2 IServiceCollection 101
4.3 服务的消费 105
4.3.1 IServiceProvider 105
4.3.2 服务实例的创建 106
4.3.3 生命周期 109

4.4 实现概览 113
4.4.1 ServiceProviderEngine 和 ServiceProviderEngineScope 113
4.4.2 ServiceProvider 115
4.4.3 注入 IServiceProvider 对象 117
4.5 扩展 119
4.5.1 适配 120
4.5.2 IServiceProviderFactory<TContainerBuilder> 120
4.5.3 整合第三方依赖注入框架 121

第 5 章 文件系统 126
5.1 抽象的文件系统 126
5.1.1 树形层次结构 126
5.1.2 读取文件内容 128
5.1.3 监控文件的变化 130
5.2 设计详解 131
5.2.1 IChangeToken 132
5.2.2 IFileProvider 133
5.2.3 PhysicalFileProvider 135
5.2.4 EmbeddedFileProvider 139
5.2.5 两个特殊的 IFileProvider 实现 144
5.3 远程文件系统 147
5.3.1 HttpFileInfo 与 HttpDirectoryContents 147
5.3.2 HttpFileProvider 150
5.3.3 FileProviderMiddleware 151
5.3.4 远程文件系统的应用 153

第 6 章 配置选项（上篇） 155
6.1 读取配置信息 155
6.1.1 配置编程模型三要素 155
6.1.2 以键值对的形式读取配置 156
6.1.3 读取结构化的配置 157
6.1.4 将结构化配置直接绑定为对象 160
6.1.5 将配置定义在文件中 161
6.2 配置模型 165
6.2.1 数据结构及其转换 166

- 6.2.2 IConfiguration 167
- 6.2.3 IConfigurationProvider 169
- 6.2.4 IConfigurationSource 171
- 6.2.5 IConfigurationBuilder 171

6.3 配置绑定 172
- 6.3.1 绑定配置项的值 173
- 6.3.2 绑定复合数据类型 175
- 6.3.3 绑定集合对象 177
- 6.3.4 绑定字典 180

6.4 配置的同步 181
- 6.4.1 配置数据流 181
- 6.4.2 ConfigurationReloadToken 182
- 6.4.3 ConfigurationRoot 183
- 6.4.4 ConfigurationSection 185

6.5 多样性的配置源 186
- 6.5.1 MemoryConfigurationSource 187
- 6.5.2 EnvironmentVariablesConfigurationSource 188
- 6.5.3 CommandLineConfigurationSource 191
- 6.5.4 FileConfigurationSource 194
- 6.5.5 StreamConfigurationSource 207
- 6.5.6 ChainedConfigurationSource 208
- 6.5.7 自定义 ConfigurationSource（S616） 210

第 7 章 配置选项（下篇） 215

7.1 Options 模式 215
- 7.1.1 将配置绑定为 Options 对象 215
- 7.1.2 提供具名的 Options 217
- 7.1.3 配置源的同步 219
- 7.1.4 直接初始化 Options 对象 221
- 7.1.5 根据依赖服务的 Options 设置 223
- 7.1.6 验证 Options 的有效性 225

7.2 Options 模型 226
- 7.2.1 OptionsManager<TOptions> 226
- 7.2.2 IOptionsFactory<TOptions> 228

		7.2.3	IOptionsMonitorCache\<TOptions>	237
		7.2.4	IOptionsMonitor\<TOptions>	238
	7.3	依赖注入		240
		7.3.1	服务注册	240
		7.3.2	IOptions\<TOptions>与IOptionsSnapshot\<TOptions>	246
		7.3.3	扩展与定制	248
		7.3.4	集成配置系统	256
第8章	诊断日志（上篇）			258
	8.1	各种诊断日志形式		258
		8.1.1	调试日志	258
		8.1.2	跟踪日志	259
		8.1.3	事件日志	262
		8.1.4	诊断日志	265
	8.2	Debugger 调试日志		268
		8.2.1	Debugger	268
		8.2.2	Debug	270
	8.3	TraceSource 跟踪日志		271
		8.3.1	跟踪日志模型三要素	271
		8.3.2	预定义 TraceListener	280
		8.3.3	Trace	284
	8.4	EventSource 事件日志		287
		8.4.1	EventSource	287
		8.4.2	EventListener	294
		8.4.3	荷载对象序列化	298
		8.4.4	活动跟踪	302
		8.4.5	性能计数	306
	8.5	DiagnosticSource 诊断日志		308
		8.5.1	标准的观察者模式	308
		8.5.2	AnonymousObserver\<T>	310
		8.5.3	强类型的事件订阅	313
		8.5.4	针对活动的跟踪	315
第9章	诊断日志（下篇）			317
	9.1	统一日志编程模式		317

	9.1.1	将日志输出到不同的渠道 317
	9.1.2	日志过滤 323
	9.1.3	日志范围 329
	9.1.4	LoggerMessage 331
9.2	日志模型详解 334	
	9.2.1	日志模型三要素 334
	9.2.2	ILogger 335
	9.2.3	日志范围 339
	9.2.4	ILoggerProvider 342
	9.2.5	ILoggerFactory 342
	9.2.6	LoggerMessage 347
9.3	依赖注入 348	
	9.3.1	服务注册 349
	9.3.2	设置日志过滤规则 351
9.4	日志输出渠道 353	
	9.4.1	控制台 353
	9.4.2	调试器 357
	9.4.3	TraceSource 日志 359
	9.4.4	EventSource 日志 362

第 10 章　承载系统 377

10.1	服务承载 377	
	10.1.1	承载长时间运行服务 377
	10.1.2	依赖注入 379
	10.1.3	配置选项 382
	10.1.4	承载环境 385
	10.1.5	日志 388
10.2	承载模型 391	
	10.2.1	IHostedService 392
	10.2.2	IHost 392
	10.2.3	IHostBuilder 397
10.3	实现原理 402	
	10.3.1	服务宿主 403
	10.3.2	针对配置系统的设置 406

	10.3.3 针对依赖注入框架的设置	407
	10.3.4 创建宿主	412
	10.3.5 静态类型 Host	418

第 11 章 管道（上篇） ... 421

11.1 管道式的请求处理 ... 421
- 11.1.1 两个承载体系 ... 421
- 11.1.2 请求处理管道 ... 423
- 11.1.3 中间件 ... 424
- 11.1.4 定义强类型中间件 ... 427
- 11.1.5 按照约定定义中间件 ... 428

11.2 依赖注入 ... 430
- 11.2.1 服务注册 ... 430
- 11.2.2 服务的消费 ... 433
- 11.2.3 生命周期 ... 437
- 11.2.4 集成第三方依赖注入框架 ... 443

11.3 配置 ... 444
- 11.3.1 初始化配置 ... 444
- 11.3.2 以键值对形式读取和修改配置 ... 446
- 11.3.3 合并配置 ... 448
- 11.3.4 注册 IConfigurationSource ... 449

11.4 承载环境 ... 450
- 11.4.1 IWebHostEnvironment ... 450
- 11.4.2 通过配置定制承载环境 ... 452
- 11.4.3 针对环境的编程 ... 454

11.5 初始化 ... 459
- 11.5.1 Startup ... 459
- 11.5.2 IHostingStartup ... 461
- 11.5.3 IStartupFilter ... 464

第 12 章 管道（中篇） ... 467

12.1 中间件委托链 ... 467
- 12.1.1 HttpContext ... 467
- 12.1.2 中间件 ... 468

XVI | ASP.NET Core 3 框架揭秘（上册）

  12.1.3 中间件管道的构建 469
 12.2 服务器 470
  12.2.1 IServer 471
  12.2.2 针对服务器的适配 471
  12.2.3 HttpListenerServer 473
 12.3 承载服务 476
  12.3.1 WebHostedService 476
  12.3.2 WebHostBuilder 476
  12.3.3 应用构建 478

第 13 章 管道（下篇） 480

 13.1 请求上下文 480
  13.1.1 HttpContext 480
  13.1.2 服务器适配 483
  13.1.3 获取上下文 487
  13.1.4 上下文的创建与释放 488
  13.1.5 RequestServices 489
 13.2 IServer + IHttpApplication 491
  13.2.1 IServer 491
  13.2.2 HostingApplication 492
  13.2.3 诊断日志 495
 13.3 中间件委托链 501
  13.3.1 IApplicationBuilder 501
  13.3.2 弱类型中间件 504
  13.3.3 强类型中间件 507
  13.3.4 注册中间件 509
 13.4 应用的承载 510
  13.4.1 GenericWebHostServiceOptions 510
  13.4.2 GenericWebHostService 512
  13.4.3 GenericWebHostBuilder 515
  13.4.4 ConfigureWebHostDefaults 530

附录 A 实例演示 1 533

第 1 章

全新的开发体验

微软于 2000 年推出 .NET 战略，并于两年后推出第一个版本的 .NET Framework 和 IDE（Visual Studio.NET 2002，后来改名为 Visual Studio）。如果读者是一个资深的 .NET 程序员，那么传统的 .NET 应用的开发方式应该已经深深地烙印在脑海中。.NET Core 带来了全新的开发体验，但开发方式的差异根本不足以成为快速跨入 .NET Core 世界的门槛，因为 .NET Core 在很多应用开发方面比传统的 .NET Framework 更加简单。为了消除尚未接触过 .NET Core 的读者对未知世界的恐惧，本章先介绍几个简单的 Hello World 应用，从而使读者提前了解 .NET Core 全新的开发体验。

1.1 Windows 平台

"跨平台"是 .NET Core 区别于传统 .NET Framework 最核心的特征。相较于传统的 .NET Framework 应用只能运行在微软的 Windows 平台上，经过全新设计的 .NET Core 在诞生的时候就被注入了跨平台的"基因"。.NET Core 应用在无须经过任何更改的情况下就可以直接运行在 Windows、macOS 以及各种 Linux（RHEL、Ubuntu、Debian、Fedora、CentOS 和 SUSE 等）平台上。除此之外，.NET Core 针对容器也提供了原生的支持，一个 .NET Core 应用可以同时运行在 Windows Container 和 Linux Container 上。下面先介绍 Windows 平台上 .NET Core 应用的开发体验。

1.1.1 构建开发环境

.NET Core 的官方站点介绍了在各种平台上构建开发环境的方式。总的来说，在不同的平台上开发 .NET Core 应用都需要安装相应的 SDK 和 IDE。成功安装 SDK 之后，我们在本地将自动拥有 .NET Core 的运行时（CoreCLR）、基础类库以及相应的开发工具。

dotnet.exe 是 .NET Core SDK 提供的一个重要的命令行工具，在进行 .NET Core 应用的开发部署时会频繁地使用 dotnet.exe。dotnet.exe 提供了很多有用的命令，此处不做系统介绍，如果后续章节涉及相关命令，我们再对它们做针对性的介绍。当 .NET Core SDK 安装结束之后，通过运行"dotnet"命令可以确认 SDK 是否安装成功。如图 1-1 所示，执行"dotnet --info"命令可以查看当前安装的 .NET Core SDK 的基本信息，显示的信息包含 SDK 的版本、运行时环境以及本机安装的所有运行时版本。

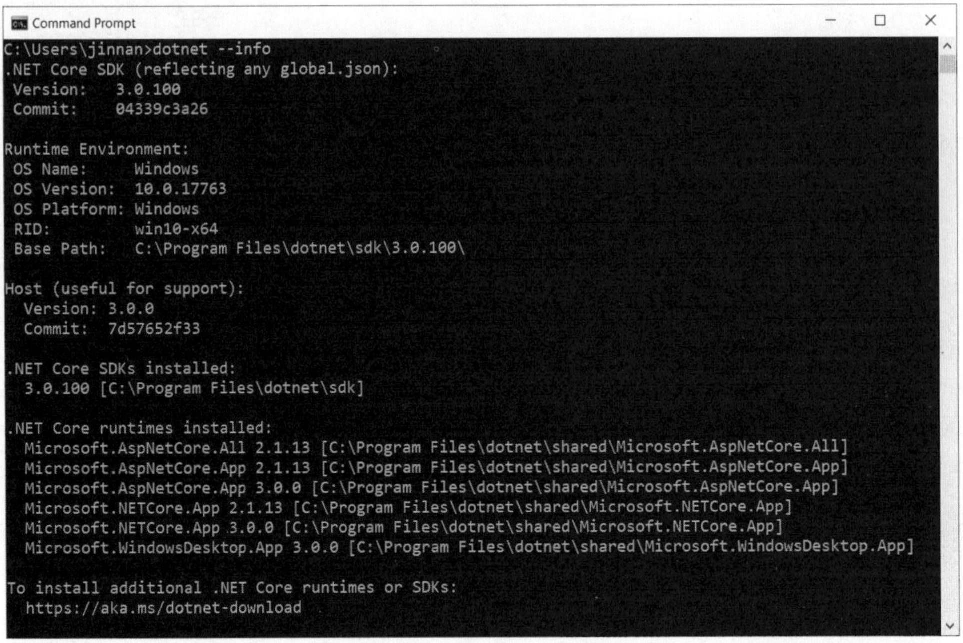

```
C:\Users\jinnan>dotnet --info
.NET Core SDK (reflecting any global.json):
 Version:   3.0.100
 Commit:    04339c3a26

Runtime Environment:
 OS Name:     Windows
 OS Version:  10.0.17763
 OS Platform: Windows
 RID:         win10-x64
 Base Path:   C:\Program Files\dotnet\sdk\3.0.100\

Host (useful for support):
  Version: 3.0.0
  Commit:  7d57652f33

.NET Core SDKs installed:
  3.0.100 [C:\Program Files\dotnet\sdk]

.NET Core runtimes installed:
  Microsoft.AspNetCore.All 2.1.13 [C:\Program Files\dotnet\shared\Microsoft.AspNetCore.All]
  Microsoft.AspNetCore.App 2.1.13 [C:\Program Files\dotnet\shared\Microsoft.AspNetCore.App]
  Microsoft.AspNetCore.App 3.0.0 [C:\Program Files\dotnet\shared\Microsoft.AspNetCore.App]
  Microsoft.NETCore.App 2.1.13 [C:\Program Files\dotnet\shared\Microsoft.NETCore.App]
  Microsoft.NETCore.App 3.0.0 [C:\Program Files\dotnet\shared\Microsoft.NETCore.App]
  Microsoft.WindowsDesktop.App 3.0.0 [C:\Program Files\dotnet\shared\Microsoft.WindowsDesktop.App]

To install additional .NET Core runtimes or SDKs:
  https://aka.ms/dotnet-download
```

图 1-1　执行 "dotnet --info" 命令获取 .NET Core SDK（Windows）的基本信息

高效的开发离不开优秀的 IDE，在这方面作为一个 .NET 开发者是幸福的，因为我们拥有 Visual Studio。虽然 Visual Studio Code 也是一个优秀的 IDE，但 Windows 依旧是主要的开发环境，所以笔者推荐使用 Visual Studio。截至 2019 年，Visual Studio 的最新版本为 2019。另外，Visual Studio 已经提供了 Mac 版本。

Visual Studio Code 是一个完全免费并且提供全平台（Windows、Mac 和 Linux）支持的 IDE，读者可以直接在其官网上下载。Visual Studio 2019 提供了社区版（Community）、专业版（Professional）和企业版（Enterprise），其中社区版是免费的，专业版和企业版需要付费购买。

除了 Visual Studio 和 Visual Studio Code，我们还可以使用一款叫作 Rider 的 IDE 开发 .NET Core 应用。Rider 是 JetBrains 公司开发的一款专门针对 .NET 的 IDE，我们可以利用它开发 ASP.NET、.NET Core、Xmarin 及 Unity 应用。和 Visual Studio Code 一样，Rider 也是一个跨平台的 IDE，我们可以同时在 Windows、Max OS 以及各种桌面版本的 Linux Distribution 上使用它。但 Rider 不是一款免费的 IDE，对其感兴趣的读者可以在官方站点下载 30 天试用版。

1.1.2　利用命令行创建 .NET Core 应用

dotnet.exe 提供了基于"脚手架"（Scaffolding）创建初始应用的 new 命令。如果需要开发某种类型的 .NET Core 应用，我们一般不会从第一行代码开始写，而是利用这个 new 命令创建一个具有初始结构的应用程序。除此之外，在开发过程中如果需要添加某种类型的文件（如各种类型的配置文件、MVC 的视图文件等），我们也可以利用该命令来完成，通过这种方式添加的文件具有预定义的初始内容。.NET Core SDK 在安装时提供了一系列预定义的脚手架模板，我们可以按照图 1-2 所示的方式执行“dotnet new --list”命令列出当前安装的脚手架模板。

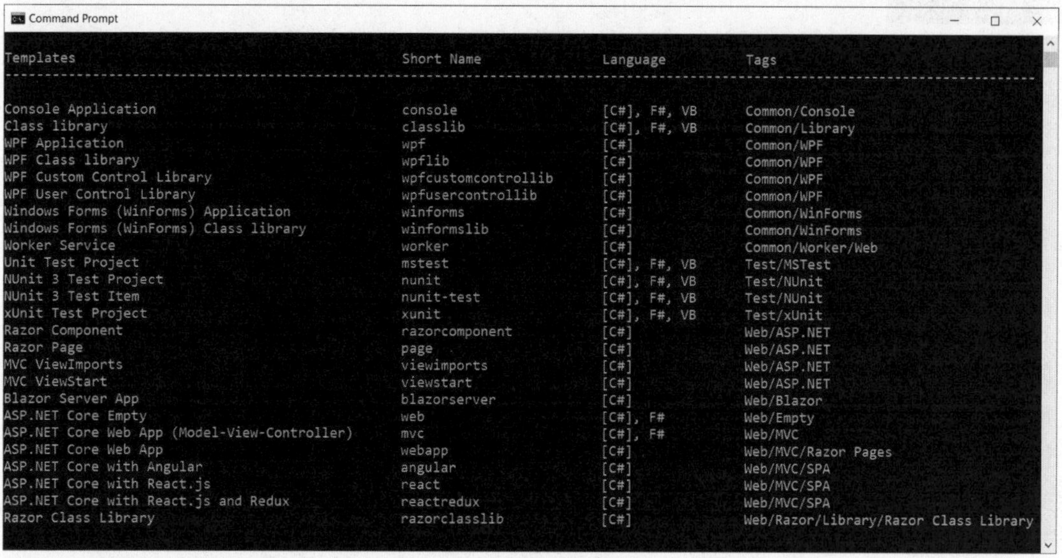

```
Command Prompt

Templates                                         Short Name               Language          Tags
----------------------------------------------------------------------------------------------------------------------------------

Console Application                               console                  [C#], F#, VB      Common/Console
Class library                                     classlib                 [C#], F#, VB      Common/Library
WPF Application                                   wpf                      [C#]              Common/WPF
WPF Class library                                 wpflib                   [C#]              Common/WPF
WPF Custom Control Library                        wpfcustomcontrollib      [C#]              Common/WPF
WPF User Control Library                          wpfusercontrollib        [C#]              Common/WPF
Windows Forms (WinForms) Application              winforms                 [C#]              Common/WinForms
Windows Forms (WinForms) Class library            winformslib              [C#]              Common/WinForms
Worker Service                                    worker                   [C#]              Common/Worker/Web
Unit Test Project                                 mstest                   [C#], F#, VB      Test/MSTest
NUnit 3 Test Project                              nunit                    [C#], F#, VB      Test/NUnit
NUnit 3 Test Item                                 nunit-test               [C#], F#, VB      Test/NUnit
xUnit Test Project                                xunit                    [C#], F#, VB      Test/xUnit
Razor Component                                   razorcomponent           [C#]              Web/ASP.NET
Razor Page                                        page                     [C#]              Web/ASP.NET
MVC ViewImports                                   viewimports              [C#]              Web/ASP.NET
MVC ViewStart                                     viewstart                [C#]              Web/ASP.NET
Blazor Server App                                 blazorserver             [C#]              Web/Blazor
ASP.NET Core Empty                                web                      [C#], F#          Web/Empty
ASP.NET Core Web App (Model-View-Controller)      mvc                      [C#], F#          Web/MVC
ASP.NET Core Web App                              webapp                   [C#]              Web/MVC/Razor Pages
ASP.NET Core with Angular                         angular                  [C#]              Web/MVC/SPA
ASP.NET Core with React.js                        react                    [C#]              Web/MVC/SPA
ASP.NET Core with React.js and Redux              reactredux               [C#]              Web/MVC/SPA
Razor Class Library                               razorclasslib            [C#]              Web/Razor/Library/Razor Class Library
```

图 1-2　执行“dotnet new --list”命令获取脚手架模板列表

图 1-2 列出的就是安装 .NET Core SDK 后提供的预定义的脚手架模板，这些模板大致分为 Project Template 和 Item Template 两类，前者为我们创建一个初始项目，后者则在一个现有项目中针对某种项目元素添加一个或者多个对应的文件。细心的读者可以从图 1-2 中看到 dotnet new 命令具有一个--type 参数，该参数具有 3 个预定义的选项（project、item 和 other），其中，前两个预定义的选项分别对应 Project 和 Item 这两种模板类型。

如果这些预定义的脚手架模板无法满足我们的需求，还可以创建自定义的 Project 或者 Item 模板，至于自定义模板应该如何定义，有兴趣的读者可以参考 .NET Core 官方文档。自定义模板最终会封装成一个 NuGet 包，我们可以通过执行“dotnet new --i”命令或者“dotnet new --install”命令对其进行安装。除此之外，对于已经安装的模板，我们可以通过执行“dotnet new --u”命令或者“dotnet new --uninstall”命令将其卸载。

下面执行“dotnet new”命令（dotnet new console -n helloworld），按照图 1-3 所示的方式创建一个名为“helloworld”的控制台程序。和传统的 .NET Framework 应用一样，一个针对 C#语言的 .NET Core 项目依然由一个对应的.csproj 文件进行定义，图 1-3 中的 helloworld.csproj 就是这样的一个文件。

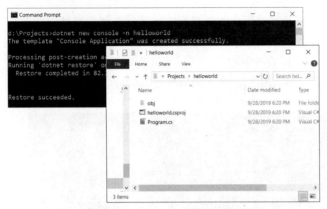

图 1-3　执行"dotnet new"命令创建一个控制台程序

对于传统的 .NET Framework 应用来说，即使是一个空的 C#语言项目，定义该项目的.csproj 文件在内容和结构上都是很复杂的，因为这个.csproj 文件的结构并不是面向开发者设计的，我们也不会直接编辑这个文件，而是利用 Visual Studio 通过设置当前项目的某些属性间接地修改它。但是对于一个 .NET Core 应用来说，这个.csproj 文件的结构变得相对简单并且更加清晰，所以开发人员经常会直接编辑它。对于前面我们执行脚手架命令创建的控制台程序，定义项目的 helloworld.csproj 文件的完整内容如下。

```xml
<Project Sdk="Microsoft.NET.Sdk">
  <PropertyGroup>
    <OutputType>Exe</OutputType>
    <TargetFramework>netcoreapp3.0</TargetFramework>
  </PropertyGroup>
</Project>
```

如上面的代码片段所示，这个 helloworld.csproj 是一个根节点为<Project>的 XML 文件，与项目相关的属性可以分组定义在相应的<PropertyGroup>节点下。这个 helloworld.csproj 文件实际上只定义了两个属性，分别是通过<OutputType>节点和<TargetFramework>节点表示的编译输出类型与目标框架。由于我们创建的是一个针对 .NET Core 3.0 的可执行控制台应用，所以目标框架为 netcoreapp3.0，编译输出为 Exe。

dotnet new 命令行除了可以创建一个空的控制台程序，还可以生成一些初始化代码，这就是项目目录下的 Program.cs 文件的内容。如下所示的代码片段给出了定义在这个文件中的整个 C#代码的定义，同时定义了代表程序入口点的 Main 方法，并且在这个方法中将字符串"Hello World!"打印在控制台上。

```csharp
using System;
namespace helloworld
{
    class Program
    {
        static void Main(string[] args)
        {
            Console.WriteLine("Hello World!");
```

```
        }
    }
}
```

通过执行脚手架命令行创建的应用程序虽然简单，但它是一个完整的 .NET Core 应用，可以在无须任何修改的情况下直接编译和运行。针对 .NET Core 应用的编译和运行同样是执行 "dotnet.exe" 命令行完成的。如图 1-4 所示，在进入当前项目所在目录之后，可以执行 "dotnet build" 命令对这个控制台应用实施编译，由于默认采用 Debug 编译模式，所以编译生成的程序集会保存在 "\bin\Debug\" 目录下。除此之外，针对不同目标框架编译生成的程序集是不同的，由于我们创建的是针对 .NET Core 3.0 的应用程序，所以最终生成的程序集被保存在 "\bin\Debug\netcoreapp3.0\" 目录下。

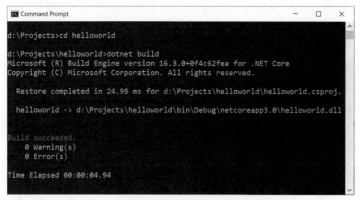

图 1-4　执行 "dotnet build" 命令编译一个控制台程序

如果查看编译的输出目录，可以发现两个同名（helloworld）的文件，一个是 helloworld.dll，另一个是 helloworld.exe，后者在尺寸上会大很多。另外，helloworld.exe 是一个可以直接运行的可执行文件；而 helloworld.dll 仅仅是一个单纯的动态链接库，需要借助命令行 dotnet 才能执行。

如图 1-5 所示，当我们在项目目录下执行 "dotnet run" 命令后，编译后的程序随即被执行，程序入口 Main 方法中指定的 "Hello World!" 字符串被直接打印在控制台上。其实，执行 "dotnet run" 命令启动程序之前无须显式执行 "dotnet build" 命令对源代码实施编译，因为该命令会自动触发编译操作。在执行 "dotnet" 命令启动应用程序集时，我们也可以直接指定启动程序集的路径（"dotnet bin\Debug\netcoreapp3.0\helloworld.dll"）。（S101）①

图 1-5　执行 "dotnet run" 命令运行一个控制台程序

---

① 解释见附录 A

1.1.3 ASP.NET Core 应用

前面利用 dotnet new 命令创建了一个简单的控制台程序,下面将其改造成一个 ASP.NET Core 应用。一个 ASP.NET Core 应用构建在 ASP.NET Core 框架之上,ASP.NET Core 框架利用一个消息处理管道完成对 HTTP 请求的监听、接收、处理和最终的响应(第 11 章至第 13 章对管道有详细的介绍)。ASP.NET Core 管道由一个服务器(Server)和若干中间件(Middleware)构成,当宿主(Host)程序启动之后,管道被构建出来,作为管道"龙头"的服务器就开始监听来自客户端的 HTTP 请求。

添加引用

下面直接利用 Visual Studio 打开项目文件 helloworld.csproj。为了能够使用 ASP.NET Core 框架提供的程序集,我们可以通过修改项目文件(.csproj)添加针对"Microsoft.AspNetCore.App"的框架引用(FrameworkReference)。在 Visual Studio 中修改项目文件非常方便,我们只需要右击选择目标项目,并从弹出的菜单中选择"Edit Project File"命令即可。如果没有特殊说明,本书后续章节提供的控制台实例演示基本上添加了针对"Microsoft.AspNetCore.App"的框架引用。下面是修改后的项目文件,针对"Microsoft.AspNetCore.App"的框架引用被添加到<ItemGroup/>节点下。

```
<Project Sdk="Microsoft.NET.Sdk">
  <PropertyGroup>
    <OutputType>Exe</OutputType>
    <TargetFramework>netcoreapp3.0</TargetFramework>
  </PropertyGroup>
  <ItemGroup>
    <FrameworkReference Include="Microsoft.AspNetCore.App"/>
  </ItemGroup>
</Project>
```

注册服务器与中间件

从应用承载或者寄宿(Hosting)方面来看,.NET Core 具有一个以 IHost/IHostBuilder 为核心的服务承载系统(第 10 章会对承载系统进行详细介绍),任何需要长时间运行的操作都可以定义成 IHostedService 服务并通过该系统来承载。IHost 对象可以视为所有承载服务的宿主(Host),而 IHostBuilder 对象则是它的构建者(Builder)。一个 ASP.NET Core 应用本质上就是一个用来监听、接收和处理 HTTP 请求的后台服务,所以它被定义成一个 GenericWebHostService(实现了 IHostedService 接口),并将它注册到承载系统中,进而实现了针对 ASP.NET Core 应用的承载。

一个运行的 ASP.NET Core 应用本质上体现为由一个服务器和若干中间件构成的消息处理管道,服务器解决针对 HTTP 请求的监听、接收和最终的响应,具体针对请求的处理则由它递交给后续的中间件来完成。这个管道是由 GenericWebHostService 服务构建的。

ASP.NET Core 提供了几种原生的服务类型,比较常用的是 KestrelServer 和 HTTP.sys。KestrelServer 是采用 libuv 创建的跨平台的 Web 服务器,可以在 Windows、macOS 和 Linux 平台

上使用。它不仅可以作为独立的 Web 服务器直接对外提供服务，还可以结合传统的 Web 服务器（如 IIS、Apache 和 Nginx）将它们作为反向代理来使用。HTTP.sys 则是一种只能在 Windows 平台上使用的 Web 服务器，由于它本质上是一个在操作系统内核模式运行的驱动，所以能够提供非常好的性能。本书所有的实例全部采用 KestrelServer。

在对项目文件 helloworld.csproj 做了上述修改之后，我们对定义在 Program.cs 中的 Main 方法做了如下改动：调用静态类型 Host 的 CreateDefaultBuilder 方法创建了一个 IHostBuilder 对象，并最终调用该对象的 Build 方法构建作为服务宿主的 IHost 对象。当我们调用 IHost 对象的 Run 扩展方法时，ASP.NET Core 应用程序将会被启动。

在调用 Build 方法构建 IHost 对象之前，可以调用 IHostBuilder 接口的 ConfigureWebHost 扩展方法，并利用指定的 Action<IWebHostBuilder>委托对象构建 ASP.NET Core 应用的请求处理管道。具体来说，我们调用 IWebHostBuilder 接口的 UseKestrel 扩展方法将 KestrelServer 注册为服务器，调用 Configure 扩展方法注册了用来处理请求的中间件。Configure 扩展方法的输入参数是一个 Action<IApplicationBuilder>对象，所需的中间件注册在 IApplicationBuilder 对象上。演示程序注册的唯一中间件是通过调用 IApplicationBuilder 接口的 Run 扩展方法注册的，该中间件利用指定的 Func<HttpContext,Task>对象将响应的主体内容设置为 "Hello World."。

```
using Microsoft.AspNetCore.Builder;
using Microsoft.AspNetCore.Hosting;
using Microsoft.AspNetCore.Http;
using Microsoft.Extensions.Hosting;

namespace helloworld
{
    class Program
    {
        static void Main()
        {
            Host.CreateDefaultBuilder()
                .ConfigureWebHost(webHostBuilder => webHostBuilder
                    .UseKestrel()
                    .Configure(app => app.Run(
                        context => context.Response.WriteAsync("Hello World."))))
                .Build()
                .Run();
        }
    }
}
```

我们可以按照上面演示的程序通过执行"dotnet"命令行启动该程序，也可以直接在 Visual Studio 中按 F5 键或者 Ctrl+F5 组合键启动该程序。执行"dotnet run"命令启动 ASP.NET Core 程序，输出结果如图 1-6 所示，这些输出其实是通过日志的形式输出的（第 8 章和第 9 章会对诊断日志进行详细介绍）。从这些输出可以看出 ASP.NET Core 应用采用的默认监听地址（"http://localhost:5000" 和 "https//localhost:5001"）和承载环境（Production）。如果需要关闭应用程序，按 Ctrl+C 组合键即可。

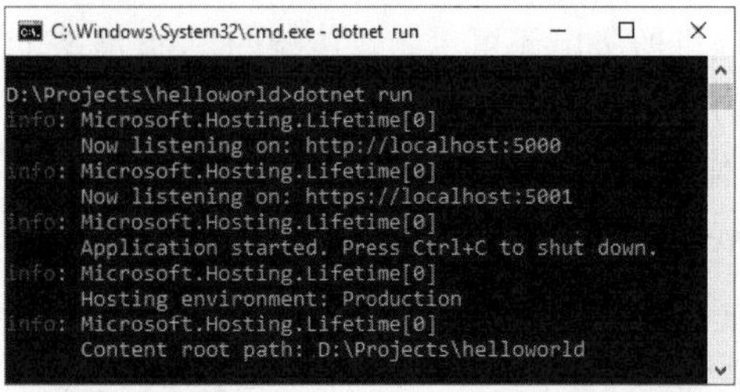

图 1-6　执行"dotnet run"命令启动 ASP.NET Core 程序

注册的 KestrelServer 会绑定到"http：//localhost:5000"和"https：//localhost:5001"这两个地址监听请求，如果利用浏览器分别对这两个地址发起请求会得到怎样的响应？如图 1-7 所示，两个请求都会得到主体内容为"Hello World."的响应（由于证书的问题，Chrome 浏览器为 HTTPS 的请求会显示"Not secure"的警告），毫无疑问，该内容就是中间件写入的。（S102）

图 1-7　"Hello World."应用的响应

修改 SDK

每个 .NET Core 应用都针对一种具体的 SDK 类型。前面展示了项目文件 helloworld.csproj 的完整定义，这是一个 XML 文件，在根节点的<Project>上通过 SDK 属性设置了当前项目采用的 SDK 类型。前面通过 dotnet new 命令工具创建的控制台应用默认采用的 SDK 类型为"Microsoft.NET.Sdk"，而 ASP.NET Core 应用通常采用另一种名为"Microsoft.NET.Sdk.Web"的 SDK 类型。

如果将 SDK 设置为"Microsoft.NET.Sdk.Web"，就可以将针对"Microsoft. AspNetCore.App"的框架引用从项目文件中删除。由于不需要利用生成的.exe 文件启动 ASP.NET Core 应用，所以应该将 XML 元素<OutputType>Exe</OutputType>从<PropertyGroup>节点中删除。因此，最终的项目文件只需要保留如下内容即可。

```xml
<Project Sdk="Microsoft.NET.Sdk.Web">
  <PropertyGroup>
    <TargetFramework>netcoreapp3.0</TargetFramework>
  </PropertyGroup>
</Project>
```

launchSettings.json

当我们通过修改项目文件 helloworld.csproj 将 SDK 类型改为"Microsoft.NET.Sdk.Web"之

后，如果使用 Visual Studio 打开这个文件，一个名为 launchSettings.json 的配置文件将自动生成并且被保存在 "\Properties" 目录下。顾名思义，launchSettings.json 是一个在应用启动时自动加载的配置文件，该配置文件可以在不同的设置下执行应用程序。下面的代码片段就是 Visual Studio 自动创建的 launchSettings.json 文件的内容。由此可以看出，该配置文件默认添加了两个节点，其中，iisSettings 节点用于设置 IIS 相关的选项，而 profiles 节点定义了一系列用于描述应用启动场景的 Profile。

```
{
  "iisSettings": {
    "windowsAuthentication": false,
    "anonymousAuthentication": true,
    "iisExpress": {
      "applicationUrl": "http://localhost:51127/",
      "sslPort": 0
    }
  },
  "profiles": {
    "IIS Express": {
      "commandName": "IISExpress",
      "launchBrowser": true,
      "environmentVariables": {
        "ASPNETCORE_ENVIRONMENT": "Development"
      }
    },
    "helloworld": {
      "commandName": "Project",
      "launchBrowser": true,
      "environmentVariables": {
        "ASPNETCORE_ENVIRONMENT": "Development"
      },
      "applicationUrl": "http://localhost:51128/"
    }
  }
}
```

初始的 launchSettings.json 文件会默认创建两个 Profile，一个被命名为 "IIS Express"，另一个则使用应用名称命名（"helloworld"）。每个 Profile 相当于定义了应用的启动场景，相关的设置包括应用启动的方式、环境变量和 URL 等，具体的设置包括以下几点。

- commandName：启动当前应用程序的命令类型，有效的选项包括 IIS、IISExpress、Executable 和 Project，前三个选项分别表示采用 IIS、IISExpress 和指定的可执行文件（.exe）来启动应用程序。如果使用 "dotnet run" 命令启动程序，那么对应 Profile 的启动命名名称应该设置为 Project。
- executablePath：如果 commandName 属性被设置为 Executable，就需要利用该属性设置启动可执行文件的路径（绝对路径或者相对路径）。
- environmentVariables：该属性用来设置环境变量。由于 launchSettings.json 文件仅仅在开发环境中使用，所以默认会添加一个名为 "ASPNETCORE_ENVIRONMENT" 的环境变量,并将它的值设置为"Development",ASP.NET Core 应用就是利用这样一个环境变量来表示当前部署环境的。

- commandLineArgs:命令行参数,即传入 Main 方法的参数列表。
- workingDirectory:启动当前应用运行的工作目录。
- applicationUrl:应用程序采用的 URL 列表,多个 URL 之间用分号(;)进行分隔。
- launchBrowser:一个布尔类型的开关,表示应用程序的时候是否自动启动浏览器。
- launchUrl:如果 launchBrowser 被设置为 True,浏览器采用的初始化路径就通过该属性进行设置。
- nativeDebugging:是否启动本地代码调试(Native Code Debugging),默认值为 False。
- externalUrlConfiguration:如果该属性被设置为 True,就意味着禁用本地的配置,默认值为 False。
- use64Bit:如果 commandName 属性被设置为 IIS Express,该属性决定采用 X64 版本还是 X86 版本,默认值为 False,意味着 ASP.NET Core 应用默认采用 X86 版本的 IIS Express。

launchSettings.json 文件中的所有设置仅仅针对开发环境,在产品环境下是不需要这个文件的,应用发布后生成的文件列表中也不会包含该文件。该文件不需要手动编辑,当前项目属性对话框(通过在解决方案对话框中右击选择"Properties"(属性)选项来打开)中"Debug"(调试)选项卡下的所有设置最终都会体现在该文件上(见图 1-8)。

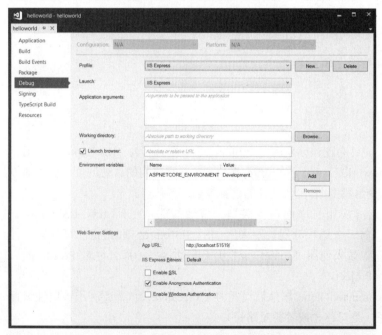

图 1-8　在 Visual Studio 中通过设置调试选项编辑 launchSettings.json 文件

如果在 launchSettings.json 文件中设置了多个 Profile，它们会以图 1-9 所示的形式出现在 Visual Studio 的工具栏中，我们可以选择任意一个 Profile 中定义的配置选项启动当前的应用程序。如果在 Profile 中通过设置 launchBrowser 属性选择启动浏览器，我们还可以选择浏览器的类型。

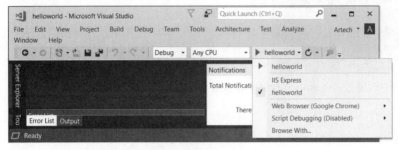

图 1-9　在 Visual Studio 中选择 Profile

如果我们在当前项目所在目录下通过执行"dotnet run"命令启动应用程序，launchSettings.json 文件就会默认被加载。我们可以通过命令行参数 --launch-profile 指定采用的 Profile。如果没有对 Profile 做显式指定，那么该配置文件中第一个 commandName 为"Project"的 Profile 会默认被使用。如图 1-10 所示，在创建应用的根目录下通过执行"dotnet run"命令启动应用程序，其中，第一次执行"dotnet run"命令时显式设置了 Profile 名称（--launch-profile helloworld）。从输出结果可以看出，两次采用的是同一个 Profile。

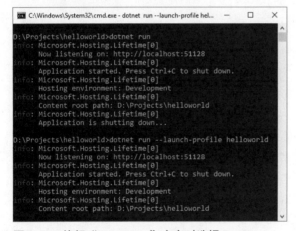

图 1-10　执行"dotnet run"命令时选择 Profile

如果在执行"dotnet run"命令时不希望加载 launchSettings.json 文件，可以通过显式指定命令行参数 --no-launch-profile 实现。如图 1-11 所示，执行"dotnet run"命令时指定了参数 --no-launch-profile，所以应用会采用 KestrelServer 默认的监听地址（"http://localhost:5000"和"https://localhost:5001"）。由于 launchSettings.json 文件没有被加载，所以当前的执行环境从 Development 变为默认的 Production。

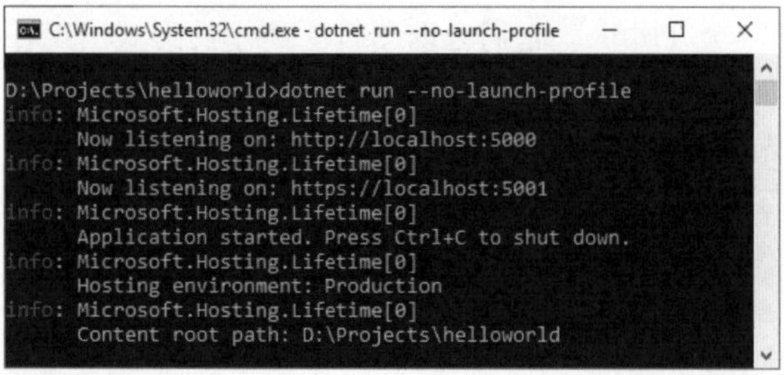

图 1-11　执行 "dotnet run" 命令时通过--no-launch-profile 参数阻止加载 launchSettings.json 文件

显式指定 URL

如果既不使用 launchSettings.json 文件中定义的 URL，也不使用 KestrelServer 默认采用的监听地址，我们可以在应用程序中显式指定应用的 URL。正如下面的代码片段所示，只需要调用 IWebHostBuilder 接口的 UseUrls 扩展方法指定一组以分号分隔的 URL 即可。（S103）

```
class Program
{
    static void Main()
    {
        Host.CreateDefaultBuilder()
            .ConfigureWebHost(webHostBuilder => webHostBuilder
                .UseKestrel()
                .UseUrls("http://0.0.0.0:3721;https://0.0.0.0:9527")
                .Configure(app => app.Run(context =>
                    context.Response.WriteAsync("Hello World."))))
            .Build()
            .Run();
    }
}
```

ConfigureWebHostDefaults

上面演示的实例，都是通过调用 IHostBuilder 接口的 ConfigureWebHost 扩展方法，借助指定的 Action<IWebHostBuilder>委托对象构建处理请求，来处理管道的。该接口还有一个 ConfigureWebHostDefaults 扩展方法，它会做一些默认设置。正如下面的代码片段所示，如果调用这个方法，KestrelServer 就不需要进行显式注册。至于 ConfigureWebHostDefaults 扩展方法究竟做了哪些默认设置，有兴趣的读者可以参阅第 13 章。（S104）

```
class Program
{
    static void Main()
    {
        Host.CreateDefaultBuilder()
            .ConfigureWebHostDefaults(webHostBuilder => webHostBuilder
                .Configure(app => app.Run(
                    context => context.Response.WriteAsync("Hello World."))))
            .Build()
            .Run();
```

            }
}

1.1.4　ASP.NET Core MVC 应用

　　由于 ASP.NET Core 框架在本质上就是由服务器和中间件构建的消息处理管道，所以在它上面构建的应用开发框架都建立在某种类型的中间件上，整个 ASP.NET Core MVC 开发框架就是建立在用来实现路由的 EndpointRoutingMiddleware 中间件和 EndpointMiddleware 中间件上的。ASP.NET Core MVC 利用路由系统为它分发请求，并在此基础上实现针对目标 Controller 的激活、Action 方法的选择和执行，以及最终对于执行结果的响应。在介绍的实例演示中，我们将对上面创建的 ASP.NET Core 做进一步改造，使之转变成一个 MVC 应用。

注册服务与中间件

　　ASP.NET Core 框架内置了一个原生的依赖注入框架（第 3 章和第 4 章会对依赖注入进行系统而深入的介绍），该框架利用一个依赖注入容器提供管道在构建及请求处理过程中所需的服务，而这些服务需要在应用启动的时候被预先注册。对于 ASP.NET Core MVC 框架来说，它在处理 HTTP 请求的过程中所需的一系列服务同样需要预先注册。对这个概念有了基本的了解之后，读者对如下所示的代码就容易理解了。

```
using Microsoft.AspNetCore.Builder;
using Microsoft.AspNetCore.Hosting;
using Microsoft.Extensions.DependencyInjection;
using Microsoft.Extensions.Hosting;

namespace helloworld
{
    class Program
    {
        static void Main()
        {
            Host.CreateDefaultBuilder()
                .ConfigureWebHostDefaults(webHostBuilder => webHostBuilder
                    .ConfigureServices(servicecs => servicecs
                        .AddRouting()
                        .AddControllersWithViews())
                    .Configure(app => app
                        .UseRouting()
                        .UseEndpoints(endpoints => endpoints.MapControllers())))
                .Build()
                .Run();
        }
    }
}
```

　　整个 ASP.NET MVC 框架建立在 EndpointRoutingMiddleware 中间件和 EndpointMiddleware 中间件构建的路由系统上，这两个中间件采用"终结点（Endpoint）映射"的方式实现针对

HTTP 请求的路由（由这两个中间件构建的路由系统在第 15 章有详细介绍）。这里所谓的终结点可以视为应用程序提供的针对 HTTP 请求的处理器，这两个中间件通过预先设置的规则将具有某些特征的请求（如路径、HTTP 方法等）映射到对应的终结点，进而实现路由的功能。对于一个 MVC 应用程序来说，我们可以以定义在 Controller 类型中的 Action 方法视为一个终结点，那么路由映射最终体现在 HTTP 请求与目标 Action 方法的映射上。

上面的代码片段先后调用 IApplicationBuilder 接口的 UseRouting 方法与 UseEndpoints 方法注册了 EndpointRoutingMiddleware 中间件和 EndpointMiddleware 中间件。在调用 UseEndpoints 方法的时候，我们利用指定的 Action<IEndpointRouteBuilder>委托对象调用了 IEndpointRouteBuilder 接口的 MapControllers 扩展方法，从而完成了针对定义在 Controller 类型中所有 Action 方法的映射。

由于注册的中间件具有对其他服务的依赖，所以需要预先将这些服务注册到依赖注入框架中。依赖服务的注册是通过调用 IWebHostBuilder 接口的 ConfigureServices 方法完成的，该方法的参数类型为 Action<IServiceCollection>，添加的服务注册就保存在 IServiceCollection 接口表示的集合中。在上面的演示程序中，两个中间件依赖的服务是通过调用 IServiceCollection 接口的 AddRouting 方法和 AddControllersWithViews 方法进行注册的。

如下所示的 HelloController 是 Controller 类型。按照约定，所有的 Controller 类型名称都应该以"Controller"字符作为后缀。与之前版本的 ASP.NET MVC 不同，ASP.NET Core MVC 下的 Controller 类型并不要求强制继承某个基类。我们在 HelloController 中定义了一个唯一的 Action 方法 SayHello，该方法直接返回一个内容为"Hello World!"的字符串。该方法通过标注的 HttpGetAttribute 特性注册了一个模板为"/hello"的路由，意味着请求地址为"/hello"的 GET 请求最终会被路由到这个 Action 方法上，而该方法执行的结果将作为请求的响应内容。所以，启动该程序后使用浏览器访问地址"http://localhost:5000/hello"得到的输出结果如图 1-7 所示。（S105）

```
public class HelloController
{
    [HttpGet("/hello")]
    public string SayHello() => "Hello World!";
}
```

引入视图

上面这个程序并没有涉及视图，所以不是一个典型的 MVC 应用，下面对其做进一步改造。为了使 HelloController 具有视图呈现的能力，我们让它派生于基类 Controller。Action 方法 SayHello 的返回类型被修改为 IActionResult 接口，它表示 Action 方法执行的结果。可以为该方法定义一个表示姓名的参数 name，通过 HttpGetAttribute 特性注册的路由模板（"/hello/{name}"）中具有与之对应的路由参数。换句话说，满足该路径模式的请求 URL 携带的姓名，将自动绑定到该 Action 方法的参数 name 上。在 SayHello 方法中，可以利用 ViewBag 将代表姓名的参数 name 的值传递给呈现的视图，该方法最终调用 View 方法返回当前 Action 方法对应的 ViewResult 对象。

```csharp
public class HelloController : Controller
{
    [HttpGet("/hello/{name}")]
    public IActionResult SayHello(string name)
    {
        ViewBag.Name = name;
        return View();
    }
}
```

由于调用 View 方法时没有显式指定视图的名称，所以视图引擎会将当前 Action 的名称（SayHello）作为视图的名称。如果该视图还没有经过编译（部署时针对 View 的预编译，或者在这之前针对该 View 的动态编译），视图引擎将从若干候选路径中读取对应的.cshtml 文件进行编译，其中首选的路径为"{ContentRoot}\Views\{ControllerName}\{ViewName}.cshtml"。为了满足视图引擎定位视图文件的规则，我们需要将 SayHello 对应的视图文件（SayHello.cshtml）定义在目录"\Views\Hello\"下（见图 1-12）。

图 1-12 View 文件所在的路径

如下代码片段就是 SayHello.cshtml 文件的内容，这是一个针对 Razor 引擎的视图文件。从文件的扩展名（.cshtml）可以看出，这样的文件可以同时包含 HTML 标签和 C#代码。总的来说，视图文件会在服务端生成最终在浏览器中呈现出来的 HTML，既可以在这个文件中直接提供原样输出的 HTML 标签，也可以内嵌一段动态执行的 C#代码。虽然 Razor 引擎对 View 文件的编写有严格的语法，但是笔者认为没有必要在 Razor 语法上花费太多精力，因为 Razor 语法的目的就是"自然"地将动态 C#代码和静态 HTML 标签结合起来，最终生成一份完整的 HTML 文档，因此它的语法和普通的思维基本上是一致的。例如，下面的 View 文件最终会生成一个完整的 HTML 文档，其主体部分只有一个<p>标签。该标签的内容是动态的，这是因为包含利用 ViewBag 从 Controller 传进来的姓名。

```html
<html>
  <head>
    <title>Hello World</title>
  </head>
  <body>
    <p>Hello, @ViewBag.Name</p>
  </body>
</html>
```

再次运行该程序后，可以利用浏览器访问地址"http://localhost:5000/hello/foobar"。由于请求地址与 Action 方法 SayHello 上的路由规则相匹配，所以路径携带的姓名（foobar）会绑定到该方法的参数 name 上，最终的输出结果如图 1-13 所示。（S106）

图 1-13　启动并访问 ASP.NET Core MVC 应用

注册 Startup 类型

任何一个 ASP.NET Core 应用在初始化的时候都会根据请求处理的需求注册对应的中间件。前面演示的实例都是直接调用 IWebHostBuilder 接口的 Configure 扩展方法注册所需的中间件，但是在大部分真实的开发场景中，我们一般会将中间件及依赖服务的注册定义在一个单独的类型中，按照约定，通常将这个类型命名为 Startup。例如，针对服务和中间件的注册就可以放在如下代码片段定义的 Startup 类型中。

```
public class Startup
{
    public void ConfigureServices(IServiceCollection services) => services
        .AddRouting()
        .AddControllersWithViews();

    public void Configure(IApplicationBuilder app) => app
        .UseRouting()
        .UseEndpoints(endpoints => endpoints.MapControllers());
}
```

如上面的代码片段所示，不需要让 Startup 类型实现某个预定义的接口或者继承某个预定义基类，所采用的完全是一种基于"约定"的定义方式。随着对 ASP.NET Core 框架认识的加深，可以发现这种"约定优于配置"的设计广泛地应用在整个框架之中。按照约定，服务注册和中间件注册分别在 ConfigureServices 方法与 Configure 方法中实现，它们的第一个参数类型分别为 IServiceCollection 接口和 IApplicationBuilder 接口。由于已经将两种核心操作转移到 Startup 类型中，所以需要注册该类型。Startup 类型可以调用 IWebHostBuilder 接口的 UseStartup<TStartup>扩展方法进行注册。（S107）

```
class Program
{
    static void Main()
    {
        Host.CreateDefaultBuilder()
            .ConfigureWebHostDefaults(
                webHostBuilder => webHostBuilder.UseStartup<Startup>())
            .Build()
            .Run();
    }
}
```

前面对 .NET Core、ASP.NET Core 及 ASP.NET Core MVC 应用的编程做了初步介绍，但是这仅仅限于我们熟悉的 Windows 平台。作为一个跨平台的开发框架，我们有必要在其他操作系统平台上体验 .NET Core 开发的乐趣。

1.2 macOS 用户

除了微软的 Windows 平台，.NET Core 针对 macOS 以及各种 Linux Distribution（RHEL、Ubuntu、Debian、Fedora、CentOS 和 SUSE 等）都提供了很好的支持。本节主要介绍使用 Mac 开发 .NET Core 应用，但需要先在 macOS 上构建开发环境。

1.2.1 构建开发环境

和 Windows 一样，如果要在 macOS 上进行 .NET Core 应用的开发，只需要安装 .NET Core SDK 和相应的 IDE 即可。.NET Core SDK 可以直接从微软官方站点下载，安装之后我们将拥有 .NET Core 针对 macOS 的运行时和相应工具，其中包含前面频繁使用的命令行工具 dotnet。另外，虽然老版本的 macOS（如 10.12 Sierra）可以安装 .NET Core 3.0 SDK，但是只有 10.13 High Sierra 或者更高版本的 macOS 才能编译和运行 .NET Core 3.0 应用程序。

对于 macOS 的 .NET Core 应用的开发人员来说，他们在 IDE 上同样具有广泛的选择。首先，Visual Studio 目前已经推出了 Mac 版本，虽然与 Windows 版本在功能和稳定性上还有一定的差距，但是对于习惯使用 Visual Studio 的 macOS 用户来说是最好的选择。除此之外，也可以选择免费的 Visual Studio Code，如果不喜欢这种风格的 IDE，还可以选择 JetBrains 公司的 Rider。

1.2.2 利用命令行创建 .NET Core 应用

对于 Windows 和 macOS 用户来说，针对他们的开发体验基本上是一致的，因为 .NET Core SDK 提供的命令行（主要是 dotnet 命令行工具）在各个平台具有一致的定义，在 IDE（Visual Studio 和 Visual Studio Code）层面也具有相同的选择。下面先介绍用于创建初始 .NET Core 项目的脚手架命令行在 macOS 上的应用，为此可以按照图 1-14 所示的方式执行"dotnet new console"命令在当前目录下创建一个控制台应用程序。

图 1-14 执行"dotnet new console"命令创建控制台程序

图 1-14 右图是执行脚手架命令行创建的控制台应用的项目结构，可以看出它与在 Windows 上执行相同命令创建的应用具有相同的文件结构。不仅如此，生成的文件内容也完全一致，如下所示的代码片段是项目文件 helloworld.csproj 和程序文件 Program.cs 的内容。

helloworld.csproj：

```xml
<Project Sdk="Microsoft.NET.Sdk">
  <PropertyGroup>
    <OutputType>Exe</OutputType>
    <TargetFramework>netcoreapp3.0</TargetFramework>
  </PropertyGroup>
</Project>
```

Program.cs：

```csharp
using System;
namespace helloworld
{
    class Program
    {
        static void Main(string[] args)
        {
            Console.WriteLine("Hello World!");
        }
    }
}
```

在无须对原文件做任何改动的情况下，可以直接执行"dotnet"命令行来启动该控制台程序。如图 1-15 所示，在将当前目录切换到控制台应用所在项目根目录之后，可以直接执行"dotnet run"命令启动创建的程序，随后可以看到作为程序入口的 Main 方法输出到控制台上的"Hello World!"文本。

图 1-15　执行"dotnet run"命令执行控制台应用

1.2.3　ASP.NET Core MVC 应用

1.1 节演示了将一个脚手架命令行创建的控制台应用改造成 ASP.NET Core MVC 应用的步骤，由于 .NET Core 提供真正的跨平台支持，所以按照相同方式改造的 ASP.NET Core MVC 应用同样可以在 macOS 上运行。验证过程如下：首先直接编辑项目文件 helloworld.csproj，将 SDK 改成"Microsoft.NET.Sdk.Web"，并移除表示输出类型的属性节点（<OutputType>Exe</OutputType>）。

```xml
<Project Sdk="Microsoft.NET.Sdk.Web">
  <PropertyGroup>
    <TargetFramework>netcoreapp3.0</TargetFramework>
  </PropertyGroup>
</Project>
```

我们选择相应的 IDE 或者纯文本编辑器对程序文件 Program.cs 进行如下所示的修改，那么应用就变成一个简单的 ASP.NET Core MVC 应用。经过前面的介绍，读者应该可以理解每行代码的含义，所以此处不再赘述。

```csharp
using Microsoft.AspNetCore.Builder;
using Microsoft.AspNetCore.Hosting;
using Microsoft.AspNetCore.Mvc;
using Microsoft.Extensions.DependencyInjection;

namespace helloworld
{
    class Program
    {
        static void Main(string[] args)
        {
            new WebHostBuilder()
                .UseKestrel()
                .UseUrls("http://0.0.0.0:5000")
                .ConfigureServices(svcs => svcs.AddMvc())
                .Configure(app => app.UseMvc())
                .Build()
                .Run();
        }
    }

    public class HelloController
    {
        [HttpGet("/hello/{name}")]
        public string SayHello(stirng name) => $"Hello {name}!";
    }
}
```

至此，所有的编程工作都已经结束，只需要按照我们熟悉的方式执行 "dotnet run" 命令就可以启动这个程序。程序启动之后利用浏览器访问 "http://localhost:5000/hello/foobar" 可以得到图 1-16 所示的输出结果。

图 1-16　启动并访问 ASP.NET Core MVC 应用

1.3　Linux

在 Linux 环境下开发 .NET Core 应用有多种选择。第一种选择是在一台物理机上安装原生的

Linux，读者可以根据自己的习惯选择某种 Linux Distribution，目前的 RHEL、Ubuntu、Debian、Fedora、CentOS 和 SUSE 这些主流的 Distribution 都是支持 .NET Core 的。第二种选择是采用虚拟机的形式安装相应的 Linux Distribution，笔者经常使用的是安装在 VirtualBox 上的 Ubuntu。对于 X64 Windows 10 的用户来说，第三种选择更加方便快捷，也就是使用 Windows 10 提供的 Linux 子系统（Windows Subsystem for Linux，WSL），目前最新版本为 WSL 2。

1.3.1　启用 Linux 子系统

　　WSL 2 要求操作系统必须是 Windows 10 build 18917 或者更高的版本。在 Windows 10 上启用 WSL 2 只需要开启"Virtual Machine Platform"和"Windows Subsystem for Linux"这两个特性即可。开启这两个特性的方式主要有两种：第一种是在 PowerShell 中以如下方式执行"Enable-WindowsOptionalFeature"命令（该命令需要以管理员身份执行）。

```
Enable-WindowsOptionalFeature -Online -FeatureName VirtualMachinePlatform
Enable-WindowsOptionalFeature -Online -FeatureName Microsoft-Windows-Subsystem-Linux
```

　　第二种是采用可视化的方式。具体来说，可以通过"Control Panel"→"Programs and Features"打开图 1-17 所示的"Programs and Features"对话框，并选择"Turn Windows features on or off"，打开"Windows Features"对话框。选中"Virtual Machine Platform"和"Windows Subsystem for Linux"复选框，确定并重启计算机之后，针对 Linux 的 Windows 子系统就会被启用。

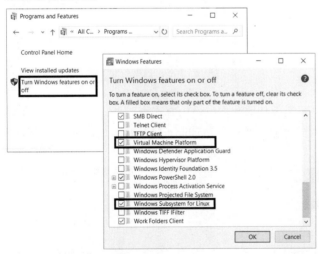

图 1-17　开启 WSL 2 相关特性

　　启用 WSL 之后，我们可以根据需要安装对应的 Linux Distribution。最方便的方式就是直接在 Microsoft Store 中下载并安装对应的 Linux Distribution，如 Microsoft Store 提供了图 1-18 所示的免费版 Ubuntu 18.04 LTS。

图 1-18　利用 Microsoft Store 安装 Ubuntu 18.04 LTS

如果 Microsoft Store 不可用（笔者的操作系统就是没有 Microsoft Store 的 Windows 10 Enterprise LTSC 版本），我们可以直接利用 curl 或者 Invoke-WebRequest 以命令行的方式下载 Linux 安装包。下面列举了几个常用的 Linux Distribution 的下载地址。

- Ubuntu 18.04：https://aka.ms/wsl-ubuntu-1804。
- Ubuntu 18.04 ARM：https://aka.ms/wsl-ubuntu-1804-arm。
- Ubuntu 16.04：https://aka.ms/wsl-ubuntu-1604。
- Debian GNU/Linux：https://aka.ms/wsl-debian-gnulinux。
- Kali Linux：https://aka.ms/wsl-kali-linux。
- OpenSUSE：https://aka.ms/wsl-opensuse-42。
- SLES：https://aka.ms/wsl-sles-12。

当 Linux 被成功安装之后，我们可以通过在 CMD 命令行中执行"bash"命令进入 Linux Bash Shell。如图 1-19 所示，在 CMD 命令行中执行"bash"命令之后会自动进入 Linux Bash Shell。在 Bash Shell 中执行"lsb_release -a"命令会输出当前安装的 Linux Distribution 的版本信息。如果想恢复到 CMD 命令行模式，只需要执行"exit"命令退出 Bash Shell 即可。

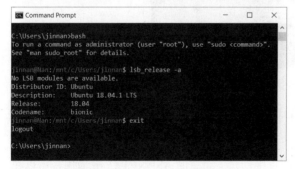

图 1-19　查看 Linux Distribution 的版本信息

作为主系统的 Windows 和 Linux 子系统可以共享网络系统与文件系统。Windows 下的文件系统直接挂载到"/mnt"目录下，所以 Windows 下的文件或者目录路径加上"/mnt"前缀就会变成基于 Linux 子系统下的路径。如图 1-19 所示，CMD 命令行环境下的当前工作目录为"C:\Users\jinnan"，当我们切换到 Bash Shell 后，当前工作目录其实并没有发生变化，只是路径变成"/mnt/c/Users/jinnan"而已。

1.3.2　构建开发环境

如果想在 Linux 下运行 .NET Core 应用，只需要安装 .NET Core SDK 即可。针对不同的 Linux Distribution，.NET Core SDK 的安装略有不同。对于主流的 Linux 发行版本（RHEL、Ubuntu、Debian、Fedora、CentOS 和 SUSE 等），.NET Core 官方网站给出了安装 .NET Core SDK 的详细教程。对于 Ubuntu 18.04 LTS，.NET Core 3.0 SDK 可以执行如下命令进行安装。

```
~$ sudo apt-get update
~$ sudo apt-get install apt-transport-https
~$ sudo apt-get install dotnet-sdk-3.0
```

安装完成之后，可以执行"dotnet --info"命令查看当前 .NET Core SDK 的基本信息，如果得到图 1-20 所示的输出结果，就证明 .NET Core 3.0 SDK 安装成功。

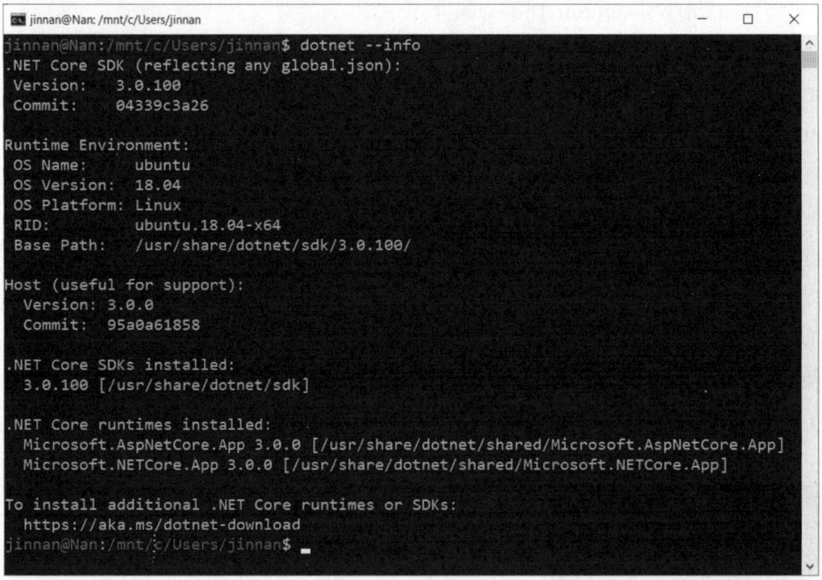

图 1-20　执行"dotnet --info"命令查看 .NET Core SDK（Linux）的基本信息

1.3.3　利用命令行创建 ASP.NET Core 应用

.NET Core SDK 同样为 Linux 提供了重要的命令行工具 dotnet，这说明利用该命令提供的脚手架模板可以创建相应的 .NET Core 应用。下面直接创建一个空的 ASP.NET Core 应用，所以在执行"dotnet new"命令的时候将模板名称指定为"web"。如图 1-21 所示，执行"dotnet new"

命令时的当前工作目录为"/mnt/c/helloworld",也就是主系统 Windows 下的"C:\helloworld",图 1-21 列出了该目录下的所有文件和子目录。

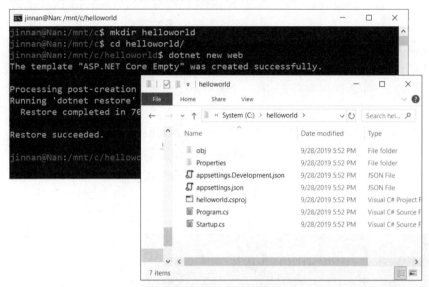

图 1-21　执行"dotnet new"命令创建一个 ASP.NET Core 应用

与在 Windows 和 macOS 下利用脚手架命令创建的应用一样,在 Linux 下可以在无须对这些命令做任何更改的情况下直接加以运行,为此我们只需要在当前应用所在的目录下执行"dotnet run"命令即可。如图 1-22 所示,启动后的 ASP.NET Core 应用默认绑定在 5000(HTTP)和 5001(HTTPS)端口上进行请求监听,由于 Linux 是当前 Windows 的子系统而非虚拟机,所以主、子系统可以共享网络。因此,利用浏览器请求地址"http://localhost:5000"可以访问这个应用的主页。

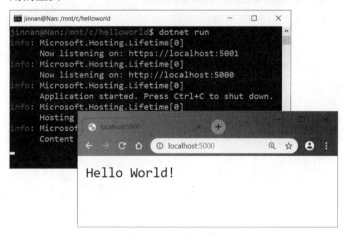

图 1-22　启动并访问 ASP.NET Core 应用

1.4 Docker

有的 .NET Core 开发人员可能没有使用过 Docker，但是应该听说过 Docker。Docker 是 GitHub 上最受欢迎的开源项目之一，有人认为 Docker 是所有云应用的基石，并且可以把互联网升级到下一代。Docker 是 DotCloud 公司开源的一款产品，从其诞生那一刻算起，在短短的两三年内就已经成为开源社区最热门的项目之一。对于完全拥抱开源的 .NET Core 来说，它自然应该对 Docker 提供完美的支持。下面的内容基于这样一个前提：读者对 Docker 有基本的了解，并且在计算机（Windows）上已经安装了 Docker。

下面将演示如何创建一个 ASP.NET Core 程序并将其编译成 Docker 镜像，Docker 环境针对该镜像创建一个容器来启动一个应用实例。简单起见，可以直接采用脚手架命令行的形式创建 ASP.NET Core 应用。如图 1-23 所示，执行"dotnet new web"命令在"D:\projects\helloworld"目录下创建一个空的 ASP.NET Core 应用。

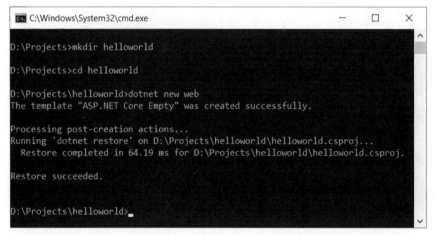

图 1-23　执行"dotnet new web"命令创建一个空的 ASP.NET Core 应用

现在需要将 ASP.NET Core 应用制作成一个 Docker 镜像，为此我们需要在项目根目录下创建一个 Dockerfile 文件（文件名就是 Dockerfile，没有扩展名），并在该文件中定义如下内容。如果读者对 Dockerfile 具有基本的了解，对于这个文件的内容应该不难理解。

```
# 1. 指定编译和发布应用的镜像
FROM mcr.microsoft.com/dotnet/core/sdk:3.0 AS build

# 2. 指定（编译和发布）工作目录
WORKDIR /app

# 3. 将.csproj 复制到工作目录"/app"，然后执行"dotnet restore"命令恢复所有安装的 NuGet 包
COPY *.csproj ./
RUN dotnet restore

# 4. 将所有文件复制到工作目录("/app")，然后执行"dotnet publish"命令将应用发布到"/app/out"目录下
```

```
COPY . ./
RUN dotnet publish -c Release -o out

# 5. 编译生成 Docker 镜像
# 5.1. 设置基础镜像
FROM mcr.microsoft.com/dotnet/core/aspnet:3.0 AS runtime

# 5.2. 设置（运行）工作目录，并将发布文件复制到 out 子目录下
WORKDIR /app
COPY --from=build /app/out .

# 5.3. 利用环境变量设置 ASP.NET Core 应用的监听地址
ENV ASPNETCORE_URLS http://0.0.0.0:3721

# 5.4. 执行 "dotnet" 命令启动 ASP.NET Core 应用
ENTRYPOINT ["dotnet", "helloworld.dll"]
```

这个 Dockerfile 文件采用一个中间层（build）暂存 ASP.NET Core MVC 应用发布后的资源，其工作目录为 "/app"。具体来说，这个层将 "microsoft/aspnetcore-build:2" 作为基础镜像，我们先将定义项目的 .csproj 文件（helloworld.csproj）复制到当前工作目录，然后运行 "dotnet restore" 命令恢复所有注册在这个项目文件中的 NuGet 包。最后将当前项目的所有文件复制到当前工作目录，并执行 "dotnet publish" 命令对整个项目进行编译发布（针对 Release 模式），发布后的资源被保存到目录 "/app/out" 中。

真正编译生成 Docker 镜像时，通常采用 "mcr.microsoft.com/dotnet/core/aspnet:3.0" 作为基础镜像，由于应用在上面进行了预先发布，所以只需要将发布后的所有文件复制到当前工作目录即可。接下来我们通过环境变量设置了 ASP.NET Core 应用的监听地址 "http://0.0.0.0:3721"。针对 ENTRYPOINT 的定义（ENTRYPOINT ["dotnet", "helloworld.dll"]），当容器被启动的时候，"dotnet helloworld.dll" 命令会被执行以启动 ASP.NET Core 应用。

定义 Dockerfile 文件之后，可以打开 CMD 命令行并切换到项目所在根目录（也就是 Dockerfile 文件所在的目录），然后执行 "docker build -t helloworldapp ." 命令，该命令会利用这个 Dockerfile 文件生成一个命名为 helloworldapp 的 Docker 镜像。图 1-24 展示了执行 "docker build" 命令过程中的输出结果。

成功创建 Docker 镜像之后，只需要针对这个镜像创建对应的容器，最终的 ASP.NET Core 应用的启动就可以直接通过启动该容器来完成。如图 1-25 所示，我们执行 "docker run -d -p 8080:3721 --name myapp helloworldapp" 命令针对前面生成的 Docker 镜像（helloworldapp）创建并启动了一个命名为 myapp（--name myapp）的容器。由于我们从外面访问这个应用，所以通过端口映射（-p 8080:3721）可以将内部监听端口 3721 映射为当前宿主机器的端口 8080，然后利用地址 "http://localhost:8080" 访问通过 Docker 容器承载的 ASP.NET Core 应用。

图 1-24　执行"docker build"命令生成 Docker 镜像（一）

图 1-25　执行"docker build"命令生成 Docker 镜像（二）

第 2 章

跨平台的奥秘

跨平台是 .NET Core 区别于传统的 .NET Framework 最显著的特征。第 1 章不仅介绍了在 Windows、macOS 和 Linux 上开发与运行 .NET Core 程序，还介绍了如何让创建的 .NET Core 应用在 Docker 容器中运行。本章将继续研究跨平台，并揭开 .NET Core 跨平台的奥秘。

2.1 历史的枷锁

对于计算机从业人员来说，"平台"（Platform）是一个司空见惯的词语。在不同的语境中，平台具有不同的语义，它可以指代操作系统环境和 CPU 指令类型，也可以表示硬件设备类型。目前，微软已经为 Windows 平台构建了一个完整的且支持多种设备的 .NET 生态系统。与此同时，通过借助 Mono 和 Xamarin，.NET 已经可以被成功移植到包括 macOS、Linux、iOS、Android 和 FreeBSD 等非 Windows 平台。

2.1.1 Windows 下的 .NET

微软于 2002 年推出的第一个版本的 .NET Framework 是一个主要面向 Windows 桌面（Windows Forms）和服务器（ASP.NET Web Forms）的基础框架。在此之后，个人计算机的地位不断受到其他设备的挑战，为此微软根据设备自身的需求对 .NET Framework 做了相应的简化和修改，不断推出针对不同设备类型的 .NET Framework，如 Windows Phone、Windows Store、Silverlight 和 .NET Micro Framework 等，它们分别对移动设备、平板设备和嵌入式设备提供支持。由于这些不同的 .NET Framework 分支是完全独立的，所以开发一款同时支持多种设备的"可移植"（Portable）应用的难度很大。

.NET Framework 的层次结构

由于不同设备 .NET Framework 的独立性，所以跨设备平台的代码重用显得异常困难。为了使读者深刻理解这个问题，本节从 .NET Framework 的层次结构开始介绍。.NET Framework 由图 2-1 所示的两个层次构成，它们分别是提供运行环境的 CLR（Common Language Runtime）和提供 API 的 FCL（Framework Class Library）。

```
┌─────────────────────────────────────────────┐
│              Framework Library              │
│  ┌──────────────┬──────────────┬──────────┐ │
│  │Windows Forms │   ASP.NET    │   WPF    │ │
│  ├──────────────┼──────────────┼──────────┤ │
│  │     WCF      │      WF      │Data Service│
│  ├──────────────┼──────────────┼──────────┤ │
│  │   ADO.NET    │Entity Framework│Linq to SQL│
│  ├──────────────┼──────────────┼──────────┤ │
│  │     ...      │     ...      │   ...    │ │
│  ├──────────────┴──────────────┴──────────┤ │
│  │         Basic Class Library (BCL)      │ │
│  └────────────────────────────────────────┘ │
└─────────────────────────────────────────────┘
┌─────────────────────────────────────────────┐
│         Common Language Runtime (CLR)       │
└─────────────────────────────────────────────┘
```

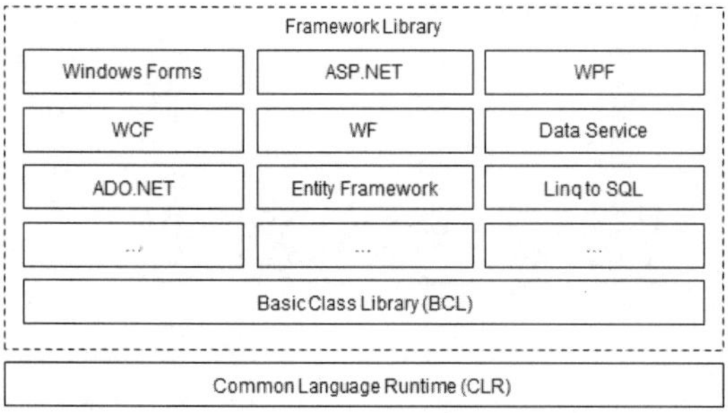

图 2-1　.NET Framework = Runtime + FCL

CLR 与 .NET 的关系等同于 JVM 与 Java 的关系，CLR 本质上就是 .NET 虚拟机。作为一个运行时（Runtime），CLR 为程序提供一个托管（Managed）的执行环境，它是 .NET Framework 的执行引擎，为托管程序的执行提供内存分配、垃圾回收、安全控制、异常处理和多线程管理等方面的服务。CLR 是 .NET Framework 的子集，但是两者具有不同的版本策略。截至 2019 年，微软仅发布了 4 个版本的 CLR，分别是 CLR 1.0、CLR 1.1、CLR 2.0 和 CLR 4.0，.NET Framework 1.0 和 .NET Framework 1.1 分别采用 CLR 1.0 与 CLR 1.1，CLR 2.0 被 .NET Framework 2.0 和 .NET Framework 3.x 共享，.NET Framework 4.x 下的运行时均为 CLR 4.0。

FCL 是一个旨在为开发人员提供 API 的类库，由它提供的 API 又可以划分为两个层次（见图 2-1）。处于最底层的部分被称为 BCL（Basic Class Library），用于描述一些基本的数据类型和数据结构（如字符串、数字、日期/时间和集合等），同时提供一些基础性的操作（如 IO、诊断、反射、文本编码、安全控制、多线程管理等）。在 BCL 之上则是面向具体应用类型的 API，大体上可以将它们划分为以下 3 种类型。

- 面向应用（如 ASP.NET、WPF 和 Windows Forms 等）。
- 面向服务（如 WCF、WF 和 Data Services 等）。
- 面向数据（如 ADO .NET、Entity Framework 和 LINQ to SQL 等）。

也可以采用另一种方式对 FCL 进行重新划分，即：将面向某种应用或者服务类型（如 Windows Forms、WPF、ASP.NET 和 WCF 等）的部分称为 AppModel，那么整个 .NET Framework 就具有三层结构（见图 2-2）。

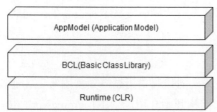

图 2-2　.NET Framework = Runtime + BCL+ AppModel

大而全的 BCL

微软的 .NET 战略是在 2000 年提出的，2002 年 .NET Framework 1.0 和 IDE（VS .NET 2002）随之问世。在之后的 10 多年中，.NET Framework 的一系列版本被先后推出。截至 2019 年，微软发布的最新 .NET Framework 版本为 4.7，图 2-3 展示了整个 .NET Framework 不断升级的演进过程，以及各个版本提供的主要特性。

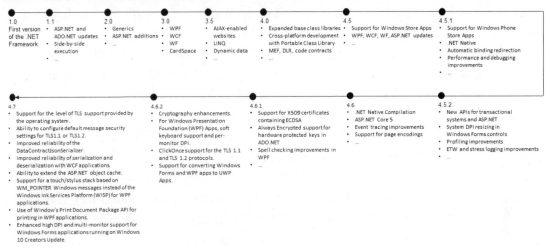

图 2-3　.NET Framework 版本升级（功能点）

图 2-3 展示了 .NET Framework 的发展历程，由此说明：作为整个 .NET 平台的基础框架，.NET Framework 在不断升级过程中虽然变得更加强大和完整，但是在另一方面也变得越来越臃肿。随着版本的不断升级，构成 .NET Framework 的 AppModel、BCL 和 CLR 都在不断膨胀（.NET Framework 2.0/3.x 和 .NET Framework 4.x 分别采用 CLR 2.0 和 CLR 4.0），图 2-4 很直观地说明了这个问题。

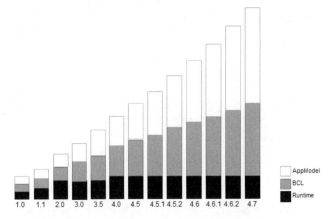

图 2-4　.NET Framework 版本升级（自身尺寸）

程序集是 .NET 最基本的部署单元，无论定义中的多少类型被使用，CLR 总是将整个程序集加载到内存中。对于上面介绍的构成 .NET Framework 的三个层次来说，AppModel 是针对具体应用/服务类型的，相应的 API 通过独立的程序集来承载（如 ASP.NET 的核心框架定义在程序集 System.Web.dll 中，承载整个 Windows Forms 框架的程序集则是 System.Windows.Forms.dll），所以 .NET Framework 的各个应用模型是相互独立的。在开发某种类型的应用时，引用应用模型对应的程序集即可。例如，如果开发的是一个 Windows Forms 应用，是不需要引用 System.Web.dll 程序集的。

将针对某种 AppModel 的 API 定义在单一程序中并不会带来很严重的问题，但是针对单程序集的 BCL 承载方式就会不一样。因为 BCL 绝大部分的核心代码都定义在 mscorlib.dll 程序集中，所以 BCL 基本上是作为一个不可分割的整体存在于 .NET Framework 之中的。由于 .NET Framework 需要对运行在本机各种类型的托管程序提供支持，所以针对所有应用类型的基础类型均需要定义在 BCL 中。在很多情况下，我们的应用可能仅仅需要使用 BCL 一个很小的子集，但是依然需要将定义整个程序集都加载到内存之中。

一方面，BCL 总是作为一个不可分割的整体被加载；另一方面，其自身的尺寸也在随着 .NET Framework 的升级而不断膨胀。对于客户端应用（如 Windows Forms/WPF 应用）来说，这个问题可以忽略，但是对于运行在移动设备和服务器上的应用（包括部署于云端应用）来说，由此带来的对性能和吞吐量的响应就成了一个不得不考虑的问题。

理想的 BCL 消费方式是"按需消费"，我们需要哪个部分就加载哪个部分。由于作为独立部署单元的程序集总是作为一个整体被 CLR 加载到内存中，所以要完全实现这种理想的 BCL 消费方式，唯一的办法就是按照图 2-5 所示的方式将其划分为若干小的单元，并分别定义到独立的程序集中。除在运行的时候减少内存占用外，按照模块化的原则对整个 BCL 进行拆分也使维护和版本升级变得更加容易，如果现有版本具有需要修复的漏洞（Bug），或者性能需要改进，那么只需要改动并升级相应的模块即可。

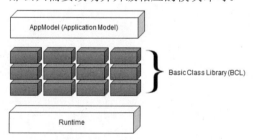

图 2-5　模块化的 BCL

多个设备平台独自为政

目前，微软已经构建了一个完整且支持多种设备的 .NET 生态系统，从最初单纯的桌面和服务器平台，逐渐扩展到移动、平板和游戏等平台。设备运行环境的差异性导致了针对它们的应用不能构建在一个统一的 .NET Framework 平台上，所以微软采用独立的 .NET Framework 平台对它们提供针对性的支持。就目前来说，除了支持 Windows 桌面和服务器设备的"全尺寸"的 .NET Framework，微软还推出了一系列压缩版 .NET Framework，包括 Windows Phone、Windows Store、Silverlight 和 .NET Micro Framework 等，它们分别对移动设备、平板设备和嵌入式设备提供支持。

这些 .NET Framework 并不是仅仅在 AppModel 层提供针对相应设备平台的开发框架，它们提供的 BCL 和 Runtime 也是不同的。换句话说，这些 .NET Framework 平台是完全独立的（见图 2-6）。由于目标平台具有独立性，所以很难编写能够在各个平台复用的代码，关于这一点笔者会在 2.2 节重点讨论。

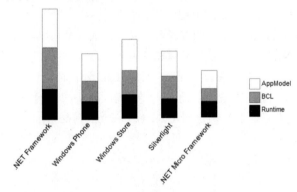

图 2-6   .NET Framework 平台之间的独立性

2.1.2 非 Windows 下的 .NET

尽管微软多年来基本上仅研发 Windows 平台下的 .NET，但是 .NET 通过 Mono 和 Xamarin 已经延伸到其他平台（macOS、Linux、iOS 和 Android 等）。虽然目前做得并不算完美，但是可以认为 .NET 具备跨平台的能力。

从 CLI 谈起

.NET 跨平台的能力建立在一种开放的标准或者规范之上，这个所谓的标准或者规范就是 CLI。制定 CLI 旨在解决这样一个问题：由不同（高级）编程语言开发的 .NET 应用能够在无须任何更改的情况下运行于不同的系统环境中。要实现这个目标，必须有效解决这里涉及的两种类型的差异，即编程语言的差异和运行时环境的差异。只有编程语言之间能够实现相互兼容，运行时环境能够得到统一，跨平台方可实现。

CLI 的英文全称为 Common Language Infrastructure，其中 Common Language 说的是语言，具体描述的是一种通用语言，旨在解决各种高级开发语言的兼容性问题。Infrastructure 指的则是运行时环境，旨在弥补不同平台之间执行方式的差异。Common Language 是对承载应用的二进制内容的静态描述，Infrastructure 则表示动态执行应用的引擎，所以 CLI 旨在为可执行代码本身和执行它的引擎确立一个统一的标准。

编程语言可分为编译型和解释型两类。前者需要通过编译器实施编译以生成可执行代码，

CLI 涉及的 Common Language 指的是编译型语言。要实现真正的跨平台，最终需要解决的是可执行代码在不同平台之间的兼容和可移植的问题，而编程语言的选择仅仅决定了应用程序源文件的原始状态，应用的兼容性和可移植性由编译后的结果决定。如果通过不同编程语言开发的应用通过相应的编译器编译后能够生成相同的目标代码，那么编程语言之间的差异就不再是一个问题。计算机领域有这样一种思维——任何一个软件设计方面的难题都可以通过增加一个抽象层来解决，CLI 就贯彻了这一方针。

按照 CLI 的规定，用来描述可执行代码的是一种被称为 CIL（Common Intermediate Language）的语言，这是一种介于高级语言和机器语言之间的中间语言。如图 2-7 所示，虽然程序源文件由不同的编程语言编写，但是我们可以借助相应的编译器将其编译成 CIL 代码。从原则上讲，设计新的编程语言并将其加入 .NET 中，只需要配以相应的编译器来生成统一的 CIL 代码即可。我们也可以为现有的某种编程语言（如 Java）设计一种以 CIL 为目标语言的编译器，使之成为 .NET 语言。CIL 是一门中间语言，也是一门面向对象的语言，所以对于一个 CIL 程序来说，类型是基本的组成单元和核心要素。微软制定的 CTS（Common Type System）为 CLI 确立了一个统一的类型系统。

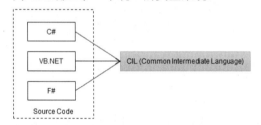

图 2-7　将不同语言编写的源代码编译成统一的中间代码

编程语言的差异通过编译器得以"同一化"，运行环境的差异则可以通过虚拟机（Virtual Machine，VM）技术来解决。虚拟机是 CIL 的执行容器，能够在执行 CIL 代码的过程中采用即时编译的方式将其动态地翻译成与当前执行环境完全匹配的机器指令。如图 2-8 所示，虚拟机屏蔽了不同操作系统之间的差异，使目标程序可以不做任何修改就能运行于不同的底层执行环境中，CIL 实际上是一种面向虚拟机的语言。

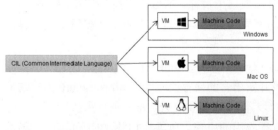

图 2-8　虚拟机采用即时编译的方式将 CIL 翻译成机器代码

从实现原理来看，让 .NET 能够跨平台其实不难，但是让各种相关的人员参与进来以构建一个健康而完善的跨平台 .NET 生态圈不是一件一蹴而就的事情，这里涉及的利益相关方包括编程语言的设计者，设计和开发编译器、虚拟机、IDE 以及其他相关工具的人员还包括广大的应用开发者。跨平台 .NET 生态环境必须建立在一个标准的规范之上，所以微软为此制定了 CLI，然后提交给欧洲计算机制造商协会（European Computer Manufacturers Association，ECMA），成为一个编号为 335 的规范，所以 CLI 又被称为 ECMA-335。另外，ECMA 还接受了微软为 C#这门编程语言制定的规范，即 ECMA-334。

Mono 与 Xamarin

CLI（ECMA-335）在 .NET 诞生的那一刻就被赋予了跨平台的"基因"，但微软似乎根本不曾想过将 .NET 推广到其他非 Windows 平台，真正完成这一使命的是一个叫作 Mono 的项目。虽然 Mono 不是一个新的项目，但是依然有很多人对其不甚了解。下面先对 Mono 进行简单介绍。

1999 年，Miguel de Icaza 创建了 Ximian 公司，该公司旨在为 GNOME 项目（这是一个为类 UNIX 系统提供桌面环境的 GNU 项目，GNOME 项目是目前 Linux 最常用的桌面环境之一）开发软件和提供支持。2000 年 6 月，微软正式发布 .NET Framework，Miguel de Icaza 被"基于互联网的全新开发平台"（.NET 在发布之初被认为是"a new platform based on Internet standards"）深深吸引。

2000 年 11 月，微软发布了 CLI 规范（ECMA-335），并为公众开发了独立实现的许可，Miguel de Icaza 认为 CLI 实际上为 .NET 走向非 Windows 平台提供了可能。Miguel de Icaza 于 2001 年 7 月开启了 Mono 项目，并将 C#作为主要的开发语言（目前支持 VB .NET），所以针对 CLI 规范和 C#语言的两个 ECMA 规范是构建 Mono 项目的理论基础，如果访问 Mono 的官方网站，可以了解 Mono 的定义："Mono is an open source implementation of Microsoft's .NET Framework based on the ECMA standards for C# and the Common Language Runtime."

Mono 的使命不仅仅局限于将 .NET 应用正常运行在其他非 Windows 平台，它还可以帮助开发人员直接在其他平台进行 .NET 应用的开发，所以 Mono 不仅根据 CLI 规范为相应的平台开发了作为虚拟机的 CLR 和编译器，还提供了 IDE 和相应的开发工具（被称为 MonoDevelop）。Mono 的第一个正式版本（Mono 1.0）发布于 2004 年 6 月。

2003 年 8 月，Ximian 公司被 Novell 公司收购。Novell 公司继续支持 Miguel de Icaza 开发 Mono 项目，在这期间 Mono 陆续推出了若干 Mono 2.x 版本。2011 年 4 月，Novell 公司被 Attachmate 公司收购，后者决定放弃 Mono 项目，于是 Miguel de Icaza 带着整个 Mono 团队成立了 Xamarin 公司。2011 年 7 月，Xamarin 公司从原来的母公司 Novell 得到了 Mono 的开发许可。之后，Xamarin 公司先后发布了 Mono 3.x、Mono 4.0 和 Mono 5.x。Mono 项目目前的目标是实现 .NET 4.5 除 WPF、WF 和部分 WCF 外的所有特性，目前缺失的部分的开发正在通过 Olive 项目（Mono 的一个子项目）进行着。

在 Mono 项目的基础之上，Xamarin 公司开始开发以新公司命名的产品，其中最重要的版本是 2013 年 2 月发布的 Xamarin 2.0。Xamarin 2.0 由 Xamarin.Android、Xamarin.iOS 和 Xamarin.Windows 组成，可以采用 C#语言开发针对 Android、iOS 和 Windows 平台的 Native 应用。除此之外，Xamarin 2.0 还携带着一个被称为 Xamarin Studio（MonoDevelop 的升级版）的 IDE 以及与一些与

Visual Studio 集成的工具。2014 年 5 月 Xamarin 3.0 发布，作为其核心的 Xamarin.Forms 为不同平台的 Native 应用提供统一的控件，也就是说，利用 Xamarin.Forms API 开发 Native 应用可以在无须做任何改变的情况下运行在 Android、iOS 和 Windows 平台上。

2016 年 2 月，微软和 Xamarin 公司签署协议并达成了前者针对后者的收购。在 2016 年 Build 大会上，微软宣布将整个 Xamarin SDK 开源，并将它作为一个免费的工具集成到 Visual Studio 中，Visual Studio 企业版的用户还可以免费使用 Xamarin 企业版的所有特性。

综上所述，由于 .NET 是建立在 CLI 规范之上的，所以它具有跨平台的"基因"。在微软发布了第一个针对桌面和服务器平台的 .NET Framework 之后，它开始对完整版的 .NET Framework 进行不同范围和层次的"阉割"，进而出现了 Windows Phone、Windows Store、Silverlight 和 .NET Micro Framework 的压缩版的 .NET Framework。从这个意义上讲，Mono 和它们并没有本质上的区别，唯一不同的是，Mono 真正突破了 Windows 平台的屏障。

包括 Mono 在内的这些分支促成了 .NET 的繁荣，但这仅仅是一种表象而已。虽然都是 .NET Framework 的子集，但是由于它们采用完全独立的运行时和基础类库，所以很难开发一个支持多种设备的"可移植"（Portable）应用，这些分支反而成为制约 .NET 发展的一道道枷锁。

2.2 复用之伤

针对不同平台的 .NET Framework 相互独立的特性，给应用的开发者带来的最大的问题就是代码难以复用。比较极端的场景就是，如果为一个现有的桌面应用提供针对移动设备的支持，我们需要重新开发一个全新的应用，现有的代码难以被新的应用重用。"代码复用"是软件设计最根本的目标，但是平台之间的差异导致跨平台代码重用确实困难重重。虽然做得不算理想，但微软在这方面确实做出了很多尝试，在某些方面甚至具有创新之处，下面介绍跨平台代码重用的一些实现方案。

2.2.1 源代码复用

对于包括 Mono 在内的各个 .NET Framework 平台的 BCL 来说，虽然在 API 定义层面存在一些共同之处，但是由于它们定义在不同的程序集之中，所以在 PCL（Portal Class Library）推出之前，针对程序集的共享是不可能实现的，只能在源代码层面实现共享。源代码的共享通过在不同项目之间共享源文件的方式实现，下面介绍 3 种不同的方式。

源文件共享

如果针对不同 .NET Framework 平台的多个项目文件存在于同一个物理目录下，那么该相同目录下的源文件可以同时包含在这些项目中以达到共享的目的。如图 2-9 所示，两个分别针对 Silverlight 和 WPF 的项目共享相同的目录，与两个项目文件在同一个目录下的 C#文件 Shared.cs 可以同时被包含在这两个项目之中。

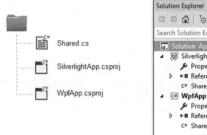

图 2-9　源文件共享

文件链接

如果采用默认的方式将一个现有的文件添加到当前项目之中，Visual Studio 会将目标文件复制到项目本地的目录下，所以无法达到共享的目的。但是针对现有文件的添加，它支持一种叫作"链接"的方式，使添加到项目中的文件指向的依然是原来的路径，我们可以采用这种方式为多个项目添加针对同一个文件的链接，以实现源文件跨项目共享。如果针对的是 Silverlight 和 WPF 这两个项目，不论项目文件和需要被共享的文件存在于哪个目录下面，我们都可以采用添加文件链接的方式分享 Shared.cs 文件（见图 2-10）。

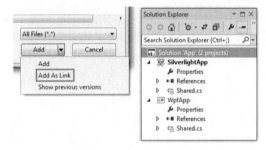

图 2-10　文件链接

Shared Project

项目一般都是为了组织源文件和其他相关资源并最终将它们编译成一个可被部署的程序集。但 Shared Project 这种项目类型比较特别，它只具有对源文件进行组织的功能，无法通过编译生成程序集，它的存在就是为了实现源文件的共享。对于上面介绍的两种源代码的共享方式来说，它们都是针对某个单一文件的共享，而 Shared Project 则可以对多个源文件进行打包以实现批量共享。

如图 2-11 所示，我们可以创建一个 Shared Project 类型的项目 Shared.shproj，并将需要共享的 3 个 C#文件（Foo.cs、Bar.cs 和 Baz.cs）添加进来。将针对这个项目的引用同时添加到 Silverlight 项目（SilverlightApp.csproj）和 Windows Phone 项目（WinPhoneApp.csproj）之中，如果需要对这两个项目实施编译，包含在项目 Shared.shproj 中的 3 个 C#文件就会自动作为当前项目的源文件参与编译。

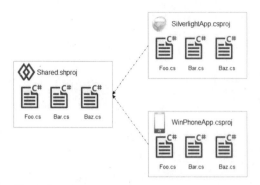

图 2-11　Shared Project

2.2.2　程序集复用

采用 C#、VB .NET 这样的编程语言编写的源文件经过编译会生成有 IL 代码和元数据构成的托管模块，一个或者多个托管模块合并生成一个程序集。程序集的文件名、版本、语言文化和签名的公钥令牌共同组成了它的唯一标识，我们将该标识称为程序集有效名称（Assembly Qualified Name）。除了包含必要的托管模块，我们还可以将其他文件作为资源内嵌到程序集中，程序集的文件构成由一个清单（Manifest）文件进行描述，这个清单文件包含在某个托管模块之中。

除描述程序集文件构成外，这个清单文件也包含描述程序集的元数据。元数据使程序集成为一个自描述性（Self-Describing）的部署单元，除了描述定义在本程序集中的所有类型，这些元数据还包括对引用的外部程序集的所有类型的描述。包含在元数据中针对外部程序集的描述，是由编译时引用的程序集决定的，引用程序集的名称（包含文件名、版本和签名的公钥令牌）会直接体现在当前程序集的元数据中。针对程序集引用的元数据采用 ".assembly" 形式被记录在清单文件中，被记录下来的不仅包含被引用的程序集文件名（Foo 和 Bar），还包括程序集的版本，对于签名的程序集（Foo）来说，公钥令牌也一并包含其中。

```
.assembly extern Foo
{
  .publickeytoken = (B7 7A 5C 56 19 34 E0 89 )
  .ver 1:0:0:0
}
.assembly extern Bar
{
  .ver 1:0:0:0
}
```

包含在当前程序集清单文件中针对引用程序集的元数据是 CLR 加载目标程序集的依据。在默认情况下，CLR 要求加载与程序集引用元数据完全一致的程序集。换句话说，如果引用的是一个未被签名的程序集（Bar），那么只要求被加载的程序集具有一致的文件名和版本就可以。如果引用的是一个经过签名的程序集，那么必须要求被加载的程序集具有一致的公钥令牌。

前面介绍了关于 .NET 多目标框架的问题。虽然不同目标框架的 BCL 在 API 层面具有很多交集，但是这些 API 实际上被定义在不同的程序集中，所以在不同的目标框架下共享同一个程序集几乎成了不可能的事情。如果要使跨目标平台程序集复用成为现实，就必须要求 CLR 在加载程序集时放宽"完全匹配"的限制，因为当前程序集清单文件中描述的某个引用程序集，在不同的目标框架下可能指向不同的程序集。实际上，确实存在一些机制或者策略可以使 CLR 加载一个与引用元数据的描述不一致的程序集。

程序集一致性

.NET Framework 是向后兼容的，也就是说，原来针对低版本 .NET Framework 编译生成的程序集是可以直接在高版本 CLR 下运行的。对于 .NET Framework 2.0 编译生成的程序集来说，所有引用的基础程序集的版本都应该是 2.0，如果这个程序集在 .NET Framework 4.0 环境下执行，CLR 在决定加载它所依赖的程序集时，应该选择 .NET Framework 2.0 还是 .NET Framework 4.0？

如果利用 Visual Studio 创建一个针对 .NET Framework 2.0 的控制台应用（将程序命名为 App），并在作为程序入口的 Main 方法上编写如下一段代码。如下面的代码片段所示，控制台上输出了 3 个基本类型（Int32、XmlDocument 和 DataSet）所在程序集的全名。

```
class Program
{
    static void Main()
    {
        Console.WriteLine(typeof(int).Assembly.FullName);
        Console.WriteLine(typeof(XmlDocument).Assembly.FullName);
        Console.WriteLine(typeof(DataSet).Assembly.FullName);
    }
}
```

直接将这段程序在默认版本的 CLR（2.0）下运行，在控制台上输出的结果如图 2-12 所示，由此可以发现上述 3 个基本类型所在程序集的版本都是 2.0.0.0。也就是说，在这种情况下，运行时加载的程序集和编译时引用的程序集是一致的。

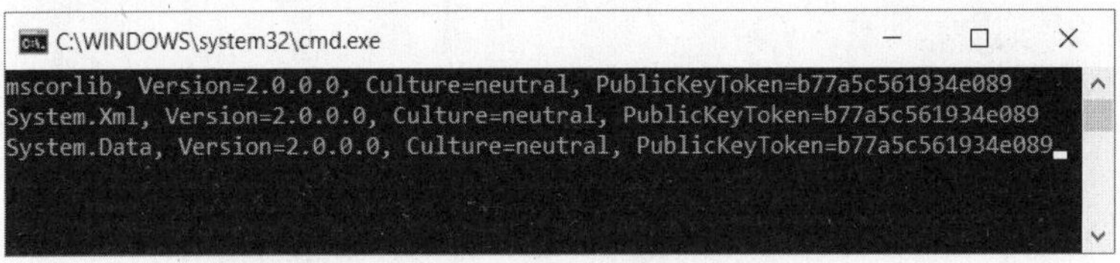

图 2-12 加载程序集与编译程序集保持一致

如果在目录 "\bin\debug" 下可以直接找到以 Debug 模式编译生成的程序集 App.exe，并按照如下形式修改对应的配置文件（App.exe.config），该配置的目的在于将启动应用时采用的运行时（CLR）版本从默认的 2.0 切换到 4.0。

```
<configuration>
  <startup>
    <supportedRuntime version="v4.0"/>
  </startup>
</configuration>
```

或者：
```
<configuration>
  <startup>
    <requiredRuntime version="v4.0"/>
  </startup>
</configuration>
```

无须重新编译（确保运行的依然是同一个程序集）直接运行 App.exe，在控制台上得到的输出结果如图 2-13 所示，可以看到 3 个程序集的版本全部变成 4.0.0.0，也就是说，真正被 CLR 加载的这些基础程序集与当前 CLR 的版本是匹配的。

图 2-13　加载程序集与运行时版本保持一致

这个简单的实例说明：运行过程中加载的 .NET Framework 程序集（承载 FCL 的程序集）是由当前运行时（CLR）决定的，这些程序集的版本总是与 CLR 的版本相匹配。包含在元数据中的程序集信息提供目标程序集的名称，而版本则由当前运行的 CLR 决定，我们将这个重要的机制称为程序集一致性（Assembly Unification），图 2-14 揭示了程序集一致性。

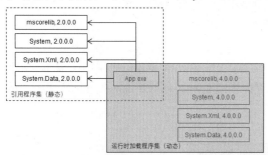

图 2-14　程序集一致性

Retargetable 程序集

在默认情况下，如果某个程序集引用了另一个具有强签名的程序集，那么 CLR 在执行的时候总是会根据程序集文件名、版本和公钥令牌定位目标程序集。如果无法找到一个与之完全匹配的程序集，通常会抛出一个 FileNotFoundException 类型的异常。如果当前引用的是一个 Retargetable 程序集，则意味着 CLR 在定位目标程序集时可以"放宽"匹配的要求，即只要求目标程序集具有相同的文件名即可。

如图 2-15 所示，应用程序（App）引用了具有强签名的程序集 "Foobar, Version=1.0.0.0, Culture=neutral, PublicKeyToken=b03f5f7f11d50a3a"，所以对于编译后生成的程序集 App.exe 来说，对应的程序集引用将包含目标程序集的文件名、版本和公钥令牌。如果在运行的时候只提供了一个有效名称为"Foobar, Version=2.0.0.0, Culture=neutral, PublicKeyToken=d7fg7asdf7asd7aer"的程序集，除了文件名，后者的版本号和公钥令牌都与程序集引用元数据的描述不一样。在默认情况下，系统此时会抛出一个 FileNotFoundException 类型的异常。倘若 Foobar 是一个 Retargetable 程序集，它将作为目标程序集被正常加载并使用。除了定义程序集的元数据多了如下一个 retargetable 标记，Retargetable 程序集与普通程序集并没有本质区别。

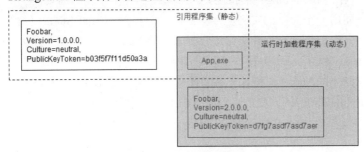

图 2-15 针对 Retargetable 程序集的引用

普通程序集：
`.assembly Lib`

Retargetable 程序集：
`.assembly retargetable Lib`

retargetable 标记可以通过在程序集上标注 AssemblyFlagsAttribute 特性设定，但这样的重定向仅仅对 .NET Framework 自身提供的基础程序集有效。虽然可以通过使用 AssemblyFlagsAttribute 特性为自定义的程序集添加一个 retargetable 标记，但是 CLR 并不会赋予它重定向的能力。

`[assembly:AssemblyFlags(AssemblyNameFlags.Retargetable)]`

如果某个程序集引用了一个 Retargetable 程序集，自身清单文件针对该程序集的引用元数据同样具有如下所示的 retargetable 标记。CLR 正是利用这个标记确定它引用的是否是一个 Retargetable 程序集，进而确定针对该程序集的加载策略，即采用针对文件名、版本和公钥令牌的完全匹配策略，还是采用只针对文件名的降级匹配策略。

针对普通程序集的引用：
```
.assembly extern Foobar
{
  .publickeytoken = (B7 7A 5C 56 19 34 E0 89 )
  .ver 1:0:0:0
}
```

针对 Retargetable 程序集的引用：
```
.assembly extern retargetable Foobar
{
  .publickeytoken = (B7 7A 5C 56 19 34 E0 89)
  .ver 1:0:0:0
}
```

类型转移

在进行框架或者产品升级过程中，经常会遇到针对程序集合并和拆分的场景。例如，在新版本添加额外的 API 之后需要对现有的 API 重新进行规划，可能会将定义在程序集 A 中的类型转移到程序集 B 中。但即使发生了这样的情况，我们依然需要为新框架或者产品提供向后兼容的能力，这就需要使用类型转移（Type Forwarding）的特性。

为了使读者了解类型转移，下面用一个简单的实例进行演示。我们利用 Visual Studio 创建一个针对 .NET Framework 3.5 的控制台应用 App，并在作为程序入口的 Main 方法中编写了两行代码，从而将两个常用的类型（String 和 Func<>）所在的程序集名打印出来。程序编译之后会在"\bin\Debug"目录下生成可执行文件 App.exe 和对应的配置文件 App.exe.config。从下面的配置文件的内容可以看出，.NET Framework 3.5 采用的运行时（CLR）版本为 v2.0.50727。

```
class Program
{
    static void Main()
    {
        Console.WriteLine(typeof(string).Assembly.FullName);
        Console.WriteLine(typeof(Func<>).Assembly.FullName);
    }
}
```

App.exe.config：

```
<configuration>
  <startup>
    <supportedRuntime version="v2.0.50727"/></startup>
  </startup>
</configuration>
```

如果直接以命令行的形式执行编译生成的 App.exe，就会呈现图 2-16 所示的输出结果。可以看出，两个基础类型（String 和 Func<>）中只有 String 类型被定义在 mscorlib.dll 程序集之中，而 Func<> 类型其实被定义在 System.Core.dll 程序集之中。其实，.NET Framework 2.0、.NET Framework 3.0 和 .NET Framework 3.5 不仅仅共享相同的运行时（CLR 2.0），对于提供基础类型的核心程序集 mscorlib.dll 也是共享的。也就是说，.NET Framework 2.0 发布时提供的程序集 mscorlib.dll 在 .NET Framework 3.x 时代没有升级。Func<>类型是在 .NET Framework 3.5 发布时提供的一个基础类型，所以不得不将其定义在一个另一个程序集之中，微软将这个程序集命名为 System.Core.dll。

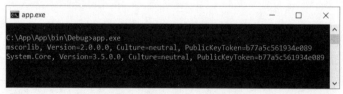

图 2-16 Func<>类型所在的程序集（.NET Framework 3.5）

下面验证在 .NET Framework 4.0（CLR 4.0）环境下运行同一个应用程序（App.exe）是否会

输出不同的结果。为此，在不对项目做重新编译的情况下直接修改配置文件 App.exe.config，并将运行时版本设置为 v4.0。

```
<configuration>
  <startup>
    <supportedRuntime version="v4.0"/>
  </startup>
</configuration>
```

图 2-17 所示是同一个 App.exe 在 .NET Framework 4.0 环境下的输出结果，可以看出，提供的两个基础类型所在的程序集都是 mscorlib.dll。也就是当 .NET Framework 升级到 4.0 之后，不仅运行时升级到了全新的 CLR 4.0，微软也对承载基础类型的 mscorlib.dll 程序集进行了重新规划，所以定义在 System.Core.dll 程序集中的基础类型也基本上重新回到了 mscorlib.dll 程序集中。

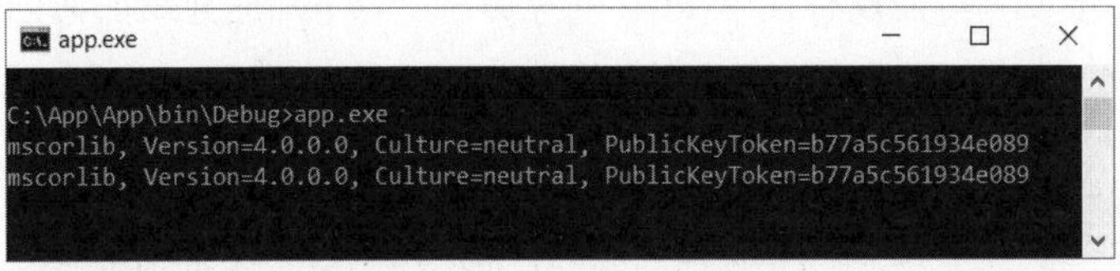

图 2-17 Func<>类型所在的程序集（.NET Framework 4.0）

下面继续分析上面演示的程序。由于 App.exe 最初是针对目标框架 .NET Framework 3.5 编译生成的，所以它的清单文件将包含针对 mscorlib.dll（2.0.0.0）和 System.Core.dll（3.5.0.0）的程序集引用。下面是针对这两个程序集引用的元数据的定义。

```
.assembly extern mscorlib
{
  .publickeytoken = (B7 7A 5C 56 19 34 E0 89 )
  .ver 2:0:0:0
}
.assembly extern System.Core
{
  .publickeytoken = (B7 7A 5C 56 19 34 E0 89 )
  .ver 3:5:0:0
}
```

当 App.exe 在 .NET Framework 4.0 环境中运行时，由于它的元数据提供的是针对 System.Core.dll 程序集的引用，所以 CLR 总是试图加载该程序集并从中定位目标类型（如演示实例中的 Func<>类型）。如果当前运行环境无法提供这个程序集，那么一个 FileNotFoundException 类型的异常就会被抛出。换句话说，虽然 Func<>类型在 .NET Framework 4.0 环境中已经转移到了新的程序集 mscorlib.dll 中，但当前环境依然需要提供一个文件名为 System.Core.dll 的程序集。

System.Core.dll 程序集存在的目的是告诉 CLR 它需要加载的类型已经发生转移，并将该类型所在的新的程序集名称告诉它，那么 .NET Framework 4.0 环境中的 System.Core.dll 程序集是如何描述 Func<>类型已经转移到 mscorlib.dll 程序集之中的？如果分析 System.Core.dll 程序集中的元数据，可以看到如下一段与此相关的代码。在程序集的清单文件中，每个被转移的类型都对应一个 .class extern forwarder 指令。

```
.class extern forwarder System.Func`1
{
  .assembly extern mscorlib
}
```

不同于上面介绍的 Retargetable 程序集，类型的转移并不是只针对 .NET Framework 提供的基础程序集，如果项目需要提供类似的向后兼容行，也可以使用这个特性。针对类型转移的编程只涉及一个类型为 TypeForwardedToAttribute 的特性。

可移植类库

在 .NET Framework 时代，创建 PCL 是实现跨多个目标框架程序集共享的唯一途径。上面介绍的内容都是在为 PCL 做铺垫，只有充分理解 Retargetable 程序集和类型转移，才可能对 PCL 的实现原理具有深刻的认识。由于读者可能没有使用 PCL 的经历，所以下面先介绍如何创建一个 PCL 项目。

如果采用 Visual Studio（2015）的 Class Library（Portal）项目模板创建一个 PCL 项目，就需要在图 2-18 所示的对话框中选择支持的目标平台及其版本。Visual Studio 会为新建的项目添加一个名为 .NET 的引用，这个引用指向一个由选定 .NET Framework 平台决定的程序集列表。由于这些程序集提供的 API 能够兼容所有选择的平台，所以在此基础上编写的程序也具有平台兼容性。

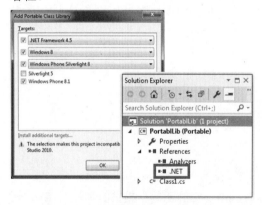

图 2-18　为 PCL 项目选择目标平台

如果查看这个特殊的 .NET 引用所在的地址，我们会发现它指向目录"%ProgramFiles%\ReferenceAssemblies\Microsoft\Framework\.NETPortable\{version}\Profile\ProfileX"。如果查看"%ProgramFiles%\ReferenceAssemblies\Microsoft\Framework\.NETPortable"目录，我们会发现它具有图 2-19 所示的结构。该目录下具有几个代表 .NET Framework 版本的子目录（v4.0、v4.5 和 v4.6）。具体到针对某个 .NET Framework 版本的目录（如 v4.6），其子目录"Profile"下具有一系列以"Profile"+"数字"（如 Profile31、Profile32 和 Profile44 等）命名的子目录，而 PCL 项目引用的就是存储在这些目录下的程序集。

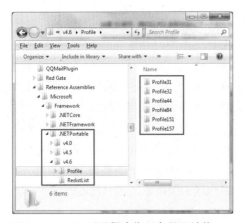

图 2-19　PCL 引用程序集所在目录结构

对于两个不同平台的 .NET Framework 来说，它们的 BCL 在 API 的定义上肯定存在交集。从理论上来说，建立在这个交集基础上的程序是可以被这两个平台共享的。如图 2-20 所示，如果我们编写的代码需要分别对 Windows Desktop/Phone、Windows Phone/Store 和 Windows Store/Desktop 平台提供支持，那么这样的代码依赖的部分仅限于两两的交集 A+B、A+C 和 A+D。如果要求这部分代码能够运行在 Windows Desktop/Phone/Store 这 3 个平台上，那么它们只能建立在三者之间的交集 A 上。

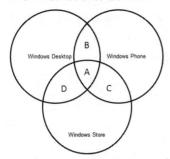

图 2-20　针对 .NET Framework 平台在 BCL 上的交集

对于所有可能的 .NET Framework 平台（包括版本）的组合，微软都会将它们在 BCL 上的交集提取出来并定义在相应的程序集中。例如，所有的 .NET Framework 平台都包含一个核心的程序集 mscorlib.dll，虽然定义的类型及其成员在各个 .NET Framework 平台不尽相同，但是它们之间肯定存在交集，微软针对不同的 .NET Framework 平台组合将这些交集提取出来并定义在一系列同名程序集中（同样命名为 mscorlib.dll）。微软按照这样的方式创建了其他针对不同 .NET Framework 平台组合的基础程序集，这些针对某个组合的所有程序集构成一系列的 Profile，并定义在上面提到过的目录下。值得注意的是，所有这些针对某个 Profile 的程序集均为 Retargetable 程序集。

创建一个 PCL 项目时，第一个必需的步骤是选择兼容的 .NET Framework 平台，Visual

Studio 会根据我们的选择确定一个具体的 Profile，并为创建的项目添加针对该 Profile 的程序集引用。由于所有引用的程序集是根据我们选择的 .NET Framework 平台"量身定制"的，所以定义在 PCL 项目中的代码才具有可移植的能力。

上面仅仅从开发的角度解释了定义在 PCL 项目中的代码本身能够确保与目标 .NET Framework 平台兼容的原因，但是从运行的角度来看，却存在如下两个问题。

- 元数据描述的引用程序集与真实加载的程序集不一致。例如，如果创建一个兼容 .NET Framework 4.5 和 Silverlight 5.0 的 PCL 项目，被引用的程序集 mscorlib.dll 的版本为 2.0.5.0，但是 Silverlight 5.0 运行时环境中的程序集 mscorlib.dll 的版本则为 5.0.5.0。
- 元数据描述的引用程序集的类型定义与运行时加载程序集类型定义不一致。例如，引用程序集中的某个类型被转移到另一个程序集中。

由于 PCL 项目在编译时引用的均为 Retargetable 程序集，所以程序集的重定向机制可以解决第一个问题。如果 CLR 在加载某个 Retargetable 程序集时找不到一个与引用程序集在文件名、版本、语言文化和公钥令牌完全匹配的程序集，通常只考虑文件名的一致性。第二个问题可以通过上面介绍的类型转移机制解决。

2.3 全新的布局

从本质上讲，按照 CLI 规范设计的 .NET 从其诞生的那一刻就具有跨平台的"基因"，它与 Java 没有本质区别。由于采用了统一的中间语言，所以只需要针对不同的平台设计不同的虚拟机（运行时）就能弥补不同操作系统与处理器架构之间的差异，但是理想与现实的差距非常大。在过去的十多年中，微软将 .NET 引入各个不同的应用领域，并且采用完全独立的多目标框架的设计思路，所以针对多目标框架的代码复用只能通过 PCL 这种"妥协"的方式来解决。如果依旧按照这条道路走下去， .NET 的触角延伸得越广，枷锁就越多，所以 .NET 已经到了不得不做出彻底改变的时刻。

2.3.1 跨平台的 .NET Core

综上所述，要真正实现 .NET 的跨平台，需要解决两个问题：一是针对不同的平台设计相应的运行时，从而为中间语言 CIL 提供一致性的执行环境；二是提供统一的 BCL，从而彻底解决代码复用的难题。对于真正跨平台的 .NET Core 来说，微软不仅为它设计了针对不同平台被称为 CoreCLR 的运行时，还重新设计了一套被称为 CoreFX 的 BCL。

如图 2-21 所示， .NET Core 目前支持的 AppModel 有 4 种（ASP.NET Core、Windows Forms、WPF 和 UWP），其中 ASP.NET Core 提供了全平台的支持，而 Windows Forms、WPF 和 UWP 只能在 Windows 上运行。CoreFX 是经过完全重写的 BCL，除自身具有跨平台执行的能力外，其提供的 API 也不再统一定义在少数几个单一的程序集中，而是经过有效划分之后被定义在各自独立的模块中。这些模块对应一个单一的程序集，并最终以 NuGet 包的形式进行分发。至于底层的虚拟机，微软则为主流的操作系统类型（Windows、macOS 和 Linux）和处理器架构（x86、 x64 和 ARM）设计了针对性的运行时，被称为 CoreCLR。

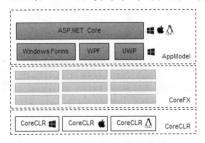

图 2-21　.NET Core 层次结构（一）

作为运行时的 CoreCLR 和提供 BCL 的 CoreFX 是 .NET Core 重要的基石，但是就开发成本来看，微软在后者投入的精力是前者无法比拟的。我们知道 .NET Core 自诞生到现在已经有很多年，目前的版本还只是到了 3.0，在发布进度上稍显缓慢，其中一个主要的原因是重写 CoreFX 提供的基础 API 确实是一件烦琐且耗时的工程，而且这项工程远未结束。为了对 CoreFX 提供的 BCL 有一个大致的了解，下面介绍常用的基础 API 在哪些命名空间定义。

- System.Collections：定义了常用的集合类型。
- System.Console：提供 API 完成基本的控制台操作。
- System.Data：提供用于访问数据库的 API，相当于原来的 ADO.NET。
- System.Diagnostics：提供基本的诊断、调试和追踪的 API。
- System.DirectoryServices：提供基于 AD（Active Directory）管理的 API。
- System.Drawing：提供 GDI 相关的 API。
- System.Globalization：提供 API 实现多语言及全球化支持。
- System.IO：提供与文件输入和输出相关的 API。
- System.NET：提供与网络通信相关的 API。
- System.Numerics：定义一些数值类型作为基元类型的补充，如 BigInteger、Complex 和 Plane 等。
- System.Reflection：提供 API 以实现与反射相关的操作。
- System.Runtime：提供与运行时相关的一些基础类型。
- System.Security：提供与数据签名和加密/解密相关的 API。
- System.Text：提供针对字符串/文本编码与解码相关的 API。
- System.Threading：提供用于管理线程的 API。
- System.Xml：提供 API 用于操作 XML 结构的数据。

对于传统的 .NET Framework 来说，承载 BCL 的 API 几乎都定义在 mscorlib.dll 程序集中，这些 API 并不是全部都转移到组成 CoreFX 的众多程序集中，那些与运行时（CoreCLR）具有紧密关系的底层 API 被定义到 System.Private.CoreLib.dll 程序集中，所以图 2-22 反映了真正的 .NET Core 层次结构。我们在编程过程中使用的基础数据类型基本上都定义在这个程序集中，目前这个程序集的尺寸已经超过了 10MB。由于该程序集提供的 API 与运行时关联较为紧密，较之 CoreFX 提供的 API，这些基础 API 具有较高的稳定性，所以它们是随着 CoreCLR 一起发布的。

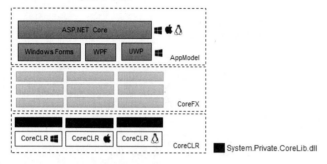

图 2-22　.NET Core 层次结构（二）

虽然我们在编程过程中使用的绝大部分基础类型都定义在 System.Private.CoreLib.dll 程序集中，但是这是一个"私有"的程序集，从其命名可以看出这一点。将 System.Private.CoreLib.dll 称为一个私有程序集，并不是因为定义其中的都是一些私有类型，而是因为在编程过程中不会真正引用这个程序集，这与 .NET Framework 下的 mscorlib.dll 程序集是一样的。不仅如此，当 .NET Core 代码被编译时，编译器也不会链接到这个私有程序集上，也就是说，编译后生成的程序集中也没有针对该程序集引用的元数据。但是当应用被真正执行的时候，所有引用的基础类型会全部自动"转移"到这个程序集中。而实现运行过程中的类型转移其实就是利用了 Type Forwarding 技术。

实例演示：针对 System.Private.CoreLib.dll 程序集的类型转移

对于上面介绍的针对 System.Private.CoreLib.dll 程序集的类型转移，很多读者可能还是难以理解，为了彻底认识这个问题，下面做一个简单的实例演示。首先利用 Visual Studio 创建一个 .NET Core 控制台应用，然后在作为程序入口的 Main 方法中编写如下几行代码，从而将常用的几个数据类型（System.String、System.Int32 和 System.Boolean）所在的程序集名称打印在控制台上。

```
class Program
{
    static void Main()
    {
        Console.WriteLine(typeof(string).Assembly.FullName);
        Console.WriteLine(typeof(int).Assembly.FullName);
        Console.WriteLine(typeof(bool).Assembly.FullName);
    }
}
```

根据上面的分析可知，程序运行过程中使用的这些基础类型全部来源于 System.Private.CoreLib.dll 程序集，这在图 2-23 所示的输出结果中得到了证实。通过图 2-23 所示的输出结果，我们不仅可以知道这个核心程序集的名称，还可以知道该程序集目前的版本（4.0.0.0）。

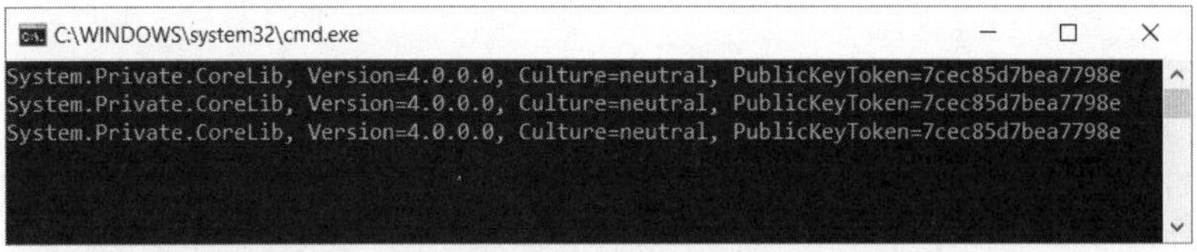

```
C:\WINDOWS\system32\cmd.exe
System.Private.CoreLib, Version=4.0.0.0, Culture=neutral, PublicKeyToken=7cec85d7bea7798e
System.Private.CoreLib, Version=4.0.0.0, Culture=neutral, PublicKeyToken=7cec85d7bea7798e
System.Private.CoreLib, Version=4.0.0.0, Culture=neutral, PublicKeyToken=7cec85d7bea7798e
```

图 2-23　基础类型来源于 System.Private.CoreLib.dll 程序集

应用程序编译后生成的程序集并不会具有针对 System.Private.CoreLib.dll 程序集引用的元数据，为了证明这一点，只需要利用 Windows SDK（在目录"%ProgramFiles(x86)%\Microsoft SDKs\Windows\{version}\Bin"下）提供的反编译工具 ildasm.exe 即可。利用 ildasm.exe 打开这个控制台应用编译生成的程序集之后，可以发现它具有如下两个程序集的引用。

```
.assembly extern System.Runtime
{
  .publickeytoken = (B0 3F 5F 7F 11 D5 0A 3A )
  .ver 4:2:0:0
}
.assembly extern System.Console
{
  .publickeytoken = (B0 3F 5F 7F 11 D5 0A 3A )
  .ver 4:1:1:0
}
```

实际上，程序只涉及 4 个类型，即 1 个 Console 类型和 3 个基础数据类型（String、Int32 和 Boolean），而程序集层面只有针对 System.Runtime 程序集和 System.Console 程序集的引用。毫无疑问，后面这 3 个基础数据类型肯定与 System.Runtime 程序集有关，要了解该程序集针对这 3 个基础数据类型的相关定义，需要先知道这个程序集究竟被保存在哪里。"%ProgramFiles%\dotnet\"是 .NET Core 应用的根目录，而 System.Runtime.dll 作为共享程序集被保存在子目录 "\shared\Microsoft.NETCore.App\3.0.0"下面，这个目录下面还保存着很多其他的共享程序集。

我们依然利用反编译工具 ildasm.exe 查看 System.Runtime.dll 程序集清单文件的元数据定义。可以发现，整个程序集除了定义少数几个核心类型（如两个重要的委托类型 Action 和 Func 就定义在这个程序集中），它的作用就是将所有基础类型采用 Type Forwarding 方式转移到 System.Private.CoreLib.dll 程序集中，下面的代码片段展示了针对程序使用的 3 个基础数据类型转移的相关定义。

```
.assembly extern System.Private.CoreLib
{
  .publickeytoken = (7C EC 85 D7 BE A7 79 8E )
  .ver 4:0:0:0
}
.class extern forwarder System.String
{
  .assembly extern System.Private.CoreLib
}
.class extern forwarder System.Int32
{
```

```
  .assembly extern System.Private.CoreLib
}
.class extern forwarder System.Boolean
{
  .assembly extern System.Private.CoreLib
}
```

演示实例体现的程序集之间的引用关系,以及上述代码片段体现的相关基础类型(System.String、System.Int32 和 System.Boolean)的转移方向基本上体现在图 2-24 所示的关系图中。

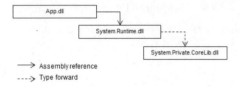

图 2-24　程序集引用关系和类型转移方向

复用 .NET Framework 程序集

上述利用 Type Forwarding 方式实现跨程序集类型转移的技术称为"垫片"(Shim),这是实现程序集跨平台复用的重要手段。除了 System.Runtime.dll 程序集,.NET Core 还提供了一些其他垫片程序集,正是源于这些垫片程序集的存在,我们可以将在 .NET Framework 环境下编译的程序集在 .NET Core 应用中使用。为了使读者对此有深刻认识,下面做一个简单的实例演示。

首先利用 Visual Studio 创建一个空的解决方案,然后添加图 2-25 所示的 3 个项目(NetApp、NetCoreApp、NetLib)。其中,NetApp 和 NetCoreApp 分别是针对 .NET Framework(4.7)和 .NET Core(3.0)的控制台程序,而 NetLib 则是针对 .NET Framework 的类库项目,该项目定义的 API 将在 NetApp 和 NetCoreApp 中被调用。

图 2-25　在 .NET Core 应用中使用 .NET Framework 程序集

我们在 NetLib 项目中定义了一个 Utils 工具类,并在其中定义了一个 PrintAssemblyNames 方法。如下面的代码片段所示,在这个方法中打印出 3 个常用的类型(Task、Uri 和 XmlWriter)所在的程序集的名称。通过在不同类型(.NET Framework 和 .NET Core)的应用程序中调用这个方法,我们可以确定它们在运行时究竟是从哪个程序集中加载的。另外,我们分别在 NetApp 和 NetCoreApp 这两个不同类型的控制台程序中调用了这个方法。

NetLib:
```
public class Utils
{
    public static void PrintAssemblyNames()
    {
        Console.WriteLine(typeof(Task).Assembly.FullName);
```

```
        Console.WriteLine(typeof(Uri).Assembly.FullName);
        Console.WriteLine(typeof(XmlWriter).Assembly.FullName);
    }
}
```

NetApp：
```
class Program
{
    static void Main()
    {
        Console.WriteLine(".NET Framework 4.7.2");
        Utils.PrintAssemblyNames();
    }
}
```

NetCoreApp：
```
class Program
{
    static void Main()
    {
        Console.WriteLine(".NET Core 3.0");
        Utils.PrintAssemblyNames();
    }
}
```

直接运行 NetApp 和 NetCoreApp 这两个控制台程序后，会输出不同的结果。如图 2-26 所示，对于指定的 3 个类型（Task、Uri 和 XmlWriter），分别在 .NET Framework 和 .NET Core 环境下承载它们的程序集是不同的。具体来说，.NET Framework 环境下的这 3 个类型分别定义在 mscorlib.dll、System.dll 和 System.Xml.dll 中。当切换到 .NET Core 环境后，运行时则会在 3 个私有的程序集 System.Private.CoreLib.dll、System.Private.Uri.dll 和 System.Private.Xml.dll 中加载这 3 个类型。

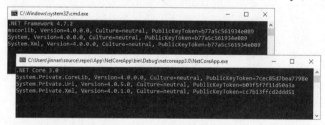

图 2-26　同一个类型来源于不同的程序集

由于 NetApp 和 NetCoreApp 这两个控制台应用使用的是同一个针对 .NET Framework 编译的程序集 NetLib.dll，所以先利用反编译工具 ildasm.exe 查看它具有怎样的程序集引用。如下面的代码片段所示，程序集 NetLib.dll 引用的程序集与控制台应用 NetApp 的输出结果是一致的。

```
.assembly extern mscorlib
{
  .publickeytoken = (B7 7A 5C 56 19 34 E0 89 )
  .ver 4:0:0:0
```

```
}
.assembly extern System
{
  .publickeytoken = (B7 7A 5C 56 19 34 E0 89 )
  .ver 4:0:0:0
}
.assembly extern System.Xml
{
  .publickeytoken = (B7 7A 5C 56 19 34 E0 89 )
  .ver 4:0:0:0
}
```

之后，核心问题就变成 Task、Uri 和 XmlWriter 这 3 个类型在 .NET Core 运行环境下是如何转移到其他程序集中的。要回答这个问题，只需要利用 ildasm.exe 查看 mscorlib.dll、System.dll 和 System.Xml.dll 反编译这 3 个程序集即可。这 3 个程序集同样存在于 "%ProgramFiles%\dotnet\shared\Microsoft.NETCore.App\3.0.0" 目录下，通过反编译与它们相关的程序集就可以得到如下所示的相关元数据。

mscorlib.dll：
```
.assembly extern System.Private.CoreLib
{
  .publickeytoken = (7C EC 85 D7 BE A7 79 8E )
  .ver 4:0:0:0
}
.class extern forwarder System.Threading.Tasks.Task
{
  .assembly extern System.Private.CoreLib
}
```

System.dll：
```
.assembly extern System.Private.Uri
{
  .publickeytoken = (B0 3F 5F 7F 11 D5 0A 3A )
  .ver 4:0:5:0
}
.class extern forwarder System.Uri
{
  .assembly extern System.Private.Uri
}
```

System.Xml.dll：
```
.assembly extern System.Private.Xml
{
  .publickeytoken = (CC 7B 13 FF CD 2D DD 51 )
  .ver 4:0:1:0
}
.class extern forwarder System.Xml.XmlWriter
{
  .assembly extern System.Private.Xml
}
```

由上述代码片段可知，针对 Task、Uri 和 XmlWriter 这 3 个类型的转移共涉及 6 个程序集，从程序集（NetLib.dll）引用元数据角度来看，这 3 个类型分别定义在 mscorlib.dll、System.dll 和 System.Xml.dll 中。当程序运行的时候，这 3 个类型分别转移到 3 个私有程序集（System.Private.CoreLib、System.Private.Uri、System.Private.Xml）中。

2.3.2 统一的 BCL

虽然 .NET Core 借助 CoreCLR 和 CoreFX 实现了真正的跨平台，但是目前的 .NET Core 仅仅提供 ASP.NET Core、UWP、WPF 和 Windows Forms 这 4 种编程模型，后面 3 种依旧专注于 Windows 平台。对于传统 .NET Framework 下面向桌面应用的 WPF、Windows Forms 和 ASP.NET，依然被保留下来作为 .NET 的一大分支。除了 .NET Framework 和 .NET Core，.NET 还具有另一个重要分支，即 Xamarin，它可以为 iOS、OS 和 Android 编写统一的应用（见图 2-27）。

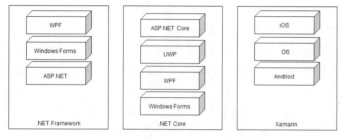

图 2-27　由 .NET Framework、.NET Core 和 Xamarin 组成的全新 .NET 平台

虽然被微软重新布局的 .NET 平台只包含 3 个分支，但是之前遇到的代码复用问题依然存在。之前解决程序集复用的方案就是 PCL，但这并不是一种理想的解决方案，由于各个目标框架具有各种独立的 BCL，所以我们创建的 PCL 项目只能建立在指定的几种兼容目标框架的 BCL 交集之上。对于全新的 .NET Core 平台来说，这个问题可以通过提供统一的 BCL 得到根本解决，这个统一的 BCL 被称为 .NET Standard。

可以将 .NET Standard 称为新一代的 PCL，PCL 提供的可移植能力仅仅限于创建时就确定下来的几种目标平台，但是 .NET Standard 做得更加彻底，因为它在设计的时候就已经考虑针对三大分支的复用。如图 2-28 所示，.NET Standard 为 .NET Framework、.NET Core 和 Xamarin 提供了统一的 API，所以在这组标准 API 基础上编写的代码能被所有类型的 .NET 应用复用。

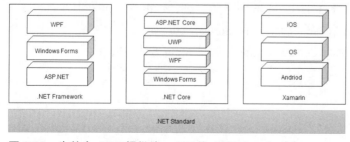

图 2-28　为整个 .NET 提供统一 API 的 .NET Standard

.NET Standard 提供的 API 主要是根据现有 .NET Framework 定义的，它的版本升级反映了其提供的 API 不断丰富的过程。Visual Studio 提供的相应的项目模板可以帮助我们创建基于 .NET Standard 的类库项目，这样的项目会采用专门的目标框架别名 netstandard{version}。一个针对 .NET Standard 2.1 的类库项目具有如下定义，同时可以看到它采用的目标框架别名为 .NET Standard 2.1。

```xml
<Project Sdk="Microsoft.NET.Sdk">
  <PropertyGroup>
    <TargetFramework>netstandard2.1</TargetFramework>
  </PropertyGroup>
</Project>
```

顾名思义，.NET Standard 仅仅是一个标准，而不提供具体的实现。.NET Standard 定义了一整套标准的接口，各个分支需要针对自身的执行环境对这套接口提供实现。对于 .NET Core 来说，它的基础 API 主要由 CoreFX 和 System.Private.CoreLib.dll 这个核心程序集来承载，这些 API 基本上就是根据 .NET Standard 设计的。但是对 .NET Framework 来说，它的 BCL 提供的 API 与 .NET Standard 存在着很大的交集，实际上，.NET Standard 基本上就是根据 .NET Framework 现有的 API 设计的，所以微软不可能在 .NET Framework 上重写一套类似于 CoreFX 的实现，只需要采用某种技术链接到现有的程序集即可。

一个针对 .NET Standard 编译生成的程序集在不同的执行环境中针对真正提供实现的程序集的所谓链接依然是通过垫片技术实现的。为了彻底搞清楚这个问题，下面做一个简单的实例演示。如图 2-29 所示，我们创建了与上面演示实例具有类似结构的解决方案，与之不同的是，分别针对 .NET Framework 和 .NET Core 的控制台应用 NetApp 与 NetCoreApp 共同引用的类库 NetStandardLib 是一个 .NET Standard 2.1 类库项目。

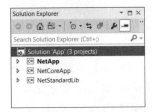

图 2-29　在 .NET Framework 和 .NET Core 应用复用的 .NET Standard 2.0 类库

与上面演示的实例一样，我们在 NetStandardLib 中定义了一个 Utils 类，并利用其静态方法 PrintAssemblyNames 将两个数据类型（Dictionary<,>和 SortedDictionary<,>）所在的程序集名称输出到控制台上，该方法分别在 NetApp 和 NetCoreApp 的入口 Main 方法中被调用。

NetStandardLib：

```
public class Utils
{
    public static void PrintAssemblyNames()
    {
        Console.WriteLine(typeof(Dictionary<,>).Assembly.FullName);
        Console.WriteLine(typeof(SortedDictionary<,>).Assembly.FullName);
```

```
    }
}
```
NetApp：
```
class Program
{
    static void Main()
    {
        Console.WriteLine(".NET Framework 4.7");
        Utils.PrintAssemblyNames();
    }
}
```
NetCoreApp：
```
class Program
{
    static void Main()
    {
        Console.WriteLine(".NET Core 2.0");
        Utils.PrintAssemblyNames();
    }
}
```

直接运行这两个分别针对 .NET Framework 和 .NET Core 的控制台应用 NetApp 与 NetCoreApp，可以发现它们会生成不同的输出结果。如图 2-30 所示，在 .NET Framework 和 .NET Core 执行环境下，Dictionary<,> 和 SortedDictionary<,> 这两个泛型字典类型其实来源于不同的程序集。具体来说，Dictionary<,> 类型在 .NET Framework 4.7 和 .NET Core 2.0 环境下分别定义在程序集 mscorlib.dll 与 System.Private.CoreLib.dll 中，而 SortedDictionary<,> 类型所在的程序集分别是 System.dll 和 System.Collection.dll。

图 2-30　同一个类型来源于不同的程序集

对于演示的这个实例来说，NetStandardLib 类库项目针对的目标框架为 .NET Standard 2.1，后者最终体现为一个名为 NetStandard.Library 的 NuGet 包，这一点其实可以从 Visual Studio 针对该项目的依赖节点看出来（见图 2-31）。这个名为 NetStandard.Library 的 NuGet 包具有一个核心程序集 netstandard.dll，上面的 .NET Standard API 就定义在该程序集中。

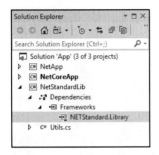

图 2-31　.NET Standard 项目对 NuGet 包 NetStandard.Library 的依赖

也就是说，所有 .NET Standard 2.1 项目都具有针对程序集 netstandard.dll 的依赖，这个依赖也会体现在编译后生成的程序集上。对于演示实例中 NetStandardLib 类库项目编译生成的同名程序集来说，它对 netstandard.dll 程序集的依赖体现在如下所示的元数据中。

```
.assembly extern netstandard
{
  .publickeytoken = (CC 7B 13 FF CD 2D DD 51 )
  .ver 2:1:0:0
}
.assembly NetStandardLib
{
  ...
}
...
```

按照我们既有的知识，原本定义在 netstandard.dll 程序集的两个类型（Dictionary<,>和SortedDictionary<,>）在不同的执行环境中需要被转移到另一个程序集中，所以完全可以在相应的环境中提供一个同名的垫片程序集，并借助类型的跨程序集转移机制来实现，实际上微软也就是这么做的。下面先介绍针对 .NET Framework 的垫片程序集 netstandard.dll 的相关定义，我们可以直接在 NetApp 编译的目标目录中找到这个程序集。借助反编译工具 ildasm.exe，可以得到与 Dictionary<,>和 SortedDictionary<,>这两个泛型字典类型转移相关的元数据，具体的代码片段如下。

```
.assembly extern mscorlib
{
  .publickeytoken = (B7 7A 5C 56 19 34 E0 89 )
  .ver 0:0:0:0
}
.assembly extern System
{
  .publickeytoken = (B7 7A 5C 56 19 34 E0 89 )
  .ver 0:0:0:0
}
.class extern forwarder System.Collections.Concurrent.ConcurrentDictionary`2
{
  .assembly extern mscorlib
}
```

```
.class extern forwarder System.Collections.Generic.SortedDictionary`2
{
  .assembly extern System
}
```

针对 .NET Core 的垫片程序集 netstandard.dll 被保存在共享目录 "%ProgramFiles%dotnet\shared\Microsoft.NETCore.App\2.0.0"下，我们采用同样的方式提取与 Dictionary<,>和 SortedDictionary<,>这两个泛型字典类型转移相关的元数据。从下面的代码片段可以看出，Dictionary<,>和 SortedDictionary<,>这两个类型都被转移到程序集 System.Collections.dll 之中。

```
.assembly extern System.Collections
{
  .publickeytoken = (B0 3F 5F 7F 11 D5 0A 3A )
  .ver 0:0:0:0
}
.class extern forwarder System.Collections.Generic.Dictionary`2
{
  .assembly extern System.Collections
}
.class extern forwarder System.Collections.Generic.SortedDictionary`2
{
  .assembly extern System.Collections
}
```

由演示实例的执行结果可知，SortedDictionary<,>确实定义在程序集 System.Collections.dll 中，但是 Dictionary<,>类型则出自核心程序集 System.Private.CoreLib.dll，所以可以断定 Dictionary<,>类型在 System.Collections.dll 中必然出现了二次转移。为了确认该推论，我们只需要采用相同的方式反编译程序集 System.Collections.dll，该程序集也被存储在共享目录 "%ProgramFiles%dotnet\shared\Microsoft.NETCore.App\3.0.0"下，该程序集中针对 Dictionary<,>类型的转移体现在如下所示的元数据中。

```
.assembly extern System.Private.CoreLib
{
  .publickeytoken = (7C EC 85 D7 BE A7 79 8E )
  .ver 4:0:0:0
}
.class extern forwarder System.Collections.Generic.Dictionary`2
{
  .assembly extern System.Private.CoreLib
}
```

Dictionary<,>和 SortedDictionary<,>这两个类型在 .NET Framework 4.7.2 与 .NET Core 3.0 环境下的跨程序集转移路径基本上体现在图 2-32 中。简单来说，.NET Framework 环境下的垫片程序集 netstandard.dll 将这两个类型分别转移到程序集 mscorlib.dll 和 System.dll 之中。如果执行环境切换到 .NET Core，这两个类型先被转移到程序集 System.Collection.dll 之中，但是 Dictionary<,>类型最终是由 System.Private.CoreLib.dll 这个基础程序集承载的，所以在程序集 System.Collection.dll 中针对该类型做了二次转移。

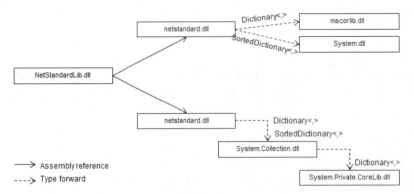

图 2-32　Dictionary<,>和 SortedDictionary<,>分别针对 .NET Framework 和 .NET Core 的类型转移

上面这个简单的实例基本上揭示了 .NET Standard 能够提供全平台可移植性的原因，下面对此进行简单总结。.NET Standard API 由 NetStandard.Library 的 NuGet 包承载，后者提供了一个名为 netstandard.dll 的程序集，保留在这个程序集中的仅仅是 .NET Standard API 的存根（Stub），而不提供具体的实现。所有 .NET Standard 类库项目编译生成的程序集保留了针对程序集 netstandard.dll 的引用。

.NET 平台的三大分支（.NET Framework、.NET Core 和 Xamarin）分别按照自己的方式实现了 .NET Standard 规定的标准的 API。由于在运行时真正承载 .NET Standard API 的类型被分布到多个程序集中，所以 .NET Standard 程序集能够被复用的前提是运行时能够将这些基础类型链接到对应的程序集上。由于 .NET Standard 程序集是针对 netstandard.dll 进行编译的，所以在各自环境中提供这个同名的程序集完成类型的转移即可。

2.3.3　展望未来

天下大势，分久必合，合久必分，技术发展亦是如此。当我们设计一个全新框架或者平台的时候，总是希望它尽可能地兼容和适配未来可能发生的变化，但是没有人能够准确地预测未来。如果现有的框架平台不能满足新需求，我们就需要在现有基础上开启另一个分支。

"开枝散叶"最初可能是权宜之计，但是随着时间的累积，我们发现这样的操作已经成为一种常见的解决方案，2.1 节描述的就是这种情况。"枝繁叶茂"其实是一种病态的繁荣，当它成为一种无法承受的负累的时候，对现有分支的整合就成为唯一的解决方案。

.NET 正走在"大一统"的道路上，.NET Core 只是这条漫长道路上的一个里程碑。2019 年 11 月，微软推出 .NET Core 3.1 LTS（Long Term Support），这将是最后一个版本的 .NET Core。今后不会再有 .NET Core 和 .NET Framework 之分，未来的 .NET 就是图 2-33 所示的统一平台，我们将其称为 .NET 5。由于所有类型的应用都是在统一的 .NET Standard 上开发的，所以不会再有移植性问题。按照微软提供的时间表，.NET 5 GA（General Available）将于 2020 年 11 月发布，而 .NET 6 LTS（Long Term Support）的发布会发生在 2021 年以后。在正常情况下，一年一

个大版本将是 .NET 未来的发展节奏，偶数年份和奇数年份发布的分别是 GA 版本与 LTS 版本，或者奇数年份发布 GA 版本，偶数年份发布 LTS 版本。

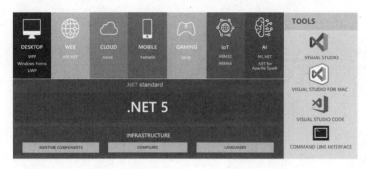

图 2-33 "大一统"的 .NET 5

第 3 章

依赖注入(上篇)

ASP.NET Core 框架建立在一些核心的基础框架之上,这些基础框架包括依赖注入、文件系统、配置选项和诊断日志等,所以本书会依次展开介绍。上述这些框架不仅仅是支撑 ASP.NET Core 框架的基础,在进行应用开发的时候同样会频繁地使用到它们。对于上述几个基础框架,依赖注入尤为重要,所以第 3 章和第 4 章会详细讲述。本章主要从理论角度介绍控制反转和依赖注入,第 4 章主要介绍 .NET Core 依赖注入框架的设计与实现。

3.1 控制反转

ASP.NET Core 应用在启动以及后续针对请求的处理过程中会依赖各种组件提供服务。为了便于定制,这些组件一般会以接口的形式进行标准化,我们将这些标准化的组件统一称为"服务"(Service)。整个 ASP.NET Core 框架建立在一个底层的依赖注入框架之上,它使用依赖注入容器提供所需的服务对象。要了解这个依赖注入容器以及它的服务提供机制,我们需要先了解什么是依赖注入(Dependence Injection,DI)。提到依赖注入,就不得不介绍控制反转(Inverse of Control,IoC)。

3.1.1 流程控制的反转

软件开发中的一些设计理念往往没有明确的定义,如 SOA、微服务(Micro Service)和无服务器(Serverless),我们无法从"内涵"方面准确定义它们,只能从"外延"上描述这些架构设计应该具有怎样的特性。由于无法给出一个明确的界定,所以针对同一个概念往往会有很多不同的理解。IoC 也是这种情况,所以本章所述只是笔者的观点,仅供读者参考。

很多人认为 IoC 是一种面向对象的设计模式,但笔者认为 IoC 不但不能算作一种设计模式,其自身也与面向对象没有直接关系。很多人之所以不能非常准确地理解 IoC,只是因为他们忽略了一个最根本的东西,那就是 IoC 本身。

IoC 的英文全称是 Inverse of Control,可译为控制反转或者控制倒置。控制反转和控制倒置体现的都是控制权的转移,即控制权原来在 A 手中,现在需要 B 来接管。对于软件设计来说,

IoC 所谓的控制权转移是如何体现的？要回答这个问题，就需要先了解 IoC 的 C（Control）究竟指的是什么。对于任何一项任务，不论其大小，基本上都可以分解成相应的步骤，所以任何一项任务的实施都有其固有的流程，而 IoC 涉及的控制可以理解为"针对流程的控制"。

下面通过一个具体实例来说明传统的设计在采用了 IoC 之后针对流程的控制是如何实现反转的。例如，如果要设计一个针对 Web 的 MVC 类库，可以将其命名为 MvcLib。简单起见，这个类库中只包含如下所示的同名的静态类。

```
public static class MvcLib
{
    public static Task ListenAsync(Uri address);
    public static Task<Request> ReceiveAsync();
    public static Task<Controller> CreateControllerAsync(Request request);
    public static Task<View> ExecuteControllerAsync(Controller controller);
    public static Task RenderViewAsync(View view);
}
```

MvcLib 提供了上述 5 个方法帮助我们完成整个 HTTP 请求流程中的 5 个核心任务。具体来说，ListenAsync 方法启动一个监听器并将其绑定到指定的地址进行 HTTP 请求的监听，抵达的请求通过 ReceiveAsync 方法进行接收，接收的请求通过一个 Request 对象来表示。CreateControllerAsync 方法根据接收的请求解析并激活目标 Controller 对象。ExecuteControllerAsync 方法执行激活的 Controller 对象并返回一个表示视图的 View 对象。RenderViewAsync 方法最终将 View 对象转换成 HTML 请求，并作为当前请求响应的内容返回请求的客户端。

下面在 MvcLib 的基础上创建一个真正的 MVC 应用。我们会发现，除了按照 MvcLib 的规范自定义具体的 Controller 对象和 View 对象，还需要自行控制包括请求的监听与接收、Controller 对象的激活与执行、View 对象的最终呈现在内的整个流程，这样一个执行流程反映在如下所示的代码片段中。

```
class Program
{
    static async Task Main()
    {
        while (true)
        {
            var address = new Uri("http://0.0.0.0:8080/mvcapp");
            await MvcLib.ListenAsync(address);
            while (true)
            {
                var request = await MvcLib.ReceiveAsync();
                var controller = await MvcLib.CreateControllerAsync(request);
                var view = await MvcLib.ExecuteControllerAsync(controller);
                await MvcLib.RenderViewAsync(view);
            }
        }
    }
}
```

上面的示例体现了图 3-1 所示的流程控制方式（应用的代码完全采用异步的方式来处理请

求,为了使流程图显得更加简单,我们在流程图中将其画成了同步的形式,读者不必纠结这个问题)。我们设计的类库(MvcLib)仅仅通过 API 的形式提供各种单一功能的实现,作为类库消费者的应用程序(App)则需要自行编排整个工作流程。如果从代码重用的角度来讲,这里被重用的仅限于实现某个环节单一功能的代码,编排整个工作流程的代码并没有得到重用。

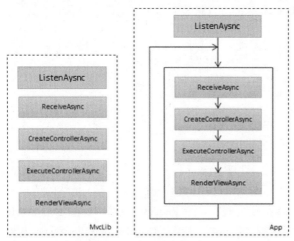

图 3-1 流程控制掌握在应用程序中

但在真实开发场景下,我们需要的不是一个仅仅能够提供单一 API 的类库,而是能够直接在上面构建应用的框架。类库(Library)和框架(Framework)的不同之处在于:前者往往只是提供实现某种单一功能的 API;而后者则针对一个目标任务对这些单一功能进行编排,以形成一个完整的流程,并利用一个引擎来驱动这个流程使之能够自动执行。

对于上面演示的 MvcLib 来说,作为消费者的应用程序需要自行控制整个 HTTP 请求的处理流程,但实际上这是一个很"泛化"的工作流程,几乎所有的 MVC 应用均采用这样的流程来监听、接收请求并最终对请求予以响应。如果将这个流程在一个 MVC 框架之中实现,由它构建的所有 MVC 应用就可以直接使用这个请求处理流程,而不需要做无谓的 DIY(Do It Yourself)。

如果将 MvcLib 从类库改造成一个框架,可以将其称为 MvcFrame。如图 3-2 所示,MvcFrame 的核心是一个被称为 MvcEngine 的执行引擎,它驱动一个编排好的工作流对 HTTP 请求进行一致性处理。如果要利用 MvcFrame 构建一个具体的 MVC 应用,除了根据业务需求定义相应的 Controller 类型和 View 文件,我们只需要初始化这个引擎并直接启动它即可。如果读者曾经开发过 ASP.NET MVC 应用,会发现 ASP.NET MVC 就是这样一个框架。

有了上面演示的这个例子作为铺垫,我们应该很容易理解 IoC 的本质。总的来说,IoC 是设计框架所采用的一种基本思想,所谓的控制反转就是将应用对流程的控制转移到框架之中。以上面的示例来说,在传统面向类库编程的时代,针对 HTTP 请求处理的流程被牢牢地控制在应用程序之中。引入框架之后,请求处理的控制权转移到了框架之中。

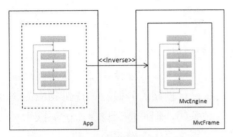

图 3-2　流程控制反转到框架之中

3.1.2　好莱坞法则

在好莱坞，演员把简历递交给电影公司后就只能回家等待消息。由于电影公司对整个娱乐项目具有完全控制权，演员只能被动地接受电影公司的邀约。"不要给我们打电话，我们会给你打电话"（Don't call us, we'll call you）——这就是著名的好莱坞法则（Hollywood Principle 或者 Hollywood Low）（见图 3-3），IoC 完美地体现了这一法则。

Don't call us, we'll call you

图 3-3　好莱坞法则

在 IoC 的应用语境中，框架就如同掌握整个电影制片流程的电影公司，由于它是整个工作流程的实际控制者，所以只有它知道哪个环节需要哪些组件。应用程序就像是演员，它只需要按照框架制定的规则注册这些组件即可，因为框架会在适当的时机自动加载并执行注册的组件。

以 ASP.NET MVC 应用开发来说，我们只需要按照约定的规则（如约定的目录结构和文件与类型命名方式等）定义相应的 Controller 类型和 View 文件即可。ASP.NET MVC 框架在处理请求的过程中会根据路由解析生成参数得到目标 Controller 的类型，然后自动创建 Controller 对象并执行。如果目标 Action 方法需要呈现一个 View，框架就会根据预定义的目录约定找到对应的 View 文件（.cshtml 文件），并对其实施动态编译以生成对应的类型。当目标 View 对象创建之后，它执行之后生成的 HTML 会作为响应回复给客户端。可以看出，整个请求流程处处体现了"框架 Call 应用"的好莱坞法则。

总的来说，在一个框架的基础上进行应用开发，就相当于在一条调试好的流水线上生产某种产品。只需要在相应的环节准备对应的原材料，最终下线的就是我们希望得到的产品。IoC 几乎是所有框架均具有的一个固有属性，从这个意义上讲，IoC 框架其实是一种错误的说法，可以说世界上本没有 IoC 框架，也可以说所有的框架都是 IoC 框架。

3.1.3 流程定制

采用 IoC 可以实现流程控制从应用程序向框架的转移，但是被转移的仅仅是一个泛化的流程，任何一个具体的应用可能都需要对该流程的某些环节进行定制。以 MVC 框架来说，默认实现的请求处理流程可以只考虑针对 HTTP 1.1 的支持，但是我们在设计框架的时候应该提供相应的扩展点来支持 HTTP 2。作为一个 Web 框架，用户认证功能是必备的，但是框架自身不能局限于某一种或者几种固定的认证方式，它应该允许我们通过扩展实现任意的认证模式。

其实也可以说得更加宽泛一些。如图 3-4 所示，如果将一个泛化的工作流程（A→B→C）定义在框架之中，建立在该框架的两个应用就需要对组成这个流程的某些环节进行定制。例如，步骤 A 和步骤 C 可以被 App1 重用，但是步骤 B 需要被定制（B1）。App2 则重用步骤 A 和步骤 B，但是需要按照自己的方式处理步骤 C。

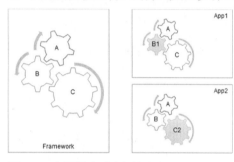

图 3-4　应用程序对流程的定制

IoC 将对流程的控制从应用程序转移到框架之中，框架利用一个引擎驱动整个流程的自动化执行。应用程序无须关心工作流程的细节，它只需要启动这个引擎即可。这个引擎一旦被启动，框架就会完全按照预先编排好的流程进行工作，如果应用程序希望整个流程按照自己希望的方式被执行，就需要在启动之前对流程进行定制。

一般来说，框架会以相应的形式提供一系列的扩展点，应用程序通过注册扩展的方式实现对流程某个环节的定制。在引擎被启动之前，应用程序将所需的扩展注册到框架之中。一旦引擎被正常启动，这些注册的扩展会自动参与整个流程的执行。

综上所述，IoC 一方面通过流程控制从应用程序向框架的反转，实现了针对流程自身的重用，另一方面通过内置的扩展机制使这个被重用的流程能够自由地被定制，这两个因素决定了框架自身的价值。重用使框架不是为应用程序提供实现单一功能的 API，而是提供一整套可执行的解决方案，并为不同的应用程序提供框架定制，从而使框架可以应用到更多的应用之中。

3.2　IoC 模式

正如前面所提到的，很多人将 IoC 理解为一种面向对象的设计模式，实际上，IoC 不仅与面向对象没有必然的联系，它自身甚至不算是一种设计模式。一般来讲，设计模式提供了一种解决某种具体问题的方案，但 IoC 既没有一个针对性的问题领域，其自身也没有提供一种可操作的解决方案，所以我们更加倾向于将 IoC 视为一种设计原则。很多设计模式都采用了 IoC 原则，下面介绍几种典型的设计模式。

3.2.1 模板方法

提到 IoC，很多人首先想到的是依赖注入，但是笔者认为与 IoC 联系最紧密的是一种被称为"模板方法"（Template Method）的设计模式。模板方法设计模式与 IoC 的意图一致，该模式主张将一个可复用的工作流程或者由多个步骤组成的算法定义成模板方法，组成这个流程或者算法的单一步骤则在相应的虚方法之中实现，模板方法根据预先编排的流程调用这些虚方法。这些方法均定义在一个类中，可以通过派生该类并重写相应的虚方法的方式达到对流程定制的目的。

对于前面演示的 MVC 示例，我们可以将整个请求处理流程在一个 MvcEngine 类中实现。如下面的代码片段所示，可以将请求的监听与接收、目标 Controller 的激活与执行、View 的呈现分别定义在 5 个受保护的虚方法中，模板方法 StartAsync 根据预定义的请求处理流程先后调用这 5 个方法。

```
public class MvcEngine
{
    public async Task StartAsync(Uri address)
    {
        await ListenAsync(address);
        while (true)
        {
            var request = await ReceiveAsync();
            var controller = await CreateControllerAsync(request);
            var view = await ExecuteControllerAsync(controller);
            await RenderViewAsync(view);
        }
    }
    protected virtual Task ListenAsync(Uri address);
    protected virtual Task<Request> ReceiveAsync();
    protected virtual Task<Controller> CreateControllerAsync(Request request);
    protected virtual Task<View> ExecuteControllerAsync(Controller controller);
    protected virtual Task RenderViewAsync(View view);
}
```

对于具体的应用程序来说，如果定义在 MvcEngine 中针对请求的处理方式完全符合要求，那么只需要创建一个 MvcEngine 对象，然后指定一个监听地址来调用模板方法 StartAsync，以开启 MVC 引擎即可。如果该 MVC 引擎对请求某个环节的处理无法满足要求，我们可以创建 MvcEngine 的派生类，然后重写实现该环节相应的虚方法即可。例如，定义在某个应用程序中的 Controller 都是无状态的，如果采用单例（Singleton）的方式重用已经激活的 Controller 对象以提高性能，那么可以按照如下方式创建一个自定义的 FoobarMvcEngine，并按照自己的方式重写 CreateControllerAsync 方法。

```
public class FoobarMvcEngine : MvcEngine
```

```
{
    protected override Task<View> CreateControllerAsync (Request request)
    {
        <<省略实现>>
    }
}
```

3.2.2 工厂方法

对于一个复杂的流程来说，我们倾向于将组成该流程的各个环节实现在相应的组件之中，所以针对流程的定制可以通过提供相应组件的形式实现。23 种设计模式之中有一种重要的类型，即创建型模式，如常用的工厂方法和抽象工厂，IoC 体现的针对流程的复用与定制同样可以通过这些设计模式来完成。

所谓的工厂方法，其实就是在某个类中定义用来提供所需服务对象的方法，这个方法可以是一个单纯的抽象方法，也可以是具有默认实现的虚方法，至于方法声明的返回类型，可以是一个接口或者抽象类，也可以是未封闭（Sealed）的具体类型。派生类型可以采用重写工厂方法的方式提供所需的服务对象。

同样，以 MVC 框架为例，我们让独立的组件完成整个请求处理流程的几个核心环节。具体来说，可以为这些核心组件定义如下几个对应的接口：IWebListener 接口用来监听、接收和响应请求（针对请求的响应由 ReceiveAsync 方法返回的 HttpContext 上下文来完成）；IControllerActivator 接口用于根据当前 HttpContext 上下文激活目标 Controller 对象，并在 Controller 对象执行后做一些释放回收工作；IControllerExecutor 接口和 IViewRenderer 接口分别用来完成针对 Controller 的执行以及针对视图的呈现。

```
public interface IWebListener
{
    Task ListenAsync(Uri address);
    Task<HttpContext> ReceiveAsync();
}

public interface IControllerActivator
{
    Task<Controller> CreateControllerAsync(HttpContext httpContext);
    Task ReleaseAsync(Controller controller);
}

public interface IControllerExecutor
{
    Task<View> ExecuteAsync(Controller controller, HttpContext httpContext);
}

public interface IViewRenderer
{
    Task RendAsync(View view, HttpContext httpContext);
}
```

我们在作为 MVC 引擎的 MvcEngine 中定义了 4 个工厂方法（GetWebListener、GetControllerActivator、GetControllerExecutor 和 GetViewRenderer），用来提供上述 4 种组件。这 4 个工厂方法均为具有默认实现的虚方法，可以利用它们提供默认的组件。在用于启动引擎的 StartAsync 方法中，可以利用这些工厂方法提供的对象完成整个请求处理流程。

```
public class MvcEngine
{
    public async Task StartAsync(Uri address)
    {
        var listener        = GetWebListener();
        var activator       = GetControllerActivator();
        var executor        = GetControllerExecutor();
        var renderer        = GetViewRenderer();

        await listener.ListenAsync(address);
        while (true)
        {
            var httpContext = await listener.ReceiveAsync();
            var controller = await activator.CreateControllerAsync(httpContext);
            try
            {
                var view = await executor.ExecuteAsync(controller, httpContext);
                await renderer.RendAsync(view, httpContext);
            }
            finally
            {
                await activator.ReleaseAsync(controller);
            }
        }
    }
    protected virtual IWebLister GetWebListener();
    protected virtual IControllerActivator GetControllerActivator();
    protected virtual IControllerExecutor GetControllerExecutor();
    protected virtual IViewRenderer GetViewRenderer();
}
```

对于具体的应用程序来说，如果需要对请求处理的某个环节进行定制，那么需要将定制的操作实现在对应接口的实现类中。在 MvcEngine 的派生类中，我们需要重写对应的工厂方法来提供被定制的对象。例如，以单例模式提供目标 Controller 对象的实现就定义在 SingletonControllerActivator 类中，可以在派生于 MvcEngine 的 FoobarMvcEngine 类中重写工厂方法 GetControllerActivator，使其返回一个 SingletonControllerActivator 对象。

```
public class SingletonControllerActivator : IControllerActivator
{
    public Task<Controller> CreateControllerAsync(HttpContext httpContext)
    {
        <<省略实现>>
    }
    public Task ReleaseAsync(Controller controller) => Task.CompletedTask;
```

```
}
public class FoobarMvcEngine : MvcEngine
{
    protected override ControllerActivator GetControllerActivator()
        => new SingletonControllerActivator();
}
```

3.2.3 抽象工厂

虽然工厂方法和抽象工厂均提供了一个"生产"对象实例的工厂，但是两者在设计上有本质上的区别。工厂方法利用定义在某个类型的抽象方法或者虚方法完成了针对"单一对象"的提供，而抽象工厂则利用一个独立的接口或者抽象类提供"一组相关的对象"。

具体来说，我们需要定义一个独立的工厂接口或者抽象工厂类，并在其中定义多个工厂方法来提供"同一系列"的多个相关对象。如果希望抽象工厂具有一组默认的"产出"，也可以将一个未被封闭的类型作为抽象工厂，以虚方法形式定义的工厂方法将默认的对象作为返回值。在具体的应用开发中，可以通过实现工厂接口或者继承抽象工厂类（不一定是抽象类）的方式来定义具体工厂类，并利用它来提供一组定制的对象系列。

可以采用抽象工厂模式改造 MVC 框架。如下面的代码片段所示，可以定义了一个名为 IMvcEngineFactory 的接口作为抽象工厂，并在其中定义了 4 个方法，用来提供请求监听和处理过程使用到的 4 种核心对象。如果 MVC 框架提供了针对这 4 种核心组件的默认实现，就可以按照如下方式为这个抽象工厂提供一个默认实现（MvcEngineFactory）。

```
public interface IMvcEngineFactory
{
    IWebLister GetWebListener();
    IControllerActivator GetControllerActivator();
    IControllerExecutor GetControllerExecutor();
    IViewRenderer GetViewRenderer();
}

public class MvcEngineFactory: IMvcEngineFactory
{
    public virtual IWebLister GetWebListener();
    public virtual IControllerActivator GetControllerActivator();
    public virtual IControllerExecutor GetControllerExecutor();
    public virtual IViewRenderer GetViewRenderer();
}
```

可以在创建 MvcEngine 对象的时候提供一个具体的 IMvcEngineFactory 对象，如果没有显式指定，MvcEngine 通常默认使用 EngineFactory 对象。在用于启动引擎的 StartAsync 方法中，MvcEngine 利用 IMvcEngineFactory 对象来获取相应的对象完成对请求的处理流程。

```
public class MvcEngine
{
    public IMvcEngineFactory EngineFactory { get; }
```

```csharp
public MvcEngine(IMvcEngineFactory engineFactory = null)
    => EngineFactory = engineFactory ?? new MvcEngineFactory();

public async Task StartAsync(Uri address)
{
    var listener           = EngineFactory.GetWebListener();
    var activator          = EngineFactory.GetControllerActivator();
    var executor           = EngineFactory.GetControllerExecutor();
    var renderer           = EngineFactory.GetViewRenderer();

    await listener.ListenAsync(address);
    while (true)
    {
        var httpContext = await listener.ReceiveAsync();
        var controller = await activator.CreateControllerAsync(httpContext);
        try
        {
            var view = await executor.ExecuteAsync(controller, httpContext);
            await renderer.RendAsync(view, httpContext);
        }
        finally
        {
            await activator.ReleaseAsync(controller);
        }
    }
}
```

如果具体的应用程序需要采用 SingletonControllerActivator 以单例的模式来激活目标 Controller 对象，就可以按照如下方式定义一个具体的工厂类 FoobarEngineFactory。最终的应用程序将利用 FoobarEngineFactory 对象来创建作为引擎的 MvcEngine 对象。

```csharp
public class FoobarEngineFactory : MvcEngineFactory
{
    public override ControllerActivator GetControllerActivator()
    {
        return new SingletonControllerActivator();
    }
}

public class App
{
    static async Task Main()
    {
        var address     = new Uri("http://0.0.0.0:8080/mvcapp");
        var engine      = new MvcEngine(new FoobarEngineFactory());
        await engine.StartAsync(address);
        ...
    }
}
```

3.3 依赖注入

IoC 主要体现了这样一种设计思想：通过将一组通用流程的控制权从应用转移到框架之中以实现对流程的复用，并按照好莱坞法则实现应用程序的代码与框架之间的交互。我们可以采用若干设计模式以不同的方式实现 IoC，如模板方法、工厂方法和抽象工厂，下面介绍一种更有价值的 IoC 模式：依赖注入（Dependency Injection，DI）。

3.3.1 由容器提供对象

与前面介绍的工厂方法和抽象工厂模式一样，依赖注入是一种"对象提供型"的设计模式，可以将提供的对象统称为"服务""服务对象""服务实例"。在一个采用依赖注入的应用中，我们定义某个类型时，只需要直接将它依赖的服务采用相应的方式注入进来即可。

在应用启动时，我们会对所需的服务进行全局注册。一般来说，服务大都是针对实现的接口或者继承的抽象类进行注册的，服务注册信息会在后续消费过程中帮助我们提供对应的服务实例。按照好莱坞法则，应用只需要定义并注册好所需的服务，服务实例的提供则完全交给框架来完成，框架利用一个独立的容器（Container）来提供所需的每个服务实例。

我们将这个被框架用来提供服务的容器称为依赖注入容器，也有很多人将其称为 IoC 容器，根据前面针对 IoC 的介绍，笔者认为后者不是一个合理的称谓。依赖注入容器之所以能够按照我们希望的方式来提供所需的服务是因为该容器是根据服务注册信息创建的，服务注册包含提供所需服务实例的所有信息。

例如，如果创建一个名为 Cat 的依赖注入容器类型，那么可以调用 GetService<T>扩展方法从某个 Cat 对象中获取指定类型的服务对象。笔者之所以将其命名为 Cat，主要源于卡通形象"机器猫"（哆啦 A 梦）。机器猫的四次元口袋就是一个理想的依赖注入容器，大熊只需要告诉机器猫相应的需求，它就能从这个口袋中得到相应的法宝。依赖注入容器亦是如此，服务消费者只需要告诉容器所需服务的类型（一般是一个服务接口或者抽象服务类），就能得到与之匹配的服务实例。

```
public static class CatExtensions
{
    public static T GetService<T>(this Cat cat);
}
```

对于 MVC 框架来说，我们在前面分别采用不同的设计模式对框架的核心类型 MvcEngine 进行了"改造"，而采用依赖注入的方式，并利用上述 Cat 容器按照如下方式对其重新实现，我们会发现 MvcEngine 变得异常简洁而清晰。

```
public class MvcEngine
{
    public Cat Cat { get; }
    public MvcEngine(Cat cat) => Cat = cat;

    public async Task StartAsync(Uri address)
    {
```

```csharp
        var listener           = Cat.GetService<IWebListener>();
        var activator          = Cat.GetService<IControllerActivator>();
        var executor           = Cat.GetService<IControllerExecutor>();
        var renderer           = Cat.GetService<IViewRenderer>();

        await listener.ListenAsync(address);
        while (true)
        {
            var httpContext = await listener.ReceiveAsync();
            var controller = await activator.CreateControllerAsync(httpContext);
            try
            {
                var view = await executor.ExecuteAsync(controller, httpContext);
                await renderer.RendAsync(view, httpContext);
            }
            finally
            {
                await activator.ReleaseAsync(controller);
            }
        }
    }
}
```

依赖注入体现了一种最直接的服务消费方式，消费者只需要告诉提供者（依赖注入容器）所需服务的类型，后者就能根据预先注册的规则提供一个匹配的服务实例。由于服务注册最终决定了依赖注入容器根据指定的服务类型会提供一个什么样的服务实例，所以我们可以通过修改服务注册的方式来实现对框架的定制。如果应用程序需要采用 SingletonControllerActivator 以单例的模式来激活目标 Controller，那么它可以在启动 MvcEngine 之前按照如下形式将 SingletonControllerActivator 注册到依赖注入容器中。

```csharp
public class App
{
    static void Main(string[] args)
    {
        var cat = new Cat()
            .Register<IControllerActivator, SingletonControllerActivator>();
        var engine    = new MvcEngine(cat);
        var address   = new Uri("http://localhost/mvcapp");
        engine.StartAsync(address);
    }
}
```

3.3.2 3种依赖注入方式

一项任务往往需要多个对象相互协作才能完成，或者说某个对象在完成某项任务的时候需要直接或者间接地依赖其他的对象来完成某些必要的步骤，所以运行时对象之间的依赖关系是由目标任务决定的，是"恒定不变"的，自然也无所谓"解耦"的说法。但是运行时对象通过对应的类来定义，类与类之间的耦合可以通过对依赖进行抽象的方式来降低或者解除。

从服务消费的角度来讲，如果借助一个接口对消费的服务进行抽象，那么服务消费程序针对具体服务类型的依赖可以转移到对服务接口的依赖上面，但是在运行时提供给消费者的总是一个针对某个具体服务类型的对象。不仅如此，要完成定义在服务接口的操作，这个对象可能需要其他相关对象的参与，换句话说，提供的这个依赖服务对象可能具有对其他服务对象的依赖。作为服务对象提供者的依赖注入容器，它会根据依赖链提供所有的依赖服务实例。

如图 3-5 所示，当应用框架调用 GetService<IFoo>方法向依赖注入容器索取一个实现了 IFoo 接口的服务对象时，该方法会根据预先注册的类型映射关系创建一个类型为 Foo 的对象。由于 Foo 对象需要 Bar 对象和 Qux 对象的参与才能完成目标操作，所以 Foo 具有针对 Bar 和 Qux 的直接依赖。而服务对象 Bar 又依赖 Baz，所以 Baz 成了 Foo 的间接依赖。对于依赖注入容器最终提供的 Foo 对象，它所直接或者间接依赖的对象 Bar、Baz 和 Qux 都会预先被初始化并自动注入该对象之中。

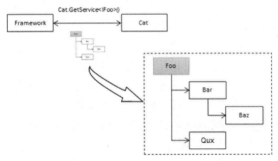

图 3-5　依赖注入容器对依赖的自动注入

从面向对象编程的角度来讲，类型中的字段或者属性是依赖的一种主要体现形式。如果类型 A 中具有一个类型 B 的字段或者属性，那么类型 A 就对类型 B 产生了依赖，所以可以将依赖注入简单地理解为一种针对依赖字段或者属性的自动化初始化方式，我们可以通过 3 种方式达到这个目的。下面着重介绍 3 种依赖注入方式。

构造器注入

构造器注入就是在构造函数中借助参数将依赖的对象注入由它创建的对象之中。如下面的代码片段所示，Foo 针对 Bar 的依赖体现在只读属性 Bar 上，针对该属性的初始化是在构造函数中实现的，具体的属性值由构造函数传入的参数提供。

```
public class Foo
{
    public IBar Bar{get;}
    public Foo(IBar bar) =>Bar = bar;
}
```

除此之外，构造器注入还体现在对构造函数的选择上。如下面的代码片段所示，Foo 类定义了两个构造函数，依赖注入容器在创建 Foo 对象之前需要先选择一个合适的构造函数。至于目标构造函数如何选择，不同的依赖注入容器可能有不同的策略，如可以选择参数最多或者最少的构造函数，也可以按照如下方式在目标构造函数上标注一个 InjectionAttribute 特性。

```
public class Foo
{
    public IBar Bar{get;}
    public IBaz Baz {get;}

    [Injection]
    public Foo(IBar bar) =>Bar = bar;
    public Foo(IBar bar, IBaz baz):this(bar)
        =>Baz = baz;
}
```

属性注入

如果依赖直接体现为类的某个属性，并且该属性不是只读的，就可以让依赖注入容器在对象创建之后自动对其进行赋值，进而达到依赖注入的目的。一般来说，在定义这种类型的时候，需要显式地将这样的属性标识为需要自动注入的依赖属性，从而与其他普通属性进行区分。如下面的代码片段所示，Foo 类中定义了两个可读写的公共属性（Bar 和 Baz），通过标注 InjectionAttribute 特性的方式可以将属性 Baz 设置为自动注入的依赖属性。由依赖注入容器提供的 Foo 对象的 Baz 属性将会自动被初始化。

```
public class Foo
{
    public IBar Bar{get; set;}

    [Injection]
    public IBaz Baz {get; set;}
}
```

方法注入

体现依赖关系的字段或者属性可以通过方法的形式初始化。如下面的代码片段所示，Foo 对 Bar 的依赖体现在只读属性上，针对该属性的初始化实现在 Initialize 方法中，具体的属性值由该方法传入的参数提供。同样，通过标注特性（InjectionAttribute）的方式可以将该方法标识为注入方法。依赖注入容器在调用构造函数创建一个 Foo 对象之后，它会自动调用 Initialize 方法对只读属性 Bar 进行赋值。

```
public class Foo
{
    public IBar Bar{get;}

    [Injection]
    public Initialize(IBar bar)=> Bar = bar;
}
```

除了通过依赖注入容器在初始化服务过程中自动调用的实现，我们还可以利用它实现另一种更加自由的方法注入，这种注入方式在 ASP.NET Core 应用中具有广泛应用。ASP.NET Core 在启动的时候会调用注册的 Startup 对象来完成中间件的注册，而定义 Startup 类型的时候不需要让它实现某个接口,所以用于注册中间件的 Configure 方法没有一个固定的声明,但可以按照如下方式将任意依赖的服务实例直接注入这个方法之中。

```csharp
public class Startup
{
    public void Configure(IApplicationBuilder app, IFoo foo, IBar bar, IBaz baz);
}
```

类似的注入方式同样可以应用到中间件类型的定义上。与用来注册中间件的 Startup 类型一样,ASP.NET Core 框架下的中间件类型同样不需要实现某个预定义的接口,用于处理请求的 InvokeAsync 方法或者 Invoke 方法同样可以按照如下方式注入任意的依赖服务。

```csharp
public class FoobarMiddleware
{
    private readonly RequestDelegate _next;
    public FoobarMiddleware(RequestDelegate next)=> _next = next;

    public Task InvokeAsync(HttpContext httpContext, IFoo foo, IBar bar, IBaz baz);
}
```

上面这种方式的方法注入促成了一种"面向约定"的编程方式。由于不再需要实现某个预定义的接口或者继承某个预定义的基类,所以需要实现或者重写方法的声明也就少了对应的限制,这样就可以采用最直接的方式将依赖的服务注入方法中。对于前面介绍的这几种注入方式,构造器注入是最理想的形式,笔者不建议使用属性注入和方法注入(前面介绍的这种基于约定的方法注入除外)。

3.3.3　Service Locator 模式

假设我们需要定义一个服务类型 Foo,它依赖于服务 Bar 和 Baz,后者对应的服务接口分别为 IBar 和 IBaz。如果当前应用中具有一个依赖注入容器(假设类似于前面定义的 Cat),那么我们就可以采用如下两种方式定义服务类型 Foo。

```csharp
public class Foo : IFoo
{
    public IBar Bar { get; }
    public IBaz Baz { get; }
    public Foo(IBar bar, IBaz baz)
    {
        Bar = bar;
        Baz = baz;
    }
    public async Task InvokeAsync()
    {
        await Bar.InvokeAsync();
        await Baz.InvokeAsync();
    }
}

public class Foo : IFoo
{
```

```
    public Cat Cat { get; }
    public Foo(Cat cat) => Cat = cat;
    public async Task InvokeAsync()
    {
        await Cat.GetService<IBar>().InvokeAsync();
        await Cat.GetService<IBaz>().InvokeAsync();
    }
}
```

从表面上看，上面提供的这两种服务类型的定义方式都可以解决针对依赖服务的耦合问题，并将针对服务实现的依赖转变成针对接口的依赖。很多人会选择第二种定义方式，因为这种定义方式不仅代码量更少，针对服务的提供也更加直接。我们直接在构造函数中"注入"了代表依赖注入容器的 Cat 对象，在任何使用到依赖服务的地方，只需要利用它来提供对应的服务实例即可。

但第二种定义方式采用的设计模式不是依赖注入，而是一种被称为 Service Locator 的设计模式。Service Locator 模式同样具有一个通过服务注册创建的全局的容器来提供所需的服务实例，该容器被称为 Service Locator。依赖注入容器和 Service Locator 实际上是同一事物在不同设计模式中的不同称谓。那么，依赖注入和 Service Locator 之间的差异主要体现在哪些方面？

笔者认为可以从依赖注入容器或者 Service Locator 被谁使用的角度来区分这两种设计模式的差别。在一个采用依赖注入的应用中，我们只需要采用标准的注入形式定义服务类型，并在应用启动之前完成相应的服务注册即可，框架自身的引擎在运行过程中会利用依赖注入容器来提供当前所需的服务实例。换句话说，依赖注入容器的使用者应该是框架而不是应用程序。Service Locator 模式显然不是这样，而是应用程序在利用它来提供所需的服务实例，所以它的使用者是应用程序。

我们也可以从另外一个角度区分两者之间的差别。由于依赖服务是以"注入"的方式来提供的，所以采用依赖注入模式的应用可以看作将服务推送给被依赖对象，Service Locator 模式下的应用则是利用 Service Locator 拉取所需的服务，这一"推"一"拉"也准确地体现了两者之间的差异。那么既然两者之间有差异，究竟孰优孰劣？

2010 年，Mark Seemann 就已经将 Service Locator 视为一种反模式（Anti-Pattern），虽然也有人对此提出不同的意见，但笔者不推荐使用这种设计模式。笔者反对使用 Service Locator 模式与前面提到的反对使用属性注入和方法注入具有类似的缘由。

本着"松耦合、高内聚"的设计原则，我们既然将一组相关的操作定义在一个能够复用的服务中，就应该尽量要求服务自身不但具有独立和自治的特性，而且要求服务之间应该具有明确的界定，服务之间的依赖关系应该是明确的而不是模糊的。不论采用属性注入或者方法注入，还是使用 Service Locator 提供当前依赖的服务，都相当于为当前的服务增添一个新的依赖，即针对依赖注入容器或者 Service Locator 的依赖。

当前服务针对另一个服务的依赖与针对依赖注入容器或者 Service Locator 的依赖具有本质

上的不同。前者是一种基于类型的依赖，不论是基于服务的接口还是实现类型，这是一种基于"契约"的依赖。这种依赖不仅是明确的，也是有保障的。但是依赖注入容器或者 Service Locator 本质上是一个黑盒，它能够提供所需服务的前提是相应的服务注册已经预先添加到容器之中，但是这种依赖不仅是模糊的也是不可靠的。

正因为如此，ASP.NET Core 框架使用的依赖注入框架只支持构造器注入，而不支持属性注入和方法注入（类似于 Startup 和中间件基于约定的方法注入除外），但是我们可能会不知不觉地按照 Service Locator 模式编写代码。从某种意义上讲，当我们在程序中使用 IServiceProvider（表示依赖注入容器）提取某个服务实例时，就意味着我们已经在使用 Service Locator 模式了，所以遇到这种情况时应该思考是否一定需要这么做。

3.4 一个简易版的依赖注入容器

前面从纯理论的角度对依赖注入进行了深入论述，第 4 章会对 ASP.NET Core 框架内部使用的依赖注入框架进行单独介绍。为了使读者能够更好地理解第 4 章的内容，我们按照类似的原理创建了一个简易版本的依赖注入框架，也就是前面多次提及的 Cat。

3.4.1 编程体验

虽然我们对这个名为 Cat 的依赖注入框架进行了最大限度的简化，但是与 ASP.NET Core 框架内部使用的真实依赖注入框架相比，Cat 不但采用了一致的设计，而且具备所有的功能特性。为了使读者对 Cat 具有感官方面的认识，下面先演示如何利用 Cat 提供所需的服务实例。

作为依赖注入容器的 Cat 对象不仅可以作为服务实例的提供者，还需要维护服务实例的生命周期。Cat 提供了 3 种生命周期模式，如果要了解它们之间的差异，就必须对多个 Cat 之间的层次关系有充分的认识。一个代表依赖注入容器的 Cat 对象用来创建其他的 Cat 对象，后者将前者视为"父容器"，所以多个 Cat 对象通过其"父子关系"维系一个树形层次化结构。但这仅仅是一个逻辑结构而已，实际上，每个 Cat 对象只会按照图 3-6 所示的方式引用整棵树的根。

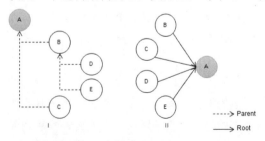

图 3-6　Cat 对象之间的关系

了解了多个 Cat 对象之间的关系之后，就很容易理解 3 种预定义的生命周期模式。如下所示的 Lifetime 枚举代表 3 种生命周期模式：Transient 代表容器针对每次服务请求都会创建一个新的服务实例；Self 将提供服务实例保存在当前容器中，它代表针对某个容器范围内的单例模式；Root 将每个容器提供的服务实例统一存放到根容器中，所以该模式能够在多个"同根"容器范围内确保提供的服务是单例的。

```
public enum Lifetime
{
    Root,
    Self,
    Transient
}
```

代表依赖注入容器的 Cat 对象之所以能够为我们提供所需的服务实例，其根本前提是相应的服务注册在此之前已经添加到容器之中。服务总是针对服务类型（接口、抽象类或者具体类型）进行注册的，Cat 通过定义的扩展方法提供了如下 3 种注册方式。除了直接提供服务实例的形式（默认采用 Root 模式），在注册服务时必须指定一个具体的生命周期模式。

- 指定具体的实现类型。
- 提供一个服务实例。
- 指定一个创建服务实例的工厂。

我们定义了如下所示的接口和对应的实现类型，来演示针对 Cat 的服务注册。其中，Foo、Bar、Baz 和 Qux 分别实现了对应的接口 IFoo、IBar、IBaz 和 IQux，其中 Qux 类型上标注的 MapToAttribute 特性，注册了与对应接口 IQux 之间的映射。为了反映 Cat 对服务实例生命周期的控制，可以让它们派生于同一个基类 Base。Base 实现了 IDisposable 接口，我们在其构造函数和实现的 Dispose 方法中输出相应的文本，以确定对应的实例何时被创建和释放。另外，我们还定义了一个泛型的接口 IFoobar<T1, T2>和对应的实现类 Foobar<T1, T2>，来演示 Cat 针对泛型服务实例的提供。

```
public interface IFoo {}
public interface IBar {}
public interface IBaz {}
public interface IQux {}
public interface IFoobar<T1, T2> {}
public class Base : IDisposable
{
    public Base()
        => Console.WriteLine($"Instance of {GetType().Name} is created.");
    public void Dispose()
        => Console.WriteLine($"Instance of {GetType().Name} is disposed.");
}

public class Foo : Base, IFoo{ }
public class Bar : Base, IBar{ }
public class Baz : Base, IBaz{ }
[MapTo(typeof(IQux), Lifetime.Root)]
public class Qux : Base, IQux { }
public class Foobar<T1, T2>: IFoobar<T1,T2>
```

```
{
    public T1Foo { get; }
    public T2Bar { get; }
    public Foobar(T1foo, T2bar)
    {
        Foo = foo;
        Bar = bar;
    }
}
```

如下所示的代码片段创建了一个 Cat 对象,并采用上面提到的方式针对接口 IFoo、IBar 和 IBaz 注册了对应的服务,它们采用的生命周期模式分别为 Transient、Self 和 Root。另外,我们还调用了另一个将当前入口程序集作为参数的 Register 方法,该方法会解析指定程序集中标注了 MapToAttribute 特性的类型并做相应的服务注册,对于我们演示的程序来说,该方法会完成针对 IQux/Qux 类型的服务注册。接下来我们利用 Cat 对象创建了它的两个子容器,并调用子容器的 GetService<T>方法来提供相应的服务实例。

```
class Program
{
    static void Main()
    {
        var root = new Cat()
            .Register<IFoo, Foo>(Lifetime.Transient)
            .Register<IBar>(_ => new Bar(), Lifetime.Self)
            .Register<IBaz, Baz>(Lifetime.Root)
            .Register(Assembly.GetEntryAssembly());
        var cat1 = root.CreateChild();
        var cat2 = root.CreateChild();

        void GetServices<TService>(Cat cat)
        {
            cat.GetService<TService>();
            cat.GetService<TService>();
        }

        GetServices<IFoo>(cat1);
        GetServices<IBar>(cat1);
        GetServices<IBaz>(cat1);
        GetServices<IQux>(cat1);
        Console.WriteLine();
        GetServices<IFoo>(cat2);
        GetServices<IBar>(cat2);
        GetServices<IBaz>(cat2);
        GetServices<IQux>(cat2);
    }
}
```

上面的程序运行之后会在控制台上输出图 3-7 所示的结果,输出结果不仅表明 Cat 能够根据添加的服务注册提供对应类型的服务实例,还体现了它对生命周期的控制。由于服务 IFoo 被注册为 Transient 服务，所以 Cat 针对该接口的服务提供的 4 次请求都会创建一个全新的 Foo 对象。IBar 服务的生命周期模式为 Self，如果利用同一个 Cat 对象提供对应的服务实例，那么该 Cat 对象只会创建一个 Bar 对象，所以整个过程中会创建两个 Bar 对象。IBaz 和 IQux 服务采用 Root 生命周期，所以具有同根的两个 Cat 对象提供的总是同一个 Baz/Qux 对象，后者只会被创建一次。（S301）

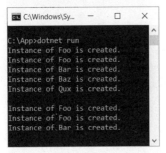

图 3-7 Cat 按照服务注册对应的生命周期模式提供服务实例

除了提供类似于 IFoo、IBar 和 IBaz 这种非泛型的服务实例，如果具有针对泛型定义（Generic Definition）的服务注册，Cat 同样可以提供泛型服务实例。如下面的代码片段所示，在为创建的 Cat 对象添加了针对 IFoo 接口和 IBar 接口的服务注册之后，我们调用 Register 方法注册了针对泛型定义 IFoobar<,>的服务注册，具体的实现类型为 Foobar<,>。当我们利用 Cat 对象提供一个类型为 IFoobar<IFoo, IBar>的服务实例时，它会创建并返回一个 Foobar<Foo, Bar>对象。（S302）

```
public class Program
{
    public static void Main()
    {
        var cat = new Cat()
            .Register<IFoo, Foo>(Lifetime.Transient)
            .Register<IBar, Bar>(Lifetime.Transient)
            .Register(typeof(IFoobar<,>), typeof(Foobar<,>), Lifetime.Transient);

        var foobar = (Foobar<IFoo, IBar>)cat.GetService<IFoobar<IFoo, IBar>>();
        Debug.Assert(foobar.Foo is Foo);
        Debug.Assert(foobar.Bar is Bar);
    }
}
```

在进行服务注册时，可以为同一个类型添加多个服务注册。虽然添加的所有服务注册均是有效的，但由于 GetService<TService>扩展方法总是返回一个唯一的服务实例，所以可以对该方法采用"后来居上"的策略，即总是采用最近添加的服务注册创建服务实例。如果调用另一个 GetServices<TService>扩展方法，该方法将返回根据所有服务注册提供的服务实例。

下面的代码片段为创建的 Cat 对象添加了 3 个针对 Base 类型的服务注册，对应的实现类型分别为 Foo、Bar 和 Baz。我们将 Base 作为泛型参数调用了 GetServices<Base>方法，该方法会返回包含 3 个 Base 对象的集合，集合元素的类型分别为 Foo、Bar 和 Baz。（S303）

```csharp
public class Program
{
    public static void Main()
    {
        var services = new Cat()
            .Register<Base, Foo>(Lifetime.Transient)
            .Register<Base, Bar>(Lifetime.Transient)
            .Register<Base, Baz>(Lifetime.Transient)
            .GetServices<Base>();
        Debug.Assert(services.OfType<Foo>().Any());
        Debug.Assert(services.OfType<Bar>().Any());
        Debug.Assert(services.OfType<Baz>().Any());
    }
}
```

如果提供的服务实例实现了 IDisposable 接口,就应该在适当的时候调用其 Dispose 方法释放该服务实例。由于服务实例的生命周期完全由作为依赖注入容器的 Cat 对象来管理,所以通过调用 Dispose 方法来释放服务实例也应该由它负责。Cat 对象针对提供服务实例的释放策略取决于采用的生命周期模式,具体的策略如下。

- **Transient 和 Self**:所有实现了 IDisposable 接口的服务实例会被当前 Cat 对象保存起来,当 Cat 对象自身的 Dispose 方法被调用的时候,这些服务实例的 Dispose 方法会随之被调用。
- **Root**:由于服务实例保存在作为根容器的 Cat 对象上,所以当这个 Cat 对象的 Dispose 方法被调用的时候,这些服务实例的 Dispose 方法会随之被调用。

上述释放策略可以通过如下演示实例来印证。如下代码片段创建了一个 Cat 对象,并添加了相应的服务注册;然后调用 CreateChild 方法创建了代表子容器的 Cat 对象,并用它提供了 4 个注册服务对应的实例。

```csharp
class Program
{
    static void Main()
    {
        using (var root = new Cat()
            .Register<IFoo, Foo>(Lifetime.Transient)
            .Register<IBar>(_ => new Bar(), Lifetime.Self)
            .Register<IBaz, Baz>(Lifetime.Root)
            .Register(Assembly.GetEntryAssembly()))
        {
            using (var cat = root.CreateChild())
            {
                cat.GetService<IFoo>();
                cat.GetService<IBar>();
                cat.GetService<IBaz>();
                cat.GetService<IQux>();
                Console.WriteLine("Child cat is disposed.");
            }
            Console.WriteLine("Root cat is disposed.");
        }
```

```
        }
}
```

  由于两个 Cat 对象的创建都是在 using 块中进行的，所以它们的 Dispose 方法都会在 using 块结束的地方被调用。为了确定方法被调用的时机，我们特意在控制台上打印了相应的文字。该程序运行之后会在控制台上输出图 3-8 所示的结果，可以看到，当作为子容器的 Cat 对象的 Dispose 方法被调用时，由它提供的两个生命周期模式分别为 Transient 和 Self 的两个服务实例（Foo 和 Bar）被正常释放。而生命周期模式为 Root 的服务实例是 Baz 和 Qux，它的 Dispose 方法会延迟到作为根容器的 Cat 对象的 Dispose 方法被调用的时候。（S304）

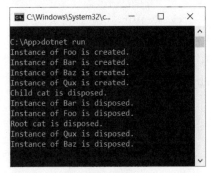

图 3-8　服务实例的释放

3.4.2　设计与实现

  在完成针对 Cat 的编程体验之后，下面介绍依赖注入容器的设计原理和具体实现。由于作为依赖注入容器的 Cat 对象总是利用预先添加的服务注册来提供对应的服务实例，所以服务注册至关重要。如下所示的代码片段就是表示服务注册的 ServiceRegistry 类型的定义，它具有 3 个核心属性（ServiceType、Lifetime 和 Factory），分别代表服务类型、生命周期模式和用来创建服务实例的工厂。最终用来创建服务实例的工厂体现为一个类型为 Func<Cat,Type[], object>的委托对象，它的两个输入分别代表当前使用的 Cat 对象以及提供服务类型的泛型参数，如果提供的服务类型并不是一个泛型类型，这个参数就会被指定为一个空的数组。

```
public class ServiceRegistry
{
    public Type                             ServiceType { get; }
    public Lifetime                         Lifetime { get; }
    public Func<Cat,Type[], object>         Factory { get; }
    internal ServiceRegistry                Next { get; set; }

    public ServiceRegistry(Type serviceType, Lifetime lifetime,
        Func<Cat,Type[], object> factory)
    {
        ServiceType     = serviceType;
        Lifetime        = lifetime;
        Factory         = factory;
    }
```

```csharp
internal IEnumerable<ServiceRegistry> AsEnumerable()
{
    var list = new List<ServiceRegistry>();
    for (var self = this; self!=null; self= self.Next)
    {
        list.Add(self);
    }
    return list;
}
```

将针对同一个服务类型(ServiceType 属性相同)的多个 ServiceRegistry 组成一个链表,那么作为相邻节点的两个 ServiceRegistry 对象将通过 Next 属性关联起来。我们为 ServiceRegistry 定义了一个 AsEnumerable 方法,使它返回由当前及后续节点组成的 ServiceRegistry 集合。如果当前 ServiceRegistry 为链表头,那么这个方法会返回链表上的所有 ServiceRegistry 对象。图 3-9 体现了服务注册的 3 个核心要素和链表结构。

图 3-9　服务注册

在了解了表示服务注册的 ServiceRegistry 之后,下面着重介绍表示依赖注入容器的 Cat 类型。如下面的代码片段所示,Cat 类型同时实现了 IServiceProvider 接口和 IDisposable 接口,定义在前者中的 GetService 方法用于提供服务实例。作为根容器的 Cat 对象通过公共构造函数创建,另一个内部构造函数则用来创建作为子容器的 Cat 对象,指定的 Cat 对象将作为父容器。

```csharp
public class Cat : IServiceProvider, IDisposable
{
    internal readonly Cat                                           _root;
    internal readonly ConcurrentDictionary<Type, ServiceRegistry>   _registries;
    private readonly ConcurrentDictionary<Key, object>              _services;
    private readonly ConcurrentBag<IDisposable>                     _disposables;
    private volatile bool _disposed;

    public Cat()
    {
        _registries     = new ConcurrentDictionary<Type, ServiceRegistry>();
        _root           = this;
        _services       = new ConcurrentDictionary<Key, object>();
        _disposables    = new ConcurrentBag<IDisposable>();
    }

    internal Cat(Cat parent)
    {
        _root           = parent._root;
```

```
    _registries     = _root._registries;
    _services       = new ConcurrentDictionary<Key, object>();
    _disposables    = new ConcurrentBag<IDisposable>();
}

private void EnsureNotDisposed()
{
    if (_disposed)
    {
        throw new ObjectDisposedException("Cat");
    }
}
...
}
```

作为根容器的 Cat 对象通过_root 字段表示。_registries 字段返回的 ConcurrentDictionary<Type, ServiceRegistry>对象用来存储所有添加的服务注册，该字典对象的 Key 和 Value 分别表示服务类型与 ServiceRegistry 链表，图 3-10 可以体现这一映射关系。由于需要负责完成对提供服务实例的释放工作，所以需要将实现了 IDisposable 接口的服务实例保存在通过_disposables 字段表示的集合中。

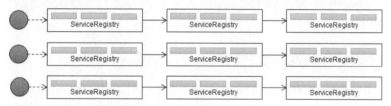

图 3-10　服务类型与服务注册链表的映射

由当前 Cat 对象提供的非 Transient 服务实例保存在由_services 字段表示的一个 ConcurrentDictionary<Key, object>对象上，该字典对象的键类型为如下代码片段中的 Key，它相当于创建服务实例所使用的 ServiceRegistry 对象和泛型参数类型数组的组合。

```
internal class Key : IEquatable<Key>
{
    public ServiceRegistry       Registry { get; }
    public Type[]                GenericArguments { get; }

    public Key(ServiceRegistry registry, Type[] genericArguments)
    {
        Registry            = registry;
        GenericArguments    = genericArguments;
    }

    public bool Equals(Key other)
    {
        if (Registry != other.Registry)
        {
            return false;
```

```csharp
        }
        if (GenericArguments.Length != other.GenericArguments.Length)
        {
            return false;
        }
        for (int index = 0; index < GenericArguments.Length; index++)
        {
            if (GenericArguments[index] != other.GenericArguments[index])
            {
                return false;
            }
        }
        return true;
    }

    public override int GetHashCode()
    {
        var hashCode = Registry.GetHashCode();
        for (int index = 0; index < GenericArguments.Length; index++)
        {
            hashCode ^= GenericArguments[index].GetHashCode();
        }
        return hashCode;
    }
    public override bool Equals(object obj) => obj is Key key ? Equals(key) : false;
}
```

虽然我们为 Cat 类型定义了若干扩展方法来提供多种不同的服务注册,但是这些方法最终都会调用如下代码片段中的 Register 方法,该方法会将提供的 ServiceRegistry 对象添加到 _registries 字段表示的字典对象中。值得注意的是,无论调用哪个 Cat 对象的 Register 方法,指定的 ServiceRegistry 对象都会被添加到作为根容器的 Cat 对象上。

```csharp
public class Cat : IServiceProvider, IDisposable
{
    public Cat Register(ServiceRegistry registry)
    {
        EnsureNotDisposed();
        if (_registries.TryGetValue(registry.ServiceType, out var existing))
        {
            _registries[registry.ServiceType] = registry;
            registry.Next = existing;
        }
        else
        {
            _registries[registry.ServiceType] = registry;
        }
        return this;
    }
    ...
}
```

用来提供服务实例的核心操作实现在如下代码片段的 GetServiceCore 方法中。如下面的代码片段所示，在调用 GetServiceCore 方法时需要指定对应的 ServiceRegistry 对象的服务类型的泛型参数。当该方法被执行的时候，对于 Transient 生命周期模式，它会直接利用 ServiceRegistry 对象提供的工厂来创建服务实例。如果服务实例的类型实现了 IDisposable 接口，它会被添加到_disposables 字段表示的待释放服务实例列表中。如果生命周期模式为 Root 和 Self，该方法会先根据提供的 ServiceRegistry 对象判断对应的服务实例是否已经存在，存在的服务实例会直接返回。

```csharp
public class Cat : IServiceProvider, IDisposable
{
    private object GetServiceCore(ServiceRegistry registry,
        Type[] genericArguments)
    {
        var key = new Key(registry, genericArguments);
        var serviceType = registry.ServiceType;

        switch (registry.Lifetime)
        {
            case Lifetime.Root: return GetOrCreate(_root._services,
                _root._disposables);
            case Lifetime.Self: return GetOrCreate(_services, _disposables);
            default:
                {
                    var service = registry.Factory(this, genericArguments);
                    if (service is IDisposable disposable && disposable != this)
                    {
                        _disposables.Add(disposable);
                    }
                    return service;
                }
        }

        object GetOrCreate(ConcurrentDictionary<Key, object> services,
            ConcurrentBag<IDisposable> disposables)
        {
            if (services.TryGetValue(key, out var service))
            {
                return service;
            }
            service = registry.Factory(this, genericArguments);
            services[key] = service;
            if (service is IDisposable disposable)
            {
                disposables.Add(disposable);
            }
            return service;
        }
    }
}
```

GetServiceCore 方法只有在指定 ServiceRegistry 对应的服务实例不存在的情况下才会利用提供的工厂来创建服务实例，创建的服务实例会根据生命周期模式保存到作为根容器的 Cat 对象或者当前 Cat 对象上。如果提供的服务实例实现了 IDisposable 接口，在采用 Root 生命周期模式时会被保存到作为根容器的 Cat 对象的待释放列表中。如果生命周期模式为 Self，那么它会被添加到当前 Cat 对象的待释放列表中。

在实现的 GetService 方法中，Cat 会根据指定的服务类型找到对应的 ServiceRegistry 对象，并最终调用 GetServiceCore 方法来提供对应的服务实例。GetService 方法还会解决一些特殊服务的供给问题：若服务类型为 Cat 或者 IServiceProvider，该方法返回的就是它自己；如果服务类型为 IEnumerable<T>，GetService 方法会根据泛型参数类型 T 找到所有的 ServiceRegistry 并利用它们来创建对应的服务实例，最终返回的是由这些服务实例组成的集合。除了这些，针对泛型服务实例的提供也是在 GetService 方法中解决的。

```
public class Cat : IServiceProvider, IDisposable
{
    public object GetService(Type serviceType)
    {
        EnsureNotDisposed();

        if (serviceType == typeof(Cat) || serviceType == typeof(IServiceProvider))
        {
            return this;
        }

        ServiceRegistry registry;
        //IEnumerable<T>
        if (serviceType.IsGenericType && serviceType.GetGenericTypeDefinition() ==
            typeof(IEnumerable<>))
        {
            var elementType = serviceType.GetGenericArguments()[0];
            if (!_registries.TryGetValue(elementType, out registry))
            {
                return Array.CreateInstance(elementType, 0);
            }

            var registries = registry.AsEnumerable();
            var services = registries.Select(it => GetServiceCore(it,
                Type.EmptyTypes)).ToArray();
            Array array = Array.CreateInstance(elementType, services.Length);
            services.CopyTo(array, 0);
            return array;
        }

        //Generic
        if (serviceType.IsGenericType && !_registries.ContainsKey(serviceType))
        {
            var definition = serviceType.GetGenericTypeDefinition();
```

```
            return _registries.TryGetValue(definition, out registry)
                ? GetServiceCore(registry, serviceType.GetGenericArguments())
                : null;
        }

        //Normal
        return _registries.TryGetValue(serviceType, out registry)
                ? GetServiceCore(registry, new Type[0])
                : null;
    }
    ...
}
```

在实现的 Dispose 方法中，由于所有待释放的服务实例已经保存到_disposables 字段表示的集合中，所以依次调用它们的 Dispose 方法即可。释放了所有服务实例并清空待释放列表之后，Dispose 方法还会清空 _services 字段表示的服务实例列表。

```
public class Cat : IServiceProvider, IDisposable
{
    public void Dispose()
    {
        _disposed = true;
        foreach(var disposable in _disposables)
        {
            disposable.Dispose();
        }
        _disposables.Clear();
        _services.Clear();
    }
    ...
}
```

3.4.3 扩展方法

为了方便服务注册，可以定义如下 6 个 Register 扩展方法。由于服务注册的添加总是需要调用 Cat 自身的 Register 扩展方法来完成，所以这些方法最终都需要创建一个代表服务注册的 ServiceRegistry 对象。对于一个 ServiceRegistry 对象来说，它最核心的元素就是表示服务实例创建工厂的 Func<Cat,Type[], object>对象，所以这 6 个扩展方法需要解决的问题就是创建一个委托对象。

```
public static class CatExtensions
{
    public static Cat Register(this Cat cat, Type from, Type to, Lifetime lifetime)
    {
        Func<Cat, Type[], object> factory =
            (_, arguments) => Create(_, to, arguments);
        cat.Register(new ServiceRegistry(from, lifetime, factory));
        return cat;
    }
```

```csharp
public static Cat Register<TFrom, TTo>(this Cat cat, Lifetime lifetime)
    where TTo:TFrom
    => cat.Register(typeof(TFrom), typeof(TTo), lifetime);

public static Cat Register(this Cat cat, Type serviceType, object instance)
{
    Func<Cat, Type[], object> factory = (_, arguments) => instance;
    cat.Register(new ServiceRegistry(serviceType, Lifetime.Root, factory));
    return cat;
}

public static Cat Register<TService>(this Cat cat, TService instance)
{
    Func<Cat, Type[], object> factory = (_, arguments) => instance;
    cat.Register(new ServiceRegistry(typeof(TService), Lifetime.Root,
        factory));
    return cat;
}

public static Cat Register(this Cat cat, Type serviceType,
    Func<Cat, object> factory, Lifetime lifetime)
{
    cat.Register(new ServiceRegistry(serviceType, lifetime,
        (_, arguments) => factory(_)));
    return cat;
}

public static Cat Register<TService>(this Cat cat,
    Func<Cat,TService> factory, Lifetime lifetime)
{
    cat.Register(new ServiceRegistry(typeof(TService), lifetime,
        (_,arguments)=>factory(_)));
    return cat;
}

private static object Create(Cat cat, Type type, Type[] genericArguments)
{
    if (genericArguments.Length > 0)
    {
        type = type.MakeGenericType(genericArguments);
    }
    var constructors = type.GetConstructors();
    if (constructors.Length == 0)
    {
        throw new InvalidOperationException($"Cannot create the instance of
            {type} which does not have a public constructor.");
    }
    var constructor = constructors.FirstOrDefault(it =>
        it.GetCustomAttributes(false).OfType<InjectionAttribute>().Any());
```

```csharp
        constructor ??= constructors.First();
        var parameters = constructor.GetParameters();
        if (parameters.Length == 0)
        {
            return Activator.CreateInstance(type);
        }
        var arguments = new object[parameters.Length];
        for (int index = 0; index < arguments.Length; index++)
        {
            arguments[index] = cat.GetService(parameters[index].ParameterType);
        }
        return constructor.Invoke(arguments);
    }
}
```

由于前两个重载指定的是服务实现类型，所以我们需要调用对应的构造函数来创建服务实例，这一逻辑在私有的 Create 方法中实现。第三个扩展方法直接指定服务实例，所以将提供的参数转换成一个 Func<Cat,Type[], object>非常容易。

我们刻意简化了构造函数的筛选逻辑。为了解决构造函数的选择问题，可以引入 InjectionAttribute 特性。如果将所有公共实例构造函数作为候选的构造函数，就会优先选择标注了该特性的构造函数。当构造函数被选择出来之后，我们需要通过分析其参数类型并利用 Cat 对象来提供具体的参数值，这实际上是一个递归的过程。最终我们将针对构造函数的调用转换成 Func<Cat,Type[], object>对象，进而创建出表示服务注册的 ServiceRegistry 对象。

```csharp
[AttributeUsage( AttributeTargets.Constructor)]
public class InjectionAttribute: Attribute {}
```

上述 6 个扩展方法可以完成针对单一服务的注册，有时项目中可能会出现非常多的服务需要注册，完成针对它们的批量注册是一个不错的选择。依赖注入框架提供了针对程序集范围的批量服务注册。为了标识带注册的服务，我们需要在服务实现类型上标注如下所示的 MapToAttribute 类型，并指定服务类型（一般为它实现的接口或者继承的基类）和生命周期。

```csharp
[AttributeUsage( AttributeTargets.Class, AllowMultiple = true)]
public sealed class MapToAttribute: Attribute
{
    public Type        ServiceType { get; }
    public Lifetime    Lifetime { get; }

    public MapToAttribute(Type serviceType, Lifetime lifetime)
    {
        ServiceType = serviceType;
        Lifetime = lifetime;
    }
}
```

针对程序集范围的批量服务注册在 Cat 的如下所示的 Register 扩展方法中实现。如下面的代码片段所示，该方法会从指定程序集中获取所有标注了 MapToAttribute 特性的类型，并提取服务类型、实现类型和生命周期模型，然后利用它们批量完成所需的服务注册。

```csharp
public static class CatExtensions
{
    public static Cat Register(this Cat cat, Assembly assembly)
    {
        var typedAttributes = from type in assembly.GetExportedTypes()
            let attribute = type.GetCustomAttribute<MapToAttribute>()
            where attribute != null
            select new { ServiceType = type, Attribute = attribute };
        foreach (var typedAttribute in typedAttributes)
        {
            cat.Register(typedAttribute.Attribute.ServiceType,
                typedAttribute.ServiceType, typedAttribute.Attribute.Lifetime);
        }
        return cat;
    }
}
```

除了上述 6 个用来注册服务的 Register 扩展方法，我们还为 Cat 类型定义了 3 个扩展方法：GetService<T>方法以泛型参数的形式指定服务类型，GetServices<T>方法会提供指定服务类型的所有实例，而 CreateChild 方法则帮助我们创建一个代表子容器的 Cat 对象。

```csharp
public static class CatExtensions
{
    public static T GetService<T>(this Cat cat) => (T)cat.GetService(typeof(T));
    public static IEnumerable<T> GetServices<T>(this Cat cat)
        => cat.GetService<IEnumerable<T>>();
    public static Cat CreateChild(this Cat cat) => new Cat(cat);
}
```

第 4 章

依赖注入（下篇）

毫不夸张地说，整个 ASP.NET Core 框架就是建立在依赖注入框架之上的。ASP.NET Core 应用在启动时构建管道，以及利用该管道处理每个请求过程中使用到的服务对象，均来源于依赖注入容器。该依赖注入容器不仅为 ASP.NET Core 框架自身提供必要的服务，还是应用程序的服务提供者，依赖注入已经成为 ASP.NET Core 应用的基本编程模式。第 3 章主要从理论层面讲述了依赖注入这种设计模式，本章主要介绍 .NET Core 的依赖注入框架。

4.1 利用容器提供服务

由于依赖注入对 ASP.NET Core 框架本身和基于该框架的应用开发都具有举足轻重的作用，所以本书的多个章节都会涉及这一主题。本章只是单纯地介绍这个独立的基础框架，不会涉及它在 ASP.NET Core 中的应用。本书在接下来讲述每个主题时都会采用"先简单体验，后深入剖析"的模式，所以我们先从编程的层面体验如何利用依赖注入容器来提供所需的服务实例。

4.1.1 服务的注册与消费

为了使读者更容易理解 .NET Core 提供的依赖注入框架，笔者在第 3 章创建了一个名为 Cat 的 Mini 版依赖注入框架。不论是编程模式还是实现原理，Cat 与我们即将介绍的依赖注入框架都非常相似。这个依赖注入框架主要涉及两个 NuGet 包，我们在编程过程中频繁使用的一些接口和基础数据类型都定义在 NuGet 包 "Microsoft.Extensions.DependencyInjection.Abstractions" 中[①]，而依赖注入的具体实现则由 NuGet 包 "Microsoft.Extensions.DependencyInjection" 来承载。

在设计 Cat 框架时，既可以将 Cat 对象作为提供服务实例的依赖注入容器，也可以将它作

[①] .NET Core 提供的很多基础框架和组件在设计的时候都会考虑"抽象和实现"的分离，并且将这种分离直接体现在部署的 NuGet 包上。例如，如果要开发一款名为 Foobar 的组件，微软会将一些抽象的接口和必要的类型定义在 NuGet 包 "Foobar.Abstractions" 中，而将具体的实现类型定义在 NuGet 包 "Foobar" 中。如果另一个组件或者框架需要使用 Foobar 组件，或者它需要对 Foobar 组件进行扩展与定制，只需要添加针对 NuGet 包 "Foobar.Abstractions" 的依赖即可。这种做法体现了"最小依赖"的设计原则，我们在进行组件或者框架开发的时候也可以采用这种做法。

为存放服务注册的集合,但是 .NET Core 依赖注入框架则将这两者分离开来。我们添加的服务注册被保存在通过 IServiceCollection 接口表示的集合之中,由这个集合创建的依赖注入容器体现为一个 IServiceProvider 对象。

作为依赖注入容器的 IServiceProvider 对象不仅具有类似于 Cat 的层次结构,两者对提供的服务实例也采用一致的生命周期管理方式。依赖注入框架利用 ServiceLifetime 来表示 Singleton、Scoped 和 Transient 这 3 种生命周期模式,笔者在 Cat 中则将其命名为 Root、Self 和 Transient,前者关注的是现象,后者关注的是内部实现。

```
public enum ServiceLifetime
{
    Singleton,
    Scoped,
    Transient
}
```

应用程序初始化过程中添加的服务注册是依赖注入容器用来提供所需服务实例的依据。由于 IServiceProvider 对象总是利用指定的服务类型来提供对应的服务实例,所以服务总是基于类型进行注册。我们倾向于利用接口来对服务进行抽象,所以这里的服务类型一般为接口,但是依赖注入框架对服务注册的类型并没有任何限制。具体的服务注册主要体现为如下 3 种形式,除了直接提供一个服务实例的注册形式(这种形式默认采用 Singleton 模式),在注册服务的时候还必须指定一个具体的生命周期模式。

- 指定具体的服务实现类型。
- 提供一个现成的服务实例。
- 指定一个创建服务实例的工厂。

我们的演示实例是一个普通的控制台应用。由于 NuGet 包 "Microsoft.Extensions.DependencyInjection" 承载了整个依赖注入框架的实现,所以应该添加该 NuGet 包的依赖。由于这是 ASP.NET Core 框架的基础 NuGet 包之一,所以可以通过修改项目文件并按照如下方式添加针对 "Microsoft.AspNetCore.App" 的框架引用(FrameworkReference)来引入该 NuGet 包。本书后续章节中采用 "Microsoft.NET.Sdk"[1] 作为 SDK 的演示实例,如果没有具体说明,则默认采用这种方式添加所需 NuGet 包的依赖。

```
<Project Sdk="Microsoft.NET.Sdk">
  <PropertyGroup>
    <OutputType>Exe</OutputType>
    <TargetFramework>netcoreapp3.0</TargetFramework>
  </PropertyGroup>
  <ItemGroup>
    <FrameworkReference Include="Microsoft.AspNetCore.App" />
  </ItemGroup>
</Project>
```

---

[1] 如果项目采用的 SDK 为 "Microsoft .NET.Sdk.Web",那么 ASP.NET Core 框架会自动添加基础 NuGet 包的依赖,所以无须指定对应的框架引用。

在添加了针对"Microsoft.Extensions.DependencyInjection"这个 NuGet 包的依赖之后,我们定义了如下接口和实现类型来表示相应的服务。如下面的代码片段所示,Foo、Bar 和 Baz 分别实现了对应的接口 IFoo、IBar 与 IBaz。为了反映 DI 框架对服务实例生命周期的控制,我们让它们派生于同一个基类 Base。Base 实现了 IDisposable 接口,我们在其构造函数和实现的 Dispose 方法中打印出相应的文字以确定对应的实例何时被创建与释放。我们还定义了一个泛型的接口 IFoobar<T1, T2>和对应的实现类 Foobar<T1, T2>来演示针对泛型服务实例的提供。

```csharp
public interface IFoo {}
public interface IBar {}
public interface IBaz {}
public interface IFoobar<T1, T2> {}
public class Base : IDisposable
{
    public Base()
        => Console.WriteLine($"An instance of {GetType().Name} is created.");
    public void Dispose()
        => Console.WriteLine($"The instance of {GetType().Name} is disposed.");
}

public class Foo : Base, IFoo, IDisposable { }
public class Bar : Base, IBar, IDisposable { }
public class Baz : Base, IBaz, IDisposable { }
public class Foobar<T1, T2>: IFoobar<T1,T2>
{
    public T1 Foo { get; }
    public T2 Bar { get; }
    public Foobar(T1 foo, T2 bar)
    {
        Foo = foo;
        Bar = bar;
    }
}
```

如下所示的代码片段创建了一个 ServiceCollection 对象(它是对 IServiceCollection 接口的默认实现),并调用相应的方法(AddTransient、AddScoped 和 AddSingleton)针对接口 IFoo、IBar 和 IBaz 注册了对应的服务,从方法命名可以看出注册的服务采用的生命周期模式分别为Transient、Scoped 和 Singleton。完成服务注册之后,我们调用 IServiceCollection 接口的BuildServiceProvider 扩展方法创建出代表依赖注入容器的 IServiceProvider 对象,并调用该对象的 GetService<T>方法来提供相应的服务实例。调试断言表明 IServiceProvider 对象提供的服务实例与预先添加的服务注册是一致的。(S401)

```csharp
class Program
{
    static void Main()
    {
        var provider = new ServiceCollection()
            .AddTransient<IFoo, Foo>()
```

```
            .AddScoped<IBar>(_ => new Bar())
            .AddSingleton<IBaz, Baz>()
            .BuildServiceProvider();
        Debug.Assert(provider.GetService<IFoo>() is Foo);
        Debug.Assert(provider.GetService<IBar>() is Bar);
        Debug.Assert(provider.GetService<IBaz>() is Baz);
    }
}
```

除了提供类似于 IFoo、IBar 和 IBaz 的服务实例，IServiceProvider 对象还能提供泛型服务实例。如下面的代码片段所示，为创建的 ServiceCollection 对象添加了针对 IFoo 接口和 IBar 接口的服务注册之后，我们调用 AddTransient 方法注册了针对泛型定义 IFoobar<,>的服务注册（实现的类型为 Foobar<,>）。当我们利用 ServiceCollection 对象创建出代表依赖注入容器的 IServiceProvider 对象并由它提供一个类型为 IFoobar<IFoo, IBar>的服务实例的时候，它会创建并返回一个 Foobar<Foo, Bar>对象。（S402）

```
public class Program
{
    public static void Main()
    {
        var provider = new ServiceCollection()
            .AddTransient<IFoo, Foo>()
            .AddTransient<IBar, Bar>()
            .AddTransient(typeof(IFoobar<,>), typeof(Foobar<,>))
            .BuildServiceProvider();

        var foobar = (Foobar<IFoo, IBar>)provider.GetService<IFoobar<IFoo, IBar>>();
        Debug.Assert(foobar.Foo is Foo);
        Debug.Assert(foobar.Bar is Bar);
    }
}
```

进行服务注册时可以为同一个类型添加多个服务注册。虽然添加的所有服务注册均是有效的，但是 GetService<T>扩展方法总是返回一个服务实例。依赖注入框架对该方法采用了"后来居上"的策略，也就是说，依赖注入容器总是采用最近添加的服务注册来创建服务实例。如果调用 GetServices<TService>扩展方法，该方法将利用指定服务类型的所有服务注册来提供一组服务实例。

下面的代码片段为创建的 ServiceCollection 对象添加了 3 个针对 Base 类型的服务注册，对应的实现类型分别为 Foo、Bar 和 Baz。我们将 Base 作为泛型参数调用了 GetServices<Base>方法，该方法会返回包含 3 个 Base 对象的集合，集合元素的类型分别为 Foo、Bar 和 Baz。（S403）

```
public class Program
{
    public static void Main()
    {
        var services = new ServiceCollection()
            .AddTransient<Base, Foo>()
            .AddTransient<Base, Bar>()
```

```
            .AddTransient<Base, Baz>()
            .BuildServiceProvider()
            .GetServices<Base>();
    Debug.Assert(services.OfType<Foo>().Any());
    Debug.Assert(services.OfType<Bar>().Any());
    Debug.Assert(services.OfType<Baz>().Any());
    }
}
```

对于 IServiceProvider 对象针对服务实例的提供还有一个细节：如果在调用 GetService 方法或者 GetService<T>方法时将服务类型设置为 IServiceProvider 接口，提供的服务实例实际上就是当前的 IServiceProvider 对象。这说明可以将代表依赖注入容器的 IServiceProvider 对象作为服务进行注入，这一特性体现在如下所示的调试断言中。第 3 章提及，一旦在应用中利用注入的 IServiceProvider 来获取其他依赖的服务实例，就意味着使用了 Service Locator 模式。这是一种反模式，当应用程序中出现了这样的代码时，应认真思考是否真的需要这么做。

```
var provider = new ServiceCollection().BuildServiceProvider();
Debug.Assert(provider.GetService<IServiceProvider>() == provider);
```

4.1.2 生命周期

代表依赖注入容器的 IServiceProvider 对象之间的层次结构创建了 3 种不同的生命周期模式。由于 Singleton 服务实例保存在作为根容器的 IServiceProvider 对象上，所以它能够在多个同根 IServiceProvider 对象之间提供真正的单例保证。Scoped 服务实例被保存在当前 IServiceProvider 对象上，所以它只能在当前范围内保证提供的实例是单例的。没有实现 IDisposable 接口的 Transient 服务则采用"即用即建，用后即弃"的策略。

下面通过对前面演示的实例稍做修改来演示 3 种不同生命周期模式的差异。如下所示的代码片段创建了一个 ServiceCollection 对象，并针对接口 IFoo、IBar 和 IBaz 注册了对应的服务，它们采用的生命周期模式分别为 Transient、Scoped 和 Singleton。利用 ServiceCollection 对象创建出代表依赖注入容器的 IServiceProvider 对象之后，我们调用其 CreateScope 方法创建了两个代表"服务范围"的 IServiceScope 对象，该对象的 ServiceProvider 属性返回一个新的 IServiceProvider 对象，它实际上是当前 IServiceProvider 对象的子容器。最后利用作为子容器的 IServiceProvider 对象来提供相应的服务实例。

```
class Program
{
    static void Main()
    {
        var root = new ServiceCollection()
            .AddTransient<IFoo, Foo>()
            .AddScoped<IBar>(_ => new Bar())
            .AddSingleton<IBaz, Baz>()
            .BuildServiceProvider();
        var provider1 = root.CreateScope().ServiceProvider;
        var provider2 = root.CreateScope().ServiceProvider;
```

```
        GetServices<IFoo>(provider1);
        GetServices<IBar>(provider1);
        GetServices<IBaz>(provider1);
        Console.WriteLine();
        GetServices<IFoo>(provider2);
        GetServices<IBar>(provider2);
        GetServices<IBaz>(provider2);

        static void GetServices<T>(IServiceProvider provider)
        {
            provider.GetService<T>();
            provider.GetService<T>();
        }
    }
}
```

运行上面的程序在控制台上输出的结果如图 4-1 所示。由于服务 IFoo 被注册为 Transient 服务，所以 IServiceProvider 对象针对该接口类型的 4 次调用都会创建一个全新的 Foo 对象。IBar 服务的生命周期模式为 Scoped，如果利用同一个 IServiceProvider 对象提供对应的服务实例，它只会创建一个 Bar 对象，所以整个程序在执行过程中会创建两个 Bar 对象。IBaz 服务采用 Singleton 生命周期，所以具有同根的两个 IServiceProvider 对象提供的总是同一个 Baz 对象，后者只会被创建一次。（S404）

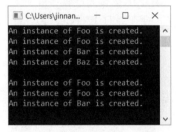

图 4-1　IServiceProvider 对象按照服务注册对应的生命周期模式提供服务实例

作为依赖注入容器的 IServiceProvider 对象不仅可以提供所需的服务实例，还可以管理这些服务实例的生命周期。如果某个服务类型实现了 IDisposable 接口，就意味着当生命周期完结的时候需要通过调用 Dispose 方法执行一些资源释放操作，这些操作同样由提供该服务实例的 IServiceProvider 对象来驱动执行。依赖注入框架针对提供服务实例的释放策略取决于对应的服务注册所采用的生命周期模式，具体的策略如下。

- Transient 和 Scoped：所有实现了 IDisposable 接口的服务实例会被当前 IServiceProvider 对象保存起来，当 IServiceProvider 对象的 Dispose 方法被调用的时候，这些服务实例的 Dispose 方法会随之被调用。
- Singleton：由于服务实例保存在作为根容器的 IServiceProvider 对象上，所以只有当后者的 Dispose 方法被调用的时候，这些服务实例的 Dispose 方法才会随之被调用。

对于一个 ASP.NET Core 应用来说，它具有一个与当前应用绑定代表全局根容器的

IServiceProvider 对象。对于处理的每一次请求，ASP.NET Core 框架都会利用这个根容器来创建基于当前请求的服务范围，并利用后者提供的 IServiceProvider 对象来提供请求处理所需的服务实例。请求处理完成之后，创建的服务范围被终结，对应的 IServiceProvider 对象也随之被释放，此时由该 IServiceProvider 对象提供的 Scoped 服务实例以及实现了 IDisposable 接口的 Transient 服务实例得以及时释放。

上述释放策略可以通过如下演示实例进行印证。如下代码片段创建了一个 ServiceCollection 对象，并针对不同的生命周期模式添加了针对 IFoo、IBar 和 IBaz 的服务注册。利用 ServiceCollection 集合创建出作为根容器的 IServiceProvider 对象之后，可以调用它的 CreateScope 方法创建出对应的服务范围。然后我们利用创建的服务范围得到代表子容器的 IServiceProvider 对象，并用它提供了 3 个注册服务对应的实例。

```
class Program
{
    static void Main()
    {
        using (var root = new ServiceCollection()
            .AddTransient<IFoo, Foo>()
            .AddScoped<IBar, Bar>()
            .AddSingleton<IBaz, Baz>()
            .BuildServiceProvider())
        {
            using (var scope = root.CreateScope())
            {
                var provider = scope.ServiceProvider;
                provider.GetService<IFoo>();
                provider.GetService<IBar>();
                provider.GetService<IBaz>();
                Console.WriteLine("Child container is disposed.");
            }
            Console.WriteLine("Root container is disposed.");
        }
    }
}
```

由于代表根容器的 IServiceProvider 对象和服务范围的创建都是在 using 块中进行的，所以所有针对它们的 Dispose 方法都会在 using 块结束的地方被调用。为了确定方法被调用的时机，可以在控制台上打印相应的文字。该程序运行之后在控制台上输出的结果如图 4-2 所示，可以看到，当作为子容器的 IServiceProvider 对象被释放的时候，由它提供的两个生命周期模式分别为 Transient 和 Scoped 的两个服务实例（Foo 和 Bar）被正常释放。而对于生命周期模式为 Singleton 的服务实例 Baz，它的 Dispose 方法会延迟到作为根容器的 IServiceProvider 对象被释放的时候才释放。（S405）

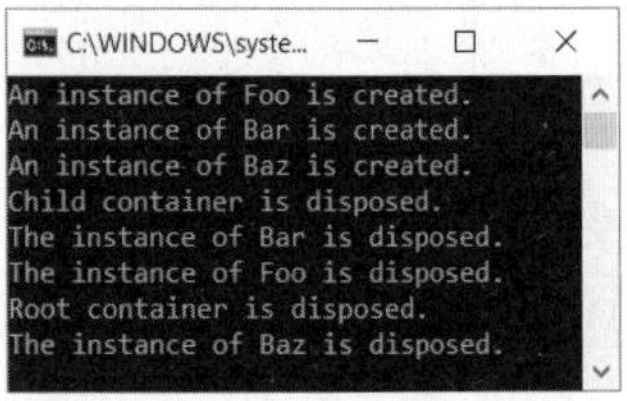

图4-2 服务实例的释放

4.1.3 针对服务注册的验证

Singleton 和 Scoped 这两种不同的生命周期是通过将提供的服务实例分别存放到作为根容器的 IServiceProvider 对象和当前 IServiceProvider 对象来实现的，这意味着作为根容器的 IServiceProvider 对象提供的 Scoped 服务实例也是单例的。如果某个 Singleton 服务依赖另一个 Scoped 服务，那么 Scoped 服务实例将被一个 Singleton 服务实例所引用，也就意味着 Scoped 服务实例成了一个 Singleton 服务实例。

在 ASP.NET Core 应用中，将某个服务注册的生命周期设置为 Scoped 的真正意图是希望依赖注入容器根据接收的每个请求来创建和释放服务实例，但是一旦出现上述这种情况，就意味着 Scoped 服务实例将变成一个 Singleton 服务实例，这样的 Scoped 服务实例直到应用关闭才会被释放，这无疑不是我们希望得到的结果。如果某个 Scoped 服务实例引用的资源（如数据库连接）需要被及时释放，这可能会造成难以估量的后果。为了避免这种情况的出现，在利用 IServiceProvider 对象提供服务的过程中可以开启针对服务范围的验证。

如果希望 IServiceProvider 对象在提供服务的过程中可以对服务范围做有效性检验，只需要在调用 IServiceCollection 接口的 BuildServiceProvider 扩展方法时，将一个布尔类型的 True 值作为参数即可。下面的演示程序定义了两个服务接口（IFoo 和 IBar）和对应的实现类型（Foo 和 Bar），其中，Foo 需要依赖 IBar。如果将 IFoo 和 IBar 分别注册为 Singleton 服务与 Scoped 服务，当调用 BuildServiceProvider 方法创建代表依赖注入容器的 IServiceProvider 对象的时候，我们将参数设置为 True 以开启针对服务范围的检验。最后分别利用代表根容器和子容器的 IServiceProvider 对象来提供这两种类型的服务实例。

```
class Program
{
    static void Main()
    {
        var root = new ServiceCollection()
            .AddSingleton<IFoo, Foo>()
            .AddScoped<IBar, Bar>()
            .BuildServiceProvider(true);
        var child = root.CreateScope().ServiceProvider;

        void ResolveService<T>(IServiceProvider provider)
        {
            var isRootContainer = root == provider ? "Yes" : "No";
            try
```

```
        {
            provider.GetService<T>();
            Console.WriteLine( $"Status: Success;
                Service Type: {typeof(T).Name}; Root: {isRootContainer}");
        }
        catch (Exception ex)
        {
            Console.WriteLine($"Status: Fail;
                Service Type: {typeof(T).Name}; Root: {isRootContainer}");
            Console.WriteLine($"Error: {ex.Message}");
        }
    }

    ResolveService<IFoo>(root);
    ResolveService<IBar>(root);
    ResolveService<IFoo>(child);
    ResolveService<IBar>(child);
    }
}

public interface IFoo {}
public interface IBar {}
public class Foo : IFoo
{
    public IBar Bar { get; }
    public Foo(IBar bar) => Bar = bar;
}
public class Bar : IBar {}
```

上面的演示实例启动之后在控制台上输出的结果如图 4-3 所示。从输出结果可以看出，4 个服务解析只有 1 次（使用代表子容器的 IServiceProvider 提供 IBar 服务实例）是成功的。这个实例充分说明：一旦开启了针对服务范围的验证，IServiceProvider 对象不可能提供以单例形式存在的 Scoped 服务。（S406）

图 4-3　IServiceProvider 针对服务范围的检验

针对服务范围的检验体现在配置选项类型 ServiceProviderOptions 的 ValidateScopes 属性上。如下面的代码片段所示，ServiceProviderOptions 还具有另一个名为 ValidateOnBuild 的属性，如果将该属性设置为 True，就意味着 IServiceProvider 对象被构建的时候会检验提供的每个 ServiceDescriptor 的有效性，即确保它们最终都具有提供对应服务实例的能力。在默认情况下，ValidateOnBuild 的属性值为 False，意味着只有利用 IServiceProvider 对象来提供我们所需的服务实例，相应的异常采用才会被抛出来。

```
public class ServiceProviderOptions
{
    public bool ValidateScopes { get; set; }
    public bool ValidateOnBuild { get; set; }
}
```

我们照例来做一个在构建 IServiceProvider 对象时检验服务注册有效性的实例。如下代码片段定义了一个接口 IFoobar 和对应的实现类型 Foobar，若采用单例的形式来使用 Foobar 对象，可以定义唯一的私有构造函数。

```
public interface IFoobar {}
public class Foobar : IFoobar
{
    private Foobar() {}
    public static readonly Foobar Instance = new Foobar();
}
```

如下所示的演示实例定义了一个内嵌的 BuildServiceProvider 方法，从而完成针对 IFoobar/Foobar 的服务注册和最终对 IServiceProvider 对象的创建。在调用 BuildServiceProvider 扩展方法创建对应 IServiceProvider 对象时指定了一个 ServiceProviderOptions 对象，而该对象的 ValidateOnBuild 属性来源于内嵌方法的同名参数。

```
class Program
{
    static void Main()
    {
        BuildServiceProvider(false);
        BuildServiceProvider(true);

        static void BuildServiceProvider(bool validateOnBuild)
        {
            try
            {
                var options = new ServiceProviderOptions
                {
                    ValidateOnBuild = validateOnBuild
                };
                new ServiceCollection()
                    .AddSingleton<IFoobar, Foobar>()
                    .BuildServiceProvider(options);
                Console.WriteLine(
                    $"Status: Success; ValidateOnBuild: {validateOnBuild}");
            }
            catch (Exception ex)
            {
                Console.WriteLine(
                    $"Status: Fail; ValidateOnBuild: {validateOnBuild}");
                Console.WriteLine($"Error: {ex.Message}");
            }
        }
    }
}
```

```
        }
    }
}
```

由于 Foobar 具有唯一的私有构造函数，而内嵌方法 BuildServiceProvider 提供的服务注册并不能提供我们所需的服务实例，所以这个服务注册是无效的。由于在默认情况下构建 IServiceProvider 对象的时候并不会对服务注册做有效性检验，所以此时无效的服务注册并不会及时被探测到。一旦将 ValidateOnBuild 选项设置为 True，IServiceProvider 对象在被构建的时候就会抛出异常，图 4-4 所示的输出结果就体现了这一点。（S407）

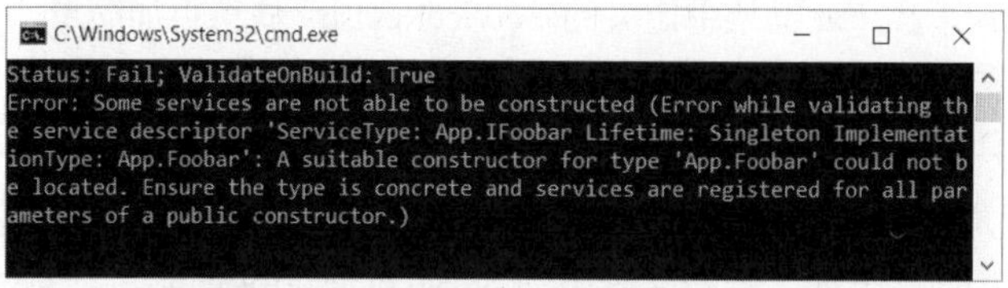

图 4-4　构建 IServiceProvider 对象针对服务注册有效性的检验

4.2　服务注册

由上面的实例演示可知，作为依赖注入容器的 IServiceProvider 对象是通过调用 IServiceCollection 接口的 BuildServiceProvider 扩展方法创建的，IServiceCollection 对象是一个存放服务注册信息的集合。第 3 章创建的 Cat 框架中的服务注册是通过类型 ServiceRegistry 表示的，在 .NET Core 依赖注入框架中，与 ServiceRegistry 对应的类型是 ServiceDescriptor。

4.2.1　ServiceDescriptor

ServiceDescriptor 是对某个服务注册项的描述，作为依赖注入容器的 IServiceProvider 对象正是利用该对象提供的描述信息才得以提供我们需要的服务实例。服务描述总是注册到通过 ServiceType 属性表示的服务类型上，ServiceDescriptor 的 Lifetime 表示采用的生命周期模式。

```csharp
public class ServiceDescriptor
{
    public Type                              ServiceType { get; }
    public ServiceLifetime                   Lifetime { get; }

    public Type                              ImplementationType { get; get; }
    public Func<IServiceProvider, object>    ImplementationFactory { get; }
    public object                            ImplementationInstance { get; }

    public ServiceDescriptor(Type serviceType, object instance);
    public ServiceDescriptor(Type serviceType,
        Func<IServiceProvider, object> factory, ServiceLifetime lifetime);
    public ServiceDescriptor(Type serviceType, Type implementationType,
        ServiceLifetime lifetime);
}
```

ServiceDescriptor 的其他 3 个属性体现了服务实例的 3 种提供方式，并且分别对应 3 个构造函数。如果指定了服务的实现类型（对应 ImplementationType 属性），那么最终的服务实例将通过调用定义在该类型中的某个构造函数来创建。如果指定的是一个 Func<IServiceProvider, object>对象（对应 ImplementationFactory 属性），那么该委托对象将作为提供服务实例的工厂。如果直接指定一个现成的对象（对应的属性为 ImplementationInstance），那么该对象就是最终提供的服务实例。

如果采用现成的服务实例来创建 ServiceDescriptor 对象，对应的服务注册就会采用 Singleton 生命周期模式。对于通过其他两个构造函数创建的 ServiceDescriptor 对象来说，需要显式指定采用的生命周期模式。相较于 ServiceDescriptor，在 Cat 框架中定义的 ServiceRegistry 显得更加简单，因为我们直接提供了一个类型为 Func<Cat,Type[], object>的对象来提供对应的服务实例。

除了可以调用上面介绍的 3 个构造函数来创建对应的 ServiceDescriptor 对象，还可以利用定义在 ServiceDescriptor 类型中的一系列静态方法来创建该对象。如下面的代码片段所示，ServiceDescriptor 提供了两个名为 Describe 的方法重载来创建对应的 ServiceDescriptor 对象。

```
public class ServiceDescriptor
{
    public static ServiceDescriptor Describe(Type serviceType,
        Func<IServiceProvider, object> implementationFactory,
        ServiceLifetime lifetime);
    public static ServiceDescriptor Describe(Type serviceType,
        Type implementationType, ServiceLifetime lifetime);
}
```

如果调用上面两个 Describe 方法来创建 ServiceDescriptor 对象，就需要指定采用的生命周期模式，为了使对象创建变得更加简单，ServiceDescriptor 类型中还定义了一系列针对具体生命周期模式的静态工厂方法。如下所示的代码片段是针对 Singleton 模式的一组静态工厂方法重载的定义，针对其他两种模式 Scoped 和 Transient 的方法具有类似的定义。

```
public class ServiceDescriptor
{
    public static ServiceDescriptor Singleton
        <TService, TImplementation>()
        where TService: class where TImplementation: class, TService;
    public static ServiceDescriptor Singleton
        <TService, TImplementation>(
        Func<IServiceProvider, TImplementation> implementationFactory)
        where TService: class where TImplementation: class, TService;
    public static ServiceDescriptor Singleton<TService>(
        Func<IServiceProvider, TService> implementationFactory)
        where TService: class;
    public static ServiceDescriptor Singleton<TService>(
        TService implementationInstance) where TService: class;
    public static ServiceDescriptor Singleton(Type serviceType,
        Func<IServiceProvider, object> implementationFactory);
    public static ServiceDescriptor Singleton(Type serviceType,
        object implementationInstance);
    public static ServiceDescriptor Singleton(Type service,
```

```
        Type implementationType);
}
```

4.2.2 IServiceCollection

依赖注入框架将服务注册存储在一个通过 IServiceCollection 接口表示的集合之中。如下面的代码片段所示，一个 IServiceCollection 对象本质上就是一个元素类型为 ServiceDescriptor 的列表。在默认情况下使用的是实现该接口的 ServiceCollection 类型。

```
public interface IServiceCollection : IList<ServiceDescriptor> {}
public class ServiceCollection : IServiceCollection {}
```

应用启动时，针对服务的注册本质上就是创建相应的 ServiceDescriptor 对象并将其添加到指定 IServiceCollection 对象中的过程。考虑到服务注册是一个高频调用的操作，所以依赖注入框架为 IServiceCollection 接口定义了一系列扩展方法来完成服务注册。如下所示的两个 Add 方法可以将指定的一个或者多个 ServiceDescriptor 对象添加到 IServiceCollection 集合中。

```
public static class ServiceCollectionDescriptorExtensions
{
    public static IServiceCollection Add(this IServiceCollection collection,
        ServiceDescriptor descriptor);
    public static IServiceCollection Add(this IServiceCollection collection,
        IEnumerable<ServiceDescriptor> descriptors);
}
```

依赖注入框架还针对具体生命周期模式为 IServiceCollection 的接口定义了一系列的扩展方法，它们会根据提供的输入创建对应的 ServiceDescriptor 对象，并将其添加到指定的 IServiceCollection 对象中。如下所示的代码片段是针对 Singleton 模式的 AddSingleton 方法重载的定义，针对其他两个生命周期模式的 AddScoped 方法和 AddTransient 方法具有类似的定义。

```
public static class ServiceCollectionServiceExtensions
{
    public static IServiceCollection AddSingleton<TService>(
        this IServiceCollection services) where TService: class;
    public static IServiceCollection AddSingleton<TService, TImplementation>(
        this IServiceCollection services)
        where TService: class
        where TImplementation: class, TService;
    public static IServiceCollection AddSingleton<TService>(
        this IServiceCollection services, TService implementationInstance)
        where TService: class;
    public static IServiceCollection AddSingleton<TService, TImplementation>(
        this IServiceCollection services,
        Func<IServiceProvider, TImplementation> implementationFactory)
        where TService: class where TImplementation: class, TService;
    public static IServiceCollection AddSingleton<TService>(
        this IServiceCollection services,
        Func<IServiceProvider, TService> implementationFactory)
        where TService: class;
    public static IServiceCollection AddSingleton(
        this IServiceCollection services, Type serviceType);
```

```csharp
    public static IServiceCollection AddSingleton(this IServiceCollection services,
        Type serviceType, Func<IServiceProvider, object> implementationFactory);
    public static IServiceCollection AddSingleton(this IServiceCollection services,
        Type serviceType, object implementationInstance);
    public static IServiceCollection AddSingleton(this IServiceCollection services,
        Type serviceType, Type implementationType);
}
```

虽然针对同一个服务类型可以添加多个 ServiceDescriptor 对象，但这种情况只有在应用需要使用同一类型的多个服务实例的情况下才有意义，所以可以注册多个 ServiceDescriptor 对象来提供同一个主题的多个订阅者。如果总是根据指定的服务类型来提取单一的服务实例，那么一个服务类型只需要一个 ServiceDescriptor 对象就够了。对于这种场景我们可能会使用如下两个名为 TryAdd 的扩展方法，该方法会根据指定 ServiceDescriptor 提供的服务类型判断对应的服务注册是否存在，只有在指定类型的服务注册不存在的情况下，ServiceDescriptor 才会被添加到指定的 IServiceCollection 对象之中。

```csharp
public static class ServiceCollectionDescriptorExtensions
{
    public static void TryAdd(this IServiceCollection collection,
        ServiceDescriptor descriptor);
    public static void TryAdd(this IServiceCollection collection,
        IEnumerable<ServiceDescriptor> descriptors);
}
```

TryAdd 扩展方法同样具有基于 3 种生命周期模式的版本，如下所示的代码片段是针对 Singleton 模式的 TryAddSingleton 方法的定义。在指定服务类型对应的 ServiceDescriptor 对象不存在的情况下，这些方法会根据提供的实现类型、服务实例或者服务实例工厂来创建生命周期模式为 Singleton 的 ServiceDescriptor 对象，并将其添加到指定的 IServiceCollection 对象之中。针对其他两种生命周期模式的 TryAddScoped 方法和 TryAddTransient 方法具有类似的定义。

```csharp
public static class ServiceCollectionDescriptorExtensions
{
    public static void TryAddSingleton<TService>(this IServiceCollection collection)
        where TService: class;
    public static void TryAddSingleton<TService, TImplementation>(
        this IServiceCollection collection)
        where TService: class
        where TImplementation: class, TService;
    public static void TryAddSingleton(this IServiceCollection collection,
        Type service);
    public static void TryAddSingleton<TService>(this IServiceCollection collection,
        TService instance) where TService: class;
    public static void TryAddSingleton<TService>(this IServiceCollection services,
        Func<IServiceProvider, TService> implementationFactory)
        where TService: class;
    public static void TryAddSingleton(this IServiceCollection collection,
        Type service, Func<IServiceProvider, object> implementationFactory);
    public static void TryAddSingleton(this IServiceCollection collection,
        Type service, Type implementationType);
```

}

除了上面介绍的扩展方法 TryAdd 和 TryAdd{Lifetime}，IServiceCollection 接口还具有如下两个名为 TryAddEnumerable 的扩展方法。当 TryAddEnumerable 方法在决定将指定的 ServiceDescriptor 添加到 IServiceCollection 对象之前，它也会做存在性检验。与 TryAdd 方法和 TryAdd{Lifetime}方法不同的是，TryAddEnumerable 方法在判断执行的 ServiceDescriptor 是否存在时同时考虑服务类型和实现类型。

```
public static class ServiceCollectionDescriptorExtensions
{
    public static void TryAddEnumerable(this IServiceCollection services,
        ServiceDescriptor descriptor);
    public static void TryAddEnumerable(this IServiceCollection services,
        IEnumerable<ServiceDescriptor> descriptors);
}
```

TryAddEnumerable 方法用来判断存在性的实现类型不是只有 ServiceDescriptor 的 ImplementationType 属性。如果 ServiceDescriptor 是通过一个指定的服务实例创建的，那么该实例的类型会用来判断对应的服务注册是否存在。如果 ServiceDescriptor 是通过提供的服务实例工厂创建的，那么代表服务实例创建工厂的 Func<in T, out TResult>对象的第二个参数类型将被用于判断 ServiceDescriptor 的存在性。TryAddEnumerable 扩展方法的实现逻辑可以通过如下代码片段进行验证。

```
var services = new ServiceCollection();

services.TryAddEnumerable(ServiceDescriptor.Singleton<IFoobarbazqux, Foo>());
Debug.Assert(services.Count == 1);

services.TryAddEnumerable(ServiceDescriptor.Singleton<IFoobarbazqux, Foo>());
Debug.Assert(services.Count == 1);

services.TryAddEnumerable(ServiceDescriptor.Singleton<IFoobarbazqux>(new Foo()));
Debug.Assert(services.Count == 1);

Func<IServiceProvider, Foo> factory4Foo = _ => new Foo();
services.TryAddEnumerable(ServiceDescriptor.Singleton<IFoobarbazqux>(factory4Foo));
Debug.Assert(services.Count == 1);

services.TryAddEnumerable(ServiceDescriptor.Singleton<IFoobarbazqux, Bar>());
Debug.Assert(services.Count == 2);

services.TryAddEnumerable(ServiceDescriptor.Singleton<IFoobarbazqux>(new Baz()));
Debug.Assert(services.Count == 3);

Func<IServiceProvider, Qux> factory4Qux = _ => new Qux();
services.TryAddEnumerable(ServiceDescriptor.Singleton<IFoobarbazqux>(factory4Qux));
Debug.Assert(services.Count == 4);
```

如果通过上述策略得到的实现类型为 Object，那么 TryAddEnumerable 方法会因为实现类型不明确而抛出一个 ArgumentException 类型的异常。这主要发生在提供的 ServiceDescriptor 对象是由服务实例工厂创建的情况下，所以上面实例中用来创建 ServiceDescriptor 的工厂类型分别为 Func<IServiceProvider, Foo>和 Func<IServiceProvider, Qux>，而不是 Func<IServiceProvider, object>。

```
var service = ServiceDescriptor.Singleton<IFoobarbazqux>(_ => new Foo());
new ServiceCollection().TryAddEnumerable(service);
```

如果采用如上所示的方式利用一个 Lamda 表达式创建一个 ServiceDescriptor 对象，对于创建的 ServiceDescriptor 来说，其服务实例工厂就是一个 Func<IServiceProvider, object>对象，所以将它作为参数调用 TryAddEnumerable 方法时会抛出图 4-5 所示的 ArgumentException 异常，并提示"Implementation type cannot be 'App.IFoobarbazqux' because it is indistinguishable from other services registered for 'App.IFoobarbazqux'."。

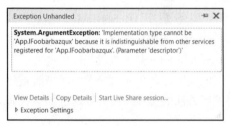

图 4-5　实现类型不明确导致的异常

上面介绍的这些方法最终的目的都是在指定的 IServiceCollection 集合中添加新的 ServiceDescriptor 对象，有时需要删除或者替换现有的某个 ServiceDescriptor 对象，这种情况通常发生在需要对当前使用框架中由某个服务提供的功能进行定制的时候。由于 IServiceCollection 实现了 IList<ServiceDescriptor>接口，所以可以调用其 Clear 方法、Remove 方法和 RemoveAt 方法来清除或者删除现有的 ServiceDescriptor 对象。除此之外，还可以选择如下这些扩展方法。

```
public static class ServiceCollectionDescriptorExtensions
{
    public static IServiceCollection RemoveAll<T>(
        this IServiceCollection collection);
    public static IServiceCollection RemoveAll(this IServiceCollection collection,
        Type serviceType);
    public static IServiceCollection Replace(this IServiceCollection collection,
        ServiceDescriptor descriptor);
}
```

RemoveAll 方法和 RemoveAll<T>方法可以帮助我们根据指定的服务类型删除现有的 ServiceDescriptor 对象。Replace 方法会使用指定的 ServiceDescriptor 替换第一个具有相同服务类型（对应 ServiceType 属性）的 ServiceDescriptor，实际操作是先删除后添加。如果从目前的 IServiceCollection 集合中找不到服务类型匹配的 ServiceDescriptor 对象，那么指定的 ServiceDescriptor 对象会直接添加到 IServiceCollection 对象中，这一逻辑也可以利用如下程序进行验证。

```
var services = new ServiceCollection();
services.Replace(ServiceDescriptor.Singleton<IFoobarbazqux, Foo>());
Debug.Assert(services.Any(it => it.ImplementationType == typeof(Foo)));

services.AddSingleton<IFoobarbazqux, Bar>();
services.Replace(ServiceDescriptor.Singleton<IFoobarbazqux, Baz>());
Debug.Assert(!services.Any(it=>it.ImplementationType == typeof(Foo)));
Debug.Assert(services.Any(it => it.ImplementationType == typeof(Bar)));
Debug.Assert(services.Any(it => it.ImplementationType == typeof(Baz)));
```

4.3 服务的消费

包含服务注册信息的 IServiceCollection 集合最终被用来创建作为依赖注入容器的 IServiceProvider 对象。当需要消费某个服务实例的时候，只需要指定服务类型调用 IServiceProvider 接口的 GetService 方法即可，IServiceProvider 对象就会根据对应的服务注册提供所需的服务实例。

4.3.1 IServiceProvider

如下面的代码片段所示，IServiceProvider 接口定义了唯一的 GetService 方法，可以根据指定的类型来提供对应的服务实例。当利用包含服务注册的 IServiceCollection 对象创建出 IServiceProvider 对象之后，我们需要将服务注册的服务类型（对应 ServiceDescriptor 的 ServiceType 属性）作为参数调用 GetService 方法，该方法能够根据服务注册信息提供对应的服务实例。

```
public interface IServiceProvider
{
    object GetService(Type serviceType);
}
```

针对 IServiceProvider 对象的创建，体现在 IServiceCollection 接口的 3 个 BuildServiceProvider 扩展方法重载上。如下面的代码片段所示，这 3 个扩展方法提供的都是一个类型为 ServiceProvider 的对象，该对象是根据提供的配置选项创建的。配置选项类型 ServiceProviderOptions 提供了两个属性：ValidateScopes 属性表示是否需要开启针对服务范围的检验；ValidateOnBuild 属性表示是否需要预先检验作为服务注册的每个 ServiceDescriptor 对象能否提供对应的服务实例。在默认情况下，ValidateScopes 和 ValidateOnBuild 描述的检验都是关闭的。

```
public class ServiceProviderOptions
{
    public bool ValidateScopes { get; set; }
    public bool ValidateOnBuild { get; set; }
    internal static readonly ServiceProviderOptions Default
        = new ServiceProviderOptions();
}
```

```csharp
public static class ServiceCollectionContainerBuilderExtensions
{
    public static ServiceProvider BuildServiceProvider(
        this IServiceCollection services)
        => BuildServiceProvider(services, ServiceProviderOptions.Default);
    public static ServiceProvider BuildServiceProvider(
        this IServiceCollection services, bool validateScopes)
        => services.BuildServiceProvider(new ServiceProviderOptions {
            ValidateScopes = validateScopes });
    public static ServiceProvider BuildServiceProvider(
        this IServiceCollection services, ServiceProviderOptions options)
        => new ServiceProvider(services, options);
}
```

虽然调用 IServiceCollection 接口的 BuildServiceProvider 扩展方法返回的总是一个 ServiceProvider 对象，但是笔者并不打算详细介绍这个类型，这是因为实现在该类型中针对服务实例的提供机制一直在不断变化，而且这个变化趋势在未来版本更替过程中可能还将继续下去。除此之外，ServiceProvider 对象还涉及一系列内部类型和接口，所以本书不涉及具体的细节，只进行总体设计。

除了定义在 IServiceProvider 接口中的 GetService 方法，该接口提供服务实例的扩展方法还有如下几种：GetService<T>方法以泛型参数的形式指定了服务类型，返回的服务实例也会做对应的类型转换。如果指定服务类型的服务注册不存在，GetService 方法会返回 Null，如果调用 GetRequiredService 方法或者 GetRequiredService<T>方法则会抛出一个 InvalidOperationException 类型的异常。如果所需的服务实例是必需的，我们一般会调用这两个 GetRequiredService 扩展方法。

```csharp
public static class ServiceProviderServiceExtensions
{
    public static T GetService<T>(this IServiceProvider provider);

    public static T GetRequiredService<T>(this IServiceProvider provider);
    public static object GetRequiredService(this IServiceProvider provider,
        Type serviceType);

    public static IEnumerable<T> GetServices<T>(this IServiceProvider provider);
    public static IEnumerable<object> GetServices(this IServiceProvider provider,
        Type serviceType);
}
```

如果针对某个类型添加了多个服务注册，那么 GetService 方法总是采用最新添加的服务注册来提供服务实例。如果希望利用所有的服务注册创建一组服务实例列表，我们既可以调用 GetServices 方法或者 GetServices<T>方法，也可以调用 GetService<IEnumerable<T>>方法。

4.3.2 服务实例的创建

对于通过调用 IServiceCollection 集合的 BuildServiceProvider 方法创建的 IServiceProvider 对象来说，如果通过指定服务类型调用其 GetService 方法以获取对应的服务实例，那么它总是会根据提供的服务类型从服务注册列表中找到对应的 ServiceDescriptor 对象，并根据它来提供所需的服务实例。

ServiceDescriptor 对象具有 3 个不同的构造函数，分别对应服务实例最初的 3 种提供方式，我们既可以提供一个 Func<IServiceProvider, object>对象作为工厂来创建对应的服务实例，也可以直接提供一个创建好的服务实例。如果提供的是服务的实现类型，最终提供的服务实例将通过调用该类型的某个构造函数来创建，那么构造函数是通过什么策略被选择出来的？

如果 IServiceProvider 对象试图通过调用构造函数的方式来创建服务实例，传入构造函数的所有参数必须先被初始化，所以最终被选择的构造函数必须具备一个基本的条件，即 IServiceProvider 对象能够提供构造函数的所有参数。为了使读者能够更加深刻地理解 IServiceProvider 对象在构造函数选择过程中采用的策略，我们会采用实例演示的方式对此进行讲述。

我们在一个控制台应用中定义了 4 个服务接口（IFoo、IBar、IBaz 和 IQux）以及实现它们的 4 个类（Foo、Bar、Baz 和 Qux）。如下面的代码片段所示，我们为 Qux 定义了 3 个构造函数，参数都定义了服务接口类型。为了确定 IServiceProvider 最终选择哪个构造函数来创建目标服务实例，我们在构造函数执行时在控制台上输出相应的指示性文字。

```csharp
public interface IFoo {}
public interface IBar {}
public interface IBaz {}
public interface IQux {}

public class Foo : IFoo {}
public class Bar : IBar {}
public class Baz : IBaz {}
public class Qux : IQux
{
    public Qux(IFoo foo)
        => Console.WriteLine("Selected constructor: Qux(IFoo)");
    public Qux(IFoo foo, IBar bar)
        => Console.WriteLine("Selected constructor: Qux(IFoo, IBar)");
    public Qux(IFoo foo, IBar bar, IBaz baz)
        => Console.WriteLine("Selected constructor: Qux(IFoo, IBar, IBaz)");
}
```

如下演示程序创建了一个 ServiceCollection 对象，并在其中添加针对 IFoo、IBar 及 IQux 这 3 个服务接口的服务注册，但未添加针对服务接口 IBaz 的注册。利用由这个 ServiceCollection 对象创建的 IServiceProvider 来提供针对服务接口 IQux 的实例，是否能够得到一个 Qux 对象？如果可以，它又是通过执行哪个构造函数创建的？

```csharp
class Program
{
    static void Main()
    {
        new ServiceCollection()
            .AddTransient<IFoo, Foo>()
            .AddTransient<IBar, Bar>()
```

```
            .AddTransient<IQux, Qux>()
            .BuildServiceProvider()
            .GetServices<IQux>();
    }
}
```

对于定义在 Qux 中的 3 个构造函数来说，由于创建 IServiceProvider 对象提供的 IServiceCollection 集合包含针对 IFoo 接口和 IBar 接口的服务注册，所以它能够提供前面两个构造函数的所有参数。由于第三个构造函数具有一个类型为 IBaz 的参数，所以无法通过 IServiceProvider 对象来提供。根据前面介绍的第一个原则（IServiceProvider 对象能够提供构造函数的所有参数），Qux 的前两个构造函数会成为合法的候选构造函数，那么 IServiceProvider 对象最终会选择哪一个构造函数？

在所有合法的候选构造函数列表中，最终被选择的构造函数具有如下特征：每个候选构造函数的参数类型集合都是这个构造函数参数类型集合的子集。如果这样的构造函数并不存在，一个 InvalidOperationException 类型的异常会被抛出来。根据这个原则，Qux 的第二个构造函数的参数类型包括 IFoo 和 IBar，而第一个构造函数只具有一个类型为 IFoo 的参数，所以最终被选择的是 Qux 的第二个构造函数，运行实例程序，控制台上产生的输出结果如图 4-6 所示。（S408）

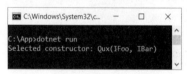

图 4-6　构造函数的选择策略

下面对实例程序略加改动。如下面的代码片段所示，我们只为 Qux 定义两个构造函数，它们都具有两个参数，参数类型分别为 IFoo & IBar 和 IBar & IBaz，并且将针对 IBaz / Baz 的服务注册添加到创建的 ServiceCollection 集合中。

```
class Program
{
    static void Main()
    {
        new ServiceCollection()
            .AddTransient<IFoo, Foo>()
            .AddTransient<IBar, Bar>()
            .AddTransient<IBaz, Baz>()
            .AddTransient<IQux, Qux>()
            .BuildServiceProvider()
            .GetServices<IQux>();
    }
}

public class Qux : IQux
{
    public Qux(IFoo foo, IBar bar) {}
    public Qux(IBar bar, IBaz baz) {}
}
```

虽然 Qux 的两个构造函数的参数都可以由 IServiceProvider 对象来提供，但是并没有一个构造函数的参数类型集合能够成为所有有效构造函数参数类型集合的超集，所以 IServiceProvider 对象无法选择一个最佳的构造函数。运行该程序后会抛出图 4-7 所示的 InvalidOperationException 类型的异常，并提示无法从两个候选的构造函数中选择一个最优的来创建服务实例。（S409）

图 4-7　构造函数的选择策略

4.3.3　生命周期

生命周期决定了 IServiceProvider 对象采用什么样的方式提供和释放服务实例。虽然不同版本的依赖注入框架针对服务实例的生命周期管理采用了不同的实现方式，但总的来说原理还是类似的。在提供的依赖注入框架 Cat 中，我们已经模拟了 3 种生命周期模式的实现原理，下面结合服务范围做进一步阐述。

服务范围

对于依赖注入框架采用的 3 种生命周期模式（Singleton、Scoped 和 Transient）来说，Singleton 和 Transient 都具有明确的语义，但是很多初学者不清楚 Scoped 代表一种什么样的生命周期模式。Scoped 指的是由 IServiceScope 接口表示的服务范围，该范围由 IServiceScopeFactory 接口表示的"服务范围工厂"来创建。如下面的代码片段所示，IServiceProvider 的 CreateScope 扩展方法正是利用提供的 IServiceScopeFactory 服务实例来创建作为服务范围的 IServiceScope 对象的。

```
public interface IServiceScope : IDisposable
{
    IServiceProvider ServiceProvider { get; }
}

public interface IServiceScopeFactory
{
    IServiceScope CreateScope();
}

public static class ServiceProviderServiceExtensions
{
   public static IServiceScope CreateScope(this IServiceProvider provider)
       => provider.GetRequiredService<IServiceScopeFactory>().CreateScope();
}
```

任何一个 IServiceProvider 对象都可以利用其注册的 IServiceScopeFactory 服务创建一个代表

服务范围的 IServiceScope 对象，后者代表的"范围"内具有一个新创建的 IServiceProvider 对象（对应 IServiceScope 接口的 ServiceProvider 属性），该对象与当前 IServiceProvider 在逻辑上具有图 4-8 所示的"父子关系"。

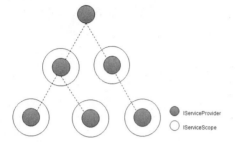

图 4-8　IServiceScope 与 IServiceProvider（逻辑结构）

图 4-8 所示的树形层次结构只是一种逻辑结构，从对象引用层面来看，通过某个 IServiceScope 封装的 IServiceProvider 对象不需要知道自己的"父亲"是谁，它只关心作为根节点的 IServiceProvider 对象在哪里。图 4-9 从物理层面揭示了 IServiceScope/IServiceProvider 对象之间的关系，任何一个 IServiceProvider 对象都具有针对根容器的引用。

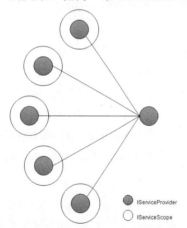

图 4-9　IServiceScope 与 IServiceProvider（物理结构）

3 种生命周期模式

只有充分了解 IServiceScope 对象的创建过程以及它与 IServiceProvider 对象之间的关系，我们才会对 3 种生命周期管理模式（Singleton、Scoped 和 Transient）具有深刻的认识。就服务实例的提供方式来说，它们之间具有如下几方面差异。

- Singleton：IServiceProvider 对象创建的服务实例保存在作为根容器的 IServiceProvider 对象中，所以多个同根的 IServiceProvider 对象提供的针对同一类型的服务实例都是同一个对象。
- Scoped：IServiceProvider 对象创建的服务实例由自己保存，所以同一个 IServiceProvider

对象提供的针对同一类型的服务实例均是同一个对象。
- Transient：针对每次服务提供请求，IServiceProvider 对象总是创建一个新的服务实例。

IServiceProvider 除了提供所需的服务实例，对于由它提供的服务实例，它还具有回收释放的作用。这里所说的回收释放与 .NET Core 自身的垃圾回收机制无关，仅仅针对自身类型实现了 IDisposable 接口或者 IAsyncDisposable 接口的服务实例（下面称为 Disposable 服务实例），针对服务实例的释放体现为调用它们的 Dispose 方法或者 DisposeAsync 方法。IServiceProvider 对象针对服务实例采用的回收释放策略取决于采用的生命周期模式，具体策略主要体现为如下两点。
- Singleton：提供 Disposable 服务实例保存在作为根容器的 IServiceProvider 对象上，只有在这个 IServiceProvider 对象被释放的时候，这些 Disposable 服务实例才能被释放。
- Scoped 和 Transient：IServiceProvider 对象会保存由它提供的 Disposable 服务实例，当自己被释放的时候，这些 Disposable 服务实例就会被释放。

综上所述，每个作为依赖注入容器的 IServiceProvider 对象都具有图 4-10 所示的两个列表来存放服务实例，其被分别命名为 Realized Services 和 Disposable Services，对于一个作为非根容器的 IServiceProvider 对象来说，由它提供的 Scoped 服务保存在自身的 Realized Services 列表中，而 Singleton 服务实例保存在根容器的 Realized Services 列表中。如果服务实现类型实现了 IDisposable 接口或者 IAsyncDisposable 接口，Scoped 服务实例和 Transient 服务实例则保存在自身的 Disposable Services 列表中，而 Singleton 服务实例保存在根容器的 Disposable Services 列表中。

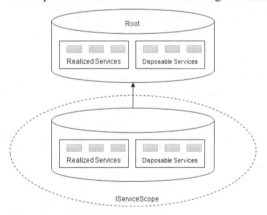

图 4-10　生命周期管理

对于作为根容器的 IServiceProvider 对象来说，Singleton 和 Scoped 对它来说是两种等效的生命周期模式，由它提供的 Singleton 服务实例和 Scoped 服务实例保存在自身的 Realized Services 列表中，而所有需要被释放的服务实例保存在 Disposable Services 列表中。当某个 IServiceProvider 对象被用于提供针对指定类型的服务实例时，它会根据服务类型提取出表示服务注册的 ServiceDescriptor 对象，并根据它得到对应的生命周期模式。
- 如果提供服务的生命周期模式为 Singleton，并且作为根容器的 Realized Services 列表中包含对应的服务实例，那么它将作为最终提供的服务实例。如果这样的服务实例尚未创建，那么新的服务将会被创建出来并作为提供的服务实例。这个服务实例会被添加到根容器的 Realized Services 列表中。如果实例类型实现了 IDisposable 接口或者 IAsyncDisposable 接口，创建的服务实例就会被添加到根容器的 Disposable Services 列表中。
- 如果提供服务的生命周期模式为 Scoped，那么 IServiceProvider 会先确定自身的 Realized Services 列表中是否存在对应的服务实例。如果 Realized Services 列表中存在对应的服务实例，那么该服务实例将作为最终的返回值；如果 Realized Services 列表中不存在对应的服务实例，那么新的服务实例会被创建出来。在作为最终的服务实例被返回之前，新创建的服务实例会被添加到自身的 Realized Services 列表中，如果实例类型实现了 IDisposable 接口或者 IAsyncDisposable 接口，新创建的服务实例会被添加到自身的 Disposable Services 列表中。
- 如果提供服务的生命周期模式为 Transient，那么 IServiceProvider 会直接创建一个新的服务实例。在作为最终的服务实例被返回之前，如果实例类型实现了 IDisposable 接口或者 IAsyncDisposable 接口，创建的服务实例会被添加到自身的 Disposable Services 列表中。

对于非根容器的 IServiceProvider 对象来说，它的生命周期是由"包裹"着它的 IServiceScope 对象控制的。从前面给出的定义可以看出，IServiceScope 实现了 IDisposable 接口，Dispose 方法的执行不仅标志着当前服务范围的终结，也意味着对应 IServiceProvider 对象生命周期的结束。

当代表服务范围的 IServiceScope 对象的 Dispose 方法被调用的时候，它会调用对应 IServiceProvider 对象的 Dispose 方法。一旦 IServiceProvider 对象因自身 Dispose 方法的调用而被释放的时候，它会从自身的 Disposable Services 列表中提取出所有需要被释放的服务实例，并调用它们的 Dispose 方法或者 DisposeAsync 方法。在这之后，Disposable Services 列表和 Realized Services 列表会被清空，列表中的服务实例和 IServiceProvider 对象自身会成为垃圾对象被 GC 回收。

ASP.NET Core 应用

依赖注入框架所谓的服务范围在 ASP.NET Core 应用中具有明确的边界，指的是针对每个 HTTP 请求的上下文，也就是说，服务范围的生命周期与每个请求的上下文绑定在一起。如图 4-11 所示，ASP.NET Core 应用中用于提供服务实例的 IServiceProvider 对象分为两种类型：一种是作为根容器并与应用具有相同生命周期的 IServiceProvider 对象，一般称为 ApplicationServices；另一种是根据请求及时创建和释放的 IServiceProvider 对象，一般称为 RequestServices。

在 ASP.NET Core 应用初始化过程（即请求管道构建过程）中使用的服务实例都是由 ApplicationServices 提供的。在具体处理每个请求时，ASP.NET Core 框架会利用注册的一个中间件来针对当前请求创建一个代表服务范围的 IServiceScope 对象，该服务范围提供的 RequestServices 用来提供当前请求处理过程中所需的服务实例。一旦服务请求处理完成，IServiceScope 对象代表的服务范围被终结，在当前请求处理过程中的 Scoped 服务会变成垃圾对象并最终被 GC 回收。对于实现了 IDisposable 接口或者 IAsyncDisposable 接口的 Scoped 服务实例或者 Transient 服务实例来说，在变成垃圾对象之前，它们的 Dispose 方法或者 DisposeAsync 方法会被调用。

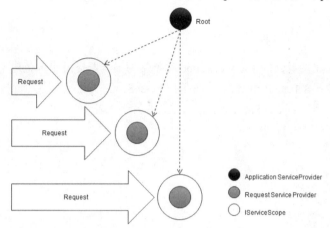

图 4-11　生命周期管理

4.4　实现概览

上面从实现原理的角度对 .NET Core 的依赖注入框架进行了介绍，下面进一步介绍该框架的总体设计和实现。在过去的多个版本更迭过程中，依赖注入框架的底层实现一直都在发生改变，并且底层涉及的大多是内容接口和类型，所以本节不涉及太过细节的层面。

4.4.1　ServiceProviderEngine 和 ServiceProviderEngineScope

对于依赖注入的底层设计和实现来说，ServiceProviderEngine 和 ServiceProviderEngineScope 是两个核心类型。顾名思义，ServiceProviderEngine 表示提供服务实例的提供引擎，容器提供的服务实例最终是通过该引擎提供的，在一个应用范围内只存在一个全局唯一的 ServiceProviderEngine 对象。ServiceProviderEngineScope 代表服务范围，它利用对提供服务实例的缓存来实现对生命周期的控制。ServiceProviderEngine 实现了 IServiceProviderEngine 接口，从如下代码片段可以看出，一个 ServiceProviderEngine 对象不仅是一个 IServiceProvider 对象，还是一个 IServiceScopeFactory 对象。

```
internal interface IServiceProviderEngine :
    IServiceProvider, IDisposable, IAsyncDisposable
{
    void ValidateService(ServiceDescriptor descriptor);
    IServiceScope RootScope { get; }
}

internal abstract class ServiceProviderEngine :
    IServiceProviderEngine, IServiceScopeFactory
{
    public IServiceScope RootScope { get; }
```

```csharp
    public IServiceScope CreateScope();
    ...
}
```

ServiceProviderEngine 的 RootScope 属性返回的 IServiceScope 对象是为根容器提供服务范围的。作为一个 IServiceScopeFactory 对象，ServiceProviderEngine 的 CreateScope 方法会创建一个新的服务范围，这两种服务范围都通过一个 ServiceProviderEngineScope 对象来表示。

```csharp
internal class ServiceProviderEngineScope : IServiceScope, IDisposable,
    IServiceProvider, IAsyncDisposable
{
    public ServiceProviderEngine Engine { get; }
    public IServiceProvider ServiceProvider { get; }

    public object GetService(Type serviceType);
}
```

如上面的代码片段所示，ServiceProviderEngineScope 对象不仅是一个 IServiceScope 对象，还是一个 IServiceProvider 对象。上面提及的表示服务范围的 IServiceScope 对象是对一个表示依赖注入容器的 IServiceProvider 对象的封装，实际上两者合并为同一个 ServiceProviderEngineScope 对象，一个 ServiceProviderEngineScope 对象的 ServiceProvider 属性返回的就是它自己。换句话说，子容器和它所在的服务范围引用的是同一个 ServiceProviderEngineScope 对象。上述这些核心类型与接口之间的关系如图 4-12 所示。

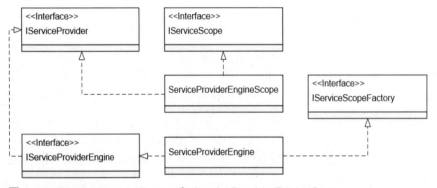

图 4-12　ServiceProviderEngine 和 ServiceProviderEngineScope

图 4-13 进一步揭示了 ServiceProviderEngine 和 ServiceProviderEngineScope 之间的关系。对于一个通过调用 ServiceProviderEngine 对象的 CreateScope 创建的 ServiceProviderEngineScope 来说，由于它也是一个 IServiceProvider 对象，所以如果调用它的 GetService<IServiceProvider>方法，该方法同样返回它自己。如果调用它的 GetService<IServiceScopeFactory>方法，它返回创建它的 ServiceProviderEngine 对象，也就是该方法和 Engine 属性返回的是同一个对象。

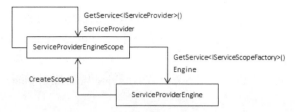

图 4-13　ServiceProviderEngine 和 ServiceProviderEngineScope 之间的关系（一）

依赖注入框架提供的服务实例最终是通过 ServiceProviderEngine 对象提供的。从上面给出的代码片段可以看出，ServiceProviderEngine 是一个抽象类，.NET Core 依赖注入框架提供了如下 4 个具体的实现类型，默认使用的是 DynamicServiceProviderEngine。

- RuntimeServiceProviderEngine：采用反射的方式提供服务实例。
- ILEmitServiceProviderEngine：采用 IL Emit 的方式提供服务实例。
- ExpressionsServiceProviderEngine：采用表达式树的方式提供服务实例。
- DynamicServiceProviderEngine：根据请求并发数量动态决定最终的服务实例提供方案（反射与 IL Emit 或者反射与表达式树，是否选择 IL Emit 取决于当前运行时是否支持 Reflection Emit）。

4.4.2　ServiceProvider

调用 IServiceCollection 集合的 BuildServiceProvider 扩展方法创建的是一个 ServiceProvider 对象。如图 4-14 所示，作为根容器的 ServiceProvider 对象，与前面介绍的 ServiceProviderEngine 和 ServiceProviderEngineScope 对象共同构建整个依赖注入框架的设计蓝图。

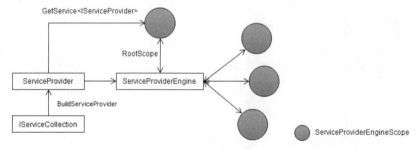

图 4-14　ServiceProviderEngine 和 ServiceProviderEngineScope 之间的关系（二）

在利用 IServiceCollection 集合创建 ServiceProvider 对象的时候，提供的服务注册将用来创建一个具体的 ServiceProviderEngine 对象。该 ServiceProviderEngine 对象的 RootScope 就是它创建的一个 ServiceProviderEngineScope 对象，子容器提供的 Singleton 服务实例由它维护。如果调用 ServiceProvider 对象的 GetService<IServiceProvider>方法，返回的其实不是它自己，而是作为 RootScope 的 ServiceProviderEngineScope 对象（调用 ServiceProviderEngineScope 对象的 GetService<IServiceProvider>方法返回的是它自己）。

ServiceProvider 和 ServiceProviderEngineScope 都实现了 IServiceProvider 接口，如果调用了

它们的 GetService<IServiceScopeFactory>方法，返回的就是同一个 ServiceProviderEngine 对象。该特性决定了调用它们的 CreateScope 扩展方法会创建一个新的 ServiceProviderEngineScope 对象作为子容器。综上所述，依赖注入框架具有如下几方面特性。

- ServiceProviderEngine 的唯一性：整个服务提供体系只存在一个 ServiceProviderEngine 对象。
- ServiceProviderEngine 与 IServiceFactory 的同一性：唯一存在的 ServiceProviderEngine 会作为创建服务范围的 IServiceFactory 工厂。
- ServiceProviderEngineScope 和 IServiceProvider 的同一性：表示服务范围的 ServiceProviderEngineScope 也作为服务提供者的依赖注入容器。

为了印证依赖注入框架的特性，我们编写了如下所示的测试代码。由于 ServiceProviderEngine 和 ServiceProviderEngineScope 都是内部类型，所以只能采用反射的方式得到它们的属性或者字段成员。上面总结的这些特征体现在如下几组调试断言中。（S410）

```
class Program
{
    static void Main()
    {
        var (engineType, engineScopeType) = ResolveTypes();
        var root = new ServiceCollection().BuildServiceProvider();
        var child1 = root.CreateScope().ServiceProvider;
        var child2 = root.CreateScope().ServiceProvider;

        var engine = GetEngine(root);
        var rootScope = GetRootScope(engine, engineType);

        //ServiceProviderEngine 的唯一性
        Debug.Assert(ReferenceEquals(
            GetEngine(rootScope, engineScopeType), engine));
        Debug.Assert(ReferenceEquals(GetEngine(child1, engineScopeType), engine));
        Debug.Assert(ReferenceEquals(GetEngine(child2, engineScopeType), engine));

        //ServiceProviderEngine 和 IServiceScopeFactory 的同一性
        Debug.Assert(ReferenceEquals(
             root.GetRequiredService<IServiceScopeFactory>(), engine));
        Debug.Assert(ReferenceEquals(
            child1.GetRequiredService<IServiceScopeFactory>(), engine));
        Debug.Assert(ReferenceEquals(
            child2.GetRequiredService<IServiceScopeFactory>(), engine));

        //ServiceProviderEngineScope 提供的 IServiceProvider 是它自己
        //ServiceProvider 提供的 IServiceProvider 是 RootScope
        Debug.Assert(ReferenceEquals(
            root.GetRequiredService<IServiceProvider>(), rootScope));
        Debug.Assert(ReferenceEquals(
            child1.GetRequiredService<IServiceProvider>(), child1));
        Debug.Assert(ReferenceEquals(
```

```csharp
        child2.GetRequiredService<IServiceProvider>(), child2));

    //ServiceProviderEngineScope 和 IServiceProvider 的同一性
    Debug.Assert(ReferenceEquals((rootScope).ServiceProvider, rootScope));
    Debug.Assert(ReferenceEquals(
        ((IServiceScope)child1).ServiceProvider, child1));
    Debug.Assert(ReferenceEquals(
        ((IServiceScope)child2).ServiceProvider, child2));
}

static (Type Engine, Type EngineScope) ResolveTypes()
{
    var assembly = typeof(ServiceProvider).Assembly;
    var engine = assembly.GetTypes().Single(
        it => it.Name == "IServiceProviderEngine");
    var engineScope = assembly.GetTypes().Single(
        it => it.Name == "ServiceProviderEngineScope");
    return (engine, engineScope);
}

static object GetEngine(ServiceProvider serviceProvider)
{
    var field = typeof(ServiceProvider)
        .GetField("_engine", BindingFlags.Instance | BindingFlags.NonPublic);
    return field.GetValue(serviceProvider);
}

static object GetEngine(object enginScope, Type engineScopeType)
{
    var property = engineScopeType.GetProperty("Engine",
        BindingFlags.Instance | BindingFlags.Public);
    return property.GetValue(enginScope);
}

static IServiceScope GetRootScope(object engine, Type engineType)
{
    var property = engineType.GetProperty("RootScope",
        BindingFlags.Instance | BindingFlags.Public);
    return (IServiceScope)property.GetValue(engine);
}
}
```

4.4.3 注入 IServiceProvider 对象

第 3 章从 Service Locator 模式是反模式的角度说明了不推荐在服务中注入 IServiceProvider 对象的原因。但反模式并不等于就是完全不能用的模式，有时直接在服务构造函数中注入作为依赖注入容器的 IServiceProvider 对象可能是最快捷、最省事的解决方案。对于 IServiceProvider 对象的注入，如下细节可能会被人们忽略或者误解。

请读者思考如下问题：如果在某个服务中注入了 IServiceProvider 对象，那么利用某个 IServiceProvider 对象提供该服务实例的时候，注入的 IServiceProvider 对象是否是它自己？如下代码片段定义了两个在构造函数中注入了 IServiceProvider 对象的服务类型 SingletonService 和 ScopedService，并按照命名所示的生命周期进行了注册。

```
class Program
{
    static void Main()
    {
        var serviceProvider = new ServiceCollection()
            .AddSingleton<SingletonService>()
            .AddScoped<ScopedService>()
            .BuildServiceProvider();
        using (var scope = serviceProvider.CreateScope())
        {
            var child = scope.ServiceProvider;
            var singletonService = child.GetRequiredService<SingletonService>();
            var scopedService = child.GetRequiredService<ScopedService>();

            Debug.Assert(ReferenceEquals(child, scopedService.RequestServices));
            Debug.Assert(ReferenceEquals(
                serviceProvider, singletonService.ApplicationServices));
        }
    }

    public class SingletonService
    {
        public IServiceProvider ApplicationServices { get; }
        public SingletonService(IServiceProvider serviceProvider)
            => ApplicationServices = serviceProvider;
    }

    public class ScopedService
    {
        public IServiceProvider RequestServices { get; }
        public ScopedService(IServiceProvider serviceProvider)
            => RequestServices = serviceProvider;
    }
}
```

我们最终利用一个作为子容器的 IServiceProvider 对象（ServiceProviderEngineScope 对象）来提供这两个服务类型的实例，并通过调试断言确定注入的 IServiceProvider 对象是否就是作为当前依赖注入容器的 ServiceProviderEngineScope 对象。如果在 Debug 模式下运行上述测试代码可以发现，第一个断言是成立的，第二个断言是不成立的。

再次回到两个服务类型的定义，SingletonService 类型和 ScopedService 类型中通过注入 IServiceProvider 对象初始化的属性分别被命名为 ApplicationServices 和 RequestServices，意味着它们希望注入的分别是针对当前应用程序的根容器和针对请求的子容器。所以，利用针对请求的子容器来提供针对这两个类型的服务实例时,如果注入的是当前子容器,就与 ApplicationServices 的意图不符。由此可知,在提供服务实例时注入的 IServiceProvider 对象取决于采用的生命周期模式,具体策略如下。

- Singleton:注入的是 ServiceProviderEngine 的 RootScope 属性表示的 ServiceProviderEngineScope 对象。
- Scoped 和 Transient:如果当前 IServiceProvider 对象的类型为 ServiceProviderEngineScope,注入的就是它自己;如果是一个 ServiceProvider 对象,注入的就是 ServiceProviderEngine 的 RootScope 属性表示的 ServiceProviderEngineScope 对象。

基于生命周期模式注入 IServiceProvider 对象的策略可以通过如下测试程序进行验证(S411)。另外,如果将调用 IServiceCollection 集合的 BuildServiceProvider 扩展方法创建的 ServiceProvider 对象作为根容器,那么它对应的 ServiceProviderEngine 对象的 RootScope 属性返回作为根服务范围的 ServiceProviderEngineScope 对象,ServiceProvider、ServiceProviderEngine 和 ServiceProviderEngineScope 这 3 个类型全部实现了 IServiceProvider 接口,可以将这 3 个对象都视为根容器。

```
class Program
{
    static void Main()
    {
        var serviceProvider = new ServiceCollection()
            .AddSingleton<SingletonService>()
            .AddScoped<ScopedService>()
            .BuildServiceProvider();
        var rootScope = serviceProvider.GetService<IServiceProvider>();
        using (var scope = serviceProvider.CreateScope())
        {
            var child = scope.ServiceProvider;
            var singletonService = child.GetRequiredService<SingletonService>();
            var scopedService = child.GetRequiredService<ScopedService>();

            Debug.Assert(ReferenceEquals(
                child, child.GetRequiredService<IServiceProvider>()));
            Debug.Assert(ReferenceEquals(child, scopedService.RequestServices));
            Debug.Assert(ReferenceEquals(
                rootScope, singletonService.ApplicationServices));
        }
    }
}
```

4.5 扩展

.NET Core 的服务承载系统无缝集成了依赖注入框架,但是目前还有很多开源的依赖注入框架,比较常用有 Castle、StructureMap、Spring.NET、AutoFac、Unity 和 Ninject 等,应如何实现与第三方依赖注入框架的整合?

4.5.1 适配

.NET Core 具有一个承载（Hosting）系统（详见第 10 章），该系统承载需要在后台长时间运行的服务，一个 ASP.NET Core 应用仅仅是该系统承载的一种服务而已。承载系统总是采用依赖注入的方式消费它在服务承载过程中所需的服务。对于承载系统来说，原始的服务注册总是体现为一个 IServiceCollection 集合，最终的依赖注入容器则体现为一个 IServiceProvider 对象，如果要将第三方依赖注入框架整合进来，就需要利用它们解决从 IServiceCollection 集合到 IServiceProvider 对象的适配问题。

具体来说，我们可以在 IServiceCollection 集合和 IServiceProvider 对象之间设置一个针对某个第三方依赖注入框架的 ContainerBuilder 对象：先利用包含原始服务注册的 IServiceCollection 集合创建一个 ContainerBuilder 对象，再利用该对象构建作为依赖注入容器的 IServiceProvider 对象（见图 4-15）。

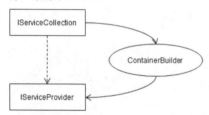

图 4-15　IServiceCollection—ContainerBuilder—IServiceProvider

4.5.2　IServiceProviderFactory<TContainerBuilder>

如图 4-15 所示，两种转换是利用一个 IServiceProviderFactory<TContainerBuilder>对象完成的。如下面的代码片段所示，IServiceProviderFactory<TContainerBuilder>接口定义了两个方法：CreateBuilder 方法利用指定的 IServiceCollection 集合创建对应的 ContainerBuilder 对象；CreateServiceProvider 方法则进一步利用这个 ContainerBuilder 对象创建作为依赖注入容器的 IServiceProvider 对象。

```
public interface IServiceProviderFactory<TContainerBuilder>
{
    TContainerBuilder CreateBuilder(IServiceCollection services);
    IServiceProvider CreateServiceProvider(TContainerBuilder containerBuilder);
}
```

.NET Core 的承载系统总是利用注册的 IServiceProviderFactory<TContainerBuilder>服务创建最终作为依赖注入容器的 IServiceProvider 对象。承载系统默认注册的是如下所示的 DefaultServiceProviderFactory 类型。如下面的代码片段所示，DefaultServiceProviderFactory 对象会直接调用指定 IServiceCollection 集合的 BuildServiceProvider 方法创建对应的 IServiceProvider 对象。

```
public class DefaultServiceProviderFactory :
    IServiceProviderFactory<IServiceCollection>
{
    public DefaultServiceProviderFactory()
        : this(ServiceProviderOptions.Default){}
```

```csharp
    public DefaultServiceProviderFactory(ServiceProviderOptions options)
        => _options = options;

    public IServiceCollection CreateBuilder(IServiceCollection services)
        => services;

    public IServiceProvider CreateServiceProvider(
        IServiceCollection containerBuilder) =>
        containerBuilder.BuildServiceProvider(_options);
}
```

4.5.3 整合第三方依赖注入框架

为了使读者对利用注册的 IServiceProviderFactory<TContainerBuilder>服务整合第三方依赖注入框架有更加深刻的理解，本节将演示一个具体的实例。第 3 章创建了一个名为 Cat 的 Mini 版依赖注入框架，下面将提供一个具体的 IServiceProviderFactory<TContainerBuilder>实现类型完成对它的整合。

首先，创建一个名为 CatBuilder 的类型作为对应的 ContainerBuilder。由于需要涉及针对服务范围的创建，所以我们在 CatBuilder 类中定义了如下两个内嵌的私有类型。其中，表示服务范围的 ServiceScope 对象实际上就是对一个 IServiceProvider 对象的封装；而 ServiceScopeFactory 类型表示创建该对象的工厂，它是对一个 Cat 对象的封装。

```csharp
public class CatBuilder
{
    private class ServiceScope : IServiceScope
    {
        public ServiceScope(IServiceProvider serviceProvider)
            => ServiceProvider = serviceProvider;
        public IServiceProvider ServiceProvider { get; }
        public void Dispose()=> (ServiceProvider as IDisposable)?.Dispose();
    }

    private class ServiceScopeFactory : IServiceScopeFactory
    {
        private readonly Cat _cat;
        public ServiceScopeFactory(Cat cat) => _cat = cat;
        public IServiceScope CreateScope() => new ServiceScope(_cat);
    }
}
```

一个 CatBuilder 对象是对一个 Cat 对象的封装，它的 BuildServiceProvider 方法会直接返回这个 Cat 对象，并作为最终提供的依赖注入容器。CatBuilder 对象在初始化过程中添加了针对 IServiceScopeFactory 接口的服务注册，具体注册的是根据作为当前子容器的 Cat 对象创建的 ServiceScopeFactory 对象。为了实现程序集范围内的批量服务注册，可以为 CatBuilder 类型定义一个 Register 方法。

```csharp
public class CatBuilder
{
```

```csharp
    private readonly Cat _cat;
    public CatBuilder(Cat cat)
    {
        _cat = cat;
        _cat.Register<IServiceScopeFactory>(
            c => new ServiceScopeFactory(c.CreateChild()), Lifetime.Transient);
    }
    public IServiceProvider BuildServiceProvider() => _cat;
    public CatBuilder Register(Assembly assembly)
    {
        _cat.Register(assembly);
        return this;
    }
    ...
}
```

如下面的代码片段所示，CatServiceProviderFactory 类型实现了 IServiceProviderFactory<CatBuilder>接口。在实现的 CreateBuilder 方法中，我们创建了一个 Cat 对象，并将指定 IServiceCollection 集合包含的服务注册（ServiceDescriptor 对象）转换成兼容 Cat 的服务注册（ServiceRegistry 对象），然后将其应用到创建的 Cat 对象上。我们最终利用这个 Cat 对象创建出返回的 CatBuilder 对象。实现的 CreateServiceProvider 方法返回的是通过调用 CatBuilder 对象的 CreateServiceProvider 方法得到的 IServiceProvider 对象。

```csharp
public class CatServiceProviderFactory : IServiceProviderFactory<CatBuilder>
{
    public CatBuilder CreateBuilder(IServiceCollection services)
    {
        var cat = new Cat();
        foreach (var service in services)
        {
            if (service.ImplementationFactory != null)
            {
                cat.Register(service.ServiceType, provider
                    => service.ImplementationFactory(provider),
                    service.Lifetime.AsCatLifetime());
            }
            else if (service.ImplementationInstance != null)
            {
                cat.Register(service.ServiceType, service.ImplementationInstance);
            }
            else
            {
                cat.Register(service.ServiceType, service.ImplementationType,
                    service.Lifetime.AsCatLifetime());
            }
        }
        return new CatBuilder(cat);
    }
    public IServiceProvider CreateServiceProvider(CatBuilder containerBuilder)
```

```
    => containerBuilder.BuildServiceProvider();
}
```

Cat 对象具有与 .NET Core 依赖注入框架一致的服务生命周期表达方式，所以将服务注册从 ServiceDescriptor 类型转化成 ServiceRegistry 类型时，可以实现直接完成两种生命周期模式的转换，具体的转换可以在 AsCatLifetime 扩展方法中实现。

```
internal static class Extensions
{
    public static Lifetime AsCatLifetime(this ServiceLifetime lifetime)
    {
        return lifetime switch
        {
            ServiceLifetime.Scoped    => Lifetime.Self,
            ServiceLifetime.Singleton => Lifetime.Root,
            _                         => Lifetime.Transient,
        };
    }
}
```

其次，演示如何利用 CatServiceProviderFactory 创建作为依赖注入容器的 IServiceProvider 对象。我们定义了如下接口和对应的实现类型，其中，Foo、Bar、Baz 和 Qux 类型分别实现了对应的接口 IFoo、IBar、IBaz 与 IQux，Qux 类型上标注了一个 MapToAttribute 特性，并注册了与对应接口 IQux 之间的映射。为了反映 Cat 框架对服务实例生命周期的控制，我们让这些类型派生于同一个基类 Base。Base 实现了 IDisposable 接口，可以在其构造函数和实现的 Dispose 方法中输出相应的文本，以确定对应的实例何时被创建和释放。

```
public interface IFoo {}
public interface IBar {}
public interface IBaz {}
public interface IQux {}
public interface IFoobar<T1, T2> {}
public class Base : IDisposable
{
    public Base()
        => Console.WriteLine($"Instance of {GetType().Name} is created.");
    public void Dispose()
        => Console.WriteLine($"Instance of {GetType().Name} is disposed.");
}

public class Foo : Base, IFoo{ }
public class Bar : Base, IBar{ }
public class Baz : Base, IBaz{ }
[MapTo(typeof(IQux), Lifetime.Root)]
public class Qux : Base, IQux { }
public class Foobar<T1, T2>: IFoobar<T1,T2>
{
    public IFoo Foo { get; }
    public IBar Bar { get; }
    public Foobar(IFoo foo, IBar bar)
```

```
        {
            Foo = foo;
            Bar = bar;
        }
}
```

在如下所示的演示程序中,我们先创建了一个 ServiceCollection 集合,并采用 3 种不同的生命周期模式分别添加了针对 IFoo 接口、IBar 接口和 IBaz 接口的服务注册。然后根据 ServiceCollection 集合创建了一个 CatServiceProviderFactory 对象,并调用其 CreateBuilder 方法创建出对应的 CatBuilder 对象。最后调用 CatBuilder 对象的 Register 方法完成了针对当前入口程序集的批量服务注册,其目的在于添加针对 IQux/Qux 的服务注册。

```
class Program
{
    static void Main()
    {
        var services = new ServiceCollection()
            .AddTransient<IFoo, Foo>()
            .AddScoped<IBar>(_ => new Bar())
            .AddSingleton<IBaz>(new Baz());

        var factory = new CatServiceProviderFactory();
        var builder = factory.CreateBuilder(services)
            .Register(Assembly.GetEntryAssembly());
        var container = factory.CreateServiceProvider(builder);

        GetServices();
        GetServices();
        Console.WriteLine("\nRoot container is disposed.");
        (container as IDisposable)?.Dispose();

        void GetServices()
        {
            using (var scope = container.CreateScope())
            {
                Console.WriteLine("\nService scope is created.");
                var child = scope.ServiceProvider;

                child.GetService<IFoo>();
                child.GetService<IBar>();
                child.GetService<IBaz>();
                child.GetService<IQux>();

                child.GetService<IFoo>();
                child.GetService<IBar>();
                child.GetService<IBaz>();
                child.GetService<IQux>();
                Console.WriteLine("\nService scope is disposed.");
            }
        }
    }
```

        }
}

在调用 CatServiceProviderFactory 对象的 CreateServiceProvider 方法来创建出作为依赖注入容器的 IServiceProvider 对象之后，我们先后两次调用了本地方法 GetServices。GetServices 方法会利用这个 IServiceProvider 对象来创建一个服务范围，并利用此服务范围内的 IServiceProvider 提供两组服务实例。利用 CatServiceProviderFactory 创建的 IServiceProvider 对象最终通过调用其 Dispose 方法进行释放。该程序运行之后在控制台上输出的结果如图 4-16 所示，输出结果体现的服务生命周期与演示程序体现的生命周期是完全一致的。（S412）

图 4-16　利用 CatServiceProviderFactory 创建 IServiceProvider 对象

第 5 章

文件系统

ASP.NET Core 应用具有很多读取文件的场景，如读取配置文件、静态 Web 资源文件（如 CSS、JavaScript 和图片文件等）、MVC 应用的 View 文件，以及直接编译到程序集中的内嵌资源文件。这些文件的读取都需要使用一个 IFileProvider 对象。IFileProvider 对象构建了一个抽象的文件系统，我们不仅可以利用该系统提供的统一 API 来读取各种类型的文件，还能及时监控目标文件的变化。

5.1 抽象的文件系统

IFileProvider 对象可以构建一个具有层次化目录结构的文件系统。由于 IFileProvider 是一个接口，所以它构建的是一个抽象的文件系统，这里所谓的目录和文件都是一个抽象的概念。具体的文件可能对应一个物理文件，也可能保存在数据库中，或者来源于网络，甚至有可能根本就不存在，其内容需要在读取时动态生成。目录也仅仅是组织文件的逻辑容器。为了使读者对这个文件系统有一个大体认识，下面先演示几个简单的实例。

5.1.1 树形层次结构

文件系统管理的所有文件以目录的形式进行组织，一个 IFileProvider 对象可以视为针对一个根目录的映射。目录除了可以存放文件，还可以包含子目录，所以目录/文件在整体上呈现出树形层次化结构。接下来我们将一个 IFileProvider 对象映射到一个物理目录，并利用它将所在目录的结构呈现出来。

下面的演示实例是一个普通的控制台程序。我们在演示实例中定义了如下一个 IFileManager 接口，它利用 ShowStructure 方法将文件系统的整体结构显示出来。该方法的 Action<int, string>中的参数可以将文件系统的节点（目录或者文件）名称呈现出来。这个 Action<int, string>对象的两个参数分别代表缩进的层级和目录/文件的名称。

```
public interface IFileManager
{
    void ShowStructure(Action<int, string> render);
}
```

我们定义如下这个 FileManager 类，作为对 IFileManager 接口的默认实现，它利用只读 _fileProvider 字段表示的 IFileProvider 对象来提取目录结构。目标文件系统的整体结构通过 Render 方法以递归的方式呈现出来，其中涉及对 IFileProvider 对象的 GetDirectoryContents 方法的调用。该方法返回一个 IDirectoryContents 对象，以表示指定目录的内容，如果对应的目录存在，就可以遍历该对象得到它的子目录和文件。目录和文件最终体现为一个 IFileInfo 对象，而 IFileInfo 对象对应的是一个目录还是一个文件，则通过其 IsDirectory 属性进行区分。

```csharp
public class FileManager : IFileManager
{
    private readonly IFileProvider _fileProvider;
    public FileManager(IFileProvider fileProvider) => _fileProvider = fileProvider;
    public void ShowStructure(Action<int, string> render)
    {
        int indent = -1;
        Render("");
        void Render(string subPath)
        {
            indent++;
            foreach (var fileInfo in _fileProvider.GetDirectoryContents(subPath))
            {
                render(indent, fileInfo.Name);
                if (fileInfo.IsDirectory)
                {
                    Render($@"{subPath}\{fileInfo.Name}".TrimStart('\\'));
                }
            }
            indent--;
        }
    }
}
```

接下来构建一个本地物理目录"c:\test\"，并按照图 5-1 中的结构在其下面创建相应的子目录和文件。我们会将这个目录映射到一个 IFileProvider 对象上，并进一步利用它创建上面的 FileManager 对象。最终调用 FileManager 对象的 ShowStructure 方法将目录结构呈现出来。

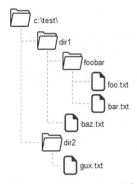

图 5-1　FileProvider 映射的物理目录结构

整个演示程序体现在如下所示的代码片段中。我们针对目录"c:\test\"创建了一个表示物理文件系统的 PhysicalFileProvider 对象，并将其注册到创建的 ServiceCollection 对象上。除此之外，ServiceCollection 对象上还添加了针对 IFileManager/FileManager 的服务注册。

```
class Program
{
    static void Main()
    {
        static void Print(int layer, string name)
          => Console.WriteLine($"{new string(' ', layer * 4)}{name}");
        new ServiceCollection()
            .AddSingleton<IFileProvider>(new PhysicalFileProvider(@"c:\test"))
            .AddSingleton<IFileManager, FileManager>()
            .BuildServiceProvider()
            .GetRequiredService<IFileManager>()
            .ShowStructure(Print);
    }
}
```

我们最终利用 ServiceCollection 生成的 IServiceProvider 对象得到 FileManager 对象，并调用该对象的 ShowStructure 方法将 PhysicalFileProvider 对象映射的目录结构呈现出来。运行该程序之后，控制台上输出的结果如图 5-2 所示，该结果展示了映射物理目录的真实结构。（S501）

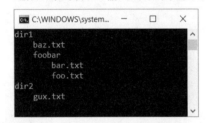

图 5-2　运行程序显示的目录结构

5.1.2　读取文件内容

前面演示了如何利用 IFileProvider 对象将文件系统的结构完整地呈现出来，接下来我们将演示如何利用它来读取一个物理文件的内容。我们为 IFileManager 定义一个 ReadAllTextAsync 方法，以异步的方式读取指定文件内容，方法的参数表示文件的路径。如下面的代码片段所示，ReadAllTextAsync 方法将指定的文件路径作为参数来调用 IFileProvider 对象的 GetFileInfo 方法，以得到一个 IFileInfo 对象。最终调用 IFileInfo 对象的 CreateReadStream 方法得到读取文件的输出流，进而得到文件的真实内容。

```
public interface IFileManager
{
    ...
    Task<string> ReadAllTextAsync(string path);
}

public class FileManager : IFileManager
```

```
{
    ...
    public async Task<string> ReadAllTextAsync(string path)
    {
        byte[] buffer;
        using (var stream = _fileProvider.GetFileInfo(path).CreateReadStream())
        {
            buffer = new byte[stream.Length];
            await stream.ReadAsync(buffer, 0, buffer.Length);
        }
        return Encoding.Default.GetString(buffer);
    }
}
```

如果依然将 FileManager 使用的 IFileProvider 映射为目录 "c:\test\"，现在我们就在该目录中创建一个名为 data.txt 的文本文件，并在该文件中任意写入一些内容。然后在 Main 方法中编写如下程序，利用依赖注入的方式得到 FileManager 对象，并读取文件 data.txt 的内容。最终的调试断言旨在确定通过 IFileProvider 读取的确实就是目标文件的真实内容。（S502）

```
class Program
{
    static async Task Main()
    {
        var content = await new ServiceCollection()
            .AddSingleton<IFileProvider>(new PhysicalFileProvider(@"c:\test"))
            .AddSingleton<IFileManager, FileManager>()
            .BuildServiceProvider()
            .GetRequiredService<IFileManager>()
            .ReadAllTextAsync("data.txt");

        Debug.Assert(content == File.ReadAllText(@"c:\test\data.txt"));
    }
}
```

我们一直强调，IFileProvider 对象构建的是一个抽象的具有目录结构的文件系统，具体文件的提供方式取决于具体的 IFileProvider 对象的类型。演示实例中定义的 FileManager 并没有限定具体使用何种类型的 IFileProvider，该对象是在应用中通过依赖注入的方式指定的。由于上面的应用程序注入的是一个 PhysicalFileProvider 对象，所以可以利用它读取对应物理目录下的某个文件。如果将 data.txt 文件直接以资源文件的形式编译到程序集中，就需要使用另一个名为 EmbeddedFileProvider 的实现类型。

可以直接将 data.txt 文件添加到控制台应用的项目根目录下。在默认情况下，编译项目的时候这样的文件并不能成为内嵌到目标程序集的资源文件，为此我们需要修改项目文件（.csproj 文件）的内容。具体来说，我们可以在项目文件中按照如下形式添加一个<EmbeddedResource>元素，从而将 data.txt 文件设置为内嵌到编译后生成的程序集的内嵌资源文件中。

```
<Project Sdk="Microsoft.NET.Sdk">
    ...
    <ItemGroup>
```

```
      <EmbeddedResource Include="data.txt"/>
  </ItemGroup>
</Project>
```

如下程序可以演示针对内嵌于程序集中的资源文件的读取。我们首先得到当前入口程序集，并利用它创建了一个 EmbeddedFileProvider 对象，用于代替原来的 PhysicalFileProvider 对象来被注册到 ServiceCollection 之中。然后采用完全一致的编程方式得到 FileManager 对象，并利用它读取内嵌文件 data.txt 的内容。为了验证读取的目标文件准确无误，可以采用直接读取资源文件的方式得到内嵌文件 data.txt 的内容，并利用一个调试断言确定两者的一致性。（S503）

```
class Program
{
    static async Task Main()
    {
        var assembly = Assembly.GetEntryAssembly();

        var content1 = await new ServiceCollection()
            .AddSingleton<IFileProvider>(new EmbeddedFileProvider(assembly))
            .AddSingleton<IFileManager, FileManager>()
            .BuildServiceProvider()
            .GetRequiredService<IFileManager>()
            .ReadAllTextAsync("data.txt");

        var stream = assembly
            .GetManifestResourceStream($"{assembly.GetName().Name}.data.txt");
        var buffer = new byte[stream.Length];
        stream.Read(buffer, 0, buffer.Length);
        var content2 = Encoding.Default.GetString(buffer);

        Debug.Assert(content1 == content2);
    }
}
```

5.1.3 监控文件的变化

在文件读取场景中，确定加载到内存中的数据与源文件的一致性并自动同步，是一个很常见的需求。例如，可以将配置定义在一个 JSON 文件中，应用启动的时候会读取该文件并将其转换成对应的 Options 对象。在很多情况下，如果改动了配置文件，最新的配置数据只有在应用重启之后才能生效。如果能够以一种高效的方式对配置文件进行监控，并在其发生改变的情况下向应用发送通知，那么应用就能在不用重启的情况下重新读取配置文件，进而实现 Options 对象承载的内容和原始配置文件完全同步。

对文件系统实施监控并在其发生改变时发送通知也是 IFileProvider 对象提供的核心功能之一。下面依然使用前面这个程序来演示如何使用 PhysicalFileProvider 对某个物理文件实施监控，并在目标文件的内容发生改变时重新读取新的内容。

```
class Program
{
```

```
static async Task Main()
{
    using (var fileProvider = new PhysicalFileProvider(@"c:\test"))
    {
        string original = null;
        ChangeToken.OnChange(() => fileProvider.Watch("data.txt"), Callback);
        while (true)
        {
            File.WriteAllText(@"c:\test\data.txt", DateTime.Now.ToString());
            await Task.Delay(5000);
        }

        async void Callback()
        {
            var stream = fileProvider.GetFileInfo("data.txt").CreateReadStream();
            {
                var buffer = new byte[stream.Length];
                await stream.ReadAsync(buffer, 0, buffer.Length);
                string current = Encoding.Default.GetString(buffer);
                if (current != original)
                {
                    Console.WriteLine(original = current);
                }
            }
        }
    }
}
```

如上面的代码片段所示，我们针对目录"c:\test"创建了一个 PhysicalFileProvider 对象，并调用其 Watch 方法对指定的 data.txt 文件实施监控。该方法会返回一个 IChangeToken 对象，我们正是利用这个对象来接收文件改变通知的。我们调用 ChangeToken 的静态方法 OnChange，针对这个对象注册了一个回调，用于实现对源文件的重新读取和显示，当源文件发生改变时，注册的回调会自动执行。我们每隔 5 秒对 data.txt 文件进行一次修改，而文件的内容为当前时间。所以，程序启动之后，每隔 5 秒当前时间就会以图 5-3 所示的方式呈现在控制台上。（S504）

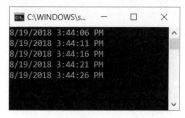

图 5-3　实时显示监控文件的内容

5.2　设计详解

5.1 节通过几个简单的实例演示从编程的角度对文件系统做了初步介绍，下面从设计的角度进一步阐述文件系统。这个抽象的文件系统以目录的形式来组织文件，我们可以利用它读取某个文件的内容，也可以对目录或者文件实施监控并及时得到变化的通知。由于 IFileProvider 对象提供了针对文件系统变换的监控功能，在 .NET Core 应用中类似的功能大都利用一个 IChangeToken 对象来实现，所以在对 IFileProvider 进行深入介绍之前应该先了解 IChangeToken。

5.2.1 IChangeToken

从字面上理解，IChangeToken 对象就是一个与某组监控数据相关联的"令牌"（Token），它能够在检测到数据改变时及时对外发出一个通知。如果 IChangeToken 对象关联的数据发生改变，那么它的 HasChanged 属性将变成 True。我们可以调用其 RegisterChangeCallback 方法注册一个在数据发生改变时可以自动执行的回调，该方法会返回一个 IDisposable 对象，可以用其 Dispose 方法解除注册的回调。IChangeToken 接口的另一个属性 ActiveChangeCallbacks 表示当数据发生变化时是否需要主动执行注册的回调操作。

```
public interface IChangeToken
{
    bool HasChanged { get; }
    bool ActiveChangeCallbacks { get; }
    IDisposable RegisterChangeCallback(Action<object> callback, object state);
}
```

.NET Core 提供了若干原生的 IChangeToken 接口实现类型，我们最常使用的是一个名为 CancellationChangeToken 的实现。CancellationChangeToken 的实现原理很简单，就是按照如下形式借助 CancellationToken 对象来发送通知的。

```
public class CancellationChangeToken : IChangeToken
{
    private readonly CancellationToken _token;
    public CancellationChangeToken(CancellationToken token)
        => _token = token;
    public bool HasChanged
        => _token.IsCancellationRequested;
    public bool ActiveChangeCallbacks
        => true;
    public IDisposable RegisterChangeCallback(Action<object> callback, object state)
        => _token.Register(callback, state);
}
```

除了 CancellationChangeToken，有时也会使用一个名为 CompositeChangeToken 的实现类型。顾名思义，CompositeChangeToken 代表由多个 IChangeToken 组合而成的复合型 IChangeToken 对象。如下面的代码片段所示，在调用构造函数创建一个 CompositeChangeToken 对象时，需要提供这些 IChangeToken 对象。对于一个 CompositeChangeToken 对象来说，只要组成它的任何一个 IChangeToken 发生改变，其 HasChanged 属性就变成 True，而注册的回调自然会被执行。只要任何一个 IChangeToken 的同名属性返回 True，ActiveChangeCallbacks 属性就会返回 True。

```
public class CompositeChangeToken : IChangeToken
{
    public bool                                      ActiveChangeCallbacks { get; }
```

```
    public IReadOnlyList<IChangeToken>        ChangeTokens { get; }
    public bool                               HasChanged { get; }

    public CompositeChangeToken(IReadOnlyList<IChangeToken> changeTokens);
    public IDisposable RegisterChangeCallback(Action<object> callback, object state);
}
```

我们可以直接调用 IChangeToken 提供的 RegisterChangeCallback 方法注册在接收到数据变化通知后的回调操作，但是更常用的方式则是直接调用静态类型 ChangeToken 提供的如下两个 OnChange 方法重载进行回调注册，这两个方法的第一个参数需要被指定为一个用来提供 IChangeToken 对象的 Func<IChangeToken>委托。

```
public static class ChangeToken
{
    public static IDisposable OnChange(Func<IChangeToken> changeTokenProducer,
        Action changeTokenConsumer) ;
    public static IDisposable OnChange<TState>(Func<IChangeToken> changeTokenProducer,
        Action<TState> changeTokenConsumer, TState state) ;
}
```

5.2.2　IFileProvider

了解 IChangeToken 对象之后，我们将关注点转移到文件系统的核心接口 IFileProvider 上，该接口定义在 NuGet 包 "Microsoft.Extensions.FileProviders.Abstractions" 中。前面做了几个简单的实例演示，体现了文件系统承载的 3 个基本功能，而这 3 个基本功能分别体现在 IFileProvider 接口如下所示的 3 个方法中。

```
public interface IFileProvider
{
    IFileInfo GetFileInfo(string subpath);
    IDirectoryContents GetDirectoryContents(string subpath);
    IChangeToken Watch(string filter);
}
```

虽然文件系统采用目录组织文件，但不论是目录还是文件都通过一个 IFileInfo 对象来表示，至于具体是目录还是文件则通过 IFileInfo 的 IsDirectory 属性来确定。对于一个 IFileInfo 对象，我们可以通过只读属性 Exists 判断指定的目录或者文件是否真实存在。而 Name 属性和 PhysicalPath 属性分别表示文件或者目录的名称与物理路径。LastModified 属性返回一个时间戳，表示目录或者文件最后一次被修改的时间。对于一个表示具体文件的 IFileInfo 对象来说，我们可以利用 Length 属性得到文件内容的字节长度。我们还可以借助 CreateReadStream 方法返回的 Stream 对象读取文件的内容。

```
public interface IFileInfo
{
    bool              Exists { get; }
    bool              IsDirectory { get; }
    string            Name { get; }
    string            PhysicalPath { get; }
    DateTimeOffset    LastModified { get; }
```

```
    long                Length { get; }
    Stream CreateReadStream();
}
```

IFileProvider 接口的 GetFileInfo 方法会根据指定的路径得到表示所在文件的 IFileInfo 对象。换句话说，虽然一个 IFileInfo 对象可以用于描述目录和文件，但是 GetFileInfo 方法的目的在于得到指定路径返回的文件而不是目录（笔者不太认同这种容易产生歧义的 API 设计）。一般来说，不论指定的文件是否存在，GetFileInfo 方法总会返回一个具体的 IFileInfo 对象，因为目标文件的存在与否是由该对象的 Exists 属性确定的。

如果希望得到某个目录的内容，如需要查看多少文件或者子目录包含在这个目录下，可以调用 IFileProvider 对象的 GetDirectoryContents 方法，并将所在目录的路径作为参数。目录内容通过该方法返回的 IDirectoryContents 对象来表示。如下面的代码片段所示，一个 IDirectoryContents 对象实际上是一组 IFileInfo 对象的集合，组成这个集合的所有 IFileInfo 就是对包含在这个目录下的所有文件和子目录的描述。和 GetFileInfo 方法一样，不论指定的目录是否存在，GetDirectoryContents 方法总是返回一个具体的 IDirectoryContents 对象，它的 Exists 属性可以确定指定目录是否存在。

```
public interface IDirectoryContents : IEnumerable<IFileInfo>
{
    bool Exists { get; }
}
```

如果要监控 IFileProvider 所在目录或者文件的变化，我们可以调用它的 Watch 方法，但前提是对应的 IFileProvider 对象提供了这样的监控功能。这个方法接受一个字符串类型的参数 filter，我们可以利用这个参数指定一个针对"文件匹配模式"（File Globbing Pattern）表达式（以下简称 Globbing Pattern 表达式）来筛选需要监控的目标目录或者文件。

Globbing Pattern 表达式比正则表达式简单，它只包含"*"一种通配符，如果认为它包含两种通配符，那么另一个通配符是"**"。Globbing Pattern 表达式体现为一个文件路径，其中，"*"代表所有不包括路径分隔符（"/"或者"\"）的所有字符，"**"代表包含路径分隔符在内的所有字符。表 5-1 列举了常见的几种 Globbing Pattern 表达式。

表 5-1　常见的几种 Globbing Pattern 表达式

Globbing Pattern 表达式	匹配的文件
src/foobar/foo/settings.*	子目录"src/foobar/foo/"（不含其子目录）下名为 settings 的所有文件，如 settings.json、settings.xml 和 settings.ini 等
src/foobar/foo/*.cs	子目录"src/foobar/foo/"（不含其子目录）下的所有 .cs 文件
src/foobar/foo/*.*	子目录"src/foobar/foo/"（不含其子目录）下的所有文件
src/**/*.cs	子目录"src"（含其子目录）下的所有 .cs 文件

一般来说，不论是调用 IFileProvider 对象的 GetFileInfo 方法或者 GetDirectoryContents 方法所指定的目标文件或者目录的路径，还是调用 Watch 方法指定的筛选表达式，都是一个针对当前 IFileProvider 对象映射根目录的相对路径。指定的这个路径可以采用"/"字符作为前缀，但

是这个前缀不是必要的。换句话说，下面两组程序是完全等效的。
```
//路径不包含前缀 "/"
var dirContents = fileProvider.GetDirectoryContents("foobar");
var fileInfo = fileProvider.GetFileInfo("foobar/foobar.txt");
var changeToken = fileProvider.Watch("foobar/*.txt");

//路径包含前缀 "/"
var dirContents = fileProvider.GetDirectoryContents("/foobar");
var fileInfo = fileProvider.GetFileInfo("/foobar/foobar.txt");
var changeToken = fileProvider.Watch("/foobar/*.txt");
```

总的来说，以 IFileProvider 对象为核心的文件系统从设计上来看是非常简单的。除了 IFileProvider 接口，文件系统还涉及其他一些对象，如 IDirectoryContents、IFileInfo 和 IChangeToken 等。文件系统涉及的接口及其相互之间的关系如图 5-4 所示。

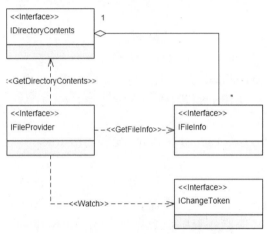

图 5-4　文件系统涉及的接口及其相互之间的关系

5.2.3　PhysicalFileProvider

ASP.NET Core 应用中使用得最多的还是具体的物理文件，如配置文件、View 文件以及作为 Web 资源的静态文件。物理文件系统由定义在 NuGet 包 "Microsoft.Extensions.FileProviders. Physical" 中的 PhysicalFileProvider 来构建。System.IO 命名空间下定义了一整套针对操作物理目录和文件的 API，但 PhysicalFileProvider 最终也是通过调用这些 API 来完成相关的 IO 操作的。

```
public class PhysicalFileProvider : IFileProvider, IDisposable
{
    public PhysicalFileProvider(string root);

    public IFileInfo GetFileInfo(string subpath);
    public IDirectoryContents GetDirectoryContents(string subpath);
    public IChangeToken Watch(string filter);

    public void Dispose();
}
```

PhysicalFileInfo

一个 PhysicalFileProvider 对象总是映射到某个具体的物理目录上，被映射的目录所在的路径通过构造函数的参数 root 来提供，该目录将作为 PhysicalFileProvider 的根目录。GetFileInfo 方法返回的 IFileInfo 对象代表指定路径对应的文件，这是一个类型为 PhysicalFileInfo 的对象。一个物理文件可以通过一个 System.IO.FileInfo 对象来表示，一个 PhysicalFileInfo 对象实际上就是对该对象的封装，定义在 PhysicalFileInfo 的所有属性都来源于 FileInfo 对象。对于创建读取文件输出流的 CreateReadStream 方法来说，它返回的是一个根据物理文件绝对路径创建的 FileStream 对象。

```
public class PhysicalFileInfo : IFileInfo
{
    ...
    public PhysicalFileInfo(FileInfo info);
}
```

对于 PhysicalFileProvider 的 GetFileInfo 方法来说，即使指定的路径指向一个具体的物理文件，也不是总返回一个 PhysicalFileInfo 对象。PhysicalFileProvider 会将一些场景视为 "目标文件不存在"，并让 GetFileInfo 方法返回一个 NotFoundFileInfo 对象。具体来说，PhysicalFileProvider 的 GetFileInfo 方法在如下场景中会返回一个 NotFoundFileInfo 对象。

- 确实没有一个物理文件与指定的路径相匹配。
- 如果指定的是一个绝对路径（如 "c:\foobar"），即 Path.IsPathRooted 方法返回 True。
- 如果指定的路径指向一个隐藏文件。

顾名思义，具有如下定义的 NotFoundFileInfo 类型表示一个 "不存在" 的文件。NotFoundFileInfo 对象的 Exists 属性总是返回 False，而其他属性则变得没有任何意义。当调用其 CreateReadStream 试图读取一个根本不存在的文件的内容时，会抛出一个 FileNotFoundException 类型的异常。

```
public class NotFoundFileInfo : IFileInfo
{
    public bool                 Exists => false;
    public long                 Length => throw new NotImplementedException();
    public string               PhysicalPath => null;
    public string               Name { get; }
    public DateTimeOffset       LastModified => DateTimeOffset.MinValue;
    public bool                 IsDirectory => false;

    public NotFoundFileInfo(string name) => this.Name = name;

    public Stream CreateReadStream()
        => throw new FileNotFoundException($"The file {Name} does not exist.");
}
```

PhysicalDirectoryInfo

PhysicalFileProvider 利用一个 PhysicalFileInfo 对象来描述某个具体的物理文件，而一个物理目录则通过一个 PhysicalDirectoryInfo 对象来描述。既然 PhysicalFileInfo 是对一个 FileInfo 对象的封装，那么 PhysicalDirectoryInfo 对象封装的就是表示目录的 DirectoryInfo 对象。如下面的代码片段所示，我们需要在创建一个 PhysicalDirectoryInfo 对象时提供 DirectoryInfo 对象，PhysicalDirectoryInfo 实现的所有属性的返回值都来源于 DirectoryInfo 对象。由于 CreateReadStream 方法的目的总是读取文件的内容，所以 PhysicalDirectoryInfo 类型的这个方法会抛出一个 InvalidOperationException 类型的异常。

```
public class PhysicalDirectoryInfo : IFileInfo
{
    ...
    public PhysicalDirectoryInfo(DirectoryInfo info);
}
```

PhysicalDirectoryContents

调用 PhysicalFileProvider 的 GetDirectoryContents 方法时，如果指定的路径指向一个具体的目录，那么该方法会返回一个类型为 PhysicalDirectoryContents 的对象。PhysicalDirectoryContents 是一个 IFileInfo 对象的集合，该集合中包括所有描述子目录的 PhysicalDirectoryInfo 对象和描述文件的 PhysicalFileInfo 对象。PhysicalDirectoryContents 的 Exists 属性取决于指定的目录是否存在。

```
public class PhysicalDirectoryContents : IDirectoryContents
{
    public bool Exists { get; }
    public PhysicalDirectoryContents(string directory);
    public IEnumerator<IFileInfo> GetEnumerator();
    IEnumerator IEnumerable.GetEnumerator();
}
```

NotFoundDirectoryContents

如果指定的路径并不指向一个存在的目录，或者指定的是一个绝对路径，GetDirectoryContents 方法都会返回一个 Exists 属性为 False 的 NotFoundDirectoryContents 对象。如下所示的代码片段展示了 NotFoundDirectoryContents 类型的定义，如果需要使用这样一个类型，就可以直接利用静态属性 Singleton 得到对应的单例对象。

```
public class NotFoundDirectoryContents : IDirectoryContents
{
    public static NotFoundDirectoryContents Singleton { get; }
        = new NotFoundDirectoryContents();
    public bool Exists => false;
    public IEnumerator<IFileInfo> GetEnumerator()
        => Enumerable.Empty<IFileInfo>().GetEnumerator();
    IEnumerator IEnumerable.GetEnumerator() => GetEnumerator();
}
```

PhysicalFilesWatcher

下面介绍 PhysicalFileProvider 的 Watch 方法。调用该方法时，PhysicalFileProvider 会通过解

析 Globbing Pattern 表达式来确定我们期望监控的文件或者目录，最终利用 FileSystemWatcher 对象对这些文件实施监控。这些文件或者目录的变化（创建、修改、重命名和删除等）都会实时地反映到 Watch 方法返回的 IChangeToken 上。

PhysicalFileProvider 的 Watch 方法中指定的 Globbing Pattern 表达式必须是针对当前根目录的相对路径，可以使用"/"或者"./"前缀，也可以不采用任何前缀。一旦使用了绝对路径（如"c:\test*.txt"）或者"../"前缀（如"../test/*.txt"），不论解析出的文件是否存在于 PhysicalFileProvider 的根目录下，这些文件都不会被监控。除此之外，如果没有指定 Globbing Pattern 表达式，PhysicalFileProvider 也不会有任何文件会被监控。

PhysicalFileProvider 针对物理文件系统变化的监控是通过下面的 PhysicalFilesWatcher 对象实现的，其 Watch 方法内部会直接调用 PhysicalFileProvider 的 CreateFileChangeToken 方法，并返回得到的 IChangeToken 对象。这是一个公共类型，如果有监控物理文件系统变化的需要，可以直接使用这个类型。

```
public class PhysicalFilesWatcher: IDisposable
{
    public PhysicalFilesWatcher(string root, FileSystemWatcher fileSystemWatcher,
        bool pollForChanges);
    public IChangeToken CreateFileChangeToken(string filter);
    public void Dispose();
}
```

从 PhysicalFilesWatcher 构造函数的定义可以看出，它最终利用一个 FileSystemWatcher 对象（对应参数 fileSystemWatcher）来完成针对指定根目录下（对应参数 root）所有子目录和文件的监控。FileSystemWatcher 的 CreateFileChangeToken 方法返回的 IChangeToken 对象会帮助我们感知到子目录或者文件的添加、删除、修改和重命名，但是它会忽略隐藏的目录和文件。需要注意的是，如果不再需要对指定目录实施监控，应调用 PhysicalFileProvider 的 Dispose 方法，该方法可以将 FileSystemWatcher 对象关闭。

小结

可以借助图 5-5 所示的 UML 对由 PhysicalFileProvider 构建物理文件系统的整体设计进行总结。首先，该文件系统使用 PhysicalDirectoryInfo 对象和 PhysicalFileInfo 对象来描述目录与文件，它们分别是对 DirectoryInfo 对象和 FileInfo（System.IO.FileInfo）对象的封装。

PhysicalFileProvider 的 GetDirectoryContents 方法返回一个 EnumerableDirectoryContents 对象（如果指定的目录存在），组成该对象的分别是根据其所有子目录和文件创建的 PhysicalDirectoryInfo 对象和 PhysicalFileInfo 对象。当调用 PhysicalFileProvider 的 GetFileInfo 方法时，如果指定的文件存在，返回的是描述该文件的 PhysicalFileInfo 对象。而 PhysicalFileProvider 的 Watch 方法则利用 FileSystemWatcher 来监控指定文件或者目录的变化。

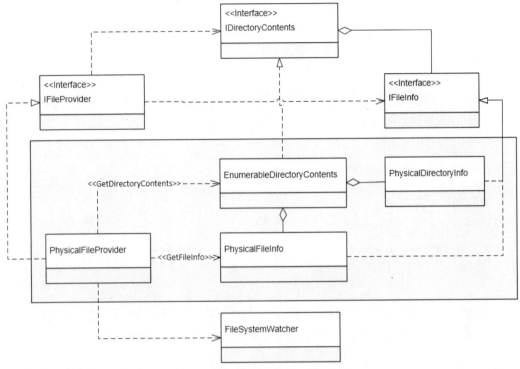

图 5-5　PhysicalFileProvider 涉及的主要类型及其相互之间的关系

5.2.4　EmbeddedFileProvider

一个物理文件可以直接作为资源内嵌到编译生成的程序集中。借助 EmbeddedFileProvider，我们可以采用统一的编程方式来读取内嵌的资源文件，该类型定义在 NuGet 包 "Microsoft.Extensions.FileProviders.Embedded" 中。在正式介绍 EmbeddedFileProvider 之前，我们必须知道如何将一个项目文件作为资源内嵌到编译生成的程序集中。

将项目文件变成内嵌资源

在默认情况下，添加到 .NET Core 项目中的静态文件并不会成为目标程序集的内嵌资源文件。如果需要将静态文件作为目标程序集的内嵌文件，就需要修改当前项目对应的.csproj 文件。具体来说，我们需要按照前面实例演示的方式在.csproj 文件中添加<ItemGroup>/<EmbeddedResource>元素，并利用 Include 属性显式地将对应的资源文件包含进来。

<EmbeddedResource>的 Include 属性可以设置多个路径，路径之间采用分号（;）作为分隔符。以图 5-6 所示的目录结构为例，如果需要将 root 目录下的 4 个文件作为程序集的内嵌文件，我们可以修改.csproj 文件，并按照如下形式将 4 个文件的路径包含进来。

图 5-6 包含资源文件的 .NET Core 项目

```
<Project Sdk="Microsoft.NET.Sdk">
    ...
    <ItemGroup>
        <EmbeddedResource
            Include="root/dir1/foobar/foo.txt;root/dir1/foobar/bar.txt;root/dir1/baz.txt;
                    root/dir2/qux.txt"></EmbeddedResource>
    </ItemGroup>
</Project>
```

除了指定每个需要内嵌的资源文件的路径，我们还可以采用基于通配符"*"和"**"的 Globbing Pattern 表达式将一组匹配的文件批量包含进来。同样是将 root 目录下的所有文件作为程序集的内嵌文件，但下面的定义方式更加简洁。

```
<Project Sdk="Microsoft.NET.Sdk">
    ...
    <ItemGroup>
        <EmbeddedResource Include="root/**"></EmbeddedResource>
    </ItemGroup>
</Project>
```

<EmbeddedResource>具有两个属性：Include 属性用来添加内嵌资源文件，Exclude 属性负责排除不符合要求的文件。还是以前面的项目为例，对于 root 目录下的 4 个文件，如果不希望 baz.txt 文件作为内嵌资源文件，也可以按照如下方式将其排除。

```
<Project Sdk="Microsoft.NET.Sdk">
    ...
    <ItemGroup>
        <EmbeddedResource
            Include="root/**"
            Exclude="root/dir1/baz.txt"></EmbeddedResource>
    </ItemGroup>
</Project>
```

读取资源文件

每个程序集都有一个清单文件（Manifest），其作用是记录组成程序集的所有文件成员。总的来说，一个程序集主要由两种类型的文件构成，即承载 IL 代码的托管模块文件和编译时内嵌的资源文件。针对图 5-6 所示的项目结构，如果将 4 个文本文件以资源文件的形式内嵌到生成的程序集（App.dll）中，程序集的清单文件将采用如下形式来记录它们。

```
.mresource public App.root.dir1.baz.txt
{
  // Offset: 0x00000000 Length: 0x0000000C
}
.mresource public App.root.dir1.foobar.bar.txt
{
  // Offset: 0x00000010 Length: 0x0000000C
}
.mresource public App.root.dir1.foobar.foo.txt
{
  // Offset: 0x00000020 Length: 0x0000000C
}
.mresource public App.root.dir2.qux.txt
{
  // Offset: 0x00000030 Length: 0x0000000C
}
```

虽然文件在原始项目中具有层次化的目录结构，但是当它们成功转移到编译生成的程序集中之后，目录结构将不复存在，所有的内嵌文件将统一存放在同一个容器中。如果通过 Reflector 打开程序集，资源文件的扁平化存储将会一目了然（见图 5-7）。为了避免命名冲突，编译器会根据原始文件所在的路径对资源文件重新命名，具体的规则是 "{BaseNamespace}.{Path}"，目录分隔符将统一转换成 "."。值得强调的是，资源文件名称的前缀不是程序集的名称，而是为项目设置的基础命名空间的名称。

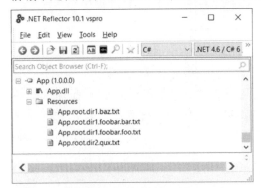

图 5-7　内嵌资源文件的扁平化存储

表示程序集的 Assembly 对象定义的如下几个方法用来提取内嵌资源文件的相关信息和读取指定资源文件的内容。GetManifestResourceNames 方法用于获取记录在程序集清单文件中的资源文件名，GetManifestResourceInfo 方法则用于获取指定资源文件的描述信息。如果需要读取某个资源文件的内容，可以将资源文件名称作为参数调用 GetManifestResourceStream 方法，该方法会返回一个读取文件内容的 Stream 对象。

```
public abstract class Assembly
{
    public virtual string[] GetManifestResourceNames();
```

```
public virtual ManifestResourceInfo GetManifestResourceInfo(string resourceName);
public virtual Stream GetManifestResourceStream(string name);
}
```

同样是针对前面这个演示项目对应的目录结构，当 4 个文件作为内嵌文件被成功转移到编译生成的程序集中后，我们可以调用程序集对象的 GetManifestResourceNames 方法获取这 4 个内嵌文件的资源名称。如果以资源名称（App.root.dir1.foobar.foo.txt）作为参数调用 GetManifestResourceStream 方法，可以读取资源文件的内容，具体的演示如下所示。

```
class Program
{
    static void Main()
    {
        var assembly = typeof(Program).Assembly;
        var resourceNames = assembly.GetManifestResourceNames();
        Debug.Assert(resourceNames.Contains("App.root.dir1.foobar.foo.txt"));
        Debug.Assert(resourceNames.Contains("App.root.dir1.foobar.bar.txt"));
        Debug.Assert(resourceNames.Contains("App.root.dir1.baz.txt"));
        Debug.Assert(resourceNames.Contains("App.root.dir2.qux.txt"));

        var stream = assembly.GetManifestResourceStream("App.root.dir1.foobar.foo.txt");
        var buffer = new byte[stream.Length];
        stream.Read(buffer, 0, buffer.Length);
        var content = Encoding.Default.GetString(buffer);
        Debug.Assert(content == File.ReadAllText("App/root/dir1/foobar/foo.txt"));
    }
}
```

EmbeddedFileProvider

对内嵌于程序集的资源文件有了大致的了解之后，针对 EmbeddedFileProvider 的实现原理就比较容易理解。由于内嵌于程序集的资源文件采用扁平化存储形式，所以通过 EmbeddedFileProvider 构建的文件系统中并没有目录层级的概念。我们可以认为所有的资源文件都保存在程序集的根目录下。对于 EmbeddedFileProvider 构建的文件系统来说，它提供的 IFileInfo 对象总是对一个具体的资源文件进行描述，这是一个具有如下定义的 EmbeddedResourceFileInfo 对象。

```
public class EmbeddedResourceFileInfo : IFileInfo
{
    private readonly Assembly        _assembly;
    private long?                    _length;
    private readonly string          _resourcePath;

    public EmbeddedResourceFileInfo(Assembly assembly, string resourcePath, string name,
        DateTimeOffset lastModified)
    {
        _assembly          = assembly;
        _resourcePath      = resourcePath;
        this.Name          = name;
        this.LastModified  = lastModified;
```

```
    }
    public Stream CreateReadStream()
    {
        Stream stream = _assembly.GetManifestResourceStream(_resourcePath);
        if (!this._length.HasValue)
        {
            this._length = new long?(stream.Length);
        }
        return stream;
    }

    public bool Exists => true;
    public bool IsDirectory => false;
    public DateTimeOffset LastModified { get; }

    public string Name { get; }
    public string PhysicalPath => null;
    public long Length
    {
        get
        {
            if (!_length.HasValue)
            {
                using (Stream stream =_assembly
                    .GetManifestResourceStream(this._resourcePath))
                {
                    _length = new long?(stream.Length);
                }
            }
            Return _length.Value;
        }
    }
}
```

如上面的代码片段所示,在创建一个 EmbeddedResourceFileInfo 对象时需要指定内嵌资源文件在清单文件中的路径(resourcePath)、所在的程序集、资源文件的名称(name),以及作为文件最后修改时间的 DateTimeOffset 对象。由于一个 EmbeddedResourceFileInfo 对象总是对应一个具体的内嵌资源文件,所以它的 Exists 属性总是返回 True,IsDirectory 属性则返回 False。由于资源文件系统并不具有层次化的目录结构,它所谓的物理路径毫无意义,所以 PhysicalPath 属性直接返回 Null。CreateReadStream 方法返回的是调用程序集的 GetManifestResourceStream 方法返回的输出流,而表示文件长度的 Length 返回的是这个 Stream 对象的长度。

下面的代码片段是 EmbeddedFileProvider 的定义。创建一个 EmbeddedFileProvider 对象时,除了指定资源文件所在的程序集,还可以指定一个基础命名空间。如果该命名空间没有显式设置,在默认情况下会将程序集的名称作为命名空间,也就是说,如果为项目指定了一个不同于程序集名称的基础命名空间,那么创建 EmbeddedFileProvider 对象时必须指定这个命名空间。

```csharp
public class EmbeddedFileProvider : IFileProvider
{
    public EmbeddedFileProvider(Assembly assembly);
    public EmbeddedFileProvider(Assembly assembly, string baseNamespace);

    public IDirectoryContents GetDirectoryContents(string subpath);
    public IFileInfo GetFileInfo(string subpath);
    public IChangeToken Watch(string pattern);
}
```

调用 EmbeddedFileProvider 的 GetFileInfo 方法并指定资源文件的逻辑名称时，该方法会将它与命名空间一起组成资源文件在程序集清单的名称（路径分隔符会被替换成"."）。如果对应的资源文件存在，那么一个 EmbeddedResourceFileInfo 会被创建并返回，否则返回的将是一个 NotFoundFileInfo 对象。对于内嵌资源文件系统来说，根本就不存在所谓的文件更新问题，所以它的 Watch 方法会返回一个 HasChanged 属性总是 False 的 IChangeToken 对象。

由于内嵌于程序集的资源文件总是只读的，它所谓的最后修改时间实际上是程序集的生成日期，所以 EmbeddedFileProvider 在提供 EmbeddedResourceFileInfo 对象时会将程序集文件的最后更新时间作为资源文件的最后更新时间。如果不能正确解析这个时间，那么 EmbeddedResourceFileInfo 的 LastModified 属性将被设置为当前 UTC 时间。

由于 EmbeddedFileProvider 构建的内嵌资源文件系统不存在层次化的目录结构，所有的资源文件可以视为全部存储在程序集的根目录下，所以它的 GetDirectoryContents 方法只有在指定一个空字符串或者"/"（空字符串和"/"都表示根目录）时才会返回一个描述这个根目录的 DirectoryContents 对象，该对象实际上是一组 EmbeddedResourceFileInfo 对象的集合。在其他情况下，EmbeddedFileProvider 的 GetDirectoryContents 方法总是返回一个 NotFoundDirectoryContents 对象。

5.2.5 两个特殊的 IFileProvider 实现

PhysicalFileProvider 和 EmbeddedFileProvider 分别构建了针对物理文件与程序集内嵌文件的文件系统，除此之外，还有两个特殊的 IFileProvider 实现类型可供选择，它们分别是 NullFileProvider 和 CompositeFileProvider。NullFileProvider 代表一个不包含任何内容的空文件系统，CompositeFileProvider 则是通过多个 IFileProvider 共同构建的组合式的文件系统。这两个特殊的 FileProvider 类型都定义在 "Microsoft.Extensions.FileProviders.Abstractions" 这个 NuGet 包中。

NullFileProvider

顾名思义，一个 NullFileProvider 对象代表一个不包含任何子目录和文件的空文件系统，所以它的 GetDirectoryContents 方法和 GetFileInfo 方法分别返回一个 NotFoundDirectoryContents 对象和 NotFoundFileInfo 对象。对于一个空的文件系统来说，并不存在所谓的目录和文件变化，所以其 Watch 方法返回一个 NullChangeToken 对象。相关的类型定义在如下所示的代码片段中。

```csharp
public class NullFileProvider : IFileProvider
```

```
{
    public IDirectoryContents GetDirectoryContents(string subpath)
        => NotFoundDirectoryContents.Singleton;
    public IFileInfo GetFileInfo(string subpath) => new NotFoundFileInfo(subpath);
    public IChangeToken Watch(string filter) => NullChangeToken.Singleton;
}

public class NullChangeToken : IChangeToken
{
    public bool HasChanged => false;
    public bool ActiveChangeCallbacks => false;
    public static NullChangeToken Singleton { get; } = new NullChangeToken();

    public IDisposable RegisterChangeCallback(Action<object> callback, object state)
        => EmptyDisposable.Instance;
}

internal class EmptyDisposable : IDisposable
{
    public static EmptyDisposable Instance { get; } = new EmptyDisposable();
    public void Dispose() { }
}
```

CompositeFileProvider

NullFileProvider 代表一个空的文件系统；CompositeFileProvider 则正好相反，代表一个由多个 IFileProvider 构建的复合型文件系统。如下面的代码片段所示，当调用构造函数创建一个 CompositeFileProvider 对象时，需要提供一组构建这个复合型文件系统的 IFileProvider 对象。

```
public class CompositeFileProvider : IFileProvider
{
    private readonly IFileProvider[] _fileProviders;
    public CompositeFileProvider(params IFileProvider[] fileProviders)
        => _fileProviders = fileProviders ?? new IFileProvider[0];
    ...
}
```

由于 CompositeFileProvider 由多个 IFileProvider 对象构成，所以当调用其 GetFileInfo 方法根据指定的路径获取对应文件时，它会遍历这些 IFileProvider 对象，直到找到一个存在的（对应 IFileInfo 的 Exists 属性返回 True）文件。如果所有的 IFileProvider 都不能提供这个文件，它会返回一个 NotFoundFileInfo 对象。由于遍历的顺序取决于构建 CompositeFileProvider 时提供的 IFileProvider 的顺序，所以如果对这些 IFileProvider 具有优先级的要求，应该将高优先级的 IFileProvider 对象放在前面。

```
public class CompositeFileProvider : IFileProvider
{
    private readonly IFileProvider[] _fileProviders;
    public IFileInfo GetFileInfo(string subpath)
    {
        foreach (var provider in _fileProviders)
```

```
            {
                var file = provider.GetFileInfo(subpath);
                if (file?.Exists == true)
                {
                    return file;
                }
            }
            return new NotFoundFileInfo(subpath);
        }
        ...
}
```

对于表示复合文件系统的 CompositeFileProvider 来说，某个目录的内容是由所有这些内部 IFileProvider 共同提供的，所以 GetDirectoryContents 方法返回的也是一个复合型的 DirectoryContents，具体的类型为具有如下定义的 CompositeDirectoryContents。与 GetFileInfo 方法一样，如果多个 IFileProvider 存在一个具有相同路径的文件，那么 CompositeFileProvider 总是从优先提供的 IFileProvider 中提取。

```
public class CompositeFileProvider : IFileProvider
{
    public IDirectoryContents GetDirectoryContents(string subpath)
        => new CompositeDirectoryContents(_fileProviders, subpath);
}

public class CompositeDirectoryContents : IDirectoryContents
{
    public CompositeDirectoryContents(IList<IFileProvider> fileProviders, string subpath);
    public bool Exists{get;}
    public IEnumerator<IFileInfo> GetEnumerator();
    IEnumerator IEnumerable.GetEnumerator() ;
}
```

CompositeFileProvider 的 Watch 方法返回的也是一个复合型的 IChangeToken 对象，其类型就是前面介绍的 CompositeChangeToken。这个 CompositeChangeToken 对象由组成 CompositeFileProvider 的所有 IFileProvider 来提供（通过调用 Watch 方法），所以它能监控任何一个 IFileProvider 对象对应的文件系统的变化，如下所示的代码片段体现了 Watch 方法的实现逻辑。

```
public class CompositeFileProvider : IFileProvider
{
    private readonly IFileProvider[] _fileProviders;

    public IChangeToken Watch(string pattern)
    {
        var tokens = _fileProviders
            .Select(it => it.Watch(pattern))
            .Where(it => it!= null)
            .ToList();

        return tokens.Count == 0
            ? (IChangeToken)NullChangeToken.Singleton
```

```
            : new CompositeChangeToken(tokens);
    }
}
```

5.3 远程文件系统

IFileProvider 构建了一个抽象文件系统，作为它的两个具体实现，PhysicalFileProvider 和 EmbeddedFileProvider 分别构建了一个物理文件系统与程序集内嵌文件系统。总的来说，它们针对的都是"本地"文件，下面通过自定义 IFileProvider 实现类型构建一个"远程"文件系统，我们可以将它视为一个只读的"云盘"。由于文件系统的目录结构和文件内容都是通过 HTTP 请求的方式读取的，所以可以将这个自定义的 IFileProvider 实现命名为 HttpFileProvider。读者可以将本节作为选读内容，跳过本节不会影响后续章节的阅读。

图 5-8 基本上体现了以 HttpFileProvider 为核心的远程文件系统的设计和实现原理。我们将真实的文件保存在文件服务器上，客户端可以通过公布的 Web API 得到指定路径所在的目录结构和描述信息，以及读取指定文件的内容。文件服务器中的每个目录都对应一个 URL，客户端可以通过指定相应的 URL 将某个目录作为本地文件系统的根目录。

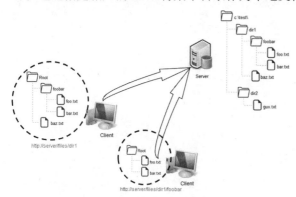

图 5-8　远程文件系统设计和实现原理

如图 5-8 所示，服务器上的文件系统实际上是直接通过指向"c:\test"目录的 PhysicalFileProvider 对象构建的，根目录对应的 URL 表示为"http://server/files/"。两个客户端采用 HttpFileProvider 将远程文件系统"挂载"到本地，采用的 URL 分别为"http://server/files/dir1"和"http://server/files/dir1/foobar"，所以它们分别映射为文件服务器上的目录"c:\dir1"和"c:\dir1\foobar"。

5.3.1　HttpFileInfo 与 HttpDirectoryContents

在以 HttpFileProvider 为核心的文件系统中，我们通过 HttpFileInfo 来表示目录和文件，包含子目录和文件的目录内容则通过 HttpDirectoryContents 类型来表示。在给出这两个类型的定义之前，我们需要介绍两个对应的描述类型：描述文件和目录的 HttpFileDescriptor，描述目录内容的 HttpDirectoryContentsDescriptor。如下面的代码片段所示，HttpFileDescriptor 的属性成员基本上是根据 IFileInfo 接口定义的。由于真实的目录或者文件存在于文件服务器上，所以 HttpFileDescriptor 的 PhysicalPath 属性表示的路径实际上是对应的 URL，该 URL 在构造时通过指定的 Func<string, string>委托计算出来。

```
public class HttpFileDescriptor
{
    public bool                Exists { get; set; }
    public bool                IsDirectory { get; set; }
    public DateTimeOffset      LastModified { get; set; }
    public long                Length { get; set; }
    public string              Name { get; set; }
    public string              PhysicalPath { get; set; }

    public HttpFileDescriptor()
    {}

    public HttpFileDescriptor(IFileInfo fileInfo,
        Func<string, string> physicalPathResolver)
    {
        Exists          = fileInfo.Exists;
        IsDirectory     = fileInfo.IsDirectory;
        LastModified    = fileInfo.LastModified;
        Length          = fileInfo.Length;
        Name            = fileInfo.Name;
        PhysicalPath    = physicalPathResolver(fileInfo.Name);
    }

    public IFileInfo ToFileInfo(HttpClient httpClient)
    {
        return this.Exists
            ? new HttpFileInfo(this, httpClient)
            : (IFileInfo)new NotFoundFileInfo(this.Name);
    }
}
```

用于描述文件和目录的 HttpFileDescriptor 对象是对一个 IFileInfo 对象的封装，与之类似，用来描述目录内容的 HttpDirectoryContentsDescriptor 对象则是对一个 IDirectoryContents 对象的封装。如下面的代码片段所示，HttpDirectoryContentsDescriptor 的 FileDescriptors 属性返回一组 HttpFileDescriptor 对象的集合，集合中的每个 HttpFileDescriptor 对象对应当前目录下的某个子目录或者文件。

```
public class HttpDirectoryContentsDescriptor
{
    public bool                                     Exists { get; set; }
    public IEnumerable<HttpFileDescriptor>          FileDescriptors { get; set; }

    public HttpDirectoryContentsDescriptor()
```

```
    {
        FileDescriptors = new HttpFileDescriptor[0];
    }

    public HttpDirectoryContentsDescriptor(IDirectoryContents directoryContents,
        Func<string, string> physicalPathResolver)
    {
        Exists = directoryContents.Exists;
        FileDescriptors = directoryContents.Select(
            _ => new HttpFileDescriptor(_, physicalPathResolver));
    }
}
```

从前面的代码片段可以看出，HttpFileDescriptor 类型具有一个 ToFileInfo 方法，可以将自己转换成一个 HttpFileInfo 对象。由于 HttpFileInfo 是通过一个 HttpFileDescriptor 对象创建的，所以它的所有属性最初都来源于这个对象。除了提供目录或者文件的描述信息，HttpFileInfo 还可以通过自身的 CreateReadStream 方法承载读取文件内容的职责。由于真正的文件保存在服务器上，所以我们需要利用构建时提供的 HttpClient 对象向目标文件所在的 URL 发送 HTTP 请求来读取文件内容。

```
public class HttpFileInfo: IFileInfo
{
    private readonly HttpClient _httpClient;

    public bool               Exists { get; }
    public bool               IsDirectory { get; }
    public DateTimeOffset     LastModified { get; }
    public long               Length { get; }
    public string             Name { get; }
    public string             PhysicalPath { get; }

    public HttpFileInfo(HttpFileDescriptor descriptor, HttpClient httpClient)
    {
        Exists        = descriptor.Exists;
        IsDirectory   = descriptor.IsDirectory;
        LastModified  = descriptor.LastModified;
        Length        = descriptor.Length;
        Name          = descriptor.Name;
        PhysicalPath  = descriptor.PhysicalPath;
        _httpClient   = httpClient;
    }

    public Stream CreateReadStream()
    {
        var message =  _httpClient.GetAsync(this.PhysicalPath).Result;
        return message.Content.ReadAsStreamAsync().Result;
    }
}
```

与 HttpFileInfo 对象类似，表示目录内容的 HttpDirectoryContents 对象依然是根据对应的描述对象（一个 HttpDirectoryContentsDescriptor 对象）创建的。一个 HttpDirectoryContents 对象本质上就是一个 IFileInfo 对象的集合，集合中的每个元素都是一个根据 HttpFileDescriptor 对象创建的 HttpFileInfo 对象。

```
public class HttpDirectoryContents : IDirectoryContents
{
    private readonly IEnumerable<IFileInfo> _fileInfos;
    public bool Exists { get; }

    public HttpDirectoryContents(HttpDirectoryContentsDescriptor descriptor,
        HttpClient httpClient)
    {
        this.Exists   = descriptor.Exists;
        _fileInfos    = descriptor.FileDescriptors.Select(
            file => file.ToFileInfo(httpClient));
    }

    public IEnumerator<IFileInfo> GetEnumerator() => _fileInfos.GetEnumerator();
    IEnumerator IEnumerable.GetEnumerator() => _fileInfos.GetEnumerator();
}
```

5.3.2　HttpFileProvider

下面介绍 HttpFileProvider 这个核心类型的实现。IFileProvider 承载了 3 项职责：通过 GetDirectoryContents 方法得到指定目录的内容，通过 GetFileInfo 方法得到指定目录或者文件的描述，通过 Watch 方法监控目录或者文件的变化。虽然可以采用某种技术手段从服务端向客户端发送通知，但是针对远程文件的监控意义不大，所以 HttpFileProvider 只提供前面两项基本功能。

```
public class HttpFileProvider : IFileProvider
{
    private readonly string         _baseAddress;
    private readonly HttpClient     _httpClient;

    public HttpFileProvider(string baseAddress)
    {
        _baseAddress  = baseAddress.TrimEnd('/');
        _httpClient   = new HttpClient();
    }

    public IDirectoryContents GetDirectoryContents(string subpath)
    {
        string url        = $"{_baseAddress}/{subpath.TrimStart('/')}?dir-meta";
        string content    = _httpClient.GetStringAsync(url).Result;
        HttpDirectoryContentsDescriptor descriptor = JsonConvert
            .DeserializeObject<HttpDirectoryContentsDescriptor>(content);
        return new HttpDirectoryContents(descriptor, _httpClient);
    }

    public IFileInfo GetFileInfo(string subpath)
    {
```

```csharp
        string url = $"{_baseAddress}/{subpath.TrimStart('/')}?file-meta";
        string content = _httpClient.GetStringAsync(url).Result;
        HttpFileDescriptor descriptor = JsonConvert
            .DeserializeObject<HttpFileDescriptor>(content);
        return descriptor.ToFileInfo(_httpClient);
    }

    public IChangeToken Watch(string filter) => NullChangeToken.Singleton;
}
```

由于文件系统由服务器托管，目录内容和目录与文件的描述信息都只能通过发送 HTTP 请求的形式来获取。HttpFileProvider 利用一个 HttpClient 对象来获取这些远程资源。HttpFileProvider 建立的本地文件系统的根目录可以指向文件服务器上的任意一个目录，我们将指向这个目录的 URL 称为基地址，该地址通过字段 _baseAddress 表示。对于任何一个目录或者文件来说，它对应的 URL 通过这个基地址和相对地址合并而成。

不论是 GetFileInfo 方法还是 GetDirectoryContents 方法，HttpFileProvider 发送 HTTP 请求的地址都是所在目录或者文件对应的 URL，但是它们返回的内容是不同的。前者返回的是目录或者文件的描述信息，后者返回的是目录内容的描述信息。因此，可以采用相应的查询字符串来区分这两种具有相同路径的 HTTP 请求，它们采用的查询字符串的名称分别是"?file-meta"和"?dir-meta"。

对于 HttpFileProvider 实现的 GetDirectoryContents 方法和 GetFileInfo 方法来说，它们会根据指定的相对路径解析出对应的 URL，然后利用 HttpClient 针对这个地址发送 HTTP 请求。响应的内容利用 JsonConvert 反序列化成一个 HttpDirectoryContentsDescriptor 对象或者 HttpFileDescriptor 对象，然后据此创建并返回一个 HttpDirectoryContents 对象或者 HttpFileInfo 对象。

5.3.3 FileProviderMiddleware

文件服务器是一个简单的 ASP.NET Core 应用，HttpFileProvider 调用的 Web API 是通过一个类型为 FileProviderMiddleware 的中间件实现的。这个自定义的 FileProviderMiddleware 需要处理如下 3 种类型的 HTTP 请求。

- 读取文件内容：请求地址指向目标文件，不含任何查询字符串，如"/files/dir1/foobar/foo.txt"。
- 读取文件或者目录的描述：请求地址指向目标文件或者目录，将"?file-meta"作为查询字符串，如"/files/dir1/foobar?file-meta"或者"/files/dir1/foobar/foo.txt?file-meta"。
- 读取目录内容：请求地址指向目标目录，将"?dir-meta"作为查询字符串，如"/files/dir1/foobar?dir-meta"。

如下所示的代码片段体现了 FileProviderMiddleware 这个中间件的完整定义。可以看出，它直接使用一个 PhysicalFileProvider 来构建本地文件系统，对应的根目录直接在构造函数中指定。针对上述 3 种 HTTP 请求的处理在 InvokeAsync 方法中实现，具体的实现逻辑其实很简单：如果请求地址携带查询字符串"dir-meta",中间件会根据请求目标目录创建一个 HttpDirectoryContentsDescriptor 对象,然后利用 JsonConvert 将其序列化后写入响应报文。如果请求地址携带查询字符串"file-meta",则根据请求的目录或者文件创建一个 HttpFileDescriptor 对象,并采用相同的方式序列化后写入响应报文。如果请求地址没有包含上面的两个查询字符串,则直接读取目标文件的内容并写入响应报文的主体。

```csharp
public class FileProviderMiddleware
{
    private readonly RequestDelegate _next;
    private readonly IFileProvider   _fileProvider;

    public FileProviderMiddleware(RequestDelegate next, string root)
    {
        _next                = next;
        _fileProvider        = new PhysicalFileProvider(root);
    }

    public async Task InvokeAsync(HttpContext context)
    {
        if (context.Request.Query.ContainsKey("dir-meta"))
        {
            var dirContents = _fileProvider.GetDirectoryContents(context.Request.Path);
            var dirDecriptor = new HttpDirectoryContentsDescriptor(dirContents,
                CreatePhysicalPathResolver(context, true));
            await context.Response.WriteAsync(JsonConvert.SerializeObject(dirDecriptor));
        }
        else if (context.Request.Query.ContainsKey("file-meta"))
        {
            var fileInfo = _fileProvider.GetFileInfo(context.Request.Path);
            var fileDescriptor = new HttpFileDescriptor(
                fileInfo, CreatePhysicalPathResolver(context, false));
            await context.Response.WriteAsync(JsonConvert.SerializeObject(fileDescriptor));
        }
        else
        {
            await context.Response.SendFileAsync(
                _fileProvider.GetFileInfo(context.Request.Path));
        }
    }

    private Func<string, string> CreatePhysicalPathResolver(HttpContext context,
        bool isDirRequest)
    {
        string schema            = context.Request.IsHttps ? "https" : "http";
        string host              = context.Request.Host.Host;
        int port                 = context.Request.Host.Port ?? 8080;
        string pathBase          = context.Request.PathBase.ToString().Trim('/');
        string path              = context.Request.Path.ToString().Trim('/');
```

```
        pathBase     = string.IsNullOrEmpty(pathBase) ? string.Empty : $"/{pathBase}";
        path         = string.IsNullOrEmpty(path) ? string.Empty : $"/{path}";

        return isDirRequest
            ? (Func<string, string>)(
                name => $"{schema}://{host}:{port}{pathBase}{path}/{name}")
            : name => $"{schema}://{host}:{port}{pathBase}{path}";
    }
}
```

5.3.4 远程文件系统的应用

整个文件系统由 FileProviderMiddleware 对象和 HttpFileProvider 对象组成，我们可以利用前者创建一个 ASP.NET Core 应用来作为文件服务器，客户端则利用后者在本地建立一个虚拟的文件系统。下面演示如何在一个具体的应用程序中使用这两个对象。首先创建一个 ASP.NET Core 应用来承载文件服务器，具体的代码如下所示。

```
public class Program
{
    public static void Main()
    {
        Host.CreateDefaultBuilder()
            .ConfigureWebHostDefaults(builder => builder.Configure(app=>app
                .UsePathBase("/files")
                .UseMiddleware<FileProviderMiddleware>(@"c:\test")))
            .Build()
            .Run();
    }
}
```

FileProviderMiddleware 中间件直接通过调用 IWebHostBuilder 的 Configure 扩展方法进行注册，并在注册的同时指定根目录的路径。下面直接利用在本章开篇创建的实例来演示如何利用 HttpFileProvider 来展示指定的目录结构和远程读取文件内容，并且对之前的程序进行了如下改写。

```
class Program
{
    static void Main()
    {
        var baseAddress = "http://localhost:5000/files/dir1";
        var fileManager = new ServiceCollection()
            .AddSingleton<IFileProvider>(new HttpFileProvider(baseAddress))
            .AddSingleton<IFileManager, FileManager>()
            .BuildServiceProvider()
            .GetRequiredService<IFileManager>();
        fileManager.ShowStructure((layer, name)
            => Console.WriteLine($"{new string('\t', layer)}{name}"));
    }
}
```

上面的代码片段创建并注册了一个 HttpFileProvider 对象,它采用的根目录的 URL 为 "http://localhost:5000/files/dir1"。由于文件服务器和客户端处于同一台主机,所以通过 HttpFileProvider 建立的本地文件系统的根目录实际上指向目录 "c:\test\dir1"。调用 FileManager 的 ShowStructure 方法之后,控制台上会以图 5-9 所示的形式呈现出本地文件系统的虚拟结构。(S505)

图 5-9 通过 HttpFileProvider 构建的文件系统结构

我们依然可以直接调用 FileManager 的 ReadAllTextAsync 方法远程读取某个文件的内容。如下面的代码片段所示,调用这个方法读取的文件路径为 "foobar/foo.txt",由于 HttpFileProvider 采用的基地址为 "/files/dir1",所以读取的这个文件在本地的路径为 "c:\test\dir1\foobar\foo.txt"。如下所示的调试断言表明,利用 HttpFileProvider 读取的文件就是这个物理文件。(S506)

```
class Program
{
    static async Task Main()
    {
        var baseAddress = "http://localhost:5000/files/dir1";
        var fileManager = new ServiceCollection()
            .AddSingleton<IFileProvider>(new HttpFileProvider(baseAddress))
            .AddSingleton<IFileManager, FileManager>()
            .BuildServiceProvider()
            .GetRequiredService<IFileManager>();

        var content1 = await fileManager.ReadAllTextAsync("foobar/foo.txt");
        var content2 = await File.ReadAllTextAsync(@"c:\test\dir1\foobar\foo.txt");
        Debug.Assert(content1 == content2);
    }
}
```

第6章
配置选项（上篇）

"配置选项"代表两个独立的基础框架："配置"（Configuration）通过构建的抽象配置模型弥补了不同配置数据源的差异，并在此基础上通过提供一致性的编程方式来读取配置数据；"选项"（Options）借助依赖注入框架实现了 Options 模式，该模式可以将应用的设置直接注入所需的服务之中。本章主要关注配置，第 7 章重点讲述选项。

6.1 读取配置信息

提及配置，大部分 .NET 开发人员会想起 app.config 和 web.config，多年以来我们已经习惯将结构化的配置定义在这两个 XML 格式的文件之中。到了 .NET Core 时代，很多我们习以为常的东西都发生了改变，其中就包括定义配置的方式。总的来说，新的配置系统具有更好的扩展性，其最大的特点就是支持多样化的数据源。既可以采用内存的变量作为配置的数据源，也可以将配置定义在持久化的文件甚至数据库中。在对配置系统进行系统介绍之前，下面先从编程的角度阐述全新的配置读取方式。

6.1.1 配置编程模型三要素

就编程层面来讲，.NET Core 的配置系统由图 6-1 中的 3 个核心对象构成。读取的配置信息最终会转换成一个 IConfiguration 对象供应用程序使用。IConfigurationBuilder 对象是 IConfiguration 对象的构建者，IConfigurationSource 对象则代表配置数据最原始的来源。

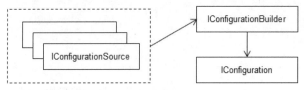

图 6-1　配置系统的 3 个核心对象

在读取配置时，可以根据配置的定义方式（数据源）创建相应的 IConfigurationSource 对象，并将其注册到 IConfigurationBuilder 对象上。提供配置的最初来源可能不止一个，我们可以在一个 IConfigurationBuilder 对象上注册多个相同或者不同类型的 IConfigurationSource 对象。IConfigurationBuilder 对象正是利用注册的这些 IConfigurationSource 对象提供的数据构建程序中使用的 IConfiguration 对象的。

这里介绍的 IConfiguration 接口、IConfigurationSource 接口和 IConfigurationBuilder 接口以及其他一些基础类型均定义在 NuGet 包 "Microsoft.Extensions.Configuration.Abstractions" 中。对这些接口的默认实现，则大多定义在 NuGet 包 "Microsoft.Extensions.Configuration" 中。

6.1.2 以键值对的形式读取配置

虽然大部分情况下的配置从整体上来说都具有结构化层次关系，但是"原子"配置项都体现为最简单的键值对形式，并且键和值通常都是字符串。下面通过一个简单的实例演示如何以键值对的形式来读取配置。

假设应用程序需要通过配置来设定日期/时间的显示格式，所以可以将相关的配置信息定义在如下所示的 DateTimeFormatOptions 类中，它的 4 个属性体现了针对 DateTime 对象的 4 种显示格式（分别为长日期/时间和短日期/时间）。

```
public class DateTimeFormatOptions
{
    ...
    public string LongDatePattern { get; set; }
    public string LongTimePattern { get; set; }
    public string ShortDatePattern { get; set; }
    public string ShortTimePattern { get; set; }
}
```

如果通过配置的形式控制由 DateTimeFormatOptions 的 4 个属性所体现的显示格式，就需要定义一个构造函数。如下面的代码片段所示，该构造函数具有一个 IConfiguration 接口类型的参数。键值对是配置的基本表现形式，所以 IConfiguration 对象提供的索引使我们可以根据配置项的 Key 得到配置项的值，下面的代码正是以索引的方式得到对应配置信息的。

```
public class DateTimeFormatOptions
{
    ...
    public DateTimeFormatOptions (IConfiguration config)
    {
        LongDatePattern      = config["LongDatePattern"];
        LongTimePattern      = config["LongTimePattern"];
        ShortDatePattern     = config["ShortDatePattern"];
        ShortTimePattern     = config["ShortTimePattern"];
    }
}
```

如果要创建一个体现当前配置的 DateTimeFormatOptions 对象，就必须提供这个承载相关配置信息的 IConfiguration 对象。正如前面提及的，IConfiguration 对象是由 IConfigurationBuilder 对象创建的，而原始的配置信息则是通过相应的 IConfigurationSource 对象提供的，所以创建一个 IConfiguration 对象的正确编程方式如下：创建一个 ConfigurationBuilder（IConfigurationBuilder 接口的默认实现类型）对象，并为之注册一个或者多个 IConfigurationSource 对象，最后利用它来创建我们需要的 IConfiguration 对象。

可以通过如下程序来读取配置并将其转换成一个 DateTimeFormatOptions 对象。简单起见，我们采用的 IConfigurationSource 实现类型为 MemoryConfigurationSource，它直接利用一个保存在内存中的字典对象作为最初的配置来源。下面的代码片段在为 MemoryConfigurationSource 提供的字典对象中设置了 4 种类型的日期/时间显示格式。

```csharp
public class Program
{
    public static void Main()
    {
        var source = new Dictionary<string, string>
        {
            ["longDatePattern"]     = "dddd, MMMM d, yyyy",
            ["longTimePattern"]     = "h:mm:ss tt",
            ["shortDatePattern"]    = "M/d/yyyy",
            ["shortTimePattern"]    = "h:mm tt"
        };

        var config = new ConfigurationBuilder()
            .Add(new MemoryConfigurationSource { InitialData = source })
            .Build();

        var options = new DateTimeFormatOptions(config);
        Console.WriteLine($"LongDatePattern: {options.LongDatePattern}");
        Console.WriteLine($"LongTimePattern: {options.LongTimePattern}");
        Console.WriteLine($"ShortDatePattern: {options.ShortDatePattern}");
        Console.WriteLine($"ShortTimePattern: {options.ShortTimePattern}");
    }
}
```

上面的代码片段创建了一个 ConfigurationBuilder 对象，并在它上面注册了一个根据内存字典创建的 MemoryConfigurationSource 对象。下面调用 ConfigurationBuilder 对象的 Build 方法创建 IConfiguration 对象，并利用 IConfiguration 对象创建了 DateTimeFormatOptions 对象。为了验证该 Options 对象是否与原始配置一致，我们将它的 4 个属性打印在控制台上。程序运行之后，控制台上的输出结果如图 6-2 所示。（S601）

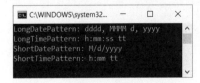

图 6-2　以键值对的形式读取配置

6.1.3　读取结构化的配置

真实项目中涉及的配置大都具有结构化的层次结构，所以 IConfiguration 对象同样具有这样的结构。由于配置具有一个树形层次结构，所以可以将其称为"配置树"，一个 IConfiguration 对象对应这棵配置树的某个节点，而整棵配置树自然可以由根节点对应的 IConfiguration 对象来表示。以键值对体现的"原子配置项"对应配置树中不具有子节点的"叶子节点"。

下面以实例来演示如何定义并读取具有层次结构的配置数据。我们依然沿用 6.1.2 节的应用场景，但现在不仅需要设置日期/时间的格式，还需要设置其他数据类型的格式，如表示货币的 Decimal 类型。因此，我们定义了一个 CurrencyDecimalFormatOptions 类，它的 Digits 属性和 Symbol 属性分别表示小数位数与货币符号，CurrencyDecimalFormatOptions 对象依然是利用 IConfiguration 对象创建的。

```
public class CurrencyDecimalFormatOptions
{
    public int          Digits { get; set; }
    public string       Symbol { get; set; }

    public CurrencyDecimalFormatOptions (IConfiguration config)
    {
        Digits = int.Parse(config["Digits"]);
        Symbol = config["Symbol"];
    }
}
```

我们定义了另一个名为 FormatOptions 的类型来表示针对不同数据类型的格式设置。如下面的代码片段所示，它的 DateTime 属性和 CurrencyDecimal 属性分别表示针对日期/时间与货币数字的格式设置。FormatOptions 依然具有一个参数类型为 IConfiguration 的构造函数，它的两个属性均在此构造函数中被初始化。值得注意的是，初始化这两个属性采用的是当前 IConfiguration 的"子配置节"，我们可以通过调用 GetSection 方法根据指定的名称（DateTime 和 CurrencyDecimal）获得这两个子配置节。

```
public class FormatOptions
{
    public DateTimeFormatOptions            DateTime { get; set; }
    public CurrencyDecimalFormatOptions     CurrencyDecimal { get; set; }

    public FormatOptions (IConfiguration config)
    {
        DateTime = new DateTimeFormatOptions (
            config.GetSection("DateTime"));
        CurrencyDecimal = new CurrencyDecimalFormatOptions (
            config.GetSection("CurrencyDecimal"));
    }
}
```

FormatOptions 类型体现的配置具有图 6-3 所示的树形层次结构。在前面演示的实例中，我们使用 MemoryConfigurationSource 对象来提供原始的配置信息。承载原始配置信息的是一个元素类型为 KeyValuePair<string, string>的集合，但是它在物理存储上并不具有树形层次结构，那么它如何提供一个结构化的 IConfiguration 对象承载的数据？

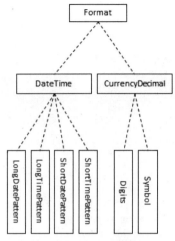

图 6-3 树形层次结构的配置

解决方案其实很简单，对于一棵完整的配置树，具体的配置信息最终是通过叶子节点来承载的，所以 MemoryConfigurationSource 对象只需要在配置字典中保存叶子节点的数据即可。除此之外，为了描述配置树的结构，配置字典还需要将对应叶子节点在配置树中的路径作为 Key。所以，MemoryConfigurationSource 对象可以采用表 6-1 列举的配置字典对配置树进行扁平化处理，作为 Key 的路径将冒号（:）作为分隔符。

表 6-1 配置的物理结构

Key	Value
Format:DateTime:LongDatePattern	dddd, MMMM d, yyyy
Format:DateTime:LongTimePattern	h:mm:ss tt
Format:DateTime:ShortDatePattern	M/d/yyyy
Format:DateTime:ShortTimePattern	h:mm tt
Format:CurrencyDecimal:Digits	2
Format:CurrencyDecimal:Symbol	$

下面的代码片段按照表 6-1 列举的结构创建了一个 Dictionary<string, string>对象，并利用它创建了 MemoryConfigurationSource 对象。在利用 ConfigurationBuilder 对象得到 IConfiguration 对象之后，我们调用其 GetSection 方法得到名称为 Format 的配置节，并利用后者创建了一个 FormatOptions。

```
public class Program
{
    public static void Main()
    {
        var source = new Dictionary<string, string>
        {
            ["format:dateTime:longDatePattern"] = "dddd, MMMM d, yyyy",
            ["format:dateTime:longTimePattern"] = "h:mm:ss tt",
            ["format:dateTime:shortDatePattern"] = "M/d/yyyy",
            ["format:dateTime:shortTimePattern"] = "h:mm tt",
```

```csharp
            ["format:currencyDecimal:digits"] = "2",
            ["format:currencyDecimal:symbol"] = "$",
        };
        var configuration = new ConfigurationBuilder()
            .Add(new MemoryConfigurationSource { InitialData = source })
            .Build();

        var options = new FormatOptions(configuration.GetSection("Format"));
        var dateTime = options.DateTime;
        var currencyDecimal = options.CurrencyDecimal;

        Console.WriteLine("DateTime:");
        Console.WriteLine($"\tLongDatePattern: {dateTime.LongDatePattern}");
        Console.WriteLine($"\tLongTimePattern: {dateTime.LongTimePattern}");
        Console.WriteLine($"\tShortDatePattern: {dateTime.ShortDatePattern}");
        Console.WriteLine($"\tShortTimePattern: {dateTime.ShortTimePattern}");

        Console.WriteLine("CurrencyDecimal:");
        Console.WriteLine($"\tDigits:{currencyDecimal.Digits}");
        Console.WriteLine($"\tSymbol:{currencyDecimal.Symbol}");
    }
}
```

在得到利用读取的配置创建的 FormatOptions 对象之后，为了验证该对象与原始配置数据是否一致，我们依然将它的相关属性打印在控制台上。这个程序运行之后在控制台上呈现的输出结果如图 6-4 所示。（S602）

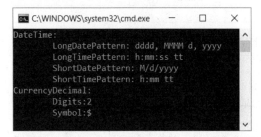

图 6-4　读取结构化的配置

6.1.4　将结构化配置直接绑定为对象

在真正的项目开发过程中，我们倾向于像前面演示的实例一样将一组相关的配置转换成一个 POCO 对象，如演示实例中的 DateTimeFormatOptions 对象、CurrencyDecimalOptions 对象和 FormatOptions 对象。在前面的演示实例中，为了创建这些封装配置的对象，我们通常采用手动读取配置的形式。如果定义的配置项太多，逐条读取配置项其实是一项非常烦琐的工作。

如果承载配置数据的 IConfiguration 对象与对应的 POCO 类型具有兼容的结构，那么利用配置的自动绑定机制可以将 IConfiguration 对象直接转换成对应的 POCO 对象。对于我们演示的实例来说，如果采用自动化配置绑定来创建对应的 Options 对象，那么这些类型中就不再需要实现手动绑定的构造函数。

在删除所有 Options 类型的构造函数之后，再修改 Options 对象的创建方式。如下面的代码片段所示，在调用 IConfigurationBuilder 对象的 Build 方法创建出对应 IConfiguration 对象之后，调用 GetSection 方法可以得到其 format 配置节，而 FormatOptions 对象不用再通过调用构造函数来创建，而是直接调用该配置节的 Get<T>方法，该方法完成了从 IConfiguration 对象到 POCO 对象之间的自动化绑定。修改后的程序运行之后，同样会得到图 6-4 所示的输出结果。（S603）

```
...
var options = new ConfigurationBuilder()
    .Add(new MemoryConfigurationSource { InitialData = source })
    .Build()
    .GetSection("format")
    .Get<FormatOptions>();
...
```

6.1.5 将配置定义在文件中

前面演示的 3 个实例都是采用 MemoryConfigurationSource 将一个字典对象作为配置源，下面演示一种更加常见的配置定义方法，那就是将原始配置的内容定义在一个 JSON 文件中。我们将原本通过一个内存字典对象承载的配置定义在一个 JSON 文件中，为此在项目的根目录下创建一个名为 appsettings.json 的配置文件，并将该文件的 "Copy to Output Directory" 属性设置为 "Copy always"[①]，其目的是促使项目在编译的时候能够将此文件复制到输出目录下。可以采用如下形式定义关于日期/时间和货币的格式配置。

```
{
    "format": {
        "dateTime": {
            "longDatePattern"       : "dddd, MMMM d, yyyy",
            "longTimePattern"       : "h:mm:ss tt",
            "shortDatePattern"      : "M/d/yyyy",
            "shortTimePattern"      : "h:mm tt"
        },
        "currencyDecimal": {
            "digits": 2,
            "symbol": "$"
        }
    }
}
```

由于配置源发生了改变，原来的 MemoryConfigurationSource 需要替换成 JsonConfigurationSource，但不需要手动创建这个 JsonConfigurationSource 对象，只需要调用 IConfiguration

---

[①] 如果项目采用的 SDK 类型为 "Microsoft .NET.Sdk"，该应用在 Visual Studio 中运行时会将编译输出目录作为当前目录。如果项目采用的 SDK 类型为 "Microsoft .NET.Sdk.Web"，那么项目根目录就是当前执行的目录，此时不需要设置配置文件的 "Copy to Output Directory" 属性。

Builder 接口的 AddJsonFile 扩展方法添加指定的 JSON 文件即可。执行修改后的程序，依然可以得到图 6-4 所示的输出结果。（S604）

```
var options = new ConfigurationBuilder()
    .AddJsonFile("appsettings.json")
    .Build()
    .GetSection("format")
    .Get<FormatOptions>();
...
```

根据环境动态加载配置文件

真实项目开发过程中使用的配置往往取决于应用当前执行的环境，也就是说，不同的执行环境（开发、测试、预发和产品等）会采用不同的配置。如果采用基于物理文件的配置，则可以为不同的环境提供对应的配置文件，具体的做法如下：除了提供一个基础配置文件（如 appsettings.json），我们还需要为相应的环境提供对应的差异化配置文件，后者通常采用环境名称作为文件扩展名（如 appsettings.production.json）。

以目前演示的程序为例，现有的配置文件 appsettings.json 可以作为基础配置文件，如果某个环境需要采用不同的配置，也可以将差异化的配置定义在对应的文件中。如图 6-5 所示，我们额外添加了两个配置文件（appsettings.staging.json 和 appsettings.production.json），从文件命名可以看出这两个配置文件分别对应预发环境和产品环境。

图 6-5 针对执行环境的配置文件

我们在 JSON 文件中定义了针对日期/时间和货币格式的配置，假设预发环境和产品环境需要采用不同的货币格式，那么就需要将差异化的配置定义在针对环境的两个配置文件中。简单起见，我们仅仅将货币的小数位数定义在配置文件中。如下面的代码片段所示，货币小数位数（默认值为 2）在预发环境和产品环境中分别被设置为 3 与 4。

appsettings.staging.json：

```
{
    "format": {
        "currencyDecimal": {
            "digits": 3
```

```
        }
    }
}
```

appsettings.production.json：
```
{
    "format": {
        "currencyDecimal": {
            "digits": 4
        }
    }
}
```

一般来说，可以采用环境变量来决定应用的执行环境，但为了在演示过程中能够灵活地进行环境切换，可以采用命令行参数（如/env staging）的形式来设置环境。到目前为止，针对某一环境的配置被分布到两个配置文件中，所以在启动文件时就应该根据当前执行环境动态地加载对应的配置文件。如果两个文件涉及同一段配置，就应该首选当前环境对应的那个配置文件。由于配置默认采用"后来居上"的原则，所以应该先加载基础配置文件，再加载针对环境的配置文件。针对执行环境的判断以及针对环境的配置加载体现在如下所示的代码片段中。

```
class Program
{
    static void Main(string[] args)
    {
        var index = Array.IndexOf(args, "/env");
        var environment = index > -1
            ? args[index + 1]
            : "Development";

        var options = new ConfigurationBuilder()
            .AddJsonFile("appsettings.json",false)
            .AddJsonFile($"appsettings.{environment}.json",true)
            .Build()
            .GetSection("format")
            .Get<FormatOptions>();
        ...
    }
}
```

如上面的代码片段所示，在利用传入的命令行参数确定了当前执行环境之后，我们先后两次调用 IConfigurationBuilder 对象的 AddJsonFile 方法将两个配置文件加载进来，所以两个文件合并后的内容将用于构建 Build 方法创建的 IConfiguration 对象。然后以命令行的形式启动这个控制台程序，并通过命令行参数指定相应的环境名称。从图 6-6 所示的输出结果可以看出，打印出的配置数据（货币的小数位数）确实来源于环境对应的配置文件。（S605）

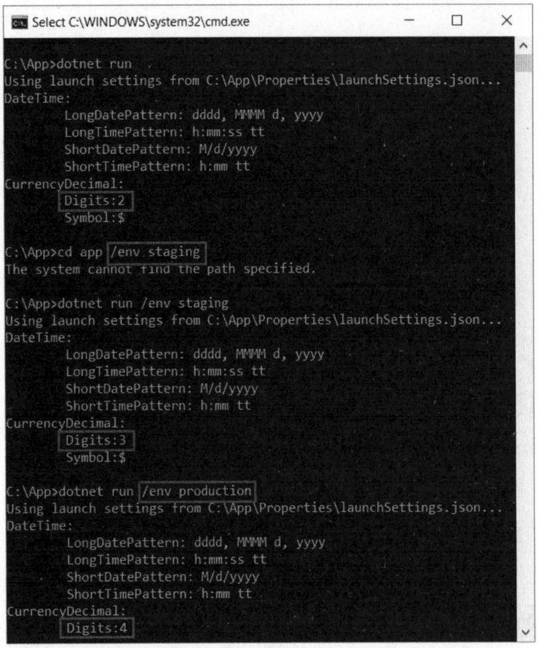

图6-6　输出与当前环境匹配的配置

配置文件的同步

在很多情况下，应用程序的配置只会在启动的时候从相应的配置源中读取，并在整个应用的生命周期中保持不变，一旦需要更新配置，就需要重新启动应用程序。.NET Core 的配置模型提供了针对配置源的监控功能，它能保证一旦原始配置改变之后应用程序能够及时接收到通知，此时我们可以利用预先注册的回调进行配置的同步。

前面演示的应用程序采用 JSON 文件作为配置源，所以我们希望应用程序能够感知该文件的改变，并在文件发生改变的时候自动加载新的配置，然后将其重新应用到程序之中。为了演示配置的同步，我们对程序做了如下改变。

```
class Program
{
    static void Main()
    {
        var config = new ConfigurationBuilder()
            .AddJsonFile(path: "appsettings.json",
                         optional:true,
                         reloadOnChange: true)
            .Build();
        ChangeToken.OnChange(() => config.GetReloadToken(), () =>
        {
            var options = config.GetSection("format").Get<FormatOptions>();
            var dateTime = options.DateTime;
            var currencyDecimal = options.CurrencyDecimal;
```

```
            Console.WriteLine("DateTime:");
            Console.WriteLine($"\tLongDatePattern: {dateTime.LongDatePattern}");
            Console.WriteLine($"\tLongTimePattern: {dateTime.LongTimePattern}");
            Console.WriteLine($"\tShortDatePattern: {dateTime.ShortDatePattern}");
            Console.WriteLine($"\tShortTimePattern: {dateTime.ShortTimePattern}");

            Console.WriteLine("CurrencyDecimal:");
            Console.WriteLine($"\tDigits:{currencyDecimal.Digits}");
            Console.WriteLine($"\tSymbol:{currencyDecimal.Symbol}\n\n");
        });
        Console.Read();
    }
}
```

表示 JSON 文件配置源的 JsonConfigurationSource 在默认情况下并不会监控源文件的变化，所以在调用 IConfigurationBuilder 的 AddJsonFile 扩展方法时，需要通过传入的 reloadOnChange 参数开启这个功能。通过 IConfigurationBuilder 的 Build 方法创建的 IConfiguration 对象具有一个返回类型为 IChangeToken 的 GetReloadToken 方法，我们正是利用它返回的 IChangeToken 对象来感知配置源的变化的。一旦配置源发生变化，IConfiguration 对象将自动加载新的内容，所以只需要通过注册的回调将同一个 IConfiguration 对象应用到程序之中即可。

上述程序会在感知到配置源发生变化后自动将新的配置内容打印出来，所以当该程序被启动之后，我们对 appsettings.json 文件[①]所做的任何修改都会触发应用对该文件的重新加载。图 6-7 中的输出结果是两次修改货币小数位数导致的。（S606）

图 6-7　配置文件更新触发配置的重新加载

6.2　配置模型

6.1 节通过实例演示了几种典型的配置读取方式，下面从设计的角度重写认识配置模型。配

[①] 由于加载的是编译后复制到输出目录下的配置文件，在做一个实验的时候，修改的是当前工作目录下的配置文件，而不是当前项目根目录下的配置文件。

置的编程模型涉及 3 个核心对象，分别通过 3 个对应的接口（IConfiguration、IConfigurationSource 和 IConfigurationBuilder）来表示。如果从设计层面审视背后的配置模型，还缺少另一个通过 IConfigurationProvider 接口表示的核心对象。总的来说，配置模型由这 4 个核心对象组成，但是要彻底了解这 4 个核心对象之间的关系，需要先了解配置的几种数据结构。

6.2.1 数据结构及其转换

相同的数据具有不同的表现形式和承载方式，同时体现出不同的数据结构。对于配置来说，它在被应用程序消费的过程中是以 IConfiguration 对象的形式来体现的，该对象在逻辑上具有一个树形层次结构，所以将其称为配置树，并将这棵树视为配置的逻辑结构。

配置具有多种原始来源，可以是内存对象、物理文件、数据库或者其他自定义的存储介质。如果采用物理文件来存储配置数据，还可以选择不同的文件格式，常见的文件类型包括 XML、JSON 和 INI 这 3 种，所以配置的原始数据结构是多种多样的。配置模型的最终目的在于提取原始的配置数据并将其转换成一个 IConfiguration 对象。换句话说，配置模型就是为了按照图 6-8 所示的方式将配置数据从原始结构转变成逻辑结构。

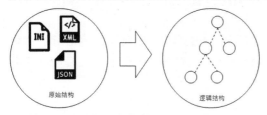

图 6-8　配置由原始结构向逻辑结构的转变

配置从原始结构向逻辑结构的转变不是一蹴而就的，在它们之间有一种中间结构。原始的配置数据被读取出来之后先统一转换成这种中间结构的数据，这种中间结构究竟是一种什么样的数据结构？

6.1 节提及，一棵配置树通过其叶子节点承载所有的原子配置项，这棵树的结构和承载的数据完全可以利用一个简单的数据字典来表达。具体来说，我们只需要将所有叶子节点在配置树中的路径作为 Key，将叶子节点承载的配置数据作为 Value 即可。所谓的中间结构指的就是这样的数据字典，可以将其称为配置字典。所以，配置模型会按照图 6-9 所示的方式将具有不同原始结构的配置数据统一转换成配置字典，然后完成针对逻辑结构的转换。

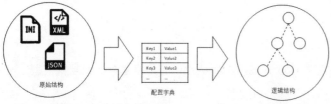

图 6-9　配置"三态"转换

对于配置模型的 4 个核心对象来说，IConfiguration 对象是对配置树的描述，其他 3 个核心对象（IConfigurationSource、IConfigurationBuilder 和 IConfigurationProvider）在配置的结构转换过程中扮演不同的角色，至于它们究竟具有什么样的作用，下面将分别进行介绍。

6.2.2 IConfiguration

配置在应用程序中总是以一个 IConfiguration 对象的形式供我们使用。一个 IConfiguration 对象具有树形层次结构并不是说对应的类型具有对应的数据成员定义，而是说它提供的 API 在逻辑上体现出树形层次结构，所以说配置树是一种逻辑结构。如下所示的代码片段是 IConfiguration 接口的完整定义，所谓的层次化逻辑结构就体现在它的 GetChildren 方法和 GetSection 方法上。

```
public interface IConfiguration
{
    IEnumerable<IConfigurationSection>      GetChildren();
    IConfigurationSection                   GetSection(string key);
    IChangeToken                            GetReloadToken();

    string this[string key] { get; set; }
}
```

一个 IConfiguration 对象表示配置树的某个配置节点。对于组成整棵树的所有配置节点来说，表示根节点的 IConfiguration 对象与表示其他配置节点的 IConfiguration 对象是不同的，所以配置模型采用不同的接口来表示它们。根节点所在的 IConfiguration 对象体现为一个 IConfigurationRoot 对象，其他节点对象则用一个 IConfigurationSection 对象来表示，IConfigurationRoot 接口和 IConfigurationSection 接口继承自 IConfiguration 接口。图 6-10 展示了由一个 IConfigurationRoot 对象和一组 IConfigurationSection 对象组成的配置树。

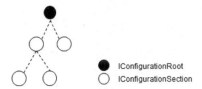

图 6-10　由一个 IConfigurationRoot 对象和一组 IConfigurationSection 对象组成的配置树

如下所示的代码片段是 IConfigurationRoot 接口的定义，它具有的唯一的方法 Reload 实现对配置数据的重新加载。IConfigurationRoot 对象表示配置树的根，所以也代表了整棵配置树，如果它被重新加载，则意味着整棵配置树承载的所有配置数据均被重新加载。

```
public interface IConfigurationRoot : IConfiguration
{
    void Reload();
}
```

表示非根配置节点的 IConfigurationSection 接口具有如下 3 个属性：只读属性 Key 用来唯一标识多个具有相同父节点的 ConfigurationSection 对象；而 Path 则表示当前配置节点在配置树中

的路径,由组成当前路径的所有IConfigurationSection对象的Key构成,Key之间采用冒号(:)作为分隔符。Path和Key的组合体现了对应配置节在整棵配置树中的位置。

```
public interface IConfigurationSection : IConfiguration
{
    string Path { get; }
    string Key { get; }
    string Value { get; set; }
}
```

IConfigurationSection 接口的 Value 属性表示配置节点承载的数据。在大部分情况下,只有配置树的叶子节点对应的 IConfigurationSection 对象才包含具体的值,非叶子节点对应的 IConfigurationSection 对象实际上仅仅表示存放所有子配置节的逻辑容器,它们的 Value 属性一般返回 Null。值得注意的是,这个 Value 属性并不是只读的,而是可读可写的,但写入的值一般不会被持久化,一旦配置树被重新加载,该值将丢失。

在对 IConfigurationRoot 接口和 IConfigurationSection 接口具有基本了解之后,下面介绍定义在 IConfiguration 接口中的成员。它的 GetChildren 方法返回的 IConfigurationSection 集合表示它的所有子配置节,而 GetSection 方法则根据指定的 Key 得到一个具体的子配置节。当执行 GetSection 方法时,指定的参数会与当前 IConfigurationSection 的 Path 进行组合,从而确定目标配置节点所在的路径,如果在调用该方法的时候指定一个当前配置节的相对路径,就可以得到子节点以下的任何一个配置节。

```
var source = new Dictionary<string, string>
{
    ["A:B:C"] = "ABC"
};

var root = new ConfigurationBuilder()
    .AddInMemoryCollection(source)
    .Build();

var section1 = root.GetSection("A:B:C");                    //A:B:C
var section2 = root.GetSection("A:B").GetSection("C");      //A:C->C
var section3 = root.GetSection("A").GetSection("B:C");      //A->B:C

Debug.Assert(section1.Value == "ABC");
Debug.Assert(section2.Value == "ABC");
Debug.Assert(section3.Value == "ABC");

Debug.Assert(!ReferenceEquals(section1, section2));
Debug.Assert(!ReferenceEquals(section1, section3));
Debug.Assert(null != root.GetSection("D"));
```

如上面的代码片段所示,我们以不同的方式调用 GetSection 方法得到的都是路径为"A:B:C"的 IConfigurationSection 对象。上面这段代码还体现了另一个有趣的现象:虽然这 3 个 IConfigurationSection 对象均指向配置树的同一个节点,但是它们却并非同一个对象。换句话说,调用 GetSection 方法时,不论配置树中是否存在一个与指定路径相匹配的配置节,它总是会创建新的 IConfigurationSection 对象。

IConfiguration 还具有一个索引，我们可以指定子配置节的 Key 或者相对当前配置节点的路径得到对应 IConfigurationSection 的值。当执行这个索引的时候，它会按照与 GetSection 方法完全一致的逻辑得到一个 IConfigurationSection 对象，并返回其 Value 属性。如果配置树中不具有匹配的配置节，该索引会返回 Null 而不会抛出异常。

6.2.3　IConfigurationProvider

在 6.1 节介绍 IConfigurationSource 对象的时候，我们说它是对原始配置源的体现。虽然每种不同类型的配置源都有一个对应的 IConfigurationSource 实现，但是针对原始数据的读取并不是由它完成的，而是委托一个与之对应的 IConfigurationProvider 对象来实现。在前面介绍的配置结构转换过程中，针对不同配置源类型的 IConfigurationProvider 对象可以按照图 6-11 所示的方式实现配置数据从原始结构向物理结构的转换。

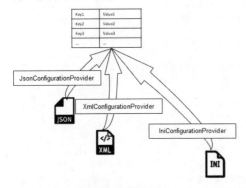

图 6-11　原始配置数据通过 IConfigurationProvider 转换成配置字典

由于 IConfigurationProvider 对象的目的在于将配置数据从原始结构转换成配置字典，所以定义在 IConfigurationProvider 接口中的方法大都体现为针对字典对象的操作。

```
public interface IConfigurationProvider
{
    void Load();
    void Set(string key, string value);
    bool TryGet(string key, out string value);

    IEnumerable<string> GetChildKeys(IEnumerable<string> earlierKeys,
        string parentPath);
    IChangeToken GetReloadToken();
}
```

配置数据的加载通过调用 IConfigurationProvider 的 Load 方法来完成。我们可以调用 TryGet 方法获取由指定的 Key 所标识的配置项的值。从数据持久化的角度来讲，IConfigurationProvider 基本上是一个只读的对象，也就是说，它只负责从持久化资源中读取配置数据，而不负责持久化更新后的配置数据，所以它提供的 Set 方法设置的配置数据一般只是保存在内存中，但是在实现该方法时对提供的值进行持久化也是可以的。

　　IConfigurationProvider 的 GetChildKeys 方法用于获取某个指定配置节点（对应参数 parentPath）的所有子节点的 Key。当 IConfiguration 的 GetChildren 方法被调用时，注册的所有 IConfigurationSource 对应的 IConfigurationProvider 的 GetChildKeys 方法会被调用。这个方法的第一个参数 earlierKeys 代表的 Key 来源于其他 IConfigurationProvider 对象，当解析出当前 IConfigurationProvider 提供的 Key 后，该方法需要将它们合并到 earlierKeys 集合中，合并后的结果将作为方法的返回值。值得注意的是，返回的 Key 的集合是经过排序的。

　　每种类型的配置源都对应一个 IConfigurationProvider 接口的实现类型，但它们一般不会直接实现 IConfigurationProvider 接口，而是选择继承另一个名为 ConfigurationProvider 的抽象类。这个抽象类的定义其实很简单，从如下代码片段可以看出，ConfigurationProvider 仅仅是对一个 IDictionary<string, string>对象（Key 不区分大小写）的封装，其 Set 方法和 TryGetValue 方法最终操作的都是这个字典对象。

```
public abstract class ConfigurationProvider : IConfigurationProvider
{
    protected IDictionary<string, string> Data { get; set; }
    protected ConfigurationProvider()=> Data =
        new Dictionary<string, string>(StringComparer.OrdinalIgnoreCase);
    public IEnumerable<string> GetChildKeys(
        IEnumerable<string> earlierKeys, string parentPath)
    {
        var prefix = parentPath == null ? string.Empty : $"{parentPath}:" ;
        return Data
            .Where(it => it.Key.StartsWith(
                prefix, StringComparison.OrdinalIgnoreCase))
            .Select(it => Segment(it.Key, prefix.Length))
            .Concat(earlierKeys)
            .OrderBy(it => it);
    }
    public virtual void Load() {}
    public void Set(string key, string value) => Data[key] = value;
    public bool TryGet(string key, out string value)
        => Data.TryGetValue(key, out value);

    private static string Segment(string key, int prefixLength)
    {
        var indexOf = key.IndexOf(
            ":", prefixLength, StringComparison.OrdinalIgnoreCase);
        return indexOf < 0
            ? key.Substring(prefixLength)
            : key.Substring(prefixLength, indexOf - prefixLength);
    }
    ...
}
```

　　抽象类 ConfigurationProvider 实现了 Load 方法并将其定义成虚方法，这个方法并没有提供具体的实现，所以它的派生类可以通过重写这个方法从相应的数据源中读取配置数据，并通过对 Data 属性的赋值完成对配置数据的加载。

6.2.4　IConfigurationSource

IConfigurationSource 对象在配置模型中代表配置源，被注册到 IConfigurationBuilder 对象上，为由它创建的 IConfiguration 对象提供原始的配置数据。由于针对原始配置数据的读取在相应的 IConfigurationProvider 对象中实现，所以 IConfigurationSource 对象的作用就在于提供相应的 IConfigurationProvider 对象。如下面的代码片段所示，IConfigurationSource 接口具有一个唯一的 Build 方法，根据指定的 IConfigurationBuilder 对象提供对应的 IConfigurationProvider 对象。

```
public interface IConfigurationSource
{
    IConfigurationProvider Build(IConfigurationBuilder builder);
}
```

6.2.5　IConfigurationBuilder

IConfigurationBuilder 对象在整个配置模型中处于核心地位，代表原始配置源的 IConfigurationSource 对象就注册在它上面。IConfigurationBuilder 对象会利用注册的 IConfigurationSource 对象提供的原始数据创建供应用程序使用的 IConfiguration 对象。如下面的代码片段所示，IConfigurationBuilder 接口定义了两个方法：Add 方法用于注册 IConfigurationSource 对象；最终的 IConfiguration 对象则由 Build 方法创建，该方法返回一个代表整棵配置树的 IConfigurationRoot 对象。注册的 IConfigurationSource 对象被保存在通过 Sources 属性表示的集合中，而 Properties 属性则以字典的形式存放任意的自定义属性。

```
public interface IConfigurationBuilder
{
    IEnumerable<IConfigurationSource>      Sources { get; }
    Dictionary<string, object>             Properties { get; }

    IConfigurationBuilder       Add(IConfigurationSource source);
    IConfigurationRoot          Build();
}
```

配置系统提供了一个名为 ConfigurationBuilder 的类作为 IConfigurationBuilder 接口的默认实现。定义在它上面的 Build 方法体现了配置系统读取原始配置数据并生成配置树的默认机制，这是下面要重点讲述的内容。ConfigurationBuilder 类型的 Build 方法返回一个类型为 ConfigurationRoot 的对象，由它表示的配置树的每个非根配置节点均是一个类型为 ConfigurationSection 的对象。

本节主要从设计和实现原理的角度对配置模型进行详细介绍。总的来说，配置模型涉及 4 个核心对象，包括承载配置逻辑结构的 IConfiguration 对象和它的创建者 IConfigurationBuilder 对象，以及与配置源相关的 IConfigurationSource 对象和 IConfigurationProvider 对象。这 4 个核心对象之间的关系简单而清晰，完全可以通过一句话进行概括：IConfigurationBuilder 对象利用注册在它上面的所有 IConfigurationSource 对象提供的 IConfigurationProvider 对象来读取原始配置数据并创建出相应的 IConfiguration 对象。图 6-12 中的 UML 展示了配置模型涉及的主要接口/类型以及它们之间的关系。

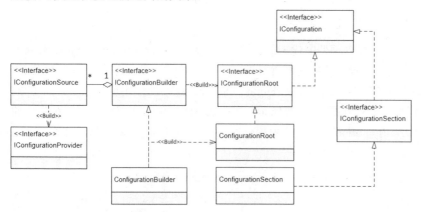

图 6-12　配置模型涉及的主要接口/类型以及它们之间的关系

6.3　配置绑定

虽然应用程序可以直接通过 IConfigurationBuilder 对象创建的 IConfiguration 对象来提取配置数据，但是我们更倾向于将其转换成一个 POCO 对象，以面向对象的方式来使用配置，我们将这个转换过程称为配置绑定。配置绑定可以通过如下几个针对 IConfiguration 的扩展方法来实现，这些扩展方法都定义在 NuGet 包 "Microsoft.Extensions.Configuration.Binder" 中。

```
public static class ConfigurationBinder
{
    public static void Bind(this IConfiguration configuration, object instance);
    public static void Bind(this IConfiguration configuration, object instance,
        Action<BinderOptions> configureOptions);
    public static void Bind(this IConfiguration configuration, string key,
        object instance);

    public static T Get<T>(this IConfiguration configuration);
    public static T Get<T>(this IConfiguration configuration,
        Action<BinderOptions> configureOptions);
    public static object Get(this IConfiguration configuration, Type type);
    public static object Get(this IConfiguration configuration, Type type,
        Action<BinderOptions> configureOptions);
}

public class BinderOptions
{
    public bool BindNonPublicProperties { get; set; }
}
```

Bind 方法将指定的 IConfiguration 对象（对应参数 configuration）绑定一个预先创建的对象（对应参数 instance），如果参数绑定的只是当前 IConfiguration 对象的某个子配置节，就需要通过参数 sectionKey 指定对应子配置节的相对路径。Get 方法和 Get<T>方法则直接将指定的 IConfiguration 对象转换成指定类型的 POCO 对象。

旨在生成 POCO 对象的配置绑定实现在 IConfiguration 接口的 Bind 扩展方法上。配置绑定的目标类型可以是一个简单的基元类型，也可以是一个自定义数据类型，还可以是一个数组、集合或者字典类型。通过前面的介绍可知，IConfigurationProvider 对象将原始配置数据读取出来之后会将其转换成 Key 和 Value 均为字符串的数据字典，那么针对这些完全不同的目标类型，原始配置数据是如何通过数据字典的形式来体现的？

6.3.1 绑定配置项的值

配置模型采用字符串键值对的形式来承载基础配置数据，我们将这组键值对称为配置字典，扁平的字典因为采用路径化的 Key 使配置项在逻辑上具有了层次结构。IConfigurationBuilder 对象将配置的层次化结构体现在由它创建的 IConfigurationRoot 对象上，我们将 IConfigurationRoot 对象看作一棵配置树。所谓的配置绑定体现为如何将映射在配置树上某个节点的 IConfiguration 对象（可以是 IConfigurationRoot 对象或者 IConfigurationSection 对象）转换成一个对应的 POCO 对象。

对于针对 IConfiguration 对象的配置绑定来说，最简单的是针对叶子节点的 IConfigurationSection 对象的绑定。表示配置树叶子节点的 IConfigurationSection 对象承载着原子配置项的值，而且这个值是一个字符串，所以针对它的配置绑定最终体现为如何将这个字符串转换成指定的目标类型，这样的操作体现在 IConfiguration 如下两个 GetValue 扩展方法上。

```
public static class ConfigurationBinder
{
    public static T GetValue<T>(IConfiguration configuration, string sectionKey);
    public static T GetValue<T>(IConfiguration configuration, string sectionKey,
        T defaultValue);
    public static object GetValue(IConfiguration configuration, Type type,
        string sectionKey);
    public static object GetValue(IConfiguration configuration, Type type,
        string sectionKey, object defaultValue);
}
```

对于给出的这 4 个重载，其中两个方法定义了一个表示默认值的参数 defaultValue，如果对应配置节的值为 Null 或者空字符串，那么指定的默认值将作为方法的返回值。对于其他的方法重载，它们实际上是将 Null 或者 Default(T)作为隐式默认值。执行上述这些 GetValue 方法时，它们会将配置节名称（对应参数 sectionKey）作为参数调用指定 IConfiguration 对象的 GetSection 方法得到表示对应配置节的 IConfigurationSection 对象，它的 Value 属性被提取出来并按照如下逻辑转换成目标类型。

- 如果目标类型为 object，那么直接返回原始值（字符串或者 Null）。
- 如果目标类型不是 Nullable<T>，那么针对目标类型的 TypeConverter 将被用来做类型转换。

- 如果目标类型为 Nullable<T>，那么在原始值不是 Null 或者空字符串的情况下会将基础类型 T 作为新的目标类型进行转换，否则直接返回 Null。

为了验证上述这些类型转化规则，我们编写了如下测试程序。如下面的代码片段所示，我们利用注册的 MemoryConfigurationSource 添加了 3 个配置项，对应的值分别为 Null、空字符串和 "123"，然后调用 GetValue 方法分别对它们进行类型转换，转换的目标类型分别是 Object、Int32 和 Nullable<Int32>，上述类型转换规则体现在对应的调试断言中。（S607）

```
public class Program
{
    public static void Main()
    {
        var source = new Dictionary<string, string>
        {
            ["foo"] = null,
            ["bar"] = "",
            ["baz"] = "123"
        };

        var root = new ConfigurationBuilder()
            .AddInMemoryCollection(source)
            .Build();

        //针对 object
        Debug.Assert(root.GetValue<object>("foo") == null);
        Debug.Assert("".Equals(root.GetValue<object>("bar")));
        Debug.Assert("123".Equals(root.GetValue<object>("baz")));

        //针对普通类型
        Debug.Assert(root.GetValue<int>("foo") == 0);
        Debug.Assert(root.GetValue<int>("baz") == 123);

        //针对 Nullable<T>
        Debug.Assert(root.GetValue<int?>("foo") == null);
        Debug.Assert(root.GetValue<int?>("bar") == null);
    }
}
```

按照前面介绍的类型转换规则，如果目标类型支持源自字符串的类型转换，就能够将配置项的原始值绑定为该类型的对象，而让某个类型支持某种类型转换规则的途径就是为之注册相应的 TypeConverter。下面的代码片段定义了一个表示二维坐标的 Point 对象，并且为它注册了一个类型为 PointTypeConverter 的 TypeConverter，PointTypeConverter 通过实现的 ConvertFrom 方法将坐标的字符串表达式（如 "123" 和 "456"）转换成一个 Point 对象。

```
[TypeConverter(typeof(PointTypeConverter))]
public class Point
{
    public double X { get; set; }
    public double Y { get; set; }
```

```csharp
public class PointTypeConverter : TypeConverter
{
    public override bool CanConvertFrom(ITypeDescriptorContext context,
        Type sourceType)
    => sourceType == typeof(string);

    public override object ConvertFrom(ITypeDescriptorContext context,
        CultureInfo culture, object value)
    {
        string[] split = value.ToString().Split(',');
        double x = double.Parse(split[0].Trim().TrimStart('('));
        double y = double.Parse(split[1].Trim().TrimEnd(')'));
        return new Point { X = x, Y = y };
    }
}
```

由于定义的 Point 类型支持源自字符串的类型转换，所以如果配置项的原始值（字符串）具有与之兼容的格式，就可以按照如下方式将其绑定为一个 Point 对象。（S608）

```csharp
public class Program
{
    public static void Main()
    {
        var source = new Dictionary<string, string>
        {
            ["point"] = "(123,456)"
        };

        var root = new ConfigurationBuilder()
            .AddInMemoryCollection(source)
            .Build();

        var point = root.GetValue<Point>("point");
        Debug.Assert(point.X == 123);
        Debug.Assert(point.Y == 456);
    }
}
```

6.3.2 绑定复合数据类型

这里所谓的复合类型就是一个具有属性数据成员的自定义类型。如果用一棵树表示一个复合对象，那么叶子节点承载所有的数据，并且叶子节点的数据类型均为基元类型。如果用数据字典来提供一个复杂对象所有的原始数据，那么这个字典中只需要包含叶子节点对应的值即可。只要将叶子节点所在的路径作为字典元素的 Key，就可以通过一个字典对象体现复合对象的结构。

```csharp
public class Profile: IEquatable<Profile>
{
    public Gender              Gender { get; set; }
```

```csharp
    public int                  Age { get; set; }
    public ContactInfo          ContactInfo { get; set; }

    public Profile() {}
    public Profile(Gender gender, int age, string emailAddress, string phoneNo)
    {
        Gender              = gender;
        Age                 = age;
        ContactInfo         = new ContactInfo
        {
            EmailAddress    = emailAddress,
            PhoneNo         = phoneNo
        };
    }
    public bool Equals(Profile other)
    {
        return other    == null
            ? false
            : Gender    == other.Gender &&
              Age       == other.Age &&
              ContactInfo.Equals(other.ContactInfo);
    }
}

public class ContactInfo: IEquatable<ContactInfo>
{
    public string EmailAddress { get; set; }
    public string PhoneNo { get; set; }
    public bool Equals(ContactInfo other)
    {
        return other == null
            ? false
            : EmailAddress      == other.EmailAddress &&
              PhoneNo           == other.PhoneNo;
    }
}

public enum Gender
{
    Male,
    Female
}
```

上面的代码片段定义了一个表示个人基本信息的 Profile 类,它的 3 个属性(Gender、Age 和 ContactInfo)分别表示性别、年龄和联系方式。由于配置绑定会调用默认无参构造函数来创建绑定的目标对象,所以需要为 Profile 类定义一个默认构造函数。表示联系方式的 ContactInfo 类型具有两个属性(EmailAddress 和 PhoneNo),它们分别表示电子邮箱地址和电话号码。一个完整的 Profile 对象可以通过图 6-13 所示的树来体现。

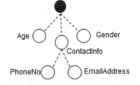

图 6-13　复杂对象的配置树

如果需要通过配置的形式表示一个完整的 Profile 对象，只需要将 4 个叶子节点（性别、年龄、电子邮箱地址和电话号码）对应的数据由配置来提供即可。对于承载配置数据的数据字典，我们需要按照表 6-2 列举的方式将这 4 个叶子节点的路径作为字典元素的 Key。

表 6-2　针对复杂对象的配置数据结构

Key	Value
Gender	Male
Age	18
ContactInfo:Email	foobar@outlook.com
ContactInfo:PhoneNo	123456789

可以通过下面的程序来验证针对复合数据类型的配置绑定。先创建一个 Configuration Builder 对象，然后为它添加了一个 MemoryConfigurationSource 对象，并按照表 6-2 列举的结构提供原始配置数据。在调用 Build 方法构建出 IConfiguration 对象之后，可以直接调用 Get<T>扩展方法将它转换成一个 Profile 对象。（S609）

```
public class Program
{
    public static void Main()
    {
        var source = new Dictionary<string, string>
        {
            ["gender"]                       = "Male",
            ["age"]                          = "18",
            ["contactInfo:emailAddress"]     = "foobar@outlook.com",
            ["contactInfo:phoneNo"]          = "123456789"
        };

        var configuration = new ConfigurationBuilder()
            .AddInMemoryCollection(source)
            .Build();

        var profile = configuration.Get<Profile>();
        Debug.Assert(profile.Equals(
            new Profile(Gender.Male, 18, "foobar@outlook.com", "123456789")));
    }
}
```

6.3.3　绑定集合对象

如果配置绑定的目标类型是一个集合（包括数组），那么当前 IConfiguration 对象的每个子配置节将绑定为集合的元素。如果将一个 IConfiguration 对象绑定为一个元素类型为 Profile 的集合，那么它表示的配置树应该具有图 6-14 所示的结构。

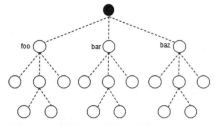

图 6-14　集合对象的配置树

既然能够正确地将集合对象通过一个合法的配置树体现出来，那么就可以将它转换成配置字典。而图 6-14 表示的这个包含 3 个元素的 Profile 集合，就可以采用表 6-3 列举的结构来定义对应的配置字典。

表 6-3　针对集合的配置数据结构

Key	Value
foo:Gender	Male
foo:Age	18
foo:ContactInfo:EmailAddress	foo@outlook.com
foo:ContactInfo:PhoneNo	123
bar:Gender	Male
bar:Age	25
bar:ContactInfo:EmailAddress	bar@outlook.com
bar:ContactInfo:PhoneNo	456
baz:Gender	Female
baz:Age	36
baz:ContactInfo:EmailAddress	baz@outlook.com
baz:ContactInfo:PhoneNo	789

下面通过一个简单的实例来演示针对集合的配置绑定。如下面的代码片段所示，我们创建了一个 ConfigurationBuilder 对象，并为它注册了一个 MemoryConfigurationSource 对象，它按照表 6-3 列举的结构提供了原始的配置数据。在得到这个 ConfigurationBuilder 对象创建的 IConfiguration 对象之后，我们两次调用其 Get<T>方法将它分别绑定为一个 IEnumerable<Profile>对象和一个 Profile[]数组。由于 IConfigurationProvider 通过 GetChildKeys 方法提供的 Key 是经过排序的，所以在绑定生成的集合或者数组中的元素的顺序与配置源是不相同的，如下的调试断言也体现了这一点。（S610）

```
public class Program
{
    public static void Main()
    {
        var source = new Dictionary<string, string>
        {
```

```
            ["foo:gender"]                          = "Male",
            ["foo:age"]                             = "18",
            ["foo:contactInfo:emailAddress"]        = "foo@outlook.com",
            ["foo:contactInfo:phoneNo"]             = "123",

            ["bar:gender"]                          = "Male",
            ["bar:age"]                             = "25",
            ["bar:contactInfo:emailAddress"]        = "bar@outlook.com",
            ["bar:contactInfo:phoneNo"]             = "456",

            ["baz:gender"]                          = "Female",
            ["baz:age"]                             = "36",
            ["baz:contactInfo:emailAddress"]        = "baz@outlook.com",
            ["baz:contactInfo:phoneNo"]             = "789"
        };

        var configuration = new ConfigurationBuilder()
            .AddInMemoryCollection(source)
            .Build();

        var profiles = new Profile[]
        {
            new Profile(Gender.Male,18,"foo@outlook.com","123"),
            new Profile(Gender.Male,25,"bar@outlook.com","456"),
            new Profile(Gender.Female,36,"baz@outlook.com","789"),
        };

        var collection = configuration.Get<IEnumerable<Profile>>();
        Debug.Assert(collection.Any(it => it.Equals(profiles[0])));
        Debug.Assert(collection.Any(it => it.Equals(profiles[1])));
        Debug.Assert(collection.Any(it => it.Equals(profiles[2])));

        var array = configuration.Get<Profile[]>();
        Debug.Assert(array[0].Equals(profiles[1]));
        Debug.Assert(array[1].Equals(profiles[2]));
        Debug.Assert(array[2].Equals(profiles[0]));
    }
}
```

在针对集合类型的配置绑定过程中，如果某个配置节绑定失败，该配置节将被忽略并选择下一个配置节继续进行绑定。但是如果目标类型为数组，最终绑定生成的数组长度与子配置节的个数总是一致的，绑定失败的元素将被设置为 Null。例如，保存原始配置的字典对象包含两个元素，如果将上面的程序进行改写，第一个元素的性别从"Male"改为"男"，那么这个值是不可能转换成 Gender 枚举对象的，所以针对这个 Profile 的配置绑定会失败。如果将目标类型设置为 IEnumerable<Profile>，那么最终生成的集合只有两个元素，倘若目标类型切换成 Profile 数组，数组的长度依然为 3，第一个元素就是 Null。（S611）

```
public class Program
```

```
{
    public static void Main()
    {
        var source = new Dictionary<string, string>
        {
            ["foo:gender"]                          = "男",
            ["foo:age"]                             = "18",
            ["foo:contactInfo:emailAddress"]        = "foo@outlook.com",
            ["foo:contactInfo:phoneNo"]             = "123",

            ["bar:gender"]                          = "Male",
            ["bar:age"]                             = "25",
            ["bar:contactInfo:emailAddress"]        = "bar@outlook.com",
            ["bar:contactInfo:phoneNo"]             = "456",

            ["baz:gender"]                          = "Female",
            ["baz:age"]                             = "36",
            ["baz:contactInfo:emailAddress"]        = "baz@outlook.com",
            ["baz:contactInfo:phoneNo"]             = "789"
        };

        var configuration = new ConfigurationBuilder()
            .AddInMemoryCollection(source)
            .Build();

        var collection = configuration.Get<IEnumerable<Profile>>();
        Debug.Assert(collection.Count() == 2);

        var array = configuration.Get<Profile[]>();
        Debug.Assert(array.Length == 3);
        Debug.Assert(array[2] == null);
        //由于配置节按照Key进行排序，所以绑定失败的配置节为最后一个
    }
}
```

6.3.4 绑定字典

能够通过配置绑定生成的字典是一个实现了 IDictionary<string,T>的类型，也就是说，配置模型对字典的 Value 类型没有任何要求，但字典对象的 Key 必须是一个字符串（或者枚举）。如果采用配置树的形式表示这样一个字典对象，就会发现它与针对集合的配置树在结构上几乎是一样的，唯一的区别是集合元素的索引直接变成字典元素的 Key。

也就是说，图 6-14 所示的配置树同样可以表示成一个具有 3 个元素的 Dictionary<string, Profile>对象，它们对应的 Key 分别是 Foo、Bar 和 Baz，所以可以按照如下方式将承载相同数据的 IConfiguration 对象绑定为一个 IDictionary<string,T>对象。（S612）

```
public class Program
{
    public static void Main()
```

```
{
    var source = new Dictionary<string, string>
    {
        ["foo:gender"]                          = "Male",
        ["foo:age"]                             = "18",
        ["foo:contactInfo:emailAddress"]        = "foo@outlook.com",
        ["foo:contactInfo:phoneNo"]             = "123",

        ["bar:gender"]                          = "Male",
        ["bar:age"]                             = "25",
        ["bar:contactInfo:emailAddress"]        = "bar@outlook.com",
        ["bar:contactInfo:phoneNo"]             = "456",

        ["baz:gender"]                          = "Female",
        ["baz:age"]                             = "36",
        ["baz:contactInfo:emailAddress"]        = "baz@outlook.com",
        ["baz:contactInfo:phoneNo"]             = "789"
    };

    var configuration = new ConfigurationBuilder()
        .AddInMemoryCollection(source)
        .Build();

    var profiles = configuration.Get<IDictionary<string, Profile>>();
    Debug.Assert(profiles["foo"].Equals(
        new Profile(Gender.Male, 18, "foo@outlook.com", "123")));
    Debug.Assert(profiles["bar"].Equals(
        new Profile(Gender.Male, 25, "bar@outlook.com", "456")));
    Debug.Assert(profiles["baz"].Equals(
        new Profile(Gender.Female, 36, "baz@outlook.com", "789")));
}
```

6.4 配置的同步

前面介绍配置模型的核心对象时，我们刻意回避了与配置同步相关的 API，本节专门介绍配置的同步。配置的同步涉及两个方面：第一，对原始的配置源实施监控并在其发生变化之后重新加载配置；第二，配置重新加载之后及时通知应用程序，进而使应用能够及时使用最新的配置。要了解配置同步机制的实现原理，需要先了解配置数据的流向。

6.4.1 配置数据流

通过前面的介绍，我们已经对配置模型有了充分的了解。处于核心地位的 IConfigurationBuilder 对象，借助注册的 IConfigurationSource 对象提供的 IConfigurationProvider 对象，从相应的配置源中加载数据，而各种针对 IConfigurationProvider 接口的实现，就是为了将形态各异的原始配置数据转换成配置字典。应用程序中使用的配置数据直接来源于 IConfigurationBuilder 对象创建的 IConfiguration 对象，那么调用定义在 IConfiguration 对象上的 API 获取配置数据时，配置数据究竟具有怎样的流向？

前面提及，由 ConfigurationBuilder（IConfigurationBuilder 接口的默认实现）的 Build 方法提供的 IConfiguration 对象是一个 ConfigurationRoot 对象，它代表整棵配置树，而组成这棵树的配置节则通过 ConfigurationSection 对象表示。这棵由 ConfigurationRoot 对象表示的配置树其实是无状态的，也就是说，无论是 ConfigurationRoot 对象还是 ConfigurationSection 对象，它们并没有利用某个字段存储任何配置数据。

ConfigurationRoot 对象保持着对所有注册的 IConfigurationSource 对象提供的 IConfigurationProvider 对象的引用，当调用 ConfigurationRoot 对象或者 ConfigurationSection 对象相应的 API 提取配置数据时，最终都会直接从这些 IConfigurationProvider 对象中提取数据。换句话说，配置数据在整个模型中只以配置字典的形式存储在 IConfigurationProvider 对象上面。

应用程序在读取配置时产生的数据流基本体现在图 6-15 中。下面从 ConfigurationRoot 和 ConfigurationSection 这两个类型的定义来对这个数据流，以及建立在此基础上的配置同步机制做进一步介绍，但在此之前需要先介绍一个名为 ConfigurationReloadToken 的类型。

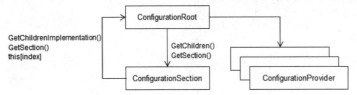

图 6-15　配置数据流

6.4.2　ConfigurationReloadToken

ConfigurationRoot 类型和 ConfigurationSection 类型的 GetReloadToken 方法返回的 IChangeToken 对象都是 ConfigurationReloadToken 类型。不仅如此，对于组成同一棵配置树的所有节点对应的 IConfiguration 对象（ConfigurationRoot 或者 ConfigurationSection）来说，它们的 GetReloadToken 方法返回的其实是同一个 ConfigurationReloadToken 对象。

另外，对于 IConfiguration 接口的 GetReloadToken 方法返回的 IChangeToken 对象，其作用不是在配置源发生变化时向应用程序发送通知，而是通知应用程序：配置源已经发生改变，并且新的数据已经被相应的 IConfigurationProvider 对象重新加载进来。由于 ConfigurationRoot 对象和 ConfigurationSection 对象都不维护任何数据，它们仅仅将 API 调用转移到 IConfigurationProvider 对象上，所以应用程序使用原来的 IConfiguration 对象就可以获取最新的配置数据。

ConfigurationReloadToken 本质上是对一个 CancellationTokenSource 对象的封装。从如下代码片段可以看出，ConfigurationReloadToken 与第 5 章介绍的 CancellationChangeToken 具有类似的定义和实现。两者唯一的不同之处在于：CancellationChangeToken 对象利用创建时提供的 CancellationTokenSource 对象对外发送通知，而 ConfigurationReloadToken 对象则通过调用 OnReload 方法利用

内置的CancellationTokenSource对象发送通知。

```csharp
public class ConfigurationReloadToken : IChangeToken
{
    private CancellationTokenSource _cts = new CancellationTokenSource();
    public IDisposable RegisterChangeCallback(Action<object> callback, object state)
        =>_cts.Token.Register(callback, state);
    public bool ActiveChangeCallbacks => True;
    public bool HasChanged =>_cts.IsCancellationRequested;

    public void OnReload() => _cts.Cancel();
}
```

6.4.3　ConfigurationRoot

下面介绍由ConfigurationBuilder对象的Build方法直接创建的ConfigurationRoot对象具有怎样的实现。正如前面提及的，一个ConfigurationRoot对象是根据一组IConfigurationProvider对象创建的，这些IConfigurationProvider对象则由注册的IConfigurationSource对象来提供。

```csharp
public class ConfigurationRoot : IConfigurationRoot
{
    private IList<IConfigurationProvider> _providers;
    private ConfigurationReloadToken     _changeToken;

    public ConfigurationRoot(IList<IConfigurationProvider> providers)
    {
        _providers    = providers;
        _changeToken = new ConfigurationReloadToken();
        foreach (var provider in providers)
        {
            provider.Load();
            ChangeToken.OnChange(
               () => provider.GetReloadToken(), () => RaiseChanged());
        }
    }
    public void Reload()
    {
        foreach (var provider in _providers)
        {
            provider.Load();
        }
        RaiseChanged();
    }
    public IChangeToken GetReloadToken() => _changeToken;

    private void RaiseChanged()
        => Interlocked.Exchange(ref _changeToken, new ConfigurationReloadToken())
        .OnReload();
    ...
}
```

ConfigurationRoot 的 GetReloadToken 方法返回的是一个 ConfigurationReloadToken 对象，该对象用字段_changeToken 表示。如果需要利用这个对象对外发送配置重新加载的通知，就需要调用其 OnReload 方法，由上面的代码片段可知，该方法会在 RaiseChanged 方法中被调用。由于一个 IChangeToken 对象只能发送一次通知，所以该方法还负责创建新的 ConfigurationReloadToken 对象并对_changeToken 字段赋值。

换句话说，一旦 ConfigurationRoot 的 RaiseChanged 方法被调用，我们就可以利用其 GetReloadToken 方法返回的 IChangeToken 对象来接收配置被重新加载的通知。通过上面提供的代码我们可以看到，RaiseChanged 方法在两个地方被调用：第一，在构造函数中调用每个 IConfigurationProvider 对象的 GetReloadToken 方法，得到对应的 IChangeToken 对象，并在它们注册的回调中调用这个方法；第二，实现的 Reload 方法在依次调用每个 IConfigurationProvider 对象的 Load 方法来重新加载配置数据之后，调用了 RaiseChanged 方法。按照这个逻辑，应用程序会在如下两个场景中利用 ConfigurationRoot 返回的 IChangeToken 接收配置被重新加载的通知。

- 某个 IConfigurationProvider 对象捕捉到对应配置源的改变后自动重新加载配置，并在加载完成后利用其 GetReloadToken 方法返回的 IChangeToken 发送通知。
- 显式调用 ConfigurationRoot 的 Reload 方法手动加载配置。

了解了 ConfigurationRoot 的 GetReloadToken 方法返回的是什么样的 IChangeToken 之后，下面介绍它的其他成员具有怎样的实现。如下面的代码片段所示，在 ConfigurationRoot 的索引定义中，它分别调用了 IConfigurationProvider 对象的 TryGet 方法和 Set 方法，根据配置字典的 Key 获取和设置对应的 Value。

```
public class ConfigurationRoot : IConfigurationRoot
{
    private IList<IConfigurationProvider> _providers;

    public string this[string key]
    {
        get
        {
            foreach (var provider in _providers.Reverse())
            {
                if (provider.TryGet(key, out var value))
                {
                    return value;
                }
            }
            return null;
        }
        set
        {
            foreach (var provider in _providers)
            {
                provider.Set(key, value);
            }
```

```csharp
        }
    }

    public IConfigurationSection GetSection(string key)
        => new ConfigurationSection(this, key);

    public IEnumerable<IConfigurationSection> GetChildren()
        => GetChildrenImplementation(null);

    internal IEnumerable<IConfigurationSection> GetChildrenImplementation(
        string path)
    {
        return _providers
            .Aggregate(Enumerable.Empty<string>(),
                (seed, source) => source.GetChildKeys(seed, path))
            .Distinct()
            .Select(key => GetSection(path == null ? key : $"{path}:{key}"));
    }

    public IEnumerable<IConfigurationProvider> Providers => _providers;
}
```

从索引的定义可以看出，ConfigurationRoot 在读取 Value 值时针对 IConfigurationProvider 列表的遍历是从后往前的，这一点非常重要，因为该特性决定了 IConfigurationSource 的注册会采用"后来居上"的原则。也就是说，如果多个 IConfigurationSource 配置源提供的 IConfigurationProvider 对象包含同名的配置项，后面注册的 IConfigurationSource 对象就具有更高的选择优先级，我们应该根据这个特性合理安排 IConfigurationSource 对象的注册顺序。在设置 Value 时，ConfigurationRoot 对象会调用每个 IConfigurationProvider 对象的 Set 方法，这意味着新的值会被保存到所有 IConfigurationProvider 对象的配置字典中。

正如前面多次提到过的，通过 ConfigurationRoot 表示的配置树的所有配置节都是一个类型为 ConfigurationSection 的对象，这一点体现在实现的 GetSection 方法上。将对应的路径作为参数，可以得到组成配置树的所有配置节。

用于获取所有子配置节的 GetChildren 方法可以通过调用内部方法 GetChildrenImplementation 来实现。GetChildrenImplementation 方法旨在获取配置树某个节点的所有子节点，该方法的参数表示指定节点针对配置树根的路径。当这个方法被执行时，它会以聚合的形式遍历所有的 IConfigurationProvider，并调用它们的 GetChildKeys 方法获取所有子节点的 Key，这些 Key 与当前节点的路径合并后代表子节点的路径，这些路径最终被作为参数调用 GetSection 方法创建对应的配置节。

6.4.4　ConfigurationSection

如下所示的代码片段大体上体现了代表配置节的 ConfigurationSection 类型的实现逻辑。如下面的代码片段所示，一个 ConfigurationSection 对象是通过代表配置树根的 ConfigurationRoot 对象和当前配置节在配置树中的路径来构建的。ConfigurationSection 的 Path 属性直接返回构建时指定的路径，而 Key 属性则是根据这个路径解析出来的。

```csharp
public class ConfigurationSection : IConfigurationSection
{
    private readonly ConfigurationRoot      _root;
    private readonly string                 _path;
    private string                          _key;

    public ConfigurationSection(ConfigurationRoot root, string path)
    {
        _root = root;
        _path = path;
    }

    public string this[string key]
    {
        get => _root[string.Join(':', new string[] { _path, _key })];
        set => _root[string.Join(':', new string[] { _path, _key })] = value;
    }

    public string Key => _key
        ?? (_key = _path.Contains(':') ? _path.Split(':').Last() : _path);
    public string Path => _path;
    public string Value
    {
        get => _root[_path];
        set => _root[_path] = value;
    }
    public IEnumerable<IConfigurationSection> GetChildren()
        => _root.GetChildrenImplementation(_path);
    public IChangeToken GetReloadToken() => _root.GetReloadToken();
    public IConfigurationSection GetSection(string key)
        => _root.GetSection(string.Join(':', new string[] { _path, key }));
}
```

如图 6-15 所示，在 ConfigurationSection 类型中实现的大部分成员都是调用 ConfigurationRoot 对象相应的 API 来实现的。ConfigurationSection 的索引直接调用 ConfigurationRoot 的索引来获取或者设置配置字典的 Value，GetChildren 方法返回的就是调用 GetChildrenImplementation 方法得到的结果，而 GetReloadToken 方法和 GetSection 方法都是通过调用同名方法实现的。

6.5 多样性的配置源

.NET Core 采用的全新的配置模型的一个主要特点就是对多种不同配置源提供支持。我们可以将内存变量、命令行参数、环境变量和物理文件作为原始配置数据的来源。如果将物理文件作为配置源，还可以选择不同的格式（如 XML、JSON 和 INI 等）。如果这些默认支持的配置源形式无法满足需求，也可以通过注册自定义 IConfigurationSource 的方式将其他形式的数据作为配置源。

6.5.1　MemoryConfigurationSource

6.1 节至 6.4 节的大部分实例演示都使用 MemoryConfigurationSource 来提供原始的配置。我们知道 MemoryConfigurationSource 配置源采用一个字典对象（具体来说应该是一个元素类型为 KeyValuePair<string, string>的集合）作为存放原始配置数据的容器。作为一个 IConfigurationSource 对象，它总是通过创建某个对应的 IConfigurationProvider 对象来完成具体的配置数据读取工作，那么 MemoryConfigurationSource 究竟会提供一个什么样的 IConfigurationProvider 对象？

```
public class MemoryConfigurationSource : IConfigurationSource
{
    public IEnumerable<KeyValuePair<string, string>> InitialData { get; set;}
    public IConfigurationProvider Build(IConfigurationBuilder builder)
        => new MemoryConfigurationProvider(this);
}
```

上面的代码片段体现了 MemoryConfigurationSource 类型的完整定义，由此可以看到，它具有一个 IEnumerable<KeyValuePair<string, string>>类型的属性 InitialData 来存放初始的配置数据。从 Build 方法的实现可以看出，真正被它用来读取原始配置数据的是一个 MemoryConfigurationProvider 类型的对象，该类型的定义如下面的代码片段所示。

```
public class MemoryConfigurationProvider : ConfigurationProvider,
    IEnumerable<KeyValuePair<string, string>>
{
    public MemoryConfigurationProvider(MemoryConfigurationSource source);
    public void Add(string key, string value);
    public IEnumerator<KeyValuePair<string, string>> GetEnumerator();
    IEnumerator IEnumerable.GetEnumerator();
}
```

从上面的代码片段可以看出，MemoryConfigurationProvider 派生于抽象类 ConfigurationProvider，同时实现了 IEnumerable<KeyValuePair<string, string>>接口。ConfigurationProvider 对象直接使用一个 Dictionary<string, string>来保存配置数据，当我们根据一个 MemoryConfigurationSource 对象调用构造函数创建 MemoryConfigurationProvider 时，它只需要将通过 InitialData 属性保存的配置数据转移到这个字典中即可。MemoryConfigurationProvider 定义的 Add 方法在任何时候都可以向配置字典中添加一个新的配置项。

通过前面对配置模型的介绍可知，IConfigurationProvider 对象在配置模型中所起的作用就是读取原始配置数据并将其转换成配置字典。在所有预定义的 IConfigurationProvider 实现类型中，MemoryConfigurationProvider 最为简单、直接，因为它对应的配置源就是一个配置字典，所以根本不需要做任何结构转换。

利用 MemoryConfigurationSource 生成配置时，我们需要将其注册到 IConfigurationBuilder 对象之上。具体来说，我们可以像前面演示的实例一样直接调用 IConfigurationBuilder 接口的 Add 方法，也可以调用如下所示的两个重载的 AddInMemoryCollection 扩展方法。

```
public static class MemoryConfigurationBuilderExtensions
```

```
{
    public static IConfigurationBuilder AddInMemoryCollection(
        this IConfigurationBuilder configurationBuilder);
    public static IConfigurationBuilder AddInMemoryCollection(
        this IConfigurationBuilder configurationBuilder,
        IEnumerable<KeyValuePair<string, string>> initialData);
}
```

6.5.2　EnvironmentVariablesConfigurationSource

顾名思义，环境变量就是描述当前执行环境并影响进程执行行为的变量。按照作用域的不同，可以将环境变量分为 3 类，即针对当前系统、当前用户和当前进程的环境变量。系统和用户级别的环境变量保存在注册表中，它们的路径分别为"HKEY_LOCAL_MACHINE\SYSTEM\ControlSet001\Control\Session Manager\Environment"和"HKEY_CURRENT_USER\Environment"。

环境变量的提取和维护可以通过静态类型 Environment 来完成。具体来说，我们可以调用它的静态方法 GetEnvironmentVariable 获得某个指定名称的环境变量的值，而 GetEnvironmentVariables 方法则会返回所有的环境变量，EnvironmentVariableTarget 枚举类型的参数代表环境变量作用域决定的存储位置。如果在调用 GetEnvironmentVariable 方法或者 GetEnvironmentVariables 方法时没有显式指定参数 target 或者将参数指定为 EnvironmentVariableTarget.Process，在进程初始化前存在的所有环境变量（包括针对系统、当前用户和当前进程）将会作为候选列表。

```
public static class Environment
{
    public static string GetEnvironmentVariable(string variable);
    public static string GetEnvironmentVariable(string variable,
        EnvironmentVariableTarget target);

    public static IDictionary GetEnvironmentVariables();
    public static IDictionary GetEnvironmentVariables(
        EnvironmentVariableTarget target);

    public static void SetEnvironmentVariable(string variable, string value);
    public static void SetEnvironmentVariable(string variable, string value,
        EnvironmentVariableTarget target);
}

public enum EnvironmentVariableTarget
{
    Process,
    User,
    Machine
}
```

环境变量的添加、修改和删除均由 SetEnvironmentVariable 方法完成，如果没有显式指定参数 target，默认采用的就是 EnvironmentVariableTarget.Process。如果希望删除指定名称的环境变量，在调用这个方法的时候将参数 value 设置为 Null 或者空字符串即可。

除了在程序中利用静态类型 Environment，还可以采用命令行的方式查看和设置环境变量。除此之外，在开发环境中还可以利用"System Properties"（系统属性）设置工具，以可视化的方式查看和设置系统与用户级别的环境变量（"This PC"→"Properties"→"Change Settings"→"Advanced"→"Environment Variables"）。如果采用 Visual Studio 调试编写的应用，则可以采用设置项目属性的方式来设置进程级别的环境变量（"Properties"→"Debug"→"Environment Variables"），如图 6-16 所示。如第 1 章中陈述，设置的环境变量会被保存到 launchSettings.json 文件中。

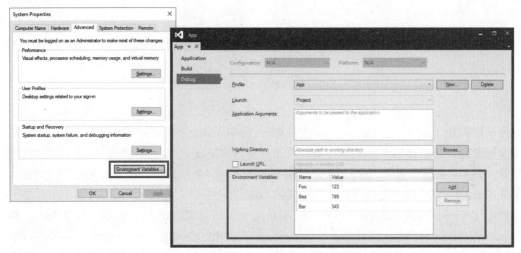

图 6-16　设置环境变量

针对环境变量的配置源可以通过 EnvironmentVariablesConfigurationSource 类型来表示，该类型定义在 NuGet 包"Microsoft.Extensions.Configuration.EnvironmentVariables"之中。该类型定义了一个字符串类型的属性 Prefix，表示环境变量名的前缀。如果设置了 Prefix 属性，系统只会选择名称作为前缀的环境变量。

```
public class EnvironmentVariablesConfigurationSource : IConfigurationSource
{
    public string Prefix { get; set; }

    public IConfigurationProvider Build(IConfigurationBuilder builder)
        => new EnvironmentVariablesConfigurationProvider(.Prefix);
}
```

由前面给出的代码片段可以看出，EnvironmentVariablesConfigurationSource 配置源会利用对应的 EnvironmentVariablesConfigurationProvider 对象来读取环境变量，此操作体现在如下所示的 Load 方法中。由于环境变量本身就是一个数据字典，所以 EnvironmentVariablesConfigurationProvider 对象无须再进行结构上的转换。当 Load 方法被执行之后，它只需要将符合条件的环境变量筛选出来并添加到自己的配置字典中即可。

```
public class EnvironmentVariablesConfigurationProvider : ConfigurationProvider
```

```csharp
{
    private readonly string _prefix;

    public EnvironmentVariablesConfigurationProvider(string prefix = null)
        => _prefix = prefix ?? string.Empty;

    public override void Load()
    {
        var dictionary = Environment.GetEnvironmentVariables()
            .Cast<DictionaryEntry>()
            .Where(it => it.Key.ToString().StartsWith(
                _prefix, StringComparison.OrdinalIgnoreCase))
            .ToDictionary(it => it.Key.ToString().Substring(_prefix.Length),
                it => it.Value.ToString());
        Data = new Dictionary<string, string>(
            dictionary, StringComparer.OrdinalIgnoreCase);
    }
}
```

值得一提的是,如果创建 EnvironmentVariablesConfigurationProvider 对象时指定了用于筛选环境变量的前缀,当符合条件的环境变量被添加到自身的配置字典之后,配置项的名称会将此前缀剔除。例如,前缀设置为"FOO_",环境变量 FOO_BAR 被添加到配置字典之后,配置项名称会变成 BAR,这个细节也体现在上面定义的 Load 方法中。

在使用 EnvironmentVariablesConfigurationSource 时,可以调用 Add 方法将它注册到指定的 IConfigurationBuilder 对象上。除此之外,EnvironmentVariablesConfigurationSource 的注册还可以直接调用 IConfigurationBuilder 接口的如下 3 个重载的 AddEnvironmentVariables 扩展方法来完成。

```csharp
public static class EnvironmentVariablesExtensions
{
    public static IConfigurationBuilder AddEnvironmentVariables(
        this IConfigurationBuilder configurationBuilder);
    public static IConfigurationBuilder AddEnvironmentVariables(
        this IConfigurationBuilder builder,
        Action<EnvironmentVariablesConfigurationSource> configureSource);
    public static IConfigurationBuilder AddEnvironmentVariables(
        this IConfigurationBuilder configurationBuilder, string prefix);
}
```

下面的实例演示了如何将环境变量作为配置源。如下面的代码片段所示,可以调用 Environment 的静态方法 SetEnvironmentVariable 设置 4 个环境变量,变量名称具有相同的前缀"TEST_"。我们调用 AddEnvironmentVariables 方法创建了一个 EnvironmentVariablesConfigurationSource 对象,并将其注册到创建的 ConfigurationBuilder 对象之上,在调用该方法时可以将环境变量名称的前缀设置为"TEST_"。最终将由 ConfigurationBuilder 构建的 IConfiguration 对象绑定成一个 Profile 对象。(S613)

```csharp
public class Program
{
    public static void Main()
```

```
{
    Environment.SetEnvironmentVariable("TEST_GENDER", "Male");
    Environment.SetEnvironmentVariable("TEST_AGE", "18");
    Environment.SetEnvironmentVariable("TEST_CONTACTINFO:EMAILADDRESS",
        "foobar@outlook.com");
    Environment.SetEnvironmentVariable("TEST_CONTACTINFO:PHONENO", "123456789");

    var profile = new ConfigurationBuilder()
        .AddEnvironmentVariables("TEST_")
        .Build()
        .Get<Profile>();

    Debug.Assert(profile.Equals(
        new Profile(Gender.Male, 18, "foobar@outlook.com", "123456789")));
}
```

6.5.3 CommandLineConfigurationSource

在很多情况下，我们会采用 Self-Host 方式将一个 ASP.NET Core 应用寄宿到一个托管进程中，此时我们倾向于采用命令行的方式来启动寄宿程序。当以命令行的形式启动一个 ASP.NET Core 应用时，我们希望直接使用命名行开关（Switch）来控制应用的一些行为，所以命令行开关自然也就成了配置常用的来源之一。配置模型针对这种配置源的支持是通过 CommandLineConfigurationSource 实现的，该类型定义在 NuGet 包 "Microsoft.Extensions.Configuration.CommandLine" 中。

以命令行的形式执行某个命令时，命令行开关（包括名称和值）体现为一个简单的字符串数组，所以 CommandLineConfigurationSource 的根本目的在于将命名行开关从字符串数组转换成配置字典。要充分理解这个转换规则，我们需要先了解 CommandLineConfigurationSource 支持的命令行开关究竟采用什么样的形式来指定。下面通过一个简单的实例来说明命令行开关的几种指定方式。假设我们有一个命令 exec，并采用如下方式执行某个托管程序（app）。

```
exec app {options}
```

在执行"exec"命令时可以通过相应的命令行开关指定多个选项。总的来说，命令行开关的指定形式大体上分为两种：单参数（Single Argument）和双参数（Double Arguments）。单参数形式就是采用等号（=）将命令行开关的名称和值通过如下方法采用一个参数来指定。

- {name}={value}。
- {prefix}{name}={value}。

对于第二种单参数命令行开关的指定形式，我们可以在开关名称前面添加一个前缀，目前的前缀支持"/"、"--"和"-"这 3 种。遵循这样的格式，我们可以采用如下 3 种方式将命令行开关 architecture 设置为 x64。下面的列表之所以没有使用前缀"-"，是因为这个前缀要求使用命令行开关映射（Switch Mapping），下面会单独进行介绍。

```
exec app architecture=x64
exec app /architecture=x64
```

```
exec app --architecture=x64
```

除了采用单参数形式,还可以采用双参数形式来指定命令行开关。双参数就是使用两个参数分别定义命令行开关的名称和值。这种形式采用的具体格式为{prefix}{name} {value},所以上述命令行开关 architecture 也可以采用如下方式来指定。

```
exec app /architecture x64
exec app --architecture x64
```

命令行开关的全名和缩写之间具有一个映射关系(Switch Mapping)。以上述两个命令行开关为例,我们可以采用首字母 "a" 来代替 "architecture"。如果使用 "-" 作为前缀,不论采用单参数还是双参数,都必须使用映射后的开关名称。值得一提的是,同一个命令行开关可以具有多个映射,如可以同时将 "architecture" 映射为 "arch"。假设 "architecture" 具有这两种映射,我们就可以按照如下两种方式指定 CPU 架构。

```
exec app -a=x64
exec app -arch=x64
exec app -a x64
exec app -arch x64
```

了解了命令行开关的指定形式之后,下面介绍 CommandLineConfigurationSource 类型和由它提供的 CommandLineConfigurationProvider 对象。由于原始的命令行参数总是体现为一个采用空格分隔的字符串,这样的字符串可以进一步转换成一个字符串集合,所以 CommandLineConfigurationSource 对象以字符串集合作为配置源。如下面的代码片段所示,CommandLineConfigurationSource 类型具有两个属性,即 Args 属性和 SwitchMappings 属性:前者代表承载原始命令行参数的字符串集合,后者则保存了命令行开关的缩写与全称之间的映射关系。CommandLineConfigurationSource 实现的 Build 方法会根据这两个属性创建并返回一个 CommandLineConfigurationProvider 对象。

```
public class CommandLineConfigurationSource : IConfigurationSource
{
    public IEnumerable<string>            Args { get; set; }
    public IDictionary<string, string>    SwitchMappings { get; set; }

    public IConfigurationProvider Build(IConfigurationBuilder builder)
        => new CommandLineConfigurationProvider( Args,SwitchMappings);
}
```

具有如下定义的 CommandLineConfigurationProvider 类型依然是抽象类 ConfigurationProvider 的继承者。CommandLineConfigurationProvider 类型的目的很明确,就是对体现为字符串集合的原始命令行参数进行解析,并将解析出的参数名称和值添加到配置字典中,这一切都是在重写的 Load 方法中完成的。

```
public class CommandLineConfigurationProvider : ConfigurationProvider
{
    protected IEnumerable<string> Args { get; }
    public CommandLineConfigurationProvider(IEnumerable<string> args,
        IDictionary<string, string> switchMappings = null);
    public override void Load();
}
```

在采用基于命令行参数作为配置源时，我们可以创建一个 CommandLineConfiguration
Source 对象，并将其注册到 ConfigurationBuilder 上。我们也可以调用 IConfigurationBuilder 接口
的如下 3 个 AddCommandLine 扩展方法将两个步骤合二为一。

```
public static class CommandLineConfigurationExtensions
{
    public static IConfigurationBuilder AddCommandLine(
        this IConfigurationBuilder builder,
        Action<CommandLineConfigurationSource> configureSource);
    public static IConfigurationBuilder AddCommandLine(
        this IConfigurationBuilder configurationBuilder, string[] args);
    public static IConfigurationBuilder AddCommandLine(
        this IConfigurationBuilder configurationBuilder, string[] args,
        IDictionary<string, string> switchMappings);
}
```

为了使读者对 CommandLineConfigurationSource/CommandLineConfigurationProvider 解析命
令行参数采用的策略有深刻的认识，下面演示一个简单的实例。如下面的代码片段所示，我们
创建了一个 ConfigurationBuilder 对象，并调用 AddCommandLine 方法注册了针对命令行参数的
配置源，Main 方法的参数 args 直接作为原始的命令行参数。

```
class Program
{
    static void Main(string[] args)
    {
        try
        {
            var mapping = new Dictionary<string, string>
            {
                ["-a"]       = "architecture",
                ["-arch"]    = "architecture"
            };
            var configuration = new ConfigurationBuilder()
                .AddCommandLine(args, mapping)
                .Build();
            Console.WriteLine($"Architecture: {configuration["architecture"]}");
        }
        catch (Exception ex)
        {
            Console.WriteLine($"Error: {ex.Message}");
        }
    }
}
```

在调用 AddCommandLine 扩展方法注册 CommandLineConfigurationSource 对象时，我们指
定了一个命令行开关映射表，它将命令行开关 "architecture" 映射为 "a" 和 "arch"。需要注意
的是，在通过字典定义命令行开关映射时，作为目标名称的 Key 应该添加前缀 "-"。下面调用
ConfigurationBuilder 对象的 Build 方法创建 IConfiguration 对象，然后从中提取 "architecture" 配
置项的值并打印出来。如图 6-17 所示，可以采用命令行的形式启动这个程序，并以不同的形式

指定"architecture"的值。（S614）

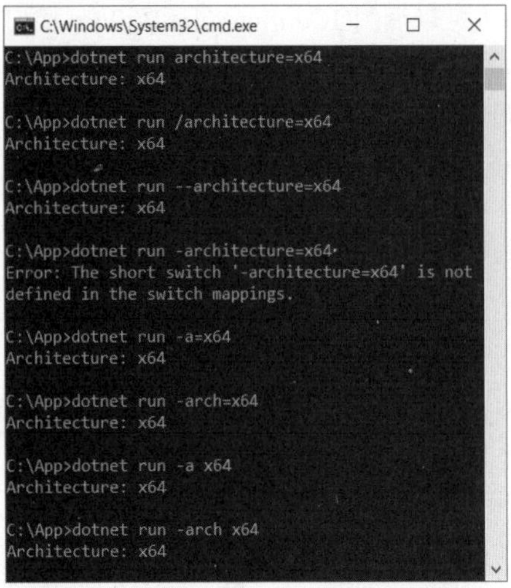

图 6-17 以命令行参数的形式提供配置

6.5.4 FileConfigurationSource

物理文件是我们最常用到的原始配置载体，而最佳的配置文件格式主要有 3 种，即 JSON、XML 和 INI，对应的配置源类型分别是 JsonConfigurationSource、XmlConfigurationSource 和 IniConfigurationSource，它们具有如下一个相同的基类 FileConfigurationSource。

```
public abstract class FileConfigurationSource : IConfigurationSource
{
    public IFileProvider                            FileProvider { get; set; }
    public string                                   Path { get; set; }
    public bool                                     Optional { get; set; }
    public int                                      ReloadDelay { get; set; }
    public bool                                     ReloadOnChange { get; set; }
    public Action<FileLoadExceptionContext>         OnLoadException { get; set; }

    public abstract IConfigurationProvider Build(IConfigurationBuilder builder);
    public void EnsureDefaults(IConfigurationBuilder builder);
    public void ResolveFileProvider();
}
```

FileConfigurationSource 对象总是利用 IFileProvider 对象来读取配置文件，我们可以利用 FileProvider 属性来设置这个对象。配置文件的路径可以用 Path 属性表示，一般来说这是一个针对 IFileProvider 对象根目录的相对路径。在读取配置文件时，这个路径将作为参数调用 IFileProvider 对象的 GetFileInfo 方法，以得到描述配置文件的 IFileInfo 对象，该对象的 CreateReadStream 方法最终会被调用来读取文件内容。

如果 FileProvider 属性并没有被显式赋值，并且指定的配置文件路径是一个绝对路径（如"c:\app\appsettings.json"），那么将创建一个针对配置文件所在目录（"c:\app"）的 Physical

FileProvider，并作为 FileProvider 的属性值，而 Path 属性将被设置成配置文件名。如果指定的仅仅是一个相对路径，FileProvider 属性就不会被自动初始化。这个逻辑可以在 ResolveFileProvider 方法中实现，并体现在如下测试程序中。

```csharp
class Program
{
    static void Main()
    {
        var source = new FakeConfigurationSource
        {
            Path = @"C:\App\appsettings.json"
        };
        Debug.Assert(source.FileProvider == null);

        source.ResolveFileProvider();
        var fileProvider = (PhysicalFileProvider)source.FileProvider;
        Debug.Assert(fileProvider.Root == @"C:\App\");
        Debug.Assert(source.Path == "appsettings.json");
    }

    private class FakeConfigurationSource : FileConfigurationSource
    {
        public override IConfigurationProvider Build(IConfigurationBuilder builder)
            => throw new NotImplementedException();
    }
}
```

除了 ResolveFileProvider 方法，FileConfigurationSource 还定义了 EnsureDefaults 方法，该方法会确保 FileConfigurationSource 总是具有一个用于加载配置文件的 IFileProvider 对象。具体来说，EnsureDefaults 方法最终会调用 IConfigurationBuilder 接口具有如下定义的 GetFileProvider 扩展方法来获取默认的 IFileProvider 对象。

```csharp
public static class FileConfigurationExtensions
{
    public static IFileProvider GetFileProvider(this IConfigurationBuilder builder)
    {
        if (builder.Properties.TryGetValue("FileProvider", out object provider))
        {
            return builder.Properties["FileProvider"] as IFileProvider;
        }
        return new PhysicalFileProvider(AppContext.BaseDirectory ?? string.Empty);
    }
}
```

从上面给出的代码片段可以看出，GetFileProvider 扩展方法实际上是将 IConfigurationBuilder 对象的 Properties 属性表示的字典作为存放 IFileProvider 对象的容器（对应的 Key 为 FileProvider）。如果这个容器中存在一个 IFileProvider 对象，那么它将作为方法的返回值。反之，该方法会将当前应用的基础目录（默认为当前应用程序域的基础目录，也就是当前执行.exe 文件所在的目录）作为根目录创建一个 PhysicalFileProvider 对象。

由于默认情况下 EnsureDefaults 方法会从 IConfigurationBuilder 对象的属性字典中提取 IFileProvider 对象，所以可以在这个属性字典中存放一个默认的 IFileProvider 对象供所有注册在它上面的 FileConfigurationSource 对象共享。实际上，IConfigurationBuilder 接口提供的 SetFileProvider 方法和 SetBasePath 方法可以实现这个功能。

```
public static class FileConfigurationExtensions
{
    public static IConfigurationBuilder SetFileProvider(
        this IConfigurationBuilder builder, IFileProvider fileProvider)
    {
        builder.Properties["FileProvider"] = fileProvider;
        return builder;
    }

    public static IConfigurationBuilder SetBasePath(
        this IConfigurationBuilder builder, string basePath)
        =>builder.SetFileProvider(new PhysicalFileProvider(basePath));
}
```

FileConfigurationSource 对象的 Optional 属性表示当前配置源是否可以缺省。如果该属性被设置成 True，即使指定的配置文件不存在也不会抛出异常。可缺省的配置文件在支持多环境的场景中具有广泛应用。正如前面的演示实例，我们可以按照如下方式加载两个配置文件：基础配置文件 appsettings.json 一般包含相对全面的配置，针对某个环境的差异化配置则定义在 appsettings.{environment}.json 文件中。前者是必需的，后者则是可以缺省的，这保证了应用程序在缺少基于当前环境的差异化配置文件的情况下依然可以使用定义在基础配置文件中的默认配置。

```
var configuration = new ConfigurationBuilder()
    .SetBasePath(Directory.GetCurrentDirectory())
    .AddJsonFile(path: "appsettings.json", optional: false)
    .AddJsonFile(path: $"appsettings.{environment}.json", optional: true)
    .Build();
```

FileConfigurationSource 借助 IFileProvider 对象提供的文件系统监控功能，实现了配置文件在更新后的自动实时加载功能，这个特性通过 ReloadOnChange 属性来开启或者关闭。在默认情况下这个特性是关闭的，我们需要通过将这个属性设置为 True 来显式地开启该特性。如果开启了配置文件的重新加载功能，一旦配置文件发生变化，IFileProvider 对象会在第一时间将通知发送给对应的 FileConfigurationProvider 对象，后者会调用 Load 方法重新加载配置文件。考虑到针对配置文件的写入此时可能尚未结束，所以 FileConfigurationSource 采用"延时加载"的方式来解决这个问题，具体的延时通过 ReloadDelay 属性来控制。该属性的单位是毫秒，默认设置的延时为 250 毫秒。

考虑到针对配置文件的加载不可能百分之百成功，所以 FileConfigurationSource 提供了相应的异常处理机制。具体来说，可以通过 FileConfigurationSource 对象的 OnLoadException 属性注册一个 Action<FileLoadExceptionContext>类型的委托作为异常处理器。作为参数的

FileLoadExceptionContext 对象代表 FileConfigurationProvider 在加载配置文件出错的情况下为异常处理器提供的执行上下文。

```
public class FileLoadExceptionContext
{
    public Exception                    Exception { get; set; }
    public FileConfigurationProvider    Provider { get; set; }
    public bool                         Ignore { get; set; }
}
```

如上面的代码片段所示，我们可以从 FileLoadExceptionContext 上下文中获取抛出的异常和当前 FileConfigurationProvider 对象。如果异常处理结束之后上下文对象的 Ignore 属性被设置为 True，FileConfigurationProvider 对象就会认为目前的异常（可能是原来抛出的异常，也可能是异常处理器设置的异常）是可以被忽略的，此时程序会继续执行，否则异常还是会抛出来。另外，最终抛出来的是原来的异常，所以通过修改上下文的 Exception 属性无法达到抛出另一个异常的目的。

就像为注册到 IConfigurationBuilder 对象上的所有 FileConfigurationSource 注册一个共享的 IFileProvider 对象一样，我们也可以调用 IConfigurationBuilder 接口的 SetFileLoadExceptionHandler 扩展方法注册一个共享的异常处理器，该方法依然是利用 IConfigurationBuilder 对象的属性字典来存放这个作为异常处理器的委托对象的。注册的这个异常处理器可以通过对应的 GetFileLoadExceptionHandler 扩展方法来获取。

```
public static class FileConfigurationExtensions
{
    public static IConfigurationBuilder SetFileLoadExceptionHandler(
        this IConfigurationBuilder builder, Action<FileLoadExceptionContext> handler)
    {
        builder.Properties["FileLoadExceptionHandler"] = handler;
        return builder;
    }

    public static Action<FileLoadExceptionContext> GetFileLoadExceptionHandler(
        this IConfigurationBuilder builder)
        => builder.Properties.TryGetValue("FileLoadExceptionHandler",
            out object handler)
            ? handler as Action<FileLoadExceptionContext>
            : null;
}
```

前面我们提到 FileConfigurationSource 的 EnsureDefaults 方法，除了在 IFileProvider 对象没有被初始化的情况下调用 IConfigurationBuilder 的 GetFileProvider 扩展方法提供一个默认的 IFileProvider 对象，EnsureDefaults 方法还会在异常处理器没有初始化的情况下调用上面的 GetFileLoadExceptionHandler 扩展方法提供一个默认的异常处理器。

对于配置系统默认提供的针对 3 种文件格式化的 FileConfigurationSource 类型来说，它们提供的 IConfigurationProvider 实现都派生于抽象基类 FileConfigurationProvider。对于自定义的

FileConfigurationSource，我们也倾向于将这个抽象类作为对应 IConfigurationProvider 实现类型的基类。

```
public abstract class FileConfigurationProvider : ConfigurationProvider
{
    public FileConfigurationSource Source { get; }
    public FileConfigurationProvider(FileConfigurationSource source);

    public override void Load();
    public abstract void Load(Stream stream);
}
```

创建一个 FileConfigurationProvider 对象时需要提供对应的 FileConfigurationSource 对象，它会赋值给 Source 属性。如果指定的 FileConfigurationSource 对象开启了配置文件更新监控和自动加载功能（其属性 OnLoadException 返回 True），FileConfigurationProvider 对象会利用 FileConfigurationSource 对象提供的 IFileProvider 对象对配置文件实施监控，并通过注册回调的方式在配置文件更新的时候调用 Load 方法重新加载配置。

由于 FileConfigurationSource 提供了 IFileProvider 对象，所以 FileConfigurationProvider 对象可以调用其 CreateReadStream 方法获取读取配置文件内容的流对象，也就是说，我们可以利用这个 Stream 对象来完成配置的加载。根据基于 Stream 加载配置的功能体现在抽象方法 Load 上，所以 FileConfigurationProvider 的派生类都需要重写这个方法。

JsonConfigurationSource

JsonConfigurationSource 代表针对基于 JSON 文件的配置源，该类型定义在 NuGet 包 "Microsoft.Extensions.Configuration.Json" 中。从下面的定义可以看出，JsonConfigurationSource 重写的 Build 方法在提供对应的 JsonConfigurationProvider 对象之前会调用 EnsureDefaults 方法，这个方法确保用于读取配置文件的 IFileProvider 对象和处理配置文件加载异常的处理器被初始化。JsonConfigurationProvider 派生于抽象类 FileConfigurationProvider，它利用重写的 Load 方法读取配置文件的内容并将其转换成配置字典。

```
public class JsonConfigurationSource : FileConfigurationSource
{
    public override IConfigurationProvider Build(IConfigurationBuilder builder)
    {
        EnsureDefaults(builder);
        return new JsonConfigurationProvider(this);
    }
}

public class JsonConfigurationProvider : FileConfigurationProvider
{
    public JsonConfigurationProvider(JsonConfigurationSource source);
    public override void Load(Stream stream);
}
```

IConfigurationBuilder 接口用如下几个 AddJsonFile 扩展方法来注册 JsonConfiguration

Source。如果调用第一个 AddJsonFile 方法重载，就可以利用指定的 Action<JsonConfiguration Source>对象对创建的 JsonConfigurationSource 进行初始化。而其他 AddJsonFile 方法重载实际上就是通过相应的参数初始化 JsonConfigurationSource 对象的 Path 属性、Optional 属性和 ReloadOnChange 属性。

```
public static class JsonConfigurationExtensions
{
    public static IConfigurationBuilder AddJsonFile(
        this IConfigurationBuilder builder,
        Action<JsonConfigurationSource> configureSource);
    public static IConfigurationBuilder AddJsonFile(
        this IConfigurationBuilder builder, string path);
    public static IConfigurationBuilder AddJsonFile(
        this IConfigurationBuilder builder, string path, bool optional);
    public static IConfigurationBuilder AddJsonFile(
        this IConfigurationBuilder builder, string path, bool optional,
        bool reloadOnChange);
    public static IConfigurationBuilder AddJsonFile(
        this IConfigurationBuilder builder, IFileProvider provider, string path,
        bool optional, bool reloadOnChange);
}
```

当使用 JSON 文件定义配置时，不论对于何种数据结构（复杂对象、集合、数组和字典），我们都能通过 JSON 文件以一种简单而自然的方式来定义它们。同样以前面定义的 Profile 类型为例，我们可以利用如下所示的 3 个 JSON 文件分别定义一个完整的 Profile 对象、一个 Profile 对象的集合，以及一个 Key 和 Value 类型分别为字符串与 Profile 的字典。

Profile 对象：

```
{
    "profile": {
        "gender"        : "Male",
        "age"           : "18",
        "contactInfo"   : {
            "email"     : "foobar@outlook.com",
            "phoneNo"   : "123456789"
        }
    }
}
```

Profile 集合或者数组：

```
{
    "profiles": [
        {
            "gender"        : "Male",
            "age"           : "18",
            "contactInfo"   : {
                "email"     : "foo@outlook.com",
                "phoneNo"   : "123"
            }
```

```
    },
    {
        "gender"           : "Male",
        "age"              : "25",
        "contactInfo"      : {
            "email"        : "bar@outlook.com",
            "phoneNo"      : "456"
        }
    },
    {
        "gender"           : "Female",
        "age"              : "40",
        "contactInfo"      : {
            "email"        : "baz@outlook.com",
            "phoneNo"      : "789"
        }
    }
  ]
}
```

Profile 字典(Dictionary<string, Profile>):

```
{
    "profiles": {
        "foo": {
            "gender"           : "Male",
            "age"              : "18",
            "contactInfo"      : {
                "email"        : "foo@outlook.com",
                "phoneNo"      : "123"
            }
        },
        "bar": {
            "gender"           : "Male",
            "age"              : "25",
            "contactInfo"      : {
                "email"        : "bar@outlook.com",
                "phoneNo"      : "456"
            }
        },
        "baz": {
            "gender"           : "Female",
            "age"              : "40",
            "contactInfo"      : {
                "email"        : "baz@outlook.com",
                "phoneNo"      : "789"
            }
        }
    }
}
```

XmlConfigurationSource

XML 也是一种常用的配置定义形式，它对数据的表达能力甚至强于 JSON，几乎所有类型的数据结构都可以用 XML 表示出来。用一个 XML 元素表示一个复杂对象时，对象的数据成员可以定义为当前 XML 元素的子元素。如果数据成员是一个简单的数据类型，我们还可以选择将其定义成当前 XML 元素的属性（Attribute）。针对一个 Profile 对象，我们可以采用如下两种不同的形式来定义。

```xml
<Profile>
    <Gender>Male</Gender>
    <Age>18</Age>
    <ContactInfo>
        <EmailAddress>foobar@outlook.com</EmailAddress>
        <PhoneNo>123456789</PhoneNo>
    </ContactInfo>
</Profile>
```

或者：

```xml
<Profile Gender="Male" Age="18">
  <ContactInfo EmailAddress ="foobar@outlook.com" PhoneNo="123456789"/>
</Profile>
```

虽然 XML 对数据结构的表达能力总体上强于 JSON，但是 XML 作为配置模型的数据来源有其局限性，如它们对集合的表现形式有点不尽如人意。例如，对于一个元素类型为 Profile 的集合，我们可以采用如下结构的 XML 来表现。

```xml
<Profiles>
    <Profile Gender="Male" Age="18">
        <ContactInfo EmailAddress ="foo@outlook.com" PhoneNo="123"/>
    </Profile>
    <Profile Gender="Male" Age="25">
        <ContactInfo EmailAddress ="bar@outlook.com" PhoneNo="456"/>
    </Profile>
    <Profile Gender="Female" Age="36">
        <ContactInfo EmailAddress ="baz@outlook.com" PhoneNo="789"/>
    </Profile>
</Profiles>
```

上述 XML 不能正确地转换成配置字典，这是因为字典的 Key 必须是唯一的，而且最终构成配置树的每个节点必须具有不同的路径。上面这段 XML 无法满足这个基本的要求，因为表示一个 Profile 对象的 3 个 XML 元素（<Profile>...</Profile>）是同质的，对于由它们表示的 3 个 Profile 对象来说，分别表示性别、年龄、电子邮箱地址和电话号码的 4 个叶子节点的路径是完全一样的，所以根本无法作为配置字典的 Key。通过前面针对配置绑定的介绍可知，如果需要用配置字典来表示一个 Profile 对象的集合，就需要按照如下方式为每个集合元素加上相应的索引（foo、bar 和 baz）。

```
foo:Gender
foo:Age
foo:ContactInfo:EmailAddress
```

```
foo:ContactInfo:PhoneNo

bar:Gender
bar:Age
bar:ContactInfo:EmailAddress
bar:ContactInfo:PhoneNo

baz:Gender
baz:Age
baz:ContactInfo:EmailAddress
baz:ContactInfo:PhoneNo
```

按照这样的结构，如果需要以 XML 的方式来表示一个 Profile 对象的集合，就不得不采用如下结构。但是这样的定义方式从语义角度来讲是不合理的，因为同一个集合的所有元素就应该是同质的，同质的 XML 元素采用不同的名称是不合理的。根据配置绑定的规则，这样的结构同样可以表示一个由 3 个元素组成的 Dictionary<string, Profile>对象，Key 分别是 Foo、Bar 和 Baz。如果用这样的 XML 来表示一个字典对象，那么语义上完全没有问题。

```xml
<Profiles>
  <Foo Gender="Male" Age="18">
    <ContactInfo EmailAddress ="foobar@outlook.com" PhoneNo="123"/>
  </Foo>
  <Bar Gender="Male" Age="25">
    <ContactInfo EmailAddress ="foobar@outlook.com" PhoneNo="123"/>
  </Bar>
  <Baz Gender="Male" Age="18">
    <ContactInfo EmailAddress ="baz@outlook.com" PhoneNo="789"/>
  </Baz>
</Profiles>
```

针对 XML 文件的配置源类型为 XmlConfigurationSource，该类型定义在 NuGet 包"Microsoft.Extensions.Configuration.Xml"中。如下面的代码片段所示，XmlConfigurationSource 通过重写的 Build 方法创建出对应的 XmlConfigurationProvider 对象。作为抽象类型 FileConfigurationProvider 的继承者，XmlConfigurationProvider 通过重写的 Load 方法完成了针对 XML 文件的读取和配置字典的初始化。

```csharp
public class XmlConfigurationSource : FileConfigurationSource
{
    public override IConfigurationProvider Build(IConfigurationBuilder builder)
    {
        EnsureDefaults(builder);
        return new XmlConfigurationProvider(this);
    }
}

public class XmlConfigurationProvider : FileConfigurationProvider
{
    public XmlConfigurationProvider(XmlConfigurationSource source);
    public override void Load(Stream stream);
```

}

JsonConfigurationSource 的注册可以通过调用针对 IConfigurationBuilder 接口的 AddJsonFile 扩展方法来完成。与之类似，IConfigurationBuilder 接口同样具有如下一系列名为 AddXmlFile 的扩展方法，这些方法可以帮助我们注册根据指定 XML 文件创建的 XmlConfigurationSource 对象。

```
public static class XmlConfigurationExtensions
{
    public static IConfigurationBuilder AddXmlFile(
        this IConfigurationBuilder builder, string path);
    public static IConfigurationBuilder AddXmlFile(
        this IConfigurationBuilder builder, string path, bool optional);
    public static IConfigurationBuilder AddXmlFile(
        this IConfigurationBuilder builder, string path, bool optional,
        bool reloadOnChange);
    public static IConfigurationBuilder AddXmlFile(
        this IConfigurationBuilder builder, IFileProvider provider, string path,
        bool optional, bool reloadOnChange);
}
```

XML 之所以不能像 JSON 格式那样可以以一种很自然的形式表示集合或者数组，是因为后者对这两种数据类型提供了明确的定义方式（采用中括号定义），但 XML 只有子元素的概念，我们无法确定它的子元素是否是一个集合。可以做这样一个假设：如果同一个 XML 元素下的所有子元素都具有相同的名称，那么我们可以将其视为集合。根据这个假设，对 XmlConfigurationSource 略加修改就可以解决 XML 难以表示集合数据结构的问题。

我们通过派生 XmlConfigurationSource 创建一个新的 IConfigurationSource 实现类型，姑且将其命名为 ExtendedXmlConfigurationSource。XmlConfigurationSource 提供的 ConfigurationProvdier 类型为 ExtendedXmlConfigurationProvider，它派生于 XmlConfigurationProvider。在重写的 Load 方法中，ExtendedXmlConfigurationProvider 对原始的 XML 结构进行相应的改动，可以使原本不合法的 XML（XML 元素具有相同的名称）转换成一个针对集合的配置字典。图 6-18 展示了 XML 结构转换所采用的规则和步骤。

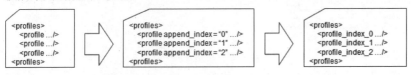

图 6-18　XML 结构转换所采用的规则和步骤

如图 6-18 所示，针对集合对原始 XML 所做的结构转换由两个步骤组成：第一步，为表示集合元素的 XML 元素添加一个名为 append_index 的属性（Attribute），我们采用零基索引作为该属性的值；第二步，根据第一步转换的结果创建一个新的 XML，同名的集合元素（如 <profile>）将根据添加的索引值重新命名（如<profile_index_0>）。毫无疑问，转换后的这个 XML 可以很好地表示一个集合对象。如下所示的代码片段是 ExtendedXmlConfigurationProvider 类型的定义，上述转换逻辑就体现在重写的 Load 方法中。（S615）

```
public class ExtendedXmlConfigurationProvider : XmlConfigurationProvider
```

```csharp
{
    public ExtendedXmlConfigurationProvider(XmlConfigurationSource source) :
        base(source)
    {}

    public override void Load(Stream stream)
    {
        //加载源文件并创建一个 XmlDocument
        var sourceDoc = new XmlDocument();
        sourceDoc.Load(stream);

        //添加索引
        AddIndexes(sourceDoc.DocumentElement);

        //根据添加的索引创建一个新的 XmlDocument
        var newDoc = new XmlDocument();
        var documentElement =
            newDoc.CreateElement(sourceDoc.DocumentElement.Name);
        newDoc.AppendChild(documentElement);

        foreach (XmlElement element in sourceDoc.DocumentElement.ChildNodes)
        {
            Rebuild(element, documentElement,
                name => newDoc.CreateElement(name));
        }

        //根据新的 XmlDocument 初始化配置字典
        using (Stream newStream = new MemoryStream())
        {
            using (XmlWriter writer = XmlWriter.Create(newStream))
            {
                newDoc.WriteTo(writer);
            }
            newStream.Position = 0;
            base.Load(newStream);
        }
    }

    private void AddIndexes(XmlElement element)
    {
        if (element.ChildNodes.OfType<XmlElement>().Count() > 1)
        {
            if (element.ChildNodes.OfType<XmlElement>()
                .GroupBy(it => it.Name).Count() == 1)
            {
                var index = 0;
                foreach (XmlElement subElement in element.ChildNodes)
                {
                    subElement.SetAttribute("append_index", (index++).ToString());
                    AddIndexes(subElement);
```

```
            }
        }
    }
}

private void Rebuild(XmlElement source, XmlElement destParent,
    Func<string, XmlElement> creator)
{
    var index = source.GetAttribute("append_index");
    var elementName = string.IsNullOrEmpty(index)
        ? source.Name : $"{source.Name}_index_{index}";
    var element = creator(elementName);
    destParent.AppendChild(element);
    foreach (XmlAttribute attribute in source.Attributes)
    {
        if (attribute.Name != "append_index")
        {
            element.SetAttribute(attribute.Name, attribute.Value);
        }
    }

    foreach (XmlElement subElement in source.ChildNodes)
    {
        Rebuild(subElement, element, creator);
    }
}
}
```

为了能够将上面的 XmlConfigurationProvider 应用到程序中，需要为它定义相应的 IConfigurationSource 类型，为此我们定义了下面的 ExtendedXmlConfigurationSource 类型。它直接继承自 XmlConfigurationSource 类型，并在重写的 Build 方法中提供 ExtendedXmlConfigurationProvider 对象。为了方便将 ExtendedXmlConfigurationSource 对象注册到 IConfigurationBuilder 对象上，也可以进一步定义如下这些扩展方法。

```
public class ExtendedXmlConfigurationSource : XmlConfigurationSource
{
    public override IConfigurationProvider Build(IConfigurationBuilder builder)
    {
        EnsureDefaults(builder);
        return new ExtendedXmlConfigurationProvider(this);
    }
}

public static class ExtendedXmlConfigurationExtensions
{
    public static IConfigurationBuilder AddExtendedXmlFile(
        this IConfigurationBuilder builder, string path)
        => builder.AddExtendedXmlFile(path, false, false);
    public static IConfigurationBuilder AddExtendedXmlFile(
        this IConfigurationBuilder builder, string path, bool optional)
```

```csharp
    => builder.AddExtendedXmlFile(path, optional, false);
public static IConfigurationBuilder AddExtendedXmlFile(
    this IConfigurationBuilder builder, string path, bool optional,
    bool reloadOnChange)
{
    builder.Add(new ExtendedXmlConfigurationSource
        { Path = path, Optional = optional, ReloadOnChange = reloadOnChange });
    return builder;
}
}
```

IniConfigurationSource

INI 是 Initialization 的缩写形式。INI 文件又被称为初始化文件,也是 Windows 操作系统普遍使用的配置文件,也被一些 Linux 操作系统和 UNIX 操作系统所支持。INI 文件直接以键值对的形式定义配置项,如下所示的代码片段体现了 INI 文件的基本格式。总的来说,INI 文件以 {Key}={Value} 的形式定义配置项,{Value} 可以定义在可选的双引号中(如果值的前后包括空白字符,就必须使用双引号,否则会被忽略)。

```ini
[Section]
key1=value1
key2 = " value2 "
; comment
# comment
/ comment
```

除了以 {Key}={Value} 的形式定义的原子配置项,我们还可以采用 [{SectionName}] 的形式定义配置节对它们进行分组。由于中括号([])是下一个配置节开始和上一个配置节结束的标志,所以采用 INI 文件定义的配置节并不存在层次化的结构,即没有子配置节的概念。除此之外,我们可以在 INI 文件中定义相应的注释,注释行前置的字符可以采用";"、"#"或者"/"。

由于 INI 文件自身就体现为一个数据字典,所以可以采用"路径化"的 Key 来定义最终绑定为复杂对象、集合或者字典的配置数据。如果采用 INI 文件定义一个 Profile 对象的基本信息,我们也可以采用如下定义形式。

```ini
Gender                  = "Male"
Age                     = "18"
ContactInfo:EmailAddress      = "foobar@outlook.com"
ContactInfo:PhoneNo           = "123456789"
```

由于 Profile 的配置信息具有两个层次(Profile/ContactInfo),所以可以按照如下形式将 EmailAddress 和 PhoneNo 定义在配置节 ContactInfo 中,这个 INI 文件在语义表达上和上面是完全等效的。

```ini
Gender = "Male"
Age    = "18"

[ContactInfo]
EmailAddress   = "foobar@outlook.com"
PhoneNo        = "123456789"
```

针对 INI 文件类型的配置源类型可以通过下面的 IniConfigurationSource 来表示，该类型定义在 NuGet 包 "Microsoft.Extensions.Configuration.Ini" 中。IniConfigurationSource 重写的 Build 方法创建的是一个 IniConfigurationProvider 对象。作为抽象类 FileConfigurationProvider 的继承者，IniConfigurationProvider 利用重写的 Load 方法，来完成 INI 文件内容的读取和配置字典的初始化。

```
public class IniConfigurationSource : FileConfigurationSource
{
    public override IConfigurationProvider Build(IConfigurationBuilder builder)
    {
        EnsureDefaults(builder);
        return new IniConfigurationProvider(this);
    }
}

public class IniConfigurationProvider : FileConfigurationProvider
{
    public IniConfigurationProvider(IniConfigurationSource source);
    public override void Load(Stream stream);
}
```

既然 JsonConfigurationSource 和 XmlConfigurationSource 的注册可以通过调用 IConfigurationBuilder 接口的 AddJsonFile 方法与 AddXmlFile 方法来完成，那么 NuGet 包 "Microsoft.Extensions.Configuration.Ini" 也会为 IniConfigurationSource 定义 AddIniFile 扩展方法。

```
public static class IniConfigurationExtensions
{
    public static IConfigurationBuilder AddIniFile(
        this IConfigurationBuilder builder, string path);
    public static IConfigurationBuilder AddIniFile(
        this IConfigurationBuilder builder, string path, bool optional);
    public static IConfigurationBuilder AddIniFile(
        this IConfigurationBuilder builder, string path, bool optional,
        bool reloadOnChange);
    public static IConfigurationBuilder AddIniFile(
        this IConfigurationBuilder builder, IFileProvider provider, string path,
        bool optional, bool reloadOnChange);
}
```

6.5.5　StreamConfigurationSource

StreamConfigurationSource 对象通过指定的 Stream 对象来读取配置内容，所以这种配置源具有更加灵活的应用。如下面的代码片段所示，StreamConfigurationSource 是一个抽象类，用于读取配置内容的输出流体现在它的 Stream 属性中。

```
public abstract class StreamConfigurationSource : IConfigurationSource
{
    public Stream Stream { get; set; }
    public abstract IConfigurationProvider Build(IConfigurationBuilder builder);
}
```

上面介绍了 3 种针对 JSON 文件、XML 文件和 INI 文件的 FileConfigurationSource 类型，在它们所在的 NuGet 包中，还定义了针对 StreamConfigurationSource 的版本。如下面的代码片段所示，这些具体的 StreamConfigurationSource 类型通过重写的 Build 方法提供对应的 IConfigurationProvider 对象，然后由它们利用指定的 Stream 对象读取对应的 JSON 文件、XML 文件和 INI 文本并转换成配置字典。针对上述 3 种具体的 StreamConfigurationSource 类型，我们可以调用如下 3 个对应的扩展方法来注册。

```csharp
public class JsonStreamConfigurationSource : StreamConfigurationSource
{
    public override IConfigurationProvider Build(IConfigurationBuilder builder)
        => new JsonStreamConfigurationProvider(this);
}

public class XmlStreamConfigurationSource : StreamConfigurationSource
{
    public override IConfigurationProvider Build(IConfigurationBuilder builder)
        => new XmlStreamConfigurationProvider(this);
}

public class IniStreamConfigurationSource : StreamConfigurationSource
{
    public override IConfigurationProvider Build(IConfigurationBuilder builder)
        => new IniStreamConfigurationProvider(this);
}

public static class JsonConfigurationExtensions
{
    public static IConfigurationBuilder AddJsonStream(
        this IConfigurationBuilder builder, Stream stream);
}

public static class XmlConfigurationExtensions
{
    public static IConfigurationBuilder AddXmlStream(
        this IConfigurationBuilder builder, Stream stream);
}

public static class IniConfigurationExtensions
{
    public static IConfigurationBuilder AddIniStream(
        this IConfigurationBuilder builder, Stream stream);
}
```

6.5.6　ChainedConfigurationSource

如下所示的 ChainedConfigurationSource 类型代表的配置源比较特殊，因为它承载的原始数据体现为一个 IConfiguration 对象。作为数据源的 IConfiguration 对象，它可以由 Chained

ConfigurationSource 类型的 Configuration 属性来表示。实现的 Build 方法会返回对应的 ChainedConfigurationProvider 对象。除此之外，ChainedConfigurationSource 类型还具有一个布尔类型的 ShouldDisposeConfiguration 属性，它决定了当提供的 ChainedConfigurationProvider 对象释放（调用其 Dispose 方法）的时候，指定的 IConfiguration 对象是否应该随之被释放。

```csharp
public class ChainedConfigurationSource : IConfigurationSource
{
    public IConfiguration      Configuration { get; set; }
    public bool                ShouldDisposeConfiguration { get; set; }

    public IConfigurationProvider Build(IConfigurationBuilder builder)
        => new ChainedConfigurationProvider(this);
}
```

虽然 IConfiguration 没有继承 IDisposable 接口，但是具体的实现类型（如表示配置树根节点的 ConfigurationRoot 类型）实现了 IDisposable 接口，所以当确定指定的 IConfiguration 对象没有使用时，我们应该调用其 Dispose 方法完成相应的释放和回收工作。如果指定的 IConfiguration 对象的生命周期由创建的 ChainedConfigurationSource 控制，那么我们应该将 ShouldDisposeConfiguration 属性设置为 True。如果该 IConfiguration 对象的释放由其他对象负责，就应该将该属性设置为 False。

```csharp
public class ChainedConfigurationProvider : IConfigurationProvider, IDisposable
{
    private readonly IConfiguration      _config;
    private readonly bool                _shouldDisposeConfig;

    public ChainedConfigurationProvider(ChainedConfigurationSource source)
    {
        _config = source.Configuration;
        _shouldDisposeConfig = source.ShouldDisposeConfiguration;
    }

    public bool TryGet(string key, out string value)
        => !string.IsNullOrEmpty(_config[key]);

    public void Set(string key, string value) => _config[key] = value;

    public IChangeToken GetReloadToken() => _config.GetReloadToken();

    public void Load() { }

    public IEnumerable<string> GetChildKeys(IEnumerable<string> earlierKeys,
        string parentPath)
    {
        var section = parentPath == null ? _config : _config.GetSection(parentPath);
        var children = section.GetChildren();
        var keys = new List<string>();
        keys.AddRange(children.Select(c => c.Key));
```

```
            return keys.Concat(earlierKeys)
                .OrderBy(k => k, ConfigurationKeyComparer.Instance);
        }

        public void Dispose()
        {
            if (_shouldDisposeConfig)
            {
                (_config as IDisposable)?.Dispose();
            }
        }
}
```

上面是 ChainedConfigurationProvider 类型的完整定义。与上面介绍的 IConfigurationProvider 实现类型有所不同，ChainedConfigurationProvider 并没有继承自基类 ConfigurationProvider，而是直接利用提供的 IConfiguration 对象实现了 IConfigurationProvider 接口的所有成员。ChainedConfigurationSource 对象的注册可以通过如下所示的两个 AddConfiguration 扩展方法重载来完成。如果调用第一个方法重载，注册 ChainedConfigurationSource 对象的 ShouldDisposeConfiguration 属性就被设置为 False，这意味着提供 IConfiguration 对象的生命周期由外部控制。

```
public static class ChainedBuilderExtensions
{
    public static IConfigurationBuilder AddConfiguration(
        this IConfigurationBuilder configurationBuilder, IConfiguration config);
    public static IConfigurationBuilder AddConfiguration(
        this IConfigurationBuilder configurationBuilder, IConfiguration config,
        bool shouldDisposeConfiguration);
}
```

6.5.7　自定义 ConfigurationSource（S616）

前面对配置模型中默认提供的各种 IConfigurationSource 实现类型进行了深入详尽的介绍，如果它们依然无法满足项目中的需求，还可以通过自定义 IConfigurationSource 实现类型来支持我们希望的配置源。就配置数据的持久化方式来说，将配置存储在数据库中是一种常见的方式。下面将创建一个针对数据库的 IConfigurationSource 实现类型，它采用 Entity Framework Core 来完成数据库的存取操作。由于篇幅所限，笔者没有对 Entity Framework Core 相关的编程进行单独介绍，如果读者对此不太熟悉，可以查阅 Entity Framework Core 的在线文档。

我们将自定义 ConfigurationSource 命名为 DbConfigurationSource。在正式介绍 DbConfigurationSource 的实现之前，下面先介绍它在项目中的应用。我们将配置保存在 SQL Server 数据库中的某个数据表中，并采用 Entity Framework Core 来读取它。可以将连接字符串作为配置定义在一个名为 appSettings.json 的 JSON 文件中。

```
{
  "connectionStrings": {
    "DefaultDb": "Server = ... ; Database=...; Uid = ...; Pwd = ..."
  }
```

如下所示的演示程序先创建了一个 ConfigurationBuilder 对象,并在它上面注册了一个指向 connectionString.json 文件的 JsonConfigurationSource 对象。针对 DbConfigurationSource 对象的注册体现在 AddDatabase 扩展方法上,这个方法具有两个参数,分别代表连接字符串的名称和初始的配置数据。前者正是 connectionString.json 设置的连接字符串名称 DefaultDb;后者是一个字典对象,它提供的原始配置正好可以构成一个 Profile 对象。利用 ConfigurationBuilder 创建出相应的 IConfiguration 对象之后,可以读取配置并将其绑定为一个 Profile 对象。

```
public class Program
{
    static void Main()
    {
        var initialSettings = new Dictionary<string, string>
        {
            ["Gender"]                    = "Male",
            ["Age"]                       = "18",
            ["ContactInfo:EmailAddress"]  = "foobar@outlook.com",
            ["ContactInfo:PhoneNo"]       = "123456789"
        };

        var profile = new ConfigurationBuilder()
            .AddJsonFile("appSettings.json")
            .AddDatabase("DefaultDb", initialSettings)
            .Build()
            .Get<Profile>();

        Debug.Assert(profile.Gender == Gender.Male);
        Debug.Assert(profile.Age == 18);
        Debug.Assert(profile.ContactInfo.EmailAddress == "foobar@outlook.com");
        Debug.Assert(profile.ContactInfo.PhoneNo == "123456789");
    }
}
```

如上面的代码片段所示,针对 DbConfigurationSource 对象的应用仅仅体现在为 IConfigurationBuilder 接口定义的 AddDatabase 扩展方法上,所以使用起来是非常方便的,那么这个扩展方法背后具有怎样的逻辑实现?DbConfigurationSource 采用 Entity Framework Core 并以 Code First 的方式进行数据操作,如下所示的 ApplicationSetting 是表示基本配置项的 POCO 类型,可以将配置项的 Key 以小写的方式存储。而 ApplicationSettingsContext 是对应的 DbContext 类型。

```
[Table("ApplicationSettings")]
public class ApplicationSetting
{
    private string key;

    [Key]
    public string Key
    {
```

```csharp
        get { return key; }
        set { key = value.ToLowerInvariant(); }
    }

    [Required]
    [MaxLength(512)]
    public string Value { get; set; }

    public ApplicationSetting()
    {}

    public ApplicationSetting(string key, string value)
    {
        Key     = key;
        Value   = value;
    }
}
public class ApplicationSettingsContext : DbContext
{
    public ApplicationSettingsContext(DbContextOptions options) : base(options)
    {}

    public DbSet<ApplicationSetting> Settings { get; set; }
}
```

如下所示的代码片段是 DbConfigurationSource 类型的定义，它的构造函数有两个参数：第一个参数的类型为 Action<DbContextOptionsBuilder>，可以用这个委托对象对创建 DbContext 采用的 DbContextOptions 进行设置；第二个可选的参数用来指定一些需要自动初始化的配置项。DbConfigurationSource 类型在重写的 Build 方法中利用这两个参数创建一个 DbConfigurationProvider 对象。

```csharp
public class DbConfigurationSource : IConfigurationSource
{
    private Action<DbContextOptionsBuilder> _setup;
    private IDictionary<string, string> _initialSettings;

    public DbConfigurationSource(Action<DbContextOptionsBuilder> setup,
        IDictionary<string, string> initialSettings = null)
    {
        _setup              = setup;
        _initialSettings    = initialSettings;
    }
    public IConfigurationProvider Build(IConfigurationBuilder builder)
    {
        return new DbConfigurationProvider(_setup, _initialSettings);
    }
}
```

DbConfigurationProvider 派生于抽象类 ConfigurationProvider。在重写的 Load 方法中，它会

根据提供的 Action<DbContextOptionsBuilder>创建 ApplicationSettingsContext 对象，利用它从数据库中读取配置数据并转换成字典对象，该字典对象最终被赋值给代表配置字典的 Data 属性。如果数据表中没有数据，Load 方法还会利用这个 DbContext 对象将提供的初始化配置添加到数据库中。

```csharp
public class DbConfigurationProvider: ConfigurationProvider
{
    private readonly IDictionary<string, string>          _initialSettings;
    private readonly Action<DbContextOptionsBuilder>      _setup;

    public DbConfigurationProvider(Action<DbContextOptionsBuilder> setup,
        IDictionary<string, string> initialSettings)
    {
        _setup              = setup;
        _initialSettings    = initialSettings?? new Dictionary<string, string>() ;
    }

    public override void Load()
    {
        var builder =
            new DbContextOptionsBuilder<ApplicationSettingsContext>();
        _setup(builder);
        using (ApplicationSettingsContext dbContext =
            new ApplicationSettingsContext(builder.Options))
        {
            dbContext.Database.EnsureCreated();
            Data = dbContext.Settings.Any()
                ? dbContext.Settings.ToDictionary(it => it.Key, it => it.Value,
                    StringComparer.OrdinalIgnoreCase)
                : Initialize(dbContext);
        }
    }

    private IDictionary<string, string> Initialize(
        ApplicationSettingsContext dbContext)
    {
        foreach (var item in _initialSettings)
        {
            dbContext.Settings.Add(new ApplicationSetting(item.Key, item.Value));
        }
        return _initialSettings.ToDictionary(it => it.Key, it => it.Value,
            StringComparer.OrdinalIgnoreCase);
    }
}
```

实例演示中用来注册 DbConfigurationSource 对象的 AddDatabase 扩展方法具有如下定义。该方法首先调用 IConfigurationBuilder 对象的 Build 方法创建一个 IConfiguration 对象，并调用该 IConfiguration 对象的 GetConnectionString 扩展方法根据指定的连接字符串名称得到完整的连接字符串。下面调用构造函数创建一个 DbConfigurationSource 对象，并将其注册到 Configuration Builder 对象上。创建 DbConfigurationSource 对象时指定的 Action<DbContextOptionsBuilder>会完成针对连接字符串的设置。

```csharp
public static class DbConfigurationExtensions
{
    public static IConfigurationBuilder AddDatabase(
        this IConfigurationBuilder builder, string connectionStringName,
        IDictionary<string, string> initialSettings = null)
    {
        var connectionString = builder.Build()
            .GetConnectionString(connectionStringName);
        var source = new DbConfigurationSource(
            optionsBuilder => optionsBuilder.UseSqlServer(connectionString),
            initialSettings);
        builder.Add(source);
        return builder;
    }
}
```

第 7 章

配置选项（下篇）

.NET Core 组件、框架和应用基本上都会将配置选项绑定为一个 POCO 对象，并以依赖注入的形式来使用它。我们将这个承载配置选项的 POCO 对象称为 Options 对象，将这种以依赖注入方式来消费它的编程方式称为 Options 模式（Options Pattern）。第 6 章介绍的配置系统是 Options 对象的主要数据来源，除了将承载配置数据的 IConfiguration 对象绑定为 Options 对象，我们还可以直接通过编程的方式来初始化 Options 对象。

7.1 Options 模式

依赖注入不仅是支撑整个 ASP.NET Core 框架的基石，也是开发 ASP.NET Core 应用采用的基本编程模式，所以依赖注入十分重要。依赖注入使我们可以将依赖的功能定义成服务，最终以一种松耦合的形式注入消费该功能的组件或者服务中。除了可以采用依赖注入的形式消费承载某种功能的服务，还可以采用相同的方式消费承载配置数据的 Options 对象。

7.1.1 将配置绑定为 Options 对象

Options 模式是一种采用依赖注入来提供 Options 对象的编程方式，但这并不意味着我们会直接利用依赖注入框架来提供 Options 对象本身，因为利用依赖注入框架获取的是一个能够提供 Options 对象的 IOptions<TOptions>对象，泛型参数 TOptions 表示的正是 Options 对象的类型。下面的演示实例利用 IOptions<TOptions>服务来提供我们需要的 Options 对象，该对象由一个承载配置数据的 IConfiguration 对象绑定而成。简单起见，我们依然沿用第 6 章定义的 Profile 作为基础的 Options 类型，下面先回顾相关类型的定义。

```
public class Profile: IEquatable<Profile>
{
    public Gender            Gender { get; set; }
    public int               Age { get; set; }
    public ContactInfo       ContactInfo { get; set; }

    public Profile() {}
```

```csharp
    public Profile(Gender gender, int age, string emailAddress, string phoneNo)
    {
        Gender          = gender;
        Age             = age;
        ContactInfo     = new ContactInfo
        {
            EmailAddress        = emailAddress,
            PhoneNo             = phoneNo
        };
    }
    public bool Equals(Profile other)
    {
        return other == null
            ? false
            :Gender == other.Gender &&
              Age == other.Age &&
              ContactInfo.Equals(other.ContactInfo);
    }
}

public class ContactInfo: IEquatable<ContactInfo>
{
    public string EmailAddress { get; set; }
    public string PhoneNo { get; set; }
    public bool Equals(ContactInfo other)
    {
        return other == null
            ? false
            : EmailAddress      == other.EmailAddress &&
              PhoneNo           == other.PhoneNo;
    }
}

public enum Gender
{
    Male,
    Female
}
```

下面通过一个简单的控制台应用来演示 Options 编程模式。在演示程序中定义了上面这些类型之后，我们创建承载一个 Profile 对象的配置文件 profile.json。如下所示的代码片段就是这个 JSON 文件的内容，它提供了构成一个完整 Profile 对象的所有数据。为了使该文件能够在编译后自动复制到输出目录，我们需要将 "Copy to Output Directory" 属性设置为 "Copy Always"。

```
{
    "gender"    : "Male",
    "age"       : "18",
    "contactInfo": {
        "emailAddress"  : "foobar@outlook.com",
        "phoneNo"       : "123456789"
    }
```

下面编写代码来演示如何采用 Options 模式获取由配置文件提供的数据绑定生成的 Profile 对象。我们调用 AddJsonFile 扩展方法将针对 JSON 配置文件（profile.json）的配置源注册到创建的 ConfigurationBuilder 对象上，并利用它创建对应的 IConfiguration 对象。

```
class Program
{
    static void Main()
    {
        var configuration = new ConfigurationBuilder()
            .AddJsonFile("profile.json")
            .Build();
        var profile = new ServiceCollection()
            .AddOptions()
            .Configure<Profile>(configuration)
            .BuildServiceProvider()
            .GetRequiredService<IOptions<Profile>>()
            .Value;
        Console.WriteLine($"Gender: {profile.Gender}");
        Console.WriteLine($"Age: {profile.Age}");
        Console.WriteLine($"Email Address: {profile.ContactInfo.EmailAddress}");
        Console.WriteLine($"Phone No: {profile.ContactInfo.PhoneNo}");
    }
}
```

下面创建一个 ServiceCollection 对象，在调用 AddOptions 扩展方法注册 Options 编程模式的核心服务后，可以将创建的 IConfiguration 对象作为参数调用 Configure<Profile>扩展方法。Configure<TOptions>扩展方法相当于将提供的 IConfiguration 对象与指定的 TOptions 类型做了一个映射，在需要提供对应的 TOptions 对象时，IConfiguration 对象承载的配置数据会被提取出来并绑定生成返回的 TOptions 对象。

在调用 IServiceCollection 的 BuildServiceProvider 扩展方法得到作为依赖注入容器的 IServiceProvider 对象之后，可以直接调用其 GetRequiredService<T>扩展方法来提供 IOptions<Profile>对象，该对象的 Value 属性返回的就是指定 IConfiguration 对象绑定生成的 Profile 对象。我们将这个 Profile 对象承载的相关数据直接打印在控制台上，输出结果如图 7-1 所示，由此可以看出，通过 Options 模式得到的 Profile 对象承载的数据完全来源于配置文件。（S701）

图 7-1　绑定配置生成的 Profile 对象

7.1.2　提供具名的 Options

针对同一个 Options 类型，IOptions<TOptions>服务在整个应用范围内只能提供一个单一的

Options 对象，但是在很多情况下我们需要利用多个同类型的 Options 对象来承载不同的配置。就演示实例中用来表示个人信息的 Profile 类型来说，应用程序中可能会使用它来表示不同用户的信息，如张三、李四和王五。为了解决这个问题，我们可以在添加 IConfiguration 对象与 Options 类型映射关系时赋予它们一个唯一标识，这个标识最终会被用来提取对应的 Options 对象。这种具名的 Options 对象由 IOptionsSnapshot<TOptions>接口表示的服务提供。

同样，针对前面的演示实例，假设应用需要采用 Options 模式提取承载不同用户信息的 Profile 对象，具体应该如何实现？由于采用 JSON 格式的配置文件来提供原始的用户信息，所以需要将针对多个用户的信息定义在 profile.json 文件中。我们通过如下形式提供了两个用户（foo 和 bar）的基本信息。

```
{
    "foo": {
        "gender"      : "Male",
        "age"         : "18",
        "contactInfo": {
            "emailAddress"   : "foo@outlook.com",
            "phoneNo"        : "123"
        }
    },
    "bar": {
        "gender"      : "Female",
        "age"         : "25",
        "contactInfo": {
            "emailAddress"   : "bar@outlook.com",
            "phoneNo"        : "456"
        }
    }
}
```

具名 Options 的注册和提取体现在如下所示的代码片段中。在调用 IServiceCollection 接口的 Configure<TOptions>扩展方法时，我们将注册的映射关系命名为 foo 和 bar，提供原始配置数据的 IConfiguration 对象也由原来的 ConfigurationRoot 对象变成它的两个子配置节。

```
class Program
{
    static void Main()
    {
        var configuration = new ConfigurationBuilder()
            .AddJsonFile("profile.json")
            .Build();

        var serviceProvider = new ServiceCollection()
            .AddOptions()
            .Configure<Profile>("foo", configuration.GetSection("foo"))
            .Configure<Profile>("bar", configuration.GetSection("bar"))
            .BuildServiceProvider();

        var optionsAccessor = serviceProvider
            .GetRequiredService<IOptionsSnapshot<Profile>>();
```

```
        Print(optionsAccessor.Get("foo"));
        Print(optionsAccessor.Get("bar"));

        static void Print(Profile profile)
        {
            Console.WriteLine($"Gender: {profile.Gender}");
            Console.WriteLine($"Age: {profile.Age}");
            Console.WriteLine($"Email Address: {profile.ContactInfo.EmailAddress}");
            Console.WriteLine($"Phone No: {profile.ContactInfo.PhoneNo}\n");
        }
    }
}
```

为了使用指定的用户名来提取对应的 Profile 对象，可以利用作为依赖注入容器的 IServiceProvider 对象得到 IOptionsSnapshot<TOptions>服务，并将用户名作为参数调用其 Get 方法得到对应的 Profile 对象。程序运行后，针对两个不同用户的基本信息将以图 7-2 所示的形式输出到控制台上。（S702）

图 7-2　根据用户名提取对应的 Profile 对象

7.1.3　配置源的同步

通过第 6 章的介绍可知，配置模型不仅支持对配置源的监控，还可以在检测到更新之后及时加载新的配置数据，并通过一个 IChangeToken 对象对外发送通知。对于前面演示的两个实例来说，提供的 Options 对象都是由配置文件提供的数据绑定生成的，如果新的配置数据被重新加载之后能够提供与之匹配的 Options 对象，那么这将是最理想的编程模式，可以通过 IOptionsMonitor<TOptions>服务来实现。

前面演示的第一个实例（S701）利用 JSON 文件定义了一个单一 Profile 对象的信息，下面对它做相应的修改来演示如何监控这个 JSON 文件，并在监测到文件改变之后及时提取新的配置信息生成新的 Profile 对象。如下面的代码片段所示，调用 AddJsonFile 扩展方法注册对应配置源时应将该方法的参数 reloadOnChange 设置为 True，从而开启对对应配置文件的监控功能。

```
class Program
{
    static void Main()
    {
        var configuration = new ConfigurationBuilder()
            .AddJsonFile(path: "profile.json", optional: false,
                reloadOnChange: true)
            .Build();
```

```
            new ServiceCollection()
                .AddOptions()
                .Configure<Profile>(configuration)
                .BuildServiceProvider()
                .GetRequiredService<IOptionsMonitor<Profile>>()
                .OnChange(profile =>
                {
                    Console.WriteLine($"Gender: {profile.Gender}");
                    Console.WriteLine($"Age: {profile.Age}");
                    Console.WriteLine(
                        $"Email Address: {profile.ContactInfo.EmailAddress}");
                    Console.WriteLine($"Phone No: {profile.ContactInfo.PhoneNo}\n");
                });
            Console.Read();
        }
    }
```

在得到作为依赖注入容器的 IServiceProvider 对象之后，可以利用它得到 IOptionsMonitor<TOptions>服务，该对象会接收到配置系统发出的关于配置被重新加载的通知，并在收到通知后重新生成 Options 对象。我们调用 IOptionsMonitor<TOptions>对象的 OnChange 方法注册了一个类型为 Action<TOptions>的委托对象，该委托对象会在接收到 Options 变化时自动执行，而作为输入的正是重新生成的 Options 对象。

由于注册的委托对象会将新 Profile 对象的相关属性打印在控制台上，所以程序启动后针对配置文件的任何修改都会导致新的数据被打印在控制台上。例如，我们先后修改了年龄（25）和性别（Female），新的数据将按照图 7-3 所示的形式反映在控制台上。（S703）

图 7-3　及时提取新的 Profile 对象并应用到程序中（一）

具名 Options 同样可以采用类似的编程模式来实现配置的同步。在前面演示的提供具名 Options 的第二个实例的基础上，我们对程序做了如下修改。与之前不同的是，在利用 IServiceProvider 对象得到 IOptionsMonitor<TOptions>服务之后，可以调用其 OnChange 方法注册的回调是一个 Action<TOptions, String>对象，该委托对象的第二个参数表示的正是在注册 IConfiguration 对象与 Options 类型应用关系时指定的名称。

```
class Program
{
    static void Main()
    {
        var configuration = new ConfigurationBuilder()
```

```
            .AddJsonFile(path: "profile.json", optional: false,
                reloadOnChange: true)
            .Build();
    new ServiceCollection()
        .AddOptions()
        .Configure<Profile>("foo", configuration.GetSection("foo"))
        .Configure<Profile>("bar", configuration.GetSection("bar"))
        .BuildServiceProvider()
        .GetRequiredService<IOptionsMonitor<Profile>>()
        .OnChange((profile, name) =>
        {
            Console.WriteLine($"Name: {name}");
            Console.WriteLine($"Gender: {profile.Gender}");
            Console.WriteLine($"Age: {profile.Age}");
            Console.WriteLine(
                $"Email Address: {profile.ContactInfo.EmailAddress}");
            Console.WriteLine($"Phone No: {profile.ContactInfo.PhoneNo}\n");
        });
    Console.Read();
}
```

由于通过调用 OnChange 方法注册的委托对象会将 Options 的名称和承载的数据打印在控制台上，所以控制台上输出的内容总是与配置文件的内容同步。例如，在程序启动后，我们分别修改了用户 foo 的年龄（25）和用户 bar 的性别（Male），新的内容将以图 7-4 所示的形式及时呈现在控制台上。（S704）

图 7-4　及时提取新的 Profile 对象并应用到程序中（二）

7.1.4　直接初始化 Options 对象

前面演示的几个实例具有一个共同的特征，即都采用配置系统来提供绑定 Options 对象的原始数据，实际上，Options 框架具有一个完全独立的模型，可以称为 Options 模型。这个独立的 Options 模型本身并不依赖于配置系统，让配置系统来提供配置数据仅仅是通过 Options 模型的一个扩展点实现的。在很多情况下，可能并不需要将应用的配置选项定义在配置文件中，在应用启动时直接初始化可能是一种更方便、快捷的方式。

我们依然沿用前面演示的应用场景，现在摒弃配置文件，转而采用编程的方式直接对用户信息进行初始化，所以需要对程序做如下改写。在调用 IServiceCollection 接口的 Configure<Profile>扩展方法时，不需要再指定一个 IConfiguration 对象，而是利用一个 Action<Profile>类型的委托对作为参数的 Profile 对象进行初始化。程序运行后会在控制台上产生图 7-1 所示的输出结果。（S705）

```
class Program
{
    static void Main()
    {
        var profile = new ServiceCollection()
            .AddOptions()
            .Configure<Profile>(it =>
            {
                it.Gender            = Gender.Male;
                it.Age               = 18;
                it.ContactInfo       = new ContactInfo
                {
                    PhoneNo          = "123456789",
                    EmailAddress     = "foobar@outlook.com"
                };
            })
            .BuildServiceProvider()
            .GetRequiredService<IOptions<Profile>>()
            .Value;

        Console.WriteLine($"Gender: {profile.Gender}");
        Console.WriteLine($"Age: {profile.Age}");
        Console.WriteLine($"Email Address: {profile.ContactInfo.EmailAddress}");
        Console.WriteLine($"Phone No: {profile.ContactInfo.PhoneNo}\n");
    }
}
```

具名 Options 同样可以采用类似的方式进行初始化。如果需要根据指定的名称对 Options 进行初始化，那么调用方法时就需要指定一个 Action<TOptions,String>类型的委托对象，该委托对象的第二个参数表示 Options 的名称。在如下所示的代码片段中，我们通过类似的方式设置了两个用户（foo 和 bar）的信息，然后利用 IOptionsSnapshot<TOptions>服务将它们分别提取出来。该程序运行后会在控制台上产生图 7-2 所示的输出结果。（S706）

```
class Program
{
    static void Main()
    {
```

```csharp
        var optionsAccessor = new ServiceCollection()
            .AddOptions()
            .Configure<Profile>("foo", it =>
            {
                it.Gender            = Gender.Male;
                it.Age               = 18;
                it.ContactInfo       = new ContactInfo
                {
                    PhoneNo          = "123",
                    EmailAddress     = "foo@outlook.com"
                };
            })
            .Configure<Profile>("bar", it =>
            {
                it.Gender            = Gender.Female;
                it.Age               = 25;
                it.ContactInfo       = new ContactInfo
                {
                    PhoneNo          = "456",
                    EmailAddress     = "bar@outlook.com"
                };
            })
            .BuildServiceProvider()
            .GetRequiredService<IOptionsSnapshot<Profile>>();

    Print(optionsAccessor.Get("foo"));
    Print(optionsAccessor.Get("bar"));

    static void Print(Profile profile)
    {
        Console.WriteLine($"Gender: {profile.Gender}");
        Console.WriteLine($"Age: {profile.Age}");
        Console.WriteLine($"Email Address: {profile.ContactInfo.EmailAddress}");
        Console.WriteLine($"Phone No: {profile.ContactInfo.PhoneNo}\n");
    };
}
```

在前面的演示中,我们利用依赖注入框架提供 IOptions<TOptions> 服务、IOptionsSnapshot <TOptions> 服务和 IOptionsMonitor<TOptions> 服务,然后进一步利用它们来提供对应的 Options 对象。既然作为依赖注入容器的 IServiceProvider 对象能够提供这 3 个对象,我们就能够将它们注入消费 Options 对象的类型中。所谓的 Options 模式就是通过注入这 3 个服务来提供对应 Options 对象的编程模式的。

7.1.5　根据依赖服务的 Options 设置

在很多情况下需要针对某个依赖的服务动态地初始化 Options 的设置,比较典型的就是根据当前的承载环境(开发、预发和产品)对 Options 做动态设置。第 6 章演示了一系列针对日期/时间输出格式的配置,下面沿用这个场景演示如何根据当前的承载环境设置对应的 Options。将

DateTimeFormatOptions 的定义进行简化，只保留如下所示的表示日期和时间格式的两个属性。

```
public class DateTimeFormatOptions
{
    public string DatePattern { get; set; }
    public string TimePattern { get; set; }
    public override string ToString()
        => $"Date: {DatePattern}; Time: {TimePattern}";
}
```

如下所示的代码片段是整个演示实例的完整定义。我们利用第 6 章介绍的配置系统来设置当前的承载环境，具体采用的是基于命令行参数的配置源。.NET Core 的承载系统（详见第 10 章）通过 IHostEnvironment 接口表示承载环境，具体实现类型为 HostingEnvironment。如下面的代码片段所示，我们利用获取的环境名称创建了一个 HostingEnvironment 对象，并针对 IHostEnvironment 接口采用 Singleton 生命周期做了相应的注册。

```
class Program
{
    public static void Main(string[] args)
    {
        var environment = new ConfigurationBuilder()
            .AddCommandLine(args)
            .Build()["env"];

        var services = new ServiceCollection();
        services
            .AddSingleton<IHostEnvironment>(
                new HostingEnvironment { EnvironmentName = environment })
            .AddOptions<DateTimeFormatOptions>().Configure<IHostEnvironment>(
        (options, env) => {
            if (env.IsDevelopment())
            {
                options.DatePattern = "dddd, MMMM d, yyyy";
                options.TimePattern = "M/d/yyyy";
            }
            else
            {
                options.DatePattern = "M/d/yyyy";
                options.TimePattern = "h:mm tt";
            }
        });

        var options = services
            .BuildServiceProvider()
            .GetRequiredService<IOptions<DateTimeFormatOptions>>().Value;
        Console.WriteLine(options);
    }
}
```

上面调用 IServiceCollection 接口的 AddOptions<DateTimeFormatOptions>扩展方法完成了针对 Options 模型核心服务的注册和针对 DateTimeFormatOptions 的设置。该方法返回的是一个封装了

IServiceCollection 集合的 OptionsBuilder<DateTimeFormatOptions>对象，可以调用其 Configure<IHostEnvironment>方法利用提供的 Action<DateTimeFormatOptions, IHostEnvironment>委托对象针对依赖的 IHostEnvironment 服务对 DateTimeFormatOptions 做相应的设置。

具体来说，我们针对开发环境和非开发环境设置了不同的日期与时间格式。如果采用命令行的方式启动这个应用程序，并利用命令行参数设置不同的环境名称，就可以在控制台上看到图 7-5 所示的针对 DateTimeFormatOptions 的不同设置。（S707）

图 7-5　针对承载环境的 Options 设置

7.1.6　验证 Options 的有效性

由于配置选项是整个应用的全局设置，为了尽可能避免错误的设置造成的影响，最好能够对内容进行有效性验证。接下来我们将上面的程序做了如下改动，从而演示如何对设置的日期和时间格式做最后的有效性验证。

```
class Program
{
    public static void Main(string[] args)
    {
        var config = new ConfigurationBuilder()
            .AddCommandLine(args)
            .Build();
        var datePattern = config["date"];
        var timePattern = config["time"];

        var services = new ServiceCollection();
        services.AddOptions<DateTimeFormatOptions>()
            .Configure(options =>
            {
                options.DatePattern = datePattern;
                options.TimePattern = timePattern;
            })
            .Validate(options => Validate(options.DatePattern)
                && Validate(options.TimePattern),"Invalid Date or Time pattern.");

        try
        {
            var options = services
                .BuildServiceProvider()
                .GetRequiredService<IOptions<DateTimeFormatOptions>>().Value;
            Console.WriteLine(options);
        }
```

```
        catch (OptionsValidationException ex)
        {
            Console.WriteLine(ex.Message);
        }

        static bool Validate(string format)
        {
            var time = new DateTime(1981, 8, 24,2,2,2);
            var formatted = time.ToString(format);
            return DateTimeOffset.TryParseExact(formatted, format,
                null, DateTimeStyles.None, out var value)
                && (value.Date == time.Date || value.TimeOfDay == time.TimeOfDay);
        }
    }
}
```

上述演示实例借助配置系统以命令行的形式提供了日期和时间格式化字符串。在创建了 OptionsBuilder<DateTimeFormatOptions>对象并对 DateTimeFormatOptions 做了相应设置之后，我们调用 Validate<DateTimeFormatOptions>方法利用提供的 Func<DateTimeFormatOptions,bool>委托对象对最终的设置进行验证。运行该程序并按照图 7-6 所示的方式指定不同的格式化字符串，系统会根据我们指定的规则来验证其有效性。（S708）

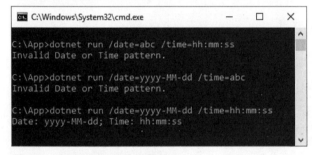

图 7-6　验证 Options 的有效性

7.2　Options 模型

通过前面演示的几个实例，我们已经对基于 Options 的编程方式有了一定程度的了解，下面从设计的角度介绍 Options 模型。我们演示的实例已经涉及 Options 模型的 3 个重要的接口，它们分别是 IOptions<TOptions>、IOptionsSnapshot<TOptions>和 IOptionsMonitor<TOptions>，最终的 Options 对象正是利用它们来提供的。在 Options 模型中，这两个接口具有同一个实现类型 OptionsManager<TOptions>。Options 模型的核心接口和类型定义在 NuGet 包"Microsoft. Extensions.Options"中。

7.2.1　OptionsManager<TOptions>

在 Options 模式的编程中，我们会利用作为依赖注入容器的 IServiceProvider 对象来提供

IOptions<TOptions>服务或者 IOptionsSnapshot<TOptions>服务，实际上，最终得到的服务实例都是一个 OptionsManager<TOptions>对象。在 Options 模型中，OptionsManager<TOptions>相关的接口和类型主要体现在图 7-7 中。

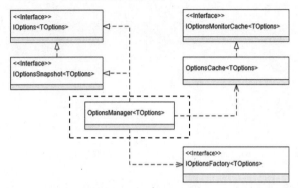

图 7-7　OptionsManager<TOptions>

下面以图 7-7 为基础介绍 OptionsManager<TOptions>对象是如何提供 Options 对象的。如下面的代码片段所示，IOptions<TOptions>接口和 IOptionsSnapshot<TOptions>接口的泛型参数的 TOptions 类型要求具有一个默认的构造函数，也就是说，Options 对象可以在无须指定参数的情况下直接采用 new 关键字进行实例化，实际上，Options 最初就是采用这种方式创建的。

```
public interface IOptions<out TOptions> where TOptions: class, new()
{
    TOptions Value { get; }
}

public interface IOptionsSnapshot<out TOptions> : IOptions<TOptions>
    where TOptions: class, new()
{
    TOptions Get(string name);
}
```

IOptions<TOptions>接口通过 Value 属性提供对应的 Options 对象，继承它的 IOptionsSnapshot<TOptions>接口则利用其 Get 方法根据指定的名称提供对应的 Options 对象。OptionsManager<TOptions>针对这两个接口成员的实现依赖其他两个对象，分别通过 IOptionsFactory<TOptions>接口和 IOptionsMonitorCache<TOptions>接口表示，这也是 Options 模型的两个核心成员。

作为 Options 对象的工厂，IOptionsFactory<TOptions>对象负责创建 Options 对象并对其进行初始化。出于性能方面的考虑，由 IOptionsFactory<TOptions>工厂创建的 Options 对象会被缓存起来，针对 Options 对象的缓存就由 IOptionsMonitorCache<TOptions>对象负责。下面会对 IOptionsFactory<TOptions>和 IOptionsMonitorCache<TOptions>进行单独讲解，在此之前需要先了解 OptionsManager<TOptions>类型是如何定义的。

```
public class OptionsManager<TOptions>
    :IOptions<TOptions>,
    IOptionsSnapshot<TOptions> where TOptions : class, new()
```

```
{
    private readonly IOptionsFactory<TOptions> _factory;
    private readonly OptionsCache<TOptions>   _cache = new OptionsCache<TOptions>();

    public OptionsManager(IOptionsFactory<TOptions> factory)
        => _factory = factory;
    public TOptions Value
        => this.Get(Options.DefaultName);
    public TOptions Get(string name)
        => _cache.GetOrAdd(name, () => _factory.Create(name));
}

public static class Options
{
    public static readonly string DefaultName = string.Empty;
}
```

OptionsManager<TOptions>对象提供 Options 对象的逻辑基本上体现在上面给出的代码中。在创建一个 OptionsManager<TOptions>对象时需要提供一个 IOptionsFactory<TOptions>工厂，而它自己还会创建一个 OptionsCache<TOptions>（该类型实现了 IOptionsMonitorCache<TOptions>接口）对象来缓存 Options 对象，也就是说，Options 对象实际上是被 OptionsManager<TOptions>对象以"独占"的方式缓存起来的，后续内容还会提到这个设计细节。

从编程的角度来讲，IOptions<TOptions>接口和 IOptionsSnapshot<TOptions>接口分别体现了非具名与具名的 Options 提供方式，但是对于同时实现这两个接口的 OptionsManager<TOptions>来说，提供的 Options 都是具名的，唯一的不同之处在于以 IOptions<TOptions>接口的名义提供 Options 对象时会采用一个空字符串作为名称。默认 Options 名称可以通过静态类型 Options 的只读字段 DefaultName 来获取。

OptionsManager<TOptions>针对 Options 对象的提供（具名或者非具名）最终体现在其实现的 Get 方法上。由于 Options 对象缓存在自己创建的 OptionsCache<TOptions>对象上，所以它只需要将指定的 Options 名称作为参数调用其 GetOrAdd 方法就能获取对应的 Options 对象。如果 Options 对象尚未被缓存，它会利用作为参数传入的 Func<TOptions>委托对象来创建新的 Options 对象，从前面给出的代码可以看出，这个委托对象最终会利用 IOptionsFactory<TOptions>工厂来创建 Options 对象。

7.2.2 IOptionsFactory<TOptions>

顾名思义，IOptionsFactory<TOptions>接口表示创建和初始化 Options 对象的工厂。如下面的代码片段所示，该接口定义了唯一的 Create 方法，可以根据指定的名称创建对应的 Options 对象。

```
public interface IOptionsFactory<TOptions> where TOptions: class, new()
{
    TOptions Create(string name);
}
```

OptionsFactory<TOptions>

OptionsFactory<TOptions>是 IOptionsFactory<TOptions>接口的默认实现。OptionsFactory<TOptions>对象针对 Options 对象的创建主要分 3 个步骤来完成，笔者将这 3 个步骤称为 Options 对象相关的"实例化"、"初始化"和"验证"。

由于 Options 类型总是具有一个公共默认的构造函数，所以 OptionsFactory<TOptions>只需要利用 new 关键字调用这个构造函数就可以创建一个空的 Options 对象。当 Options 对象被实例化之后，OptionsFactory<TOptions>对象会根据注册的一些服务对其进行初始化。Options 模型中针对 Options 对象初始化的工作由如下 3 个接口表示的服务负责。

```
public interface IConfigureOptions<in TOptions> where TOptions: class
{
    void Configure(TOptions options);
}

public interface IConfigureNamedOptions<in TOptions> :
    IConfigureOptions<TOptions> where TOptions : class
{
    void Configure(string name, TOptions options);
}

public interface IPostConfigureOptions<in TOptions> where TOptions : class
{
    void PostConfigure(string name, TOptions options);
}
```

上述 3 个接口分别通过定义的 Configure 方法和 PostConfigure 方法对指定的 Options 对象进行初始化，其中，IConfigureNamedOptions<TOptions>和 IPostConfigureOptions<TOptions>还指定了 Options 的名称。由于 IConfigureOptions<TOptions>接口的 Configure 方法没有指定 Options 的名称，意味着该方法仅仅用来初始化默认的 Options 对象，而这个默认的 Options 就是以空字符串命名的 Options 对象。从接口命名就可以看出定义中的 3 个方法的执行顺序：定义在 IPostConfigureOptions<TOptions>中的 PostConfigure 方法会在 IConfigureOptions<TOptions>和 IConfigureNamedOptions<TOptions>的 Configure 方法之后执行。

当注册的 IConfigureNamedOptions<TOptions>服务和 IPostConfigureOptions<TOptions>服务完成了对 Options 对象的初始化之后，IOptionsFactory<TOptions>对象还应该验证最终得到的 Options 对象是否有效。针对 Options 对象有效性的验证由 IValidateOptions<TOptions>接口表示的服务对象来完成。如下面的代码片段所示，IValidateOptions<TOptions>接口定义的唯一的方法 Validate 用来对指定的 Options 对象（参数 options）进行验证，而参数 name 则代表 Options 的名称。

```
public interface IValidateOptions<TOptions> where TOptions: class
{
    ValidateOptionsResult Validate(string name, TOptions options);
}
```

```csharp
public class ValidateOptionsResult
{
    public static readonly ValidateOptionsResult Success;
    public static readonly ValidateOptionsResult Skip;
    public static ValidateOptionsResult Fail(string failureMessage);

    public bool          Succeeded { get; protected set; }
    public bool          Skipped { get; protected set; }
    public bool          Failed { get; protected set; }
    public string        FailureMessage { get; protected set; }
}
```

Options 的验证结果由 ValidateOptionsResult 类型表示。总的来说，针对 Options 对象的验证会产生 3 种结果，即成功、失败和忽略，它们分别通过 3 个对应的属性来表示（Succeeded、Failed 和 Skipped）。一个表示验证失败的 ValidateOptionsResult 对象会通过其 FailureMessage 属性来描述具体的验证错误。可以调用两个静态只读字段 Success 和 Skip 以及静态方法 Fail 得到或者创建对应的 ValidateOptionsResult 对象。

Options 模型提供了一个名为 OptionsFactory<TOptions>的类型作为 IOptionsFactory<TOptions>接口的默认实现。对上述 3 个接口有了基本了解后，对实现在 OptionsFactory<TOptions>类型中用来创建并初始化 Options 对象的实现逻辑就会比较容易理解。下面的代码片段基本体现了 OptionsFactory<TOptions>类型的完整定义。

```csharp
public class OptionsFactory<TOptions> :
    IOptionsFactory<TOptions> where TOptions : class, new()
{
    private readonly IEnumerable<IConfigureOptions<TOptions>>       _setups;
    private readonly IEnumerable<IPostConfigureOptions<TOptions>>   _postConfigures;
    private readonly IEnumerable<IValidateOptions<TOptions>>        _validations;

    public OptionsFactory(IEnumerable<IConfigureOptions<TOptions>> setups,
        IEnumerable<IPostConfigureOptions<TOptions>> postConfigures)
        : this(setups, postConfigures, null)
    {}

    public OptionsFactory(IEnumerable<IConfigureOptions<TOptions>> setups,
        IEnumerable<IPostConfigureOptions<TOptions>> postConfigures,
        IEnumerable<IValidateOptions<TOptions>> validations)
    {
        _setups             = setups;
        _postConfigures     = postConfigures;
        _validations        = validations;
    }

    public TOptions Create(string name)
    {
        //步骤1：实例化
        var options = new TOptions();
```

```csharp
//步骤2-1: 针对IConfigureNamedOptions<TOptions>的初始化
foreach (var setup in _setups)
{
    if (setup is IConfigureNamedOptions<TOptions> namedSetup)
    {
        namedSetup.Configure(name, options);
    }
    else if (name == Options.DefaultName)
    {
        setup.Configure(options);
    }
}

//步骤2-2: 针对IPostConfigureOptions<TOptions>的初始化
foreach (var post in _postConfigures)
{
    post.PostConfigure(name, options);
}

//步骤3: 有效性验证
var failedMessages = new List<string>();
foreach (var validator in _validations)
{
    var reusult = validator.Validate(name, options);
    if (reusult.Failed)
    {
        failedMessages.Add(reusult.FailureMessage);
    }
}
if (failedMessages.Count > 0)
{
    throw new OptionsValidationException(name, typeof(TOptions),
        failedMessages);
}
return options;
```

如上面的代码片段所示，调用构造函数创建 OptionsFactory<TOptions>对象时需要提供 IConfigureOptions<TOptions>对象、IPostConfigureOptions<TOptions>对象和 IValidateOptions <TOptions>对象。在实现的 Create 方法中，它首先调用默认构造函数创建一个空 Options 对象，再先后利用 IConfigureOptions<TOptions>对象和 IPostConfigureOptions<TOptions>对象对这个 Options 对象进行"再加工"。

这一切完成之后，指定的 IValidateOptions<TOptions>会被逐个提取出来对最终生成的 Options 对象进行验证，如果没有通过验证，就会抛出一个 OptionsValidationException 类型的异常。图 7-8 所示的 UML 展示了 OptionsFactory<TOptions>针对 Options 对象的初始化。

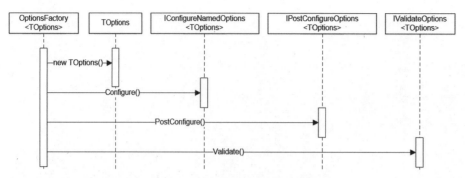

图 7-8　OptionsFactory<TOptions>针对 Options 对象的初始化

ConfigureNamedOptions<TOptions>

对于上述 3 个用来初始化 Options 对象的接口，Options 模型均提供了默认实现，其中，ConfigureNamedOptions<TOptions>类同时实现了 IConfigureOptions<TOptions>接口和 IConfigureNamedOptions<TOptions>接口。当我们创建这样一个对象时，需要指定 Options 的名称和一个用来初始化 Options 对象的 Action<TOptions>委托对象。如果指定了一个非空的名称，那么提供的委托对象将用于初始化与该名称相匹配的 Options 对象；如果指定的名称为 Null（不是空字符串），就意味着提供的初始化操作适用于所有同类的 Options 对象。

```
public class ConfigureNamedOptions<TOptions> :
    IConfigureNamedOptions<TOptions>,
    IConfigureOptions<TOptions> where TOptions : class
{
    public string              Name { get; }
    public Action<TOptions>    Action { get; }

    public ConfigureNamedOptions(string name, Action<TOptions> action)
    {
        Name   = name;
        Action = action;
    }

    public void Configure(string name, TOptions options)
    {
        if (Name == null || name == Name)
        {
            Action?.Invoke(options);
        }
    }

    public void Configure(TOptions options)
        => Configure(Options.DefaultName, options);
}
```

有时针对某个 Options 的初始化工作需要依赖另一个服务，比较典型的就是根据当前承载环境（开发、预发和产品）对某个 Options 对象做动态设置。为了解决这个问题，Options 模型提供了一个 ConfigureNamedOptions<TOptions, TDep>,其中,第二个泛型参数代表依赖的服务类型。如下面的代码片段所示,ConfigureNamedOptions<TOptions, TDep>依然是 IConfigureNamedOptions<TOptions>接口的实现类型,它利用 Action<TOptions, TDep>对象针对指定的依赖服务对 Options 做针对性初始化。

```
public class ConfigureNamedOptions<TOptions, TDep> :
    IConfigureNamedOptions<TOptions>
    where TOptions : class
    where TDep : class
{
    public string                        Name { get; }
    public Action<TOptions, TDep>        Action { get; }
    public TDep                          Dependency { get; }

    public ConfigureNamedOptions(string name, TDep dependency,
        Action<TOptions, TDep> action)
    {
        Name        = name;
        Action      = action;
        Dependency  = dependency;
    }

    public virtual void Configure(string name, TOptions options)
    {
        if (Name == null || name == Name)
        {
            Action?.Invoke(options, Dependency);
        }
    }

    public void Configure(TOptions options)
        => Configure(Options.DefaultName, options);
}
```

ConfigureNamedOptions<TOptions, TDep>仅仅实现了针对单一服务的依赖,针对 Options 的初始化可能依赖多个服务,Options 模型为此定义了如下所示的一系列类型。这些类型都实现了 IConfigureNamedOptions<TOptions>接口,并采用类似于 ConfigureNamedOptions<TOptions, TDep>类型的方式实现了 Configure 方法。

```
public class ConfigureNamedOptions<TOptions, TDep1, TDep2> :
    IConfigureNamedOptions<TOptions>
    where TOptions: class
    where TDep1: class
    where TDep2: class
{
    public string                        Name { get; }
```

```csharp
    public TDep1                                   Dependency1 { get; }
    public TDep2                                   Dependency2 { get; }
    public Action<TOptions, TDep1, TDep2>          Action { get; }

    public ConfigureNamedOptions(string name, TDep1 dependency, TDep2 dependency2,
        Action<TOptions, TDep1, TDep2> action);
    public void Configure(TOptions options);
    public virtual void Configure(string name, TOptions options);
}

public class ConfigureNamedOptions<TOptions, TDep1, TDep2, TDep3> :
    IConfigureNamedOptions<TOptions>
    where TOptions: class
    where TDep1: class
    where TDep2: class
    where TDep3: class
{
    public string                                  Name { get; }
    public TDep1                                   Dependency1 { get; }
    public TDep2                                   Dependency2 { get; }
    public TDep3                                   Dependency3 { get; }
    public Action<TOptions, TDep1, TDep2, TDep3>   Action { get; }

    public ConfigureNamedOptions(string name, TDep1 dependency, TDep2 dependency2,
        TDep3 dependency3, Action<TOptions, TDep1, TDep2, TDep3> action);
    public void Configure(TOptions options);
    public virtual void Configure(string name, TOptions options);
}

public class ConfigureNamedOptions<TOptions, TDep1, TDep2, TDep3, TDep4> :
    IConfigureNamedOptions<TOptions>
    where TOptions: class
    where TDep1: class
    where TDep2: class
    where TDep3: class
    where TDep4: class
{
    public string                                        Name { get; }
    public TDep1                                         Dependency1 { get; }
    public TDep2                                         Dependency2 { get; }
    public TDep3                                         Dependency3 { get; }
    public TDep4                                         Dependency4 { get; }
    public Action<TOptions, TDep1, TDep2, TDep3, TDep4>  Action { get; }

    public ConfigureNamedOptions(string name, TDep1 dependency, TDep2 dependency2,
        TDep3 dependency3, TDep4 dependency4,
        Action<TOptions, TDep1, TDep2, TDep3, TDep4> action);
    public void Configure(TOptions options);
```

```
    public virtual void Configure(string name, TOptions options);
}
public class ConfigureNamedOptions<TOptions, TDep1, TDep2, TDep3, TDep4, TDep5> :
    IConfigureNamedOptions<TOptions>
    where TOptions: class
    where TDep1: class
    where TDep2: class
    where TDep3: class
    where TDep4: class
    where TDep5: class
{
    public string                                                    Name { get; }
    public TDep1                                                     Dependency1 { get; }
    public TDep2                                                     Dependency2 { get; }
    public TDep3                                                     Dependency3 { get; }
    public TDep4                                                     Dependency4 { get; }
    public TDep5                                                     Dependency5 { get; }
    public Action<TOptions, TDep1, TDep2, TDep3, TDep4, TDep5>       Action { get; }

    public ConfigureNamedOptions(string name, TDep1 dependency, TDep2 dependency2,
        TDep3 dependency3, TDep4 dependency4, TDep5 dependency5
        Action<TOptions, TDep1, TDep2, TDep3, TDep4, TDep5> action);
    public void Configure(TOptions options);
    public virtual void Configure(string name, TOptions options);
}
```

PostConfigureOptions<TOptions>

默认实现 IPostConfigureOptions<TOptions>接口的是 PostConfigureOptions<TOptions>类型。从给出的代码片段可以看出，它针对 Options 对象的初始化实现方式与 ConfigureNamedOptions<TOptions>类型并没有本质上的差别。

```
public class PostConfigureOptions<TOptions> :
    IPostConfigureOptions<TOptions> where TOptions : class
{
    public string              Name { get; }
    public Action<TOptions>    Action { get; }

    public PostConfigureOptions (string name, Action<TOptions> action)
    {
        Name      = name;
        Action    = action;
    }

    public void PostConfigure(string name, TOptions options)
    {
        if (Name == null || name == Name)
        {
            Action?.Invoke(options);
```

```
    }
}
```

Options 模型同样定义了如下这一系列针对依赖服务的 IPostConfigureOptions<TOptions>接口实现。如果针对 Options 对象的后置初始化操作依赖于其他服务，就可以根据服务的数量选择对应的类型。这些类型针对 PostConfigure 方法的实现与 ConfigureNamedOptions<TOptions, TDep>类型实现 Configure 方法并没有本质区别。

- PostConfigureOptions<TOptions, TDep>。
- PostConfigureOptions<TOptions, TDep1, TDep2>。
- PostConfigureOptions<TOptions, TDep1, TDep2, TDep3>。
- PostConfigureOptions<TOptions, TDep1, TDep2, TDep3, TDep4>。
- PostConfigureOptions<TOptions, TDep1, TDep2, TDep3, TDep4, TDep5>。

ValidateOptions<TOptions>

ValidateOptions<TOptions>是对 IValidateOptions<TOptions>接口的默认实现。如下面的代码片段所示，创建一个 ValidateOptions<TOptions>对象时，需要提供 Options 的名称和验证错误消息，以及真正用于对 Options 进行验证的 Func<TOptions, bool>对象。

```
public class ValidateOptions<TOptions> : IValidateOptions<TOptions>
    where TOptions: class
{
    public string                    Name { get; }
    public string                    FailureMessage { get; }
    public Func<TOptions, bool>      Validation { get; }

    public ValidateOptions(string name, Func<TOptions, bool> validation,
        string failureMessage);

    public ValidateOptionsResult Validate(string name, TOptions options);
}
```

对 Options 的验证同样可能具有对其他服务的依赖，比较典型的依然是针对不同的承载环境（开发、预发和产品）具有不同的验证规则，所以 IValidateOptions<TOptions>接口同样具有如下 5 个针对不同依赖服务数量的实现类型。

- ValidateOptions<TOptions, TDep>。
- ValidateOptions<TOptions, TDep1, TDep2>。
- ValidateOptions<TOptions, TDep1, TDep2, TDep3>。
- ValidateOptions<TOptions, TDep1, TDep2, TDep3, TDep4>。
- ValidateOptions<TOptions, TDep1, TDep2, TDep3, TDep4, TDep5>。

前面介绍了 OptionsFactory<TOptions>类型针对 Options 对象的创建和初始化的实现原理，以及涉及的一些相关的接口和类型，图 7-9 基本上反映了这些接口与类型的关系。

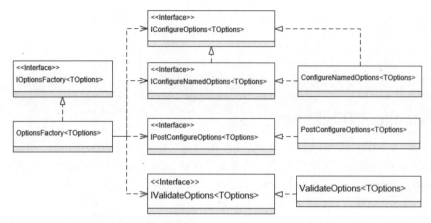

图 7-9　OptionsFactory<TOptions>

7.2.3　IOptionsMonitorCache<TOptions>

　　IOptionsFactory<TOptions>解决了 Options 的创建与初始化问题，但由于它自身是无状态的，所以 Options 模型对 Options 对象实施缓存可以获得更好的性能。Options 模型中针对 Options 对象的缓存由 IOptionsMonitorCache<TOptions>对象来完成，如下所示的代码片段是该接口的定义。

```
public interface IOptionsMonitorCache<TOptions> where TOptions : class
{
    TOptions GetOrAdd(string name, Func<TOptions> createOptions);
    bool TryAdd(string name, TOptions options);
    bool TryRemove(string name);
    void Clear();
}
```

　　由于 Options 模型总是根据名称来提供对应的 Options 对象，所以 IOptionsMonitorCache<TOptions>对象也根据名称来缓存 Options 对象。如上面的代码片段所示，IOptionsMonitorCache<TOptions>接口提供了 4 个方法，分别实现针对 Options 缓存的获取、添加、移除和清理。

　　IOptionsMonitorCache<TOptions>接口的默认实现是前面提到的 OptionsCache<TOptions>类型，OptionsManager 对象会将其作为自身的"私有"缓存。实现在 OptionsCache<TOptions>类型中针对 Options 对象的缓存逻辑其实很简单：它仅仅使用一个 ConcurrentDictionary<string, Lazy<TOptions>>对象作为缓存 Options 的容器而已。如下所示的代码片段基本上体现了 OptionsCache<TOptions>类型的实现逻辑。

```
public class OptionsCache<TOptions> :
    IOptionsMonitorCache<TOptions>
    where TOptions : class
{
    private readonly ConcurrentDictionary<string, Lazy<TOptions>> _cache =
        new ConcurrentDictionary<string, Lazy<TOptions>>(StringComparer.Ordinal);
    public void Clear() => _cache.Clear();
    public virtual TOptions GetOrAdd(string name, Func<TOptions> createOptions)
```

```csharp
        => _cache.GetOrAdd(name, new Lazy<TOptions>(createOptions)).Value;
    public virtual bool TryAdd(string name, TOptions options)
        => _cache.TryAdd(name, new Lazy<TOptions>(() => options));
    public virtual bool TryRemove(string name)
        => _cache.TryRemove(name, out var ignored);
}
```

7.2.4　IOptionsMonitor<TOptions>

Options 模型之所以将表示缓存的接口命名为 IOptionsMonitorCache<TOptions>，是因为缓存最初是为 IOptionsMonitor<TOptions>对象服务的，该对象旨在实现针对承载 Options 对象的原始数据源的监控，并在检测到数据更新后及时替换缓存的 Options 对象。

```csharp
public interface IOptionsMonitor<out TOptions>
{
    TOptions CurrentValue { get; }
    TOptions Get(string name);
    IDisposable OnChange(Action<TOptions, string> listener);
}
```

除了直接调用定义在 IOptionsMonitor<TOptions>接口中的 OnChange 方法注册应用新 Options 对象的回调，还可以调用如下这个同名的扩展方法。通过 OnChange 方法注册的回调是一个类型为 Action<TOptions>的委托对象，由于缺少输出参数来区分 Options 的名称，所以注册的回调适用于所有的 Options 对象。值得一提的是，这两个 OnChange 方法的返回类型为 IDisposable，实际上代表了针对回调的注册，我们可以调用返回对象的 Dispose 方法解除注册。

```csharp
public static class OptionsMonitorExtensions
{
    public static IDisposable OnChange<TOptions>(
        this IOptionsMonitor<TOptions> monitor, Action<TOptions> listener)
        => monitor.OnChange((o, _) => listener(o));
}
```

.NET Core 应用在进行数据变化监控时总是使用一个 IChangeToken 对象来发送通知，用于监控 Options 数据变化的 IOptionsMonitor<TOptions>对象自然也不例外。IOptionsMonitor<TOptions>对象在检测到数据变化后用于对外发送通知的 IChangeToken 对象是由一个 IOptionsChangeTokenSource<TOptions>对象完成的。IOptionsChangeTokenSource<TOptions>接口的 Name 属性表示 Options 的名称，而前面所说的 IChangeToken 对象由其 GetChangeToken 方法来提供。

```csharp
public interface IOptionsChangeTokenSource<out TOptions>
{
    string Name { get; }
    IChangeToken GetChangeToken();
}
```

Options 模型定义了如下这个 OptionsMonitor<TOptions>类型作为对 IOptionsMonitor<TOptions>接口的默认实现。当调用构造函数创建一个 OptionsMonitor<TOptions>对象时需要提供一个用来创建和初始化 Options 对象的 IOptionsFactory<TOptions>对象，一个用来对提供的 Options 对象实施缓存的IOptionsMonitorCache<TOptions>对象，以及一组用来检测配置选项数据变化并对外发送通知的IOptionsChangeTokenSource<TOptions>对象。

```csharp
public class OptionsMonitor<TOptions> :
    IOptionsMonitor<TOptions> where TOptions : class, new()
{
    private readonly IOptionsMonitorCache<TOptions>                       _cache;
    private readonly IOptionsFactory<TOptions>                            _factory;
    private readonly IEnumerable<IOptionsChangeTokenSource<TOptions>>     _sources;
    internal event Action<TOptions, string>                               _onChange;

    public OptionsMonitor(
        IOptionsFactory<TOptions> factory,
        IEnumerable<IOptionsChangeTokenSource<TOptions>> sources,
        IOptionsMonitorCache<TOptions> cache)
    {
        _factory    = factory;
        _sources    = sources;
        _cache      = cache;

        foreach (var source in _sources)
        {
            ChangeToken.OnChange<string>(
                () => source.GetChangeToken(),
                (name) => InvokeChanged(name),
                source.Name);
        }
    }

    private void InvokeChanged(string name)
    {
        name = name ?? Options.DefaultName;
        _cache.TryRemove(name);
        var options = Get(name);
        if (_onChange != null)
        {
            _onChange.Invoke(options, name);
        }
    }

    public TOptions CurrentValue { get => Get(Options.DefaultName);}

    public virtual TOptions Get(string name)
        =>_cache.GetOrAdd(name, () => _factory.Create(name));

    public IDisposable OnChange(Action<TOptions, string> listener)
    {
        var disposable = new ChangeTrackerDisposable(this, listener);
        _onChange += disposable.OnChange;
        return disposable;
```

```csharp
}

internal class ChangeTrackerDisposable : IDisposable
{
    private readonly Action<TOptions, string> _listener;
    private readonly OptionsMonitor<TOptions> _monitor;

    public ChangeTrackerDisposable(OptionsMonitor<TOptions> monitor,
        Action<TOptions, string> listener)
    {
        _listener = listener;
        _monitor = monitor;
    }

    public void OnChange(TOptions options, string name)
        => _listener.Invoke(options, name);
    public void Dispose() => _monitor._onChange -= OnChange;
}
}
```

由于 OptionsMonitor<TOptions>对象提供的 Options 对象总是来源于 IOptionsMonitorCache<TOptions>对象表示的缓存容器，所以它只需要利用提供的 IOptionsChangeTokenSource 对象来监控 Options 数据的变化，并在检测到变化之后及时删除缓存中对应的 Options 对象，这样就能保证其 CurrentValue 属性和 Get 方法返回的总是最新的 Options 数据，这样的逻辑反映在上面给出的代码片段中。

7.3 依赖注入

上面介绍了组成 Options 模型的 4 个核心对象以及它们之间的交互关系，读者对如何得到 Options 对象的实现原理可能不太了解，本节主要介绍依赖注入的相关内容。既然我们能够利用 IServiceProvider 对象提供的 IOptions<TOptions>服务、IOptionsSnapshot<TOptions>服务和 IOptionsMonitorCache<TOptions>服务来获取对应的 Options 对象，那么在这之前必然需要注册相应的服务。

7.3.1 服务注册

回顾 7.1 节演示的几个实例可以发现，Options 模式涉及的 API 其实不是很多，大都集中在相关服务的注册上。Options 模型的核心服务实现在 IServiceCollection 接口的 AddOptions 扩展方法中。

AddOptions

AddOptions 扩展方法的完整定义如下所示，由此可知，该方法将 Options 模型中的几个核心类型作为服务注册到了指定的 IServiceCollection 对象之中。由于它们都是调用 TryAdd 方法进行服务注册的，所以我们可以在需要 Options 模式支持的情况下调用 AddOptions 方法，而不需要担心是否会添加太多重复服务注册的问题。
```
public static class OptionsServiceCollectionExtensions
{
    public static IServiceCollection AddOptions(this IServiceCollection services)
    {
        services.TryAdd(ServiceDescriptor.Singleton(
            typeof(IOptions<>), typeof(OptionsManager<>)));
        services.TryAdd(ServiceDescriptor.Scoped(
            typeof(IOptionsSnapshot<>), typeof(OptionsManager<>)));
        services.TryAdd(ServiceDescriptor.Singleton(
            typeof(IOptionsMonitor<>), typeof(OptionsMonitor<>)));
        services.TryAdd(ServiceDescriptor.Transient(
            typeof(IOptionsFactory<>), typeof(OptionsFactory<>)));
        services.TryAdd(ServiceDescriptor.Singleton(
            typeof(IOptionsMonitorCache<>), typeof(OptionsCache<>)));
        return services;
    }
}
```

从给出的代码片段可以看出，**AddOptions** 扩展方法实际上注册了 5 个服务。由于这 5 个服务注册非常重要，所以笔者采用表格的形式列出了它们的 Service Type（服务接口）、Implementation（实现类型）和 Lifetime（生命周期）（见表 7-1）。虽然服务接口 IOptions<TOptions>和 IOptionsSnapshot<TOptions>映射的实现类型都是 OptionsManager<TOptions>，但是它们具有不同的生命周期。具体来说，前者的生命周期为 Singleton，后者的生命周期则是 Scoped，后续内容会单独讲述不同生命周期对 Options 对象产生什么样的影响。

表 7-1　常见的几种文件匹配模式表达式

Service Type	Implementation	Lifetime
IOptions<TOptions>	OptionsManager<TOptions>	Singleton
IOptionsSnapshot<TOptions>	OptionsManager<TOptions>	Scoped
IOptionsMonitor<TOptions>	OptionsMonitor<TOptions>	Singleton
IOptionsFactory<TOptions>	OptionsFactory<TOptions>	Transient
IOptionsMonitorCache<TOptions>	OptionsCache<TOptions>	Singleton

按照表 7-1 列举的服务注册，如果以 IOptions<TOptions>和 IOptionsSnapshot<TOptions>作为服务类型从 IServiceProvider 对象中提取对应的服务实例，得到的都是 OptionsManager<TOptions>对象。当 OptionsManager<TOptions>对象被创建时，OptionsFactory<TOptions>对象会被自动创建出来并以构造器注入的方式提供给它并且被用来创建 Options 对象。但是由于表 7-1 中并没有针对服务 IConfigureOptions<TOptions>和 IPostConfigureOptions<TOptions>的注册，所以创建的 Options 对象无法被初始化。

Configure<TOptions>与 PostConfigure<TOptions>

针对 IConfigureOptions<TOptions>和 IPostConfigureOptions<TOptions>的服务注册是通过如下这些扩展方法来完成的。具体来说，针对 IConfigureOptions<TOptions>服务的注册实现在 Configure<TOptions>方法中，而 PostConfigure<TOptions>扩展方法则帮助我们完成针对 IPost

ConfigureOptions<TOptions>的注册。

```
public static class OptionsServiceCollectionExtensions
{
    public static IServiceCollection Configure<TOptions>(
        this IServiceCollection services, Action<TOptions> configureOptions)
        where TOptions : class
        => services.Configure(Options.Options.DefaultName, configureOptions);

    public static IServiceCollection Configure<TOptions>(
        this IServiceCollection services, string name,
        Action<TOptions> configureOptions) where TOptions : class
        => services.AddSingleton<IConfigureOptions<TOptions>>(
        new ConfigureNamedOptions<TOptions>(name, configureOptions));
        return services;

    public static IServiceCollection PostConfigure<TOptions>(
        this IServiceCollection services, Action<TOptions> configureOptions)
        where TOptions : class
        => services.PostConfigure(Options.Options.DefaultName, configureOptions);

    public static IServiceCollection PostConfigure<TOptions>(
        this IServiceCollection services, string name,
        Action<TOptions> configureOptions) where TOptions : class
        => services.AddSingleton<IPostConfigureOptions<TOptions>>(
        new PostConfigureOptions<TOptions>(name, configureOptions));
}
```

从上述代码可以看出，这些方法注册的服务实现类型为 ConfigureNamedOptions<TOptions>和 PostConfigureOptions<TOptions>，采用的生命周期模式均为 Singleton。不论是 ConfigureNamedOptions<TOptions>还是 PostConfigureOptions<TOptions>，都需要指定一个具体的名称，对于没有指定具体 Options 名称的 Configure<TOptions>和 PostConfigure<TOptions>方法重载来说，最终指定的是代表默认名称的空字符串。

ConfigureAll<TOptions>与 PostConfigureAll<TOptions>

虽然 ConfigureAll<TOptions>和 PostConfigureAll<TOptions>扩展方法注册的同样是 ConfigureNamedOptions<TOptions>类型与 PostConfigureOptions<TOptions>类型，但是它们会将名称设置为 Null。通过前面介绍的内容可知，OptionsFactory 对象在 Options 对象进行初始化的过程中会将名称为 Null 的 IConfigureNamedOptions<TOptions>对象和 IPostConfigureOptions<TOptions>对象作为公共的配置对象，并且无条件执行。

```
public static class OptionsServiceCollectionExtensions
{
    public static IServiceCollection ConfigureAll<TOptions>(
        this IServiceCollection services, Action<TOptions> configureOptions)
        where TOptions : class
        => services.Configure(name: null, configureOptions: configureOptions);
```

```csharp
    public static IServiceCollection PostConfigureAll<TOptions>(
        this IServiceCollection services, Action<TOptions> configureOptions)
        where TOptions : class
        => services.PostConfigure(name: null, configureOptions: configureOptions);
}
```

ConfigureOptions

对于上面这几个将 Options 类型作为泛型参数的方法来说，它们总是利用指定的 Action<Options>对象来创建注册的 ConfigureNamedOptions<TOptions>对象和 PostConfigureOptions <TOptions>对象。对于自定义的实现了 IConfigureOptions<TOptions>接口或者 IPostConfigureOptions <TOptions>接口的类型，我们可以调用如下所示的 3 个 ConfigureOptions 扩展方法来对它们进行注册。笔者在如下所示的代码片段中通过简化的代码描述了这 3 个扩展方法的实现逻辑。

```csharp
public static class OptionsServiceCollectionExtensions
{
    public static IServiceCollection ConfigureOptions(
        this IServiceCollection services, object configureInstance)
    {
        Array.ForEach(FindIConfigureOptions(configureInstance.GetType()),
            it => services.AddSingleton(it, configureInstance));
        return services;
    }

    public static IServiceCollection ConfigureOptions(
        this IServiceCollection services, Type configureType)
    {
        Array.ForEach(FindIConfigureOptions(configureType),
            it => services.AddTransient(it, configureType));
        return services;
    }

    public static IServiceCollection ConfigureOptions<TConfigureOptions>(
        this IServiceCollection services) where TConfigureOptions : class
        => services.ConfigureOptions(typeof(TConfigureOptions));

    private static Type[] FindIConfigureOptions(Type type)
    {
        Func<Type, bool> valid = it =>
            it.IsGenericType &&
            (it.GetGenericTypeDefinition() == typeof(IConfigureOptions<>) ||
            it.GetGenericTypeDefinition() == typeof(IPostConfigureOptions<>));
        var types = type.GetInterfaces()
            .Where(valid)
            .ToArray();
        if (types.Any())
        {
            throw new InvalidOperationException();
        }
        return types;
```

```
        }
}
```

OptionsBuilder<TOptions>

Options 模式涉及非常多的服务注册，并且这些服务都是针对具体某个 Options 类型的，为了避免定义过多针对 IServiceCollection 接口的扩展方法，最新版本的 Options 模型采用 Builder 模式来完成相关的服务注册。具体来说，可以将用来存储服务注册的 IServiceCollection 集合封装到下面的 OptionsBuilder<TOptions>对象中，并利用它提供的方法间接地完成所需的服务注册。

```
public class OptionsBuilder<TOptions> where TOptions: class
{
    public string Name { get; }
    public IServiceCollection Services { get; }
    public OptionsBuilder(IServiceCollection services, string name);

    public virtual OptionsBuilder<TOptions> Configure(
        Action<TOptions> configureOptions);
    public virtual OptionsBuilder<TOptions> Configure<TDep>(
        Action<TOptions, TDep> configureOptions) where TDep: class;
    public virtual OptionsBuilder<TOptions> Configure<TDep1, TDep2>(
        Action<TOptions, TDep1, TDep2> configureOptions)
        where TDep1: class where TDep2: class;
    public virtual OptionsBuilder<TOptions>
        Configure<TDep1, TDep2, TDep3>(
        Action<TOptions, TDep1, TDep2, TDep3> configureOptions)
        where TDep1: class where TDep2: class where TDep3: class;
    public virtual OptionsBuilder<TOptions>
        Configure<TDep1, TDep2, TDep3, TDep4>(
        Action<TOptions, TDep1, TDep2, TDep3, TDep4> configureOptions)
        where TDep1: class where TDep2: class where TDep3: class where TDep4: class;
    public virtual OptionsBuilder<TOptions>
        Configure<TDep1, TDep2, TDep3, TDep4, TDep5>(
        Action<TOptions, TDep1, TDep2, TDep3, TDep4, TDep5> configureOptions)
        where TDep1: class where TDep2: class where TDep3: class
        where TDep4: class where TDep5: class;

    public virtual OptionsBuilder<TOptions> PostConfigure (
        Action<TOptions> configureOptions);
    public virtual OptionsBuilder<TOptions> PostConfigure <TDep>(
        Action<TOptions, TDep> configureOptions) where TDep: class;
    public virtual OptionsBuilder<TOptions> PostConfigure <TDep1, TDep2>(
        Action<TOptions, TDep1, TDep2> configureOptions)
        where TDep1: class where TDep2: class;
    public virtual OptionsBuilder<TOptions>
        PostConfigure <TDep1, TDep2, TDep3>(
        Action<TOptions, TDep1, TDep2, TDep3> configureOptions)
        where TDep1: class where TDep2: class where TDep3: class;
    public virtual OptionsBuilder<TOptions>
        PostConfigure <TDep1, TDep2, TDep3, TDep4>(
```

```
        Action<TOptions, TDep1, TDep2, TDep3, TDep4> configureOptions)
        where TDep1: class where TDep2: class where TDep3: class where TDep4: class;
    public virtual OptionsBuilder<TOptions>
        PostConfigure <TDep1, TDep2, TDep3, TDep4, TDep5>(
        Action<TOptions, TDep1, TDep2, TDep3, TDep4, TDep5> configureOptions)
        where TDep1: class where TDep2: class where TDep3: class
        where TDep4: class where TDep5: class;

    public virtual OptionsBuilder<TOptions> Validate(
        Func<TOptions, bool> validation);
    public virtual OptionsBuilder<TOptions> Validate<TDep>(
        Func<TOptions, TDep, bool> validation);
    public virtual OptionsBuilder<TOptions> Validate<TDep1, TDep2>(
        Func<TOptions, TDep1, TDep2, bool> validation);
    public virtual OptionsBuilder<TOptions>
        Validate<TDep1, TDep2, TDep3>(Func<TOptions, TDep1, TDep2, TDep3, bool>
        validation);
    public virtual OptionsBuilder<TOptions>
        Validate<TDep1, TDep2, TDep3, TDep4>(
        Func<TOptions, TDep1, TDep2, TDep3, TDep4, bool> validation);
    public virtual OptionsBuilder<TOptions>
        Validate<TDep1, TDep2, TDep3, TDep4, TDep5>(
        Func<TOptions, TDep1, TDep2, TDep3, TDep4, TDep5, bool> validation);

    public virtual OptionsBuilder<TOptions> Validate<TDep>(
        Func<TOptions, TDep, bool> validation, string failureMessage);
    public virtual OptionsBuilder<TOptions> Validate<TDep1, TDep2>(
        Func<TOptions, TDep1, TDep2, bool> validation, string failureMessage);
    public virtual OptionsBuilder<TOptions>
        Validate<TDep1, TDep2, TDep3>(
        Func<TOptions, TDep1, TDep2, TDep3, bool> validation,
        string failureMessage);
    public virtual OptionsBuilder<TOptions>
        Validate<TDep1, TDep2, TDep3, TDep4>(
        Func<TOptions, TDep1, TDep2, TDep3, TDep4, bool> validation,
        string failureMessage);
    public virtual OptionsBuilder<TOptions>
        Validate<TDep1, TDep2, TDep3, TDep4, TDep5>(
        Func<TOptions, TDep1, TDep2, TDep3, TDep4, TDep5, bool> validation,
        string failureMessage);
}
```

如下面的代码片段所示，OptionsBuilder<TOptions>对象不仅通过泛型参数关联对应的 Options 类型，还利用 Name 属性提供了 Options 的名称。从上面的代码片段可以看出，OptionsBuilder<TOptions>类型提供的 3 组方法分别提供了针对 IConfigureOptions<TOptions>接口、IPostConfigureOptions<TOptions>接口和 IValidateOptions<TOptions>接口的 18 个实现类型的注册。

当利用 Builder 模式来注册这些服务的时候，只需要调用 IServiceCollection 接口的如下这两个 AddOptions<TOptions>扩展方法根据指定的名称（默认名称为空字符串）创建出对应的 OptionsBuilder<TOptions>对象即可。从如下所示的代码片段可以看出，这两个方法最终都需要调用非泛型的 AddOptions 方法，由于该方法调用 TryAdd 扩展方法注册 Options 模式的 5 个核心服务，所以不会导致服务的重复注册。

```csharp
public static class OptionsServiceCollectionExtensions
{
    public static OptionsBuilder<TOptions> AddOptions<TOptions>(
        this IServiceCollection services) where TOptions: class =>
        services.AddOptions<TOptions>(Options.DefaultName);

    public static OptionsBuilder<TOptions> AddOptions<TOptions>(
        this IServiceCollection services, string name) where TOptions: class
    {
        services.AddOptions();
        return new OptionsBuilder<TOptions>(services, name);
    }
}
```

7.3.2　IOptions<TOptions>与 IOptionsSnapshot<TOptions>

通过对注册服务的分析可知，服务接口 IOptions<TOptions>和 IOptionsSnapshot<TOptions>的默认实现类型都是 OptionsManager<TOptions>，两者的不同之处体现在生命周期上，前者采用的生命周期模式为 Singleton，后者采用的生命周期模式则是 Scoped。

对于一个 ASP.NET Core 应用来说，Singleton 和 Scoped 对应的是针对当前应用和当前请求的生命周期，所以通过 IOptions<TOptions>接口获取的 Options 对象在整个应用的生命周期内都是一致的，而通过 IOptionsSnapshot<TOptions>接口获取的 Options 对象则只能在当前请求上下文中保持一致。这也是后者命名的由来，它表示针对当前请求的 Options 快照。

下面通过一个实例来演示 IOptions<TOptions>和 IOptionsSnapshot<TOptions>之间的差异。如下代码片段定义了 FoobarOptions 类型，简单起见，我们仅仅为它定义了两个整型的属性（Foo 和 Bar），并重写了 ToString 方法。

```csharp
public class FoobarOptions
{
    public int Foo { get; set; }
    public int Bar { get; set; }
    public override string ToString() => $"Foo:{Foo}, Bar:{Bar}";
}
```

整个演示程序体现在如下所示的代码片段中。我们创建了一个 ServiceCollection 对象，在调用 AddOptions 扩展方法注册 Options 模型的基础服务之后，调用 Configure<FoobarOptions>方法利用定义的本地函数 Print 将 FoobarOptions 对象的 Foo 属性和 Bar 属性设置为一个随机数。

```csharp
class Program
{
    static void Main()
    {
```

```csharp
var random = new Random();
var serviceProvider = new ServiceCollection()
    .AddOptions()
    .Configure<FoobarOptions>(foobar =>
    {
        foobar.Foo = random.Next(1, 100);
        foobar.Bar = random.Next(1, 100);
    })
    .BuildServiceProvider();

Print(serviceProvider);
Print(serviceProvider);

static void Print(IServiceProvider provider)
{
    var scopedProvider = provider
        .GetRequiredService<IServiceScopeFactory>()
        .CreateScope()
        .ServiceProvider;

    var options = scopedProvider
        .GetRequiredService<IOptions<FoobarOptions>>()
        .Value;
    var optionsSnapshot1 = scopedProvider
        .GetRequiredService<IOptionsSnapshot<FoobarOptions>>()
        .Value;
    var optionsSnapshot2 = scopedProvider
        .GetRequiredService<IOptionsSnapshot<FoobarOptions>>()
        .Value;
    Console.WriteLine($"options:{options}");
    Console.WriteLine($"optionsSnapshot1:{optionsSnapshot1}");
    Console.WriteLine($"optionsSnapshot2:{optionsSnapshot2}\n");
}
```

我们并没有直接利用 ServiceCollection 对象创建的 IServiceProvider 对象来提供服务，而是利用它创建了一个代表子容器的 IServiceProvider 对象，该对象就相当于 ASP.NET Core 应用中针对当前请求创建的 IServiceProvider 对象（RequestServices）。在利用这个 IServiceProvider 对象分别针对 IOptions<TOptions>接口和 IOptionsSnapshot<TOptions>接口得到对应的 FoobarOptions 对象之后，我们将配置选项输出到控制台上。上述操作先后执行了两次，相当于 ASP.NET Core 应用分别处理了两次请求。

图 7-10 展示了该演示程序执行后的输出结果，由此可知，只有从同一个 IServiceProvider 对象获取的 IOptionsSnapshot<TOptions>服务才能提供一致的 Options 对象，但是对于所有源自同一个根的所有 IServiceProvider 对象来说，从中提取的 IOptions<TOptions>服务都能提供一致的 Options 对象。（S709）

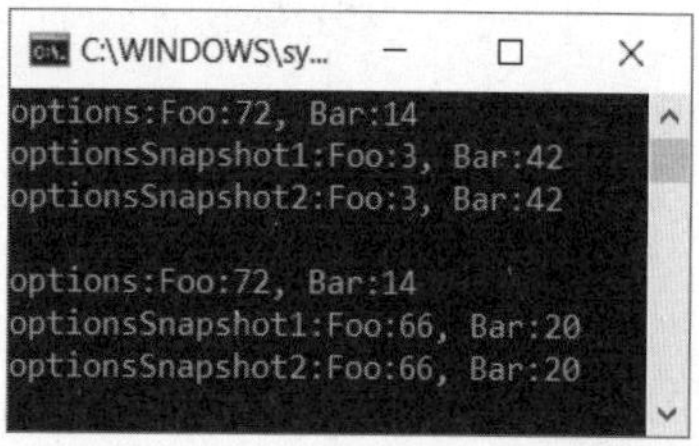

图7-10　IOptions<TOptions>和IOptionsSnapshot<TOptions>的差异

OptionsManager<Options>会利用一个自行创建的 OptionsCache<TOptions>对象来缓存Options 对象，也就是说，OptionsManager<Options>提供的 Options 对象存放在其私有缓存中。虽然 OptionsCache<TOptions>提供了清除缓存的能力，但是 OptionsManager<Options>自身无法感知原始Options 数据是否发生变化，所以不会清除缓存的 Options 对象。

这个特性决定了在一个 ASP.NET Core 应用中，以 IOptions<TOptions>服务的形式提供的Options 在整个应用的生命周期内不会发生改变，但是若使用 IOptionsSnapshot<TOptions>服务，提供的 Options 对象只能在同一个请求上下文中提供一致的保障。如果希望即使在同一个请求处理周期内也能及时应用最新的 Options 属性，就只能使用 IOptionsMonitor<TOptions>服务来提供Options 对象。

7.3.3　扩展与定制

由于 Options 模型涉及的核心对象最终都注册为相应的服务，所以从原则上讲这些对象都是可以定制的，下面提供几个这样的实例。由于 Options 模型提供了针对配置系统的集成，所以可以采用配置文件的形式来提供原始的 Options 数据，可以直接采用反序列化的方式将配置文件的内容转换成 Options 对象。

自定义 IConfigureOptions

在介绍 IConfigureOptions 扩展的实现之前，下面先演示如何在应用中使用它。首先在演示实例中定义一个 Options 类型。简单起见，我们沿用前面使用的包含两个成员的 FoobarOptions类型，从而实现 IEquatable<FoobarOptions>接口。最终绑定生成的是一个 FakeOptions 对象，为了演示针对复合类型、数组、集合和字典类型的绑定，可以为其定义相应的属性成员。

```
public class FakeOptions
{
    public FoobarOptions                        Foobar{ get; set; }
    public FoobarOptions[]                      Array { get; set; }
    public IList<FoobarOptions>                 List { get; set; }
    public IDictionary<string, FoobarOptions>   Dictionary { get; set; }
}

public class FoobarOptions: IEquatable<FoobarOptions>
{
    public int Foo { get; set; }
    public int Bar { get; set; }
```

```
    public FoobarOptions() {}
    public FoobarOptions(int foo, int bar)
    {
        Foo = foo;
        Bar = bar;
    }

    public override string ToString() => $"Foo:{Foo}, Bar:{Bar}";
    public bool Equals(FoobarOptions other)
        => this.Foo == other?.Foo && this.Bar == other?.Bar;
}
```

可以在项目根目录添加一个 JSON 文件（命名为 fakeoptions.json），如下所示的代码片段表示该文件的内容，可以看出文件的格式与 FakeOptions 类型的数据成员是兼容的，也就是说，这个文件的内容能够被反序列化成一个 FakeOptions 对象。

```
{
    "Foobar": {
        "Foo": 1,
        "Bar": 1
    },
    "Array": [{
            "Foo": 1,
            "Bar": 1
        },
        {
            "Foo": 2,
            "Bar": 2
        },
        {
            "Foo": 3,
            "Bar": 3
        }],
    "List": [{
            "Foo": 1,
            "Bar": 1
        },
        {
            "Foo": 2,
            "Bar": 2
        },
        {
            "Foo": 3,
            "Bar": 3
        }],
    "Dictionary": {
        "1": {
            "Foo": 1,
            "Bar": 1
        },
```

```
        "2": {
            "Foo": 2,
            "Bar": 2
        },
        "3": {
            "Foo": 3,
            "Bar": 3
        }
    }
}
```

下面按照 Options 模式直接读取该配置文件，并将文件内容绑定为一个 FakeOptions 对象。如下面的代码片段所示，在调用 IServiceCollection 接口的 AddOptions 扩展方法之后，我们调用了另一个自定义的 Configure<FakeOptions>扩展方法，该方法的参数表示承载原始 Options 数据的 JSON 文件的路径。这个演示程序提供的一系列调试断言表明：最终获取的 FakeOptions 对象与原始的 JSON 文件具有一致的内容。（S710）

```
class Program
{
    static void Main()
    {
        var foobar1 = new FoobarOptions(1, 1);
        var foobar2 = new FoobarOptions(2, 2);
        var foobar3 = new FoobarOptions(3, 3);

        var options = new ServiceCollection()
            .AddOptions()
            .Configure<FakeOptions>("fakeoptions.json")
            .BuildServiceProvider()
            .GetRequiredService<IOptions<FakeOptions>>()
            .Value;

        Debug.Assert(options.Foobar.Equals(foobar1));

        Debug.Assert(options.Array[0].Equals(foobar1));
        Debug.Assert(options.Array[1].Equals(foobar2));
        Debug.Assert(options.Array[2].Equals(foobar3));

        Debug.Assert(options.List[0].Equals(foobar1));
        Debug.Assert(options.List[1].Equals(foobar2));
        Debug.Assert(options.List[2].Equals(foobar3));

        Debug.Assert(options.Dictionary["1"].Equals(foobar1));
        Debug.Assert(options.Dictionary["2"].Equals(foobar2));
        Debug.Assert(options.Dictionary["3"].Equals(foobar3));
    }
}
```

Options 模型中针对 Options 对象的初始化是通过 IConfigureOptions<TOptions>对象实现的，演示程序中调用的 Configure<TOptions>方法实际上就是注册了这样一个服务。我们采用

Newtonsoft.Json 来完成针对 JSON 的序列化，并且使用基于物理文件系统的 IFileProvider 来读取文件。Configure<TOptions>方法注册的实际上就是如下这个 JsonFileConfigureOptions<TOptions>类型。JsonFileConfigureOptions<TOptions>实现了 IConfigureNamedOptions<TOptions>接口，在调用构造函数创建一个 JsonFileConfigureOptions<TOptions>对象的时候，我们指定了 Options 名称、JSON 文件的路径以及用于读取该文件的 IFileProvider 对象。

```
public class JsonFileConfigureOptions<TOptions> :
    IConfigureNamedOptions<TOptions> where TOptions : class, new()
{
    private readonly IFileProvider    _fileProvider;
    private readonly string           _path;
    private readonly string           _name;

    public JsonFileConfigureOptions(string name, string path,
        IFileProvider fileProvider)
    {
        _fileProvider = fileProvider;
        _path = path;
        _name = name;
    }

    public void Configure(string name, TOptions options)
    {
        if (name != null && _name != name)
        {
            return;
        }

        byte[] bytes;
        using (var stream = _fileProvider.GetFileInfo(_path).CreateReadStream())
        {
            bytes = new byte[stream.Length];
            stream.Read(bytes, 0, bytes.Length);
        }

        var contents = Encoding.Default.GetString(bytes);
        contents = contents.Substring(contents.IndexOf('{'));
        var newOptions = JsonConvert.DeserializeObject<TOptions>(contents);
        Bind(newOptions, options);
    }

    public void Configure(TOptions options) => Configure(Options.DefaultName,
        options);

    private void Bind(object from, object to)
    {
        var type = from.GetType();
        if (type.IsDictionary())
        {
```

```csharp
    var dest = (IDictionary)to;
    var src = (IDictionary)from;
    foreach (var key in src.Keys)
    {
        dest.Add(key, src[key]);
    }
    return;
}

if (type.IsCollection())
{
    var dest = (IList)to;
    var src = (IList)from;
    foreach (var item in src)
    {
        dest.Add(item);
    }
}

foreach (var property in type.GetProperties())
{
    if (property.IsSpecialName || property.GetMethod == null ||
        property.Name == "Item" || property.DeclaringType != type)
    {
        continue;
    }

    var src = property.GetValue(from);
    var propertyType = src?.GetType() ?? property.PropertyType;

    if ((propertyType.IsValueType || src is string || src == null)
        && property.SetMethod != null)
    {
        property.SetValue(to, src);
        continue;
    }

    var dest = property.GetValue(to);
    if (null != dest && !propertyType.IsArray())
    {
        Bind(src, dest);
        continue;
    }

    if (property.SetMethod != null)
    {
        var destType = propertyType.IsDictionary()
            ? typeof(Dictionary<,>)
                .MakeGenericType(propertyType.GetGenericArguments())
            : propertyType.IsArray()
```

```
                    ? typeof(List<>).MakeGenericType(propertyType.GetElementType())
                    : propertyType.IsCollection()
                    ? typeof(List<>)
                        .MakeGenericType(propertyType.GetGenericArguments())
                    : propertyType;

                dest = Activator.CreateInstance(destType);
                Bind(src, dest);

                if (propertyType.IsArray())
                {
                    IList list = (IList)dest;
                    dest = Array.CreateInstance(propertyType.GetElementType(),
                        list.Count);
                    list.CopyTo((Array)dest, 0);
                }
                property.SetValue(to, src);
            }
        }
    }
}

internal static class Extensions
{
    public static bool IsDictionary(this Type type)
        => type.IsGenericType && typeof(IDictionary).IsAssignableFrom(type) &&
        type.GetGenericArguments().Length == 2;
    public static bool IsCollection(this Type type)
        => typeof(IEnumerable).IsAssignableFrom(type) && type != typeof(string);
    public static bool IsArray(this Type type)
        => typeof(Array).IsAssignableFrom(type);
}
```

在实现的 Configure 方法中，JsonFileConfigureOptions<TOptions>利用提供的 IFileProvider 对象读取了指定 JSON 文件的内容，并将其反序列化成一个新的 Options 对象。由于 Options 模型最终提供的总是 IOptionsFactory<TOptions>对象最初创建的那个 Options 对象，所以针对 Options 的初始化只能针对这个 Options 对象。因此，不能使用新的 Options 对象替换现有的 Options 对象，只能将新的 Options 对象承载的数据绑定到现有的这个 Options 对象上，针对 Options 对象的绑定实现在上面提供的 Bind 方法中。如下所示的代码片段是注册 JsonFileConfigureOptions<TOptions>对象的 Configure<TOptions>扩展方法的定义。

```
public static class ServiceCollectionExtensions
{
    public static IServiceCollection Configure<TOptions>(
        this IServiceCollection services, string filePath, string basePath = null)
        where TOptions : class, new()
        => services.Configure<TOptions>(Options.DefaultName, filePath, basePath);

    public static IServiceCollection Configure<TOptions>(
```

```
    this IServiceCollection services, string name, string filePath,
    string basePath = null) where TOptions : class, new()
{
    var fileProvider = string.IsNullOrEmpty(basePath)
        ? new PhysicalFileProvider(Directory.GetCurrentDirectory())
        : new PhysicalFileProvider(basePath);

    return services.AddSingleton<IConfigureOptions<TOptions>>(
        new JsonFileConfigureOptions<TOptions>(name, filePath, fileProvider));
}
```

自定义 OptionsChangeTokenSource

通过对 IOptionsMonitor<Options>的介绍可知，它通过 IOptionsChangeTokenSource<TOptions>对象来感知 Options 数据的变化。到目前为止，我们尚未涉及针对这个服务的注册，下面演示如何通过注册该服务来实现定时刷新 Options 数据。

对于如何同步 Options 数据，最理想的场景是在数据源发生变化的时候及时将通知"推送"给应用程序。如果采用本地文件，采用这种方案是很容易实现的。但是在很多情况下，实时监控数据变化的成本很高，消息推送在技术上也不一定可行，此时需要退而求其次，使应用定时获取并更新 Options 数据。这样的应用场景可以通过注册一个自定义的 IOptionsChangeTokenSource<TOptions>实现类型来完成。

在讲述自定义 IOptionsChangeTokenSource<TOptions>类型的具体实现之前，先演示针对Options 数据的定时刷新。我们依然沿用前面定义的 FoobarOptions 作为绑定的目标 Options 类型，而具体的演示程序则体现在如下所示的代码片段中。

```
class Program
{
    static void Main()
    {
        var random = new Random();
        var optionsMonitor = new ServiceCollection()
            .AddOptions()
            .Configure<FoobarOptions>(TimeSpan.FromSeconds(1))
            .Configure<FoobarOptions>(foobar =>
            {
                foobar.Foo = random.Next(10, 100);
                foobar.Bar = random.Next(10, 100);
            })
            .BuildServiceProvider()
            .GetRequiredService<IOptionsMonitor<FoobarOptions>>();

        optionsMonitor.OnChange(foobar
            => Console.WriteLine($"[{DateTime.Now}]{foobar}"));
        Console.Read();
    }
}
```

如上面的代码片段所示，针对自定义 IOptionsChangeTokenSource<TOptions>对象的注册实现在我们为 IServiceCollection 接口定义的 Configure<FoobarOptions>扩展方法中，该方法具有一个 TimeSpan 类型的参数，表示定时刷新 Options 数据的时间间隔。在演示程序中，我们将这个时间间隔设置为 1 秒。为了模拟数据的实时变化，可以调用 Configure<FoobarOptions>扩展方法注册一个 Action<FoobarOptions>对象来更新 Options 对象的两个属性值。

利用 IServiceProvider 对象得到 IOptionsMonitor<FoobarOptions>对象，并调用其 OnChange 方法注册了一个 Action<FoobarOptions>对象，从而将 FoobarOptions 承载的数据和当前时间打印出来。由于我们设置的自动刷新时间为 1 秒，所以程序会以这个频率定时将新的 Options 数据以图 7-11 所示的形式打印在控制台上。（S711）

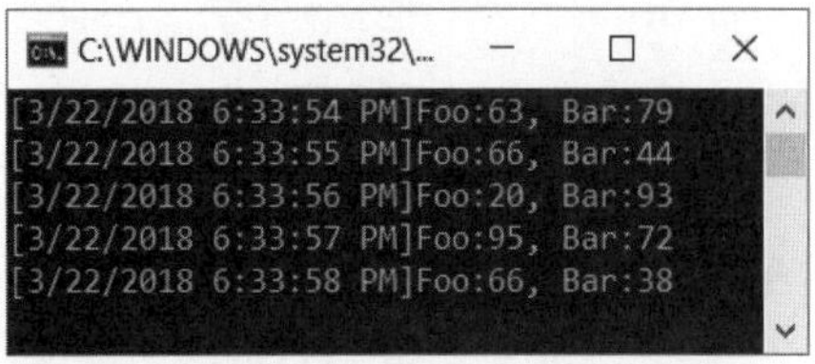

图 7-11　定时刷新 Options 数据

前面演示程序中的 Configure<TOptions>扩展方法注册了一个 TimedRefreshTokenSource<TOptions>对象，下面的代码片段给出了该类型的完整定义。从给出的代码片段可以看出，实现的 OptionsChangeToken 方法返回的 IChangeToken 对象是通过字段_changeToken 表示的 OptionsChangeToken 对象，它与第 6 章介绍的 ConfigurationReloadToken 类型具有完全一致的实现。

```
public class TimedRefreshTokenSource<TOptions> : IOptionsChangeTokenSource<TOptions>
{
    private OptionsChangeToken           _changeToken;
    public string                        Name { get; }

    public TimedRefreshTokenSource( TimeSpan interval, string name)
    {
        this.Name = name ?? Options.DefaultName;
        _changeToken = new OptionsChangeToken();
        ChangeToken.OnChange(
            () => new CancellationChangeToken(
                new CancellationTokenSource(interval).Token),
            () => {
                var previous = Interlocked.Exchange(ref _changeToken,
                    new OptionsChangeToken());
                previous.OnChange();
            });
    }

    public IChangeToken GetChangeToken() => _changeToken;

    private class OptionsChangeToken : IChangeToken
```

```
{
    private readonly CancellationTokenSource _tokenSource;

    public OptionsChangeToken()
        => _tokenSource = new CancellationTokenSource();
    public bool HasChanged
        => _tokenSource.Token.IsCancellationRequested;
    public bool ActiveChangeCallbacks
        => true;
    public IDisposable RegisterChangeCallback(Action<object> callback,
        object state)
        => _tokenSource.Token.Register(callback, state);
    public void OnChange()
        => _tokenSource.Cancel();
}
```

通过调用构造函数创建一个 TimedRefreshTokenSource<TOptions>对象时，除了需要指定 Options 的名称，还需要提供一个 TimeSpan 对象来控制 Options 自动刷新的时间间隔。在构造函数中，可以通过调用 ChangeToken 的 OnChange 方法以这个间隔定期地创建新的 OptionsChangeToken 对象并赋值给 _changeToken。与此同时，我们通过调用前一个 OptionsChangeToken 对象的 OnChange 方法对外通知 Options 已经发生变化。

```
public static class ServiceCollectionExtensions
{
    public static IServiceCollection Configure<TOptions>(
        this IServiceCollection services, string name, TimeSpan refreshInterval)
        => services.AddSingleton<IOptionsChangeTokenSource<TOptions>>(
            new TimedRefreshTokenSource<TOptions>(refreshInterval, name));
    public static IServiceCollection Configure<TOptions>(
        this IServiceCollection services, TimeSpan refreshInterval)
        => services.Configure<TOptions>(Options.DefaultName, refreshInterval);
}
```

7.3.4 集成配置系统

Options 模型本身与配置系统完全没有关系，但是配置在大部分情况下会作为绑定 Options 对象的数据源，所以有必要将两者结合在一起。与 7.3.3 节演示的两个例子一样，针对配置系统的集成同样是通过定制 Options 模型相应的对象来实现的。具体来说，集成配置系统需要解决如下两个问题。

- 将承载配置数据的 IConfiguration 对象绑定为 Options 对象。
- 自动感知配置数据的变化。

第一个问题涉及针对 Options 对象的初始化问题，这自然是通过自定义 IConfigureOptions<TOptions>实现类型来解决的，具体来说就是下面的 NamedConfigureFromConfigurationOptions<TOptions>类型，它定义在 NuGet 包 "Microsoft.Extensions.Options.ConfigurationExtensions" 中。如下面的代码片段所示，NamedConfigureFromConfigurationOptions<TOptions>通过调用 ConfigurationBinder 的静态方法 Bind 利用配置绑定机制来实现配置数据向 Options 对象的转换。

```csharp
public class NamedConfigureFromConfigurationOptions<TOptions> :
    ConfigureNamedOptions<TOptions> where TOptions : class
{
    public NamedConfigureFromConfigurationOptions(string name,
        IConfiguration config)
        : base(name, options => ConfigurationBinder.Bind(config, options))
    {}
}
```

第二个问题则采用自定义的 IOptionsChangeTokenSource<TOptions>实现类型来解决，具体提供的就是下面的 ConfigurationChangeTokenSource<TOptions>。从给出的代码片段可以看出，GetChangeToken 方法直接调用 IConfiguration 对象的 GetReloadToken 方法得到返回的 IChangeToken 对象。

```csharp
public class ConfigurationChangeTokenSource<TOptions> :
    IOptionsChangeTokenSource<TOptions>
{
    private IConfiguration   _config;
    public string            Name { get; }

    public ConfigurationChangeTokenSource(IConfiguration config) :
        this(Options.DefaultName, config) {}
    public ConfigurationChangeTokenSource(string name, IConfiguration config)
    {
        _config = config;
        Name = name ?? Options.DefaultName;
    }

    public IChangeToken GetChangeToken()
        =>_config.GetReloadToken()
}
```

将 IConfiguration 对象绑定为 Options 对象的 NamedConfigureFromConfigurationOptions<TOptions>和用来检测配置数据变化的 ConfigurationChangeTokenSource<TOptions>都是通过下面的 Configure<TOptions>扩展方法来注册的。

```csharp
public static class OptionsConfigurationServiceCollectionExtensions
{
    public static IServiceCollection Configure<TOptions>(
        this IServiceCollection services, IConfiguration config)
        where TOptions : class
        => services.Configure<TOptions>(Options.Options.DefaultName, config);

    public static IServiceCollection Configure<TOptions>(
        this IServiceCollection services, string name, IConfiguration config)
        where TOptions : class
        => services
            .AddSingleton<IOptionsChangeTokenSource<TOptions>>(
                new ConfigurationChangeTokenSource<TOptions>(name, config))
            .AddSingleton<IConfigureOptions<TOptions>>(
                new NamedConfigureFromConfigurationOptions<TOptions>(name, config));
}
```

第 8 章

诊断日志（上篇）

在整个软件开发维护生命周期内，最难的不是如何将软件系统开发出来，而是在系统上线之后及时解决遇到的问题。一个好的程序员能够在系统出现问题之后马上定位错误的根源并找到正确的解决方案，一个更好的程序员能够根据当前的运行状态预知未来可能发生的问题，并将问题扼杀在摇篮中。诊断日志能够帮助我们有效地纠错和排错。大部分程序员可能很少涉及针对诊断日志的编程，但笔者希望程序员能够改变这个习惯。

8.1 各种诊断日志形式

按照惯例，在介绍各种诊断日志方案的设计和实现原理之前，需要先介绍一些相关的编程体验。本章主要介绍 4 种典型的诊断日志记录手段，由于写入日志的对象分别为 Debugger、TraceSource、EventSource 和 DiagnosticSource（这些类型都定义在 System.Diagnostics 命名空间下），所以可以将对应的日志形式称为调试日志、跟踪日志、事件日志和诊断日志。下面对各种日志形式进行简单介绍。

8.1.1 调试日志

Debugger 这个静态类型是 .NET Core 托管代码与调试器进行通信的媒介，我们可以利用它启动调试器并将其附加到当前进程上。在调试器被附加（Attach）到当前进程之后，我们可以调用 Debugger 的 Log 方法触发一个日志事件，并提供一条日志消息。

```
class Program
{
    static void Main()
    {
        Debugger.Log(0, null, "This is a test debug message.\n");
        Console.Read();
    }
}
```

我们在演示实例中采用如上形式调用 Debugger 的 Log 方法记录了一条测试消息。Log 方法的最后一个参数代表消息内容，它的前两个参数分别代表日志的等级和类别。由于 Log 方法针对调试日志的写入只有在调试器附加在当前进程上才有效，所以需要按 F5 键（Start Debugging）启动这个程序。当程序以 Debug 模式启动之后，Visual Studio 调试输出窗口显示的日志消息如图 8-1 所示。倘若应用是通过按 Ctrl+F5 组合键（Start Without Debuging）启动的，就不会发生任何日志写入操作。

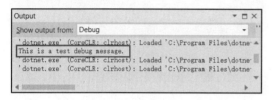

图 8-1　Visual Studio 调试输出窗口显示的日志消息

其实，直接调用 Debugger 的 Log 方法来记录调试日志的情况很少，更多的时候我们调用的是静态类型 Debug 的 Write 方法、WriteLine 方法、WriteIf 方法和 WriteLineIf 方法，上面的代码可以改写成如下形式。虽然静态类型 Debug 的 WriteLine 方法最终还是会调用 Debugger 的 Log 方法来完成日志的写入，但是两者之间还是有区别的。由于 Debug 类型上的所有方法通过条件编译的形式被设置为仅针对 Debug 模式有效，但是如果程序是在 Debug 模式下进行编译的，就意味着针对 WriteLine 方法的调用不会出现在编译生成的程序集中。

```
class Program
{
    static void Main()
    {
        Debug.WriteLine("This is a test debug message.");
        Console.Read();
    }
}
```

8.1.2　跟踪日志

从设计的角度来讲，接下来介绍的 3 种诊断日志框架采用的都是观察者模式或者发布订阅模式。在这种模式下，作为发布主题的日志事件通过 Source 进行发布，相应的 Listener 或者 Observer 作为订阅者会接收到满足订阅条件的日志事件，并将得到的日志消息写入各自的输出渠道。

日志基本上都是针对某一事件记录的，日志事件一般具有一个 ID。按照日志事件的重要程度和反映问题的严重级别，可以赋予日志事件一个等级或者类型。对于以 TraceSource 为核心的日志框架来说，日志事件等级或者类型通过 TraceEventType 枚举类型表示。除此之外，它还提供了一个 SourceLevels 类型的枚举来完成针对等级对日志事件进行的过滤。

利用 TraceSource 来记录日志时，首先需要做的就是根据名称和最低日志等级创建一个 TraceSource 对象，然后指定事件 ID、事件类型和日志消息作为参数调用 TraceEvent 方法。在如下所示的演示代码中，我们创建了一个 TraceSource 对象，并将名称和最低日志等级分别设置为 Foobar 与 SourceLevels.All，后者决定了所有等级的日志都会被记录下来。然后针对每种事件类型记录了一条日志消息，而事件 ID 被设置为一个自增的整数。

```
class Program
{
    static void Main()
    {
        var source = new TraceSource("Foobar", SourceLevels.All);
        var eventTypes = (TraceEventType[])Enum.GetValues(typeof(TraceEventType));
        var eventId = 1;
        Array.ForEach(eventTypes, it => source.TraceEvent(it, eventId++,
            $"This is a {it} message."));
        Console.Read();
    }
}
```

虽然以 TraceSource 为核心的日志框架采用订阅发布模式来记录日志，但是上面的程序只涉及作为发布者的 TraceSource 对象，作为真正完成日志写入的订阅者（监听器）没有出现。如果以 Debug 的方式执行程序，我们会发现相应的日志以图 8-2 所示的形式被输出到 Visual Studio 的输出窗口中，这是因为日志框架会默认注册一个类型为 DefaultTraceListener 的监听器。（S801）

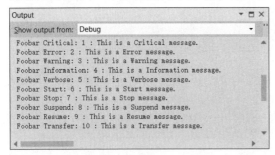

图 8-2　通过默认注册的 DefaultTraceListener 写入的日志

由于 TraceEventType 枚举类型共定义了 10 种事件类型，并且对应的枚举项是从高到低排列的（Critical 最高，Transfer 最低），所以上面的演示实例会按照等级的高低输出 10 条日志。如果我们只希望部分事件类型的日志被记录下来应如何做？一般来说，对于通过同一个 TraceSource 对象发出的日志消息，它是否应该被写入取决于对应的事件等级，等级越高越应该被记录下来。创建 TraceSource 对象时指定的 SourceLevels 枚举表示需要被记录下来的最低日志等级。对于我们的演示实例来说，如果只希望记录 Warning 等级以上的日志，做如下改写即可。

```
class Program
{
    static void Main()
    {
        var source = new TraceSource("Foobar", SourceLevels.Warning);
        var eventTypes = (TraceEventType[])Enum.GetValues(typeof(TraceEventType));
        var eventId = 1;
```

```
        Array.ForEach(eventTypes,
            it => source.TraceEvent(it, eventId++, $"This is a {it} message."));
        Console.Read();
    }
}
```

由于上面的代码在创建 TraceSource 对象时将最低日志等级设置成 SourceLevels.Warning，所以以 Debug 模式启动程序后，只有等级不低于 Warning（Warning、Error 和 Critical）的 3 条日志消息以图 8-3 所示的形式被输出到 Visual Studio 的输出窗口。（S802）

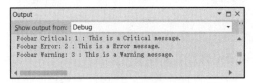

图 8-3　记录下来的被过滤的日志

到目前为止，我们都在使用系统默认注册的 DefaultTraceListener 监听器来完成对日志消息的输出（到 Visual Studio 的输出窗口）。DefaultTraceListener 除了调用 Debug 的 Write 方法将指定的消息针对调试器输出，它还支持针对物理文件的输出方式。以 TraceSource 为核心的日志框架的可扩展性体现在可以针对期望的日志输出渠道定义并注册对应的 TraceListener。

```
public class ConsoleTraceListener : TraceListener
{
    public override void Write(string message) => Console.Write(message);
    public override void WriteLine(string message) => Console.WriteLine(message);
}
```

在如上所示的代码片段中，我们通过继承抽象类 TraceListener 自定义了一个 ConsoleTraceListener 类型，它通过重写的 Write 方法和 WriteLine 方法将分发给它的日志消息输出到控制台上。下面通过对演示的程序做如下修改来注册这个自定义的针对控制台的 ConsoleTraceListener。

```
class Program
{
    static void Main()
    {
        var source = new TraceSource("Foobar", SourceLevels.Warning);
        source.Listeners.Add(new ConsoleTraceListener());
        var eventTypes = (TraceEventType[])Enum.GetValues(typeof(TraceEventType));
        var eventId = 1;
        Array.ForEach(eventTypes, it => source.TraceEvent(it, eventId++,
            $"This is a {it} message."));
    }
}
```

如上面的代码片段所示，TraceSource 类型的 Listeners 属性维护了一组注册的 TraceListener 对象，只需要将创建的 ConsoleTraceListener 添加到这个列表中即可。由于 TraceSource 注册了一个针对控制台的 TraceListener 对象，所以程序启动之后，满足过滤条件的 3 条日志消息将以图 8-4 所示的形式输出到控制台上。（S803）

图 8-4　通过注册的 ConsoleTraceListener 写入的日志

8.1.3　事件日志

　　EventSource 最初是微软为 Windows 操作系统自身的日志框架 ETW（Event Tracing for Windows）设计的，但即使在 Linux 操作系统下，我们也可以通过创建 EventListener 对象监听当前进程内所有 EventSource 发出的日志事件。从性能上讲，这是一种非常高效的记录日志的方式。从编程模式来看，它提供的强类型的编程方式可以使记录日志变得很"优雅"。

　　EventSource 所谓的强类型编程模式主要体现在如下两个方面：其一，我们需要通过继承抽象类 EventSource 定义一个具体的 EventSource 类型，并将发送日志事件的操作实现在它的某个方法中；其二，日志消息的内容可以通过一个自定义的数据类型来承载。下面介绍 EventSource 这种强类型的日志记录模式。

　　我们在演示程序中编写了如下一段简单的代码。可以认为 DatabaseSource 是为某个数据库组件定义的 EventSource。如果没有编程方面的限制，我们倾向于将其定义成一个封闭（Sealed）的类型，并且采用 Singleton 模式来访问它，所以可以定义静态只读字段 Instance 来获取这个单例的 DatabaseSource 对象。

```
class Program
{
    static void Main()
        => DatabaseSource.Instance.OnCommandExecute(CommandType.Text, "SELECT * FROM T_USER");
}

public sealed class DatabaseSource : EventSource
{
    public static readonly DatabaseSource Instance = new DatabaseSource();
    private DatabaseSource() {}

    public void OnCommandExecute(CommandType commandType, string commandText)
        => WriteEvent(1, commandType, commandText);
}
```

　　我们针对"SQL 命令执行"这一日志事件定义了 OnCommandExecute 方法，该方法定义的两个参数分别表示命令的类型和文本。如果命令类型为存储过程，那么命令文本就是对应的存储过程名称，否则就是执行的 SQL 语句。OnCommandExecute 方法最终会调用继承的 WriteEvent 方法来发送日志事件。该方法的第一个参数 1 代表日志事件的 ID。我们在作为程序入口的 Main 方法中调用了 OnCommandExecute 方法。

　　当我们创建一个 TraceSource 的时候，需要为它指定一个名称，一个 EventSource 同样需要一个确定的名称，从 ETW 层面来讲，EventSource 的名称实际上就是 ETW Provider 的名称。在默认情况下，自定义类型的名称会自动作为 EventSource 的名称，所以演示实例采用的

EventSource 名称为 DatabaseSource。

日志事件需要有一个唯一的整数作为 ID，如果没有对事件 ID 做显式设置，系统会采用从 1 开始自增的方式为每个日志方法分配一个 ID。由于 DatabaseSource 中只定义了一个唯一的日志方法 OnCommandExecute，该方法对应的日志事件 ID 自然是 1。事件方法在调用 WriteEvent 方法发送日志事件时，需要指定一个匹配的事件 ID，这就是 OnCommandExecute 方法在调用 WriteEvent 方法时将第一个参数设置为 1 的原因。

由于 EventSource 具有向 ETW 日志系统发送日志事件的能力，所以可以利用一些工具来收集这些事件。笔者习惯使用的是一款叫作 PerfView 的 GUI 工具，这是一款可以在网上直接下载的性能分析工具，解压缩后就是一个可执行文件。笔者倾向于将该工具所在的目录添加到环境变量 PATH 中，这样就可以采用命令行的形式进行启动。我们可以采用 Run 和 Collect 这两种模式启动 PerfView：前者利用 PerfView 启动和检测某个指定的应用，后者则独立启动 PerfView 并检测当前运行的所有应用进程。

我们可以将应用所在根目录作为工作目录，并以 Run 执行命令 "PerfView /onlyproviders=*DatabaseSource run dotnet run" 启动 PerfView。为了将自定义的 Trace Provider 纳入 PerfView 的检测列表中，可以将命令行开关 onlyproviders 设置为 "*DatabaseSource"。PerfView run 命令执行的应用程序为 "dotnet run"，这就意味着我们的演示程序将作为监测程序被启动。

PerfView 会将捕获到的日志打包到当前目录下一个名为 PerfViewData.etl.zip 的压缩文件中，它左侧的目录结构会以图 8-5 所示的形式列出该文件。双击该文件展开其子节点后会看到一个 Events 节点，PerfView 捕捉到的日志就可以通过它来查看。双击 Events 节点后，图 8-5 所示的事件视图将会列出捕获到的所有日志事件，我们可以输入 "DatabaseSource" 筛选由 DatabaseSource 发送的事件。可以看到，DatabaseSource 共发送了两个事件，其中一个就是 OnCommandExecute。双击事件视图左侧的 "OnCommandExecute" 可以查看该事件的详细信息，调用对应日志方法时提供的数据会包含在 Rest 列中，具体的内容如下所示。（S804）

```
ThreadID="17,608" commandType="Text" commandText="SELECT * FROM T_USER"
```

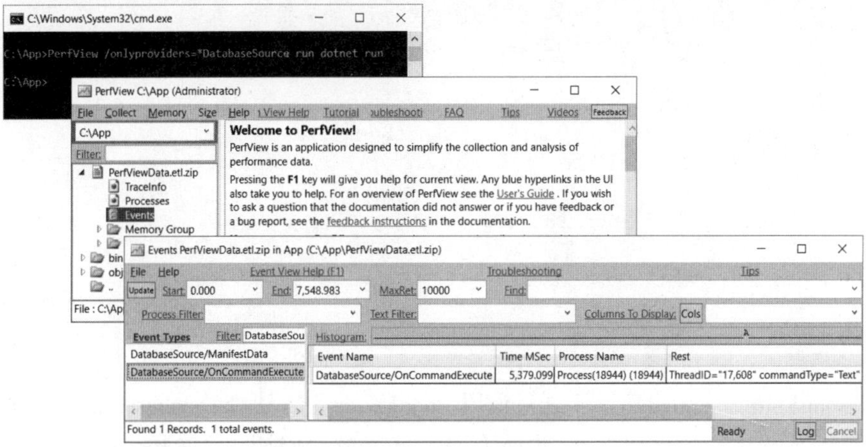

图 8-5　利用 PerfView 启动并检测应用程序

在定义具体的 EventSource 类型时，虽然系统会根据默认的规则来命名 EventSource 的名称和指定日志方法对应的事件 ID，但是对这两个属性进行显式设置会更好。所以，对上面定义的 DatabaseSource 类型进行改写，如下面的代码片段所示，我们在类型上通过标注的 EventSourceAttribute 将 EventSource 的名称设置为 Artech-Data-SqlClient。在日志方法 OnCommandExecute 上，我们利用标注的 EventAttribute 特性将事件 ID 显式设置为 1。

```
[EventSource(Name ="Artech-Data-SqlClient")]
public sealed class DatabaseSource : EventSource
{
    ...
    [Event(1)]
    public void OnCommandExecute(CommandType  commandType, string commandText)
    => WriteEvent(1, commandType, commandText);
}
```

除利用 PerfView 捕捉 EventSource 针对 ETW 发送的日志事件外，我们还可以通过创建 EventListener 对象达到相同的目的。为了专门捕捉 DatabaseSource 发出的事件，我们定义了对应的 DatabaseSourceListener 类型。如下面的代码片段所示，DatabaseSourceListener 继承自抽象类 EventListener。创建的 EventListener 对象不需要显式注册，它的 OnEventSourceCreated 方法能够感知到当前进程中所有 EventSource 对象的创建，所以我们重写了该方法，并根据名称筛选目标 EventSource。

```
public class DatabaseSourceListener : EventListener
{
    protected override void OnEventSourceCreated(EventSource eventSource)
    {
        if (eventSource.Name == "Artech-Data-SqlClient")
        {
            EnableEvents(eventSource, EventLevel.LogAlways);
        }
    }

    protected override void OnEventWritten(EventWrittenEventArgs eventData)
    {
        Console.WriteLine($"EventId: {eventData.EventId}");
        Console.WriteLine($"EventName: {eventData.EventName}");
        Console.WriteLine($"Payload");
        var index = 0;
        foreach (var payloadName in eventData.PayloadNames)
        {
            Console.WriteLine($"\t{payloadName}:{eventData.Payload[index++]}");
        }
    }
}
```

由于我们在 OnEventSourceCreated 方法中通过调用 EnableEvents 方法对 DatabaseSource（准确来说应该是名称为 Artech-Data-SqlClient 的所有 EventSource）进行了关联，所以只有 DatabaseSource 对象发出的日志事件能够被捕捉到。针对日志事件的捕捉实现在重写的

OnEventWritten 方法中，作为该方法唯一的参数，EventWrittenEventArgs 对象承载了日志事件的所有信息。我们在该方法中将事件的 ID、名称和荷载数据（Payload）输出到控制台上。

```
class Program
{
    static void Main()
    {
        var listener = new DatabaseSourceListener();
        DatabaseSource.Instance.OnCommandExecute(CommandType.Text, "SELECT * FROM T_USER");
    }
}
```

由于 EventListener 并不需要显式注册，所以只需要按照如上所示的方式在程序启动的时候创建 DatabaseSourceListener 对象即可。程序运行之后，由 DatabaseSourceListener 对象捕获的日志事件信息会以图 8-6 所示的形式输出到控制台上。（S805）

图 8-6　利用自定义的 EventListener 捕捉日志事件

8.1.4　诊断日志

对于以 TraceSource 和 EventSource 为核心的日志框架来说，它们主要关注的是日志荷载内容在进程外的持久化或者传输问题，所以被 TraceSource 对象作为内容荷载的对象必须是一个字符串，EventSource 对象虽然可以使用一个对象作为内容荷载，但是最终输出的其实还是序列化之后的结果。基于 DiagnosticSource 的日志框架与它们具有本质的不同，它的设计思路如下：作为发布者的 DiagnosticSource 对象将原始的日志荷载对象直接分发给订阅者，并完全由订阅者自行决定应该对日志内容做什么样的处理，事件的触发和监听处理是同步执行的。

同样是采用观察者模式，以 DiagnosticSource 为核心的日志框架做得似乎更加彻底，因为作为发布者和订阅者的类型显式地实现了 IObservable<T>接口与 IObserver<T>接口（泛型参数 T 代表订阅的主题类型）。IObservable<T>接口代表可被观察的对象，也就是被观察者，即发布者。IObserver<T>接口代表观察者，也就是订阅者。IObservable<T>接口定义了唯一的方法 Subscribe，用来进行订阅注册；IObserver<T>接口则提供了 3 个方法，作为接收到发布者通知后自动执行的回调，其中核心方法 OnNext 用于处理发布者发布的主题，OnCompleted 方法会在所有主题完成发布后被执行，OnError 方法则作为发布过程中出现错误时采用的异常处理器。

```
public interface IObservable<out T>
{
    IDisposable Subscribe(IObserver<T> observer);
}

public interface IObserver<in T>
{
```

```
    void OnCompleted();
    void OnError(Exception error);
    void OnNext(T value);
}
```

为了便于演示，可以先定义一个通用的观察者 Observer<T>。如下面的代码片段所示，Observer<T>实现了 IObserver<T>接口，我们利用初始化时提供的 Action<T>对象来实现其 OnNext 方法，而 OnError 方法和 OnCompleted 方法则不执行任何操作。

```
public class Observer<T> : IObserver<T>
{
    private Action<T> _onNext;
    public Observer(Action<T> onNext) => _onNext = onNext;

    public void OnCompleted() { }
    public void OnError(Exception error) { }
    public void OnNext(T value) => _onNext(value);
}
```

可以采用 DiagnosticSource 诊断日志来实现上面演示的针对数据库命令执行的日志输出场景，具体的代码如下所示。由于 DiagnosticSource 仅仅是一个抽象类型，所以用来发送日志事件的是它的子类 DiagnosticListener。换句话说，在整个日志事件分发过程中，DiagnosticListener 的角色是发布者，而不是订阅者，这一点和它的命名不太相符。如下面的代码片段所示，我们创建了一个 DiagnosticListener 对象，并将其命名为 Artech-Data-SqlClient，然后调用其 Write 方法完成了日志事件的发送。日志事件被命名为 CommandExecution，荷载内容是我们创建的一个匿名对象，其中包含两个成员，即 CommandType 和 CommandText。

```
class Program
{
    static void Main()
    {
        DiagnosticListener.AllListeners.Subscribe(new Observer<DiagnosticListener>(
            listener =>
            {
                if (listener.Name == "Artech-Data-SqlClient")
                {
                    listener.Subscribe(new Observer<KeyValuePair<string, object>>(eventData =>
                    {
                        Console.WriteLine($"Event Name: {eventData.Key}");
                        dynamic payload = eventData.Value;
                        Console.WriteLine($"CommandType: {payload.CommandType}");
                        Console.WriteLine($"CommandText: {payload.CommandText}");
                    }));
                }
            }));

        var source = new DiagnosticListener("Artech-Data-SqlClient");
        if(source.IsEnabled("CommandExecution"))
        {
            source.Write("CommandExecution",
```

```
            new { CommandType = CommandType.Text,
            CommandText = "SELECT * FROM T_USER" });
        }
    }
}
```

DiagnosticListener 的静态属性 AllListeners 以一个 IObservable<DiagnosticListener>对象的形式提供当前进程内创建的所有 DiagnosticListener 对象，我们调用它的 Subscribe 方法注册了一个 Observer<DiagnosticListener>对象。在根据名称（Artech-Data-SqlClient）筛选出目标 DiagnosticListener 之后，我们调用其 Subscribe 方法注册了一个 Observer<KeyValuePair<string, object>>对象，用来捕捉由它发出的日志事件。

日志事件的所有信息体现在作为泛型参数的 KeyValuePair<string, object>对象中，它的 Key 和 Value 分别表示事件的名称与荷载内容。由于我们已经知道了作为荷载内容的数据结构，所以可以采用动态类型的方式将成员的值提取出来。该程序执行之后，DiagnosticListener 对象记录的日志内容会以图 8-7 所示的形式输出到控制台上。（S806）

图 8-7　捕捉 DiagnosticListener 发出的日志事件

上面演示的实例通过为 DiagnosticListener 对象显式注册一个 IObserver<KeyValuePair<string, object>>对象的方式来捕捉由它发出的日志事件，实际上还有另外一种更加简便的编程方式。由于每个 DiagnosticListener 对象发出的日志事件都有一个确定的名称，并且总是将提供的荷载对象原封不动地分发给注册的订阅者，如果能够解决事件名称与方法之间的映射，以及日志事件内容荷载对象成员与方法参数之间的映射，就能够使用一个 POCO 类型作为某一个或者多个 DiagnosticListener 对象的订阅者。这种强类型的日志记录方式实现在 NuGet 包"Microsoft.Extensions.DiagnosticAdapter"中，我们需要添加针对它的依赖。

我们可以采用强类型的方式将这个作为订阅者的 POCO 类型定义成 DatabaseSourceCollector 类型。如下面的代码片段所示，如果不需要让 DatabaseSourceCollector 类型实现某个预定义的接口或者继承某个预定义的基类，那么它就是一个普通的实例类型。OnCommandExecute 方法通过标注的 DiagnosticNameAttribute 特性实现了与日志事件 CommandExecution 的关联。我们为 OnCommandExecute 方法定义了两个参数（commandType 和 commandText），它们的类型和名称需要与日志荷载对象对应的成员相匹配，借助与日志荷载对象成员与参数之间的映射，在该方法被调用之前，这两个参数会被自动绑定为匹配的内容。

```
public class DatabaseSourceCollector
{
    [DiagnosticName("CommandExecution")]
    public void OnCommandExecute(CommandType commandType, string commandText)
    {
        Console.WriteLine($"Event Name: CommandExecution");
```

```
            Console.WriteLine($"CommandType: {commandType}");
            Console.WriteLine($"CommandText: {commandText}");
        }
    }
```

为了使用上面的 DatabaseSourceCollector 对象来捕捉发出的日志事件，可以对演示程序进行改写。如下面的代码片段所示，我们不再烦琐地创建并注册一个 IObserver<KeyValuePair<string, object>>对象，而是调用 SubscribeWithAdapter 扩展方法将创建的 DatabaseSourceCollector 对象注册为日志订阅者。由于捕捉到的日志事件的相关信息在 OnCommandExecute 方法中采用与上面完全一致的输出结构，所以应用程序启动之后同样会在控制台上呈现出与图 8-7 完全一致的内容。（S807）

```
class Program
{
    static void Main()
    {
        DiagnosticListener.AllListeners.Subscribe(
            new Observer<DiagnosticListener>(listener=> {
                if (listener.Name == "Artech-Data-SqlClient")
                {
                    listener.SubscribeWithAdapter(new DatabaseSourceCollector());
                }
            }));

        var source = new DiagnosticListener("Artech-Data-SqlClient");
        source.Write("CommandExecution",
            new { CommandType = CommandType.Text, CommandText = "SELECT * FROM T_USER"
});
    }
}
```

8.2　Debugger 调试日志

静态类型 Debugger 是 .NET Core 应用与调试器进行通信的媒介，我们可以利用它人为地启动调试器，还可以利用它在某行代码上触发一个断点，本章主要利用它来向调试器发送日志消息。虽然在编程过程中涉及该类型的机会不多，但是我们也有必要了解这个类型。

8.2.1　Debugger

.NET Core 采用的 JIT Debugging 会在应用程序遇到错误或者调用到某些方法时提示加载调试器。由于调试器集成到 IDE 上，如果应用程序是直接在 Visual Studio 中以 Start Debugging 模式（F5）启动的，那么调试器会直接附加到进程上。如果采用 Start Without Debugging（Ctrl+F5）模式或者命令行的方式启动应用，那么调试器将不会参与进来，但是我们可以调用 Debugger 相应的方法来启动调试器。如下所示的代码片段展示的是静态类型 Debugger 的定义。

```
public static class Debugger
{
```

```
public static readonly string DefaultCategory;
public static bool IsAttached { get; }

public static bool Launch();
public static void Break();
public static extern bool IsLogging();
public static extern void Log(int level, string category, string message);
...
}
```

Debugger 的 IsAttached 属性表示调试器是否被附加到当前进程上，如果该属性返回 False，就可以调用 Launch 方法人为地启动调试器。当该方法被执行的时候，系统弹出的对话框如图 8-8 所示，我们可以选择一个 JIT Debugger 对当前应用进行调试。本机开启的 Visual Studio 进程（devenv.exe）会全部出现在列表框中，我们可以选择任意一个开启的 Visual Studio 进程或者启动新的 Visual Studio 进程对当前应用进行调试。

图 8-8　JIT Debugger 选择对话框

Debugger 的 Break 方法的作用就是在当前调用的地方触发一个断点，调试中的程序可以停在该位置，这与我们利用 Visual Studio 在某行代码上设置断点的作用是一致的。Debugger 的 IsLogging 方法和 Log 方法与日志有关，前者表示调试器是否已经附加到当前进程并且日志功能是否被开启，后者就是前面演示实例中用来向调试器发送日志消息的方法。我们在调用 Log 方法的时候需要指定日志消息的等级、类别和内容，如果应用程序没有对调试消息进行具体的类别划分，就可以将类别设置为 Null，实际上，表示默认类型的字段 DefaultCategory 返回的就是 Null。

由于本章的主题是"诊断日志"，所以定义在 Debugger 中的 Log 方法就是我们关注的重点。在具体的项目开发中，其实我们很少直接调用这个方法来记录调试日志，因为还有更好的选择，那就是直接调用 System.Diagnostics 命名空间下的另一个静态类型 Debug。

8.2.2 Debug

静态类型 Debug 提供了一系列方法，用于执行一些与调试相关的操作。由于这些方法全部标注了 ConditionalAttribute 特性，并将条件编译符设置为 Debug，所以针对这些方法的调用只存在于针对 Debug 模式编译生成的程序集中。由于默认情况下针对 Release 编译配置的条件编译符列表中并不包含 DEBUG，所以编译生成的程序集中不存在针对这些方法的调用，对程序执行的性能也就不会造成任何影响。

对于定义在静态类型 Debug 中的众多方法，其中有一半以上与调试日志的写入有关，最常使用的就是 Write 方法和 WriteLine 方法，它们最终都会调用 Debugger 的 Log 方法来完成对指定调试日志的写入。调用这些方法的时候需要指定日志消息（必需）和日志类别（可选）。如果没有显式设置，日志类别将被默认设置为 Null。

至于日志消息，既可以直接指定一个完整的字符串，也可以指定一个包含占位符（{0}，{1}，…，{n}）的字符串模板和对应的参数列表进行格式化。如果指定的是一个普通的对象，那么最终作为日志消息的将是该对象 ToString 方法返回的结果。另外，这些方法均没有涉及针对日志等级的指定，所以在调用 Debugger 的 Log 方法时会将等级指定为 0。

```
public static class Debug
{
    ...
    [Conditional("DEBUG")]
    public static void Write(object value);
    [Conditional("DEBUG")]
    public static void Write(string message);
    [Conditional("DEBUG")]
    public static void Write(object value, string category);
    [Conditional("DEBUG")]
    public static void Write(string message, string category);

    [Conditional("DEBUG")]
    public static void WriteLine(object value);
    [Conditional("DEBUG")]
    public static void WriteLine(string message);
    [Conditional("DEBUG")]
    public static void WriteLine(object value, string category);
    [Conditional("DEBUG")]
    public static void WriteLine(string format, params object[] args);
    [Conditional("DEBUG")]
    public static void WriteLine(string message, string category);
    [Conditional("DEBUG")]
}
```

除了上面定义的 Write 方法和 WriteLine 方法，Debug 类型还定义了对应的 WriteIf 方法和 WriteLineIf 方法。顾名思义，这些方法会提供一个布尔值作为前置条件，只有在满足指定条件的前提下，它们才会调用对应的 Write 方法和 WriteLine 方法记录调试日志。除了与调试日志相关的方法，Debug 类型中还有一些其他与调试相关的方法，由于篇幅有限，本节不再一一介绍。

```csharp
public static class Debug
{
    [Conditional("DEBUG")]
    public static void WriteIf(bool condition, object value);
    [Conditional("DEBUG")]
    public static void WriteIf(bool condition, string message);
    [Conditional("DEBUG")]
    public static void WriteIf(bool condition, object value, string category);
    [Conditional("DEBUG")]
    public static void WriteIf(bool condition, string message, string category);

    [Conditional("DEBUG")]
    public static void WriteLineIf(bool condition, object value);
    [Conditional("DEBUG")]
    public static void WriteLineIf(bool condition, string message);
    [Conditional("DEBUG")]
    public static void WriteLineIf(bool condition, object value, string category);
    [Conditional("DEBUG")]
    public static void WriteLineIf(bool condition, string message, string category);
}
```

8.3 TraceSource 跟踪日志

以 TraceSource 为核心的跟踪日志系统，以及后续即将介绍的针对 EventSource 和 DiagnosticSource 的日志框架都采用观察者模式进行设计。观察者模式又被称为发布订阅模式，这些日志框架将针对某一事件的日志消息作为发布的主题，日志消息由作为发布者的某个 Source 发出，并且被作为订阅者的一个或者多个 Listener 接收并消费。在基于 TraceSource 的跟踪日志系统中，日志消息发布者和订阅者被称为 TraceSource 与 TraceListener，除此之外，跟踪日志系统还包含一个被称为 SourceSwitch 的核心对象，可以将它们称为跟踪日志模型三要素。

8.3.1 跟踪日志模型三要素

图 8-9 揭示了以 TraceSource、TraceListener 和 SourceSwitch 为核心的跟踪日志模型的运行机制。针对某个事件的日志消息最初由某个具体的 TraceSource 发出，但是针对它们的消费（如针对不同输出渠道的持久化）则完全实现在相应的 TraceListener 中，一个 TraceSource 可以注册多个 TraceListener。在日志消息从 TraceSource 到 TraceListener 的分发过程中，SourceSwitch 起到了日志过滤的作用。每个 TraceSource 都有一个对应的 SourceSwitch，后者提供了相应的过滤策略，从而帮助前者决定是否应该将日志消息分发给注册的 TraceListener。

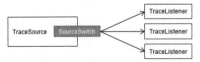

图 8-9 TraceSource、TraceListener 和 SourceSwitch 之间的关系

SourceSwitch

在介绍 TraceSource、TraceListener 和 SourceSwitch 这 3 个核心对象之前，需要先了解与跟踪日志事件类型和等级有关的两个枚举类型。记录的日志基本上是针对某一确定的事件，日志事件一般具有一个 ID。按照日志事件的重要程度和反映问题的严重级别，可以赋予日志事件一个等级或者类型。日志事件等级或者类型可以通过 TraceEventType 枚举类型表示。

```
public enum TraceEventType
{
    Critical        = 1,
    Error           = 2,
    Warning         = 4,
    Information     = 8,
    Verbose         = 16,

    Start           = 256,
    Stop            = 512,
    Suspend         = 1024,
    Resume          = 2048,
    Transfer        = 4096
}
```

如上面的代码片段所示，定义在 TraceEventType 中的每个枚举项都被赋予了一个值，它体现了对应日志事件的等级，数值越小，等级越高。为了更好地理解 TraceEventType 枚举体现了怎样的跟踪事件类型，可以将它们分为两组，其中，从 Critical 到 Verbose 是对某个独立事件的描述。等级最高的 Critical 表示致命的错误，这样的错误可能会导致程序崩溃。对于 Error 和 Warning 来说，前者表示不会影响程序继续运行的错误，后者则表示等级次之的警告。如果需要记录一些"仅供参考"之类的消息，就可以选择使用 Information 类型，如果这类消息仅供调试使用，那么最好选择等级更低的 Verbose。而从 Start 到 Transfer，它们对应的事件关联的是某个功能性的活动（Activity），这 5 种跟踪事件类型分别表示获取的 5 种状态变换，即开始、结束、中止、恢复和转换。

与跟踪事件类型或者等级相关的还包括下面的 SourceLevels 枚举。标注了 FlagsAttribute 特性的 SourceLevels 被 TraceSource 用来表示最低跟踪事件等级，它与上面介绍的 TraceEventType 具有密切的关系。如果给定一个具体的 SourceLevels，就能够确定定义在 TraceEventType 中的哪些类型是与之匹配的，判定的结果由 TraceEventType 的值与 SourceLevels 的值进行"逻辑与"运算决定。对于 All 和 Off 来说，前者表示所有的跟踪事件类型都匹配，而后者的含义则与此相反。Critical、Error、Warning、Information 和 Verbose 则表示不低于指定等级的跟踪事件类型，而 ActivityTracing 则表示只选择 5 种针对活动的事件类型。

```
[Flags]
public enum SourceLevels
{
    All         = -1,
    Off         = 0,
    Critical    = 1,
```

```
    Error            = 3,
    Warning          = 7
    Information      = 15,
    Verbose          = 31,
    ActivityTracing  = 65280
}
```

  了解了上面这两个与跟踪事件类型与等级相关的枚举类型之后，下面先介绍作为跟踪系统核心三要素之一的 SourceSwitch。当利用 TraceSource 对象触发一个跟踪事件时，与之关联的 SourceSwitch 会利用指定的 SourceLevels 来确定该跟踪事件的类型是否满足最低等级的要求。只有在当前事件类型对应的等级不低于指定等级的情况下，跟踪事件才会被分发给注册的 TraceListener 对象。

  如下面的代码片段所示，SourceSwitch 派生于抽象类 Switch，它表示一个一般意义的 "开关"，其字符串属性 DisplayName、Value 和 Description 分别表示开关的名称、值与描述。Switch 具有一个整型的 SwitchSetting 属性，用于承载具体的开关设置。当开关的值发生改变时，它的 OnValueChanged 方法会作为回调被调用。

```
public class SourceSwitch : Switch
{
    public SourceLevels Level
    {
        get => (SourceLevels)base.SwitchSetting;
        set => base.SwitchSetting = (int)value
    }

    public SourceSwitch(string name) : base(name, string.Empty){}

    public SourceSwitch(string displayName, string defaultSwitchValue)
        : base(displayName, string.Empty, defaultSwitchValue){}

    protected override void OnValueChanged()
        => base.SwitchSetting = (int)Enum.Parse(typeof(SourceLevels), base.Value, true);

    public bool ShouldTrace(TraceEventType eventType)
        => (base.SwitchSetting & (int)eventType) > 0;
}

public abstract class Switch
{
    public string            DisplayName { get; }
    protected string         Value { get; set; }
    public string            Description { get; }
    protected int            SwitchSetting { get; set; }

    protected Switch(string displayName, string description);
    protected Switch(string displayName, string description, string defaultSwitchValue);
    protected virtual void   OnSwitchSettingChanged();
```

```
    ...
}
```

一个 SourceSwitch 对象会根据指定的 SourceLevels 确定某个跟踪事件类型是否满足设定的最低等级要求，但是创建一个 SourceSwitch 对象时，我们并不会指定具体的 SourceLevels 枚举值，而是指定该枚举值的字符串表达式，所以重写的 OnValueChanged 方法会将字符串表示的等级转换成整数，并保存在 SwitchSetting 属性上，其 Level 属性再将它转换成 SourceLevels 类型。SourceSwitch 针对跟踪事件类型的过滤最终体现在它的 ShouldTrace 方法上。

TraceListener

作为跟踪日志事件的订阅者，TraceListener 对象最终会接收到 TraceSource 分发给它的跟踪日志消息，然后做进一步的处理。我们可以利用注册的 TraceListener 完成对日志消息的持久化，也可以利用它将日志消息发给某个远程服务做进一步处理。在对 TraceListener 做进一步介绍之前，需要先介绍几个与之相关的类型。

日志消息除了承载我们显式指定的内容，往往还需要包含一些与当前执行环境相关的上下文信息，如当前进程和线程的 ID、当前时间戳及调用堆栈等。是否需要输出这些上下文信息，以及具体输出哪些信息是由下面的 TraceOptions 枚举控制的。由于该枚举类型上标注了 FlagsAttribute 特性，所以枚举项是可以组合使用的。

```
[Flags]
public enum TraceOptions
{
    None                        = 0,
    LogicalOperationStack       = 1,
    DateTime                    = 2,
    Timestamp                   = 4,
    ProcessId                   = 8,
    ThreadId                    = 16,
    Callstack                   = 32,
}
```

TraceOptions 枚举涉及的这些上下文信息究竟是如何收集的？这就涉及下面的 TraceEventCache 类型，可以作为日志荷载内容的上下文信息就保存在这个对象上。从给出的代码片段可以看出，定义在 TraceOptions 中除 None 之外的每个枚举项在 TraceEventCache 中都有一个属性与之对应。为了使读者深刻了解这些上下文信息的来源，我们给出了每个属性的实现逻辑。

```
public class TraceEventCache
{
    private DateTime                        _dateTime = DateTime.MinValue;
    private string                          _stackTrace;
    private long                            _timeStamp = -1L;
    private static volatile bool            s_hasProcessId;
    private static volatile int             s_processId;

    public string Callstack
        => _stackTrace ?? (_stackTrace = Environment.StackTrace);
```

```
    public DateTime DateTime
        => _dateTime == DateTime.MinValue
            ? _dateTime = DateTime.UtcNow
            : _dateTime;
    public Stack LogicalOperationStack
        => Trace.CorrelationManager.LogicalOperationStack;
    public int ProcessId
        => GetProcessId();
    public string ThreadId
        => Environment.CurrentManagedThreadId.ToString(CultureInfo.CurrentCulture);
    public long Timestamp => _timeStamp == -1L
        ? _timeStamp = Stopwatch.GetTimestamp()
        : _timeStamp;

    internal static int GetProcessId()
    {
        if (!s_hasProcessId)
        {
            s_processId = (int)GetCurrentProcessId();
            s_hasProcessId = true;
        }
        return s_processId;
    }

    [DllImport("kernel32.dll")]
    internal static extern uint GetCurrentProcessId();
}
```

上面介绍了与跟踪日志上下文相关的两个类型，下面介绍 TraceFilter 类型。如果 SourceSwitch 是 TraceSource 用来针对所有注册 TraceListener 的全局开关，那么 TraceFilter 就是隶属于某个具体 TraceListener 的私有开关。当 TraceSource 将包含上下文信息的跟踪日志消息推送给 TraceListener 之后，后者会利用 TraceFilter 对其做进一步过滤，不满足过滤条件的日志消息会被忽略。图 8-10 展示了引入 TraceListener 对象后跟踪日志模型的完整结构。

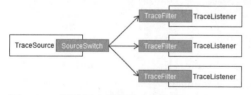

图 8-10　跟踪日志模型的完整结构

如下面的代码片段所示，TraceFilter 是一个抽象类，针对跟踪日志的过滤实现在唯一的方法 ShouldTrace 中。从方法的定义可以看出，用来检验是否满足过滤条件的输入包括承载当前上下文信息的 TraceEventCache 对象、TraceSource 的名称、跟踪日志事件类型、跟踪事件 ID、用于格式化消息内容的模板与参数列表、提供的原始数据（data1 和 data）。

```
public abstract class TraceFilter
{
```

```csharp
    public abstract bool ShouldTrace(TraceEventCache cache, string source,
        TraceEventType eventType, int id, string formatOrMessage, object[] args,
        object data1, object[] data);
}
```

系统提供了两个具体的预定义 TraceFilter 类型（EventTypeFilter 和 SourceFilter），它们分别针对事件类型和 TraceSource 名称对跟踪日志进行过滤。如下所示的代码片段展示了这两个类型的成员定义。

```csharp
public class EventTypeFilter : TraceFilter
{
    public SourceLevels EventType { get; set; }
    public EventTypeFilter(SourceLevels level) => EventType = level;

    public override bool ShouldTrace(TraceEventCache cache, string source,
        TraceEventType eventType, int id, string formatOrMessage, object[] args,
        object data1, object[] data)
        => ((int)eventType & (int)EventType) > 0;
}

public class SourceFilter : TraceFilter
{
    private string _source;
    public string Source
    {
        get => _source;
        set => _source = value ?? throw new ArgumentNullException("Source");
    }
    public SourceFilter(string source)
        => _source = source?? throw new ArgumentNullException(nameof(source));

    public override bool ShouldTrace(TraceEventCache cache, string source,
        TraceEventType eventType, int id, string formatOrMessage, object[] args,
        object data1, object[] data)
        => Source == source;
}
```

下面将关注点重新转移到 TraceListener 上。如下面的代码片段所示，TraceListener 是一个抽象类型，具有一系列的属性成员，其中包括作为名称的 Name、用于过滤跟踪日志的 Filter，以及用于控制上下文信息收集的 TraceOutputOptions。NeedIndent 属性、IndentLevel 属性和 IndentSize 属性与格式化日志输出内容采用的缩进设置有关，分别表示是否需要缩进、当前缩进的等级及每次缩进的步长。布尔类型的属性 IsThreadSafe 表示针对跟踪日志的处理是否是线程安全的，该属性默认返回 False，如果某个具体的 TraceListener 采用线程安全的方式处理跟踪日志，就应该重写该属性。TraceListener 还具有一个名为 Attributes 的只读属性，它利用一个 StringDictionary 对象作为容器来存放任意添加的属性。

```csharp
public abstract class TraceListener : MarshalByRefObject, IDisposable
{
```

```csharp
    public virtual string            Name { get; set; }
    public TraceFilter               Filter { get; set; }
    public TraceOptions              TraceOutputOptions { get; set; }

    protected bool                   NeedIndent { get; set; }
    public int                       IndentLevel { get; set; }
    public int                       IndentSize { get; set; }

    public virtual bool              IsThreadSafe { get; }
    public StringDictionary          Attributes { get; }

    protected TraceListener();
    protected TraceListener(string name);

    public abstract void Write(string message);
    public abstract void WriteLine(string message);
    ...
}
```

TraceListener 提供了很多方法来处理 TraceSource 分发给它的跟踪日志，这些方法最终都会利用 TraceFilter 判断跟踪日志是否满足过滤条件，不满足过滤条件的跟踪日志不会被处理。满足过滤规则的跟踪日志会被格式化成字符串，该字符串最终通过两个抽象方法（Write 和 WriteLine）输出到对应的目标渠道，所以一个具体的 TraceListener 类型往往只需要重写这两个方法即可。

下面介绍 TraceListener 提供的几组用于处理跟踪日志消息的方法，其中最核心的是 3 个 TraceEvent 方法重载，调用它们会发送相应的跟踪事件。这 3 个方法发出的跟踪事件包含如下信息：承载上下文信息的 TraceEventCache 对象、TraceSource 的名称、跟踪日志事件的类型和 ID，以及以两种不同的方式（消息文本或者消息模板和参数）提供日志消息的荷载内容。

```csharp
public abstract class TraceListener : MarshalByRefObject, IDisposable
{
    public virtual void TraceEvent(TraceEventCache eventCache, string source,
        TraceEventType eventType, int id);
    public virtual void TraceEvent(TraceEventCache eventCache, string source,
        TraceEventType eventType, int id, string message);
    public virtual void TraceEvent(TraceEventCache eventCache, string source,
        TraceEventType eventType, int id, string format, params object[] args);
    ...
}
```

除了上面定义的 Write 方法和 WriteLine 方法，TraceListener 还定义了如下这些方法重载，它们在根据指定的内容荷载和日志类别（Category）生成格式化的消息文本后，会直接调用上面介绍的抽象方法 Write 或者 WriteLine 将其分发到对应的输出渠道。值得注意的是，针对 TraceListener 的跟踪事件过滤对于这些方法同样有效，并且进行过滤规则检验时采用的事件类型为 Verbose。另外，这些方法并不涉及跟踪事件的 ID，但是这个 ID 对于其他方法则是必需的。

```csharp
public abstract class TraceListener : MarshalByRefObject, IDisposable
{
```

```csharp
    public virtual void Write(object o);
    public virtual void Write(object o, string category);
    public virtual void Write(string message, string category);
    public virtual void WriteLine(object o);
    public virtual void WriteLine(object o, string category);
    public virtual void WriteLine(string message, string category);
    ...
}
```

下面的 TraceTransfer 方法针对的是"活动转移"跟踪事件，所以我们需要提供关联的活动 ID。从如下所示的代码片段可以看出，TraceTransfer 方法最终还是调用 TraceEvent 方法来处理跟踪日志消息。除了将跟踪事件类型设置为 Transfer，TraceTransfer 方法还会将提供的关联活动 ID 作为后缀添加到指定的消息上。

```csharp
public abstract class TraceListener : MarshalByRefObject, IDisposable
{
    public virtual void TraceTransfer(TraceEventCache eventCache, string source, int id,
        string message, Guid relatedActivityId)
    {
        TraceEvent(eventCache, source, TraceEventType.Transfer, id,
            message + ", relatedActivityId=" + relatedActivityId.ToString());
    }
}
```

除了提供一个字符串作为日志消息的内容荷载，还可以调用如下两个 TraceData 方法重载，它们会将一个对象或者对象数组作为内容荷载，TraceFilter 的 ShouldTrace 方法最后的两个参数分别对应这两个对象。当这两个方法被调用的时候，它们会直接调用内容荷载对象的 ToString 方法将对象转换成字符串。TraceListener 还定义了其他一些方法，由于篇幅有限，本节不再一一介绍，有兴趣的读者可以参阅相关的在线文档。

```csharp
public abstract class TraceListener : MarshalByRefObject, IDisposable
{
    public virtual void TraceData(TraceEventCache eventCache, string source,
        TraceEventType eventType, int id, object data);
    public virtual void TraceData(TraceEventCache eventCache, string source,
        TraceEventType eventType, int id, params object[] data);
    ...
}
```

由于大部分 TraceListener 类型都涉及对跟踪日志内容的持久化或者网络传输，为了提供更好的性能，缓冲机制一般都会被采用。对于 TraceSource 分发的跟踪日志，TraceListener 通常先暂存于缓冲区内，等累积到一定数量之后再进行批量输出。如果希望立即输出缓存区内的跟踪日志，则可以调用它的 Flush 方法。除此之外，TraceListener 还实现了 IDisposable 接口，实现的 Dispose 方法可以帮助我们释放相应的资源。TraceListener 还定义了一个 Close 方法，按照一般的约定，该方法与 Dispose 方法是等效的。一般来说，针对 Dispose 方法或者 Close 方法的调用会强制输出缓存区中保存的跟踪日志。

```csharp
public abstract class TraceListener : MarshalByRefObject, IDisposable
{
```

```
    public virtual void Flush();
    public virtual void Close();
    public void Dispose();
}
```

TraceSource

跟踪日志事件最初都是由某个 TraceSource 对象发出的。如下面的代码片段所示，每个 TraceSource 对象都有一个通过 Name 属性表示的名称，它的 Switch 属性用于获取和设置作为全局开关的 SourceSwitch 对象，而 Listeners 属性则保存所有注册的 TraceListener 对象。创建一个 TraceSource 对象时需要指定其名称（必需）和用于创建 SourceSwitch 的 SourceLevels 枚举（可选），如果后者没有显式指定，则默认提供的等级为 Off，意味着该 TraceSource 处于完全关闭的状态。

```
public class TraceSource
{
    public TraceListenerCollection    Listeners { get; }
    public string                     Name { get; }
    public SourceSwitch               Switch { get; set; }

    public TraceSource(string name);
    public TraceSource(string name, SourceLevels defaultLevel);
    ...
}
```

定义在 TraceSource 类型中的所有公共方法全部标注了 ConditionalAttribute 特性，对应的条件编译符被设置为 TRACE。如果目标程序集是在 TRACE 条件编译符不存在的情况下编译生成的，则意味着所有与跟踪日志相关的代码都不复存在。下面定义的 TraceEvent 方法重载是 TraceSource 最核心的 3 个方法，我们可以调用它们发送指定类型的跟踪事件。调用这些方法时除了可以指定跟踪事件类型和事件 ID，我们还可以采用两种不同的形式提供日志内容荷载。如果满足过滤规则，这些方法最终会调用注册的 TraceListener 对象的同名方法。

```
public class TraceSource
{
    [Conditional("TRACE")]
    public void TraceEvent(TraceEventType eventType, int id);
    [Conditional("TRACE")]
    public void TraceEvent(TraceEventType eventType, int id, string message);
    [Conditional("TRACE")]
    public void TraceEvent(TraceEventType eventType, int id, string format,
        params object[] args);
    ...
}
```

TraceSource 还提供了如下两个名为 TraceInformation 的方法重载，它们最终还是调用 TraceEvent 方法并将跟踪事件类型设置为 Information。而 TraceTransfer 方法会发送一个类型为 Transfer 的跟踪事件，它最终调用的是注册 TraceListener 对象的同名方法。TraceSource 具有两个名为 TraceData 的方法，它们分别使用一个对象和对象数组作为跟踪日志的内容荷载，

TraceSource 类型同样具有对应的方法定义。

```
public class TraceSource
{
    [Conditional("TRACE")]
    public void TraceInformation(string message);
    [Conditional("TRACE")]
    public void TraceInformation(string format, params object[] args);
    [Conditional("TRACE")]
    public void TraceTransfer(int id, string message, Guid relatedActivityId);

    [Conditional("TRACE")]
    public void TraceData(TraceEventType eventType, int id, object data);
    [Conditional("TRACE")]
    public void TraceData(TraceEventType eventType, int id, params object[] data);

    ...
}
```

由于 TraceSource 最终会驱动注册的 TraceListener 对象来对由它发出的跟踪日志做最后的输出，这个过程可能涉及对跟踪日志的缓存，所以 TraceSource 定义 Flush 方法来驱动所有的 TraceListener 来"冲洗"它们的缓冲区。由于 TraceListener 实现了 IDispose 接口，所以 TraceSource 同样需要用下面的 Close 方法来释放它们持有的资源。

```
public class TraceSource
{
    public void Close();
    public void Flush();
    ...
}
```

8.3.2 预定义 TraceListener

在跟踪日志框架中，TraceListener 被用来捕捉由 TraceSource 对象发出的跟踪日志事件，并对接收到的日志消息进行消费。一般来说，我们会利用注册的 TraceListener 对象对跟踪日志消息进行持久化存储（如将格式化的日志消息保存在文件或者数据库中）或者可视化显示（如输出到控制台上），或者发送到某个服务做进一步处理。跟踪日志系统定义了几个原生的 TraceListener 类型。

DefaultTraceListener

创建的 TraceSource 对象的 Listeners 属性会自动添加一个 DefaultTraceListener 对象。由本章的实例演示可知，DefaultTraceListener 对象会将日志消息作为调试信息发送给 Debugger，这是通过它重写的 Write 方法和 WriteLine 方法完成的。其实，DefaultTraceListener 对象还可以帮助我们将日志内容写入指定的物理文本文件中，日志文件的路径可以通过 LogFileName 属性来指定。

```
public class DefaultTraceListener: TraceListener
{
    public string LogFileName { get; set; }
```

```csharp
public override void Write(string message);
public override void WriteLine(string message);
...
}
```

我们可以通过一个简单的程序来演示如何利用 DefaultTraceListener 对象将跟踪日志写入指定的文件中。如下面的代码片段所示，创建一个 TraceSource 对象之后，我们将默认注册的 TraceListener 清除，然后显式创建并注册了一个 DefaultTraceListener 对象，并将日志文件的路径设置为 trace.log。下面利用 TraceSource 对象针对每种事件类型分别记录一条跟踪日志消息。

```csharp
class Program
{
    static void Main()
    {
        var source = new TraceSource("Foobar", SourceLevels.All);
        source.Listeners.Clear();
        source.Listeners.Add(new DefaultTraceListener { LogFileName = "trace.log" });
        var eventTypes = (TraceEventType[])Enum.GetValues(typeof(TraceEventType));
        var eventId = 1;
        Array.ForEach(eventTypes, it =>
            source.TraceEvent(it, eventId++, $"This is a {it} message."));
    }
}
```

运行程序后可以发现，编译输出目录下有一个名为 trace.log 的日志文件，程序中生成的 10 条跟踪日志会逐条写入该文件中（见图 8-11）。DefaultTraceListener 在进行针对文件的日志输出的时候，仅仅是将格式化的日志消息以追加（Append）的形式写入指定的文件而已。但在实际的项目中，我们会考虑日志文件的大小，为了避免日志文件变得过大，要么需要对现有的日志进行存档，要么需要每隔一定的时间或者文件大小超过指定的阈值时换用新的日志文件，所以 DefaultTraceListener 针对文件的日志持久化并没有太大的使用价值。（S808）

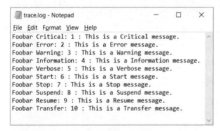

图 8-11　针对静态类型 Trace 的跟踪事件分发处理机制

TextWriterTraceListener

由于跟踪日志的内容荷载最终都会格式化成一个字符串，所以针对跟踪日志的输出体现为字符串文本的输出，这个过程可以通过一个抽象的 TextWriter 来完成。一个 TextWriterTraceListener 对象本质上就是对一个 TextWriter 对象的封装，并且最终由后者完成针对跟踪日志文本内容的输出工作。如下面的代码片段所示，这个 TextWriter 对象体现在 TextWriterTrace

Listener 的 Writer 属性上。

```csharp
public class TextWriterTraceListener : TraceListener
{
    public TextWriter Writer { get; set; }

    public TextWriterTraceListener();
    public TextWriterTraceListener(Stream stream);
    public TextWriterTraceListener(TextWriter writer);
    public TextWriterTraceListener(Stream stream, string name);
    public TextWriterTraceListener(TextWriter writer, string name);
    public TextWriterTraceListener(string fileName);
    public TextWriterTraceListener(string fileName, string name);

    public override void Write(string message);
    public override void WriteLine(string message);

    public override void Close();
    protected override void Dispose(bool disposing);
    public override void Flush();
}
```

TextWriterTraceListener 具有 3 个派生类型：ConsoleTraceListener、DelimitedListTraceListener 和 XmlWriterTraceListener。根据类型命名可以看出，它们将日志消息分别写入控制台、分隔符列表文件和 XML 文件。下面着重介绍 DelimitedListTraceListener 类型。

DelimitedListTraceListener

DelimitedListTraceListener 是 TextWriterTraceListener 的子类，这是一个很有用的 TraceListener。正如它的名称所表述的一样，DelimitedListTraceListener 对象在对跟踪日志信息进行格式化的时候会采用指定的分隔符。如下面的代码片段所示，DelimitedListTraceListener 的 Delimiter 代表的就是这个分隔符，默认情况下采用分号（;）作为分隔符。

```csharp
public class DelimitedListTraceListener : TextWriterTraceListener
{
    public string Delimiter { get; set; }

    public DelimitedListTraceListener(Stream stream);
    public DelimitedListTraceListener(TextWriter writer);
    public DelimitedListTraceListener(string fileName);
    public DelimitedListTraceListener(Stream stream, string name);
    public DelimitedListTraceListener(TextWriter writer, string name);
    public DelimitedListTraceListener(string fileName, string name);

    public override void TraceData(TraceEventCache eventCache, string source,
        TraceEventType eventType, int id, object data);
    public override void TraceData(TraceEventCache eventCache, string source,
        TraceEventType eventType, int id, params object[] data);
    public override void TraceEvent(TraceEventCache eventCache, string source,
        TraceEventType eventType, int id, string message);
    public override void TraceEvent(TraceEventCache eventCache, string source,
```

```
        TraceEventType eventType, int id, string format, params object[] args);
}
```

基于分隔符的格式化实现在重写的 TraceData 方法和 TraceEvent 方法中，所以调用 TraceSource 对象的 Write 方法或者 WriteLine 方法时输出的内容不会采用分隔符进行分隔。对于第二个 TraceData 方法重载，如果传入的内容荷载对象是一个数组，那么每个元素之间同样会采用分隔符进行分隔。在默认情况下采用的分隔符为逗号（,），但是如果 Delimiter 属性表示的主分隔符为逗号，此分隔符就会选择分号。如下所示的代码片段展示了在选用默认分隔符的情况下分别通过 TraceData 方法和 TraceEvent 方法输出的文本格式。

```
TraceData 1:
{SourceName};{EventType};{EventId};{Data};{ProcessId};{LogicalOperationStack};{ThreadId};{DateTime};{Timestamp};
TraceData 2:
{SourceName};{EventType};{EventId};{Data1},{Data2},...,{DataN};{ProcessId};{LogicalOperationStack};{ThreadId};{DateTime};{Timestamp};

TraceEvent
{SourceName};{EventType};{EventId};{Message};;{ProcessId};{LogicalOperationStack};{ThreadId};{DateTime};{Timestamp};
```

上面展示的跟踪日志输出格式中的占位符{LogicalOperationStack}表示当前逻辑操作的堆栈或者调用链。上述代码片段还揭示了另一个细节，那就是对 TraceEvent 方法的输出格式来说，在表示日志消息主体内容的{Message}和表示进程 ID 的{ProcessId}之间会出现两个分隔符，这可能是一个漏洞（Bug）。

如果我们采用逗号而不是默认的分号作为分隔符，那么最终输出的就是一个合法的 CSV（Comma Separated Value）文件。例如，在如下所示的实例演示中，如果将当前目录下一个名为 trace.csv 的文件作为日志文件，那么在这个文件中写入日志之前，需要添加一行文本作为每列的标题，由此可以看出，输出的内容就是根据针对 TraceEvent 方法的输出格式来指定的。

```
class Program
{
    static void Main()
    {
        var fileName = "trace.csv";
        File.AppendAllText(fileName,
            $"SourceName,EventType,EventId,Message,N/A,ProcessId,LogicalOperationStack,ThreadId,DateTime,Timestamp,{Environment.NewLine}");

        using (var fileStream = new FileStream(fileName, FileMode.Append))
        {
            TraceOptions options = TraceOptions.Callstack | TraceOptions.DateTime |
                TraceOptions.LogicalOperationStack | TraceOptions.ProcessId |
                TraceOptions.ThreadId | TraceOptions.Timestamp;
            var listener = new DelimitedListTraceListener(fileStream)
                { TraceOutputOptions = options, Delimiter = "," };
            var source = new TraceSource("Foobar", SourceLevels.All);
            source.Listeners.Add(listener);
```

```
            var eventTypes = (TraceEventType[])Enum.GetValues(typeof(TraceEventType));
            for (int index = 0; index < eventTypes.Length; index++)
            {
                var enventType = eventTypes[index];
                var eventId = index + 1;
                Trace.CorrelationManager.StartLogicalOperation($"Op{eventId}");
                source.TraceEvent(enventType, eventId, $"This is a {enventType} message.");
            }
            source.Flush();
        }
    }
}
```

下面针对日志文件创建一个 DelimitedListTraceListener 对象，并根据日志文件的路径创建一个 FileStream 对象，然后以此来创建 DelimitedListTraceListener 对象。为了能够捕获所有上下文信息，我们为 DelimitedListTraceListener 对象设置了包含所有选项的 TraceOptions。

我们将这个 DelimitedListTraceListener 注册到创建的 TraceSource 对象上，并利用它针对每种事件类型发出 10 个跟踪事件。为了演示上面提到的逻辑操作堆栈，可以通过 Trace 类型得到一个 CorrelationManager 对象，并调用其 StartLogicalOperation 方法启动一个以 Op{EventId}格式命名的逻辑操作。由于 DelimitedListTraceListener 对象内部采用了缓冲机制，所以我们人为地调用了 TraceSource 对象的 Flush 方法强制输出缓冲区中的跟踪日志。程序运行之后，输出的 10 条跟踪日志将全部记录在 trace.csv 文件中，如果直接利用 Excel 打开这个文件，就会看到图 8-12 所示的内容。（S809）

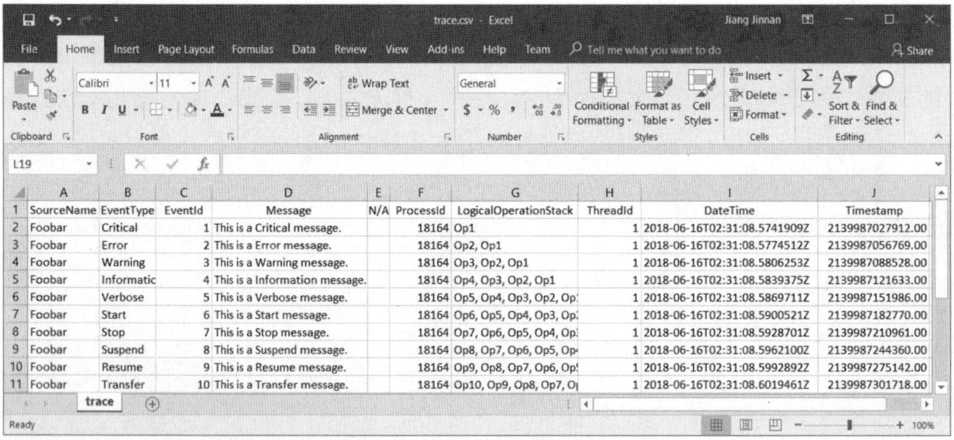

SourceName	EventType	EventId	Message	N/A	ProcessId	LogicalOperationStack	ThreadId	DateTime	Timestamp
Foobar	Critical	1	This is a Critical message.		18164	Op1	1	2018-06-16T02:31:08.5741909Z	2139987027912.00
Foobar	Error	2	This is a Error message.		18164	Op2, Op1	1	2018-06-16T02:31:08.5774512Z	2139987056769.00
Foobar	Warning	3	This is a Warning message.		18164	Op3, Op2, Op1	1	2018-06-16T02:31:08.5806253Z	2139987088528.00
Foobar	Informatic	4	This is a Information message.		18164	Op4, Op3, Op2, Op1	1	2018-06-16T02:31:08.5839375Z	2139987121633.00
Foobar	Verbose	5	This is a Verbose message.		18164	Op5, Op4, Op3, Op2, Op	1	2018-06-16T02:31:08.5869711Z	2139987151986.00
Foobar	Start	6	This is a Start message.		18164	Op6, Op5, Op4, Op3, Op	1	2018-06-16T02:31:08.5900521Z	2139987182770.00
Foobar	Stop	7	This is a Stop message.		18164	Op7, Op6, Op5, Op4, Op	1	2018-06-16T02:31:08.5928701Z	2139987210961.00
Foobar	Suspend	8	This is a Suspend message.		18164	Op8, Op7, Op6, Op5, Op	1	2018-06-16T02:31:08.5962100Z	2139987244360.00
Foobar	Resume	9	This is a Resume message.		18164	Op9, Op8, Op7, Op6, Op	1	2018-06-16T02:31:08.5992892Z	2139987275142.00
Foobar	Transfer	10	This is a Transfer message.		18164	Op10, Op9, Op8, Op7, O	1	2018-06-16T02:31:08.6019461Z	2139987301718.00

图 8-12　通过 DelimitedListTraceListener 输出的日志文件

8.3.3　Trace

除了创建一个 TraceSource 对象来记录跟踪日志，我们还可以直接使用 Trace 类型来完成类似的工作。Trace 是一个本应定义成静态类型的实例类型，因为定义其中的所有成员都是静态的。我们可以将 Trace 类型理解为一个单例的 TraceSource 对象，TraceListener 对象以全局的形式直接注册在这个 Trace 类型之上。与 TraceSource 不同的是，Trace 并不存在一个 SourceSwitch 作为全局开关对发出的跟踪事件进行过滤。也就是说，调用 Trace 相应方法发出的跟踪事件会直接分发给注册的 TraceListener 对象，后者可以通过自身的 TraceFilter 对接收的跟踪事件进行过滤，图 8-13 体现了针对静态类型 Trace 的跟踪事件分发处理机制。

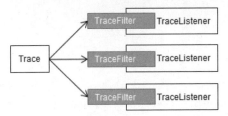

图 8-13　针对静态类型 Trace 的跟踪事件分发处理机制

下面介绍定义在 Trace 类型中与跟踪日志相关的一些成员。与 TraceSource 一样，Trace 的所有公共方法全部标注了 ConditionalAttribute 特性，并将条件编译符设置为 TRACE。如下面的代码片段所示，全局注册的 TraceListener 保存在通过静态属性 Listeners 表示的集合对象中。定义其中的 3 组方法会发送类型分别为 Error、Information 和 Warning 的跟踪事件，这些方法最终调用的是所有注册 TraceListener 对象的 TraceEvent 方法。

```
public sealed class Trace
{
    public static TraceListenerCollection Listeners { get; }

    [Conditional("TRACE")]
    public static void TraceError(string message);
    [Conditional("TRACE")]
    public static void TraceError(string format, params object[] args);

    [Conditional("TRACE")]
    public static void TraceInformation(string message);
    [Conditional("TRACE")]
    public static void TraceInformation(string format, params object[] args);

    [Conditional("TRACE")]
    public static void TraceWarning(string message);
    [Conditional("TRACE")]
    public static void TraceWarning(string format, params object[] args);
    ...
}
```

Trace 同样定义了如下所示的一系列 Write 方法和 WriteLine 方法，以及携带条件的 WriteIf 方法和 WriteLineIf 方法，这些方法最终都会调用注册的 TraceListener 对象的 Write 方法和 WriteLine 方法。

```
public sealed class Trace
{
    [Conditional("TRACE")]
    public static void Write(object value);
```

```csharp
[Conditional("TRACE")]
public static void Write(string message);
[Conditional("TRACE")]
public static void Write(object value, string category);
[Conditional("TRACE")]
public static void Write(string message, string category);

[Conditional("TRACE")]
public static void WriteIf(bool condition, object value);
[Conditional("TRACE")]
public static void WriteIf(bool condition, string message);
[Conditional("TRACE")]
public static void WriteIf(bool condition, object value, string category);
[Conditional("TRACE")]
public static void WriteIf(bool condition, string message, string category);

[Conditional("TRACE")]
public static void WriteLine(object value);
[Conditional("TRACE")]
public static void WriteLine(string message);
[Conditional("TRACE")]
public static void WriteLine(object value, string category);
[Conditional("TRACE")]
public static void WriteLine(string message, string category);

[Conditional("TRACE")]
public static void WriteLineIf(bool condition, object value);
[Conditional("TRACE")]
public static void WriteLineIf(bool condition, string message);
[Conditional("TRACE")]
public static void WriteLineIf(bool condition, object value, string category);
[Conditional("TRACE")]
public static void WriteLineIf(bool condition, string message, string category);
...
}
```

为了使被持久化的日志消息的内容更具结构化和可读性，可以利用定义在 Trace 类型中如下这些与缩进相关的成员。IndentSize 属性表示一次缩进的字符数，默认值为 4，而 IndentLevel 属性则用于返回或者设置当前的缩进层级。除了直接使用 IndentLevel 属性来控制缩进层级，我们还可以调用 Indent 方法和 Unindent 方法以递增或者递减的形式来设置缩进层级。

```csharp
public sealed class Trace
{
    public static int IndentLevel { get; set; }
    public static int IndentSize { get; set; }

    [Conditional("TRACE")]
    public static void Indent();
    [Conditional("TRACE")]
    public static void Unindent();
```

```
...
}
```

Trace 类型中同样定义了与日志输出缓冲机制相关的 API。其中，Flush 方法用来强制输出存储在 TraceListener 中的跟踪日志。如果不希望跟踪日志在 TraceListener 的缓冲区停留，就应该将 AutoFlush 属性设置为 True，在这种情况下，TraceListener 对象的 Flush 方法会在接收到分发的跟踪事件后立即被调用。除了这两个与缓存输出有关的成员，Trace 类型还定义了 Close 方法，它会调用 TraceListener 对象的同名方法来完成对资源的释放。除此之外，Trace 还有一个 CorrelationManager 属性，它返回的 CorrelationManager 对象可以帮助我们开启和关闭一个逻辑操作，以便于记录当前的调用链。

```
public sealed class Trace
{
    public static bool                      AutoFlush { get; set; }
    public static CorrelationManager        CorrelationManager { get; }

    [Conditional("TRACE")]
    public static void Close();
    [Conditional("TRACE")]
    public static void Flush();
}
```

8.4 EventSource 事件日志

基于 EventSource 的日志框架最初是为 Windows 操作系统的事件跟踪系统 ETW（Event Tracing for Windows）设计的，由于采用了发布订阅的设计思想，所以这个日志框架在逻辑上是独立的。换句话说，对于 EventSource 发出的日志事件，ETW 并非唯一的消费者，我们可以通过创建 EventListener 对象来订阅感兴趣的日志事件，并对得到的日志消息做针对性处理。本节主要将 EventSource 视为一个通用的日志框架，同时考虑到 ETW 自身的复杂性，所以本章并不会涉及 ETW，而是将关注点放到针对 EventSource 的编程上。

8.4.1 EventSource

在大部分情况下，我们倾向于定义一个派生于 EventSource 的子类型，并将日志事件的发送实现在对应的方法中，但日志事件的发送也可以直接通过创建的 EventSource 对象来完成。鉴于这两种编程模式，EventSource 定义了两组（Protected 和 Public）构造函数。

```
public class EventSource : IDisposable
{
    public string                       Name { get; }
    public Guid                         Guid { get; }
    public Exception                    ConstructionException { get; }
    public EventSourceSettings          Settings { get; }

    protected EventSource();
    protected EventSource(bool throwOnEventWriteErrors);
```

```
    protected EventSource(EventSourceSettings settings);
    protected EventSource(EventSourceSettings settings, params string[] traits);

    public EventSource(string eventSourceName);
    public EventSource(string eventSourceName, EventSourceSettings config);
    public EventSource(string eventSourceName, EventSourceSettings config,
        params string[] traits);
}
```

一个 EventSource 对象要求具有一个明确的名称,所以调用公共的构造函数创建一个 EventSource 对象时必须显式指定该名称。对于 EventSource 的派生类来说,如果类型上没有通过标注的 EventSourceAttribute 特性对名称进行显式设置,类型的名称将会作为 EventSource 的名称。一个 EventSource 对象还要求被赋予一个 GUID 作为唯一标识,对于直接创建的 EventSource 对象来说,该标识是指定的名称计算出来的,EventSource 的派生类型可以通过标注的 EventSourceAttribute 特性来对这个标识进行显式设置。

EventSource 的 Settings 属性返回一个 EventSourceSettings 对象,这是一个标注了 FlagsAttribute 特性的枚举。该枚举对象针对 EventSource 的"设置"(Settings)主要体现在两个方面:指定输出日志的格式(Manifest 或者 Manifest Self-Describing),以及决定是否应该抛出日志写入过程中出现的异常。如果没有显式设置,EventSource 的派生类型采用的设置应为 EtwManifestEventFormat,而对于直接调用公共构造函数创建的 EventSource 对象来说,它的 Settings 属性默认返回 EtwSelfDescribingEventFormat。除了上述 3 个只读属性,EventSource 还有一个 ConstructionException 属性返回构造函数执行过程中抛出的异常。

```
[Flags]
public enum EventSourceSettings
{
    Default                         = 0,
    ThrowOnEventWriteErrors         = 1,
    EtwManifestEventFormat          = 4,
    EtwSelfDescribingEventFormat    = 8
}
```

当我们在定义 EventSource 派生类时,虽然类型名称会默认作为 EventSource 的名称,但是出现命名冲突的可能性较大,所以通常在类型上通过标注 EventSourceAttribute 特性对名称进行显式设置。这样不但可以定义一个标准的名称(如采用公司和项目或者组件名称作为前缀),而且类型名称修改之后也不会造成任何影响。如下面的代码片段所示,除了指定 EventSource 的名称,通过标注 EventSourceAttribute 特性还可以设置作为唯一标识的 GUID,以及用于本地化的字符串资源。

```
[AttributeUsage(AttributeTargets.Class)]
public sealed class EventSourceAttribute : Attribute
{
    public string Name { get; set; }
    public string Guid { get; set; }
    public string LocalizationResources { get; set; }
}
```

对于 EventSource 派生类来说，它可以通过调用基类的 WriteEvent 方法来发送日志事件。所以调用 WriteEvent 方法时必须指定日志事件的 ID（必需）和内容荷载（可选）。如下面的代码片段所示，EventSource 定义了一系列重载的 WriteEvent 方法来提供具有成员结构的内容荷载。

```
public class EventSource : IDisposable
{
    protected void WriteEvent(int eventId);
    protected void WriteEvent(int eventId, int arg1);
    protected void WriteEvent(int eventId, long arg1);
    protected void WriteEvent(int eventId, string arg1);
    protected void WriteEvent(int eventId, byte[] arg1);
    protected void WriteEvent(int eventId, int arg1, int arg2);
    protected void WriteEvent(int eventId, int arg1, string arg2);
    protected void WriteEvent(int eventId, long arg1, long arg2);
    protected void WriteEvent(int eventId, long arg1, string arg2);
    protected void WriteEvent(int eventId, long arg1, byte[] arg2);
    protected void WriteEvent(int eventId, string arg1, int arg2);
    protected void WriteEvent(int eventId, string arg1, long arg2);
    protected void WriteEvent(int eventId, string arg1, string arg2);
    protected void WriteEvent(int eventId, int arg1, int arg2, int arg3);
    protected void WriteEvent(int eventId, long arg1, long arg2, long arg3);
    protected void WriteEvent(int eventId, string arg1, int arg2, int arg3);
    protected void WriteEvent(int eventId, string arg1, string arg2, string arg3);

    protected void WriteEvent(int eventId, params object[] args);
    ...
}
```

由于有最后一个 WriteEvent 方法重载的存在，所以可以采用一个或者多个 POCO 对象作为日志的内容荷载，这个特性被称为 Rich Event Payload。虽然这个特性可以使针对 EventSource 的日志编程变得很简洁，但是在高频日志写入的应用场景下应该尽可能避免调用这个 WriteEvent 方法，因为涉及针对一个对象数组的创建以及对值类型对象的装箱。

虽然没有做硬性规定，但是我们应该尽可能将 EventSource 派生类型定义在 Sealed 类型，并且采用单例的方式来使用它。在默认情况下，除非显式标注了 NonEventAttribute 特性，否则定义在 EventSource 派生类中返回类型为 void 的公共实例方法会被视为日志事件方法。每个日志事件方法具有对应的事件 ID，如果没有通过在方法上标注 EventAttribute 特性对事件 ID 进行显式设置，EventSource 就会将该方法在类型成员中的序号作为 ID。在调用上面这些 WriteEvent 方法时指定的事件 ID 要求与当前日志事件方法对应的 ID 保持一致。

除了通过在日志事件方法上标注 EventAttribute 特性显式指定事件 ID，我们还可以利用这个特性指定描述当前日志事件的其他一些元数据。如下面的代码片段所示，我们可以通过 EventAttribute 特性设置日志事件的 ID、等级、消息、版本等信息。

```
[AttributeUsage(AttributeTargets.Method)]
public sealed class EventAttribute : Attribute
{
    public int                          EventId { get; private set; }
```

```
    public EventLevel              Level { get; set; }
    public string                  Message { get; set; }
    public byte                    Version { get; set; }
    public EventOpcode             Opcode { get; set; }
    public EventTask               Task { get; set; }
    public EventKeywords           Keywords { get; set; }
    public EventTags               Tags { get; set; }
    public EventChannel            Channel { get; set; }
    public EventActivityOptions    ActivityOptions { get; set; }

    public EventAttribute(int eventId);
}
```

由 EventSource 对象发出的日志事件同样具有等级之分，具体的日志等级可以通过 EventLevel 枚举来表示。针对 EventSource 的日志框架将日志事件从高到低划分为 Critical、Error、Warning、Informational 和 Verbose 这 5 个等级。如果没有通过 EventAttribute 特性对日志等级做显式设置，日志事件方法采用的默认等级就会被设置为 Verbose。

```
public enum EventLevel
{
    LogAlways,
    Critical,
    Error,
    Warning,
    Informational,
    Verbose
}
```

如果对日志内容具有可读性要求，那么最好提供一个完整的消息文本来描述当前事件，EventAttribute 的 Message 提供了一个生成此消息文本的模板。消息模板可以采用{0}，{1}，…，{n}这样的占位符，而调用日志事件方法传入的参数将作为替换它们的参数。

EventSource 的日志采用发布订阅模式，并且日志事件的发布者和订阅者并没有耦合在一起，两者之间采用基于"契约"的交互方式，EventAttribute 特性和 EventSourceAttribute 特性提供了描述交互契约的元数据。为了确保升级之后能提供向后兼容，EventSource 为每个日志事件赋予了通过 Version 属性表示的版本号。

EventAttribute 特性的 Opcode 属性返回一个 EventOpcode 类型的枚举，该枚举表示日志事件对应操作的代码（Code）。我们将定义在 EventOpcode 枚举中的操作代码分成如下几组：第一组的 4 个操作代码与活动（Activity）有关，分别表示活动的开始（Start）、结束（Stop）、中止（Suspend）和恢复（Resume）；第二组的两个操作代码则针对基于数据收集的活动；第三组的 3 个操作代码描述的是与消息交换相关的事件，分别表示针对消息的发送（Send）、接收（Receive）和回复（Reply）；第四组是 Info 和 Extension，前者表示一般性的信息的输出，后者表示一个扩展事件。

```
public enum EventOpcode
{
    Start                       = 1,
    Stop                        = 2,
```

```
    Resume                      = 7,
    Suspend                     = 8,

    DataCollectionStart         = 3,
    DataCollectionStop          = 4,

    Send                        = 9,
    Receive                     = 240,
    Reply                       = 6,

    Info                        = 0,
    Extension                   = 5,
}
```

如果日志事件关联某项任务（Task），就可以用 Task 属性对它进行描述。我们还可以通过 Keywords 属性和 Tags 属性为日志事件关联一些关键字（Keyword）与标签（Tag）。一般来说，事件是根据订阅发送的，如果待发送的日志事件没有订阅者，那么该事件是不应该被发送的，所以日志事件的订阅原则是尽可能缩小订阅的范围，这样就可以将日志导致的性能损耗降到最低。如果为某个日志事件定义了关键字，我们就可以针对该关键字进行精准的订阅。如果当前事件的日志具有特殊的输出渠道，就可以利用其 Channel 属性来承载输出渠道信息。这些属性的返回类型都是枚举，如下所示的代码片段展示了这些枚举类型的定义。

```
[Flags]
public enum EventKeywords : long
{
    All                         = -1L,
    None                        = 0L,
    AuditFailure                = 0x10000000000000L,
    AuditSuccess                = 0x20000000000000L,
    CorrelationHint             = 0x10000000000000L,
    EventLogClassic             = 0x80000000000000L,
    MicrosoftTelemetry          = 0x2000000000000L,
    Sqm                         = 0x8000000000000L,
    WdiContext                  = 0x2000000000000L,
    WdiDiagnostic               = 0x4000000000000L
}

public enum EventChannel : byte
{
    None                        = 0,
    Admin                       = 10,
    Operational                 = 11,
    Analytic                    = 12,
    Debug                       = 13
}

[Flags]
public enum public enum EventTask
{
```

```
    None
}

[Flags]
public enum EventTags
{
    None
}
```

从上面的代码片段可以看出，枚举类型 EventTask 和 EventTags 并没有定义任何有意义的枚举选项（只定义了一个 None 选项）。因为枚举的基础类型都是整型，所以日志事件订阅者最终接收的这些数据都是相应的数字，至于不同的数值具有什么样的语义则完全可以由具体的应用来决定，所以我们可以完全不用关心这些枚举的预定义选项。EventAttribute 特性之所以将这些属性定义成枚举，并不是要求我们使用预定义的选项，只是提供一种强类型的编程方式而已，如可以采用如下形式定义 4 个 EventTags 常量来表示 4 种数据库类型。除了 EventTask 和 EventTags，表示关键字的 EventKeywords 也可以采用这种方式进行自由定义。

```
public class Tags
{
    public const EventTags MSSql    = (EventKeywords)1;
    public const EventTags Oracle   = (EventKeywords)2;
    public const EventTags Redis    = (EventKeywords)4;
    public const EventTags Mongodb  = (EventKeywords)8;
}
```

在大部分情况下，针对单一事件的日志数据往往没有实际意义，只有将在某个上下文中记录下来的一组相关日志进行聚合分析才能得到有价值的结果，采用基于"活动"（Activity）的跟踪是关联单一日志事件的常用手段。一个所谓的活动具有严格的开始边界和结束边界，并且需要耗费一定时间才能完成的操作。活动具有标准的状态机，状态之间的转换可以通过对应的事件来表示，一个活动的生命周期介于开始事件和结束事件之间。

EventSource 日志框架对基于活动的追踪（Activity Tracking）提供了很好的支持，EventAttribute 特性的 ActivityOptions 属性正是用来做这方面设置的。该属性返回的 EventActivityOptions 枚举具有如下选项：Disable 用于关闭活动追踪特性，Recursive 和 Detachable 则分别表示是否允许活动以递归或者重叠的方式运行。我们将在后续部分详细讨论这个话题。

```
[Flags]
public enum EventActivityOptions
{
    None        = 0,
    Disable     = 2,
    Recursive   = 4,
    Detachable  = 8
}
```

由上面介绍的内容可知，定义在 EventSource 派生类中的日志方法是通过调用基类的

WriteEvent 方法来发送日志事件的，但是对于一个直接调用公共构造函数创建的 EventSource 对象来说，这个受保护的 WriteEvent 方法是无法直接调用的，只能调用下面几个公共的 Write 方法和 Write<T>方法来发送日志事件。

```
public class EventSource : IDisposable
{
    public void Write(string eventName);
    public void Write(string eventName, EventSourceOptions options);
    public void Write<T>(string eventName, T data);
    public void Write<T>(string eventName, EventSourceOptions options, T data);
    public void Write<T>(string eventName, ref EventSourceOptions options, ref T data);
    public void Write<T>(string eventName, ref EventSourceOptions options,
        ref Guid activityId, ref Guid relatedActivityId, ref T data);
    ...
}
```

在调用 Write 方法和 Write<T>方法时需要设置事件的名称（必需）和其他相关设置（可选），还可以通过一个对象的形式来提供日志内容荷载。EventSource 相关的选项可以通过下面的 EventSourceOptions 结构来表示，我们可以利用它设置事件等级、操作代码、关键字、标签和活动追踪的其他选项。

```
[StructLayout(LayoutKind.Sequential)]
public struct EventSourceOptions
{
    public EventLevel              Level { get; set; }
    public EventOpcode             Opcode { get; set; }
    public EventKeywords           Keywords { get; set; }
    public EventTags               Tags { get; set; }
    public EventActivityOptions    ActivityOptions { get; set; }
}
```

另外，前面提到的 ETW 的两种格式（Manifest 和 Manifest Self-Describing）分别对应枚举类型 EventSourceSettings 的 EtwManifestEventFormat 选项和 EtwSelfDescribingEventFormat 选项，后者提供了针对 Rich Event Payload 的支持。由于泛型的 Write<T>方法采用的正是对 Rich Event Payload 特性的体现，所以通过公共构造函数创建的 EventSource 对象的 Settings 属性会被自动设置为 EtwSelfDescribingEventFormat。

只有在 EventSource 对象被订阅的前提下，针对它发送日志事件才有意义，所以为了避免无谓操作造成的性能损失，在发送日志事件之前应该调用如下几个 IsEnabled 方法，它们可以帮助我们确认 EventSource 对象针对指定的日志等级、关键字或者输出渠道是否具有订阅者。

```
public class EventSource : IDisposable
{
    public bool IsEnabled();
    public bool IsEnabled(EventLevel level, EventKeywords keywords);
    public bool IsEnabled(EventLevel level, EventKeywords keywords, EventChannel channel);
    ...
}
```

8.4.2 EventListener

EventListener 提供了一种在进程内（In-Process）订阅和消费日志事件的手段。EventListener 对象能够接收由 EventSource 分发的日志事件的前提是它向 EventSource 订阅了它感兴趣的日志事件。EventListener 对象针对 EventSource 就某种日志事件类型的订阅可以通过如下几个 EnableEvents 方法重载来完成，我们在调用这几个方法的时候可以通过指定日志等级、关键字和命令参数的方式对分发的日志事件进行过滤。除了这几个方法重载，EventListener 还定义了 DisableEvents 方法来解除订阅。

```
public abstract class EventListener : IDisposable
{
    public void EnableEvents(EventSource eventSource, EventLevel level);
    public void EnableEvents(EventSource eventSource, EventLevel level,
        EventKeywords matchAnyKeyword);
    public void EnableEvents(EventSource eventSource, EventLevel level,
        EventKeywords matchAnyKeyword, IDictionary<string, string> arguments);

    public void DisableEvents(EventSource eventSource);
    ...
}
```

虽然 EventListener 需要通过调用 EnableEvents 方法显式地就感兴趣的日志事件向 EventSource 发起订阅，但是由于 EventListener 能够感知到当前进程内任意一个 EventSource 对象的创建，所以订阅变得异常容易。如下面的代码片段所示，EventListener 类型定义了两个受保护的回调方法：OnEventSourceCreated 方法会在当前进程创建了任何一个 EventSource 对象时被调用，作为参数的正是被创建的那个 EventSource 对象，自定义的 EventListener 类型可以通过重写这个方法完成日志事件的订阅。

```
public class EventListener : IDisposable
{
    public event EventHandler<EventSourceCreatedEventArgs> EventSourceCreated;
    public event EventHandler<EventWrittenEventArgs> EventWritten;

    protected internal virtual void OnEventSourceCreated(EventSource eventSource);
    protected internal virtual void OnEventWritten(EventWrittenEventArgs eventData);
}
```

对于某个 EventSource 对象发出的日志事件，如果事件满足作为订阅者的 EventListener 设置的订阅条件，日志事件相关的信息会被封装成一个 EventWrittenEventArgs 对象，并作为参数调用 EventListener 的 OnEventWritten 方法，自定义的 EventListener 类型可以通过重写这个方法完成日志事件的消费。定义在 EventListener 的这两个方法会触发 EventSourceCreated 事件和 EventWritten 事件，所以重写这两个方法时最好能够调用基类的同名方法。

如下所示的代码片段是 EventWrittenEventArgs 类型的定义，我们不仅可以通过它获取包括荷载内容在内的用于描述当前日志事件的所有信息，还可以得到作为事件源的 EventSource 对象。EventSource 对象在发送日志事件时可以采用不同的形式来指定内容荷载，它们最终会转换成一个 ReadOnlyCollection<object> 对象分发给作为订阅者的 EventListener 对象。除了表示荷载对象集合的 Payload 属性，EventWrittenEventArgs 类型还提供了 PayloadNames 属性来表示每个荷载对象对应的名称。

```csharp
public class EventWrittenEventArgs : EventArgs
{
    public EventSource                          EventSource { get; }
    public int                                  EventId { get; }
    public string                               EventName { get; }
    public EventLevel                           Level { get; }
    public string                               Message { get; }
    public byte                                 Version { get; }
    public EventKeywords                        Keywords { get; }
    public EventOpcode                          Opcode { get; }
    public EventTags                            Tags { get; }
    public EventTask                            Task { get; }
    public EventChannel                         Channel { get; }
    public Guid                                 ActivityId { get; }
    public Guid                                 RelatedActivityId { get; }

    public ReadOnlyCollection<object>           Payload { get; }
    public ReadOnlyCollection<string>           PayloadNames { get; }
}
```

8.1 节已经演示了如何使用 EventSource 和 EventListener，下面做一个更加完整的演示。首先以常量的形式预定义一个 EventTask 对象和 EventTags 对象，前者表示操作执行所在的应用层次，后者表示 3 种典型的关系数据库类型。

```csharp
public class Tasks
{
    public const EventTask UI           = (EventTask)1;
    public const EventTask Business     = (EventTask)2;
    public const EventTask DA           = (EventTask)3;
}

public class Tags
{
    public const EventTags MSSQL        = (EventTags)1;
    public const EventTags Oracle       = (EventTags)2;
    public const EventTags Db2          = (EventTags)3;
}
```

我们依然沿用执行 SQL 命令的应用场景，为此需要将之前定义的 DatabaseSource 进行改写。首先在日志方法 OnCommandExecute 上标注的 EventAttribute 特性对它的所有属性都做了相应的设置，其中 Task 属性和 Tags 属性使用的是上面定义的常量。值得注意的是，我们为 Message 属性设置了一个包含占位符的模板，希望最终能够得到一个完整的文本信息来描述当前的事件。在 OnCommandExecute 方法内部，在调用 WriteEvent 方法发送日志事件之前，我们调用 IsEnabled 方法确认 EventSource 对象针对指定的等级和输出渠道已经被订阅。

```csharp
[EventSource(Name = "Artech-Data-SqlClient")]
public sealed class DatabaseSource : EventSource
```

```csharp
{
    public static DatabaseSource Instance = new DatabaseSource();
    private DatabaseSource() {}

    [Event(1, Level = EventLevel.Informational, Keywords = EventKeywords.None,
        Opcode = EventOpcode.Info, Task = Tasks.DA, Tags = Tags.MSSQL, Version = 1,
        Message = "Execute SQL command. Type: {0}, Command Text: {1}")]
    public void OnCommandExecute(CommandType commandType, string commandText)
    {
        if (IsEnabled(EventLevel.Informational, EventKeywords.All, EventChannel.Debug))
        {
            WriteEvent(1, (int)commandType, commandText);
        }
    }
}
```

如下所示的代码片段是作为日志事件订阅者的 DatabaseSourceListener 类型的重新定义。为了验证接收的日志事件的描述信息是否与 OnCommandExecute 方法的定义一致，我们在重写的 OnEventWritten 方法中输出了 EventWrittenEventArgs 对象的所有属性。

```csharp
public class DatabaseSourceListener : EventListener
{
    protected override void OnEventSourceCreated(EventSource eventSource)
    {
        if (eventSource.Name == "Artech-Data-SqlClient")
        {
            EnableEvents(eventSource, EventLevel.LogAlways);
        }
    }

    protected override void OnEventWritten(EventWrittenEventArgs eventData)
    {
        Console.WriteLine($"EventId: {eventData.EventId}");
        Console.WriteLine($"EventName: {eventData.EventName}");
        Console.WriteLine($"Channel: {eventData.Channel}");
        Console.WriteLine($"Keywords: {eventData.Keywords}");
        Console.WriteLine($"Level: {eventData.Level}");
        Console.WriteLine($"Message: {eventData.Message}");
        Console.WriteLine($"Opcode: {eventData.Opcode}");
        Console.WriteLine($"Tags: {eventData.Tags}");
        Console.WriteLine($"Task: {eventData.Task}");
        Console.WriteLine($"Version: {eventData.Version}");
        Console.WriteLine($"Payload");
        var index = 0;
        foreach (var payloadName in eventData.PayloadNames)
        {
            Console.WriteLine($"\t{payloadName}:{eventData.Payload[index++]}");
        }
    }
}
```

```
class Program
{
    static void Main()
    {
        var listener = new DatabaseSourceListener();
        DatabaseSource.Instance.OnCommandExecute(CommandType.Text, "SELECT * FROM T_USER");
    }
}
```

在作为程序入口的 Main 方法中，我们创建了 DatabaseSourceListener 对象，并且调用 DatabaseSource 的 OnCommandExecute 发送了一个关于 SQL 命令执行的日志事件。程序运行之后，控制台输出的日志事件的所有信息的形式如图 8-14 所示。（S810）

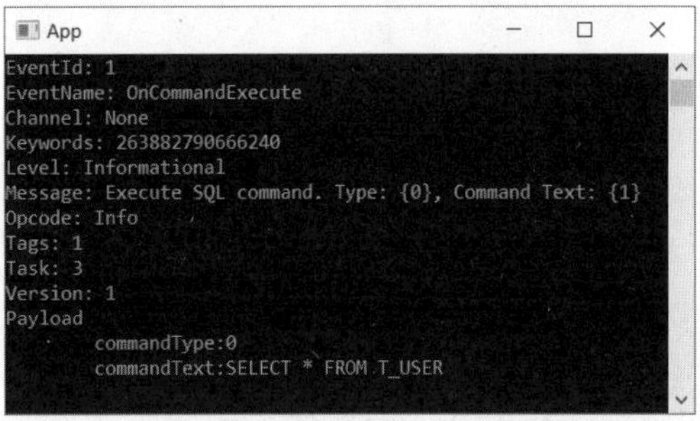

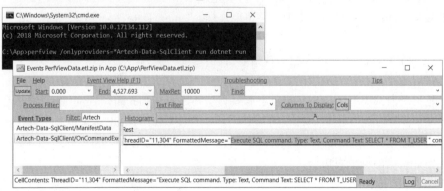

图 8-14　针对静态类型 Trace 的跟踪事件分发处理机制

由图 8-14 所示的输出结果可以发现，EventWrittenEventArgs 的 Message 属性返回的依然是没有做任何格式化的原始信息，笔者认为这是值得改进的地方。如果将日志事件发送给 ETW，它就会对消息进行有效的格式化（见图 8-15）。为了确认这一点，可以用命令行的形式利用 PerfView 来启动该应用，具体的命名为 perfview /onlyproviders=*Artech-Data-SqlClient run dotnet run。在利用 PerfView 查看对应事件的相关信息时，格式化的文本消息可以通过 FormattedMessage 属性保存起来，格式化文本内容为 "Execute SQL command. Type: Text, Command Text: SELECT * FROM T_USER"。

图 8-15　ETW 针对消息的格式化

8.4.3 荷载对象序列化

EventSource 针对 ETW 的日志事件发送属于跨进程通信，提供的原始荷载对象需要经过序列化才能实现跨进程传递。针对 EventListener 的日志事件订阅虽然没有跨越进程的边界，但是为了解除发布者和订阅者之间的耦合，最终传递给 EventListener 的并不是原始的荷载对象，而是经过序列化之后的结果。正是因为需要对原始的荷载对象进行序列化，所以并不是任何类型的对象都能成为 EventSource 的荷载对象。总的来说，能够作为日志事件内容荷载的对象主要包括以下几点。

- 所有原生类型（Primitive Type），包括 Boolean、Byte、Int16、Int32、Int64、Float 和 Double。
- 字符串、DateTime、DateTimeOffset 和 Guid。
- 类型上标注了 EventDataAttribute 特性或者 CompilerGeneratedAttribute 特性的自定义类型。
- IEnumerable<T>和 KeyValuePair<string, T>，其中，T 为有效的荷载类型。

也就是说，如果需要将一个自定义类型作为日志荷载对象，就应该在该类型上标注一个 EventDataAttribute 特性。按照上面的规则，我们也可以通过标注 CompilerGeneratedAttribute 特性来定义合法的日志荷载类型，但这个特性主要是用来识别匿名类型的。匿名类型是常用的荷载类型，匿名类型在被编译后会自动标注一个 CompilerGeneratedAttribute 特性。

将使用一个 POCO 对象作为内容荷载的做法称为 Rich Event Payload，这个特性仅对 EtwSelfDescribingEventFormat 格式才有效，但 EventSource 的派生类采用的默认格式为 EtwManifestEventFormat。如果我们希望支持这个特性，就需要通过标注的 EventSourceAttribute 特性将格式类型显式设置为 EtwSelfDescribingEventFormat。直接创建的 EventSource 对象的格式默认就是 EtwSelfDescribingEventFormat，所以可以直接使用一个有效的荷载作为参数调用其 Write<T>方法来发送相应的日志事件。

对于一个 EventListener 对象来说，它接收的日志内容荷载表示一个 ReadOnlyCollection<object>对象，组成这个集合的元素主要包含如下 3 种类型。

- 标量值（Scalar Value）：对应原生类型、字符串、DateTime、DateTimeOffset 和 Guid 类型的数据成员。
- 对象数组：对应 IEnumerable<T>类型的数据成员。
- EventPayload 对象：对应其他类型的数据成员。

EventPayload 是一个内部类型，自定义的日志荷载类型和 KeyValuePair<string, T>最终都会转换成这样一个对象。从下面给出的代码片段可以看出，一个 EventPayload 对象本质上就是一个 Key 类型和 Value 类型分别为 String 与 Object 的字典，也可以认为是一个 Key 类型和 Value 类型分别为 String 与 Object 的键值对集合。

```
internal class EventPayload : IDictionary<string, object>
{
    ...
}
```

下面介绍作为内容荷载的 POCO 对象在序列化之后具有什么样的结构。先从最简单的结构开始介绍，对于下面这个具有两个属性成员的 Foobar 对象，当它转换成 EventPayload 对象后数据结构是什么样的。

```
[EventData]
public class Foobar
{
    public int Foo { get; set; }
    public int Bar { get; set; }

    public Foobar(int foo, int bar)
    {
        Foo = foo;
        Bar = bar;
    }
}
```

标注 EventDataAttribute 特性的目的在于签署一份双方认同的序列化协议，该协议主要体现在会将公共属性作为有效的数据成员进行序列化。所以，对于一个 Foobar 对象来说，假设它的 Foo、Bar 的属性值分别为 1 和 2，由它序列化之后的 EventPayload 字典将由两个键值对组成，它们的 Key 和 Value 就是对应属性成员的名称与值，序列化之后生成的 EventPayload 对象将具有如图 8-16 所示的结构。

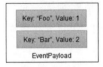

图 8-16　简单荷载对象序列化之后的结构

上面介绍了面向 EventPayload 的序列化规则，但是如果 EventSource 将具有如下定义的 Payload 对象作为荷载对象，那么当它转换成 EventListener 对象接收 ReadOnlyCollection<object>之后，这个集合对象会具有什么样的元素。

```
[EventData]
public class Payload
{
    public Foobar                        Foobar     { get; set; }
    public IEnumerable<Foobar>           Collection { get; set; }
    public IDictionary<int, Foobar>      Dictionary { get; set; }
}
```

虽然一个 Payload 对象包含 3 个公共属性成员，但它是作为一个单一的对象被序列化的，所以最终生成的 ReadOnlyCollection<object>只包含一个元素，该元素对应的是一个 EventPayload 对象。如图 8-17 所示，这个 EventPayload 对象由 3 个键值对组成，分别对应 Payload 对象的 3 个属性。由于 Foobar 属性的属性值就是一个 Foobar 对象，所以该对象将采用上述规则转换成一个 EventPayload 对象。假设 Payload 对象的 Collection 属性包含两个元素，那么对应键值对的 Value 将是一个包含两个元素的数组，数组元素为通过 Foobar 对象转换而成的

EventPayload 对象。

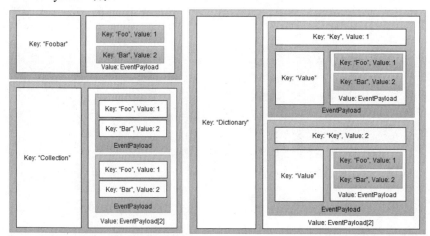

图 8-17　Payload 对象序列化之后的结构

虽然 Payload 对象的 Dictionary 属性是一个字典类型，但是我们应该将它视为一个键值对的集合。假设该属性同样由两个元素构成，对应的 Key 分别为 1 和 2，由于值为集合，所以对应键值对的 Value 应该是一个包含两个元素的数组，每个数组元素对应通过一个 KeyValuePair<Int32, Foobar> 对象转换而成的 EventPayload 对象。可以将一个键值对视为一个包含 Key 和 Value 两个属性成员的普通荷载类型，所以由它转换而成的 EventPayload 对象将由两个键值对构成，它们分别承载 Key 和 Value 的信息。

为了证实上述序列化规则，下面通过一个具体的实例按照图 8-17 所示的结构对 EventListener 对象接收的日志荷载进行解析。首先定义 FoobarSource 的自定义 EventSource 类型，日志事件方法 Test 在调用 WriteEvent 方法时将传入的 Payload 对象作为内容荷载。派生于 EventListener 的 FoobarListener 并没有重写任何方法。

```
public sealed class FoobarSource : EventSource
{
    public static readonly FoobarSource Instance = new FoobarSource();
    private FoobarSource() : base(EventSourceSettings.EtwSelfDescribingEventFormat) { }

    [Event(1)]
    public void Test(Payload payload) => WriteEvent(1, payload);
}
public class FoobarListener : EventListener {}
```

接下来我们编写了如下这段程序来发送日志事件。首先创建 FoobarListener 对象并对 FoobarSource 进行了订阅。在注册的 EventWritten 事件中，应严格按照图 8-17 所示的结构对接收的日志内容荷载进行解析，并将解析结果输出到控制台上。然后创建一个 Payload 对象，并将其作为荷载对象调用了 FoobarSource 对象的 Test 方法将对应的日志事件发送出去。

```
class Program
{
```

```csharp
static void Main()
{
    var listener = new FoobarListener();
    listener.EnableEvents(FoobarSource.Instance, EventLevel.LogAlways);
    listener.EventWritten += (sender, args) => {
        string ToString(object foobar)
        {
            var dictionary = (IDictionary<string, object>)foobar;
            return $"(Foo={dictionary["Foo"]}, Bar={dictionary["Bar"]})";
        }
        var eventData = (IDictionary<string, object>)args.Payload.Single();
        Console.WriteLine($"Foobar: {ToString(eventData["Foobar"])}");

        var array = (object[])eventData["Collection"];
        Console.WriteLine("Collection:");
        for (int index = 0; index < array.Length; index++)
        {
            Console.WriteLine($"\t[{index}]: {ToString(array[index])}");
        }

        Console.WriteLine("Dictionary:");
        array = (object[])eventData["Dictionary"];
        foreach (IDictionary<string, object> eventPayload in array)
        {
            var key = eventPayload["Key"];
            Console.WriteLine(
                $"\tKey: {key}: Value: {ToString(eventPayload["Value"])}");
        }
    };

    var payload = new Payload
    {
        Foobar = new Foobar(1, 2),
        Collection = new Foobar[] { new Foobar(11, 12), new Foobar(21, 22),
            new Foobar(31, 32) },
        Dictionary = new Dictionary<int, Foobar>
        {
            [1] = new Foobar(11, 12),
            [2] = new Foobar(21, 22),
            [3] = new Foobar(31, 32)
        }
    };

    FoobarSource.Instance.Test(payload);
}
```

对于上面创建的 Payload 对象来说，它的 Foobar 属性被赋予了一个完整的对象，而 Collection 属性则被设置为一个包含 3 个元素的数组。Dictionary 属性被添加了 3 个 Foobar 对象作为键值对的 Value，对应的 Key 分别为 1、2 和 3。程序运行后在控制台上输出的结果如

图8-18所示，该结果与 Payload 对象具有一致的结构。（S811）

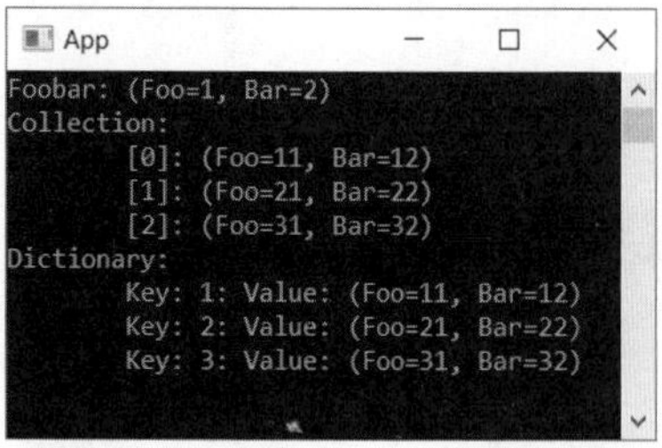

图 8-18　EventListener 对 Payload 的解析

8.4.4　活动跟踪

基于 TraceSource 的跟踪日志框架可以利用 CorrelationManager 人为地开始和结束一个逻辑操作，进而描绘出一条完整的调用链，我们可以将这个特性称为针对活动的跟踪（Activity Tracking）。代表一个逻辑操作的活动具有一个唯一的标识，其生命周期由开始事件和结束事件决定。由于一个逻辑操作可能由多个子操作协作完成，所以多个活动构成一个树形的层次化结构，逻辑调用链正是由当前活动的路径来体现的。

对于作为日志事件订阅者的 EventListener 来说，它利用 EventWrittenEventArgs 对象得到描述订阅日志事件的所有信息。如果开启了针对活动的跟踪，那么当前活动的标识也会包含其中，如下所示的代码片段中的 ActivityId 属性表示的就是这个标识，而 RelatedActivityId 属性则代表"父操作"的活动 ID。

```
public class EventWrittenEventArgs : EventArgs
{
    public Guid ActivityId { get; }
    public Guid RelatedActivityId { get; }
    ...
}
```

由于一个活动的生命周期可以由开始事件和结束事件来界定，所以可以通过发送两个关联的事件来定义一个代表逻辑操作的活动，针对 EventSource 的日志框架正是采用这种方式来实现针对活动的跟踪的。具体来说，假设需要定义一个名为 Foobar 的活动，就需要利用 EventSource 发送两个名称分别为 FoobarStart 和 FoobarStop 的事件。

要构建一个完整的调用链，必须将代表当前逻辑操作的活动流转信息保存下来，由于涉及异步调用，当前活动标识会保存在 AsyncLocal<T>对象中。针对当前活动流转信息的保存是通过 TplEtwProvider 的 EventSource 来实现的。Tpl 代表 Task Parallel Library，所以 TplEtwProvider 是为 Task 的并行编程框架创建的 ETW Provider。日志活动流转信息的保存涉及如下两个事件方法，它们具有相同的日志事件等级（Informational）和关键字（8L）。如果要利用一个自定义的 EventListener 来记录完整的调用链，就需要订阅这两个事件。

```
[EventSource(Name="System.Threading.Tasks.TplEventSource",
```

```
    Guid="2e5dba47-a3d2-4d16-8ee0-6671ffdcd7b5")]
internal sealed class TplEtwProvider : EventSource
{
    [Event(14, Version=1, Level=EventLevel.Informational, Keywords=8L)]
    public void TraceOperationBegin(int TaskID, string OperationName, long RelatedContext);

    [Event(15, Version=1, Level=EventLevel.Informational, Keywords=8L)]
    public void TraceOperationEnd(int TaskID, AsyncCausalityStatus Status);
    ...
}
```

下面通过一个简单的实例来演示如何利用自定义的 EventSource 和 EventListener 来完成针对活动的跟踪。假设一个完整的调用链由 Foo、Bar、Baz 和 Qux 这 4 个活动组成，为此我们定义了 FoobarSource。针对 4 个活动分别定义了 4 组对应的事件方法，其中 XxxStart 方法和 XxxStop 方法分别对应活动的开始事件与结束事件，前者的荷载信息包含活动开始的时间戳，后者的荷载信息包含操作耗时。

```
[EventSource(Name = "Foobar")]
public sealed class FoobarSource : EventSource
{
    public static FoobarSource Instance = new FoobarSource();

    [Event(1)]
    public void FooStart(long timestamp) => WriteEvent(1, timestamp);
    [Event(2)]
    public void FooStop(double elapsed) => WriteEvent(2, elapsed);

    [Event(3)]
    public void BarStart(long timestamp) => WriteEvent(3, timestamp);
    [Event(4)]
    public void BarStop(double elapsed) => WriteEvent(4, elapsed);

    [Event(5)]
    public void BazStart(long timestamp) => WriteEvent(5, timestamp);
    [Event(6)]
    public void BazStop(double elapsed) => WriteEvent(6, elapsed);

    [Event(7)]
    public void QuxStart(long timestamp) => WriteEvent(7, timestamp);
    [Event(8)]
    public void QuxStop(double elapsed) => WriteEvent(8, elapsed);
}
```

如下所示的 FoobarListener 会订阅上面的这些事件，并将接收的调用链的信息保存到一个 .csv 文件中（log.csv）。在重写的 OnEventSourceCreated 方法中，除了根据 EventSource 的名称注册由 FoobarSource 对象发出的 8 个事件，还需要订阅前面介绍的由 TplEtwProvider 发出的用于保存活动流转信息的事件。本着尽量缩小订阅范围的原则，我们在调用 EnableEvents 方法时采用日志等级和关键字进行了过滤。在重写的 OnEventWritten 方法中，我们将捕捉到的日志事件的相关信息（名称、活动开始时间戳和耗时、ActivityId 和 RelatedActivityId）进行格式化后写入指定的.csv 文件中。

```csharp
public sealed class FoobarListener : EventListener
{
    protected override void OnEventSourceCreated(EventSource eventSource)
    {
        if (eventSource.Name == "System.Threading.Tasks.TplEventSource")
        {
            EnableEvents(eventSource, EventLevel.Informational, (EventKeywords)0x80);
        }

        if (eventSource.Name == "Foobar")
        {
            EnableEvents(eventSource, EventLevel.LogAlways);
        }
    }

    protected override void OnEventWritten(EventWrittenEventArgs eventData)
    {
        if (eventData.EventSource.Name == "Foobar")
        {
            var timestamp = eventData.PayloadNames[0] == "timestamp"
                ? eventData.Payload[0]
                : "";
            var elapsed = eventData.PayloadNames[0] == "elapsed"
                ? eventData.Payload[0]
                : "";
            var relatedActivityId = eventData.RelatedActivityId == default
                ? ""
                : eventData.RelatedActivityId.ToString();
            var line = $"{eventData.EventName},{timestamp},{elapsed}
                ,{eventData.ActivityId},{relatedActivityId}";
            File.AppendAllLines("log.csv", new string[] { line });
        }
    }
}
```

如下所示的代码片段可以模拟由 Foo、Bar、Baz 和 Qux 这 4 个活动组成的调用链。针对这些活动的控制实现在 InvokeAsync 方法中，该方法的参数 start 和 stop 提供的委托对象分别用来发送活动的开始事件与结束事件，至于参数 body 返回的 Task 对象则代表了活动自身的操作。为了模拟活动耗时，我们人为地等待了一个随机的时间。

```csharp
public class Program
{
    static Random _random = new Random();
    public static async Task Main()
    {
```

```
        File.AppendAllLines("log.csv",
            new string[] { "EventName,StartTime,Elapsed,ActivityId,RelatedActivityId" });
        var listener = new FoobarListener();
        await FooAsync();
    }

    static Task FooAsync() => InvokeAsync(FoobarSource.Instance.FooStart,
        FoobarSource.Instance.FooStop, async () =>
    {
        await BarAsync();
        await QuxAsync();
    });
    static Task BarAsync() => InvokeAsync(FoobarSource.Instance.BarStart,
        FoobarSource.Instance.BarStop, BazAsync);
    static Task BazAsync() => InvokeAsync(FoobarSource.Instance.BazStart,
        FoobarSource.Instance.BazStop, () => Task.CompletedTask);
    static Task QuxAsync() => InvokeAsync(FoobarSource.Instance.QuxStart,
        FoobarSource.Instance.QuxStop, () => Task.CompletedTask);

    static async Task InvokeAsync(Action<long> start, Action<double> stop, Func<Task> body)
    {
        start(Stopwatch.GetTimestamp());
        var sw = Stopwatch.StartNew();
        await Task.Delay(_random.Next(10, 100));
        await body();
        stop(sw.ElapsedMilliseconds);
    }
}
```

4 个活动分别实现在 4 个对应的方法中（FooAsync、BarAsync、BazAsync 和 QuxAsync），为了模拟基于 Task 的异步编程，可以使这 4 个方法统一返回一个 Task 对象。从这 4 个方法的定义可以看出，它们体现的调用链如图 8-19 所示。

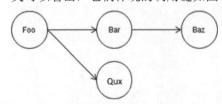

图 8-19　由相关活动构建的调用链

在作为程序入口的 Main 方法中，我们调用了 FooAsync 方法，并在这之前创建了一个 FoobarListener 对象来订阅日志事件，从而将格式化的事件信息写入指定的.csv 文件中。程序运行之后，可以在.csv 文件中看到 8 条对应的日志事件记录。如图 8-20 所示，Start 事件和 Stop 事件分别记录了活动的开始时间戳与耗时，而 ActivityId 和 RelatedActivityId 可以清晰地反映整个调用链的流转。（S812）

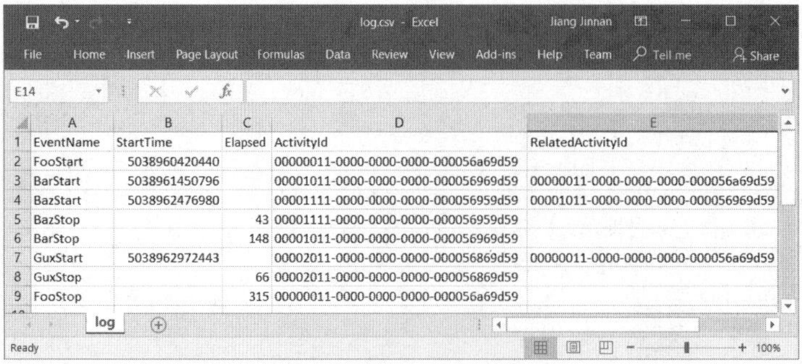

	A	B	C	D	E
1	EventName	StartTime	Elapsed	ActivityId	RelatedActivityId
2	FooStart	5038960420440		00000011-0000-0000-0000-000056a69d59	
3	BarStart	5038961450796		00001011-0000-0000-0000-000056969d59	00000011-0000-0000-0000-000056a69d59
4	BazStart	5038962476980		00001111-0000-0000-0000-000056959d59	00001011-0000-0000-0000-000056969d59
5	BazStop		43	00001111-0000-0000-0000-000056959d59	
6	BarStop		148	00001011-0000-0000-0000-000056969d59	
7	GuxStart	5038962972443		00002011-0000-0000-0000-000056869d59	00000011-0000-0000-0000-000056a69d59
8	GuxStop		66	00002011-0000-0000-0000-000056869d59	
9	FooStop		315	00000011-0000-0000-0000-000056a69d59	

图 8-20　记录在.csv 文件中的调用链信息

8.4.5　性能计数

传统的 .NET Framework 提供的 System.Diagnostics.PerformanceCounter 类型可以帮助我们收集 Windows 操作系统下物理机或者进程的性能指标，基于 PerformanceCounter 类型的性能计数 API 在 .NET Core 下被彻底放弃。但是 .NET Core 程序的很多核心性能指标都会采用事件的方式发出来，具体使用的就是如下所示的这个名为 RuntimeEventSource 的内部类型。

```
[EventSource(Guid="49592C0F-5A05-516D-AA4B-A64E02026C89", Name="System.Runtime")]
internal sealed class RuntimeEventSource : EventSource
{
    ...
}
```

我们可以利用 EventListener 对象监听由 RuntimeEventSource 发送的事件，进而得到当前的性能指标。如下所示的代码片段就是用来获取性能计数的 PerformanceCounterListener 类型的定义。在重写的 OnEventSourceCreated 方法中，可以根据名称订阅针对 RuntimeEventSource 的事件。在具体调用 EnableEvents 方法时，我们提供了一个字典作为参数，该参数利用一个名为 EventCounterIntervalSec 的元素将取样的时间间隔设置为 5 秒。

```
public class PerformanceCounterListener: EventListener
{
    private static HashSet<string> _keys = new HashSet<string>
        { "Count", "Min", "Max", "Mean", "Increment" };
    private static DateTimeOffset? _lastSampleTime;

    protected override void OnEventSourceCreated(EventSource eventSource)
    {
        base.OnEventSourceCreated(eventSource);
        if (eventSource.Name == "System.Runtime")
        {
            EnableEvents(eventSource, EventLevel.Critical, (EventKeywords)(-1),
                new Dictionary<string, string> { ["EventCounterIntervalSec"] = "5" });
        }
    }

    protected override void OnEventWritten(EventWrittenEventArgs eventData)
```

```csharp
    {
        if (_lastSampleTime != null &&
            DateTimeOffset.UtcNow - _lastSampleTime.Value > TimeSpan.FromSeconds(1))
        {
            Console.WriteLine();
        }
        _lastSampleTime = DateTimeOffset.UtcNow;
        var metrics = (IDictionary<string, object>)eventData.Payload[0];
        var name = metrics ["Name"];
        var values = metrics
            .Where(it=>_keys.Contains(it.Key))
            .Select(it => $"{it.Key} = {it.Value}");
        var timestamp = DateTimeOffset.UtcNow.ToString("yyyy-MM-dd hh:mm::ss");
        Console.WriteLine(
            $"[{timestamp}]{name, -32}: {string.Join("; ", values.ToArray())}");
    }
}

class Program
{
    static void Main()
    {
        _ = new PerformanceCounterListener();
        Console.Read();
    }
}
```

在重写的 OnEventWritten 方法中，可以得到性能计数时间的内容载荷（体现为一个字典对象），并从中提取出性能指标的名称（Name）和相关的采样值（Max、Min、Count、Mean 和 Increment）。提取出的性能指标数据连同当前时间戳经过格式化后直接输出到控制台上。在作为入口的 Main 方法中，我们直接创建了 PerformanceCounterListener 对象，它会以 5 秒的间隔收集当前的性能指标，并以图 8-21 所示的形式输出到控制台上。

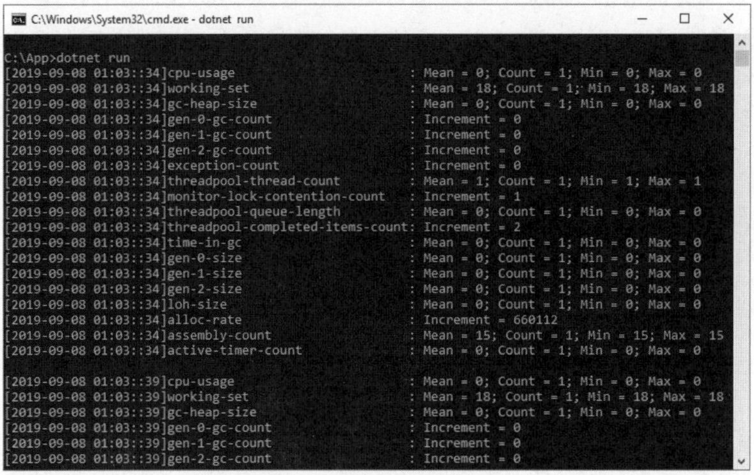

图 8-21　利用 EventListener 对象收集性能指标

如图 8-21 所示，利用 PerformanceCounterListener 对象几乎可以收集到 .NET Core 程序所在进程以及物理机的绝大部分核心指标，其中包括 CPU、内存、GC、线程池相关的指标。如果需要开发 APM（Application Performance Management）框架，或者直接集成第三方 APM（如 Elastic APM 和 SkyWalking 等），就可以直接利用这种方式采集所需的性能指标。(S813)

8.5　DiagnosticSource 诊断日志

对于以 TraceSource 和 EventSource 为核心的日志框架来说，它们关注的主要是日志荷载内容在进程外的持久化或者传输问题，所以被 TraceSource 对象作为日志内容荷载的仅限于一个字符串。EventSource 对象虽然可以将一个 POCO 对象作为内容荷载，作为订阅者的 EventListener 对象也可以在同一进程内完成对日志事件的处理，但是最终输出的其实还是序列化之后的结果。基于 DiagnosticSource 对象的日志框架与它们具有本质的不同，作为发布者的 DiagnosticSource 对象会将原始的日志荷载对象直接分发给处于同一进程内的订阅者。

8.5.1　标准的观察者模式

日志框架应该是观察者模式（或者发布订阅模式）最典型的应用，前面介绍的基于 TraceSource 和 EventSource 的日志框架采用的都是这种模式，DiagnosticSource 诊断日志框架也不例外。观察者模式在 DiagnosticSource 诊断日志框架中的使用更加标准，因为作为日志事件的发布者和订阅者类型需要显式实现 IObservable<T>接口和 IObserver<T>接口。

观察者模式体现为观察者（订阅者）预先就某个主题向被观察者（发布者）注册一个订阅，被观察者在进行主题发送时会选择匹配的观察者实施定点发送。上面提到的 IObservable<T>接口和 IObserver<T>接口分别对应主题发布涉及的这两种角色，即发布者和订阅者，而泛型参数 T 代表发布主题的类型。如下面的代码片段所示，IObservable<T>接口定义了唯一的方法 Subscribe，用来注册订阅者；表示订阅者的 IObserver<T>接口通过核心方法 OnNext 作为回调处理被观察者发布的主题，而 OnCompleted 方法和 OnError 方法则分别在所有主题全部发布结束与出现错误时被调用。

```
public interface IObservable<out T>
{
    IDisposable Subscribe(IObserver<T> observer);
}

public interface IObserver<in T>
{
    void OnCompleted();
    void OnError(Exception error);
    void OnNext(T value);
}
```

DiagnosticSource 诊断日志框架中作为发布者的是一个名为 DiagnosticListener 的类型，虽然从命名来看它更像是一个订阅者，但是它的确是一个实现了 IObservable<T>接口的日志事件发布者。抽象类型 DiagnosticSource 是 DiagnosticListener 的基类，与日志事件发送相关的方法基本上都定义在这个抽象类中。

```
public abstract class DiagnosticSource
{
    protected DiagnosticSource();

    public abstract bool IsEnabled(string name);
    public virtual bool IsEnabled(string name, object arg1, object arg2 = null);

    public abstract void Write(string name, object value);

    public Activity StartActivity(Activity activity, object args);
    public void StopActivity(Activity activity, object args);
}
```

如上面的代码片段所示，DiagnosticListener 提供的 Write 方法用来发送诊断日志事件，该方法的两个参数分别代表日志事件的名称和内容荷载。两个 IsEnabled 方法用来确定指定名称和对应参数的订阅是否存在，在发送日志事件之前先通过这两个方法进行有效性确认绝对是必要甚至可以认为是必需的。StartActivity 方法和 StopActivity 方法旨在实现针对活动的跟踪，后面会对此做单独介绍。

DiagnosticListener 实现了 IObservable<KeyValuePair<string, object>>接口，这意味着由它发布的主题内容体现为一个键值对，该键值对的 Key 表示事件名称，Value 则表示内容荷载。与创建一个 TraceSource 对象或者 EventSource 对象类似，创建 DiagnosticListener 对象时同样需要指定一个确定的名称。

```
public class DiagnosticListener : DiagnosticSource,
    IObservable<KeyValuePair<string, object>>, IDisposable
{
    public string Name { get; }
    public DiagnosticListener(string name);

    public bool IsEnabled();
    public override bool IsEnabled(string name);
    public override bool IsEnabled(string name, object arg1, object arg2 = null);

    public virtual IDisposable Subscribe(IObserver<KeyValuePair<string, object>> observer);
    public virtual IDisposable Subscribe(IObserver<KeyValuePair<string, object>> observer,
        Predicate<string> isEnabled);
    public virtual IDisposable Subscribe(IObserver<KeyValuePair<string, object>> observer,

        Func<string, object, object, bool> isEnabled);

    public override void Write(string name, object value);
    public virtual void Dispose();
}
```

既然作为发布者体现为一个 IObservable<KeyValuePair<string, object>>对象，那么日志事件

的订阅者就是一个 IObserver<KeyValuePair<string, object>>对象，通过调用 Subscribe 方法可以将这样的对象作为订阅者注册到指定的 DiagnosticListener 对象上。在调用 Subscribe 方法注册订阅者时，我们还可以指定一个额外的过滤条件。如果指定的是一个 Predicate<string>对象，就意味着针对事件名称进行过滤；如果指定的是一个 Func<string, object, object, bool>对象，就意味着作为过滤规则的输入除了需要包含事件名称，还包括两个额外的上下文参数。

DiagnosticListener 具有 3 个 IsEnabled 方法重载，用来确定当前是否具有指定类型的订阅者。如果具有任何一个订阅者，无参的 IsEnabled 方法就会返回 True，另外两个方法重载则对应两个指定过滤条件的 Subscribe 方法重载，这个方法会利用注册时指定的过滤条件来确定是否具有匹配的订阅者。

作为诊断日志事件的发布者和订阅者虽然采用进程内同步通信，但是两者在逻辑上属于相互独立的个体，如果要实现针对诊断日志事件的订阅，就需要先发现并获取作为事件发布者的 DiagnosticListener 对象。为了解决这个问题，DiagnosticListener 定义的 AllListeners 属性可以存储当前进程创建的所有 DiagnosticListener 对象。

```
public class DiagnosticListener :
    DiagnosticSource,
    IObservable<KeyValuePair<string, object>>, IDisposable
{
    public static IObservable<DiagnosticListener> AllListeners { get; }
    ...
}
```

由于代表所有 DiagnosticListener 集合的 AllListeners 属性对外体现为一个 IObservable<DiagnosticListener>对象，所以可以将诊断日志事件的注册封装成一个 IObserver<DiagnosticListener>对象来完成。如果调用 Subscribe 方法将一个 IObserver<DiagnosticListener>对象作为订阅者注册到 AllListeners 属性上，指定的订阅不仅会应用到目前创建的 DiagnosticListener 对象上，对于后续创建的 DiagnosticListener 对象，这些订阅者依然会按照定义的规则注册到它们上面。

8.5.2 AnonymousObserver<T>

不论是代表诊断日志事件发布者的 DiagnosticListener，还是定义在类型上代表所有 DiagnosticListener 对象的静态属性 AllListeners，它们都体现为一个 IObservable<T>对象。要完成针对它们的订阅，需要创建一个对应的 IObserver<T>对象，而 AnonymousObserver<T>就是对 IObserver<T>接口的一个简单的实现。

```
public abstract class ObserverBase<T> : IObserver<T>, IDisposable
{
    public void OnNext(T value);
    protected abstract void OnNextCore(T value);

    public void OnCompleted();
    protected abstract void OnCompletedCore();

    public void OnError(Exception error);
```

```
    protected abstract void OnErrorCore(Exception error);

    public void Dispose();
    protected virtual void Dispose(bool disposing);
}
public sealed class AnonymousObserver<T> : ObserverBase<T>
{
    public AnonymousObserver(Action<T> onNext);
    public AnonymousObserver(Action<T> onNext, Action onCompleted);
    public AnonymousObserver(Action<T> onNext, Action<Exception> onError);
    public AnonymousObserver(Action<T> onNext, Action<Exception> onError,
        Action onCompleted);

    protected override void OnNextCore(T value);
    protected override void OnCompletedCore();
    protected override void OnErrorCore(Exception error);
}
```

AnonymousObserver<T>定义在 NuGet 包 "System.Reactive.Core" 中，它采用与演示实例提供的 Observer<T>一样的实现方式，即通过指定的委托对象（类型分别为 Action<T>和 Action<Exception>）实现 IObservable<T>接口的 3 个方法。除了 AnonymousObserver<T>类型，该 NuGet 包还提供如下这些扩展方法来订阅注册。

```
public static class ObservableExtensions
{
    public static IDisposable Subscribe<T>(this IObservable<T> source);
    public static IDisposable Subscribe<T>(this IObservable<T> source, Action<T> onNext);
    public static IDisposable Subscribe<T>(this IObservable<T> source, Action<T> onNext,
        Action onCompleted);
    public static IDisposable Subscribe<T>(this IObservable<T> source, Action<T> onNext,
        Action<Exception> onError);
    public static IDisposable Subscribe<T>(this IObservable<T> source, Action<T> onNext,
        Action<Exception> onError, Action onCompleted);

    public static void Subscribe<T>(this IObservable<T> source, CancellationToken token);
    public static void Subscribe<T>(this IObservable<T> source, Action<T> onNext,
        CancellationToken token);
    public static void Subscribe<T>(this IObservable<T> source, IObserver<T> observer,
        CancellationToken token);
    public static void Subscribe<T>(this IObservable<T> source, Action<T> onNext,
        Action onCompleted, CancellationToken token);
    public static void Subscribe<T>(this IObservable<T> source, Action<T> onNext,
        Action<Exception> onError, CancellationToken token);
}
```

调用上面的这些方法会使日志事件的订阅注册变得非常简单。如下所示的代码片段体现了 Web 服务器针对一次 HTTP 请求处理的日志输出，服务器在接收请求后以日志的方式输出请求上下文信息和当前时间戳，在成功发送响应之后输出响应消息和整个请求处理的耗时。

```
public class Program
```

```csharp
public static void Main()
{
    DiagnosticListener.AllListeners.Subscribe(listener =>
    {
        if (listener.Name == "Web")
        {
            listener.Subscribe(eventData =>
            {
                if (eventData.Key == "ReceiveRequest")
                {
                    dynamic payload = eventData.Value;
                    var request = (HttpRequestMessage)(payload.Request);
                    var timestamp = (long)payload.Timestamp;
                    Console.WriteLine($"Receive request. Url: {request.RequestUri};
                        Timstamp:{timestamp}");
                }
                if (eventData.Key == "SendReply")
                {
                    dynamic payload = eventData.Value;
                    var response = (HttpResponseMessage)(payload.Response);
                    var elaped = (TimeSpan)payload.Elaped;
                    Console.WriteLine($"Send reply. Status code: {response.StatusCode};
                        Elaped: {elaped}");
                }
            });
        }
    });

    var source = new DiagnosticListener("Web");
    var stopwatch = Stopwatch.StartNew();
    if (source.IsEnabled("ReceiveRequest"))
    {
        var request = new HttpRequestMessage(HttpMethod.Get, "https://www.artech.top");
        source.Write("ReceiveRequest", new { Request = request,
            Timestamp = Stopwatch.GetTimestamp() });
    }
    Task.Delay(100).Wait();
    if (source.IsEnabled("SendReply"))
    {
        var response = new HttpResponseMessage(HttpStatusCode.OK);
        source.Write("SendReply", new { Response = response,
            Elaped = stopwatch.Elapsed});
    }
}
```

对于诊断日志发布的部分，需要先创建一个名为 Web 的 DiagnosticListener 对象，并利用开启的 Stopwatch 来计算请求处理耗时。我们利用手动创建的 HttpRequestMessage 对象来模拟接收到的请求，在调用 Write 方法发送一个名为 ReceiveRequest 的日志事件时，该 HttpRequest

Message 对象连同当前时间戳以一个匿名对象的形式作为日志的内容荷载对象。在人为地等待 100 毫秒以模拟请求处理耗时之后，我们调用 DiagnosticListener 对象的 Write 方法发出名为 SendReply 的日志事件，标志着针对当前请求的处理已经结束，作为内容荷载的匿名对象包含手动创建的一个 HttpResponseMessage 对象和模拟耗时。

程序前半段针对日志事件的订阅是通过调用 Subscribe 扩展方法实现的，在指定的 Action<DiagnosticListener>委托对象中，根据名称筛选出作为订阅目标的 DiagnosticListener 对象，然后订阅它的 ReceiveRequest 事件和 SendReply 事件。对于订阅的 ReceiveRequest 事件，我们采用动态类型（dynamic）的方式得到了代表当前请求的 HttpRequestMessage 对象和当前时间戳，并将请求 URL 和当前时间戳打印出来。SendReply 事件可以用相同的方法提取代表响应消息的 HttpResponseMessage 对象和耗时，并将响应状态码和耗时打印出来。程序运行之后，在控制台上看到的输出结果如图 8-22 所示。（S814）

图 8-22　针对请求的跟踪

8.5.3　强类型的事件订阅

为了降低日志事件发布者和订阅者之间的耦合度，日志事件的内容荷载在很多情况下都会采用匿名类型对象来表示，所以在本章开篇和前面演示的实例中，我们只能采用 dynamic 关键字将荷载对象转换成动态类型后才能提取出所需的成员。由于匿名类型并非公共类型，所以上述方式仅限于发布程序和订阅程序都在同一个程序集中才使用，但是在绝大部分情况下这个条件是不满足的。在不能使用动态类型提取数据成员的情况下，我们不得不采用反射或者表达式树的方式来解决这个问题，虽然可行但会变得很烦琐。

强类型日志事件订阅以一种很"优雅"的方式解决了这个问题。简单来说，所谓的强类型日志事件订阅就是将日志订阅处理逻辑定义在某个类型对应的方法中，这个方法可以按照日志内容荷载对象的成员结构来定义对应的参数。实现强类型的日志事件订阅需要实现两个绑定，即日志事件与方法之间的绑定，以及荷载的数据成员与订阅方法参数之间的绑定。

参数绑定利用荷载成员的属性名与参数名之间的映射来实现，所以订阅方法只需要根据荷载对象的属性成员来决定对应的参数的类型和名称。日志事件与方法之间的映射则可以利用下面的 DiagnosticNameAttribute 特性来实现，我们只需要在订阅方法上标注这个方法并指定映射的日志事件的名称即可。

```
public class DiagnosticNameAttribute : Attribute
{
    public string Name { get; }
    public DiagnosticNameAttribute(string name);
}
```

强类型诊断日志事件的订阅对象可以通过 DiagnosticListener 的如下几个扩展方法来完成，它们定义在 NuGet 包 "Microsoft.Extensions.DiagnosticAdapter" 中。顾名思义，这些 SubscribeWithAdapter 方法重载在指定对象和标准订阅对象之间做了一个适配，它将订阅对象转换成一个 IObserver<KeyValuePair<string, object>>对象。

```csharp
public static class DiagnosticListenerExtensions
{
    public static IDisposable SubscribeWithAdapter(this DiagnosticListener diagnostic,
        object target);
    public static IDisposable SubscribeWithAdapter(this DiagnosticListener diagnostic,
        object target, Func<string, bool> isEnabled);
    public static IDisposable SubscribeWithAdapter(this DiagnosticListener diagnostic,
        object target, Func<string, object, object, bool> isEnabled);
}
```

下面将前面演示的实例改造成强类型日志事件订阅的方式。首先定义如下这个 DiagnosticCollector 作为日志事件订阅类型，可以看出，这仅仅是一个没有实现任何接口或者继承任何基类的普通 POCO 类型。我们定义了 OnReceiveRequest 和 OnSendReply 两个日志事件方法，应用在它们上面的 DiagnosticNameAttribute 特性设置了对应的事件名称。为了自动获取日志内容荷载，可以根据荷载对象的数据结构为这两个方法定义参数。

```csharp
public sealed class DiagnosticCollector
{
    [DiagnosticName("ReceiveRequest")]
    public void OnReceiveRequest(HttpRequestMessage request, long timestamp)
        => Console.WriteLine(
            $"Receive request. Url: {request.RequestUri}; Timstamp:{timestamp}");

    [DiagnosticName("SendReply")]
    public void OnSendReply(HttpResponseMessage response, TimeSpan elaped)
        => Console.WriteLine(
            $"Send reply. Status code: {response.StatusCode}; Elaped: {elaped}");
}
```

接下来只需要改变之前的日志事件订阅方式即可。如下面的代码片段所示，根据名称找到作为订阅目标的 DiagnosticListener 对象之后，可以直接创建 DiagnosticCollector 对象，并将其作为参数调用 SubscribeWithAdapter 扩展方法进行注册即可。改动后的程序运行之后，同样会在控制台上输出图 8-22 所示的结果。（S815）

```csharp
public class Program
{
    public static void Main()
    {
        DiagnosticListener.AllListeners.Subscribe(listener =>
        {
            if (listener.Name == "Web")
            {
                listener.SubscribeWithAdapter(new DiagnosticCollector());
            }
```

```
        });
        ...
    }
}
```

8.5.4 针对活动的跟踪

针对活动的跟踪可以反映完整的调用链信息，通过前面的介绍可知，针对 TraceSource 和 EventSource 的日志框架都提供了针对活动的跟踪的支持，基于 DiagnosticSource 的日志框架也不例外。在介绍 DiagnosticSource 这个抽象类定义时，我们已经提到如下两个与活动跟踪相关的方法（StartActivity 和 StopActivity），它们分别用来发送活动的开始事件和结束事件。

```
public abstract class DiagnosticSource
{
    public Activity StartActivity(Activity activity, object args);
    public void StopActivity(Activity activity, object args);
    ...
}
```

从上面的代码片段可以看出，DiagnosticSource 涉及的活动通过一个 Activity 对象来表示。一个 Activity 对象与一个逻辑操作进行关联，不仅可以反映调用链的流转，还保存了操作开始的时间戳和耗时。如下面的代码片段所示，Activity 对象的属性 Id 和 OperationName 分别代表活动的唯一标识与操作名称。Activity 对象的属性 StartTimeUtc 和 Duration 表示活动的开始时间（UTC 时间）与耗时。调用链的流转信息可以通过表示"父活动"的 Parent 属性来体现。除此之外，还可以利用其 ParentId 和 RootId 获取"父活动"与"根活动"的标识。

```
public class Activity
{
    public string       Id { get; }
    public string       OperationName { get; }

    public DateTime     StartTimeUtc { get; }
    public TimeSpan     Duration { get; }


    public string       RootId { get; }
    public Activity     Parent { get; }
    public string       ParentId { get; }

    public Activity(string operationName);
    ...
}
```

我们可以利用 Activity 的静态属性 Current 得到以 AsyncLocal<Activity>对象形式保存的当前活动。活动的开始和结束可以通过调用 Start 方法与 Stop 方法来完成。当调用 Start 方法时，Activity 对象会被分配一个唯一标识，并将当前时间作为开始时间戳。与此同时，当前的活动会作为"父活动"赋予 Parent 属性，而自身将被作为当前活动赋予静态属性 Current。当调用 Stop 方法结束活动时，整个耗时会被计算出来并赋予 Duration 属性，然后将 Parent 属性表示的"父活动"作为当前活动。除了目前介绍的这些成员，Activity 还定义了其他若干方法和属性，由于篇幅有限，此处不再赘述。

```
public class Activity
{
    public static Activity Current { get; }

    public Activity Start();
    public void Stop();
}
```

下面介绍 DiagnosticSource 是如何实现活动跟踪的。从如下所示的代码片段可以看出，DiagnosticSource 的 StartActivity 方法和 StopActivity 方法并没有什么特别之处，它除了调用 Start 方法和 Stop 方法开始与结束指定活动对象之外，只是调用 Write 方法发送了一个日志事件。发送日志事件的名称为指定 Activity 对象的操作名称分别加上对应后缀 ".Start" 和 ".Stop"，所以需要根据此命名规则来订阅活动的开始事件和结束事件。

```
public abstract class DiagnosticSource
{
    public Activity StartActivity(Activity activity, object args)
    {
        activity.Start();
        Write(activity.OperationName + ".Start", args);
        return activity;
    }

    public void StopActivity(Activity activity, object args)
    {
        if (activity.Duration == TimeSpan.Zero)
        {
            activity.SetEndTime(Activity.GetUtcNow());
        }
        Write(activity.OperationName + ".Stop", args);
        activity.Stop();
    }
    ...
}
```

第 9 章

诊断日志（下篇）

第 8 章介绍了 4 种常用的诊断日志框架。除了微软提供的这些日志框架，还有很多第三方日志框架可供选择，如 Log4Net、NLog 和 Serilog 等。虽然这些框架大都采用类似的设计（观察者模式或者发布/订阅模式），但是它们采用的编程模式具有很大的差异。为了对这些日志框架进行整合，微软创建了一个用来提供统一的日志编程模式的日志框架。

9.1 统一日志编程模式

本章主要从设计和实现的角度对 .NET Core 框架提供的统一日志框架进行深入剖析，但在此之前必须先了解由它提供的编程模式，所以下面运用实例来演示如何触发相应等级的日志事件，并最终将日志消息写入我们期望的输出渠道。

9.1.1 将日志输出到不同的渠道

统一日志编程模型主要涉及由 ILogger 接口、ILoggerFactory 接口、ILoggerProvider 接口表示的 3 个核心对象，这 3 个核心对象以及它们之间的关系是 9.2 节着重介绍的内容，此处只需要大致了解：应用程序通过 ILoggerFactory 创建的 ILogger 对象来记录日志，而 ILoggerProvider 则完成针对相应渠道的日志输出。

EventId & LogLevel

一般来说，写入的每条日志消息总是针对某个具体的事件（Event），所以每条日志消息（Log Entry 或者 Log Message）都有一个标识事件的 ID。日志事件本身的重要程度或者反映的问题严重性不尽相同，这一点则通过日志消息的等级来标识，英文的"日志等级"可以表示成 "Log Level" "Log Verbosity Level" "Log Severity Level" 等。日志事件 ID 和日志等级可以通过如下所示的两个类型来表示。

```
public struct EventId
{
    public int          Id { get; }
```

```csharp
    public string      Name { get; }
    public EventId(int id, string name = null);

    public static implicit operator EventId(int i);
    public override string ToString();
}

public enum LogLevel
{
    Trace,
    Debug,
    Information,
    Warning,
    Error,
    Critical,
    None
}
```

表示 EventId 的结构体分别通过只读属性 Id 和 Name 表示事件的 ID（必需）与名称（可选）。EventId 重写了 ToString 方法，如果表示事件名称的 Name 属性存在，那么该方法会将事件名称作为返回值，否则这个方法会返回其 Id 属性。从上面提供的代码片段还可以看出，EventId 定义了针对整型的隐式转化器，所以任何涉及使用 EventId 的地方都可以直接用表示事件 ID 的整数来替换。

如果忽略选项 None，枚举 LogLevel 实际上定义了 6 种日志等级，枚举成员的顺序体现了等级的高低，Trace 最低，Critical 最高。表 9-1 给出了这 6 种日志等级的事件描述，我们可以在发送日志事件时根据它来决定当前日志消息应该采用何种等级。

表 9-1　日志等级

日志等级	事件描述
Trace	用于记录一些相对详细的消息，以辅助开发人员针对某个问题进行代码跟踪调试。由于这样的日志消息往往包含一些相对敏感的信息，所以在默认情况下不应该开启此等级
Debug	用于记录一些辅助调试的日志，这样的日志内容往往具有较短的时效性，如记录针对某个方法的调用及其返回值
Information	向管理员传达非关键信息，类似于"供您参考"之类的注释。这样的消息可以用来跟踪一个完整的处理流程，相应日志记录的消息往往具有相对较长的时效性，如记录当前请求的目标 URL
Warning	应用出现不正常行为，或者出现非预期的结果。尽管不是对实际错误做出的响应，但是警告指示组件或者应用程序未处于理想状态，并且一些进一步操作可能会导致关键错误，如用户登录时没有通过认证
Error	应用当前的处理流程因出现未被处理的异常而终止，但是整个应用不至于崩溃。这样的事件主要针对当前活动或者操作遇到的异常，而不是针对整个应用级别的错误，如添加记录时出现主键冲突
Critical	系统或者应用出现难以恢复的崩溃，或者需要引起足够重视的灾难性事件

针对控制台和 Debugger 的日志输出

对日志的基本编程模型有了大致的了解之后，接下来我们通过一个简单的实例来演示如何将具有不同等级的日志消息输出到两种不同的渠道：一种是直接将格式化的日志消息输出到当前控制台，另一种则是将日志作为调试信息提供给附加到当前进程的调试器（Debugger）。支持这两种日志输出渠道的 ILoggerProvider 实现类型分别为 ConsoleLoggerProvider 和 DebugLoggerProvider。

依赖注入已经变成一种基本编程模式，所以用来创建 ILogger 对象的 ILoggerFactory 工厂基本上是采用依赖注入的方式来提供的，接下来的演示实例将利用依赖注入框架提供 ILoggerFactory 对象。如下面的代码片段所示，我们创建了一个 ServiceCollection 对象，并调用 AddLogging 扩展方法注册了与日志相关的服务，然后利用由它创建的 IServiceProvider 对象来提供 ILoggerFactory 工厂。

```
namespace App
{
    public class Program
    {
        public static void Main()
        {
            var logger = new ServiceCollection()
                .AddLogging(builder => builder
                    .AddConsole()
                    .AddDebug())
                .BuildServiceProvider()
                .GetRequiredService<ILoggerFactory>()
                .CreateLogger("App.Program");

            var levels = (LogLevel[])Enum.GetValues(typeof(LogLevel));
            levels = levels.Where(it => it != LogLevel.None).ToArray();
            var eventId = 1;
            Array.ForEach(levels, level
                => logger.Log(level, eventId++,"This is a/an {0} log message.", level));
            Console.Read();
        }
    }
}
```

在调用 IServiceCollection 接口的 AddLogging 扩展方法注册日志框架相关服务时，我们利用提供的 Action<ILoggingBuilder>对象完成了针对 ConsoleLoggerProvider 和 DebugLoggerProvider 的注册。具体来说，我们调用 ILoggingBuilder 接口的 AddConsole 扩展方法注册了 ConsoleLoggerProvider 对象，而 DebugLoggerProvider 对象则是通过调用 AddDebug 扩展方法注册的。

在调用 ILoggerFactory 接口的 CreateLogger 方法创建 ILogger 对象时，我们需要提供用来对日志进行归类的类别（Category），每个 ILogger 对象都对应一个确定的类别。我们倾向于将当前写入日志的组件、服务或者类型名称作为日志类别，所以需要指定的是当前类型的全名（App.Program）。

我们通过调用 ILogger 的 Log 方法针对每个有效的日志等级分发了 6 个日志事件，事件的 ID 分别被设置成 1~6 的整数。与第 8 章介绍的基于 TraceSource 的跟踪日志框架类似，分发的日志事件的内容荷载最终体现为一个格式化的字符串，所以调用 Log 方法时可以通过指定一个包含占位符（{0}）的消息模板和对应参数的方式来格式化最终输出的消息内容。

由于注册了 ConsoleLoggerProvider 对象，所以启动该演示实例后可以直接在控制台上看到输出的日志内容。如图 9-1 所示，格式化的日志消息不仅包含格式化的消息内容，还包含日志的等级、类别和事件 ID。表示日志等级的文字还会采用不同的前景色和背景色来显示。由于涉及针对分发日志事件的过滤，对于由 ILogger 对象发出的针对不同等级的 6 个日志事件，只有 4 条日志被真正输出到控制台上。由于 LoggerFactory 上还注册了另一个 DebugLoggerProvider 对象，它会完成针对调试器的日志输出，所以 Visual Studio 的调试输出窗口也会输出这 4 条日志。（S901）

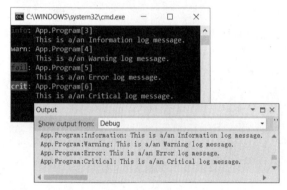

图 9-1　针对控制台和 Debugger 的日志输出

注入 ILogger<T>服务

前面演示的实例通过指定日志类别（App.Program）调用 ILoggerFactory 对象的 CreateLogger 方法创建的是一个 ILogger 对象，实际上还可以调用泛型的 CreateLogger<T>方法创建一个 ILogger<T>对象来分发日志事件。如果调用这个方法，就不需要额外提供日志类别，这是因为创建的 ILogger<T>对象会使用泛型类型的全名（命名空间+类型名称）作为日志类别。既然作为依赖注入容器的 IServiceProvider 对象能够提供 ILogger<T>服务，这就意味着可以将它作为依赖服务注入消费类型的构造函数中。

```
namespace App
{
    public class Program
    {
        public static void Main()
        {
            var logger = new ServiceCollection()
                .AddLogging(builder => builder
                    .AddConsole()
                    .AddDebug())
                .BuildServiceProvider()
                .GetRequiredService<ILogger<Program>>();
```

```
            var levels = (LogLevel[])Enum.GetValues(typeof(LogLevel));
            levels = levels.Where(it => it != LogLevel.None).ToArray();
            var eventId = 1;
            Array.ForEach(levels, level => logger.Log(level, eventId++,
                "This is a/an {level} log message.", level));
            Console.Read();
        }
    }
}
```

作为日志负载内容的消息模板除了可以采用{0}，{1}，…，{n}这样的占位符，还可以使用任意字符串（{level}）来表示。启动改写的程序之后，输出到控制台和调试输出窗口的内容与图 9-1 是完全一致的。（S902）

针对 TraceSource 和 EventSource 的输出

除了控制台和调试器这两种日志输出渠道，日志框架还可以通过一系列预定义的 ILoggerProvider 实现类型提供针对其他输出渠道的支持。第 8 章重点介绍了针对 TraceSource 和 EventSource 的日志框架，实际上它们也可以直接作为日志消息的输出渠道。

下面演示如何利用注册的 TraceSourceLoggerProvider 和 EventSourceLoggerProvider 来记录日志，为此需要将上面的代码改写成如下形式。为了捕捉由 EventSource 分发的日志事件，可以自定义一个 FoobarEventListener 类型。在程序启动时，我们创建了 FoobarEventListener 对象并注册了 EventSourceCreated 事件和 EventWritten 事件。被 EventSourceLoggerProvider 用来处理日志的 EventSource 被命名为 Microsoft-Extensions-Logging，所以我们可以根据这个名称来过滤作为注册目标的 EventSource。

```
public class Program
{
    public static void Main()
    {
        var listener = new FoobarEventListener();
        listener.EventSourceCreated += (sender, args) =>
        {
            if (args.EventSource.Name == "Microsoft-Extensions-Logging")
            {
                listener.EnableEvents(args.EventSource, EventLevel.LogAlways);
            }
        };
        listener.EventWritten += (sender, args) =>
        {
            if (args.EventName == "FormattedMessage")
            {
                var payload = args.Payload;
                var payloadNames = args.PayloadNames;
                var indexOfLevel = payloadNames.IndexOf("Level");
                var indexOfCategory = args.PayloadNames.IndexOf("LoggerName");
```

```
                    var indexOfEventId = args.PayloadNames.IndexOf("EventId");
                    var indexOfMessage = args.PayloadNames.IndexOf("FormattedMessage");
                    Console.WriteLine($"{(LogLevel)payload[indexOfLevel],-11}:
                        {payload[indexOfCategory]}[{payload[indexOfEventId]}]");
                    Console.WriteLine($"{"",-13}{payload[indexOfMessage]}");
                }
            };

            var logger = new ServiceCollection()
                .AddLogging(builder => builder
                    .AddTraceSource(new SourceSwitch("default", "All"),
                        new DefaultTraceListener { LogFileName = "trace.log" })
                    .AddEventSourceLogger())
                .BuildServiceProvider()
                .GetRequiredService<ILogger<Program>>();

            var levels = (LogLevel[])Enum.GetValues(typeof(LogLevel));
            levels = levels.Where(it => it != LogLevel.None).ToArray();
            var eventId = 1;
            Array.ForEach(levels, level => logger.Log(level, eventId++,
                "This is a/an {level} log message.", level));
        }
        private class FoobarEventListener : EventListener {}
    }
}
```

在日志事件分发处理过程中，EventSource 会针对负载内容的不同输出结构分发多个日志事件，上面的代码注册的是一个名为 FormattedMessage 的事件，它的内容荷载包含格式化之后的日志消息。对于该日志事件的内容荷载，除了作为针对格式化消息的负载成员，还包括日志等级、事件 ID 和日志类别等，我们在 EventWritten 事件处理程序中将它们从负载对象中提取出来，经过相应的格式化之后，这些数据最终被打印到控制台上。

针对 TraceSourceLoggerProvider 和 EventSourceLoggerProvider 的注册通过调用 ILoggingBuilder 的两个对应的扩展方法 AddTraceSource 和 AddEventSourceLogger 来完成。调用 AddTraceSource 方法注册 TraceSourceLoggerProvider 时，我们提供了两个参数，前者是作为全局过滤器的 SourceSwitch 对象，后者则是注册的 TraceListener。由于 DefaultTraceListener 指定了日志文件的路径，所以输出的日志消息最终会被写入指定的文件中。

日志事件的分发与前面演示的实例没有本质区别，我们依然分发针对 6 种不同等级的日志事件。程序运行后，由 TraceSourceLoggerProvider 提供的 TraceSource 会将日志消息输出到指定的文件中（trace.log），而 EventSourceLoggerProvider 会利用 EventSource 对象对原始的日志事件做相应处理之后按照自己的方式对日志事件进行继续分发，分发的日志事件会被我们注册的 FoobarEventListener 捕获，格式化的内容进而被打印到控制台上。输出到日志文件和控制台上的日志内容如图 9-2 所示。（S903）

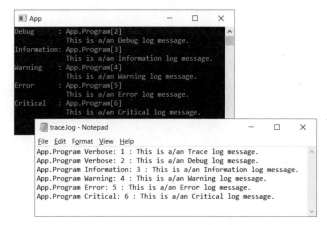

图 9-2 针对 TraceSource 和 EventSource 的日志输出

9.1.2 日志过滤

在应用程序中使用 ILogger 对象分发的日志事件，并不能保证都会进入最终的输出渠道，因为注册的 ILoggerProvider 对象会对日志进行过滤，只有符合过滤条件的日志消息才会被真正地输出到对应的渠道。我们可以采用多种方式为注册的 ILoggerProvider 对象定义过滤规则。

每一个分发的日志事件都具有一个确定的等级。一般来说，日志消息的等级越高，表明对应的日志事件越重要或者反映的问题越严重，自然就越应该被记录下来，所以在很多情况下我们指定的过滤条件只需要一个最低等级，所有不低于（等于或者高于）该等级的日志都会被记录下来。

最低日志等级在默认情况下被设置为 Information，这就是前面演示实例中等级为 Trace 和 Debug 的两条日志没有被真正输出的根源所在。如果需要将这个作为输出"门槛"的日志等级设置得更高或者更低，我们只需要将指定的等级作为参数调用 ILoggingBuilder 接口的 SetMinimumLevel 方法即可。

```
public class Program
{
    public static void Main()
    {
        var logger = new ServiceCollection().AddLogging(builder => builder
            .SetMinimumLevel(LogLevel.Trace)
            .AddConsole())
            .BuildServiceProvider()
            .GetRequiredService<ILoggerFactory>()
            .CreateLogger<Program>();

        var levels = (LogLevel[])Enum.GetValues(typeof(LogLevel));
        levels = levels.Where(it => it != LogLevel.None).ToArray();
        var eventId = 1;
        Array.ForEach(levels, level => logger.Log(level, eventId++,
            "This is a/an {level} log message.", level));
```

```
            Console.Read();
        }
}
```

如上面的代码片段所示，在调用 IServiceCollection 接口的 AddLogging 扩展方法注册日志相关服务时，我们调用 ILoggingBuilder 接口的 SetMinimumLevel 方法将最低日志等级设置为 Trace。由于设置的是最低等级，所以所有的日志消息都会以图 9-3 所示的形式被注册的 ConsoleLoggerProvider 输出到控制台上。（S904）

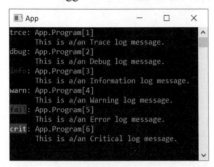

图 9-3　通过设置最低等级控制输出的日志

虽然"过滤不低于指定等级的日志消息"是常用的日志过滤规则，但可用的过滤规则并不限于此，过滤条件并不仅仅限于日志等级，很多时候还会同时考虑日志的类别。在利用 ILoggerFactory 创建对应的 ILogger 时需要指定日志类别，由于一般将当前组件、服务或者类型的名称作为日志类别，所以日志类别基本上体现了日志消息来源，如果我们只希望输出由某个组件或者服务发出的日志事件，就需要针对类别对日志事件实施过滤。

综上可知，日志过滤条件其实可以通过一个类型为 Func<string, LogLevel, bool>的委托对象来表示，它的两个输入参数分别代表日志事件的类别和等级。下面通过提供这样一个委托对象对日志消息做更细粒度的过滤，所以需要对演示程序做如下修改。

```
public class Program
{
    public static void Main()
    {
        var loggerFactory = new ServiceCollection()
            .AddLogging(builder => builder
                .AddFilter(Filter)
                .AddConsole())
            .BuildServiceProvider()
            .GetRequiredService<ILoggerFactory>();

        var fooLogger = loggerFactory.CreateLogger("Foo");
        var barLogger = loggerFactory.CreateLogger("Bar");
        var bazLogger = loggerFactory.CreateLogger("Baz");

        var levels = (LogLevel[])Enum.GetValues(typeof(LogLevel));
        levels = levels.Where(it => it != LogLevel.None).ToArray();
```

```csharp
        var eventId = 1;
        Array.ForEach(levels, level => fooLogger.Log(level, eventId++,
            "This is a/an {0} log message.", level));

        eventId = 1;
        Array.ForEach(levels, level => barLogger.Log(level, eventId++,
            "This is a/an {0} log message.", level));

        eventId = 1;
        Array.ForEach(levels, level => bazLogger.Log(level, eventId++,
            "This is a/an {0} log message.", level));

        Console.Read();

        static bool Filter(string category, LogLevel level)
        {
            return category switch
            {
                "Foo" => level >= LogLevel.Debug,
                "Bar" => level >= LogLevel.Warning,
                "Baz" => level >= LogLevel.None,
                _     => level >= LogLevel.Information,
            };
        }
}
```

如上面的代码片段所示，作为日志过滤器的 Func<string, LogLevel, bool> 对象定义的过滤规则如下：对于日志类别 Foo 和 Bar，我们只会选择等级不低于 Debug 和 Warning 的日志事件，而对于日志类别 Baz，任何等级的日志事件都不会被选择。至于其他日志类别，我们可以采用默认的最低等级 Information。在执行 AddLogging 扩展方法时，我们调用 ILoggerBuilder 接口的 AddFilter 方法将 Func<string, LogLevel, bool> 对象注册为全局过滤器。

我们利用依赖注入框架提供的 ILoggerFactory 工厂创建了 3 个 ILogger 对象，它们对应的类别分别为 Foo、Bar 和 Baz。下面利用这 3 个 ILogger 对象分发针对不同等级的 6 次日志事件，满足过滤条件的日志消息会以图 9-4 所示的形式输出到控制台上。（S905）

不论是通过调用 ILoggerBuilder 接口的 SetMinimumLevel 方法设置的最低日志等级，还是通过调用 AddFilter 扩展方法提供的过滤器，设置的日志过滤规则针对的都是所有注册的 ILoggerProvider 对象，但是有时需要将过滤规则应用到某个具体的 ILoggerProvider 对象上。如果将 ILoggerProvider 对象引入日志过滤规则中，那么日志过滤器就应该表示成一个类型为 Func<string, string, LogLevel, bool>的委托对象，该委托的 3 个输入参数分别表示 ILoggerProvider 类型的全名、日志类别和等级。为了演示针对 LoggerProvider 的日志过滤，可以将演示程序做如下改动。

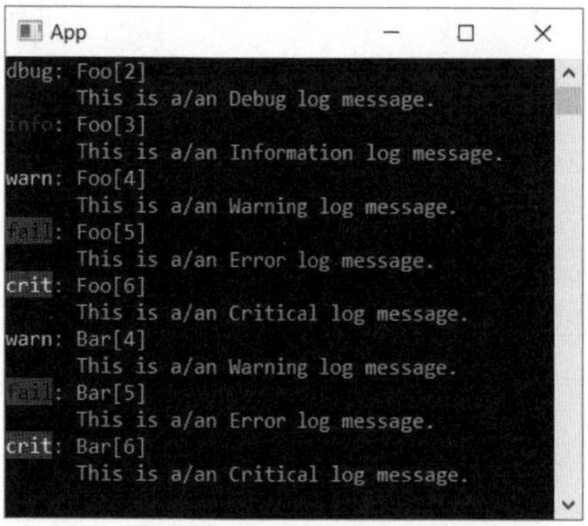

```
App
dbug: Foo[2]
      This is a/an Debug log message.
info: Foo[3]
      This is a/an Information log message.
warn: Foo[4]
      This is a/an Warning log message.
fail: Foo[5]
      This is a/an Error log message.
crit: Foo[6]
      This is a/an Critical log message.
warn: Bar[4]
      This is a/an Warning log message.
fail: Bar[5]
      This is a/an Error log message.
crit: Bar[6]
      This is a/an Critical log message.
```

图 9-4　针对类别和等级的日志过滤

```csharp
public class Program
{
    public static void Main()
    {
        var logger = new ServiceCollection()
            .AddLogging(builder => builder
                .AddFilter(Filter)
                .AddConsole()
                .AddDebug())
            .BuildServiceProvider()
            .GetRequiredService<ILogger<Program>>();

        var levels = (LogLevel[])Enum.GetValues(typeof(LogLevel));
        levels = levels.Where(it => it != LogLevel.None).ToArray();
        var eventId = 1;
        Array.ForEach(levels, level => logger.Log(level, eventId++,
            "This is a/an {0} log message.", level));
        Console.Read();

        static bool Filter(string provider, string category, LogLevel level)
        {
            if (provider == typeof(ConsoleLoggerProvider).FullName)
            {
                return level >= LogLevel.Debug;
            }
            else if (provider == typeof(DebugLoggerProvider).FullName)
            {
                return level >= LogLevel.Warning;
            }
            else
            {
                return true;
            }
        }
    }
}
```

如上面的代码片段所示，我们注册的过滤器体现的过滤规则如下：对于 Console LoggerProvider，写入的日志等级应不低于 Debug；而 DebugLoggerProvider 对应的日志等级则被设置为 Warning，至于其他的 ILoggerProvider 类型则不做任何的过滤。如图 9-5 所示，由于演示程序注册了 ConsoleLoggerProvider 和 DebugLoggerProvider，对于分发的 12 条日志消息，5 条会在控制台上输出，3 条会出现在 Visual Studio 的调试输出窗口中。（S906）

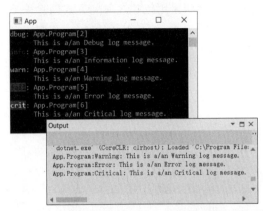

图 9-5　针对 ILoggerProvider 类型的日志过滤

通过 Func<string, string, LogLevel, bool>对象表示的日志过滤规则除了可以采用编程的形式来设置，还可以采用配置的形式来提供。以配置的形式定义的过滤规则最终都体现为一个设定的最低等级，设定的这个最低日志等级可以是一个全局的默认设置，也可以专门针对某个日志类别或者 ILoggerProvider 类型。

下面演示针对配置形式的日志过滤规则。我们先创建一个名为 logging.json 的文件，并在其中定义如下这段配置，然后将"Copy to Output Directory"的属性设置为"Copy Always"。这段配置定义了两组日志过滤规则，第一组是默认规则，第二组则是专门为 ConsoleLoggerProvider（别名为 Console）定义的过滤规则。

```
{
    "LogLevel": {
        "Default"        : "Error",
        "Foo"            : "Debug"
    },
    "Console": {
        "LogLevel": {
            "Default"    : "Information",
            "Foo"        : "Warning",
            "Bar"        : "Error"
        }
    }
}
```

前面提及，以配置形式定义的日志过滤规则最终会落实到对最低日志等级的设置上，其中，Default 表示默认设置，其他的则是针对具体日志类别的设置。上面定义的这段配置体现的过滤规则如下：对于 ConsoleLoggerProvider 来说，在默认情况下只有等级不低于 Information 的日志事件会被输出，而对日志类别 Foo 和 Bar 来说，对应的最低日志等级分别为 Warning 和 Error。对于其他 ILoggerProvider 类型来说，如果日志类别为 Foo，那么只有等级不低于 Debug 的日志才会被输出，其他日志类别则采用默认的等级 Error。

检验注册的 ILoggerProvider 对象是否会采用配置定义的规则对日志消息进行过滤的操作如下：首先加载配置文件并生成对应的 IConfiguration 对象，然后采用依赖注入的方式创建一个 ILoggerFactory 对象。除了分别注册 ConsoleLoggerProvider 和 DebugLoggerProvider，我们还将 IConfiguration 对象作为参数调用 ILoggingBuilder 接口的 AddConfiguration 扩展方法将配置承载的过滤规则应用到配置模型上。接下来我们采用不同的类别（Foo、Bar 和 Baz）创建了 3 个 Logger，并利用它们分发了 6 条具有不同等级的日志事件。

```csharp
public class Program
{
    public static void Main()
    {
        var configuration = new ConfigurationBuilder()
            .SetBasePath(Directory.GetCurrentDirectory())
            .AddJsonFile("logging.json")
            .Build();

        var loggerFactory = new ServiceCollection()
            .AddLogging(builder => builder
                .AddConfiguration(configuration)
                .AddConsole()
                .AddDebug())
            .BuildServiceProvider()
            .GetRequiredService<ILoggerFactory>();

        var fooLogger = loggerFactory.CreateLogger("Foo");
        var barLogger = loggerFactory.CreateLogger("Bar");
        var bazLogger = loggerFactory.CreateLogger("Baz");

        var levels = (LogLevel[])Enum.GetValues(typeof(LogLevel));
        levels = levels.Where(it => it != LogLevel.None).ToArray();

        var eventId = 1;
        Array.ForEach(levels, level => fooLogger.Log(level, eventId++,
            "This is a/an {0} log message.", level));

        eventId = 1;
        Array.ForEach(levels, level => barLogger.Log(level, eventId++,
            "This is a/an {0} log message.", level));

        eventId = 1;
        Array.ForEach(levels, level => bazLogger.Log(level, eventId++,
            "This is a/an {0} log message.", level));
        Console.Read();
```

```
    }
}
```

由于我们注册了 2 个不同的 ILoggerProvider 类型,创建了 3 种基于不同日志类别的 ILogger 对象,所以这里面涉及分发的 36 条日志消息。按照定义在配置文件中的过滤规则,它们能否被真正地输出 ILoggerProvider 对应的渠道体现在表 9-2 中。

表 9-2 针对配置的日志过滤

等级	ConsoleLoggerProvider			DebugLoggerProvider		
	Foo	Bar	Baz	Foo	Bar	Baz
Trace	No	No	No	No	No	No
Debug	No	No	No	Yes	No	No
Information	No	No	Yes	Yes	No	No
Warning	Yes	No	Yes	Yes	No	No
Error	Yes	Yes	Yes	Yes	Yes	Yes
Critical	Yes	Yes	Yes	Yes	Yes	Yes

表 9-2 是针对配置定义的过滤规则进行分析的结果,而图 9-6 是程序执行(以 Debug 模式进行编译)之后控制台和 Visual Studio 调试输出窗口的输出结果,可以看出两者是完全一致的。(S907)

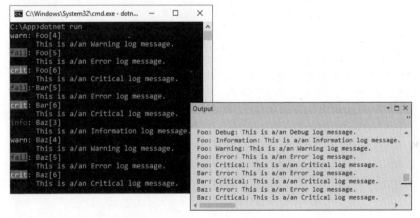

图 9-6 针对配置文件的日志过滤

9.1.3 日志范围

日志可以为针对某种目的(如纠错查错、系统优化和安全审核等)进行数据分析提供原始数据,所以孤立存在的一条日志消息对数据分析往往毫无用处,很多问题只有将多条相关的日志消息综合起来分析才能找到答案。例如,如果希望通过日志反映一个 ASP.NET Core 应用处理每次请求所花费的时间,就必须分别在请求最初被接收和最终被响应时记录当前的时间戳,但这并不足以帮助我们计算出整个请求处理的时间,因为无法确定某两条日志是否是针对同一个请求的。

为了解决上述问题,日志模型引入了日志范围(Log Scope)。所谓的日志范围是为日志记录创建的一个具有唯一标识的上下文，如果注册的 ILoggerProvider 对象支持这个特性，那么它提供的 ILogger 对象会感知到当前日志范围的存在，并将其标识连同日志消息的内容荷载一并记录下来。在进行数据分析时，就可以根据日志范围上下文标识将相关的日志消息串联起来，下面进行简单的演示。

我们在演示实例中将一个包含多个处理步骤的事务作为日志范围，这样做的目的在于利用记录的日志分析整个事务和各个步骤执行耗时。如下面的代码片段所示，可以利用依赖注入框架创建 ILogger 对象，但是在调用 ILoggerBuilder 接口的 AddConsole 扩展方法注册对应的 ConsoleLoggerProvider 对象时，我们将代表配置选项的 ConsoleLoggerOptions 对象的 IncludeScopes 属性设置为 True，这个操作相当于为注册的 ConsoleLoggerProvider 对象开启了记录日志范围上下文的特性。

```csharp
public class Program
{
    public static async Task Main()
    {
        var logger = new ServiceCollection()
            .AddLogging(builder => builder
                .AddConsole(options => options.IncludeScopes = true))
                .BuildServiceProvider()
            .GetRequiredService<ILogger<Program>>();

        using (logger.BeginScope($"Foobar Transaction[{Guid.NewGuid()}]"))
        {
            var stopwatch = Stopwatch.StartNew();
            await Task.Delay(500);
            logger.LogInformation("Operation foo completes at {0}",
                stopwatch.Elapsed);

            await Task.Delay(300);
            logger.LogInformation("Operation bar completes at {0}",
                stopwatch.Elapsed);

            await Task.Delay(800);
            logger.LogInformation("Operation baz completes at {0}",
                stopwatch.Elapsed);
        }
        Console.Read();
    }
}
```

所谓的日志范围是通过调用 ILogger 对象的 BeginScope 方法创建的，我们在调用这个方法时指定一个字符串来描述并标识创建的上下文，具体的唯一标识体现在通过 GUID 标识的事务 ID 上。创建的日志范围上下文体现为一个 IDisposable 对象，它的生命周期终止于 Dispose 方法的调用。对于任何一个支持日志范围的 ILoggerProvider 对象来说，它提供的 ILogger 对象都能感知到当前上下文的存在，所以具体的日志编程方式与之前并没有什么不同。

在我们演示的程序中，执行的事务包含 3 个操作（Foo、Bar 和 Baz）。我们将事务开始的那一刻作为基础，记录每个操作完成的时间。该程序执行后会将日志消息以图 9-7 所示的形式输出到控制台上，从输出结果可以看出，包含事务 ID 的日志范围上下文描述信息一并被记录下来。如果可以利用它将某个事务相关的日志消息筛选出来，就能清楚地计算整个事务及其每个步骤的执行时间。（S908）

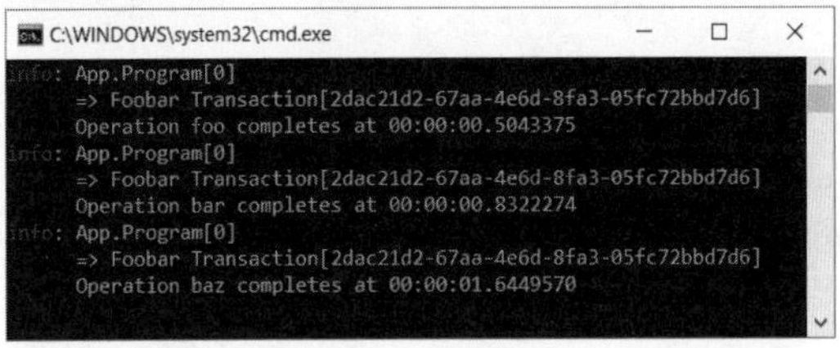

图 9-7　记录日志范围上下文

9.1.4　LoggerMessage

前面演示的应用总是调用针对具体日志等级的 Log 扩展方法来记录日志，在调用这些方法时我们总是会提供一个包含占位符的消息模板。为了提供针对语义化日志（Semantic Logging）或者结构化日志（Structured Logging）的支持，我们可以采用一个具有明确语义的字符串作为占位符。也正是因为如此，调用 Log 方法时，该方法每次都需要对提供的消息模板进行解析。如果每次提供的都是相同的消息模板，那么这种对消息模板的重复解析就会显得多余。

除此之外，这些方法总是以一个对象数组的形式来提供用于填充占位符的参数，但是对于同一个消息模板，它的每个占位符对应的参数类型在大部分情况下是明确的，所以我们更期望的是一种强类型的参数提供方式。为了解决这个问题，日志系统提供了一个名为 LoggerMessage 的静态类型，我们可以利用它根据某个具体的消息模板创建一个 Action<ILogger,...>对象来分发日志事件。

下面通过实例来演示如何利用 LoggerMessage 进行强类型的日志编程。在这个演示程序中，我们记录的日志是针对某个方法的调用，可以将调用该方法的输入参数、返回值和执行时间通过日志保存下来。如下面的代码片段所示，根据目标方法 Foobar 的定义，我们调用 LoggerMessage 的静态 Define 方法创建了一个 Action<ILogger, int, long, double, TimeSpan, Exception>类型的委托对象。

```
class Program
{
    private static Random      _random;
    private static string      _template;
    private static ILogger     _logger;
    private static Action<ILogger, int, long, double, TimeSpan, Exception> _log;

    static async Task Main()
    {
        _random = new Random();
```

```
        _template = "Method FoobarAsync is invoked." +
            "\n\t\tArguments: foo={foo}, bar={bar}" +
            "\n\t\tReturn value: {returnValue}" +
            "\n\t\tTime:{time}";
        _log = LoggerMessage.Define
            <int, long, double, TimeSpan>(LogLevel.Trace, 3721, _template);
        _logger = new ServiceCollection()
            .AddLogging(builder => builder
                .SetMinimumLevel(LogLevel.Trace)
                .AddConsole())
            .BuildServiceProvider()
            .GetRequiredService<ILogger<Program>>();
        await FoobarAsync(_random.Next(), _random.Next());
        await FoobarAsync(_random.Next(), _random.Next());
        Console.Read();
    }

    static async Task<double> FoobarAsync(int foo, long bar)
    {
        var stopwatch = Stopwatch.StartNew();
        await Task.Delay(_random.Next(100, 900));
        var result = _random.Next();
        _log(_logger, foo, bar, result, stopwatch.Elapsed, null);
        return result;
    }
}
```

在调用 Define 方法时，我们指定了日志等级（Trace）、EventId 和日志消息模板，由于在默认情况下日志系统采用的最低日志等级为 Information，为了写入等级为 Trace 的日志，可以在采用依赖注入方式创建 ILogger 对象时调用 ILoggingBuilder 接口的 SetMinimumLevel 方法将最低等级设置为 Trace。我们可以在 FoobarAsync 中利用创建的这个委托对象将当前方法的参数、返回值和执行时间通过日志记录下来。FoobarAsync 方法总共被调用了两次，所以程序执行后在控制台上输出的两组数据如图 9-8 所示。（S909）

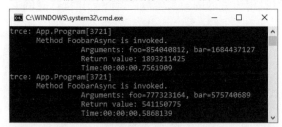

图 9-8　利用 LoggerMessage 记录日志

利用 LoggerMessage 创建的委托对象来记录日志的一个主要的目的就是避免对相同消息模板重复解析，这种基于模板的字符串解析过程不仅针对具体的日志消息，还针对日志范围上下文。调用 ILogger 对象的 BeginScope 方法时，同样是提供一个包含占位符的模板和对应的参数，针对相同模板的重复解析依然存在，所以 LoggerMessage 定义了一系列 DefineScope 方法，我们可以利用它们提供的委托对象来创建日志范围上下文。

我们在前面演示了基于事务的日志上下文，下面采用 LoggerMessage 来演示如何实现相同的功能。如下面的代码片段所示，我们调用 LoggerMessage 的 DefineScope<Guid>方法根据提供的模板创建了一个 Func<ILogger, Guid, IDisposable>类型的委托对象，它执行的结果代表的就是创建的日志范围上下文。

```csharp
class Program
{
    static async Task Main()
    {
        var logger = new ServiceCollection()
                .AddLogging(builder => builder
                    .SetMinimumLevel(LogLevel.Trace)
                    .AddConsole(options => options.IncludeScopes = true))
                .BuildServiceProvider()
                .GetRequiredService<ILogger<Program>>();

        var scopeFactory = LoggerMessage.DefineScope<Guid>(
            "Foobar Transaction[{TransactionId}]");
        var operationCompleted = LoggerMessage
            .Define<string, TimeSpan>(LogLevel.Trace, 3721,
            "Operation {operation} completes at {time}");

        using (scopeFactory(logger, Guid.NewGuid()))
        {
            await InvokeAsync();
        }

        using (scopeFactory(logger, Guid.NewGuid()))
        {
            await InvokeAsync();
        }

        Console.Read();

        async Task InvokeAsync()
        {
            var stopwatch = Stopwatch.StartNew();
            await Task.Delay(500);
            operationCompleted(logger, "foo", stopwatch.Elapsed, null);

            await Task.Delay(300);
            operationCompleted(logger, "bar", stopwatch.Elapsed, null);

            await Task.Delay(800);
            operationCompleted(logger, "baz", stopwatch.Elapsed, null);
        }
    }
}
```

我们定义了一个本地函数 InvokeAsync 来代表一个完整事务，该方法同样采用 LoggerMessage

创建的Action<ILogger, Guid, TimeSpan, Exception>来记录组成该事务的每个步骤正常结束的时间。通过提供代表事务 ID 的 Guid 对象，我们利用 Func<ILogger, Guid, IDisposable>创建了两个日志范围上下文，并在该上下文中调用了 InvokeAsync 方法。由于 InvokeAsync 方法是在代表事务的日志范围上下文中执行的，所以携带事务 ID 的日志范围信息会以图 9-9 所示的形式输出到控制台上。（S910）

图 9-9　利用 LoggerMessage 创建日志范围上下文

9.2　日志模型详解

通过前面演示的这些实例，笔者相信读者已经对 .NET Core 应用下如何记录日志有了充分的了解。按照惯例，下面从设计角度对日志系统做进一步的讲解。总的来说，日志模型由 ILogger、ILoggerFactory 和 ILoggerProvider 这 3 个核心对象构成，可以将它们称为日志模型三要素。这些接口都定义在 NuGet 包"Microsoft.Extensions.Logging.Abstractions"中，而具体的实现则由另一个 NuGet 包"Microsoft.Extensions.Logging"来提供。

9.2.1　日志模型三要素

日志模型三要素之间的关系如图 9-10 所示。其中，ILoggerFactory 对象和 ILoggerProvider 对象都是 ILogger 对象的创建者，而 ILoggerProvider 对象会注册到 ILoggerFactory 对象上。有人认为这 3 个对象之间的关系很混乱，这主要体现在 ILogger 对象具有两个不同的创建者。

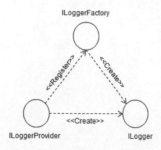

图 9-10　日志模型三要素之间的关系（一）

ILoggerProvider 和 ILoggerFactory 创建的其实是不同的 ILogger 对象。ILoggerFactory 创建的 ILogger 对象被应用程序用来分发日志事件，而 ILoggerProvider 提供的 ILogger 对象则是日志事件的真正消费者，所以从发布订阅模式的角度来讲，前者属于发布者，后者属于订阅者。一个具体的 ILoggerProvider 会利用提供的 ILogger 对象将接收的日志实现输出到对应的渠道，而 ILoggerFactory 提供的 ILogger 对象则是由这组面向具体输出渠道的 ILogger 对象组合而成的。

如果进一步引入两个实现类型可以绘制成图 9-11 所示的类图，由此可以更好地理解日志模型三要素之间的关系。LoggerFactory 类型是对 ILoggerFactory 接口的默认实现，而由它创建的是一个类型为 Logger 的对象，该对象由注册到 LoggerFactory 对象上的所有 ILoggerProvider 对象提供的一组 ILogger 对象组合而成。虽然 Logger 类型只是一个实现了 ILogger 接口的内部类型，但是应用程序用来分发日志事件的就是这个对象，正是该对象将日志事件分发给它的 ILogger 对象成员，后者最终将经过过滤的日志消息输出到对应的渠道。

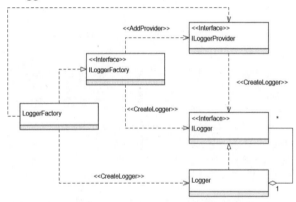

图 9-11　日志模型三要素之间的关系（二）

9.2.2　ILogger

ILogger 接口中定义了如下 3 个方法：Log、IsEnabled 和 BeginScope。一般来说，当 ILogger 对象在接收到分发给它的日志事件之后，它会将日志等级作为参数调用其 IsEnabled 方法来确定当前日志是否应该分发下去。针对日志事件的分发实现在 Log 方法中，除了提供日志等级，在调用 Log 方法时还需要提供一个 EventId 来标识当前的日志事件。

```
public interface ILogger
{
    bool IsEnabled(LogLevel logLevel);
    void Log(LogLevel logLevel, EventId eventId, object state, Exception exception,
        Func<object, Exception, string> formatter);
    IDisposable BeginScope<TState>(TState state);
}
```

日志事件的内容荷载通过 Log 方法的参数 state 来提供，由于该方法并没有对此做任何限制，所以可以提供一个任意类型的对象来承载描述日志事件的内容。日志的一个重要作用就是帮助我们更好地排错和纠错，所以这类日志承载的大部分信息会用来描述抛出的异常，该异常由

Log 方法的参数 exception 来提供。

一般来说，日志输出要么体现为日志内容的可视化呈现（如直接显示在控制台上），要么体现为持久化存储（如写入文件或者数据库中），最终的内容基本上都体现为一个格式化的字符串，因此日志在被写入之前需要先进行格式化。所谓的日志格式化就是将利用参数 state 和 exception 表示的原始内容荷载转换成一个字符串的过程，所以格式化器可以通过一个 Func<object, Exception, string>对象来表示，调用 Log 方法提供的最后一个参数就是这样一个对象。

除了定义在 ILogger 接口中的 Log 方法，还可以调用如下这些扩展的 Log 方法来分发日志事件。这些扩展方法会帮助我们完成针对日志消息的格式化。由于这些方法默认采用针对模板的格式化方式，所以调用这些方法时需要以字符串的形式提供日志消息模板和填充占位符的参数列表。除此之外，作为日志事件标识的 EventId 也不是必须提供的参数，因为这些方法会默认提供一个 ID 属性为 0 的 EventId。

```
public static class LoggerExtensions
{
    public static IDisposable BeginScope(this ILogger logger, string messageFormat,
        params object[] args);
    public static void Log(this ILogger logger, LogLevel logLevel, string message,
        params object[] args);
    public static void Log(this ILogger logger, LogLevel logLevel, EventId eventId,
        string message, params object[] args);
    public static void Log(this ILogger logger, LogLevel logLevel,
        Exception exception, string message, params object[] args);
    public static void Log(this ILogger logger, LogLevel logLevel, EventId eventId,
        Exception exception, string message, params object[] args);
}
```

对于提供的日志消息模板来说，我们可以采用两种方式来定义占位符：一种是采用连续的零基整数（如{0}、{1}和{2}等），另一种是采用具有语义的字符串（如{Id}、{Name}和{Version}等）。对于后者来说，这些以任意字符串定义的占位符会按照在模板中出现的顺序被转换成连续的数字，所以无论采用何种模板定义形式，最终的格式化日志消息都是调用 String 的 Format 方法生成的。

基于模板的日志消息格式化实现在下面的 FormattedLogValues 内部结构中。只要提供消息模板和对应的参数创建一个 FormattedLogValues 对象，我们就能利用它重写的 ToString 方法得到格式化后的消息文本。如果采用语义字符串形式的占位符，占位符的名称其实对于最终格式化后的内容没有任何影响，具有决定性作用的是占位符出现在模板消息中的顺序。另外，前面介绍的 BeginScope 扩展方法也是采用同样的方式进行消息格式化的。

```
internal struct FormattedLogValues : IReadOnlyList<KeyValuePair<string, object>>
{
    public FormattedLogValues(string format, params object[] values);
    public override string ToString();
    ...
}
```

由 ILogger 对象分发的日志事件必须具有一个明确的等级，所以调用 ILogger 对象的 Log 方法记录日志时必须显式指定日志消息采用的等级。除此之外，我们也可以调用 6 种日志等级对应的扩展方法 Log{Level}（LogDebug、LogTrace、LogInformation、LogWarning、LogError 和 LogCritical）。下面的代码片段列出了针对日志等级 Debug 的 3 个 LogDebug 方法重载的定义，针对其他日志等级的扩展方法的定义与之类似。

```csharp
public static class LoggerExtensions
{
    public static void LogDebug(this ILogger logger, EventId eventId,
        Exception exception, string message, params object[] args);
    public static void LogDebug(this ILogger logger, EventId eventId,
        string message, params object[] args);
    public static void LogDebug(this ILogger logger, string message,
        params object[] args);
    ...
}
```

每条日志消息都关联一个具体的类别（Category），这个类型实际上可以表示创建这条日志消息的"源"。日志类别指明日志消息是被谁写入的，我们一般将日志分发所在的组件、服务或者类型名称作为日志类别。日志类别是 ILogger 对象自身的属性，在利用 ILoggerFactory 工厂创建一个 ILogger 对象时，我们必须提供对应的日志类别。由同一个 ILogger 对象分发的日志事件具有相同的类别。

这种以字符串形式定义日志类别的编程方式不太"友好"，也容易出错。日志类别是日志过滤规则的重要组成部分，一旦在创建 ILogger 对象时指定了错误的日志类别，就可能导致日志不能被正常输出。为了提供一种强类型的日志类别指定方式，我们可以将日志类别与一个具体的类型进行关联，为此如下这个泛型的 ILogger<TCategoryName>被定义出来。ILogger<TCategoryName>派生于 ILogger 接口，自身并没有定义任何成员。

```csharp
public interface ILogger<out TCategoryName> : ILogger {}

public class Logger<T> : ILogger<T>
{
    private readonly ILogger _logger;

    public Logger(ILoggerFactory factory)
        => _logger = factory.CreateLogger(
           TypeNameHelper.GetTypeDisplayName(typeof(T)));
    IDisposable ILogger.BeginScope<TState>(TState state)
        => _logger.BeginScope(state);
    bool ILogger.IsEnabled(LogLevel logLevel)
        => _logger.IsEnabled(logLevel);
    void ILogger.Log<TState>(LogLevel logLevel, EventId eventId, TState state,
        Exception exception, Func<TState, Exception, string> formatter)
        => _logger.Log(logLevel, eventId, state, exception, formatter);
}
```

Logger<T>是 ILogger<TCategoryName>接口的默认实现类型。Logger<T>能够根据泛型类型解析日志类别，从上面提供的代码片段可以看出，具体的解析过程实现在 TypeNameHelper 的静态方法 GetTypeDisplayName 中，该方法采用的解析规则主要包括以下几点。
- 对于一般的类型来说，日志类别名称就是该类型的全名（命名空间+类型名）。
- 如果该类型内嵌于另一个类型之中（如 Foo.Bar+Baz），表示内嵌的"+"需要替换成"."（如 Foo.Bar.Baz）。
- 对于泛型类型，泛型参数部分将不包含在日志类型名称中（如 Foobar<T>对应的日志类别为 Foobar）。

除了可以调用构造函数创建一个 Logger<T>对象，还可以调用针对 ILoggerFactory 接口的 CreateLogger<T>扩展方法来创建。如下面的代码片段所示，除了这个 CreateLogger<T>方法，另一个 CreateLogger 方法直接指定一个 Type 类型的参数，虽然返回类型不同，但是两个方法创建的 Logger 在日志记录行为上是等效的。

```
public static class LoggerFactoryExtensions
{
    public static ILogger<T> CreateLogger<T>(this ILoggerFactory factory)
    public static ILogger CreateLogger(this ILoggerFactory factory, Type type);
}
```

为了使读者对 Logger<T>根据类型解析日志类别的逻辑有更加深刻的认识，下面做一个简单的实例演示。我们在如下所示的代码片段中定义了 3 个需要被映射为日志类别的类型（Foo、Bar 和 Baz），其中，Bar 是一个内嵌类型，而 Baz 则是一个泛型类型。在 Main 方法中，我们采用依赖注入的形式将这 3 个类型作为泛型参数创建了 3 个对应的 Logger<T>对象。

```
namespace App
{
    static class Program
    {
        static void Main()
        {
            void Log<T>()
            {
                new ServiceCollection()
                    .AddLogging(builder=>builder.AddConsole())
                    .BuildServiceProvider()
                    .GetRequiredService<ILogger<T>>()
                    .LogInformation($"{typeof(T).FullName}");
            }

            Log<Foo>();
            Log<Foo.Bar>();
            Log<Baz<Foo>>();
            Console.Read();
        }
    }

    public class Foo
    {
        public class Bar {}
```

```
    }
    public class Baz<T> {}
}
```

我们利用这 3 个 Logger<T>对象记录了 3 条 Information 等级的日志，为了更好地对比类型与日志类别的差异，可以将泛型类型的全名作为日志消息的内容。由于我们注册了 ConsoleLoggerProvider，所以记录的日志消息会以图 9-12 所示的形式输出到控制台上，可以看出输出结果与上述解析规则是一致的。（S911）

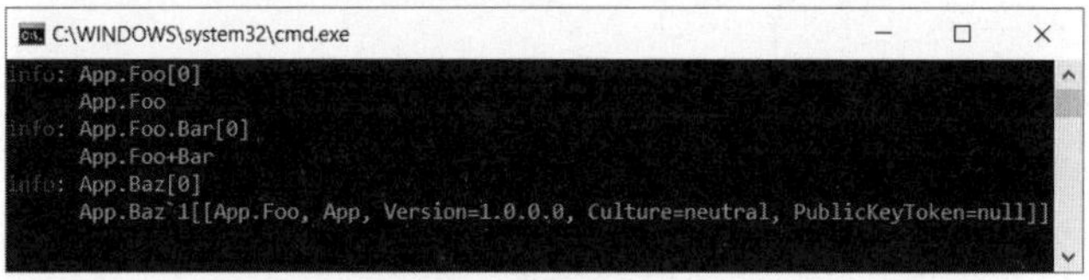

```
info: App.Foo[0]
      App.Foo
info: App.Foo.Bar[0]
      App.Foo+Bar
info: App.Baz[0]
      App.Baz`1[[App.Foo, App, Version=1.0.0.0, Culture=neutral, PublicKeyToken=null]]
```

图 9-12　Logger<T>解析日志类别

9.2.3　日志范围

ILogger 接口的泛型方法 BeginScope<TState>会为我们建立一个日志范围，调用该方法指定的参数将作为这个日志范围的标识。对于在一个日志范围中分发的日志事件，日志范围的标识将会作为一个重要的内容荷载被记录下来。日志范围最终体现为一个 IDisposable 对象，其 Dispose 方法的调用会导致日志范围的终结。除了调用 BeginScope<TState>方法，我们也可以调用下面这个扩展方法来创建日志范围，该方法会将提供的消息模板和参数格式化成一个完整的字符串作为日志范围的内容荷载。

```
public static class LoggerExtensions
{
    public static IDisposable BeginScope(this ILogger logger, string messageFormat,
        params object[] args)
}
```

逻辑调用链

第 8 章多次涉及关于记录调用链信息，基于 TraceSource、EventSource 和 DiagnosticSource 的日志框架都提供了一种被称为活动跟踪（Activity Tracking）的特性，该特性使记录的日志能够反映当前调用链路径。其实本章介绍的日志范围与这个特性在本质上是类似的，具有嵌套结构的日志范围同样可以体现调用链的流转关系。

下面通过实例演示如何利用日志范围来记录调用链信息。如下面的代码片段所示，利用依赖注入框架创建出 ILogger 对象之后，我们创建了具有内嵌关系的 3 个日志范围，并将它们命名为 Foo、Bar 和 Baz，可以将它们视为 3 个逻辑操作，所以调用链就体现为 Foo→Bar→Baz。我们在这 3 个日志范围中分别分发了一个 Information 等级的日志事件。

```
class Program
{
    static void Main()
    {
```

```
        var logger = new ServiceCollection()
            .AddLogging(builder => builder
            .AddConsole(options => options.IncludeScopes = true))
            .BuildServiceProvider()
            .GetRequiredService<ILogger<Program>>();

        using (logger.BeginScope("Foo"))
        {
            logger.Log(LogLevel.Information, "This is a log written in scope Foo.");
            using (logger.BeginScope("Bar"))
            {
                logger.Log(LogLevel.Information,
                    "This is a log written in scope Bar.");
                using (logger.BeginScope("Baz"))
                {
                    logger.Log(LogLevel.Information,
                    "This is a log written in scope Baz.");
                }
            }
        }
        Console.Read();
    }
}
```

由于我们在注册 ConsoleLoggerProvider 时开启了 IncludeScopes 特性，所以当前的日志范围信息和日志负载内容会以图 9-13 所示的形式输出到控制台上。可以看到，每条日志消息携带的日志范围都可以清晰地反映当前的调用链流转关系。（S912）

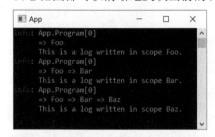

图 9-13　利用日志范围记录调用链信息

IExternalScopeProvider

日志范围是通过提供的 IExternalScopeProvider 对象提供的。日志范围呈现出具有父子关系的堆栈（Stack）结构，新的日志范围被创建之后，当前的日志范围会成为它的"父亲"。新创建的日志范围会以图 9-14 所示的方式压入堆栈并成为当前的日志范围，这一操作实现在 IExternalScopeProvider 接口的 Push 方法中。

```
public interface IExternalScopeProvider
{
    IDisposable Push(object state);
    void ForEachScope<TState>(Action<object, TState> callback, TState state);
}
```

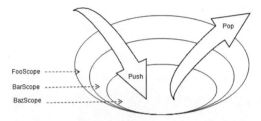

图 9-14　日志范围的堆栈结构

通过 Push 方法创建的日志范围体现为一个 IDisposable 对象，当该对象的 Dispose 方法被调用之后，它会以图 9-14 所示的形式从堆栈中弹出来，它的"父亲"会重新成为当前的日志范围。日志范围可以用来描述逻辑调用链。

IExternalScopeProvider 接口还定义了 ForEachScope 方法，该方法会利用我们调用提供的 Action<object, TState>对象处理堆栈路径上的所有日志范围。如下所示的 LoggerExternalScopeProvider 类型是对 IExternalScopeProvider 接口的默认实现。

```
public class LoggerExternalScopeProvider : IExternalScopeProvider
{
    private readonly AsyncLocal<Scope> _currentScope = new AsyncLocal<Scope>();
    public void ForEachScope<TState>(Action<object, TState> callback, TState state)
    {
        void Report(Scope current)
        {
            if (current == null)
            {
                return;
            }
            Report(current.Parent);
            callback(current.State, state);
        }
        Report(_currentScope.Value);
    }

    public IDisposable Push(object state)
    {
        var parent = _currentScope.Value;
        var newScope = new Scope(this, state, parent);
        _currentScope.Value = newScope;
        return newScope;
    }

    private class Scope : IDisposable
    {
        private readonly LoggerExternalScopeProvider      _provider;
        private bool                                      _isDisposed;
        public Scope                Parent { get; }
        public object               State { get; }
        internal Scope(LoggerExternalScopeProvider provider, object state,
```

```
                Scope parent)
        {
            _provider           = provider;
            State               = state;
            Parent              = parent;
        }

        public override string ToString()=> State?.ToString();
        public void Dispose()
        {
            if (!_isDisposed)
            {
                _provider._currentScope.Value = Parent;
                _isDisposed = true;
            }
        }
    }
}
```

ISupportExternalScope

一般来说,如果某种日志输出渠道对日志范围提供支持,那么对应的 ILoggerProvider 类型会同时实现 ISupportExternalScope 接口。如下面的代码片段所示,ISupportExternalScope 接口定义了唯一的方法 SetScopeProvider,该方法用来设置创建日志范围的 IExternalScopeProvider 对象。

```
public interface ISupportExternalScope
{
    void SetScopeProvider(IExternalScopeProvider scopeProvider);
}
```

9.2.4 ILoggerProvider

ILoggerProvider 对象在日志模型中的作用在于"提供"真正具有日志输出功能的 ILogger 对象。如下面的代码片段所示,ILoggerProvider 继承了 IDisposable 接口,如果某个具体的 ILoggerProvider 对象需要释放某种资源,就可以将相关操作实现在 Dispose 方法中。ILoggerProvider 针对 ILogger 对象的提供实现在它的 CreateLogger 方法中,该方法的参数 categoryName 代表前面提及的日志类别。

```
public interface ILoggerProvider : IDisposable
{
    ILogger CreateLogger(string categoryName);
}
```

9.2.5 ILoggerFactory

从命名的角度来讲,ILoggerProvider 和 ILoggerFactory 最终都是为了提供或者创建 ILogger 对象,但是两者提供的 ILogger 对象在本质上是不同的。针对某种日志输出渠道的 ILoggerProvider 对象提供的是真正具有日志输出功能的 ILogger 对象,而 ILoggerFactory 对象仅仅创建用于分发日志事件的 ILogger 对象,后者正是应用程序中使用的 ILogger 对象。

如下面的代码片段所示，ILoggerFactory 接口定义了两个简单的方法：针对 ILogger 对象的创建实现在 CreateLogger 方法中，而 AddProvider 方法用来注册 ILoggerProvider 对象。本章开篇演示实例中使用的 LoggerFactory 类型是对 ILoggerFactory 接口的默认实现，由它创建的是一个类型为 Logger 的对象，该对象由所有注册的 ILoggerProvider 对象提供的 ILogger 对象组合而成。

```
public interface ILoggerFactory : IDisposable
{
    ILogger CreateLogger(string categoryName);
    void AddProvider(ILoggerProvider provider);
}
```

复合型的 Logger

前面多次强调 LoggerFactory 工厂创建的是一个复合型的 ILogger 对象，它是对一组面向输出渠道的 ILogger 对象的封装。当应用程序利用这个复合型 ILogger 对象来分发日志事件时，它仅仅是将日志事件进一步分发给这些 ILogger 对象成员而已。Logger 对象在进行日志消息分发之前，它需要进行日志过滤，ILogger 对象成员只有在满足过滤规则的前提下才会接收 Logger 分发给它的日志消息，所以这个复合型的 Logger 对象包括这些 ILogger 对象成员对应的日志过滤规则。

LoggerFactory 不仅仅是上面这个 Logger 对象的创建者，还是 Logger 对象的维护者，因为它创建的是一个动态的 Logger 对象，该对象总是与日志模型的当前状态保持同步。假设具有图 9-15 所示的一个 LoggerFactory 对象，并且为它注册了两个 ILoggerProvider 对象（FooLoggerProvider 和 BarLoggerProvider）。我们利用这个 LoggerFactory 对象针对日志类别 A 和 B 创建了两个 Logger 对象，它们由注册的这两个 ILoggerProvider 对象提供的 ILogger 对象（FooLogger 和 BarLogger）组合而成。

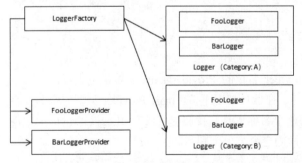

图 9-15　LoggerFactory 对象保持对所有创建的 Logger 对象的引用

LoggerFactory 对象保持由它创建的所有 Logger 对象的引用。如果后续过程有一个新的 ILoggerProvider（BazLoggerProvider）注册进来（见图 9-16），LoggerFactory 对象就会利用它创建一个新的 ILogger 对象（BazLogger），并将它们作为成员添加到现有的两个 Logger 对象之中。

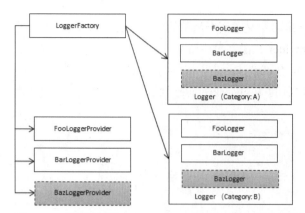

图9-16 将新注册的ILoggerProvider对象应用到现有的Logger对象之中

LoggerFactory 对象提供的是一个动态变化的 Logger 对象，这不仅体现在 LoggerFactory 对象能够将新注册的 ILoggerProvider 对象提供的 ILogger 对象应用到现有的 Logger 对象中，还体现在实时变化的日志过滤规则同样可以及时应用到现有的所有 Logger 对象上。日志模型通过 LoggerFilterRule 类型来表示日志过滤规则。

```
public class LoggerFilterRule
{
    public string                                   ProviderName { get; }
    public string                                   CategoryName { get; }
    public LogLevel?                                LogLevel { get; }
    public Func<string, string, LogLevel, bool>     Filter { get; }

    public LoggerFilterRule(string providerName, string categoryName,
        LogLevel? logLevel, Func<string, string, LogLevel, bool> filter)
    {
        ProviderName    = providerName;
        CategoryName    = categoryName;
        LogLevel        = logLevel;
        Filter          = filter;
    }
}
```

LoggerFilterRule 具有一个表示 ILoggerProvider 名称的 ProviderName 属性，毫无疑问，它对应的是 LoggerInformation 的 ProviderType 属性。在默认情况下，ILoggerProvider 类型的全名（命名空间+类型名称）就是它的名称。如果认为这个名称太长，我们也可以在 ILoggerProvider 实现类型上标注如下这个 ProviderAliasAttribute 特性为它命名一个别名。

```
[AttributeUsage(AttributeTargets.Class, AllowMultiple = false, Inherited = false)]
public class ProviderAliasAttribute: Attribute
{
    public string Alias { get; }
    public ProviderAliasAttribute(string alias) => this.Alias = alias;
}
```

在前面演示基于配置的日志规则定义时，我们之所以能够在配置文件中采用 Console 表示

ConsoleLoggerProvider，是因为 ConsoleLoggerProvider 在定义时采用这种方式将 Console 作为其别名。除了 ConsoleLoggerProvider，其他预定义的 ILoggerProvider 同样采用如下形式定义了相应的别名。

```
[ProviderAlias("Console")]
public class ConsoleLoggerProvider : ILoggerProvider {...}

[ProviderAlias("Debug")]
public class EventLogLoggerProvider : ILoggerProvider {...}

[ProviderAlias("EventLog")]
public class DebugLoggerProvider : ILoggerProvider {...}

[ProviderAlias("TraceSource")]
public class TraceSourceLoggerProvider : ILoggerProvider {...}

[ProviderAlias("EventSource")]
public class EventSourceLoggerProvider : ILoggerProvider {...}
```

日志模型采用基于"前缀"的类别匹配规则。假设我们针对日志类别 Foo.Bar 和 Foo.Baz 创建了两个 Logger 对象，如果表示过滤规则的 LoggerFilterRule 将 CategoryName 属性设置为 Foo，那么两个 Logger 对象都满足过滤条件。过滤规则是通过 LoggerFilterOptions 这个 Options 类型进行设置的。如下面的代码片段所示，除了通过属性 Rules 表示一组日志过滤规则，LoggerFilterOptions 还可以用 MinLevel 属性来设置一个全局默认的最低日志等级。

```
public class LoggerFilterOptions
{
    public LogLevel                      MinLevel { get; set; }
    public IList<LoggerFilterRule>       Rules { get; } = new List<LoggerFilterRule>();
}
```

对于一个通过 IOptionsMonitor<LoggerFilterOptions>对象创建的 LoggerFactory 对象，如果它能够感知到过滤规则的变化，那么过滤规则一旦发生变化，LoggerFactory 只需要将 LoggerFilterOptions 对象表示的新规则应用到所有 Logger 对象上，就能实现日志过滤规则的实时同步。同一个 Logger 对象可能会与多个 LoggerFilterRule 相匹配，所以 LoggerFactory 需要为每个 Logger 对象选择一个匹配度最高的 LoggerFilterRule 对象。

在选择 LoggerFilterRule 时，对于 ILoggerProvider 名称，LoggerFactory 对象会采用精确匹配的方式，对日志类别则采用最长前缀一致性原则。例如，提供的 LoggerFilterOptions 对象具有类型分别为 Foo.Bar 和 Foo 的规律规则，以及针对待过滤的类别 Foo.Bar.Baz，前者具有更高的匹配程度。

LoggerFactory

下面从代码层面进一步介绍 LoggerFactory 对象是如何创建 Logger 对象并且总是将它们的状态与当前设置保持实时同步的。如下面的代码片段所示，调用构造函数创建 LoggerFactory 对象时需要提供一组 ILoggerProvider 对象和一个 IOptionsMonitor<LoggerFilterOptions>对象，前者用于提供面向输出渠道的ILogger对象，后者则用来提供实时的日志过滤规则。

```
public class LoggerFactory : ILoggerFactory, IDisposable
{
    public LoggerFactory(IEnumerable<ILoggerProvider> providers,
        LoggerFilterOptions filterOptions);
    public LoggerFactory(IEnumerable<ILoggerProvider> providers,
        IOptionsMonitor<LoggerFilterOptions> filterOption);

    public void AddProvider(ILoggerProvider provider);
    public ILogger CreateLogger(string categoryName);
    public void Dispose();
    ...
}
```

ILoggerProvider接口派生于IDisposable接口，所以所有的ILoggerProvider对象最终都应该通过调用Dispose方法的形式被释放。由于ILoggerProvider对象是被注册到LoggerFactory对象上的，所以针对ILoggerProvider对象的释放理应由LoggerFactory负责，但是它只负责释放通过调用AddProvider方法注册的ILoggerProvider对象，在构造时指定的ILoggerProvider对象则不由它负责。这是因为后者是通过依赖注入的方式创建的，采用这种方式创建的服务对象都应该由作为依赖注入容器的IServiceProvider对象来负责释放。

对于每次针对CreateLogger方法的调用，LoggerFactory并不总是创建一个新的Logger对象，它会对提供的Logger对象进行缓存。也就是说，如果指定相同的日志类别，LoggerFactory提供的实际上是同一个Logger对象。

由于构造LoggerFactory对象时指定一个IOptionsMonitor<LoggerFilterOptions>对象来提供注册的日志过滤规则，所以它能够检测到规则是否发生改变。LoggerFactory通过调用该对象的OnChange方法注册一个回调，从而将新的规则应用到所有被缓存的Logger对象上。当LoggerFactory因为Dispose方法被回收时，它除了需要释放添加的ILoggerProvider对象，还需要解除这个回调注册，以免造成内存泄漏。

日志范围

前面提到，如果某种日志输出渠道提供针对日志范围的支持，那么对应的ILoggerProvider实现类型会同时实现ISupportExternalScope接口。ISupportExternalScope接口定义了一个SetScopeProvider方法，用来设置用于创建日志范围的IExternalScopeProvider对象，最终调用该方法为对应的ILoggerProvider对象提供IExternalScopeProvider对象的正是LoggerFactory。

```
public class LoggerFactory : ILoggerFactory
{
    private LoggerExternalScopeProvider      _scopeProvider;
    public LoggerFactory(IEnumerable<ILoggerProvider> providers,
        IOptionsMonitor<LoggerFilterOptions> filterOption)
    {
        ...
        _scopeProvider = new LoggerExternalScopeProvider();
        SetScopeProviders(providers.ToArray());
```

```
}
public void AddProvider(ILoggerProvider provider)
{
    SetScopeProviders(provider);
    ...
}
private void SetScopeProviders(params ILoggerProvider[] providers)
{
    foreach (var provider in providers.OfType<ISupportExternalScope>())
    {
        provider.SetScopeProvider(_scopeProvider);
    }
    ...
}
```

如上面的代码片段所示，LoggerFactory 在初始化过程中会创建一个 LoggerExternalScopeProvider 对象。当构造函数被调用时，它会选择所有支持日志范围（对应类型实现了 ISupportExternalScope 接口）的 ILoggerProvider 对象，并调用其 SetScopeProvider 方法将 LoggerExternalScopeProvider 对象提供给它们。类似的操作同样实现在用户注册的 ILoggerProvider 对象的 AddProvider 方法中。

9.2.6 LoggerMessage

通过 9.2.5 节的实例演示可知，利用静态类型 LoggerMessage 创建的委托对象不仅可以实现强类型的日志编程，更重要的是它能够避免针对消息模板的重复解析，所以对于涉及频繁日志写入或者对性能具有较高要求的应用最好使用这种编程方式。

如果利用 LoggerMessage 进行日志编程，就可以通过指定的消息模板、日志事件 ID 和最低日志等级调用 Define 方法或者 DefineScope 方法创建的 Action<ILogger,…, Exception >或者 Func<ILogger,…, IDisposable>对象来分发日志事件或者创建日志范围。根据模板中定义的占位符数量的不同，日志系统提供了如下这些重载方法供我们选择。

```
public static class LoggerMessage
{
    public static Action<ILogger, Exception> Define(LogLevel logLevel,
        EventId eventId, string formatString);
    public static Action<ILogger, T1, Exception> Define<T1>(LogLevel logLevel,
        EventId eventId, string formatString);
    public static Action<ILogger, T1, T2, Exception> Define<T1, T2>(
        LogLevel logLevel, EventId eventId, string formatString);
    public static Action<ILogger, T1, T2, T3, Exception> Define<T1, T2, T3>(
        LogLevel logLevel, EventId eventId, string formatString);
    public static Action<ILogger, T1, T2, T3, T4, Exception> Define
        <T1, T2, T3, T4>(LogLevel logLevel, EventId eventId, string formatString);
    public static Action<ILogger, T1, T2, T3, T4, T5, Exception> Define
        <T1, T2, T3, T4, T5>(LogLevel logLevel, EventId eventId,
        string formatString);
```

```csharp
    public static Action<ILogger, T1, T2, T3, T4, T5, T6, Exception> Define
        <T1, T2, T3, T4, T5, T6>(LogLevel logLevel, EventId eventId,
        string formatString);

    public static Func<ILogger, IDisposable> DefineScope(string formatString);
    public static Func<ILogger, T1, IDisposable> DefineScope<T1>(
        string formatString);
    public static Func<ILogger, T1, T2, IDisposable> DefineScope<T1, T2>(
        string formatString);
    public static Func<ILogger, T1, T2, T3, IDisposable> DefineScope
        <T1, T2, T3>(string formatString);
}
```

日志系统在默认情况下会利用一个 LogValuesFormatter 对象来对指定的消息模板进行解析，Define 方法或者 DefineScope 方法之所以能够避免对同一个消息模板进行重复解析，是因为由它们创建的委托对象实现了对 LogValuesFormatter 对象的重复使用，这一点可以从如下所示的代码看出来。

```csharp
public static class LoggerMessage
{
    public static Func<ILogger, T1, IDisposable> DefineScope<T1>(
        string formatString)
    {
        var formatter = new LogValuesFormatter(formatString);
        return (logger, arg1) => logger.BeginScope(new LogValues<T1>(
            formatter, arg1));
    }

    public static Action<ILogger, T1, Exception> Define<T1>(LogLevel logLevel,
        EventId eventId, string formatString)
    {
        var formatter = new  LogValuesFormatter(formatString);
        return (logger, arg1, exception) =>
        {
            if (logger.IsEnabled(logLevel))
            {
                logger.Log(logLevel, eventId, new LogValues<T1>(formatter, arg1),
                    exception, LogValues<T1>.Callback);
            }
        };
    }
    ...
}
```

9.3 依赖注入

我们总是采用依赖注入的方式来提供用于分发日志事件的 ILogger 对象。具体来说，有两种方式可供选择：一种是先利用作为依赖注入容器的 IServiceProvider 对象来提供一个

ILoggerFactory 工厂，然后利用它根据指定日志类别创建 ILogger 对象；另一种则是直接利用 IServiceProvider 对象提供一个泛型的 ILogger<TCategoryName>对象。IServiceProvider 对象能够提供期望服务对象的前提是预先添加了相应的服务注册，下面介绍用来注册服务的 API，以及它们究竟注册了哪些服务。

9.3.1 服务注册

构成日志模型的核心服务是通过 IServiceCollection 接口的 AddLogging 扩展方法进行注册的。由于可以直接利用作为依赖注入容器的 IServiceProvider 对象提供 ILoggerFactory 和 ILogger<TCategoryName>对象，AddLogging 方法自然提供了针对这两个类型的服务注册，这一点从如下所示的代码片段可以看出来。

```
public static class LoggingServiceCollectionExtensions
{
    public static IServiceCollection AddLogging(this IServiceCollection services)
        => AddLogging(services, builder => {});

    public static IServiceCollection AddLogging(this IServiceCollection services,
        Action<ILoggingBuilder> configure)
    {
        services.AddOptions();
        services.TryAdd(ServiceDescriptor.Singleton
            <ILoggerFactory, LoggerFactory>());
        services.TryAdd(ServiceDescriptor.Singleton(
            typeof(ILogger<>), typeof(Logger<>)));

        services.TryAddEnumerable(ServiceDescriptor.Singleton
            <IConfigureOptions<LoggerFilterOptions>>(
            new DefaultLoggerLevelConfigureOptions(LogLevel.Information)));

        configure(new LoggingBuilder(services));
        return services;
    }
}

internal class DefaultLoggerLevelConfigureOptions :
    ConfigureOptions<LoggerFilterOptions>
{
    public DefaultLoggerLevelConfigureOptions(LogLevel level)
        : base(options => options.MinLevel = level)
    {}
}
```

除了添加针对 LoggerFactory 和 Logger<TCategoryName>类型的服务注册，AddLogging 扩展方法还调用 IServiceCollection 接口的 AddOptions 扩展方法注册了 Options 模式的核心服务。这个扩展方法还以 Singleton 模式添加了一个针对 IConfigureOptions<LoggerFilterOptions>接口的服务注册，具体的服务实例是一个 DefaultLoggerLevelConfigureOptions 对象，它将默认的最低日

志等级设置为 Information，这正是在默认情况下等级为 Trace 和 Debug 的日志事件会被忽略的根源所在。

除了上述这些核心服务，日志事件的分发与输出还涉及其他服务，如针对 ILoggerProvider 的注册就是必需的。从上面给出的代码片段可以看出，AddLogging 方法的第二个重载提供了一个类型为 Action<ILoggingBuilder>的参数[1]，我们可以利用它来注册额外的服务。ILoggingBuilder 接口和默认实现类型 LoggingBuilder 的定义如下，与其他用来注册服务的 Builder 类型一样，LoggingBuilder 仅仅是对一个 IServiceCollection 对象的封装而已。

```
public interface ILoggingBuilder
{
    IServiceCollection Services { get; }
}

internal class LoggingBuilder : ILoggingBuilder
{
    public LoggingBuilder(IServiceCollection services) => Services = services;
    public IServiceCollection Services { get; }
}
```

ILoggingBuilder 接口具有如下几个常用的扩展方法：SetMinimumLevel 方法用来设置最低日志等级，Add 方法用来注册 ILoggerProvider 对象，注册的 ILoggerProvider 对象可以通过调用 ClearProviders 方法进行清除。

```
public static class LoggingBuilderExtensions
{
    public static ILoggingBuilder SetMinimumLevel(this ILoggingBuilder builder,
        LogLevel level)
    {
        builder.Services.Add(ServiceDescriptor.Singleton
            <IConfigureOptions<LoggerFilterOptions>>(
            new DefaultLoggerLevelConfigureOptions(level)));
        return builder;
    }

    public static ILoggingBuilder AddProvider(this ILoggingBuilder builder,
        ILoggerProvider provider)
    {
        builder.Services.AddSingleton(provider);
        return builder;
    }

    public static ILoggingBuilder ClearProviders(this ILoggingBuilder builder)
    {
```

---

[1] 为了避免将所有服务注册的扩展方法直接定义在 IServiceCollection 接口上，造成 "API 爆炸"，可以采用 Builder 模式来进行服务注册。具体来说，如果某个组件或者框架具有多个待注册的服务，可以定义一个用于封装 IServiceCollection 对象的 Builder 类型，并将服务注册的方法（含扩展方法）定义在 Builder 类型上。我们可以将 Builder 定义成接口（如 ILoggingBuilder），但是在绝大部分情况下直接定义成类型即可。

```
        builder.Services.RemoveAll<ILoggerProvider>();
        return builder;
    }
}
```

9.3.2 设置日志过滤规则

虽然 LoggerFactory 定义了若干构造函数重载，但是根据第 4 章介绍的关于构造函数筛选规则，作为依赖注入容器的 IServiceProvider 对象总是会选择如下这个构造函数重载。也就是说，代表日志过滤规则的 LoggerFilterOptions 对象是由注入的 IOptionsMonitor<LoggerFilterOptions> 对象提供的。

```
public class LoggerFactory : ILoggerFactory
{
    public LoggerFactory(IEnumerable<ILoggerProvider> providers,
        IOptionsMonitor<LoggerFilterOptions> filterOption);
    ...
}
```

如果希望对过滤规则做相应的设置，只需要注册相应的 IConfigureOptions<LoggerFilterOptions>或者 IPostConfigureOptions<LoggerFilterOptions>服务对承载过滤规则的 LoggerFilterOptions 进行针对性设置即可。ILoggingBuilder 接口定义了如下这些名为 AddFilter 的扩展方法，从给出的代码片段可以看出，它们最终都是通过调用 IServiceCollection 接口的 Configure<LoggerFilterOptions>方法注册的 ConfigureOptions<LoggerFilterOptions>对象实现对日志过滤规则的定制的。

```
public static class FilterLoggingBuilderExtensions
{
    public static ILoggingBuilder AddFilter(this ILoggingBuilder builder,
        Func<string, string, LogLevel, bool> filter)
        => builder.ConfigureFilter(options => options.AddFilter(filter));

    public static ILoggingBuilder AddFilter(this ILoggingBuilder builder,
        Func<string, LogLevel, bool> categoryLevelFilter)
        => builder.ConfigureFilter(options => options.AddFilter(categoryLevelFilter));

    public static ILoggingBuilder AddFilter<T>(this ILoggingBuilder builder,
        Func<string, LogLevel, bool> categoryLevelFilter) where T : ILoggerProvider
        => builder.ConfigureFilter(
            options => options.AddFilter<T>(categoryLevelFilter));

    public static ILoggingBuilder AddFilter(this ILoggingBuilder builder,
        Func<LogLevel, bool> levelFilter)
        => builder.ConfigureFilter(options => options.AddFilter(levelFilter));

    public static ILoggingBuilder AddFilter<T>(this ILoggingBuilder builder,
        Func<LogLevel, bool> levelFilter) where T : ILoggerProvider =>
        builder.ConfigureFilter(options => options.AddFilter<T>(levelFilter));
```

```csharp
public static ILoggingBuilder AddFilter(this ILoggingBuilder builder,
    string category, LogLevel level)
    => builder.ConfigureFilter(options => options.AddFilter(category, level));

public static ILoggingBuilder AddFilter<T>(this ILoggingBuilder builder,
    string category, LogLevel level) where T: ILoggerProvider
    => builder.ConfigureFilter(options => options.AddFilter<T>(
    category, level));

public static ILoggingBuilder AddFilter(this ILoggingBuilder builder,
    string category, Func<LogLevel, bool> levelFilter)
    => builder.ConfigureFilter(options =>
    options.AddFilter(category, levelFilter));

public static ILoggingBuilder AddFilter<T>(this ILoggingBuilder builder,
    string category, Func<LogLevel, bool> levelFilter) where T : ILoggerProvider
    => builder.ConfigureFilter(options =>
    options.AddFilter<T>(category, levelFilter));

private static ILoggingBuilder ConfigureFilter(this ILoggingBuilder builder,
    Action<LoggerFilterOptions> configureOptions)
{
    builder.Services.Configure(configureOptions);
    return builder;
}
```

除了可以调用上面这些 AddFilter<T>扩展方法来定制日志过滤规则，还可以采用配置形式来提供这些过滤规则。由本章开篇演示的实例可知，基于日志过滤规则的配置大体上分为两种形式：一种是与具体 ILoggerProvider 类型无关的，另一种则是针对某种类型的 ILoggerProvider。如果采用 JSON 格式的配置，这两种配置采用如下结构。对于第二种配置形式，我们可以将 ILoggerProvider 的全名或者别名作为配置节名称。

```
{
    "LogLevel": {
        "Default"            : "{MinLogLelel}",
        "{CategoryName}"     : "{MinLogLelel}"
    },
    "{LoggerProviderName|LoggerProviderAlias}": {
        "LogLevel": {
            "Default"            : "{MinLogLelel}",
            "{CategoryName}"     : "{MinLogLelel}"
        }
    }
}
```

如果采用配置形式来定义日志过滤规则，我们可以将承载日志过滤规则配置的 IConfiguration 对象作为参数调用 ILoggingBuilder 接口的 AddConfiguration 扩展方法。从提供的实现可以看出，该方法注册的两个接口类型分别为 IConfigureOptions<TOptions>和 IOptions

ChangeTokenSource<TOptions>的服务。前者的实现类型为 LoggerFilterConfigureOptions，它会解析配置的内容并将得到的过滤规则应用到待初始化的 LoggerFilterOptions 对象上；后者则是一个 ConfigurationChangeTokenSource<LoggerFilterOptions>对象，我们可以利用它感知配置源的变化，并及时将新的规则应用到日志系统中。

```
public static class LoggingBuilderExtensions
{
    public static ILoggingBuilder AddConfiguration(this ILoggingBuilder builder,
        IConfiguration configuration)
    {
        builder.Services.AddSingleton<IConfigureOptions<LoggerFilterOptions>>(
            new LoggerFilterConfigureOptions(configuration));
        builder.Services.AddSingleton<IOptionsChangeTokenSource
            <LoggerFilterOptions>>(
            new ConfigurationChangeTokenSource<LoggerFilterOptions>(configuration));
        return builder;
    }
}

internal class LoggerFilterConfigureOptions : IConfigureOptions<LoggerFilterOptions>
{
    public LoggerFilterConfigureOptions(IConfiguration configuration) ;
    public void Configure(LoggerFilterOptions options) ;
}
```

9.4 日志输出渠道

可以利用 LoggerFactory 工厂创建一个 Logger 对象并利用它来分发日志事件，针对日志事件的输出取决于注册到该 LoggerFactory 对象上的 ILoggerProvider 对象会提供怎样的 ILogger 对象。日志系统提供了一系列针对不同输出渠道的 ILoggerProvider 实现类型，下面依次进行介绍。

9.4.1 控制台

一个控制台应用，如采用控制台应用作为宿主的 ASP.NET Core 应用，可以将日志直接输出到控制台上。针对控制台的 ILogger 对象的实现类型为 ConsoleLogger，对应的 ILoggerProvider 对象的实现类型为 ConsoleLoggerProvider，这两个类型都定义在 NuGet 包 "Microsoft.Extensions.Logging.Console" 中。

ConsoleLogger

如下所示的代码片段展示了 ConsoleLogger 类型的定义。构造函数的参数 name 代表的名称实际上就是 ILoggerFactory 对象在创建 ILogger 对象时指定的日志类别。作为另一个参数的 ConsoleLoggerProcessor 对象可以帮助我们完成控制台的输出。

```
internal class ConsoleLogger : ILogger
{
    internal ConsoleLogger(string name, ConsoleLoggerProcessor loggerProcessor);
```

```csharp
public bool IsEnabled(LogLevel logLevel) => (logLevel != LogLevel.None);
public IDisposable BeginScope<TState>(TState state);
public void Log<TState>(LogLevel logLevel, EventId eventId, TState state,
    Exception exception, Func<TState, Exception, string> formatter);
public virtual void WriteMessage(LogLevel logLevel, string logName, int eventId,
    string message, Exception exception);
}
```

作为 ConsoleLogger 输出渠道的控制台可以通过 IConsole 接口表示，之所以没有直接用 System.Console 向控制台输出格式化的日志消息，是因为需要提供跨平台的支持，IConsole 接口表示的就是这样一个与具体平台无关的抽象控制台。如下面的代码片段所示，IConsole 接口具有如下 3 个方法。在调用 Write 方法和 WriteLine 方法向控制台输出内容时，除了指定写入的消息文本，还可以控制消息在控制台上的背景色和前景色。Flush 方法与数据输出缓冲机制有关，如果采用缓冲机制，通过 Write 方法或者 WriteLine 方法写入的消息并不会立即输出到控制台，而是先被保存到缓冲区，Flush 方法被执行时会将缓冲区的所有日志消息批量输出到控制台上。

```csharp
public interface IConsole
{
    void Write(string message, ConsoleColor? background, ConsoleColor? foreground);
    void WriteLine(string message, ConsoleColor? background,
        ConsoleColor? foreground);
    void Flush();
}
```

微软默认提供了两种类型的 Console：基于 Windows 平台的 WindowsLogConsole 类型，非 Windows 平台则通过 AnsiLogConsole 类型来表示。它们之间的不同之处主要体现在设置控制台显示颜色（前景色和背景色）的差异。对于 Windows 平台来说，消息显示在控制台上的颜色是通过显式设置 System.Console 类型的静态属性 ForegroundColor 和 BackgroundColor 实现的，但是对于非 Windows 平台来说，颜色信息会直接以基于 ANSI 标准的转意字符序列（ANSI Esacpe Sequences）的形式内嵌在消息文本之中。

当 ConsoleLogger 对象的 Log 方法被调用时，它先将指定的日志等级作为参数调用 IsEnabled 方法。如果这个方法返回 True（指定的日志等级不为 None），ConsoleLogger 对象会调用 WriteMessage 方法将提供的日志消息输出到由 Console 属性表示的控制台上。WriteMessage 方法是一个虚方法，如果它输出的消息格式和样式无法满足要求，我们可以定义 ConsoleLogger 的子类，并通过重写这个方法按照我们希望的方式输出日志消息。

```
{LogLevel} : {Category}[{EventId}]
      {Message}
```

在默认情况下，被 ConsoleLogger 对象输出到控制台上的日志消息会采用上面的格式，这也可以通过前面的演示实例进行印证。对于输出到控制台表示日志等级的部分，输出的文字与对应的日志等级的映射关系如表 9-3 所示，可以看出，日志等级在控制台上均会显示为仅包含 4 个字母的简写形式。日志等级同时决定了该部分内容在控制台上显示的前景色。如果不希望对控制台输出进行着色，就可以将 ConsoleLogger 对象的 DisableColors 属性设置为 True。

表 9-3　输出的文字与对应的日志等级的映射关系

日志等级	显示文字	前景颜色	背景颜色
Trace	trce	Gray	Black
Debug	dbug	Gray	Black
Information	info	DarkGreen	Black
Warning	warn	Yellow	Black
Error	fail	Red	Black
Critical	crit	White	Red

ConsoleLogger 对象采用异步的方式将日志消息写入控制台，所以它不会阻塞当前线程。这个异步写入的特性使我们无法确定记录的日志消息何时出现在控制台上。针对控制台的日志消息异步输出实现在 ConsoleLoggerProcessor 类型中，它会利用一个独立的线程将日志输出到控制台上。由于待输出的日志消息被存放在一个队列中，所以在控制台上的输出顺序可以保证与应用程序发出的日志事件顺序保持一致。

ConsoleLoggerProvider

ConsoleLogger 由 ConsoleLoggerProvider 对象提供，如下所示的代码片段是对应配置选项 ConsoleLoggerOptions 的定义。ConsoleLoggerOptions 对象的 Format 属性会影响生成日志消息的格式，之前的演示实例采用的都是默认格式，日志内容会格式化成多行输出，如果采用 Systemd 格式，输出的内容就只有一行。该配置选项的 DisableColors 属性用来关闭控制台针对不同日志等级的着色功能。TimestampFormat 属性表示的是格式化时间戳采用的格式化字符串，而 IncludeScopes 属性则用来开启针对日志范围的支持。

```
public class ConsoleLoggerOptions
{
    public ConsoleLoggerFormat      Format { get; set; }
    public bool                     DisableColors { get; set; }
    public string                   TimestampFormat { get; set; }
    public LogLevel                 LogToStandardErrorThreshold { get; set; }
    public bool                     IncludeScopes { get; set; }
}

public enum ConsoleLoggerFormat
{
    Default,
    Systemd
}
```

控制台输出具有两种类型：标准输出（Out）和标准错误输出（Error）。以 Console 类型表示的控制台为例，这两种输出分别体现为通过如下所示的静态属性 Out 和 Error 返回的 TextWriter 对象。ConsoleLoggerOptions 类型的 LogToStandardErrorThreshold 属性返回采用 Error 输出的最低等级，该属性的默认值为 None，所以在默认情况下总是采用 Out 输出。

```
public static class Console
{
    public static TextWriter Out { get; }
```

```
    public static TextWriter Error { get; }
}
```

如下所示的代码片段是 ConsoleLoggerProvider 类型的定义。该类型标注了 ProviderAlias Attribute 特性，并将别名设置为 Console，所以在使用配置方式设置过滤规则时，可以使用这个别名。由于创建 ConsoleLoggerProvider 对象提供的配置选项是通过 IOptionsMonitor<ConsoleLoggerOptions>对象提供的，所以配置的改变能够及时应用到由它创建的 ConsoleLogger 对象上。ConsoleLoggerProvider 对象会对创建的 ConsoleLogger 对象根据名称进行缓存，也就是说，针对同一个名称得到的 ConsoleLogger 对象其实是同一个对象。

```
[ProviderAlias("Console")]
public class ConsoleLoggerProvider : ILoggerProvider, ISupportExternalScope
{
    public ConsoleLoggerProvider(IOptionsMonitor<ConsoleLoggerOptions> options);

    public ILogger CreateLogger(string name);
    public void Dispose();
    public void SetScopeProvider(IExternalScopeProvider scopeProvider);
}
```

由于 ConsoleLoggerProvider 类型实现了 ISupportExternalScope 接口，所以当它注册到 LoggerFactory 对象时，后者会调用其 SetScopeProvider 方法为它指定一个 IExternalScopeProvider 对象，该对象最终会提供给由它创建的 ConsoleLogger 对象。如果通过配置选项的 IncludeScopes 属性开启了日志范围，ConsoleLogger 会利用 IExternalScopeProvider 对象得到当前范围的信息，并将其作为格式化日志消息的一部分。

针对 ConsoleLoggerProvider 的注册可以通过如下两个针对 ILoggingBuilder 接口的 AddConsole 扩展方法来完成，相关配置选项可以利用提供的 Action<ConsoleLoggerOptions>委托进行定制。

```
public static class ConsoleLoggerExtensions
{
    public static ILoggingBuilder AddConsole(this ILoggingBuilder builder)
    {
        builder.AddConfiguration();
        builder.Services.TryAddEnumerable(
            ServiceDescriptor.Singleton<ILoggerProvider, ConsoleLoggerProvider>());
        LoggerProviderOptions.RegisterProviderOptions
            <ConsoleLoggerOptions, ConsoleLoggerProvider>(builder.Services);
        return builder;
    }
    public static ILoggingBuilder AddConsole(this ILoggingBuilder builder,
        Action<ConsoleLoggerOptions> configure)
    {

        builder.AddConsole();
        builder.Services.Configure(configure);
        return builder;
    }
}
```

```csharp
public static class LoggerProviderOptions
{
    public static void RegisterProviderOptions<TOptions, TProvider>(
        IServiceCollection services) where TOptions: class
    {
        services.TryAddEnumerable(
            ServiceDescriptor.Singleton<IConfigureOptions<TOptions>,
            LoggerProviderConfigureOptions<TOptions, TProvider>>());
        services.TryAddEnumerable(
            ServiceDescriptor.Singleton<IOptionsChangeTokenSource<TOptions>,
            LoggerProviderOptionsChangeTokenSource<TOptions, TProvider>>());
    }
}
```

9.4.2 调试器

定义在 NuGet 包"Microsoft.Extensions.Logging.Debug"中的 DebugLogger 会直接调用 Debug 的 WriteLine 方法来写入分发给它的日志消息,写入的消息可以被附加到当前进程上的调试器(Debugger)捕获。对于 Linux 操作系统,日志消息会以文件的形式被写入,默认的目录一般为"/var/logs/messages"。由于定义在 Debug 类型中的所有方法都是针对 Debug 编译模式的,所以只有在针对 Debug 模式编译的应用中使用 DebugLogger 才有意义。

DebugLogger

我们采用如下一段相对简洁的代码模拟了 DebugLogger 的实现,DebugLogger 对象根据指定的名称(日志类别)创建。由于 DebugLogger 是面向调试器的,所以它实现的 IsEnabled 方法会验证调试器是否附加到了当前进程上。DebugLogger 不支持日志范围,所以它的 BeginScope<TState>方法会返回一个 NullScope 对象。

```csharp
internal class DebugLogger : ILogger
{
    private readonly string _name;
    public DebugLogger(string name) => _name = name;

    public IDisposable BeginScope<TState>(TState state)
    => NullScope.Instance;

    public bool IsEnabled(LogLevel logLevel)
    => (Debugger.IsAttached && (logLevel != LogLevel.None));

    public void Log<TState>(LogLevel logLevel, EventId eventId, TState state,
        Exception exception, Func<TState, Exception, string> formatter)
    {
        if (IsEnabled(logLevel))
        {
            string str = formatter(state, exception);
            if (!string.IsNullOrEmpty(str))
```

```
            {
                str = $"{logLevel}: {str}";
                if (exception != null)
                {
                    if (exception != null)
                    {
                        str = str + Environment.NewLine + Environment.NewLine +
                            exception.ToString();
                    }
                }
                Debug.WriteLine(str, _name);
            }
        }
    }
}
internal class NullScope : IDisposable
{
    public static NullScope Instance { get; } = new NullScope();
    private NullScope() { }
    public void Dispose() { }
}
```

在实现的 Log<TState>方法中，DebugLogger 会调用 IsEnabled 方法确定指定等级的日志是否需要被输出。在这个方法返回 True 的情况下，它会利用提供的格式化器生成字符串形式的日志消息。如果调用 Log<TState>方法时指定了异常对象，那么该方法还会在格式化的字符串附加上异常消息。笔者认为这是一个漏洞（Bug），因为既然作为格式化器的 Func<TState, Exception, string>委托已经考虑了异常的存在，就应该完全使用它格式化的消息。

DebugLoggerProvider

DebugLogger 对应的 ILoggerProvider 类型为 DebugLoggerProvider。如下面的代码片段所示，DebugLoggerProvider 提供 DebugLogger 对象的逻辑非常简单，它只需要在实现的 CreateLogger 方法中调用构造函数创建并返回一个 DebugLogger 对象即可。与 ConsoleLoggerProvider 不同的是，DebugLoggerProvider 的 CreateLogger 方法总是创建一个新的 DebugLogger 对象，并不会针对日志类别对其缓存，笔者认为这是一个可以改进的地方。DebugLoggerProvider 对象的注册由如下所示的 AddDebug 扩展方法来完成。

```
[ProviderAlias("Debug")]
public class DebugLoggerProvider : ILoggerProvider
{
    public ILogger CreateLogger(string name) => new DebugLogger(name);
    public void Dispose(){}
}

public static class DebugLoggerFactoryExtensions
{
```

```csharp
public static ILoggingBuilder AddDebug(this ILoggingBuilder builder)
{
    builder.Services.TryAddEnumerable(
        ServiceDescriptor.Singleton<ILoggerProvider, DebugLoggerProvider>());
    return builder;
}
```

9.4.3 TraceSource 日志

第 8 章对基于 TraceSource 的跟踪日志框架进行了详细介绍，为了整合该日志框架，可以注册一个 TraceSourceLoggerProvider，并由它提供的 TraceSourceLogger 将日志事件"转发"给 TraceSource 日志框架。

TraceSourceLogger

.NET Core 的日志系统利用一个定义在 NuGet 包 "Microsoft.Extensions.Logging.TraceSource" 中的 TraceSourceLogger 类型实现与 TraceSource 日志框架的整合。从下面的代码片段可以看出，一个 TraceSourceLogger 对象实际上就是对一个 TraceSource 对象的封装，在实现的 Log<TState> 方法中，它会调用 TraceSource 对象的 TraceEvent 方法来完成针对日志消息的写入工作。

```csharp
internal class TraceSourceLogger : ILogger
{
    private readonly TraceSource _traceSource;
    public TraceSourceLogger(TraceSource traceSource)
        => _traceSource = traceSource;
    public IDisposable BeginScope<TState>(TState state)
        => new TraceSourceScope(state);

    private static TraceEventType GetEventType(LogLevel logLevel)
    {
        return logLevel switch
        {
            LogLevel.Information => TraceEventType.Information,
            LogLevel.Warning => TraceEventType.Warning,
            LogLevel.Error => TraceEventType.Error,
            LogLevel.Critical => TraceEventType.Critical,
            _ => TraceEventType.Verbose,
        };
    }

    public bool IsEnabled(LogLevel logLevel)
    {
        if (logLevel == LogLevel.None)
        {
            return false;
        }
        TraceEventType eventType = GetEventType(logLevel);
        return _traceSource.Switch.ShouldTrace(eventType);
```

```csharp
}
public void Log<TState>(LogLevel logLevel, EventId eventId, TState state,
    Exception exception, Func<TState, Exception, string> formatter)
{
    if (!IsEnabled(logLevel))
    {
        return;
    }
    var message = string.Empty;
    if (formatter != null)
    {
        message = formatter(state, exception);
    }
    else
    {
        if (state != null)
        {
            message += state;
        }
        if (exception != null)
        {
            message += Environment.NewLine + exception;
        }
    }
    if (!string.IsNullOrEmpty(message))
    {
        _traceSource.TraceEvent(GetEventType(logLevel), eventId.Id, message);
    }
}
```

调用 TraceSource 对象的 TraceEvent 方法输出跟踪日志时，需要指定跟踪日志的事件类型，该类型由当前的日志等级来决定，表 9-4 展示了日志等级与跟踪事件类型之间的映射关系。由于 TraceSource 对象利用 SourceSwitch 来实施日志过滤，所以当 TraceSourceLogger 对象的 IsEnabled 方法被调用时，它也将指定的日志等级转换成追踪事件类型，并将其作为参数调用 SourceSwitch 对象的 ShouldTrace 方法，这个方法的返回值就是 IsEnabled 方法的返回值。

表 9-4　日志等级与跟踪事件类型之间的映射关系

日 志 等 级	跟踪事件类型
Trace	Verbose
Debug	Verbose
Information	Information
Warning	Warning
Error	Error
Critical	Critical

TraceSourceLogger 的 BeginScope<TState>方法会返回一个 TraceSourceScope 对象。如下面

的代码片段所示，在创建 TraceSourceScope 对象和调用其 Dispose 方法时，当前 Correlation Manager 对象的 StartLogicalOperation 和 StopLogicalOperation 会被调用，也就是说，它实际上是利用 Trace 系统的逻辑操作来模拟同样具有层级关系的日志范围的。

```
internal class TraceSourceScope : IDisposable
{
    private bool _isDisposed;

    public TraceSourceScope(object state)
        =>Trace.CorrelationManager.StartLogicalOperation(state);

    public void Dispose()
    {
        if (!_isDisposed)
        {
            Trace.CorrelationManager.StopLogicalOperation();
            _isDisposed = true;
        }
    }
}
```

TraceSourceLoggerProvider

TraceSourceLogger 对应的 ILoggerProvider 类型为 TraceSourceLoggerProvider。如下面的代码片段所示，创建一个 TraceSourceLoggerProvider 对象时需要提供一个 SourceSwitch 对象和 TraceListener 对象（可选）。在实现的 CreateLogger 方法中，它会根据指定的名称和 SourceSwitch 创建一个 TraceSource 对象，预先指定的 TraceListener 对象也会注册到这个 TraceSource 对象上。CreateLogger 方法最终返回的将是根据这个 TraceSource 对象创建的 TraceSourceLogger。

```
[ProviderAlias("TraceSource")]
public class TraceSourceLoggerProvider : ILoggerProvider
{
    public TraceSourceLoggerProvider(SourceSwitch rootSourceSwitch);
    public TraceSourceLoggerProvider(SourceSwitch rootSourceSwitch,
        TraceListener rootTraceListener);

    public ILogger CreateLogger(string name);
    public void Dispose();
}
```

值得注意的是，创建的 TraceSourceLogger 对象并不会根据指定的日志类型进行缓存，但是创建的 TraceSource 对象会根据指定的名称进行缓存。也就是说，针对相同的日志类别，CreateLogger 方法总是创建一个新的 TraceSourceLogger 对象，但是对应的 TraceSource 对象却是同一个。

针对 TraceSourceLoggerProvider 对象的注册可以通过调用如下这些 AddTraceSource 扩展方法重载来完成。这些 AddTraceSource 方法分为两种，分别是 ILoggerFactory 和 ILoggingBuilder

的扩展方法。ILoggerFactory 方法将创建的 TraceSourceLoggerProvider 直接注册到指定的 LoggerFactory 对象上，ILoggingBuilder 方法则采用依赖注入的方式注册 TraceSourceLoggerProvider。在调用这些方法时不仅可以指定 SourceSwitch（或者它的名称），还可以显式注册一个指定的 TraceListener 对象。

```csharp
public static class TraceSourceFactoryExtensions
{
    public static ILoggingBuilder AddTraceSource(
        this ILoggingBuilder builder,
        string switchName)
        => builder.AddTraceSource(new SourceSwitch(switchName));

    public static ILoggingBuilder AddTraceSource(
        this ILoggingBuilder builder,
        string switchName,
        TraceListener listener)
        => builder.AddTraceSource(new SourceSwitch(switchName), listener);

    public static ILoggingBuilder AddTraceSource(
        this ILoggingBuilder builder,
        SourceSwitch sourceSwitch)
    {
        builder.Services.AddSingleton<ILoggerProvider>(
            _ => new TraceSourceLoggerProvider(sourceSwitch));
        return builder;
    }

    public static ILoggingBuilder AddTraceSource(
        this ILoggingBuilder builder,
        SourceSwitch sourceSwitch,
        TraceListener listener)
    {
        builder.Services.AddSingleton<ILoggerProvider>(
            _ => new TraceSourceLoggerProvider(sourceSwitch, listener));
        return builder;
    }
}
```

9.4.4　EventSource 日志

与针对 TraceSource 日志框架的整合类似，我们也可以通过注册对应的 ILoggerProvider 并利用它提供的 ILogger 对象将日志事件转发给 EventSource 日志框架。这个用于整合 EventSource 日志框架的 ILoggerProvider 类型为 EventSourceLoggerProvider，它提供的 EventSourceLogger 会利用一个 EventSource 对象将接收的日志事件做进一步转发。

LoggingEventSource

被 EventSourceLogger 用来转发日志事件的 EventSource 类型为 LoggingEventSource。如下

面的代码片段所示，LoggingEventSource 只是一个定义在 NuGet 包 "Microsoft.Extensions.Logging.EventSource" 的内部类型。从标注到类型上的 EventSourceAttribute 特性可以看出，它采用的名称为 Microsoft-Extensions-Logging，本章开篇演示的实例正是利用此名称完成针对 LoggingEventSource 的订阅的。

```
[EventSource(Name="Microsoft-Extensions-Logging")]
internal class LoggingEventSource : EventSource
{
    public class Keywords
    {
        public const EventKeywords Message          = 2L;
        public const EventKeywords FormattedMessage = 4L;
        public const EventKeywords JsonMessage      = 8L;
    }

    [Event(1, Keywords=4L, Level=EventLevel.LogAlways)]
    internal void FormattedMessage(LogLevel Level, int FactoryID, string LoggerName,
        string EventId, string FormattedMessage);

    [Event(2, Keywords=2L, Level=EventLevel.LogAlways)]
    internal void Message(LogLevel Level, int FactoryID, string LoggerName,
        string EventId, ExceptionInfo Exception,
        IEnumerable<KeyValuePair<string, string>> Arguments);

    [Event(5, Keywords=8L, Level=EventLevel.LogAlways)]
    internal void MessageJson(LogLevel Level, int FactoryID, string LoggerName,
        string EventId, string ExceptionJson, string ArgumentsJson);
    ...
}

[EventData(Name ="ExceptionInfo")]
internal class ExceptionInfo
{
    public string       TypeName       { get; set; }
    public string       Message        { get; set; }
    public int          HResult        { get; set; }
    public string       VerboseMessage { get; set; }
}
```

当 EventSourceLogger 用于输出日志的 Log 方法被调用时，它会调用定义在 LoggingEventSource 中相应的日志事件方法。LoggingEventSource 支持 3 种不同的内容荷载，它们对应具有如下定义的 3 个 EventKeywords 常量，与这 3 个预定义关键字映射的是 3 个对应的日志事件方法。从上面的代码片段可以看出，这 3 个日志事件具有一些共同的描述信息，如日志等级、ILoggerFactory 的 ID、日志类别和事件 ID，但主要的内容荷载存在如下几方面差异。

- FormattedMessage：格式化之后生成的完整的消息文本作为日志事件的内容荷载。
- Message：指定的异常会转换成一个 ExceptionInfo 对象，指定的 State 对象将会转换成一个元素类型为 KeyValuePair<string, string>的集合对象。

- MessageJson：按照 Message 事件进行格式转换，并将最终生成的 ExceptionInfo 对象和 KeyValuePair<string, string>集合序列化成 JSON 格式。

对于采用 Message 和 JsonMessage 格式的日志事件，当 EventSourceLogger 在对指定的 State 对象进行格式转换时，它并没有通过反射的方式提取该对象的属性成员，而是尝试将 State 对象针对目标类型 IEnumerable<KeyValuePair<string, object>>进行类型转换而已。

EventSourceLogger

EventSourceLoggerProvider 创建的 ILogger 对象是一个 EventSourceLogger 对象，该对象利用上面介绍的 LoggingEventSource 对象将接收的日志事件进行二度转发。由同一个 EventSourceLoggerProvider 提供的所有的 EventSourceLogger 对象会构成一个链表，而 EventSourceLoggerProvider 对象总是保持着对链表表头的 EventSourceLogger 对象的引用。

以图 9-17 为例，由 EventSourceLoggerProvider 创建的第一个 EventSourceLogger（Foo）会直接被它引用。创建第二个 EventSourceLogger（Bar）时，Foo 将作为它的 Next，对应的引用则从 Foo 换成 Bar。后续 EventSourceLogger（如 Baz）的创建逻辑与之类似。与 LoggerFactory 会引用所有创建的 Logger 一样，这样做的好处就是 EventSourceLoggerProvider 可以通过该引用关系得到所有由它创建的 EventSourceLogger 对象，并在过滤规则发生改变时及时地将新规则应用到它们之上。

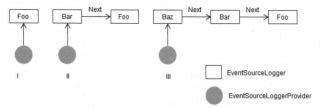

图 9-17　EventSourceLogger 链表

对 EventSourceLogger 的链表结构有了基本了解之后，下面介绍其定义。如下面的代码片段所示，EventSourceLogger 是一个并未被公开的内部类型，它的 3 个属性（CategoryName、Level 和 Next）分别表示日志类别、最低日志等级和当前创建的 EventSourceLogger 对象（它将作为链表中的下一个 EventSourceLogger 对象）。在调用构造函数创建一个 EventSourceLogger 对象时，除了需要指定日志类别和当前 EventSourceLogger，还需要指定用于发送日志事件的 LoggingEventSource 对象和当前 ILoggerFactory 的标识。

```
internal class EventSourceLogger : ILogger
{
    public string                       CategoryName { get; }
    public LogLevel                     Level { get; set; }
    public EventSourceLogger            Next { get; }

    public EventSourceLogger(string categoryName, int factoryID,
        LoggingEventSource eventSource, EventSourceLogger next);
```

```
    public bool IsEnabled(LogLevel logLevel);
    public void Log<TState>(LogLevel logLevel, EventId eventId, TState state,
        Exception exception, Func<TState, Exception, string> formatter);
    ...
}
```

  EventSourceLogger 对象针对日志事件的过滤体现在代表最低日志等级的 Level 属性上，只有不低于此等级的日志事件才会被它分发出去，这样的过滤逻辑实现在 IsEnabled 方法中。由于 EventSourceLogger 利用提供的 LoggingEventSource 对象来分发日志事件，所以它的最低日志等级取决于当前应用程序针对 LoggingEventSource 对象的订阅。由于针对 LoggingEventSource 对象的日志订阅是动态的，所以 Level 属性也应该随着当前订阅状态动态地改变，这就是 EventSourceLoggerProvider 对象一定需要通过链表的形式间接引用所有由它创建的 EventSourceLogger 对象的根源所在，因为只有这样它才能动态地改变它们的最低日志等级。

  当 EventSourceLogger 对象的 Log<TState>方法被执行时，它会先调用 IsEnabled 方法判断指定的日志等级是否低于设定的最低等级，在不低于最低等级的情况下它才会利用 LoggingEventSource 对象来分发日志事件。在具体的日志事件分发过程中，由于具有 3 种不同的内容荷载结构（对应 3 个预定义的关键字 Message、FormattedMessage 和 JsonMessage）可供选择，所以 EventSourceLogger 对象会根据当前的订阅状态选择 LoggingEventSource 对象相应的日志事件方法，具体的实现逻辑体现在如下所示的代码片段中。

```
internal class EventSourceLogger : ILogger
{
    public void Log<TState>(LogLevel logLevel, EventId eventId, TState state,
        Exception exception, Func<TState, Exception, string> formatter)
    {
        if (!IsEnabled(logLevel))
        {
            return;
        }

        if (_eventSource.IsEnabled(
            EventLevel.Critical, LoggingEventSource.Keywords.FormattedMessage))
        {
            string message = formatter(state, exception);
            _eventSource.FormattedMessage(logLevel, _factoryID,CategoryName,
                eventId.ToString(),message);
        }

        if (_eventSource.IsEnabled(
            EventLevel.Critical, LoggingEventSource.Keywords.Message))
        {
            ExceptionInfo exceptionInfo = GetExceptionInfo(exception);
            IEnumerable<KeyValuePair<string, string>> arguments =
                GetProperties(state);

            _eventSource.Message(logLevel, _factoryID, CategoryName,
```

```
                eventId.ToString(),exceptionInfo, arguments);
        }

        if (_eventSource.IsEnabled(
            EventLevel.Critical, LoggingEventSource.Keywords.JsonMessage))
        {
            string exceptionJson = "{}";
            if (exception != null)
            {
                var exceptionInfo = GetExceptionInfo(exception);
                var exceptionInfoData = new KeyValuePair<string, string>[]
                {
                    new KeyValuePair<string, string>(
                        "TypeName", exceptionInfo.TypeName),
                    new KeyValuePair<string, string>(
                        "Message", exceptionInfo.Message),
                    new KeyValuePair<string, string>(
                        "HResult", exceptionInfo.HResult.ToString()),
                    new KeyValuePair<string, string>(
                        "VerboseMessage", exceptionInfo.VerboseMessage),
                };
                exceptionJson = ToJson(exceptionInfoData);
            }
            IEnumerable<KeyValuePair<string, string>> arguments =
                GetProperties(state);
            _eventSource.MessageJson(logLevel, _factoryID, CategoryName,
                eventId.ToString(),exceptionJson, ToJson(arguments));
        }
    }
}
```

在具有针对相应关键字（Message、FormattedMessage 和 JsonMessage）的事件订阅的情况下，EventSourceLogger 对象会针对接收的同一个日志事件转发 3 次。本章开篇的实例主要演示了内容荷载的数据结构为 FormattedMessage 的场景，下面通过一个简单的实例来比较 3 次转发的日志荷载具体包含怎样的内容和结构。

为了接收 EventSourceLogger 对象分发的日志事件，可以定义如下这个名为 LoggingEventListener 的 EventListener。我们在重写的 OnEventSourceCreated 方法中针对 Microsoft-Extensions-Logging 对前面介绍的 LoggingEventSource 进行了订阅。在重写的 OnEventWritten 方法中，我们将接收的事件名称和负载成员打印在控制台上。为了反映完整的数据内容，我们针对 Object[]和 IDictionary<string, object>这两种类型的成员进行了序列化。

```
public class LoggingEventListener : EventListener
{
    protected override void OnEventSourceCreated(EventSource eventSource)
    {
        if (eventSource.Name == "Microsoft-Extensions-Logging")
        {
            EnableEvents(eventSource, EventLevel.LogAlways);
```

```csharp
        }
    }

    protected override void OnEventWritten(EventWrittenEventArgs eventData)
    {
        Console.WriteLine($"Event: {eventData.EventName}");
        for (int index = 0; index <eventData.Payload.Count; index++)
        {
            var element = eventData.Payload[index];
            if (element is object[] || element is IDictionary<string, object>)
            {
                Console.WriteLine($"{eventData.PayloadNames[index],-16}:
                    {JsonConvert.SerializeObject(element)}");
                continue;
            }
            Console.WriteLine($"{eventData.PayloadNames[index],-16}:
                {eventData.Payload[index]}");
        }
        Console.WriteLine();
    }
}
```

如下所示是日志分发的程序代码。在创建了上述这个 LoggingEventListener 对象之后，我们利用依赖注入框架添加了日志模型的核心服务，并将 EventSourceLoggerProvider 注册为唯一的 ILoggerProvider，最终利用创建的 ILogger<Program> 对象发送了一个 Error 等级的日志事件。在调用 Log 方法时，除了提供一个创建的 InvalidOperationException 对象，我们还提供一个 Dictionary<string, object> 对象作为内容荷载。作为格式化器的委托对象直接返回指定异常对象的 Message 属性。

```csharp
class Program
{
    static void Main()
    {
        var listener = new LoggingEventListener();
        var logger = new ServiceCollection()
            .AddLogging(builder => builder.AddEventSourceLogger())
            .BuildServiceProvider()
            .GetRequiredService<ILogger<Program>>();

        var state = new Dictionary<string, object>
        {
            ["ErrorCode"]   = 100,
            ["Message"]     = "Unhandled exception"
        };

        logger.Log(LogLevel.Error, 1, state, new InvalidOperationException(
            "This is a manually thrown exception."), (_, ex) => ex.Message);
        Console.Read();
    }
}
```

该程序运行之后在控制台上呈现的输出结果如图 9-18 所示。由于 EventSourceLogger Provider 提供的 EventSourceLogger 会 3 次转发日志事件,对应的事件名称分别是 FormattedMessage、Message 和 MessageJson。这 3 个日志事件承载具有 4 个相同的负载成员,它们分别是日志事件等级、ID、类别、ILoggerFactory 的 ID。3 个事件的不同之处在于,FormattedMessage 的负载成员会包含格式化的消息文本,而 Message 和 MessageJson 则会包含负载对象和异常信息,只是前者是以对象的形式(EventPayload)体现的,后者则体现为序列化后的 JSON 字符串。(S913)

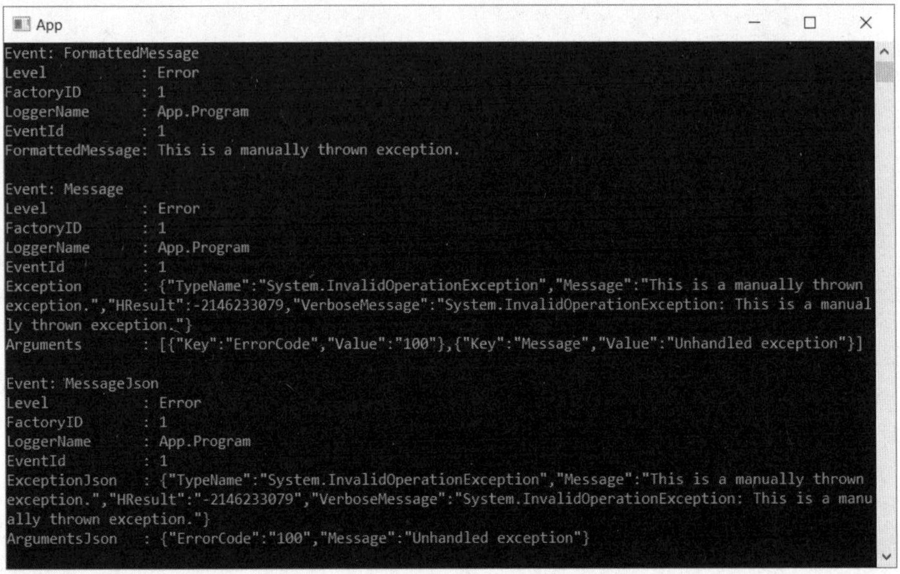

```
■ App                                                    —  □  ×
Event: FormattedMessage
Level             : Error
FactoryID         : 1
LoggerName        : App.Program
EventId           : 1
FormattedMessage: This is a manually thrown exception.

Event: Message
Level             : Error
FactoryID         : 1
LoggerName        : App.Program
EventId           : 1
Exception         : {"TypeName":"System.InvalidOperationException","Message":"This is a manually thrown
 exception.","HResult":-2146233079,"VerboseMessage":"System.InvalidOperationException: This is a manual
ly thrown exception."}
Arguments         : [{"Key":"ErrorCode","Value":"100"},{"Key":"Message","Value":"Unhandled exception"}]

Event: MessageJson
Level             : Error
FactoryID         : 1
LoggerName        : App.Program
EventId           : 1
ExceptionJson     : {"TypeName":"System.InvalidOperationException","Message":"This is a manually thrown
 exception.","HResult":"-2146233079","VerboseMessage":"System.InvalidOperationException: This is a manu
ally thrown exception."}
ArgumentsJson     : {"ErrorCode":"100","Message":"Unhandled exception"}
```

图 9-18　由 EventSourceLogger 发出的 3 种具有不同负载结构的日志事件

EventSourceLoggerProvider

　　EventSourceLogger 对应的 ILoggerProvider 实现类型为 EventSourceLoggerProvider。如下面的代码片段所示,EventSourceLoggerProvider 根据提供的 LoggingEventSource 对象创建而成。由它创建的 EventSourceLogger 对象体现在_loggers 字段上,由于 EventSourceLogger 对象采用基于链表的设置,该字段保持由当前 EventSourceLoggerProvider 对象提供的 EventSourceLogger 对象的引用,所以在实现的 Dispose 方法中,EventSourceLoggerProvider 对象可以释放所有由它创建的 EventSourceLogger 对象。

　　在实现的 CreateLogger 方法中,EventSourceLoggerProvider 对象会根据作为参数的日志类别、初始化提供的 LoggingEventSource 对象,以及当前的 EventSourceLogger 创建一个新的 EventSourceLogger 对象。EventSourceLogger 类型的构造函数具有一个参数 factoryID,从给出的代码片段可以看出,它实际上代表 EventSourceLoggerProvider 对象的全局 ID。之所以将这个参数命名为 factoryID,是因为在之前的版本中,这个参数代表的其实是 LoggerFactory 的全局 ID。

从前面给出的 LoggingEventSource 类型实现代码可以看出，这个 ID 最终会作为事件内容荷载的一部分，对应的成员名称为 FactoryID。

```csharp
[ProviderAlias("EventSource")]
public class EventSourceLoggerProvider : ILoggerProvider
{
    private static int                  _globalFactoryID;
    private readonly int                _factoryID;
    private EventSourceLogger           _loggers;
    private readonly LoggingEventSource _eventSource;

    public EventSourceLoggerProvider(LoggingEventSource eventSource)
    {
        _eventSource = eventSource;
        _factoryID = Interlocked.Increment(ref _globalFactoryID);
    }

    public ILogger CreateLogger(string categoryName)
        =>_loggers = new EventSourceLogger(categoryName, _factoryID, _eventSource,
        _loggers);

    public void Dispose()
    {
        for (EventSourceLogger logger = _loggers; logger != null;
            logger = logger.Next)
        {
            logger.Level = LogLevel.None;
        }
    }
}
```

如下所示的代码片段是用于注册 EventSourceLoggerProvider 对象的 AddEventSourceLogger 扩展方法的定义。除了注册 EventSourceLoggerProvider 和 LoggingEventSource 这两个服务，该方法还涉及与日志过滤规则相关的其他两个服务注册，我们会在后续部分对它们进行单独介绍。另外，AddEventSourceLogger 扩展方法其实应该命名为 AddEventSource，从而与其他同类方法（AddConsole、AddDebug 和 AddTraceSource）保持一致。

```csharp
public static class EventSourceLoggerFactoryExtensions
{
    public static ILoggingBuilder AddEventSourceLogger(this ILoggingBuilder builder)
    {
        builder.Services.TryAddSingleton(LoggingEventSource.Instance);
        builder.Services.TryAddEnumerable(ServiceDescriptor
            .Singleton<ILoggerProvider, EventSourceLoggerProvider>());
        builder.Services.TryAddEnumerable(ServiceDescriptor
            .Singleton<IConfigureOptions<LoggerFilterOptions>,
            EventLogFiltersConfigureOptions>());
        builder.Services.TryAddEnumerable(ServiceDescriptor
            .Singleton<IOptionsChangeTokenSource<LoggerFilterOptions>,
            EventLogFiltersConfigureOptionsChangeSource>());
```

```
        return builder;
    }
}
```

日志范围

日志范围可以体现日志事件当前的调用链，而针对 EventSource 的日志框架则提供了针对活动的跟踪功能，这在本质上也是对调用链的体现，所以两者可以很自然地合在一起。被 EventSourceLogger 用来分发日志事件的 LoggingEventSource 对象提供了如下两组针对活动跟踪的日志事件方法：如果具有针对 MessageJson 事件的订阅，EventSourceLogger 对象会调用 ActivityJsonStart 方法和 ActivityJsonStop 方法来开始与结束一个活动；如果针对 MessageJson 事件的订阅不存在，ActivityStart 方法和 ActivityStop 方法才会被调用。

```
[EventSource(Name="Microsoft-Extensions-Logging")]
internal class LoggingEventSource : EventSource
{
    [Event(3, Keywords=6L, Level=EventLevel.LogAlways,
        ActivityOptions=EventActivityOptions.Recursive)]
    internal void ActivityStart(int ID, int FactoryID, string LoggerName,
        IEnumerable<KeyValuePair<string, string>> Arguments);
    [Event(4, Keywords=6L, Level=EventLevel.LogAlways)]
    internal void ActivityStop(int ID, int FactoryID, string LoggerName);

    [Event(6, Keywords=12L, Level=EventLevel.LogAlways,
        ActivityOptions=EventActivityOptions.Recursive)]
    internal void ActivityJsonStart(int ID, int FactoryID, string LoggerName,
        string ArgumentsJson);
    [Event(7, Keywords=12L, Level=EventLevel.LogAlways)]
    internal void ActivityJsonStop(int ID, int FactoryID, string LoggerName);
    ...
}
```

如下面的代码片段所示，EventSourceLogger 对象的 BeginScope<TState>方法被调用时，它会根据是否具有针对关键字 MessageJson（8L）的日志订阅来决定调用 ActivityJsonStart 方法还是 ActivityStart 方法来启动活动。BeginScope<TState>方法最终会创建一个 ActivityScope 对象作为日志范围，当 ActivityScope 对象的 Dispose 方法被调用时，LoggingEventSource 对象的 ActivityJsonStop 方法或者 ActivityStop 方法会被调用。

```
internal class EventSourceLogger : ILogger
{
    private static int                          _activityIds;
    private readonly LoggingEventSource         _eventSource;
    private readonly int                        _factoryID;

    public IDisposable BeginScope<TState>(TState state)
    {
        if (!IsEnabled(LogLevel.Critical))
        {
            return NoopDisposable.Instance;
```

```csharp
    }
    int id = Interlocked.Increment(ref _activityIds);
    if (_eventSource.IsEnabled((EventLevel) EventLevel.Critical, 8L))
    {
        _eventSource.ActivityJsonStart(iD, _factoryID, CategoryName,
            ToJson(GetProperties(state)));
        return new ActivityScope(_eventSource, CategoryName, id, _factoryID,
            true);
    }
    _eventSource.ActivityStart(id, _factoryID, CategoryName,
        GetProperties(state));
    return new ActivityScope(teventSource, CategoryName, id, _factoryID, false);
}

private class ActivityScope : IDisposable
{
    private readonly int                    _activityID;
    private readonly string                 _categoryName;
    private readonly LoggingEventSource     _eventSource;
    private readonly int                    _factoryID;
    private readonly bool                   _isJsonStop;

    public ActivityScope(LoggingEventSource eventSource, string categoryName,
        int activityID, int factoryID, bool isJsonStop)
    {
        _categoryName   = categoryName;
        _activityID     = activityID;
        _factoryID      = factoryID;
        _isJsonStop     = isJsonStop;
        _eventSource    = eventSource;
    }

    public void Dispose()
    {
        if (_isJsonStop)
        {
            _eventSource.ActivityJsonStop(_activityID, _factoryID,
                _categoryName);
        }
        else
        {
            _eventSource.ActivityStop(_activityID, _factoryID, _categoryName);
        }
    }
}

private class NoopDisposable : IDisposable
{
    public static readonly EventSourceLogger.NoopDisposable Instance
        = new EventSourceLogger.NoopDisposable();
```

```
        public void Dispose(){}
    }
}
```

在调用 LoggingEventSource 对象的 XxxStart/XxxStop 方法时，EventSourceLogger 对象会采用一个自增长的整数作为活动的标识，但是只有这个标识并不足以描述调用链的流转，由于 EventSource 日志框架自身就能够唯一标识一个活动，所以这个负载成员并不具有多大的意义。和 MessageJson 方法一样，ActivityJsonStart 方法会将 IEnumerable<KeyValuePair<<string, object>>类型的负载对象序列化成 JSON 字符串并作为负载成员，成员名称为 ArgumentsJson，对于 ActivityStart 方法来说，对应的负载成员被命名为 Arguments，其值为转换成 IEnumerable <KeyValuePair<<string,string>>类型的对象。

下面的实例利用日志范围进行调用链跟踪。如下面的代码片段所示，用来捕捉日志事件的 LoggingEventListener 类型通过重写的 OnEventSourceCreated 订阅了针对 LoggingEventSource 的日志事件。由于需要借助 TplEtwProvider 提供针对当前活动的保存，所以 OnEventSource Created 方法还针对这个 EventSource 对象做了订阅。在另一个重写的 OnEventWritten 方法中，我们将日志事件的名称、承载内容、ActivityId 和 RelatedActivityId 写入一个指定的.csv 文件中。

```
class Program
{
    static void Main()
    {
        File.AppendAllLines("log.csv", new string[] {
            $"EventName, Payload, ActivityId, RelatedActivityId" });
        var listener = new LoggingEventListener();
        var logger = new ServiceCollection()
            .AddLogging(builder => builder.AddEventSourceLogger())
            .BuildServiceProvider()
            .GetRequiredService<ILogger<Program>>();

        using (logger.BeginScope(new Dictionary<string, object>
            { ["Operation"] = "Foo" }))
        {
            logger.LogInformation("This is a test log written in scope 'Foo'");
            using (logger.BeginScope(new Dictionary<string, object>
                { ["Operation"] = "Bar" }))
            {
                logger.LogInformation("This is a test log written in scope 'Bar'");
            }
            using (logger.BeginScope(new Dictionary<string, object>
                { ["Operation"] = "Baz" }))
            {
                logger.LogInformation("This is a test log written in scope 'Baz'");
            }
        }
    }
}
```

```csharp
public class LoggingEventListener : EventListener
{
    protected override void OnEventSourceCreated(EventSource eventSource)
    {
        if (eventSource.Name == "System.Threading.Tasks.TplEventSource")
        {
            EnableEvents(eventSource, EventLevel.Informational,
                (EventKeywords)0x80);
        }

        if (eventSource.Name == "Microsoft-Extensions-Logging")
        {
            EnableEvents(eventSource, EventLevel.LogAlways, (EventKeywords)8);
        }
    }

    protected override void OnEventWritten(EventWrittenEventArgs eventData)
    {
        int index;
        var payload =(index = eventData.PayloadNames.IndexOf("ArgumentsJson")) == -1
            ? null
            : eventData.Payload[index];
        var relatedActivityId = eventData.RelatedActivityId == default(Guid)
            ? ""
            : eventData.RelatedActivityId.ToString();

        File.AppendAllLines("log.csv", new string[] { $"{eventData.EventName},
            {payload}, {eventData.ActivityId}, {relatedActivityId}" });
    }
}
```

在作为程序入口的 Main 方法中，我们利用依赖注入框架创建了一个 ILogger<Program>对象，并进一步利用它创建了 3 个嵌套的日志范围，作为负载对象的是一个 Dictionary<string, object>对象。在这 3 个日志范围中，我们分发了 3 个 Information 等级的日志事件。程序运行之后，捕捉到的 9 个日志事件的相关信息会被写入指定的.csv 文件中，图 9-19 展示了该文件的内容。（S914）

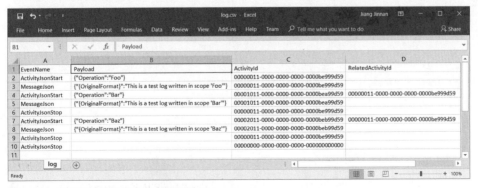

图 9-19　利用日志范围记录调用链信息

过滤规则

EventSourceLoggerProvider/EventSourceLogger 具有特殊的日志过滤规则设置方式。我们可以通过执行 EventSource 命令（通过调用静态方法 SendCommand）来设置过滤规则，设置的过滤规则以命令参数（字符串）的形式来指定。LoggingEventSource 类型通过重写的 OnEventCommand 方法获取设置的过滤规则，并调用 SetFilterSpec 方法根据默认的日志等级（取决于当前的事件订阅）将其解析为 LoggerFilterRule 对象集合。

```csharp
[EventSource(Name = "Microsoft-Extensions-Logging")]
public sealed class LoggingEventSource : EventSource
{
    private LoggerFilterRule[] _filterSpec = new LoggerFilterRule[0];
    protected override void OnEventCommand(EventCommandEventArgs command)
    {
        if (command.Command == EventCommand.Update || command.Command ==
            EventCommand.Enable)
        {
            if (!command.Arguments.TryGetValue("FilterSpecs", out var filterSpec))
            {
                filterSpec = string.Empty;
            }

            SetFilterSpec(filterSpec);
        }
        else if (command.Command == EventCommand.Disable)
        {
            SetFilterSpec(null);
        }
    }

    [NonEvent]
    private void SetFilterSpec(string filterSpec)
    {
        _filterSpec = ParseFilterSpec(filterSpec, GetDefaultLevel());
        FireChangeToken();
    }

    [NonEvent]
    internal IChangeToken GetFilterChangeToken()
    {
        var cts = LazyInitializer.EnsureInitialized(ref _cancellationTokenSource,
            () => new CancellationTokenSource());
        return new CancellationChangeToken(cts.Token);
    }

    [NonEvent]
    private void FireChangeToken()
    {
        var tcs = Interlocked.Exchange(ref _cancellationTokenSource, null);
        tcs?.Cancel();
```

```csharp
}

[NonEvent]
private static LoggerFilterRule[] ParseFilterSpec(string filterSpec,
    LogLevel defaultLevel);

[NonEvent]
private LogLevel GetDefaultLevel()
{
    var allMessageKeywords = Keywords.Message | Keywords.FormattedMessage |
      Keywords.JsonMessage;
    if (IsEnabled(EventLevel.Verbose, allMessageKeywords))
    {
        return LogLevel.Debug;
    }
    if (IsEnabled(EventLevel.Informational, allMessageKeywords))
    {
        return LogLevel.Information;
    }
    if (IsEnabled(EventLevel.Warning, allMessageKeywords))
    {
        return LogLevel.Warning;
    }
    if (IsEnabled(EventLevel.Error, allMessageKeywords))
    {
        return LogLevel.Error;
    }
    return LogLevel.Critical;
}

[NonEvent]
internal LoggerFilterRule[] GetFilterRules()=>_filterSpec;
}
```

为了使日志系统及时感知日志过滤规则的改变，并及时应用新的规则，LoggingEventSource 定义了上面所示的 GetFilterChangeToken 方法，并返回一个 CancellationChangeToken 对象。在用于设置规律规则的 SetFilterSpec 方法中，这个 CancellationChangeToken 对应 CancellationTokenSource 对象的 Cancel 方法会被调用，以此来通知规则的变化和应用新规则（具体实现在 FireChangeToken 方法中）。

由于日志规则是用注入 LoggerFactory 的 IOptionsMonitor<LoggerFilterOptions>对象提供的，所以可以通过注册对应 IOptionsChangeTokenSource<LoggerFilterOptions>服务的方式来应用新的配置选项。在 ILoggingBuilder 接口的 AddEventSourceLogger 扩展方法中注册的就是下面的 EventLogFiltersConfigureOptionsChangeSource 服务，它返回的正是 LoggingEventSource 的 GetFilterChangeToken 方法返回的 CancellationChangeToken 对象。

```csharp
internal class EventLogFiltersConfigureOptionsChangeSource:
    IOptionsChangeTokenSource<LoggerFilterOptions>
{
```

```csharp
    private readonly LoggingEventSource _eventSource;
    public EventLogFiltersConfigureOptionsChangeSource(
        LoggingEventSource eventSource)
        => _eventSource = eventSource;

    public IChangeToken GetChangeToken() => _eventSource.GetFilterChangeToken();
    public string Name { get; }
}
```

第 10 章

承载系统

借助 .NET Core 提供的承载（Hosting）系统，我们可以将任意一个或者多个长时间运行（Long-Running）的服务寄宿或者承载于托管进程中。ASP.NET Core 应用仅仅是该承载系统的一种典型的服务类型而已，任何需要在后台长时间运行的操作都可以定义成标准化的服务并利用该系统来承载。本章主要介绍"泛化"（Generic）的服务承载系统，不会涉及任何关于 ASP.NET Core 的内容。

10.1 服务承载

一个 ASP.NET Core 应用本质上是一个需要长时间运行的服务，开启这个服务是为了启动一个网络监听器。当监听到抵达的 HTTP 请求之后，该监听器会将请求传递给应用提供的管道进行处理。管道完成了对请求的处理之后会生成 HTTP 响应，并通过监听器返回客户端。除了这种最典型的承载服务，我们还有很多其他的服务承载需求，下面通过一个简单的实例来演示如何承载一个服务来收集当前执行环境的性能指标。

10.1.1 承载长时间运行服务

我们演示的承载服务会定时采集并分发当前进程的性能指标。简单起见，我们只关注处理器使用率、内存使用量和网络吞吐量这 3 种典型的性能指标，为此定义了下面的 PerformanceMetrics 类型。我们并不会实现真正的性能指标收集，所以定义静态方法 Create 利用随机生成的指标数据创建一个 PerformanceMetrics 对象。

```
public class PerformanceMetrics
{
    private static readonly Random _random = new Random();

    public int       Processor { get; set; }
    public long      Memory { get; set; }
    public long      Network { get; set; }

    public override string ToString() => $"CPU: {Processor * 100}%; Memory: {Memory / (1024
```

```
            * 1024)}M; Network: {Network / (1024 * 1024)}M/s";

    public static PerformanceMetrics Create() => new PerformanceMetrics
    {
        Processor      = _random.Next(1,8),
        Memory         = _random.Next(10, 100) * 1024 * 1024,
        Network        = _random.Next(10, 100) * 1024 * 1024
    };
}
```

承载服务通过 IHostedService 接口表示，该接口定义的 StartAsync 方法和 StopAsync 方法可以启动与关闭服务。我们将性能指标采集服务定义成如下这个实现了该接口的 PerformanceMetricsCollector 类型。在实现的 StartAsync 方法中，我们利用 Timer 创建了一个调度器，每隔 5 秒它会调用 Create 方法创建一个 PerformanceMetrics 对象，并将它承载的性能指标输出到控制台上。这个 Timer 对象会在实现的 StopAsync 方法中被释放。

```
public sealed class PerformanceMetricsCollector : IHostedService
{
    private IDisposable _scheduler;
    public Task StartAsync(CancellationToken cancellationToken)
    {
        _scheduler = new Timer(Callback, null, TimeSpan.FromSeconds(5),
            TimeSpan.FromSeconds(5));
        return Task.CompletedTask;

        static void Callback(object state)
        {
            Console.WriteLine($"[{DateTimeOffset.Now}]{PerformanceMetrics.Create()}");
        }
    }

    public Task StopAsync(CancellationToken cancellationToken)
    {
        _scheduler?.Dispose();
        return Task.CompletedTask;
    }
}
```

承载系统通过 IHost 接口表示承载服务的宿主，该对象在应用启动过程中采用 Builder 模式由对应的 IHostBuilder 对象来创建。HostBuilder 类型是对 IHostBuilder 接口的默认实现，所以可以采用如下方式创建一个 HostBuilder 对象，并调用其 Build 方法来提供作为宿主的 IHost 对象。

```
class Program
{
    static void Main()
    {
        new HostBuilder()
            .ConfigureServices(svcs => svcs
                .AddSingleton<IHostedService, PerformanceMetricsCollector>())
            .Build()
```

```
            .Run();
    }
}
```

在调用 Build 方法之前，可以调用 IHostBuilder 接口的 ConfigureServices 方法将 PerformancceMetricsCollector 注册成针对 IHostedService 接口的服务，并将生命周期模式设置成 Singleton。除了采用普通的依赖服务注册方式，针对 IHostedService 服务的注册还可以调用 IServiceCollection 接口的 AddHostedService<THostedService>扩展方法来完成，如下所示的编程方式与上面是完全等效的。

```
class Program
{
    static void Main()
    {
        new HostBuilder()
            .ConfigureServices(svcs => svcs
                .AddHostedService<PerformanceMetricsCollector>())
            .Build()
            .Run();
    }
}
```

最后调用 Run 方法启动通过 IHost 对象表示的承载服务宿主，进而启动由它承载的 PerformancceMetricsCollector 服务，该服务将以图 10-1 所示的形式每隔 5 秒显示由它"采集"的性能指标。（S1001）

图 10-1　承载指标采集服务

10.1.2　依赖注入

服务承载系统无缝整合了第 3 章和第 4 章介绍的依赖注入框架。从上面给出的代码可以看出，针对承载服务的注册实际上就是将它注册到依赖注入框架中。既然承载服务实例最终是通过依赖注入框架提供的，那么它自身所依赖的服务当然也可以注册到依赖注入框架中。

下面将 PerformanceMetricsCollector 承载的性能指标收集功能分解到由 4 个接口表示的服务中，其中 IProcessorMetricsCollector 接口、IMemoryMetricsCollector 接口和 INetworkMetricsCollector 接口代表的服务分别用于收集 3 种对应的性能指标，而 IMetricsDeliverer 接口表示的服务则负责将收集的性能指标发送出去。

```
public interface IProcessorMetricsCollector
{
```

```csharp
    int GetUsage();
}
public interface IMemoryMetricsCollector
{
    long GetUsage();
}
public interface INetworkMetricsCollector
{
    long GetThroughput();
}

public interface IMetricsDeliverer
{
    Task DeliverAsync(PerformanceMetrics counter);
}
```

我们定义的 FakeMetricsCollector 类型实现了 3 个性能指标采集接口，它们采集的性能指标直接来源于通过静态方法 Create 创建的 PerformanceMetrics 对象。FakeMetricsDeliverer 类型实现了 IMetricsDeliverer 接口，在实现的 DeliverAsync 方法中，它直接将 PerformanceMetrics 对象承载的性能指标输出到控制台上。

```csharp
public class FakeMetricsCollector :
    IProcessorMetricsCollector,
    IMemoryMetricsCollector,
    INetworkMetricsCollector
{
    long INetworkMetricsCollector.GetThroughput()
    => PerformanceMetrics.Create().NETwork;

    int IProcessorMetricsCollector.GetUsage()
    => PerformanceMetrics.Create().Processor;

    long IMemoryMetricsCollector.GetUsage()
    => PerformanceMetrics.Create().Memory;
}

public class FakeMetricsDeliverer : IMetricsDeliverer
{
    public Task DeliverAsync(PerformanceMetrics counter)
    {
        Console.WriteLine($"[{DateTimeOffset.UtcNow}]{counter}");
        return Task.CompletedTask;
    }
}
```

由于整个性能指标的采集工作被分解到 4 个接口表示的服务之中，所以可以采用如下所示的方式重新定义承载服务类型 PerformanceMetricsCollector。如下面的代码片段所示，可以直接在构造函数中注入 4 个依赖服务。对于在 StartAsync 方法创建的调用器来说，它会利用 3 个对应的服务采集 3 种类型的性能指标，并利用 IMetricsDeliverer 服务将其发送出去。

```csharp
public sealed class PerformanceMetricsCollector : IHostedService
{
    private readonly IProcessorMetricsCollector   _processorMetricsCollector;
    private readonly IMemoryMetricsCollector      _memoryMetricsCollector;
    private readonly INetworkMetricsCollector     _networkMetricsCollector;
    private readonly IMetricsDeliverer            _MetricsDeliverer;
    private IDisposable                           _scheduler;

    public PerformanceMetricsCollector(
        IProcessorMetricsCollector processorMetricsCollector,
        IMemoryMetricsCollector memoryMetricsCollector,
        INetworkMetricsCollector networkMetricsCollector,
        IMetricsDeliverer MetricsDeliverer)
    {
        _processorMetricsCollector  = processorMetricsCollector;
        _memoryMetricsCollector     = memoryMetricsCollector;
        _networkMetricsCollector    = networkMetricsCollector;
        _MetricsDeliverer           = MetricsDeliverer;
    }

    public Task StartAsync(CancellationToken cancellationToken)
    {
        _scheduler = new Timer(Callback, null, TimeSpan.FromSeconds(5),
            TimeSpan.FromSeconds(5));
        return Task.CompletedTask;

        async void Callback(object state)
        {
            var counter = new PerformanceMetrics
            {
                Processor   = _processorMetricsCollector.GetUsage(),
                Memory      = _memoryMetricsCollector.GetUsage(),
                Network     = _networkMetricsCollector.GetThroughput()
            };
            await _MetricsDeliverer.DeliverAsync(counter);
        }
    }

    public Task StopAsync(CancellationToken cancellationToken)
    {
        _scheduler?.Dispose();
        return Task.CompletedTask;
    }
}
```

在调用 IHostBuilder 接口的 Build 方法创建作为宿主的 IHost 对象之前，包括承载服务在内的所有服务都可以通过它的 ConfigureServices 方法进行注册，我们采用如下方式注册了作为承载服务的 PerformanceMetricsCollector 和它依赖的 4 个服务。修改后的程序启动之后同样会在控制台上看到图 10-1 所示的输出结果。（S1002）

```
class Program
{
    static void Main()
    {
        var collector = new FakeMetricsCollector();
        new HostBuilder()
            .ConfigureServices(svcs => svcs
                .AddSingleton<IProcessorMetricsCollector>(collector)
                .AddSingleton<IMemoryMetricsCollector>(collector)
                .AddSingleton<INetworkMetricsCollector>(collector)
                .AddSingleton<IMetricsDeliverer, FakeMetricsDeliverer>()
                .AddSingleton<IHostedService, PerformanceMetricsCollector>())
            .Build()
            .Run();
    }
}
```

10.1.3 配置选项

真正的应用开发总是会使用到配置选项，如演示程序中性能指标采集的时间间隔就应该采用配置选项的方式来指定。由于涉及对性能指标数据的发送，所以最好将发送的目标地址定义在配置选项中。如果有多种传输协议可供选择，就可以定义相应的配置选项。.NET Core 应用推荐采用 Options 模式来使用配置选项，所以可以定义如下这个 MetricsCollectionOptions 类型来承载 3 种配置选项。

```
public class MetricsCollectionOptions
{
    public TimeSpan            CaptureInterval { get; set; }
    public TransportType       Transport { get; set; }
    public Endpoint            DeliverTo { get; set; }
}

public enum TransportType
{
    Tcp,
    Http,
    Udp
}

public class Endpoint
{
    public string          Host { get; set; }
    public int             Port { get; set; }
    public override string ToString() => $"{Host}:{Port}";
}
```

传输协议和目标地址使用在 FakeMetricsDeliverer 服务中，所以我们对它进行了相应的改写。如下面的代码片段所示，我们在构造函数中通过注入的 IOptions<MetricsCollectionOptions>服务来提供上面的两个配置选项。在实现的 DeliverAsync 方法中，可以将采用的传输协议和目标地址输出到控制台上。

```csharp
public class FakeMetricsDeliverer : IMetricsDeliverer
{
    private readonly TransportType  _transport;
    private readonly Endpoint       _deliverTo;

    public FakeMetricsDeliverer(IOptions<MetricsCollectionOptions> optionsAccessor)
    {
        var options  = optionsAccessor.Value;
        _transport   = options.Transport;
        _deliverTo   = options.DeliverTo;
    }

    public Task DeliverAsync(PerformanceMetrics counter)
    {
        Console.WriteLine($"[{DateTimeOffset.Now}]Deliver performance counter {counter} to {_deliverTo} via {_transport}");
        return Task.CompletedTask;
    }
}
```

与 FakeMetricsDeliverer 提取配置选项类似,在承载服务类型 PerformanceMetricsCollector 中同样可以采用 Options 模式来提供表示性能指标采集频率的配置选项。如下所示的代码片段是 PerformanceMetricsCollector 采用配置选项后的完整定义。

```csharp
public sealed class PerformanceMetricsCollector : IHostedService
{
    private readonly IProcessorMetricsCollector  _processorMetricsCollector;
    private readonly IMemoryMetricsCollector     _memoryMetricsCollector;
    private readonly INetworkMetricsCollector    _networkMetricsCollector;
    private readonly IMetricsDeliverer           _metricsDeliverer;
    private readonly TimeSpan                    _captureInterval;
    private IDisposable                          _scheduler;

    public PerformanceMetricsCollector(
        IProcessorMetricsCollector processorMetricsCollector,
        IMemoryMetricsCollector memoryMetricsCollector,
        INetworkMetricsCollector networkMetricsCollector,
        IMetricsDeliverer metricsDeliverer,
        IOptions<MetricsCollectionOptions> optionsAccessor)
    {
        _processorMetricsCollector  = processorMetricsCollector;
        _memoryMetricsCollector     = memoryMetricsCollector;
        _networkMetricsCollector    = networkMetricsCollector;
        _metricsDeliverer           = metricsDeliverer;
        _captureInterval            = optionsAccessor.Value.CaptureInterval;
    }

    public Task StartAsync(CancellationToken cancellationToken)
```

```
    {
        _scheduler = new Timer(Callback, null, TimeSpan.FromSeconds(5),
            _captureInterval);
        return Task.CompletedTask;

        async void Callback(object state)
        {
            var counter = new PerformanceMetrics
            {
                Processor       = _processorMetricsCollector.GetUsage(),
                Memory          = _memoryMetricsCollector.GetUsage(),
                Network         = _networkMetricsCollector.GetThroughput()
            };
            await _metricsDeliverer.DeliverAsync(counter);
        }
    }

    public Task StopAsync(CancellationToken cancellationToken)
    {
        _scheduler?.Dispose();
        return Task.CompletedTask;
    }
}
```

使用配置文件可以提供上述 3 个配置选项，所以我们在根目录下添加了一个名为 appSettings.json 的配置文件。由于演示的应用程序采用的 SDK 类型为 "Microsoft.NET.Sdk"，程序运行过程中会将编译程序集的目标目录作为当前目录，所以需要将配置文件的 "Copy to output directory" 属性设置为 "Copy always"，这样可以确保它在编译时总是被复制到目标目录。我们通过在配置文件中定义如下内容来提供上述 3 个配置选项。

```
{
  "MetricsCollection": {
    "CaptureInterval": "00:00:05",
    "Transport": "Udp",
    "DeliverTo": {
      "Host": "192.168.0.1",
      "Port": 3721
    }
  }
}
```

下面针对配置选项的使用对演示程序做相应的改动。如下面的代码片段所示，我们调用了 IHostBuilder 对象的 ConfigureAppConfiguration 方法，并利用提供的 Action<IConfigurationBuilder>对象注册了指向配置文件 appsettings.json 的 JsonConfigurationSource 对象。从名称可以看出，ConfigureAppConfiguration 方法的目的在于初始化应用程序所需的配置。

```
class Program
{
    static void Main()
    {
```

```
            var collector = new FakeMetricsCollector();
            new HostBuilder()
                .ConfigureAppConfiguration(builder=>builder.AddJsonFile("appsettings.json"))
                .ConfigureServices((context,svcs) => svcs
                    .AddSingleton<IProcessorMetricsCollector>(collector)
                    .AddSingleton<IMemoryMetricsCollector>(collector)
                    .AddSingleton<INetworkMetricsCollector>(collector)
                    .AddSingleton<IMetricsDeliverer, FakeMetricsDeliverer>()
                    .AddSingleton<IHostedService, PerformanceMetricsCollector>()

                    .AddOptions()
                    .Configure<MetricsCollectionOptions>(
                        context.Configuration.GetSection("MetricsCollection")))
                .Build()
                .Run();
        }
```

之前针对依赖服务的注册是通过调用 IHostBuilder 对象的 ConfigureServices 方法利用作为参数的 Action<IServiceCollection>对象完成的，IHostBuilder 接口还有一个 ConfigureServices 方法重载，它的参数类型为 Action<HostBuilderContext, IServiceCollection>，作为上下文的 HostBuilderContext 对象可以提供应用的配置，我们在上面调用的就是 ConfigureServices 方法重载。

如上面的代码片段所示，我们利用提供的 Action<HostBuilderContext, IServiceCollection>对象通过调用 IServiceCollection 接口的 AddOptions 扩展方法注册了 Options 模式所需的核心服务，然后调用 Configure<TOptions>扩展方法从提供的 HostBuilderContext 对象中提取出当前应用的配置，并将它和对应的配置选项类型 MetricsCollectionOptions 做了绑定。我们修改后的程序运行之后在控制台上输出的结果如图 10-2 所示，可以看出，输出的结果与配置文件的内容是匹配的。（S1003）

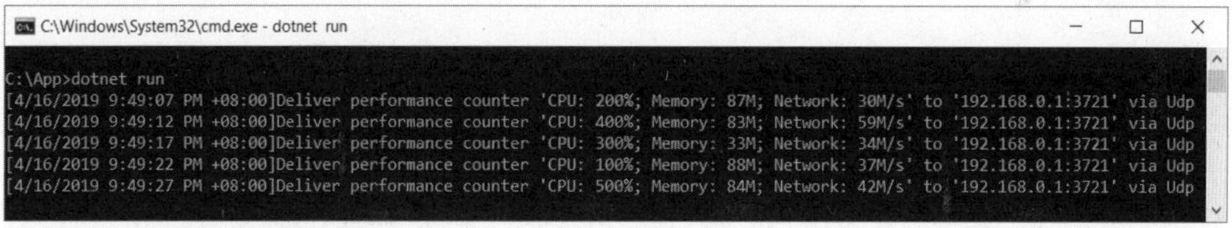

```
C:\App>dotnet run
[4/16/2019 9:49:07 PM +08:00]Deliver performance counter 'CPU: 200%; Memory: 87M; Network: 30M/s' to '192.168.0.1:3721' via Udp
[4/16/2019 9:49:12 PM +08:00]Deliver performance counter 'CPU: 400%; Memory: 83M; Network: 59M/s' to '192.168.0.1:3721' via Udp
[4/16/2019 9:49:17 PM +08:00]Deliver performance counter 'CPU: 300%; Memory: 33M; Network: 34M/s' to '192.168.0.1:3721' via Udp
[4/16/2019 9:49:22 PM +08:00]Deliver performance counter 'CPU: 100%; Memory: 88M; Network: 37M/s' to '192.168.0.1:3721' via Udp
[4/16/2019 9:49:27 PM +08:00]Deliver performance counter 'CPU: 500%; Memory: 84M; Network: 42M/s' to '192.168.0.1:3721' via Udp
```

图 10-2　引入配置选项

10.1.4　承载环境

应用程序总是针对某个具体环境进行部署的，开发（Development）、预发（Staging）和产品（Production）是 3 种典型的部署环境。这里的部署环境在承载系统中统称为承载环境（Hosting Environment）。一般来说，不同的承载环境往往具有不同的配置选项，下面演示如何为不同的承载环境提供相应的配置选项。

第 6 章已经演示了如何提供针对具体环境的配置文件，具体的做法很简单：将共享或者默认的配置定义在基础配置文件（如 appsettings.json）中，将差异化的部分定义在针对具体承载环境的配置文件（如 appsettings.staging.json 和 appsettings.production.json）中。对于我们演示的实例来说，可以采用图 10-3 所示的方式添加额外的两个配置文件来提供针对预发环境和产品环境的差异化配置。

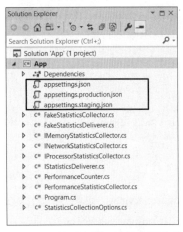

图 10-3　针对承载环境的配置文件

对于演示实例提供的 3 个配置选项来说，假设针对承载环境的差异化配合仅限于发送的目标终结点（IP 地址和端口），就可以采用如下方式将它们定义在针对预发环境的 appsettings.staging.json 和针对产品环境的 appsettings.production.json 中。

appsettings.staging.json：

```
{
  "MetricsCollection": {
    "DeliverTo": {
      "Host": "192.168.0.2",
      "Port": 3721
    }
  }
}
```

appsettings.production.json：

```
{
  "MetricsCollection": {
    "DeliverTo": {
      "Host": "192.168.0.3",
      "Port": 3721
    }
  }
}
```

在提供了针对具体承载环境的配置文件之后，还需要解决两个问题：第一，如何将它们注册到应用采用的配置框架中；第二，如何确定当前的承载环境。前者可以调用 IHostBuilder 接口的 ConfigureAppConfiguration 方法来完成，从命名可以看出，这个方法注册的是针对"应用"

层面的配置。我们可以将这里所谓的"应用"理解为承载的服务，也就是说，采用这种方式注册的配置是供承载的服务使用的。实际上，IHostBuilder 接口还有一个 ConfigureHostConfiguration 方法，它注册的服务是供服务宿主（Host）自身使用的，而当前的承载环境就可以利用此配置来指定。

我们将上述这两个问题的解决方案实现在改写的程序中。如下面的代码片段所示，为了使演示的应用程序可以采用命令行的形式来指定承载环境，可以调用 HostBuilder 接口的 ConfigureHostConfiguration 方法，并利用提供的 Action<IConfigurationBuilder>对象注册了针对命令行的配置源。为了注册针对承载环境的配置，可以调用类型为 Action<HostBuilderContext, IConfigurationBuilder>的 ConfigureAppConfiguration 方法，因为我们需要利用 HostBuilderContext 上下文对象得到当前的承载环境。

```
class Program
{
    static void Main(string[] args)
    {
        var collector = new FakeMetricsCollector();
        new HostBuilder()
            .ConfigureHostConfiguration(builder => builder.AddCommandLine(args))
            .ConfigureAppConfiguration((context, builder) => builder
                .AddJsonFile(path: "appsettings.json", optional: false)
                .AddJsonFile(
                    path: $"appsettings.{context.HostingEnvironment.EnvironmentName}.json",
                    optional: true))
            .ConfigureServices((context, svcs) => svcs
                .AddSingleton<IProcessorMetricsCollector>(collector)
                .AddSingleton<IMemoryMetricsCollector>(collector)
                .AddSingleton<INetworkMetricsCollector>(collector)
                .AddSingleton<IMetricsDeliverer, FakeMetricsDeliverer>()
                .AddSingleton<IHostedService, PerformanceMetricsCollector>()

                .AddOptions()
                .Configure<MetricsCollectionOptions>(
                    context.Configuration.GetSection("MetricsCollection")))
            .Build()
            .Run();
    }
}
```

我们调用 ConfigureAppConfiguration 方法注册了两个配置文件：一个是承载基础或者默认配置的 appsettings.json 文件，另一个是针对当前承载环境的 appsettings.{environment}.json 文件。前者是必需的，后者是可选的，这样做的目的在于确保即使当前承载环境不存在对应配置文件的情况也不会抛出异常（此时应用只会使用 appsettings.json 文件中定义的配置）。

下面以命令行的形式运行修改后的应用程序，承载环境通过命令行参数 environment 来指定。图 10-4 是先后 4 次运行演示实例得到的输出结果，从输出的 IP 地址可以看出，应用程序确实是根据当前承载环境加载对应的配置文件的。输出结果还体现了另一个细节：应用程序默认使用

的是产品（Production）环境。（S1004）

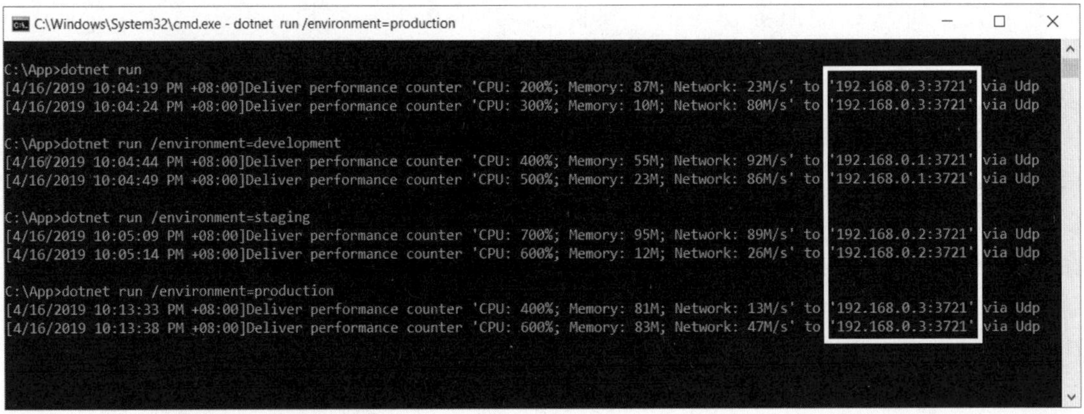

图 10-4　针对承载环境加载配置文件

10.1.5　日志

在具体的应用开发中不可避免地会涉及很多针对"诊断日志"的编程，第 8 章和第 9 章对这一主题进行了系统而详细的介绍，下面演示在通过承载系统承载的应用中如何记录日志。对于演示实例来说，它用于发送性能指标的 FakeMetricsDeliverer 对象会将收集的指标数据输出到控制台上，下面将这段文字以日志的形式进行输出，为此我们将这个类型进行了如下改写。

```
public class FakeMetricsDeliverer : IMetricsDeliverer
{
    private readonly TransportType      _transport;
    private readonly Endpoint           _deliverTo;
    private readonly ILogger            _logger;
    private readonly Action<ILogger, DateTimeOffset, PerformanceMetrics, Endpoint,
        TransportType, Exception> _logForDelivery;

    public FakeMetricsDeliverer(
        IOptions<MetricsCollectionOptions>          optionsAccessor,
        ILogger<FakeMetricsDeliverer>               logger)
    {
        var options         = optionsAccessor.Value;
        _transport          = options.Transport;
        _deliverTo          = options.DeliverTo;
        _logger             = logger;
        _logForDelivery     = LoggerMessage.Define<DateTimeOffset, PerformanceMetrics,
            Endpoint, TransportType>(LogLevel.Information, 0,
            "[{0}]Deliver performance counter {1} to {2} via {3}");
    }

    public Task DeliverAsync(PerformanceMetrics counter)
    {
        _logForDelivery(_logger, DateTimeOffset.Now, counter, _deliverTo, _transport,
```

```
            null);
        return Task.CompletedTask;
    }
}
```

如上面的代码片段所示,我们直接在构造函数中注入了 ILogger<FakeMetricsDeliverer>对象并利用它来记录日志。第 9 章提出:为了避免对同一个消息模板的重复解析,可以使用静态类型 LoggerMessage 提供的委托对象来输出日志,这也是 FakeMetricsDeliverer 中采用的编程模式。

为了将日志框架引入应用程序,我们需要在初始化应用时注册相应的服务,为此需要将应用程序做相应的改写。如下面的代码片段所示,我们调用 IHostBuilder 接口的 ConfigureLogging 扩展方法注册了日志框架的核心服务,并利用提供的 Action<ILoggingBuilder>对象注册了针对控制台作为输出渠道的 ConsoleLoggerProvider。

```
class Program
{
    static void Main(string[] args)
    {
        var collector = new FakeMetricsCollector();
        new HostBuilder()
            .ConfigureHostConfiguration(builder => builder.AddCommandLine(args))
            .ConfigureAppConfiguration((context, builder) => builder
                .AddJsonFile(path: "appsettings.json", optional: false)
                .AddJsonFile(
                    path: $"appsettings.{context.HostingEnvironment.EnvironmentName}.json",
                    optional: true))
            .ConfigureServices((context, svcs) => svcs
                .AddSingleton<IProcessorMetricsCollector>(collector)
                .AddSingleton<IMemoryMetricsCollector>(collector)
                .AddSingleton<INetworkMetricsCollector>(collector)
                .AddSingleton<IMetricsDeliverer, FakeMetricsDeliverer>()
                .AddSingleton<IHostedService, PerformanceMetricsCollector>()

                .AddOptions()
                .Configure<MetricsCollectionOptions>(
                    context.Configuration.GetSection("MetricsCollection")))
            .ConfigureLogging (builder=>builder.AddConsole())
            .Build()
            .Run();
    }
}
```

再次运行修改后的程序,控制台上的输出结果如图 10-5 所示。由输出结果可以看出,这些文字是由我们注册的 ConsoleLoggerProvider 提供的 ConsoleLogger 对象输出到控制台上的。由于承载系统自身在进行服务承载过程中也会输出一些日志,所以它们也会输出到控制台上。(S1005)

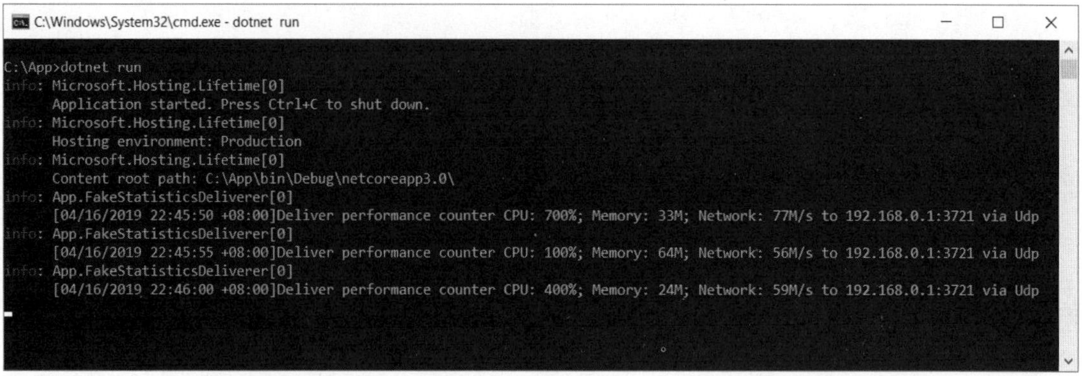

```
C:\Windows\System32\cmd.exe - dotnet run

C:\App>dotnet run
info: Microsoft.Hosting.Lifetime[0]
      Application started. Press Ctrl+C to shut down.
info: Microsoft.Hosting.Lifetime[0]
      Hosting environment: Production
info: Microsoft.Hosting.Lifetime[0]
      Content root path: C:\App\bin\Debug\netcoreapp3.0\
info: App.FakeStatisticsDeliverer[0]
      [04/16/2019 22:45:50 +08:00]Deliver performance counter CPU: 700%; Memory: 33M; Network: 77M/s to 192.168.0.1:3721 via Udp
info: App.FakeStatisticsDeliverer[0]
      [04/16/2019 22:45:55 +08:00]Deliver performance counter CPU: 100%; Memory: 64M; Network: 56M/s to 192.168.0.1:3721 via Udp
info: App.FakeStatisticsDeliverer[0]
      [04/16/2019 22:46:00 +08:00]Deliver performance counter CPU: 400%; Memory: 24M; Network: 59M/s to 192.168.0.1:3721 via Udp
```

图 10-5　将日志输出到控制台上

如果对输出的日志进行过滤，可以将过滤规则定义在配置文件中。假设对于类别以"Microsoft."为前缀的日志，我们只希望等级不低于 Warning 的才会被输出，这样会避免太多的消息被输出到控制台上造成对性能的影响，所以可以将产品环境对应的 appsettings.production.json 文件的内容做如下修改。

```
{
  "MetricsCollection": {
    "DeliverTo": {
      "Host": "192.168.0.3",
      "Port": 3721
    }
  },
  "Logging": {
    "LogLevel": {
      "Microsoft": "Warning"
    }
  }
}
```

为了应用日志配置，我们还需要对应用程序做相应的修改。如下面的代码片段所示，在对 ConfigureLogging 扩展方法的调用中，可以利用 HostBuilderContext 上下文对象得到当前配置，进而得到名为 Logging 的配置节。我们将这个配置节作为参数调用 ILoggingBuilder 对象的 AddConfiguration 扩展方法将承载的过滤规则应用到日志框架上。

```
class Program
{
    static void Main(string[] args)
    {
        var collector = new FakeMetricsCollector();
        new HostBuilder()
            .ConfigureHostConfiguration(builder => builder.AddCommandLine(args))
            .ConfigureAppConfiguration((context, builder) => builder
                .AddJsonFile(path: "appsettings.json", optional: false)
                .AddJsonFile(
                    path: $"appsettings.{context.HostingEnvironment.EnvironmentName}.json",
```

```
                    optional: true))
                .ConfigureServices((context, svcs) => svcs
                    .AddSingleton<IProcessorMetricsCollector>(collector)
                    .AddSingleton<IMemoryMetricsCollector>(collector)
                    .AddSingleton<INetworkMetricsCollector>(collector)
                    .AddSingleton<IMetricsDeliverer, FakeMetricsDeliverer>()
                    .AddSingleton<IHostedService, PerformanceMetricsCollector>()

                    .AddOptions()
                    .Configure<MetricsCollectionOptions>(
                        context.Configuration.GetSection("MetricsCollection")))
                .ConfigureLogging((context,builder) => builder
                    .AddConfiguration(context.Configuration.GetSection("Logging"))
                    .AddConsole())
                .Build()
                .Run();
        }
}
```

如果此时分别针对开发（Development）环境和产品（Production）环境以命令行的形式启动修改后的应用程序，就会发现针对开发环境控制台会输出类型前缀为"Microsoft."的日志，但是在针对产品环境的控制台上却找不到它们的踪影（见图10-6）。（S1006）

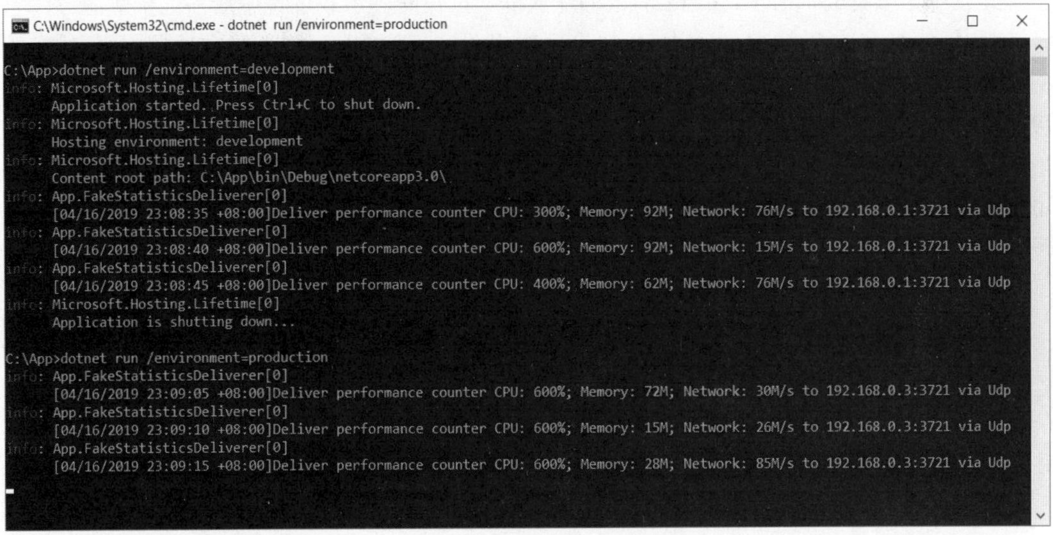

图 10-6　根据承载环境过滤日志

10.2　承载模型

前面的实例演示了服务承载的基本编程模式，下面从设计的角度重新介绍服务承载模型。总的来说，服务承载模型主要由 3 个核心对象组成（见图 10-7）：多个通过 IHostedService 接口表示的服务被承载于通过 IHost 接口表示的宿主上，IHostBuilder 接口表示 IHost 对象的构建者。

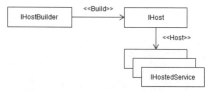

图 10-7　服务承载模型

10.2.1　IHostedService

承载的服务总是会被定义成 IHostedService 接口的实现类型。如下面的代码片段所示,该接口仅定义了两个用来启动和关闭自身服务的方法。当作为宿主的 IHost 对象被启动时,它会利用依赖注入框架激活每个注册的 IHostedService 服务,并通过调用 StartAsync 方法来启动它们。当服务承载应用程序关闭时,作为服务宿主的 IHost 对象会被关闭,由它承载的每个 IHostedService 服务对象的 StopAsync 方法也随之被调用。

```
public interface IHostedService
{
    Task StartAsync(CancellationToken cancellationToken);
    Task StopAsync(CancellationToken cancellationToken);
}
```

承载系统无缝集成了 .NET Core 的依赖注入框架,在服务承载过程中所需的依赖服务,包括承载服务自身和它所依赖的服务均由此框架提供,承载服务注册的本质就是将对应的 IHostedService 实现类型或者实例注册到依赖注入框架中。由于承载服务大都需要长时间运行直到应用被关闭,所以针对承载服务的注册一般采用 Singleton 生命周期模式。承载系统为承载服务的注册定义了 AddHostedService<THostedService>扩展方法,由于该方法通过调用 TryAddEnumerable 扩展方法来注册服务,所以不会出现服务重复注册的问题。

```
public static class ServiceCollectionHostedServiceExtensions
{
    public static IServiceCollection AddHostedService<THostedService>(
        this IServiceCollection services) where THostedService: class, IHostedService
    {
        services.TryAddEnumerable(
            ServiceDescriptor.Singleton<IHostedService, THostedService>());
        return services;
    }
}
```

10.2.2　IHost

通过 IHostedService 接口表示的承载服务最终被承载于通过 IHost 接口表示的宿主上。一般来说,一个服务承载应用在整个生命周期内只会创建一个 IHost 对象,启动和关闭应用程序本质上就是启动和关闭作为宿主的 IHost 对象。如下面的代码片段所示,IHost 接口派生于 IDisposable 接口,所以当它在关闭之后,应用程序还会调用其 Dispose 方法做一些额外的资源释放工作。

```
public interface IHost : IDisposable
{
    IServiceProvider Services { get; }
    Task StartAsync(CancellationToken cancellationToken = default);
    Task StopAsync(CancellationToken cancellationToken = default);
}
```

IHost 接口的 Services 属性返回作为依赖注入容器的 IServiceProvider 对象，该对象提供了服务承载过程中所需的服务实例，其中就包括需要承载的 IHostedService 服务。定义在 IHost 接口中的 StartAsync 方法和 StopAsync 方法完成了针对服务宿主的启动与关闭。

IHostApplicationLifetime

在前面演示的实例中，利用 HostBuilder 对象构建出 IHost 对象之后，我们并没有调用其 StartAsync 方法启动它，而是调用另一个名为 Run 的扩展方法。Run 方法涉及服务承载应用生命周期管理，如果想了解该方法的本质，需要先了解 IHostApplicationLifetime 接口。顾名思义，IHostApplicationLifetime 接口体现了服务承载应用程序的生命周期。如下面的代码片段所示，该接口除了提供了 3 个 CancellationToken 类型的属性来检测应用何时开启与关闭，还提供了一个 StopApplication 方法来关闭应用程序。

```
public interface IHostApplicationLifetime
{
    CancellationToken ApplicationStarted { get; }
    CancellationToken ApplicationStopping { get; }
    CancellationToken ApplicationStopped { get; }

    void StopApplication();
}
```

如下所示的 ApplicationLifetime 类型是对 IHostApplicationLifetime 接口的默认实现。我们可以看到它实现的 3 个属性返回的 CancellationToken 对象来源于 3 个对应的 CancellationTokenSource 对象，后者对应 3 个不同的方法（NotifyStarted、StopApplication 和 NotifyStopped）。我们可以利用 IHostApplicationLifetime 服务的 3 个属性提供的 CancellationToken 对象得到关于应用被启动和关闭的通知，这些通知最初就是由这 3 个对应的方法发出的。

```
public class ApplicationLifetime : IHostApplicationLifetime
{
    private readonly ILogger<ApplicationLifetime>      _logger;
    private readonly CancellationTokenSource           _startedSource;
    private readonly CancellationTokenSource           _stoppedSource;
    private readonly CancellationTokenSource           _stoppingSource;

    public ApplicationLifetime(ILogger<ApplicationLifetime> logger)
    {
        _startedSource      = new CancellationTokenSource();
        _stoppedSource      = new CancellationTokenSource();
        _stoppingSource     = new CancellationTokenSource();
        _logger             = logger;
    }
```

```csharp
private void ExecuteHandlers(CancellationTokenSource cancel)
{
    if (!cancel.IsCancellationRequested)
    {
        cancel.Cancel(false);
    }
}

public void NotifyStarted()
{
    try
    {
        ExecuteHandlers(_startedSource);
    }
    catch (Exception exception)
    {
        _logger.ApplicationError(6, "An error occurred starting the application",
            exception);
    }
}

public void NotifyStopped()
{
    try
    {
        ExecuteHandlers(_stoppedSource);
    }
    catch (Exception exception)
    {
        _logger.ApplicationError(8, "An error occurred stopping the application",
            exception);
    }
}

public void StopApplication()
{
    lock (_stoppingSource)
    {
        try
        {
            ExecuteHandlers(_stoppingSource);
        }
        catch (Exception exception)
        {
            _logger.ApplicationError(7,
                "An error occurred stopping the application", exception);
        }
    }
}
```

```csharp
    public CancellationToken ApplicationStarted      => _startedSource.Token;
    public CancellationToken ApplicationStopped      => _stoppedSource.Token;
    public CancellationToken ApplicationStopping     => _stoppingSource.Token;
}
```

下面通过一个简单的实例演示如何利用 IHostApplicationLifetime 服务来关闭整个承载应用程序。我们在一个控制台应用程序中定义了如下这个承载服务 FakeHostedService，并且在 FakeHostedService 类型的构造函数中注入了 IHostApplicationLifetime 服务。在得到其 3 个属性返回的 CancellationToken 对象之后，我们在它们上面分别注册了一个回调，通过回调操作，控制台上会输出相应的文字，由此我们可以知道应用程序何时被启动和关闭。

```csharp
public sealed class FakeHostedService : IHostedService
{
    private readonly IHostApplicationLifetime   _lifetime;
    private IDisposable                         _tokenSource;

    public FakeHostedService(IHostApplicationLifetime lifetime)
    {
        _lifetime = lifetime;
        _lifetime.ApplicationStarted.Register(() => Console.WriteLine(
            "[{0}]Application started", DateTimeOffset.Now));
        _lifetime.ApplicationStopping.Register(() => Console.WriteLine(
            "[{0}]Application is stopping.", DateTimeOffset.Now));
        _lifetime.ApplicationStopped.Register(() => Console.WriteLine(
            "[{0}]Application stopped.", DateTimeOffset.Now));
    }

    public Task StartAsync(CancellationToken cancellationToken)
    {
        _tokenSource = new CancellationTokenSource(TimeSpan.FromSeconds(5))
            .Token.Register(_lifetime.StopApplication);
        return Task.CompletedTask;
    }

    public Task StopAsync(CancellationToken cancellationToken)
    {
        _tokenSource?.Dispose();
        return Task.CompletedTask;
    }
}
```

在实现的 StartAsync 方法中，我们采用上面的方式在等待 5 秒之后调用 IHostApplicationLifetime 服务的 StopApplication 方法来关闭整个应用程序。FakeHostedService 服务最后采用如下所示的方式承载于当前应用程序中。

```csharp
class Program
{
    static void Main()
    {
```

```
        new HostBuilder()
            .ConfigureServices(svcs => svcs.AddHostedService<FakeHostedService>())
            .Build()
            .Run();
    }
}
```

该程序运行之后在控制台上输出的结果如图 10-8 所示,从 3 条消息输出的时间间隔可以确定当前应用程序正是承载 FakeHostedService 通过调用 IHostApplicationLifetime 服务的 StopApplication 方法关闭的。(S1007)

图 10-8　调用 IHostApplicationLifetime 服务关闭应用程序

Run 扩展方法

如果调用 IHost 对象的 Run 扩展方法,它会在内部调用 StartAsync 方法并持续等待。直到接收到来自 IHostApplicationLifetime 服务发出的关闭应用通知后,IHost 对象才会调用自身的 StopAsync 方法,针对 Run 扩展方法的调用此时才会返回。启动 IHost 对象直到应用关闭体现在下面的 WaitForShutdownAsync 扩展方法上。

```
public static class HostingAbstractionsHostExtensions
{
    public static async Task WaitForShutdownAsync(this IHost host,
        CancellationToken token = default)
    {
        var applicationLifetime = host.Services.GetService<IHostApplicationLifetime>();
        token.Register(state =>((IHostApplicationLifetime)state).StopApplication(),
            applicationLifetime);

        var waitForStop = new TaskCompletionSource<object>(
            TaskCreationOptions.RunContinuationsAsynchronously);
        applicationLifetime.ApplicationStopping.Register(state =>
        {
            var tcs = (TaskCompletionSource<object>)state;
            tcs.TrySetResult(null);
        }, waitForStop);

        await waitForStop.Task;
        await host.StopAsync();
    }
}
```

如下所示的 WaitForShutdown 方法是上面 WaitForShutdownAsync 方法的同步版本。同步的 Run 方法和异步的 RunAsync 方法的实现也体现在下面的代码片段中。除此之外,下面的代码片段还提供了 Start 方法和 StopAsync 方法，前者可以视为 StartAsync 方法的同步版本，后者可以在关闭 IHost 对象时指定一个超时时限。

```csharp
public static class HostingAbstractionsHostExtensions
{
    public static void WaitForShutdown(this IHost host)
        => host.WaitForShutdownAsync().GetAwaiter().GetResult();

    public static void Run(this IHost host)
        => host.RunAsync().GetAwaiter().GetResult();
    public static async Task RunAsync(this IHost host, CancellationToken token = default)
    {
        try
        {
            await host.StartAsync(token);
            await host.WaitForShutdownAsync(token);
        }
        finally
        {
            host.Dispose();
        }
    }

    public static void Start(this IHost host)
        => host.StartAsync().GetAwaiter().GetResult();

    public static Task StopAsync(this IHost host, TimeSpan timeout)
        => host.StopAsync(new CancellationTokenSource(timeout).Token);
}
```

10.2.3　IHostBuilder

在了解了作为服务宿主的 IHost 接口之后，下面介绍作为宿主构建者的 IHostBuilder 接口。如下面的代码片段所示，IHostBuilder 接口的核心方法 Build 用来提供由它构建的 IHost 对象。除此之外，它还有一个字典类型的只读属性 Properties，我们可以将它视为一个共享的数据字典。

```csharp
public interface IHostBuilder
{
    IDictionary<object, object> Properties { get; }
    IHost Build();
    ...
}
```

作为一个典型的设计模式，Builder 模式在最终提供给由它构建的对象之前，一般会允许做相应的前期设置，IHostBuilder 针对 IHost 对象的构建也不例外。IHostBuilder 接口提供了一系列的方法，我们可以利用它们为最终构建的 IHost 对象做相应的设置，具体的设置主要涵盖两个方面：针对配置系统的设置和针对依赖注入框架的设置。

针对配置系统的设置

IHostBuilder 接口针对配置系统的设置体现在 ConfigureHostConfiguration 方法和 ConfigureAppConfiguration 方法上。由前面的实例演示可知，ConfigureHostConfiguration 方法涉及的配置主要在服务承载过程中使用，所以是针对服务宿主的配置。ConfigureAppConfiguration 方法涉及的配置则是供承载的 IHostedService 服务使用的，所以是针对应用的配置。但前者最终会合并到后者之中，应用程序最终得到的配置实际上是两者合并的结果。

```
public interface IHostBuilder
{
    IHostBuilder ConfigureHostConfiguration(
        Action<IConfigurationBuilder> configureDelegate);
    IHostBuilder ConfigureAppConfiguration(
        Action<HostBuilderContext, IConfigurationBuilder> configureDelegate);
    ...
}
```

从上面的代码片段可以看出，ConfigureHostConfiguration 方法提供了一个 Action<IConfigurationBuilder>类型的委托作为参数，我们可以利用它注册不同的配置源或者做相应的设置（如设置配置文件所在目录的基础路径）。ConfigureAppConfiguration 方法的参数类型则是 Action<HostBuilderContext, IConfigurationBuilder>，作为第一个参数的 HostBuilderContext 对象携带了与服务承载相关的上下文信息，我们可以利用该上下文对配置系统做针对性设置。

HostBuilderContext 携带的上下文主要包含两个部分：一是通过调用 ConfigureHostConfiguration 方法设置的针对宿主的配置，对应其 Configuration 属性；二是当前的承载环境，对应其 HostingEnvironment 属性。除此之外，HostBuilderContext 类型同样具有一个作为共享数据字典的 Properties 属性。

```
public class HostBuilderContext
{
    public IConfiguration                    Configuration { get; set; }
    public IHostEnvironment                  HostingEnvironment { get; set; }
    public IDictionary<object, object>       Properties { get; }

    public HostBuilderContext(IDictionary<object, object> properties);
}
```

ConfigureAppConfiguration 方法使我们可以就当前承载上下文对应用配置做针对性配置，如针对前期提供承载配置，或者之前添加到 Properties 字典中的某个属性，以及最常见的针对当前的承载环境。如果针对配置系统的设置与当前承载上下文无关，就可以调用如下这个同名的扩展方法，该方法提供的参数依旧是一个 Action<IConfigurationBuilder>对象。

```
public static class HostingHostBuilderExtensions
{
    public static IHostBuilder ConfigureAppConfiguration(this IHostBuilder hostBuilder,
        Action<IConfigurationBuilder> configureDelegate)
    => hostBuilder.ConfigureAppConfiguration((context, builder) =>
        configureDelegate(builder));
}
```

承载环境

任何一个应用总是针对某个具体的环境进行部署的，我们将承载服务的部署环境称为承载环境。承载环境通过 IHostEnvironment 接口表示，HostBuilderContext 类型的 HostingEnvironment 属性返回的就是一个 IHostEnvironment 对象。如下面的代码片段所示，除了表示环境名称的 EnvironmentName 属性，IHostEnvironment 接口还定义了一个表示当前应用名称的 ApplicationName 属性。

```
public interface IHostEnvironment
{
    string              EnvironmentName { get; set; }
    string              ApplicationName { get; set; }
    string              ContentRootPath { get; set; }
    IFileProvider       ContentRootFileProvider { get; set; }
}
```

编译某个 .NET Core 项目时，提供的代码文件（.cs）会转换成元数据和 IL 指令保存到生成的程序集中，其他一些文件还可以作为程序集的内嵌资源。除了这些面向程序集的文件，一些文件还会以静态文件的形式供应用程序使用，如 Web 应用 3 种典型的静态文件（JavaScript、CSS 和图片），我们将这些静态文件称为内容文件（Content File）。IHostEnvironment 接口的 ContentRootPath 表示的是存放这些内容文件的根目录所在的路径，ContentRootFileProvider 属性对应的则是指向该路径的 IFileProvider 对象，我们可以利用它获取目录的层次结构，也可以直接利用它来读取文件的内容。

开发、预发和产品是 3 种典型的承载环境，如果采用 Development、Staging 和 Production 来对它们进行命名，我们针对这 3 种承载环境的判断就可以利用如下 3 个扩展方法（IsDevelopment、IsStaging 和 IsProduction）来完成。如果需要判断指定的 IHostEnvironment 对象是否属于某个指定的环境，可以直接调用 IsEnvironment 扩展方法。从给出的代码片段可以看出针对环境名称的比较是不区分大小写的。

```
public static class HostEnvironmentEnvExtensions
{
    public static bool IsDevelopment(this IHostEnvironment hostEnvironment)
        =>hostEnvironment.IsEnvironment(Environments.Development);
    public static bool IsStaging(this IHostEnvironment hostEnvironment)
        =>hostEnvironment.IsEnvironment(Environments.Staging);
    public static bool IsProduction(this IHostEnvironment hostEnvironment)
        =>hostEnvironment.IsEnvironment(Environments.Production);

    public static bool IsEnvironment(this IHostEnvironment hostEnvironment,
        string environmentName)
        =>string.Equals(hostEnvironment.EnvironmentName, environmentName,
        StringComparison.OrdinalIgnoreCase);
}

public static class Environments
{
```

```csharp
    public static readonly string Development       = "Development";
    public static readonly string Production        = "Production";
    public static readonly string Staging           = "Staging";
}
```

IHostEnvironment 对象承载的 3 个属性都是通过配置的形式提供的,对应的配置项名称为 environment、contentRoot 和 applicationName,它们对应 HostDefaults 类型中的 3 个静态只读字段。我们可以调用针对 IHostBuilder 接口的 UseEnvironment 方法和 UseContentRoot 方法来设置环境名称与内容文件的根目录路径。从给出的代码片段可以看出,UseEnvironment 方法调用的依旧是 ConfigureHostConfiguration 方法。如果没有对应用名称做显示设置,入口程序集的名称就会作为当前应用名称。由于一些组件或者框架会假定当前应用名称就是应用所在项目编译后的程序集名称,所以我们一般不会对应用名称进行设置。

```csharp
public static class HostDefaults
{
    public static readonly string EnvironmentKey = "environment";
    public static readonly string ContentRootKey = "contentRoot";
    public static readonly string ApplicationKey = "applicationName";
}

public static class HostingHostBuilderExtensions
{
    public static IHostBuilder UseEnvironment(this IHostBuilder hostBuilder,
        string environment)
    {
        return hostBuilder.ConfigureHostConfiguration(configBuilder =>
        {
            configBuilder.AddInMemoryCollection(new[]
            {
                new KeyValuePair<string, string>(HostDefaults.EnvironmentKey,environment)
            });
        });
    }
    public static IHostBuilder UseContentRoot(this IHostBuilder hostBuilder,
        string contentRoot)
    {
        return hostBuilder.ConfigureHostConfiguration(configBuilder =>
        {
            configBuilder.AddInMemoryCollection(new[]
            {
                new KeyValuePair<string, string>(HostDefaults.ContentRootKey,
                    contentRoot)
            });
        });
    }
}
```

针对依赖注入框架的设置

由于包括承载服务(IHostedService)在内的所有依赖服务都由依赖注入框架提供,所以

IHostBuilder 接口提供了更多的方法来帮助我们注册所需的依赖服务。绝大部分用来注册服务的方法最终都会调用 ConfigureServices 方法，由于该方法提供的参数是一个 Action<HostBuilder Context, IServiceCollection>类型的委托，这就意味着服务可以就当前的承载上下文进行针对性注册。如果注册的服务与当前承载上下文无关，就可以调用如下所示的这个同名的扩展方法，该方法提供的参数是一个类型为 Action<IServiceCollection>的委托对象。

```csharp
public interface IHostBuilder
{
    IHostBuilder ConfigureServices(
        Action<HostBuilderContext, IServiceCollection> configureDelegate);
    ...
}

public static class HostingHostBuilderExtensions
{
    public static IHostBuilder ConfigureServices(this IHostBuilder hostBuilder,
        Action<IServiceCollection> configureDelegate)
        => hostBuilder.ConfigureServices((context, collection) =>
            configureDelegate(collection));
}
```

IHostBuilder 接口提供了如下两个 UseServiceProviderFactory<TContainerBuilder>方法重载，我们可以利用它注册的 IServiceProviderFactory<TContainerBuilder>对象实现对第三方依赖注入框架的整合。除此之外，IHostBuilder 接口还定义了 ConfigureContainer<TContainerBuilder>方法，从而对提供的依赖注入容器做进一步设置。

```csharp
public interface IHostBuilder
{
    IHostBuilder UseServiceProviderFactory<TContainerBuilder>(
        IServiceProviderFactory<TContainerBuilder> factory);
    IHostBuilder UseServiceProviderFactory<TContainerBuilder>(
        Func<HostBuilderContext, IServiceProviderFactory<TContainerBuilder>> factory);
    IHostBuilder ConfigureContainer<TContainerBuilder>(
        Action<HostBuilderContext, TContainerBuilder> configureDelegate);
}
```

笔者认为 .NET Core 依赖注入框架已经能够满足绝大部分应用开发的需求，所以真正与第三方依赖注入框架的整合其实并没有太大的必要。由于原生的依赖注入框架使用 DefaultServiceProviderFactory 来提供作为依赖注入容器的 IServiceProvider 对象，所以针对它的注册由如下这两个 UseDefaultServiceProvider 扩展方法来完成。

```csharp
public static class HostingHostBuilderExtensions
{
    public static IHostBuilder UseDefaultServiceProvider(this IHostBuilder hostBuilder,
        Action<ServiceProviderOptions> configure)
        => hostBuilder.UseDefaultServiceProvider((context, options) => configure(options));

    public static IHostBuilder UseDefaultServiceProvider(this IHostBuilder hostBuilder,
        Action<HostBuilderContext, ServiceProviderOptions> configure)
```

```
            return hostBuilder.UseServiceProviderFactory(context =>
            {
                var options = new ServiceProviderOptions();
                configure(context, options);
                return new DefaultServiceProviderFactory(options);
            });
        }
}
```

对于定义在 IHostBuilder 接口的 ConfigureContainer<TContainerBuilder>方法来说，它提供的参数是一个类型为 Action<HostBuilderContext, TContainerBuilder>的委托对象，如果针对 TContainerBuilder 的设置与当前承载上下文无关，我们也可以调用如下这个简化的 ConfigureContainer<TContainerBuilder>扩展方法，它只需要提供一个 Action<TContainerBuilder>对象作为参数即可。

```
public static class HostingHostBuilderExtensions
{
    public static IHostBuilder ConfigureContainer<TContainerBuilder>(
        this IHostBuilder hostBuilder, Action<TContainerBuilder> configureDelegate)
    {
        return hostBuilder.ConfigureContainer<TContainerBuilder>((context, builder) =>
            configureDelegate(builder));
    }
}
```

创建并启动宿主

IHostBuilder 接口还有如下这个 StartAsync 扩展方法，它同时完成了针对 IHost 对象的创建和启动工作，它的另一个 Start 方法是 StartAsync 方法的同步版本。

```
public static class HostingAbstractionsHostBuilderExtensions
{
    public static async Task<IHost> StartAsync(this IHostBuilder hostBuilder,
        CancellationToken cancellationToken = default)
    {
        var host = hostBuilder.Build();
        await host.StartAsync(cancellationToken);
        return host;
    }

    public static IHost Start(this IHostBuilder hostBuilder)
        => hostBuilder.StartAsync().GetAwaiter().GetResult();
}
```

10.3 实现原理

10.2 节介绍了 IHostedService、IHost 和 IHostBuilder 这 3 个接口，由此可以使读者对服务承载模型有大致的了解。下面从抽象转向具体，介绍承载系统针对该模型的实现是如何实施的。要了解承载模型的默认实现，只需要了解 IHost 接口和 IHostBuilder 接口的默认实现类型即可。由图 10-9 所示的 UML 可以看出，IHost 接口和 IHostBuilder 接口的默认实现类型分别是 Host 与 HostBuilder，本节将着重介绍这两个类型。

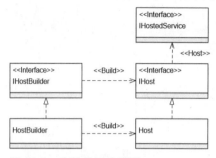

图 10-9　完整的承载模型

10.3.1　服务宿主

Host 类型是对 IHost 接口的默认实现，它仅仅是定义在 NuGet 包 "Microsoft.Extensions.Hosting" 中的一个内部类型，由于承载系统还提供了另一个同名的公共静态类型，在容易出现混淆的地方，我们会将它称为 "实例类型 Host" 以示区别。在正式介绍 Host 类型的具体实现之前，需要先认识两个与之相关的类型，其中一个是承载相关配置选项的 HostOptions，另一个是 IHostLifetime。如下面的代码片段所示，HostOptions 仅包含唯一的 ShutdownTimeout 属性，它表示关闭 Host 对象的超时时限，该属性的默认值为 5 秒。

```
public class HostOptions
{
    public TimeSpan ShutdownTimeout { get; set; } = TimeSpan.FromSeconds(5);
}
```

前面已经介绍了一个与承载应用生命周期相关的 IHostApplicationLifetime 接口，Host 类型还涉及另一个与生命周期相关的 IHostLifetime 接口。调用 Host 对象的 StartAsync 方法将 Host 对象启动之后，该方法会先调用 IHostLifetime 服务的 WaitForStartAsync 方法。Host 对象的 StopAsync 方法在执行过程中，如果它成功关闭了所有承载的服务，注册 IHostLifetime 服务的 StopAsync 方法就会被调用。

```
public interface IHostLifetime
{
    Task WaitForStartAsync(CancellationToken cancellationToken);
    Task StopAsync(CancellationToken cancellationToken);
}
```

在前面提供的针对日志的实例演示中，程序启动后控制台上会输出 3 条级别为 Information 的日志，其中第一条日志的内容为 "Application started. Press Ctrl+C to shut down."，后面两条则会输出当前承载环境的信息和存放内容文件的根目录路径。当应用程序关闭之前，控制台上还会出现一条内容为 "Application is shutting down..." 的日志。上述这 4 条日志在控制台上输出的效果体现在图 10-10 中。

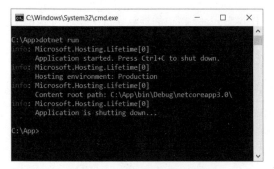

图 10-10 由 ConsoleLifetime 输出的日志

图 10-10 所示的 4 条日志都是如下这个 ConsoleLifetime 对象输出的，ConsoleLifetime 类型是对 IHostLifetime 接口的实现。除了以日志的形式输出与当前承载应用程序相关的状态信息，针对 Cancel 按键（Ctrl+C）的捕捉以及随后关闭当前应用的功能也实现在 ConsoleLifetime 类型中。ConsoleLifetime 采用的配置选项定义在 ConsoleLifetimeOptions 类型中，该类型唯一的属性成员 SuppressStatusMessages 用来决定上述 4 条日志是否需要被输出，如果不在控制台上输出这些消息，可以显式将此属性设置为 True。

```
public class ConsoleLifetime : IHostLifetime, IDisposable
{
    public ConsoleLifetime(IOptions<ConsoleLifetimeOptions> options,
        IHostEnvironment environment, IHostApplicationLifetime applicationLifetime);
    public ConsoleLifetime(IOptions<ConsoleLifetimeOptions> options,
        IHostEnvironment environment, IHostApplicationLifetime applicationLifetime,
        ILoggerFactory loggerFactory);

    public Task StopAsync(CancellationToken cancellationToken);
    public Task WaitForStartAsync(CancellationToken cancellationToken);
    public void Dispose();
}

public class ConsoleLifetimeOptions
{
    public bool SuppressStatusMessages { get; set; }
}
```

下面的代码片段展示的是经过简化的 Host 类型的定义。Host 类型的构造函数中注入了一系列依赖服务，包括作为依赖注入容器的 IServiceProvider 对象、用来记录日志的 ILogger<Host> 对象和提供配置选项 HostOptions 的 IOptions<HostOptions> 对象，以及两个与生命周期相关的 IHostApplicationLifetime 对象和 IHostLifetime 对象。值得注意的是，这里提供的 IHostApplicationLifetime 对象的类型必须是 ApplicationLifetime，因为它需要调用 NotifyStarted 方法和 NotifyStopped 方法在应用程序启动与关闭之后向订阅者发出通知，但是这两个方法并没有定义在 IHostApplicationLifetime 接口中。

```
internal class Host : IHost
{
```

```csharp
    private readonly ILogger<Host>              _logger;
    private readonly IHostLifetime              _hostLifetime;
    private readonly ApplicationLifetime        _applicationLifetime;
    private readonly HostOptions                _options;
    private IEnumerable<IHostedService>         _hostedServices;

    public IServiceProvider Services { get; }

    public Host(IServiceProvider services, IHostApplicationLifetime applicationLifetime,
        ILogger<Host> logger, IHostLifetime hostLifetime, IOptions<HostOptions> options)
    {
        Services                = services;
        _applicationLifetime    = (ApplicationLifetime)applicationLifetime;
        _logger                 = logger;
        _hostLifetime           = hostLifetime;
        _options                = options.Value);
    }

    public async Task StartAsync(CancellationToken cancellationToken = default)
    {
        await _hostLifetime.WaitForStartAsync(cancellationToken);
        cancellationToken.ThrowIfCancellationRequested();
        _hostedServices = Services.GetService<IEnumerable<IHostedService>>();
        foreach (var hostedService in _hostedServices)
        {
            await hostedService.StartAsync(cancellationToken).ConfigureAwait(false);
        }
        _applicationLifetime?.NotifyStarted();
    }

    public async Task StopAsync(CancellationToken cancellationToken = default)
    {
        using (var cts = new CancellationTokenSource(_options.ShutdownTimeout))
        using (var linkedCts = CancellationTokenSource.CreateLinkedTokenSource(
            cts.Token, cancellationToken))
        {
            var token = linkedCts.Token;
            _applicationLifetime?.StopApplication();
            foreach (var hostedService in _hostedServices.Reverse())
            {
                await hostedService.StopAsync(token).ConfigureAwait(false);
            }

            token.ThrowIfCancellationRequested();
            await _hostLifetime.StopAsync(token);
            _applicationLifetime?.NotifyStopped();
        }
    }

    public void Dispose()=>(Services as IDisposable)?.Dispose();
```

在实现的 StartAsync 方法中，Host 对象率先调用了 IHostLifetime 对象的 WaitForStartAsync 方法。如果注册的服务类型为 ConsoleLifetime，它会输出前面提及的 3 条日志。与此同时，ConsoleLifetime 对象还会注册控制台的按键事件，其目的在于确保在用户按下取消组合键（Ctrl+C）后应用能够被正常关闭。

Host 对象会利用作为依赖注入容器的 IServiceProvider 对象提取出代表承载服务的所有 IHostedService 对象，并通过 StartAsync 方法来启动它们。当所有承载的服务正常启动之后，ApplicationLifetime 对象的 NotifyStarted 方法会被调用，此时订阅者会接收到应用程序启动的通知。需要着重指出：代表承载服务的 IHostedService 对象是"逐个"（不是并发）被启动的，而且只有等待所有承载服务全部被启动之后，应用程序才算是启动成功。在整个启动过程中，如果利用作为参数的 CancellationToken 接收到取消请求，启动操作会中止。

当 Host 对象的 StopAsync 方法被调用时，它会调用 ApplicationLifetime 对象的 StopApplication 方法对外发出应用程序即将被关闭的通知，此后它会调用每个 IHostedService 对象的 StopAsync 方法。当所有承载服务被成功关闭之后，Host 对象会先后调用 IHostLifetime 对象的 StopAsync 方法和 ApplicationLifetime 对象的 NotifyStopped 方法。在 Host 关闭过程中，如果超出了通过 HostOptions 配置选项设定的超时时限，或者利用作为参数的 CancellationToken 对象接收到取消请求，整个过程会立即中止。

10.3.2　针对配置系统的设置

作为服务宿主的 IHost 对象总是通过对应的 IHostBuilder 对象构建出来的，Host 类型对应的 IHostBuilder 实现类型为 HostBuilder。除了用于构建 IHost 对象的 Build 方法，IHostBuilder 接口定义的一系列方法还可以对最终提供的 IHost 对象做相应的前期设置，这些设置会被缓存起来，最后应用到 Build 方法上。下面先介绍 HostBuilder 针对配置系统的设置。如下面的代码片段所示，调用 ConfigureHostConfiguration 方法提供的针对面向宿主配置和调用 ConfigureAppConfiguration 方法提供的面向应用配置指定的委托对象都暂存在对应的集合中，对应的字段分别是_configureHostConfigActions 和_configureAppConfigActions。

```
public class HostBuilder : IHostBuilder
{
    private List<Action<IConfigurationBuilder>> _configureHostConfigActions
        = new List<Action<IConfigurationBuilder>>();
    private List<Action<HostBuilderContext, IConfigurationBuilder>>
        _configureAppConfigActions = new
        List<Action<HostBuilderContext, IConfigurationBuilder>>();

    public IDictionary<object, object> Properties { get; }
        = new Dictionary<object, object>();

    public IHostBuilder ConfigureHostConfiguration(
        Action<IConfigurationBuilder> configureDelegate)
    {
```

```csharp
        _configureHostConfigActions.Add(configureDelegate);
        return this;
    }

    public IHostBuilder ConfigureAppConfiguration(
        Action<HostBuilderContext, IConfigurationBuilder> configureDelegate)
    {
        _configureAppConfigActions.Add(configureDelegate);
        return this;
    }
    ...
}
```

10.3.3　针对依赖注入框架的设置

针对依赖注入框架的设置主要体现在两个方面：其一，利用 ConfigureServices 方法添加服务注册；其二，利用两个 UseServiceProviderFactory<TContainerBuilder>方法注册 IServiceProviderFactory<TContainerBuilder>工厂，以及利用 ConfigureContainer<TContainerBuilder>方法对依赖注入容器做进一步设置。

注册依赖服务

与针对配置系统的设置一样，ConfigureServices 方法中用来注册依赖服务的 Action<HostBuilderContext, IServiceCollection>委托对象同样被暂存在对应的字段_configureServicesActions 表示的集合中，它们最终会在 Build 方法中被使用。

```csharp
public class HostBuilder : IHostBuilder
{
    private List<Action<HostBuilderContext, IServiceCollection>> _configureServicesActions
        = new List<Action<HostBuilderContext, IServiceCollection>>();

    public IHostBuilder ConfigureServices(
        Action<HostBuilderContext, IServiceCollection> configureDelegate)
    {
        _configureServicesActions.Add(configureDelegate);
        return this;
    }
    ...
}
```

除了直接调用 IHostBuilder 接口的 ConfigureServices 方法进行服务注册，我们还可以调用如下这些扩展方法完成针对某些特殊服务的注册。两个 ConfigureLogging 扩展方法重载帮助我们注册针对日志框架相关的服务，两个 UseConsoleLifetime 扩展方法重载添加的是针对 ConsoleLifetime 的服务注册，两个 RunConsoleAsync 扩展方法重载则在注册 ConsoleLifetime 服务的基础上，进一步构建并启动作为宿主的 IHost 对象。

```csharp
public static class HostingHostBuilderExtensions
{
    public static IHostBuilder ConfigureLogging(this IHostBuilder hostBuilder,
```

```csharp
        Action<HostBuilderContext, ILoggingBuilder> configureLogging)
{
    return hostBuilder.ConfigureServices((context, collection) =>
        collection.AddLogging(builder => configureLogging(context, builder)));
}

public static IHostBuilder ConfigureLogging(this IHostBuilder hostBuilder,
    Action<ILoggingBuilder> configureLogging)
{
    return hostBuilder.ConfigureServices((context, collection) =>
        collection.AddLogging(builder => configureLogging(builder)));
}

public static IHostBuilder UseConsoleLifetime(this IHostBuilder hostBuilder)
{
    return hostBuilder.ConfigureServices((context, collection) =>
        collection.AddSingleton<IHostLifetime, ConsoleLifetime>());
}

public static IHostBuilder UseConsoleLifetime(this IHostBuilder hostBuilder,
    Action<ConsoleLifetimeOptions> configureOptions)
{
    return hostBuilder.ConfigureServices((context, collection) =>
    {
        collection.AddSingleton<IHostLifetime, ConsoleLifetime>();
        collection.Configure(configureOptions);
    });
}

public static Task RunConsoleAsync(this IHostBuilder hostBuilder,
    CancellationToken cancellationToken = default)
{
    return hostBuilder.UseConsoleLifetime().Build().RunAsync(cancellationToken);
}

public static Task RunConsoleAsync(this IHostBuilder hostBuilder,
    Action<ConsoleLifetimeOptions> configureOptions,
    CancellationToken cancellationToken = default)
{
    return hostBuilder.UseConsoleLifetime(configureOptions).Build()
        .RunAsync(cancellationToken);
}
}
```

注册 IServiceProviderFactory<TContainerBuilder>

作为依赖注入容器的 IServiceProvider 对象总是由注册的 IServiceProviderFactory<TContainerBuilder>工厂创建的。由于 UseServiceProviderFactory<TContainerBuilder>方法注册的 IServiceProviderFactory<TContainerBuilder>是一个泛型对象，所以 HostBuilder 会将它转换成 IService

FactoryAdapter 接口类型作为适配。如下面的代码片段所示,它仅仅是将 ContainerBuilder 转换成 Object 类型而已。ServiceFactoryAdapter<TContainerBuilder>类型是对 IServiceFactoryAdapter 接口的默认实现。

```
internal interface IServiceFactoryAdapter
{
    object CreateBuilder(IServiceCollection services);
    IServiceProvider CreateServiceProvider(object containerBuilder);
}

internal class ServiceFactoryAdapter<TContainerBuilder> : IServiceFactoryAdapter
{
    private IServiceProviderFactory<TContainerBuilder>          _serviceProviderFactory;
    private readonly Func<HostBuilderContext>                    _contextResolver;
    private Func<HostBuilderContext, IServiceProviderFactory<TContainerBuilder>>
        _factoryResolver;

    public ServiceFactoryAdapter(
        IServiceProviderFactory<TContainerBuilder> serviceProviderFactory)
        =>_serviceProviderFactory = serviceProviderFactory;

    public ServiceFactoryAdapter(Func<HostBuilderContext> contextResolver,
        Func<HostBuilderContext, IServiceProviderFactory<TContainerBuilder>>
        factoryResolver)
    {
        _contextResolver = contextResolver;
        _factoryResolver = factoryResolver;
    }

    public object CreateBuilder(IServiceCollection services)
        => _serviceProviderFactory?? _factoryResolver(_contextResolver())
            .CreateBuilder(services);

    public IServiceProvider CreateServiceProvider(object containerBuilder)
        => _serviceProviderFactory
        .CreateServiceProvider((TContainerBuilder)containerBuilder);
}
```

如下所示的代码片段是两个 UseServiceProviderFactory<TContainerBuilder>重载的定义,第一个方法重载提供的 IServiceProviderFactory<TContainerBuilder>对象和第二个方法重载提供的 Func<HostBuilderContext, IServiceProviderFactory<TContainerBuilder>>会被转换成一个 ServiceFactoryAdapter<TContainerBuilder>对象,并通过_serviceProviderFactory 字段暂时保存。如果 UseServiceProviderFactory<TContainerBuilder>方法并没有被调用,那么_serviceProviderFactory 字段返回的将是根据 DefaultServiceProviderFactory 对象创建的 ServiceFactoryAdapter<IServiceCollection>对象, .NET Core 原生依赖注入框架之所以默认被承载系统使用就体现在这里。

```
public class HostBuilder : IHostBuilder
{
```

```csharp
    private List<IConfigureContainerAdapter> _configureContainerActions =
        new List<IConfigureContainerAdapter>();
    private IServiceFactoryAdapter _serviceProviderFactory =
        new ServiceFactoryAdapter<IServiceCollection>(new DefaultServiceProviderFactory());

    public IHostBuilder UseServiceProviderFactory<TContainerBuilder>(
        IServiceProviderFactory<TContainerBuilder> factory)
    {
        _serviceProviderFactory = new ServiceFactoryAdapter<TContainerBuilder>(factory);
        return this;
    }

    public IHostBuilder UseServiceProviderFactory<TContainerBuilder>(
        Func<HostBuilderContext, IServiceProviderFactory<TContainerBuilder>> factory)
    {
        _serviceProviderFactory = new ServiceFactoryAdapter<TContainerBuilder>(
            () => _hostBuilderContext, factory );
        return this;
    }
}
```

注册 IServiceProviderFactory<TContainerBuilder>工厂提供的 TContainerBuilder 对象可以通过 ConfigureContainer<TContainerBuilder>方法做进一步设置，具体的设置由提供的 Action<HostBuilderContext, TContainerBuilder>对象来完成。这个泛型的委托对象同样需要做类似的适配才能被暂时保存，它最终转换成如下所示的 IConfigureContainerAdapter 接口类型，这个适配本质上也是将 TContainerBuilder 对象转换成 Object 类型。如下所示的 ConfigureContainerAdapter<TContainerBuilder>类型是对这个接口的默认实现。

```csharp
internal interface IConfigureContainerAdapter
{
    void ConfigureContainer(HostBuilderContext hostContext, object containerBuilder);
}

internal class ConfigureContainerAdapter<TContainerBuilder> : IConfigureContainerAdapter
{
    private Action<HostBuilderContext, TContainerBuilder> _action;

    public ConfigureContainerAdapter(Action<HostBuilderContext, TContainerBuilder> action)
        => _action = action;
    public void ConfigureContainer(HostBuilderContext hostContext, object containerBuilder)
        => _action(hostContext, (TContainerBuilder)containerBuilder);
}
```

如下所示的代码片段是 ConfigureContainer<TContainerBuilder>方法的定义，该方法会将提供的 Action<HostBuilderContext, TContainerBuilder>对象转换成 ConfigureContainerAdapter<TContainerBuilder>对象，并添加到_configureContainerActions 字段表示的集合中。

```csharp
public class HostBuilder : IHostBuilder
{
    private List<IConfigureContainerAdapter> _configureContainerActions =
```

```
        new List<IConfigureContainerAdapter>();

    public IHostBuilder ConfigureContainer<TContainerBuilder>(
        Action<HostBuilderContext, TContainerBuilder> configureDelegate)
    {
        _configureContainerActions.Add(
            new ConfigureContainerAdapter<TContainerBuilder>(configureDelegate));
        return this;
    }
    ...
}
```

第 3 章创建了一个名为 Cat 的简易版依赖注入框架,第 4 章为其创建了一个 IServiceProviderFactory<TContainerBuilder>实现类型,具体类型为 CatServiceProvider,下面演示如何通过注册 CatServiceProvider 实现与 Cat 这个第三方依赖注入框架的整合。如果使用 Cat 框架,我们可以在服务类型上标注 MapToAttribute 特性来定义服务注册信息。在创建的演示程序中,我们采用这样的方式定义了 3 个服务(Foo、Bar 和 Baz)和对应的接口(IFoo、IBar 和 IBaz)。

```
public interface IFoo { }
public interface IBar { }
public interface IBaz { }

[MapTo(typeof(IFoo), Lifetime.Root)]
public class Foo : IFoo { }

[MapTo(typeof(IBar), Lifetime.Root)]
public class Bar : IBar { }

[MapTo(typeof(IBaz), Lifetime.Root)]
public class Baz : IBaz { }
```

如下所示的 FakeHostedService 类型表示应用程序承载的服务。我们在构造函数中注入了上面定义的 3 个服务,构造函数提供的调试断言用于验证上述 3 个服务被成功注入。

```
public sealed class FakeHostedService: IHostedService
{
    public FakeHostedService(IFoo foo, IBar bar, IBaz baz)
    {
        Debug.Assert(foo != null);
        Debug.Assert(bar != null);
        Debug.Assert(baz != null);
    }
    public Task StartAsync(CancellationToken cancellationToken) => Task.CompletedTask;
    public Task StopAsync(CancellationToken cancellationToken) => Task.CompletedTask;
}
```

在如下所示的服务承载程序中,我们创建了一个 HostBuilder 对象,并通过调用 ConfigureServices 方法注册了需要承载的 FakeHostedService 服务。然后调用 UseServiceProviderFactory 方法完成了对 CatServiceProvider 的注册,并在随后调用 CatBuilder 的 Register 方法完成了针对入口程序集的批量服务注册。调用 HostBuilder 的 Build 方法构建出作为宿主的 Host 对象并启动它

之后，承载的 FakeHostedService 服务将自动被创建并启动。（S1008）

```
class Program
{
    static void Main()
    {
        new HostBuilder()
            .ConfigureServices(svcs => svcs.AddHostedService<FakeHostedService>())
            .UseServiceProviderFactory(new CatServiceProviderFactory())
            .ConfigureContainer<CatBuilder>(builder=>builder.Register(
                Assembly.GetEntryAssembly()))
            .Build()
            .Run();
    }
}
```

10.3.4　创建宿主

实际上，HostBuilder 对象并没有在实现的 Build 方法中调用构造函数来创建 Host 对象，该对象是利用作为依赖注入容器的 IServiceProvider 对象创建的。为了可以采用依赖注入框架来提供构建的 Host 对象，HostBuilder 对象必须完成前期的服务注册工作。总的来说，HostBuilder 对象针对 Host 对象的构建大体可以划分为如下 4 个步骤。

- **创建 HostBuilderContext 上下文**：创建针对宿主配置的 IConfiguration 对象和表示承载环境的 IHostEnvironment 对象，然后利用二者创建出代表承载上下文的 HostBuilderContext 对象。
- **创建针对应用的配置**：创建针对应用配置的 IConfiguration 对象，并用它替换 HostBuilderContext 对象承载的配置。
- **注册依赖服务**：注册所需的依赖服务，包括应用程序通过调用 ConfigureServices 方法提供的服务注册和其他一些确保服务承载正常执行的默认服务注册。
- **创建 IServiceProvider 对象，并利用它提供 Host 对象**：利用注册的 IServiceProviderFactory<TContainerBuilder>工厂（系统默认注册或者应用程序显式注册）创建出用来提供所有依赖服务的 IServiceProvider 对象。最后利用 IServiceProvider 对象提供作为宿主的 Host 对象。

步骤一，创建 HostBuilderContext 上下文

由于很多依赖服务都是针对当前承载上下文进行注册的，所以 Build 方法首要的任务就是创建作为承载上下文的 HostBuilderContext 对象。一个 HostBuilderContext 对象由承载针对宿主配置的 IConfiguration 对象和描述当前承载环境的 IHostEnvironment 对象组成，后者提供的环境名称、应用名称和内容文件根目录路径可以通过前者来指定，具体的配置项名称定义在如下这个静态类型 HostDefaults 中。

```
public static class HostDefaults
{
```

```csharp
    public static readonly string EnvironmentKey = "environment";
    public static readonly string ContentRootKey = "contentRoot";
    public static readonly string ApplicationKey = "applicationName";
}
```

下面通过一个简单的实例演示如何利用配置的方式来指定上述 3 个与承载环境相关的属性。我们定义了如下一个名为 FakeHostedService 的承载服务，并在构造函数中注入 IHostEnvironment 对象。在 StartAsync 方法中，我们将与承载环境相关的环境名称、应用名称和内容文件根目录路径输出到控制台上。

```csharp
public class FakeHostedService : IHostedService
{
    private readonly IHostEnvironment _environment;
    public FakeHostedService(IHostEnvironment environment)
        => _environment = environment;

    public Task StartAsync(CancellationToken cancellationToken)
    {
        Console.WriteLine("{0,-15}:{1}", nameof(_environment.EnvironmentName),
            _environment.EnvironmentName);
        Console.WriteLine("{0,-15}:{1}", nameof(_environment.ApplicationName),
            _environment.ApplicationName);
        Console.WriteLine("{0,-15}:{1}", nameof(_environment.ContentRootPath),
            _environment.ContentRootPath);
        return Task.CompletedTask;
    }

    public Task StopAsync(CancellationToken cancellationToken) => Task.CompletedTask;
}
```

FakeHostedService 采用如下形式承载于当前应用程序中。如下面的代码片段所示，在创建了作为宿主构建者的 HostBuilder 之后，我们调用它的 ConfigureHostConfiguration 方法注册基于命令行的参数作为配置源，这意味着可以利用命令行参数来初始化相应的配置。

```csharp
class Program
{
    static void Main(string[] args)
    {
        new HostBuilder()
            .ConfigureHostConfiguration(builder => builder.AddCommandLine(args))
            .ConfigureServices(svcs => svcs.AddHostedService<FakeHostedService>())
            .Build()
            .Run();
    }
}
```

我们采用命令行的方式启动这个演示程序，并利用传入的命令行参数指定环境名称、应用名称和内容文件根目录路径（确保路径确实存在）。图 10-11 所示的输出结果表明，应用程序当前的承载环境与基于宿主的配置是一致的。（S1009）

图 10-11　利用配置来初始化承载环境

HostBuilder 针对 HostBuilderContext 上下文的创建体现在如下所示的 CreateBuilderContext 方法中。如下面的代码片段所示，该方法创建了一个 ConfigurationBuilder 对象，并调用 AddInMemoryCollection 扩展方法注册了针对内存变量的配置源。接下来 HostBuilder 会将这个 ConfigurationBuilder 对象作为参数调用 ConfigureHostConfiguration 方法注册的所有 Action<IConfigurationBuilder>委托。ConfigurationBuilder 对象生成的 IConfiguration 对象将作为 HostBuilderContext 上下文对象的配置。

```
public class HostBuilder: IHostBuilder
{
    private List<Action<IConfigurationBuilder>> _configureHostConfigActions ;

    public IHost Build()
    {
        var buildContext = CreateBuilderContext();
        ...
    }

    private HostBuilderContext CreateBuilderContext()
    {
        //Create Configuration
        var configBuilder = new ConfigurationBuilder().AddInMemoryCollection();
        foreach (var buildAction in _configureHostConfigActions)
        {
            buildAction(configBuilder);
        }
        var hostConfig = configBuilder.Build();

        //Create HostingEnvironment
        var contentRoot =  hostConfig [HostDefaults.ContentRootKey];
        var contentRootPath = string.IsNullOrEmpty(contentRoot)
            ? AppContext.BaseDirectory
            : Path.IsPathRooted(contentRoot)
            ? contentRoot
            : Path.Combine(Path.GetFullPath(AppContext.BaseDirectory), contentRoot);
        var hostingEnvironment = new HostingEnvironment()
        {
            ApplicationName = hostConfig [HostDefaults.ApplicationKey],
            EnvironmentName = hostConfig [HostDefaults.EnvironmentKey]
                ?? Environments.Production,
```

```
            ContentRootPath    = contentRootPath,
        };
        if (string.IsNullOrEmpty(hostingEnvironment.ApplicationName))
        {
            hostingEnvironment.ApplicationName =
                Assembly.GetEntryAssembly()?.GetName().Name;
        }
        hostingEnvironment.ContentRootFileProvider =
            new PhysicalFileProvider(hostingEnvironment.ContentRootPath);

        //Create HostBuilderContext
        return new HostBuilderContext(Properties)
        {
            HostingEnvironment      = hostingEnvironment,
            Configuration           = hostConfig
        };
    }
    ...
}
```

在创建出 HostBuilderContext 上下文的配置对象之后，HostBuilder 会根据该配置创建出代表承载环境的 HostingEnvironment 对象。如果不存在针对应用名称的配置项，应用名称就会设置为当前入口程序集的名称。如果内容文件根目录路径对应的配置项不存在，当前应用的基础路径（AppContext.BaseDirectory）就会作为内容文件根目录路径。如果指定的是一个相对路径，HostBuilder 就会根据基础路径生成一个绝对路径作为内容文件根目录路径。CreateBuilderContext 方法最终会根据创建的 HostingEnvironment 对象和之前创建的 IConfiguration 创建出代表承载上下文的 BuilderContext 对象。

步骤二，构建针对应用的配置

截至目前，作为承载上下文的 BuilderContext 对象携带的是通过调用 ConfigureHostConfiguration 方法初始化的配置。接下来，通过调用 ConfigureAppConfiguration 方法初始化的配置将与之合并，具体的逻辑体现在如下所示的 BuildAppConfiguration 方法上。

如下面的代码片段所示，BuildAppConfiguration 方法会创建一个 ConfigurationBuilder 对象，并调用其 AddConfiguration 方法合并现有的配置。与此同时，内容文件根目录的路径会作为配置文件所在目录的基础路径。HostBuilder 最后会将之前创建的 HostBuilderContext 对象和这个 ConfigurationBuilder 对象作为参数调用在 ConfigureAppConfiguration 方法注册的每一个 Action<HostBuilderContext, IConfigurationBuilder>委托。这个 ConfigurationBuilder 创建的 IConfiguration 对象会重新赋值给 HostBuilderContext 对象的 Configuration 属性，我们自此就可以从承载上下文中得到完整的配置。

```
public class HostBuilder: IHostBuilder
{
    private List<Action<HostBuilderContext, IConfigurationBuilder>>
        _configureAppConfigActions;
```

```csharp
public IHost Build()
{
    var buildContext = CreateBuilderContext();
    buildContext.Configuration = BuildAppConfiguration(buildContext);
    ...
}

private IConfiguration BuildAppConfiguration(HostBuilderContext buildContext)
{
    var configBuilder = new ConfigurationBuilder()
        .SetBasePath(buildContext.HostingEnvironment.ContentRootPath)
        .AddConfiguration(buildContext.Configuration,true);
    foreach (var action in _configureAppConfigActions)
    {
        action(_hostBuilderContext, configBuilder);
    }
    return configBuilder.Build();
}
```

步骤三，注册依赖服务

当作为承载上下文的 HostBuilderContext 对象创建出来并且被初始化后，HostBuilder 需要完成服务注册，这一实现体现在如下所示的 ConfigureAllServices 方法中。如下面的代码片段所示，ConfigureAllServices 方法在将代表承载上下文的 HostBuilderContext 对象和创建的 Service Collection 对象作为参数调用 ConfigureServices 方法中注册的每一个 Action<HostBuilderContext, IServiceCollection>委托对象之前，它会注册一些额外的系统服务。ConfigureAllServices 方法最终返回包含所有服务注册的 IServiceCollection 对象。

```csharp
public class HostBuilder: IHostBuilder
{
    private List<Action<HostBuilderContext, IServiceCollection>> _configureServicesActions;

    public IHost Build()
    {
        var buildContext = CreateBuilderContext();
        buildContext.Configuration = BuildAppConfiguration(buildContext);
        var services = ConfigureAllServices (buildContext);
        ...
    }

    private IServiceCollection ConfigureAllServices(HostBuilderContext buildContext)
    {
        var services = new ServiceCollection();
        services.AddSingleton(buildContext);
        services.AddSingleton(buildContext.HostingEnvironment);
        services.AddSingleton(_ => buildContext.Configuration);
        services.AddSingleton<IHostApplicationLifetime, ApplicationLifetime>();
        services.AddSingleton<IHostLifetime, ConsoleLifetime>();
```

```
    services.AddSingleton<IHost,Host>();
    services.AddOptions();
    services.AddLogging();

    foreach (var configureServicesAction in _configureServicesActions)
    {
        configureServicesAction(_hostBuilderContext, services);
    }
    return services;
}
```

对于 ConfigureAllServices 方法默认注册的这些服务，如果自定义的承载服务需要使用它们，就可以直接采用构造器注入的方式对它们进行消费。由于其中包含了针对 IHost/Host 的服务注册，所以由所有服务注册构建的 IServiceProvider 对象可以提供最终作为服务宿主的 Host 对象。

步骤四，创建 IServiceProvider 对象，并利用它提供 Host 对象

目前我们已经拥有了所有的服务注册，接下来的任务就是利用它们创建出作为依赖注入容器的 IServiceProvider 对象并由它提供构建的 Host 对象。针对 IServiceProvider 对象的创建体现在如下所示的 CreateServiceProvider 方法中。如下面的代码片段所示，CreateServiceProvider 方法会先得到 _serviceProviderFactory 字段表示的 IServiceFactoryAdapter 对象，该对象是根据 UseServiceProviderFactory<TContainerBuilder>方法注册的 IServiceProviderFactory<TContainerBuilder>对象创建的，调用它的 CreateBuilder 方法可以得到由注册的 IServiceProviderFactory<TContainerBuilder>对象创建的 TContainerBuilder 对象。

```
public class HostBuilder: IHostBuilder
{
    private List<IConfigureContainerAdapter> _configureContainerActions;
    private IServiceFactoryAdapter _serviceProviderFactory

    public IHost Build()
    {
        var buildContext             = CreateBuilderContext();
        buildContext.Configuration   = BuildAppConfiguration(buildContext);
        var services                 = ConfigureServices(buildContext);
        var serviceProvider          = CreateServiceProvider(buildContext, services);
         return serviceProvider.GetRequiredService<IHost>();
    }

    private IServiceProvider CreateServiceProvider(
        HostBuilderContext builderContext,IServiceCollection services)
    {
        var containerBuilder = _serviceProviderFactory.CreateBuilder(services);
        foreach (var containerAction in _configureContainerActions)
        {
            containerAction.ConfigureContainer(builderContext, containerBuilder);
        }
        return _serviceProviderFactory.CreateServiceProvider(containerBuilder);
```

```
    }
}
```

然后将这个 TContainerBuilder 对象作为参数调用 _configureContainerActions 集合中每个 IConfigureContainerAdapter 对象的 ConfigureContainer 方法，这里的每个 IConfigureContainerAdapter 对象都是根据 ConfigureContainer<TContainerBuilder>方法提供的 Action<HostBuilderContext, TContainerBuilder>对象创建的。在完成了用户针对 TContainerBuilder 对象的设置之后，CreateServiceProvider 方法会将该对象作为参数调用 IServiceFactoryAdapter 的 CreateServiceProvider 方法创建出代表依赖注入容器的 IServiceProvider 对象，Build 方法正是利用它来提供构建的 Host 对象的。

10.3.5　静态类型 Host

截至目前，我们演示的实例都是直接创建 HostBuilder 对象来创建作为服务宿主的 IHost 对象的。如果直接利用模板来创建一个 ASP.NET Core 应用，就会发现生成的程序会采用如下所示的服务承载方式。具体来说，用来创建宿主的 IHostBuilder 对象是间接地调用静态类型 Host 的 CreateDefaultBuilder方法创建的，那么这个方法究竟会提供一个什么样的 IHostBuilder 对象？

```
public class Program
{
    public static void Main(string[] args)
    {
        CreateHostBuilder(args).Build().Run();
    }

    public static IHostBuilder CreateHostBuilder(string[] args) =>
        Host.CreateDefaultBuilder(args)
            .ConfigureWebHostDefaults(webBuilder =>
            {
                webBuilder.UseStartup<Startup>();
            });
}
```

如下所示的代码片段是定义在静态类型 Host 中的两个 CreateDefaultBuilder 方法重载的定义，可以发现，它们最终提供的仍旧是一个 HostBuilder 对象，但是在返回该对象之前，该方法会帮助我们做一些初始化工作。如下面的代码片段所示，当 CreateDefaultBuilder 方法创建出 HostBuilder 对象之后，它会自动将当前目录所在的路径作为内容文件根目录的路径。接下来，该方法还会调用 HostBuilder 对象的 ConfigureHostConfiguration 方法注册针对环境变量的配置源，对应环境变量名称的前缀被设置为"DOTNET_"。如果提供了代表命令行参数的字符串数组，CreateDefaultBuilder方法还会注册针对命令行参数的配置源。

```
public static class Host
{
    public static IHostBuilder CreateDefaultBuilder()
        => CreateDefaultBuilder(args: null);

    public static IHostBuilder CreateDefaultBuilder(string[] args)
```

```csharp
{
    var builder = new HostBuilder();

    builder.UseContentRoot(Directory.GetCurrentDirectory());
    builder.ConfigureHostConfiguration(config =>
    {
        config.AddEnvironmentVariables(prefix: "DOTNET_");
        if (args != null)
        {
            config.AddCommandLine(args);
        }
    });

    builder.ConfigureAppConfiguration((hostingContext, config) =>
    {
        var env = hostingContext.HostingEnvironment;

        config
            .AddJsonFile("appsettings.json", optional: true, reloadOnChange: true)
            .AddJsonFile($"appsettings.{env.EnvironmentName}.json",
                optional: true, reloadOnChange: true);

        if (env.IsDevelopment() && !string.IsNullOrEmpty(env.ApplicationName))
        {
            var appAssembly = Assembly.Load(new AssemblyName(env.ApplicationName));
            if (appAssembly != null)
            {
                config.AddUserSecrets(appAssembly, optional: true);
            }
        }

        config.AddEnvironmentVariables();

        if (args != null)
        {
            config.AddCommandLine(args);
        }
    })
    .ConfigureLogging((hostingContext, logging) =>
    {
        logging.AddConfiguration(hostingContext.Configuration.GetSection("Logging"));
        logging.AddConsole();
        logging.AddDebug();
        logging.AddEventSourceLogger();
    })
    .UseDefaultServiceProvider((context, options) =>
    {
        options.ValidateScopes = context.HostingEnvironment.IsDevelopment();
    });
```

```
        return builder;
    }
}
```

在设置了针对宿主的配置之后，CreateDefaultBuilder 调用 HostBuilder 的 ConfigureAppConfiguration 方法设置针对应用的配置，具体的配置源包括 JSON 文件（appsettings.json 和 appsettings.{environment}.json）、环境变量（没有前缀限制）和命令行参数（如果提供了表示命令行参数的字符串数组）。

在完成了针对配置的设置之后，CreateDefaultBuilder 方法还会调用 HostBuilder 的 ConfigureLogging 扩展方法做一些与日志相关的设置，其中包括与应用日志相关的配置（对应配置节名称为 Logging）以及注册针对控制台、调试器和 EventSource 的日志输出渠道。在此之后，CreateDefaultBuilder 方法还会调用 UseDefaultServiceProvider 方法使针对服务范围的验证在开发环境下被自动开启。

第 11 章

管道（上篇）

ASP.NET Core 是一个 Web 开发平台，而不是一个单纯的开发框架。这是因为 ASP.NET Core 有一个极具扩展能力的请求处理管道，我们可以通过对这个管道的定制来满足各种场景下的 HTTP 处理需求。ASP. NET Core 应用的很多特性（如路由、会话、缓存、认证、授权等）都是通过对管道的定制来实现的，我们可以通过管道定制在 ASP.NET Core 平台上创建自己的 Web 框架。由于这部分内容是本书的核心，所以分为 3 章（第 11 章至第 13 章）对请求处理管道进行全方面讲解。

11.1 管道式的请求处理

HTTP 协议自身的特性决定了任何一个 Web 应用的工作模式都是监听、接收并处理 HTTP 请求，并且最终对请求予以响应。HTTP 请求处理是管道式设计典型的应用场景：可以根据具体的需求构建一个管道，接收的 HTTP 请求像水一样流入这个管道，组成这个管道的各个环节依次对其做相应的处理。虽然 ASP.NET Core 的请求处理管道从设计上来讲是非常简单的，但是具体的实现则涉及很多细节，为了使读者对此有深刻的理解，需要从编程的角度先了解 ASP.NET Core 管道式的请求处理方式。

11.1.1 两个承载体系

ASP.NET Core 框架目前存在两个承载（Hosting）系统。ASP.NET Core 最初提供了一个以 IWebHostBuilder/IWebHost 为核心的承载系统，其目的很单纯，就是通过图 11-1 所示的形式承载以服务器和中间件管道构建的 Web 应用。ASP.NET Core 3 依然支持这样的应用承载方式，但是本书不会涉及这种"过时"的承载方式。

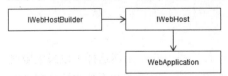

图 11-1　基于 IWebHostBuilder/IWebHost 应用的承载方式

除了承载 Web 应用本身，我们还有针对后台服务的承载需求，为此微软推出了以 IHostBuilder/IHost 为核心的承载系统，我们在第 10 章中已经对该系统做了详细的介绍。实际上，Web 应用本身就是一个长时间运行的后台服务，我们完全可以定义一个承载服务，从而将 Web 应用承载于这个系统中。如图 11-2 所示，这个用来承载 ASP.NET Core 应用的承载服务类型为 GenericWebHostService，这是一个实现了 IHostedService 接口的内部类型。

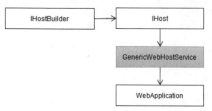

图 11-2 基于 IHostBuilder/IHost 应用的承载方式

IHostBuilder 接口上定义了很多方法（其中很多是扩展方法），这些方法的目的主要包括以下两点：第一，为创建的 IHost 对象及承载的服务在依赖注入框架中注册相应的服务；第二，为服务承载和应用提供相应的配置。其实 IWebHostBuilder 接口同样定义了一系列方法，除了这里涉及的两点，支撑 ASP.NET Core 应用的中间件也是由 IWebHostBuilder 注册的。

即使采用基于 IHostBuilder/IHost 的承载系统，我们依然会使用 IWebHostBuilder 接口。虽然我们不再使用 IWebHostBuilder 的宿主构建功能，但是定义在 IWebHostBuilder 上的其他 API 都是可以使用的。具体来说，可以调用定义在 IHostBuilder 接口和 IWebHostBuilder 接口的方法（大部分为扩展方法）来注册依赖服务与初始化配置系统，两者最终会合并在一起。利用 IWebHostBuilder 接口注册的中间件会提供给 GenericWebHostService，用于构建 ASP.NET Core 请求处理管道。

在基于 IHostBuilder/IHost 的承载系统中复用 IWebHostBuilder 的目的是通过如下所示的 ConfigureWebHost 扩展方法达成的，GenericWebHostService 服务也是在这个方法中被注册的。ConfigureWebHostDefaults 扩展方法则会在此基础上做一些默认设置（如 KestrelServer），后续章节的实例演示基本上会使用这个方法。

```
public static class GenericHostWebHostBuilderExtensions
{
    public static IHostBuilder ConfigureWebHost(this IHostBuilder builder,
        Action<IWebHostBuilder> configure);
}

public static class GenericHostBuilderExtensions
{
    public static IHostBuilder ConfigureWebHostDefaults(this IHostBuilder builder,
        Action<IWebHostBuilder> configure);
}
```

对 IWebHostBuilder 接口的复用导致很多功能都具有两种编程方式，虽然这样可以最大限度地复用和兼容定义在 IWebHostBuilder 接口上众多的应用编程接口，但笔者并不喜欢这样略显混乱的编程模式，这一点在下一个版本中也许会得到改变。

11.1.2 请求处理管道

下面创建一个最简单的 Hello World 程序。这是一个控制台应用，之前演示的大部分实例的 SDK 都采用 "Microsoft.NET.Sdk"，作为一个 ASP.NET Core Web 应用，对应的项目的 SDK 一般采用 "Microsoft.NET.Sdk.Web"。由于这种 SDK 会自动将常用的依赖或者引用添加进来，所以不需要在项目文件中显式添加针对 "Microsoft.AspNetCore.App" 的框架引用。

```xml
<Project Sdk="Microsoft.NET.Sdk.Web">
  <PropertyGroup>
    <TargetFramework>netcoreapp3.0</TargetFramework>
  </PropertyGroup>
</Project>
```

这个程序由如下所示的几行代码组成。运行这个程序之后，一个名为 KestrelServer 的服务器将会启动并绑定到本机上的 5000 端口进行请求监听。针对所有接收到的请求，我们都会采用 "Hello World" 字符串作为响应的主体内容。

```csharp
class Program
{
    static void Main()
    {
        Host.CreateDefaultBuilder()
            .ConfigureWebHost(builder => builder
                .Configure(app => app.Run(context =>
                    context.Response.WriteAsync("Hello World"))))
            .Build()
            .Run();
    }
}
```

从如上所示的代码片段可以看出，我们利用第 10 章介绍的承载系统来承载一个 ASP.NET Core 应用。在调用 Host 类型的静态方法 CreateDefaultBuilder 创建了一个 IHostBuilder 对象之后，我们调用它的 ConfigureWebHost 方法对 ASP.NET Core 应用的请求处理管道进行定制。HTTP 请求处理流程始于对请求的监听与接收，终于对请求的响应，这两项工作均由同一个对象来完成，我们称之为服务器（Server）。ASP.NET Core 请求处理管道必须有一个服务器，它是整个管道的"龙头"。在演示程序中，我们调用 IWebHostBuilder 接口的 UseKestrel 扩展方法为后续构建的管道注册了一个名为 KestrelServer 的服务器。

当承载服务 GenericWebHostService 被启动之后，定制的请求处理管道会被构建出来，管道的服务器随后会绑定到一个预设的端口（如 KestrelServer 默认采用 5000 作为监听端口）开始监听请求。HTTP 请求一旦抵达，服务器会将其标准化，并分发给管道后续的节点，我们将位于服务器之后的节点称为中间件（Middleware）。

每个中间件都具有各自独立的功能，如专门实现路由功能的中间件、专门实施用户认证和授权的中间件。所谓的管道定制主要体现在根据具体需求选择对应的中间件来构建最终的管道。在演示程序中，我们调用 IWebHostBuilder 接口的 Configure 方法注册了一个中间件，用于响应 "Hello World" 字符串。具体来说，这个用来注册中间件的 Configure 方法具有一个类型为

Action<IApplicationBuilder>的参数，我们提供的中间件就注册到提供的 IApplicationBuilder 对象上。由服务器和中间件组成的请求处理管道如图 11-3 所示。

图 11-3　由服务器和中间件组成的请求处理管道

建立在 ASP.NET Core 之上的应用基本上是根据某个框架开发的。一般来说，开发框架本身就是通过某一个或者多个中间件构建起来的。以 ASP.NET Core MVC 开发框架为例，它借助"路由"中间件实现了请求与 Action 之间的映射，并在此基础之上实现了激活（Controller）、执行（Action）及呈现（View）等一系列功能。应用程序可以视为某个中间件的一部分，如果一定要将它独立出来，由服务器、中间件和应用组成的管道如图 11-4 所示。

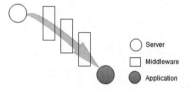

图 11-4　由服务器、中间件和应用组成的管道

11.1.3　中间件

ASP.NET Core 的请求处理管道由一个服务器和一组中间件组成，位于"龙头"的服务器负责请求的监听、接收、分发和最终的响应，而针对该请求的处理则由后续的中间件来完成。如果读者希望对请求处理管道具有深刻的认识，就需要对中间件有一定程度的了解。

RequestDelegate

从概念上可以将请求处理管道理解为"请求消息"和"响应消息"流通的管道，服务器将接收的请求消息从一端流入管道并由相应的中间件进行处理，生成的响应消息反向流入管道，经过相应中间件处理后由服务器分发给请求者。但从实现的角度来讲，管道中流通的并不是所谓的请求消息与响应消息，而是一个针对当前请求创建的上下文。这个上下文被抽象成如下这个 HttpContext 类型，我们利用 HttpContext 不仅可以获取针对当前请求的所有信息，还可以直接完成针对当前请求的所有响应工作。

```
public abstract class HttpContext
{
    public abstract HttpRequest        Request { get; set; }
    public abstract HttpResponse       Response { get; }
    ...
}
```

既然流入管道的只有一个共享的 HttpContext 上下文，那么一个 Func<HttpContext,Task>对象就可以表示处理 HttpContext 的操作，或者用于处理 HTTP 请求的处理器。由于这个委托对象非常重要，所以 ASP.NET Core 专门定义了如下这个名为 RequestDelegate 的委托类型。既然有这样一个专门的委托对象来表示"针对请求的处理"，那么中间件是否能够通过该委托对象来表示？

```
public delegate Task RequestDelegate(HttpContext context);
```

Func<RequestDelegate, RequestDelegate>

实际上，组成请求处理管道的中间件可以表示为一个类型为 Func<RequestDelegate, RequestDelegate>的委托对象，但初学者很难理解这一点，所以下面对此进行简单的解释。由于 RequestDelegate 可以表示一个 HTTP 请求处理器，所以由一个或者多个中间件组成的管道最终也体现为一个 RequestDelegate 对象。对于图 11-5 所示的中间件 Foo 来说，后续中间件（Bar 和 Baz）组成的管道体现为一个 RequestDelegate 对象，该对象会作为中间件 Foo 输入，中间件 Foo 借助这个委托对象将当前 HttpContext 分发给后续管道做进一步处理。

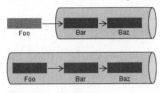

图 11-5　中间件

表示中间件的 Func<RequestDelegate, RequestDelegate>对象的输出依然是一个 RequestDelegate 对象，该对象表示将当前中间件与后续管道进行"对接"之后构成的新管道。对于表示中间件 Foo 的委托对象来说，返回的 RequestDelegate 对象体现的就是由 Foo、Bar 和 Baz 组成的请求处理管道。

既然原始的中间件是通过一个 Func<RequestDelegate, RequestDelegate>对象表示的，就可以直接注册这样一个对象作为中间件。中间件的注册可以通过调用 IWebHostBuilder 接口的 Configure 扩展方法来完成，该方法的参数是一个 Action<IApplicationBuilder>类型的委托对象，可以通过调用 IApplicationBuilder 接口的 Use 方法将表示中间件的 Func<RequestDelegate, RequestDelegate>对象添加到当前中间件链条上。

```
public static class WebHostBuilderExtensions
{
    public static IWebHostBuilder Configure(this IWebHostBuilder hostBuilder,
        Action<IApplicationBuilder> configureApp);
}

public interface IApplicationBuilder
{
    IApplicationBuilder Use(Func<RequestDelegate, RequestDelegate> middleware);
}
```

在如下所示的代码片段中，我们创建了两个 Func<RequestDelegate, RequestDelegate>对象，它们会在响应中写入两个字符串（"Hello"和"World!"）。在针对 IWebHostBuilder 接口的

Configure 方法的调用中，可以调用 IApplicationBuilder 接口的 Use 方法将这两个委托对象注册为中间件。

```
class Program
{
    static void Main()
    {
        static RequestDelegate Middleware1(RequestDelegate next)
            => async context =>
            {
                await context.Response.WriteAsync("Hello");
                await next(context);
            };
        static RequestDelegate Middleware2(RequestDelegate next)
            => async context =>
            {
                await context.Response.WriteAsync(" World!");
            };

        Host.CreateDefaultBuilder().ConfigureWebHostDefaults(builder => builder
            .Configure(app => app
                .Use(Middleware1)
                .Use(Middleware2)))
            .Build()
            .Run();
    }
}
```

由于我们注册了如上所示的两个中间件，所以它们会按照注册的顺序对分发给它们的请求进行处理。运行该程序后，如果利用浏览器对监听地址（"http://localhost:5000"）发送请求，那么两个中间件写入的字符串会以图 11-6 所示的形式呈现出来。（S1101）

图 11-6　利用注册的中间件处理请求

虽然可以直接采用原始的 Func<RequestDelegate, RequestDelegate>对象来定义中间件，但是在大部分情况下，我们依然倾向于将自定义的中间件定义成一个具体的类型。至于中间件类型的定义，ASP.NET Core 提供了如下两种不同的形式可供选择。

- **强类型定义**：自定义的中间件类型显式实现预定义的 IMiddleware 接口，并在实现的方法中完成针对请求的处理。
- **基于约定的定义**：不需要实现任何接口或者继承某个基类，只需要按照预定义的约定来定义中间件类型。

Run 方法的本质

在演示的 Hello World 应用中,我们调用 IApplicationBuilder 接口的 Run 扩展方法注册了一个 RequestDelegate 对象来处理请求,实际上,该方法仅仅是按照如下方式注册了一个中间件。由于注册的中间件并不会将请求分发给后续的中间件,如果调用 IApplicationBuilder 接口的 Run 方法后又注册了其他的中间件,后续中间件的注册将毫无意义。

```
public static class RunExtensions
{
    public static void Run(this IApplicationBuilder app, RequestDelegate handler)
        => app.Use(_ => handler);
}
```

11.1.4 定义强类型中间件

如果采用强类型的中间件类型定义方式,只需要实现如下这个 IMiddleware 接口,该接口定义了唯一的 InvokeAsync 方法,用于实现中间件针对请求的处理。这个 InvokeAsync 方法定义了两个参数:第一个参数是代表当前请求上下文的 HttpContext 对象,第二个参数是代表后续中间件组成的管道的 RequestDelegate 对象,如果当前中间件最终需要将请求分发给后续中间件进行处理,只需要调用这个委托对象即可,否则应用针对请求的处理就到此为止。

```
public interface IMiddleware
{
    Task InvokeAsync(HttpContext context, RequestDelegate next);
}
```

在如下所示的代码片段中,我们定义了一个实现了 IMiddleware 接口的 StringContent Middleware 中间件类型,在实现的 InvokeAsync 方法中,它将构造函数中指定的字符串作为响应的内容。由于中间件最终是采用依赖注入的方式来提供的,所以需要预先对它们进行服务注册,针对 StringContentMiddleware 的服务注册是通过调用 IHostBuilder 接口的 ConfigureServices 方法完成的。

```
class Program
{
    static void Main()
    {
        Host.CreateDefaultBuilder()
            .ConfigureServices(svcs => svcs.AddSingleton(
                new StringContentMiddleware("Hello World!")))
            .ConfigureWebHost(builder => builder
                .Configure(app => app.UseMiddleware<StringContentMiddleware>()))
            .Build()
            .Run();
    }

    private sealed class StringContentMiddleware : IMiddleware
    {
        private readonly string _contents;
        public StringContentMiddleware(string contents)
```

```
        => _contents = contents;
    public Task InvokeAsync(HttpContext context, RequestDelegate next)
        => context.Response.WriteAsync(_contents);
    }
}
```

针对中间件自身的注册则体现在针对 IWebHostBuilder 接口的 Configure 方法的调用上，最终通过调用 IApplicationBuilder 接口的 UseMiddleware<TMiddleware>方法来注册中间件类型。如下面的代码片段所示，在注册中间件类型时，可以以泛型参数的形式来指定中间件类型，也可以调用另一个非泛型的方法重载，直接通过 Type 类型的参数来指定中间件类型。值得注意的是，这两个方法均提供了一个参数 params，它是为针对"基于约定的中间件"注册设计的，当我们注册一个实现了 IMiddleware 接口的强类型中间件的时候是不能指定该参数的。启动该程序后利用浏览器访问监听地址，依然可以得到图 11-6 所示的输出结果。（S1102）

```
public static class UseMiddlewareExtensions
{
    public static IApplicationBuilder UseMiddleware<TMiddleware>(
        this IApplicationBuilder app, params object[] args);
    public static IApplicationBuilder UseMiddleware(this IApplicationBuilder app,
        Type middleware, params object[] args);
}
```

11.1.5 按照约定定义中间件

可能我们已经习惯了通过实现某个接口或者继承某个抽象类的扩展方式，但是这种方式有时显得约束过重，不够灵活，所以可以采用另一种基于约定的中间件类型定义方式。这种定义方式比较自由，因为它并不需要实现某个预定义的接口或者继承某个基类，而只需要遵循一些约定即可。自定义中间件类型的约定主要体现在如下几个方面。

- 中间件类型需要有一个有效的公共实例构造函数，该构造函数要求必须包含一个 RequestDelegate 类型的参数，当前中间件利用这个委托对象实现针对后续中间件的请求分发。构造函数不仅可以包含任意其他参数，对于 RequestDelegate 参数出现的位置也不做任何约束。
- 针对请求的处理实现在返回类型为 Task 的 InvokeAsync 方法或者 Invoke 方法中，它们的第一个参数表示当前请求上下文的 HttpContext 对象。对于后续的参数，虽然约定并未对此做限制，但是由于这些参数最终由依赖注入框架提供，所以相应的服务注册必须存在。

采用这种方式定义的中间件类型同样是调用前面介绍的 UseMiddleware 方法和 UseMiddleware<TMiddleware>方法进行注册的。由于这两个方法会利用依赖注入框架来提供指定类型的中间件对象，所以它会利用注册的服务来提供传入构造函数的参数。如果构造函数的参数没有对应的服务注册，就必须在调用这个方法的时候显式指定。

在如下所示的代码片段中，我们定义了一个名为 StringContentMiddleware 的中间件类型，在执行这个中间件时，它会将预先指定的字符串作为响应内容。StringContentMiddleware 的构造函数具有两个额外的参数：contents 表示响应内容，forewardToNext 则表示是否需要将请求分发给后续中间件进行处理。在调用 UseMiddleware<TMiddleware>扩展方法对这个中间件进行注册时，我们显式指定了响应的内容，至于参数 forewardToNext，我们之所以没有每次都显式指定，是因为这是一个具有默认值的参数。（S1103）

```
class Program
{
    static void Main()
    {
        Host.CreateDefaultBuilder()
            .ConfigureWebHostDefaults(builder => builder
            .Configure(app => app
                .UseMiddleware<StringContentMiddleware>("Hello")
                .UseMiddleware<StringContentMiddleware>(" World!", false)))
            .Build()
            .Run();
    }

    private sealed class StringContentMiddleware
    {
        private readonly RequestDelegate   _next;
        private readonly string            _contents;
        private readonly bool              _forewardToNext;

        public StringContentMiddleware(RequestDelegate next, string contents,
            bool forewardToNext = true)
        {
            _next           = next;
            _forewardToNext = forewardToNext;
            _contents       = contents;
        }

        public async Task Invoke(HttpContext context)
        {
            await context.Response.WriteAsync(_contents);
            if (_forewardToNext)
            {
                await _next(context);
            }
        }
    }
}
```

启动该程序后，利用浏览器访问监听地址依然可以得到图 11-6 所示的输出结果。对于前面介绍的两个中间件，它们的不同之处除了体现在定义和注册方式上，还体现在自身生命周期的差异上。具体来说，强类型方式定义的中间件可以注册为任意生命周期模式的服务，但是按照约定定义的中间件则总是一个 Singleton 服务。

11.2 依赖注入

基于 IHostBuilder/IHost 的服务承载系统建立在依赖注入框架之上，它在服务承载过程中依赖的服务（包括作为宿主的 IHost 对象）都由代表依赖注入容器的 IServiceProvider 对象提供。在定义承载服务时，也可以采用依赖注入方式来消费它所依赖的服务。作为依赖注入容器的 IServiceProvider 对象能否提供我们需要的服务实例，取决于相应的服务注册是否预先添加到依赖注入框架中。服务注册可以通过调用 IHostBuilder 接口或者 IWebHostBuilder 接口相应的方法来完成，前者在第 10 章已经有详细介绍，下面介绍基于 IWebHostBuilder 接口的服务注册。

11.2.1 服务注册

ASP.NET Core 应用提供了两种服务注册方式，一种是调用 IWebHostBuilder 接口的 ConfigureServices 方法。如下面的代码片段所示，IWebHostBuilder 定义了两个 ConfigureServices 方法重载，它们的参数类型分别是 Action<IServiceCollection>和 Action<WebHostBuilderContext, IServiceCollection>，我们注册的服务最终会被添加到作为这两个委托对象输入的 IServiceCollection 集合中。WebHostBuilderContext 代表当前 IWebHostBuilder 在构建 WebHost 过程中采用的上下文，我们可以利用它得到当前应用的配置和与承载环境相关的信息。

```
public interface IWebHostBuilder
{
    IWebHostBuilder ConfigureServices(Action<IServiceCollection> configureServices);
    IWebHostBuilder ConfigureServices(Action<WebHostBuilderContext, IServiceCollection>
        configureServices);
    ...
}

public class WebHostBuilderContext
{
    public IConfiguration            Configuration { get; set; }
    public IWebHostEnvironment       HostingEnvironment { get; set; }
}
```

除了直接调用 IWebHostBuilder 接口的 ConfigureServices 方法注册服务，还可以利用注册的 Startup 类型来完成服务的注册。所谓的 Startup 类型就是通过调用如下两个扩展方法注册到 IWebHostBuilder 接口上用来对应用程序进行初始化的。由于 ASP.NET Core 应用针对请求的处理能力与方式完全取决于注册的中间件，所以这里所谓的针对应用程序的初始化主要体现在针对中间件的注册上。

```
public static class WebHostBuilderExtensions
{
    public static IWebHostBuilder UseStartup<TStartup>(this IWebHostBuilder hostBuilder)
        where TStartup: class;
    public static IWebHostBuilder UseStartup(this IWebHostBuilder hostBuilder,
        Type startupType);
}
```

对于注册的中间件来说，它往往具有针对其他服务的依赖。当 ASP.NET Core 框架在创建具体的中间件对象时，会利用依赖注入框架来提供注入的依赖服务。中间件依赖的这些服务自然需要被预先注册，所以中间件和服务注册成为 Startup 对象的两个核心功能。与中间件类型类似，我们在大部分情况下会采用约定的形式来定义 Startup 类型。如下所示的代码片段就是一个典型的 Startup 的定义，中间件和服务的注册分别实现在 Configure 方法和 ConfigureServices 方法中。由于并不是在任何情况下都有服务注册的需求，所以 ConfigureServices 方法并不是必需的。Startup 对象的 ConfigureServices 方法的调用发生在整个服务注册的最后阶段，在此之后，ASP.NET Core 应用就会利用所有的服务注册来创建作为依赖注入容器的 IServiceProvider 对象。

```
public class Startup
{
    public void ConfigureServices(IServiceCollection servives);
    public void Configure(IApplicationBuidler app);
}
```

除了可以采用上述两种方式为应用程序注册所需的服务，ASP.NET Core 框架本身在构建请求处理管道之前也会注册一些服务，这些公共服务除了供框架自身消费，也可以供应用程序使用。那么 ASP.NET Core 框架究竟预先注册了哪些服务？为了得到这个问题的答案，我们编写了如下这段简单的程序。

```
class Program
{
    static void Main()
    {
        Host.CreateDefaultBuilder().ConfigureWebHostDefaults(builder => builder
            .UseStartup<Startup>())
        .Build()
        .Run();
    }
}

public class Startup
{
    public void ConfigureServices(IServiceCollection services)
    {
        var provider = services.BuildServiceProvider();
        foreach (var service in services)
        {
            var serviceTypeName = GetName(service.ServiceType);
            var implementationType = service.ImplementationType
                    ?? service.ImplementationInstance?.GetType()
                    ?? service.ImplementationFactory?.Invoke(provider)?.GetType();
            if (implementationType != null)
            {
                Console.WriteLine($"{service.Lifetime,-15}
                    {GetName(service.ServiceType),-50}{GetName(implementationType)}");
            }
        }
    }
}
```

```csharp
public void Configure(IApplicationBuilder app) { }

private string GetName(Type type)
{
    if (!type.IsGenericType)
    {
        return type.Name;
    }
    var name = type.Name.Split('`')[0];
    var args = type.GetGenericArguments().Select(it => it.Name);
    return $"{name}<{string.Join(",", args)}>";
}
```

在如上所示的 Startup 类型的 ConfigureServices 方法中，我们从作为参数的 IServiceCollection 对象中获取当前注册的所有服务，并打印每个服务对应的声明类型、实现类型和生命周期。这段程序执行之后，系统注册的所有公共服务会以图 11-7 所示的方式输出到控制台上，我们可以从这个列表中发现很多熟悉的类型。（S1104）

图 11-7　ASP.NET Core 框架注册的公共服务

11.2.2 服务的消费

ASP.NET Core 框架中的很多核心对象都是通过依赖注入方式提供的，如用来对应用进行初始化的 Startup 对象、中间件对象，以及 ASP.NET Core MVC 应用中的 Controller 对象和 View 对象等，所以我们可以在定义它们的时候采用注入的形式来消费已经注册的服务。下面简单介绍几种服务注入的应用场景。

在 Startup 中注入服务

构成 HostBuilderContext 上下文的两个核心对象（表示配置的 IConfiguration 对象和表示承载环境的 IHostEnvironment 对象）可以直接注入 Startup 构造函数中进行消费。由于 ASP.NET Core 应用中的承载环境通过 IWebHostEnvironment 接口表示，IWebHostEnvironment 接口派生于 IHostEnvironment 接口，所以也可以通过注入 IWebHostEnvironment 对象的方式得到当前承载环境相关的信息。

我们可以通过一个简单的实例来验证针对 Startup 的构造函数注入。如下面的代码片段所示，我们在调用 IWebHostBuilder 接口的 Startup\<TStartup\>方法时注册了自定义的 Startup 类型。在定义 Startup 类型时，我们在其构造函数中注入上述 3 个对象，提供的调试断言不仅证明了 3 个对象不为 Null，还表明采用 IHostEnvironment 接口和 IWebHostEnvironment 接口得到的其实是同一个实例。（S1105）

```
class Program
{
    static void Main()
    {
        Host.CreateDefaultBuilder().ConfigureWebHostDefaults(builder => builder
            .UseStartup<Startup>())
        .Build()
        .Run();
    }
}

public class Startup
{
    public Startup(IConfiguration configuration, IHostEnvironment hostingEnvironment,
        IWebHostEnvironment webHostEnvironment)
    {
        Debug.Assert(configuration != null);
        Debug.Assert(hostingEnvironment != null);
        Debug.Assert(webHostEnvironment != null);
        Debug.Assert(ReferenceEquals(hostingEnvironment, webHostEnvironment));
    }
    public void Configure(IApplicationBuilder app) { }
}
```

依赖服务还可以直接注入用于注册中间件的 Configure 方法中。如果构造函数注入还可以对注入的服务有所选择，那么对于 Configure 方法来说，通过任意方式注册的服务都可以注入其中，包括通过调用 IHostBuilder、IWebHostBuilder 和 Startup 自身的 ConfigureServices 方法注册的服务，还包括框架自行注册的所有服务。

如下面的代码片段所示，我们分别调用 IWebHostBuilder 和 Startup 的 ConfigureServices 方法注册了针对 IFoo 接口与 IBar 接口的服务，这两个服务直接注入 Startup 的 Configure 方法中。另外，Configure 方法要求提供一个用来注册中间件的 IApplicationBuilder 对象作为参数，但是对该参数出现的位置并未做任何限制。（S1106）

```
class Program
{
    static void Main()
    {
        Host.CreateDefaultBuilder().ConfigureWebHostDefaults(builder => builder
            .UseStartup<Startup>()
            .ConfigureServices(svcs => svcs.AddSingleton<IFoo, Foo>()))
        .Build()
        .Run();
    }
}

public class Startup
{
    public void ConfigureServices(IServiceCollection services)
        => services.AddSingleton<IBar, Bar>();
    public void Configure(IApplicationBuilder app, IFoo foo, IBar bar)
    {
        Debug.Assert(foo != null);
        Debug.Assert(bar != null);
    }
}
```

在中间件中注入服务

ASP.NET Core 请求处理管道最重要的对象是真正用来处理请求的中间件。由于 ASP.NET Core 在创建中间件对象并利用它们构建整个请求处理管道时，所有的服务都已经注册完毕，所以注册的任何一个服务都可以注入中间件类型的构造函数中。如下所示的代码片段体现了针对中间件类型的构造函数注入。（S1107）

```
class Program
{
    static void Main()
    {
        Host.CreateDefaultBuilder().ConfigureWebHostDefaults(builder => builder
            .ConfigureServices(svcs => svcs
                .AddSingleton<FoobarMiddleware>()
                .AddSingleton<IFoo, Foo>()
                .AddSingleton<IBar, Bar>())
            .Configure(app => app.UseMiddleware<FoobarMiddleware>()))
        .Build()
        .Run();
```

```csharp
    }
}
public class FoobarMiddleware: IMiddleware
{
    public FoobarMiddleware(IFoo foo, IBar bar)
    {
        Debug.Assert(foo != null);
        Debug.Assert(bar != null);
    }

    public Task InvokeAsync(HttpContext context, RequestDelegate next)
    {
        Debug.Assert(next != null);
        return Task.CompletedTask;
    }
}
```

如果采用基于约定的中间件类型定义方式，注册的服务还可以直接注入真正用于处理请求的 InvokeAsync 方法或者 Invoke 方法中。另外，将方法命名为 InvokeAsync 更符合 TAP（Task-based Asynchronous Pattern）编程模式，之所以保留 Invoke 方法命名，主要是出于版本兼容的目的。如下所示的代码片段展示了针对 InvokeAsync 方法的服务注入。（S1108）

```csharp
class Program
{
    static void Main()
    {
        Host.CreateDefaultBuilder().ConfigureWebHostDefaults(builder => builder
            .ConfigureServices(svcs => svcs
                .AddSingleton<IFoo, Foo>()
                .AddSingleton<IBar, Bar>())
            .Configure(app => app.UseMiddleware<FoobarMiddleware>()))
            .Build()
            .Run();
    }
}

public class FoobarMiddleware
{
    private readonly RequestDelegate _next;

    public FoobarMiddleware(RequestDelegate next) => _next = next;
    public Task InvokeAsync(HttpContext context, IFoo foo, IBar bar)
    {
        Debug.Assert(context != null);
        Debug.Assert(foo != null);
        Debug.Assert(bar != null);
        return _next(context);
    }
}
```

虽然约定定义的中间件类型和 Startup 类型采用了类似的服务注入方式，它们都支持构造函数注入和方法注入，但是它们之间有一些差别。中间件类型的构造函数、Startup 类型的 Configure 方法和中间件类型的 Invoke 方法或者 InvokeAsync 方法都具有一个必需的参数，其类型分别为 RequestDelegate、IApplicationBuilder 和 HttpContext，对于该参数在整个参数列表的位置，前两者都未做任何限制，只有后者要求表示当前请求上下文的参数 HttpContext 必须作为方法的第一个参数。按照上述约定，如下这个中间件类型 FoobarMiddleware 的定义是不合法的，但是 Startup 类型的定义则是合法的。对于这一点，笔者认为可以将这个限制放开，这样不仅可以使中间件类型的定义更加灵活，还能保证注入方式的一致性。

```csharp
public class FoobarMiddleware
{
    public FoobarMiddleware(RequestDelegate next);
    public Task InvokeAsync(IFoo foo, IBar bar, HttpContext context);
}

public class Startup
{
    public void Configure(IFoo foo, IBar bar, IApplicationBuilder app);
}
```

对于基于约定的中间件，构造函数注入与方法注入存在一个本质区别。由于中间件被注册为一个 Singleton 对象，所以我们不应该在它的构造函数中注入 Scoped 服务。Scoped 服务只能注入中间件类型的 InvokeAsync 方法中，因为依赖服务是在针对当前请求的服务范围中提供的，所以能够确保 Scoped 服务在当前请求处理结束之后被释放。

MVC 应用的依赖注入

在一个 ASP.NET Core MVC 应用中，我们主要在两个地方注入注册的服务：一是在定义的 Controller 中以构造函数注入的方式注入所需的服务；二是在呈现的 View 中使用@inject 指令实现服务注入。下面通过一个实例演示这两种服务注入方式。下面是 ASP.NET Core MVC 程序涉及的相关代码，我们调用 IWebHostBuilder 的 ConfigureServices 方法注册了针对 IFoo 接口和 IBar 接口的两个服务，前者注入 HomeController 的构造函数中。

```csharp
class Program
{
    static void Main()
    {
        Host.CreateDefaultBuilder().ConfigureWebHostDefaults(builder => builder
            .ConfigureServices(svcs => svcs
                .AddSingleton<IFoo, Foo>()
                .AddSingleton<IBar, Bar>()
                .AddControllersWithViews())
            .Configure(app => app
                .UseRouting()
                .UseEndpoints(endpoints => endpoints.MapControllers())))
            .Build()
            .Run();
```

```
    }
}
public class HomeController : Controller
{
    private readonly IFoo _foo;

    public HomeController(IFoo foo) => _foo = foo;

    [HttpGet("/")]
    public IActionResult Index()
    {
        ViewBag.Foo = _foo;
        return View();
    }
}
```

我们为 HomeController 定义了一个路由指向根路径（"/"）的 Action 方法 Index，该方法在调用 View 方法呈现默认的 View 之前，将注入的 IFoo 服务以 ViewBag 的形式传递到 View 中。如下所示的代码片段是这个 Action 方法对应 View（"/Views/Home/Index.cshtml"）的定义，我们通过@inject 指令注入了 IBar 服务，并将属性名设置为 Bar，这意味着当前 View 对象将添加一个 Bar 属性来引用注入的服务。

```
@inject App.IBar Bar
Foo: @ViewBag.Foo.GetType().AssemblyQualifiedName <br/>
Bar: @Bar.GetType().AssemblyQualifiedName
```

在上面这个 View 中，我们将通过 ViewBag 传递的 IFoo 服务和 IBar 服务的类型名称呈现出来。当程序启动后，如果利用浏览器访问这个 ASP.NET Core MVC 应用的基地址，这个 View 的内容最终将会以图 11-8 所示的形式呈现在浏览器中。（S1109）

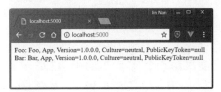

图 11-8　在 Controller 和 View 中注入服务

11.2.3　生命周期

当我们调用 IServiceCollection 相关方法注册服务的时候，总是会指定一种生命周期。由第 3 章和第 4 章的介绍可知，作为依赖注入容器的多个 IServiceProvider 对象通过 ServiceScope 构成一种层次化结构。Singleton 服务实例保存在作为根容器的 IServiceProvider 对象上，而 Scoped 服务实例以及需要回收释放的 Transient 服务实例则保存在当前 IServiceProvider 对象中，只有不需要回收的 Transient 服务才会用完就被丢弃。

至于服务实例是否需要回收释放，取决于服务实现类型是否实现 IDisposable 接口，服务实例的回收释放由保存它的 IServiceProvider 对象负责。具体来说，当 IServiceProvider 对象因自身的 Dispose 方法被调用而被回收释放时，它会调用自身维护的所有服务实例的 Dispose 方法。对于一个非根容器的 IServiceProvider 对象来说，其生命周期决定于对应的 IServiceScope 对象，调用 ServiceScope 的 Dispose 方法会导致对封装 IServiceProvider 对象的回收释放。

两个 IServiceProvider 对象

如果在一个具体的 ASP.NET Core 应用中讨论服务生命周期会更加易于理解：Singleton 是针对应用程序的生命周期，而 Scoped 是针对请求的生命周期。换句话说，Singleton 服务的生命周期会一直延续到应用程序关闭，而 Scoped 服务的生命周期仅仅与当前请求上下文绑定在一起，那么这样的生命周期模式是如何实现的？

ASP.NET Core 应用针对服务生命周期管理的实现原理其实也很简单。在应用程序正常启动后，它会利用注册的服务创建一个作为根容器的 IServiceProvider 对象，我们可以将它称为 ApplicationServices。如果应用在处理某个请求的过程中需要采用依赖注入的方式激活某个服务实例，那么它会利用这个 IServiceProvider 对象创建一个代表服务范围的 IServiceScope 对象，后者会指定一个 IServiceProvider 对象作为子容器，请求处理过程中所需的服务实例均由它来提供，我们可以将它称为 RequestServices。

在处理完当前请求后，这个 IServiceScope 对象的 Dispose 方法会被调用，与它绑定的这个 IServiceProvider 对象也随之被回收释放，由它提供的实现了 IDisposable 接口的 Transient 服务实例也会随之被回收释放，最终由它提供的 Scoped 服务实例变成可以被 GC 回收的垃圾对象。表示当前请求上下文的 HttpContext 类型具有如下所示的 RequestServices 属性，它返回的就是这个针对当前请求的 IServiceProvider 对象。

```
public abstract class HttpContext
{
    public abstract IServiceProvider RequestServices { get; set; }
    ...
}
```

为了使读者对注入服务的生命周期有深刻的认识，下面演示一个简单的实例。这是一个 ASP.NET Core MVC 应用，我们在该应用中定义了 3 个服务接口（IFoo、IBar 和 IBaz）和对应的实现类（Foo、Bar 和 Baz），后者派生于实现了 IDisposable 接口的基类 Base。我们分别在 Base 的构造函数和实现的 Dispose 方法中输出相应的文字，以确定服务实例被创建和释放的时间。

```
class Program
{
    static void Main()
    {
        Host.CreateDefaultBuilder().ConfigureWebHostDefaults(builder => builder
            .ConfigureServices(svcs => svcs
                .AddSingleton<IFoo, Foo>()
                .AddScoped<IBar, Bar>()
                .AddTransient<IBaz, Baz>()
                .AddControllersWithViews())
```

```csharp
            .Configure(app => app
                .Use(next => httpContext => {
                    Console.WriteLine($"Receive request to {httpContext.Request.Path}");
                    return next(httpContext);
                })
                .UseRouting()
                .UseEndpoints(endpoints => endpoints.MapControllers())))
        .ConfigureLogging(builder=>builder.ClearProviders())
        .Build()
        .Run();
    }
}

public class HomeController: Controller
{
    private readonly IHostApplicationLifetime _lifetime;

    public HomeController(IHostApplicationLifetime lifetime, IFoo foo,
        IBar bar1, IBar bar2, IBaz baz1, IBaz baz2)
        =>_lifetime = lifetime;

    [HttpGet("/index")]
    public void Index() {}

    [HttpGet("/stop")]
    public void Stop() => _lifetime.StopApplication();
}
public interface IFoo {}
public interface IBar {}
public interface IBaz {}
public class Base : IDisposable
{
    public Base()=> Console.WriteLine($"{this.GetType().Name} is created.");
    public void Dispose() => Console.WriteLine($"{this.GetType().Name} is disposed.");
}
public class Foo : Base, IFoo {}
public class Bar : Base, IBar {}
public class Baz : Base, IBaz {}
```

在注册 ASP.NET Core MVC 框架相关的服务之前，我们采用不同的生命周期对这 3 个服务进行了注册。为了确定应用程序何时开始处理接收的请求，可以利用注册的中间件打印出当前请求的路径。我们在 HomeController 的构造函数中注入了上述 3 个服务和 1 个用来远程关闭应用的 IHostApplicationLifetime 服务，其中 IBar 和 IBaz 被注入了两次。HomeController 包含 Index 和 Stop 两个 Action 方法，它们的路由指向的路径分别为 "/index" 和 "/stop"，Stop 方法利用注入的 IHostApplicationLifetime 服务关闭当前应用。

我们先采用命令行的形式来启动该应用程序，然后利用浏览器依次向该应用发送 3 个请求，前两个请求指向 Action 方法 Index（"/index"），后一个指向 Action 方法 Stop（"/stop"），此时控制台上出现的输出结果如图 11-9 所示。由输出结果可知：由于 IFoo 服务采用的生命周期模式为 Singleton，所以在整个应用的生命周期中只会创建一次。对于每个接收的请求，虽然 IBar 和 IBaz 都被注入了两次，但是采用 Scoped 模式的 Bar 对象只会被创建一次，而采用 Transient 模式的 Baz 对象则被创建了两次。再来看释放服务相关的输出，采用 Singleton 模式的 IFoo 服务会在应用被关闭的时候被释放，而生命周期模式分别为 Scoped 和 Transient 的 IBar 服务与 IBaz 服务都会在应用处理完当前请求之后被释放。（S1110）

图 11-9　服务的生命周期

基于服务范围的验证

由第 4 章的介绍可知，Scoped 服务既不应该由作为根容器的 ApplicationServices 来提供，也不能注入一个 Singleton 服务中，否则它将无法在请求结束之后释放。如果忽视了这个问题，就容易造成内存泄漏，下面是一个典型的例子。

如下所示的实例程序使用了一个名为 FoobarMiddleware 的中间件。在该中间件初始化过程中，它需要从数据库中加载由 Foobar 类型表示的数据。在这里我们采用 Entity Framework Core 提供的基于 SQL Server 的数据访问，所以可以为实体类型 Foobar 定义对应的 FoobarDbContext，它以服务的形式通过调用 IServiceCollection 的 AddDbContext<TDbContext>扩展方法进行注册，注册的服务默认采用 Scoped 生命周期。

```
class Program
{
    static void Main()
    {
        Host.CreateDefaultBuilder().ConfigureWebHostDefaults(builder => builder
            .UseDefaultServiceProvider(options=>options.ValidateScopes = false)
            .ConfigureServices(svcs => svcs.AddDbContext<FoobarDbContext>(
                options=>options.UseSqlServer("connection string")))
            .Configure(app =>app.UseMiddleware<FoobarMiddleware>()))
        .Build()
```

```
        .Run();
    }
}
public class FoobarMiddleware
{
    private readonly RequestDelegate     _next;
    private readonly Foobar              _foobar;
    public FoobarMiddleware(RequestDelegate next, FoobarDbContext dbContext)
    {
        _next           = next;
        _foobar         = dbContext.Foobar.SingleOrDefault();
    }

    public Task InvokeAsync(HttpContext context)
    {
        ...
        return _next(context);
    }
}

public class Foobar
{
    [Key]
    public string Foo { get; set; }
    public string Bar { get; set; }
}

public class FoobarDbContext : DbContext
{
    public DbSet<Foobar> Foobar { get; set; }
    public FoobarDbContext(DbContextOptions options) : base(options){}
}
```

采用约定方式定义的中间件实际上是一个 Singleton 对象，而且它是在应用初始化过程中由根容器的 IServiceProvider 对象创建的。由于 FoobarMiddleware 的构造函数中注入了 FoobarDbContext 对象，所以该对象自然也由同一个 IServiceProvider 对象来提供。这就意味着 FoobarDbContext 对象的生命周期会延续到当前应用程序被关闭的那一刻，造成的后果就是数据库连接不能及时地被释放。

在一个 ASP.NET Core 应用中，如果将服务的生命周期注册为 Scoped 模式，那么我们希望服务实例真正采用基于请求的生命周期模式。由第 4 章的介绍可知，我们可以通过启用针对服务范围的验证来避免采用作为根容器的 IServiceProvider 对象来提供 Scoped 服务实例。我们只需要调用 IWebHostBuilder 接口的两个 UseDefaultServiceProvider 方法重载将 ServiceProviderOptions 的 ValidateScopes 属性设置为 True 即可。

```
public static class WebHostBuilderExtensions
{
    public static IWebHostBuilder UseDefaultServiceProvider(
        this IWebHostBuilder hostBuilder, Action<ServiceProviderOptions> configure);
```

```csharp
public static IWebHostBuilder UseDefaultServiceProvider(
    this IWebHostBuilder hostBuilder, Action<WebHostBuilderContext,
    ServiceProviderOptions> configure);
}

public class ServiceProviderOptions
{
    public bool ValidateScopes { get; set; }
    public bool ValidateOnBuild { get; set; }
}
```

出于性能方面的考虑，如果在 Development 环境下调用 Host 的静态方法 CreateDefaultBuilder 来创建 IHostBuilder 对象，那么该方法会将 ValidateScopes 属性设置为 True。在上面演示的实例中，我们刻意关闭了针对服务范围的验证，如果将这行代码删除，在开发环境下启动该程序之后会出现图 11-10 所示的异常。

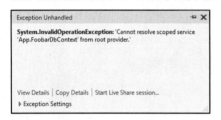

图 11-10　针对 Scoped 服务的验证

如果确实需要在中间件中注入 Scoped 服务，可以采用强类型（实现 IMiddleware 接口）的中间件定义方式，并将中间件以 Scoped 服务进行注册即可。如果采用基于约定的中间件定义方式，我们有两种方案来解决这个问题：第一种解决方案就是按照如下所示的方式在 InvokeAsync 方法中利用 HttpContext 的 RequestServices 属性得到基于当前请求的 IServiceProvider 对象，并利用它来提供依赖的服务。

```csharp
public class FoobarMiddleware
{
    private readonly RequestDelegate _next;
    public FoobarMiddleware(RequestDelegate next)=> _next = next;
    public Task InvokeAsync(HttpContext context)
    {
        var dbContext = context.RequestServices.GetRequiredService<FoobarDbContext>();
        Debug.Assert(dbContext != null);
        return _next(context);
    }
}
```

第二种解决方案则是按照如下所示的方式直接在 InvokeAsync 方法中注入依赖的服务。我们在上面介绍两种中间件定义方式时已经提及：InvokeAsync 方法注入的服务就是由基于当前请求的 IServiceProvider 对象提供的，所以这两种解决方案其实是等效的。

```csharp
public class FoobarMiddleware
{
```

```csharp
    private readonly RequestDelegate _next;
    public FoobarMiddleware(RequestDelegate next) => _next = next;
    public Task InvokeAsync(HttpContext context, FoobarDbContext dbContext)
    {
        Debug.Assert(dbContext != null);
        return _next(context);
    }
}
```

11.2.4　集成第三方依赖注入框架

由第 10 章的介绍可知，通过调用 IHostBuilder 接口的 UseServiceProviderFactory<TContainerBuilder> 方法注册 IServiceProviderFactory<TContainerBuilder>工厂的方式可以实现与第三方依赖注入框架的整合。该接口定义的 ConfigureContainer<TContainerBuilder>方法可以对提供的依赖注入容器做进一步设置，这样的设置同样可以定义在注册的 Startup 类型中。

第 3 章创建了一个名为 Cat 的简易版依赖注入框架，并在第 4 章为其创建了一个 IServiceProviderFactory<TContainerBuilder>实现，具体类型为 CatServiceProvider，下面演示如何通过注册这个 CatServiceProvider 实现与第三方依赖注入框架 Cat 的整合。如果使用 Cat 框架，我们可以通过在服务类型上标注 MapToAttribute 特性的方式来定义服务注册信息。在创建的演示程序中，我们采用如下方式定义了 3 个服务（Foo、Bar 和 Baz）和对应的接口（IFoo、IBar 和 IBaz）。

```csharp
public interface IFoo { }
public interface IBar { }
public interface IBaz { }

[MapTo(typeof(IFoo), Lifetime.Root)]
public class Foo : IFoo { }

[MapTo(typeof(IBar), Lifetime.Root)]
public class Bar : IBar { }

[MapTo(typeof(IBaz), Lifetime.Root)]
public class Baz : IBaz { }
```

在如下所示的代码片段中，我们调用 IHostBuilder 接口的 UseServiceProviderFactory 方法注册了 CatServiceProviderFactory 工厂。我们将针对 Cat 框架的服务注册实现在注册 Startup 类型的 ConfigureContainer 方法中，这是除 Configure 方法和 ConfigureServices 方法外的第三个约定的方法。我们将 CatBuilder 对象作为该方法的参数，并调用它的 Register 方法实现了针对当前程序集的批量服务注册。

```csharp
class Program
{
    static void Main()
    {
        Host.CreateDefaultBuilder()
            .ConfigureWebHostDefaults(builder => builder
                .UseStartup<Startup>())
```

```csharp
            .UseServiceProviderFactory(new CatServiceProviderFactory())
            .Build()
            .Run();
    }
}

public class Startup
{
    public void Configure(IApplicationBuilder app, IFoo foo, IBar bar, IBaz baz)
    {
        app.Run(async context =>
        {
            var response = context.Response;
            response.ContentType = "text/html";
            await response.WriteAsync($"foo: {foo}<br/>");
            await response.WriteAsync($"bar: {bar}<br/>");
            await response.WriteAsync($"baz: {baz}<br/>");
        });
    }
    public void ConfigureContainer(CatBuilder container)
        => container.Register(Assembly.GetEntryAssembly());
}
```

为了检验 ASP.NET Core 能否利用 Cat 框架来提供所需的服务，我们将注册的 3 个服务直接注入 Startup 类型的 Configure 方法中。我们在该方法中利用注册的中间件将这 3 个注入的服务实例的类型写入相应的 HTML 文档中。如果利用浏览器访问该应用，得到的输出结果如图 11-11 所示。（S1111）

图 11-11　与第三方依赖注入框架的整合

11.3　配置

通过第 10 章的介绍可知，IHostBuilder 接口中定义了 ConfigureHostConfiguration 方法和 ConfigureAppConfiguration 方法，它们可以帮助我们设置面向宿主（IHost 对象）和应用（承载服务）的配置。针对配置的初始化也可以借助 IWebHostBuilder 接口来完成。

11.3.1　初始化配置

当 IWebHostBuilder 对象被创建的时候，它会将当前的环境变量作为配置源来创建承载最初配置数据的 IConfiguration 对象，但它只会选择名称以 "ASPNETCORE_" 为前缀的环境变量

（通过静态类型 Host 的 CreateDefaultBuilder 方法创建的 HostBuilder 默认选择的是前缀为"DOTNET_"的环境变量）。在演示针对环境变量的初始化配置之前，需要先解决配置的消费问题，即如何获取配置数据。

11.2 节演示了针对 Startup 类型的构造函数注入，表示配置的 IConfiguration 对象是能够注入 Startup 类型构造函数中的两个服务对象之一。接下来我们采用 Options 模式来消费以环境变量形式提供的配置，如下所示的 FoobarOptions 是我们定义的 Options 类型。在注册的 Startup 类型中，可以直接在构造函数中注入 IConfiguration 服务，并在 ConfigureServices 方法中将其映射为 FoobarOptions 类型。在 Configure 方法中，可以通过注入的 IOptions<FoobarOptions>服务得到通过配置绑定的 FoobarOptions 对象，并将其序列化成 JSON 字符串。在通过调用 IApplicationBuilder 的 Run 方法注册的中间件中，这个 JSON 字符串直接作为请求的响应内容。

```
class Program
{
    static void Main()
    {
        Environment.SetEnvironmentVariable("ASPNETCORE_FOOBAR:FOO", "Foo");
        Environment.SetEnvironmentVariable("ASPNETCORE_FOOBAR:BAR", "Bar");
        Environment.SetEnvironmentVariable("ASPNETCORE_Baz", "Baz");

        Host.CreateDefaultBuilder().ConfigureWebHostDefaults(builder => builder
            .UseStartup<Startup>())
            .Build()
            .Run();
    }

    public class Startup
    {
        private readonly IConfiguration _configuration;

        public Startup(IConfiguration configuration)
            => _configuration = configuration;

        public void ConfigureServices(IServiceCollection services)
            => services.Configure<FoobarOptions>(_configuration);

        public void Configure(IApplicationBuilder app,
            IOptions<FoobarOptions> optionsAccessor)
        {
            var options = optionsAccessor.Value;
            var json = JsonConvert.SerializeObject(
                options, Formatting.Indented);
            app.Run(async context =>
            {
                context.Response.ContentType = "text/html";
                await context.Response.WriteAsync($"<pre>{json}</pre>");
            });
        }
    }
```

```
    }

    public class FoobarOptions
    {
        public Foobar Foobar { get; set; }
        public string Baz { get; set; }
    }

    public class Foobar
    {
        public string Foo { get; set; }
        public string Bar { get; set; }
    }
}
```

为了能够提供绑定为 FoobarOptions 对象的原始配置，我们在 Main 方法中设置了 3 个对应的环境变量，这些环境变量具有相同的前缀"ASPNETCORE_"。应用程序启动之后，如果利用浏览器访问该应用，得到的输出结果如图 11-12 所示。（S1112）

图 11-12　由环境变量提供的原始配置

11.3.2　以键值对形式读取和修改配置

第 6 章对配置模型进行了深入分析，由此可知，IConfiguration 对象是以字典的结构来存储配置数据的，该接口定义的索引可供我们以键值对的形式来读取和修改配置数据。在 ASP.NET Core 应用中，我们可以通过调用定义在 IWebHostBuilder 接口的 GetSetting 方法和 UseSetting 方法达到相同的目的。

```
public interface IWebHostBuilder
{
    string GetSetting(string key);
    IWebHostBuilder UseSetting(string key, string value);
    ...
}
```

上面演示的实例采用环境变量来提供最终绑定为 FoobarOptions 对象的原始配置，这样的配置数据也可以通过如下所示的方式调用 IWebHostBuilder 接口的 UseSetting 方法来提供。修改后的应用程序启动之后，如果利用浏览器访问该应用，同样可以得到图 11-12 所示的输出结果。（S1113）

```
class Program
{
    static void Main()
    {
```

```csharp
        Host.CreateDefaultBuilder().ConfigureWebHostDefaults(builder => builder
            .UseSetting("Foobar:Foo", "Foo")
            .UseSetting("Foobar:Bar", "Bar")
            .UseSetting("Baz", "Baz")
            .UseStartup<Startup>())
        .Build()
        .Run();
    }
}
```

配置不仅仅供应用程序来使用，ASP.NET Core 框架自身的很多特性也都可以通过配置进行定制。如果希望通过修改配置来控制 ASP.NET Core 框架的某些行为，就需要先知道对应的配置项的名称是什么。例如，ASP.NET Core 应用的服务器默认使用 launchSettings.json 文件定义的监听地址，但是我们可以通过修改配置采用其他的监听地址。包括端口在内的监听地址是通过名称为 urls 的配置项来控制的，如果记不住这个配置项的名称，也可以直接使用定义在 WebHostDefaults 中对应的只读属性 ServerUrlsKey，该静态类型中还提供了其他一些预定义的配置项名称，所以这也是一个比较重要的类型。

```csharp
public static class WebHostDefaults
{
    public static readonly string ServerUrlsKey = "urls";
    ...
}
```

针对上面演示的这个实例，如果希望为服务器设置不同的监听地址，直接调用 IWebHostBuilder 接口的 UseSetting 方法将新的地址作为 urls 配置项的内容即可。既然配置项被命名为 urls，就意味着服务器的监听地址不仅限于一个，如果希望设置多个监听地址，我们可以采用分号作为分隔符。

```csharp
class Program
{
    static void Main()
    {
        Host.CreateDefaultBuilder().ConfigureWebHostDefaults(builder => builder
            .UseSetting("Foobar:Foo", "Foo")
            .UseSetting("Foobar:Bar", "Bar")
            .UseSetting("Baz", "Baz")
            .UseSetting("urls", "http://0.0.0.0:8888;http://0.0.0.0:9999")
            .UseStartup<Startup>())
        .Build()
        .Run();
    }
}
```

为了使实例程序采用不同的监听地址，可以采用如上所示的方式调用 IWebHostBuilder 接口的 UseSetting 方法设置两个针对 8888 和 9999 端口号的监听地址。由图 11-13 所示的程序启动后的输出结果可以看出，服务器确实采用我们指定的两个地址监听请求，通过浏览器针对这两个地址发送的请求能够得到相同的结果。（S1114）

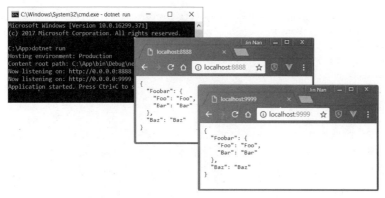

图 11-13　通过配置控制服务器的监听地址

除了调用 UseSetting 方法设置 urls 配置项来修改服务器的监听地址，直接调用 IWebHostBuilder 接口的 UseUrls 扩展方法也可以达到相同的目的。另外，我们提供的监听地址只能包含主机名称/IP 地址（Host/IP）和端口号，不能包含基础路径（PathBase）。如果我们提供"http://0.0.0.0/3721/foobar"这样一个 URL，系统会抛出一个 InvalidOperationException 类型的异常。基础路径可以通过注册中间件的方式进行设置，相关内容会在第 24 章进行介绍。

```
public static class HostingAbstractionsWebHostBuilderExtensions
{
    public static IWebHostBuilder UseUrls(this IWebHostBuilder hostBuilder,
        params string[] urls);
}
```

11.3.3　合并配置

在启动一个 ASP.NET Core 应用时，我们可以自行创建一个承载配置的 IConfiguration 对象，并通过调用 IWebHostBuilder 接口的 UseConfiguration 扩展方法将它与应用自身的配置进行合并。如果应用自身存在重复的配置项，那么该配置项的值会被指定的 IConfiguration 对象覆盖。

```
public static class HostingAbstractionsWebHostBuilderExtensions
{
    public static IWebHostBuilder UseConfiguration(this IWebHostBuilder hostBuilder,
        IConfiguration configuration);
}
```

如果前面演示的实例需要采用这种方式来提供配置，我们可以对程序代码做如下修改。如下面的代码片段所示，我们创建了一个 ConfigurationBuilder 对象，并通过调用 AddInMemoryCollection 扩展方法注册了一个 MemoryConfigurationSource 对象，它提供了绑定 FoobarOptions 对象所需的所有配置数据。我们最终利用 ConfigurationBuilder 创建出一个 IConfiguration 对象，并通过调用上述 UseConfiguration 方法将提供的配置数据合并到当前应用中。修改后的应用程序启动之后，如果利用浏览器访问该应用，同样会得到图 11-12 所示的输出结果。（S1115）

```
class Program
{
    static void Main()
```

```
{
    var configuration = new ConfigurationBuilder()
        .AddInMemoryCollection(new Dictionary<string, string>
        {
            ["Foobar:Foo"]          = "Foo",
            ["Foobar:Bar"]          = "Bar",
            ["Baz"]                 = "Baz"
        })
        .Build();

    Host.CreateDefaultBuilder().ConfigureWebHostDefaults(builder => builder
        .UseConfiguration(configuration)
        .UseStartup<Startup>())
    .Build()
    .Run();
}
```

11.3.4 注册 IConfigurationSource

配置系统最大的特点是可以注册不同的配置源。借助 IWebHostBuilder 接口的 UseConfiguration 扩展方法，虽然可以将利用配置系统提供的 IConfiguration 对象应用到 ASP.NET Core 程序中，但是这样的整合方式总显得不够彻底，更加理想的方式应该是可以直接在 ASP.NET Core 应用中注册 IConfigurationSource 对象。

针对 IConfigurationSource 的注册可以调用 IWebHostBuilder 接口的 ConfigureAppConfiguration 方法来完成，该方法与在 IHostBuilder 接口上定义的同名方法基本上是等效的。如下面的代码片段所示，这个方法的参数是一个类型为 Action<WebHostBuilderContext, IConfigurationBuilder>的委托对象，这意味着我们可以就承载上下文对配置做针对性设置。如果设置与当前承载上下文无关，我们还可以调用 ConfigureAppConfiguration 方法重载，该方法的参数类型为 Action<IConfigurationBuilder>。

```
public interface IWebHostBuilder
{
    IWebHostBuilder ConfigureAppConfiguration(Action<WebHostBuilderContext,
        IConfigurationBuilder> configureDelegate);
}

public static class WebHostBuilderExtensions
{
    public static IWebHostBuilder ConfigureAppConfiguration(
        this IWebHostBuilder hostBuilder, Action<IConfigurationBuilder> configureDelegate);
}
```

对于上面演示的这个程序来说，如果将针对 IWebHostBuilder 接口的 UseConfiguration 方法的调用替换成如下所示的针对 ConfigureAppConfiguration 方法的调用，依然可以达到相同的目的。修改后的应用程序启动之后，如果利用浏览器访问该应用，同样会得到图 11-12 所示的输

出结果。（S1116）

```
class Program
{
    static void Main()
    {
        Host.CreateDefaultBuilder().ConfigureWebHostDefaults(builder => builder
            .ConfigureAppConfiguration(config => config
                .AddInMemoryCollection(new Dictionary<string, string>
                {
                    ["Foobar:Foo"] = "Foo",
                    ["Foobar:Bar"] = "Bar",
                    ["Baz"] = "Baz"
                }))
            .UseStartup<Startup>())
            .Build()
            .Run();
    }
}
```

11.4 承载环境

基于 IHostBuilder/IHost 的承载系统通过 IHostEnvironment 接口表示承载环境，我们利用它不仅可以得到当前部署环境的名称，还可以获知当前应用的名称和存放内容文件的根目录路径。对于一个 Web 应用来说，我们需要更多的承载环境信息，额外的信息定义在 IWebHostEnvironment 接口中。

11.4.1 IWebHostEnvironment

如下面的代码片段所示，派生于 IHostEnvironment 接口的 IWebHostEnvironment 接口定义了两个属性：WebRootPath 和 WebRootFileProvider。WebRootPath 属性表示用于存放 Web 资源文件根目录的路径，WebRootFileProvider 属性则返回该路径对应的 IFileProvider 对象。如果我们希望外部可以采用 HTTP 请求的方式直接访问某个静态文件（如 JavaScript、CSS 和图片文件等），只需要将它存放于 WebRootPath 属性表示的目录之下即可。

```
public interface IWebHostEnvironment : IHostEnvironment
{
    string          WebRootPath { get; set; }
    IFileProvider   WebRootFileProvider { get; set; }
}
```

下面简单介绍与承载环境相关的 6 个属性（包含定义在 IHostEnvironment 接口中的 4 个属性）是如何设置的。IHostEnvironment 接口的 ApplicationName 代表当前应用的名称，它的默认值取决于注册的 IStartup 服务。IStartup 服务旨在完成中间件的注册，不论是调用 IWebHostBuilder 接口的 Configure 方法，还是调用它的 UseStartup/UseStartup<TStartup>方法，最终都是为了注册 IStartup 服务，所以这两个方法是不能被重复调用的。如果多次调用这两个方法，最后一次调用针对 IStartup 的服务注册会覆盖前面的注册。

如果 IStartup 服务是通过调用 IWebHostBuilder 接口的 Configure 方法注册的，那么应用的名称由调用该方法提供的 Action<IApplicationBuilder>对象来决定。具体来说，每个委托对象都会绑定到一个方法上，而方法是定义在某个类型中的，该类型所在程序集的名称会默认作为应用的名称。如果通过调用 IWebHostBuilder 接口的 UseStartup/UseStartup<TStartup>方法来注册 IStartup 服务，那么注册的 Startup 类型所在的程序集名称就是应用名称。在默认情况下，针对应用名称的设置体现在如下所示的代码片段中。

```
public static IWebHostBuilder Configure(this IWebHostBuilder hostBuilder,
    Action<IApplicationBuilder> configure)
{
    var applicationName = configure.GetMethodInfo().DeclaringType
        .GetTypeInfo().Assembly.GetName().Name;
    ...
}

public static IWebHostBuilder UseStartup(this IWebHostBuilder hostBuilder,
    Type startupType)
{
    var applicationName = startupType.GetTypeInfo().Assembly.GetName().Name;
    ...
}
```

EnvironmentName 表示当前应用所处部署环境的名称，其中开发（Development）、预发（Staging）和产品（Production）是 3 种典型的部署环境。根据不同的目的可以将同一个应用部署到不同的环境中，在不同环境中部署的应用往往具有不同的设置。在默认情况下，环境的名称为 Production。

当我们编译发布一个 ASP.NET Core 项目时，项目的源代码文件会被编译成二进制并打包到相应的程序集中，而另外一些文件（如 JavaScript、CSS 和表示 View 的.cshtml 文件等）会复制到目标目录中，我们将这些文件称为内容文件（Content File）。ASP.NET Core 应用会将所有的内容文件存储在同一个目录下，这个目录的绝对路径通过 IWebHostEnvironment 接口的 ContentRootPath 属性来表示，而 ContentRootFileProvider 属性则返回针对这个目录的 PhysicalFileProvider 对象。部分内容文件可以直接作为 Web 资源（如 JavaScript、CSS 和图片等）供客户端以 HTTP 请求的方式获取，存放此种类型内容文件的绝对目录通过 IWebHostEnvironment 接口的 WebRootPath 属性来表示，而针对该目录的 PhysicalFileProvider 自然可以通过对应的 WebRootFileProvider 属性来获取。

在默认情况下，由 ContentRootPath 属性表示的内容文件的根目录就是当前应用程序域的基础目录，也就是表示当前应用程序域的 AppDomain 对象的 BaseDirectory 属性返回的目录，静态类 AppContext 的 BaseDirectory 属性返回的也是这个目录。对于一个通过 Visual Studio 创建的 .NET Core 项目来说，该目录就是编译后保存生成的程序集的目录（如 "\bin\Debug\netcoreapp3.0" 或者 "\bin\Release\netcoreapp3.0"）。如果该目录下存在一个名为 "wwwroot" 的子目录，那么它将用来存放 Web 资源，WebRootPath 属性将返回这个目录；如果这样的子目录不存在，那么 WebRootPath 属性会返回 Null。针对这两个目录的默认设置体现在如下所示的代码片段中。

```csharp
class Program
{
    static void Main()
    {
        Host.CreateDefaultBuilder().ConfigureWebHostDefaults(builder => builder
            .UseStartup<Startup>())
        .Build()
        .Run();
    }
}
public class Startup
{
    public Startup(IWebHostEnvironment environment)
    {
        Debug.Assert(environment.ContentRootPath == AppDomain.CurrentDomain.BaseDirectory);
        Debug.Assert(environment.ContentRootPath == AppContext.BaseDirectory);

        var wwwRoot = Path.Combine(AppContext.BaseDirectory, "wwwroot");
        if (Directory.Exists(wwwRoot))
        {
            Debug.Assert(environment.WebRootPath == wwwRoot);
        }
        else
        {
            Debug.Assert(environment.WebRootPath == null);
        }
    }
    public void Configure(IApplicationBuilder app) {}
}
```

11.4.2 通过配置定制承载环境

IWebHostEnvironment 对象承载的 4 个与承载环境相关的属性（ApplicationName、EnvironmentName、ContentRootPath 和 WebRootPath）可以通过配置的方式进行定制，对应配置项的名称分别为 applicationName、environment、contentRoot 和 webroot。如果记不住这些配置项的名称也没有关系，因为我们可以利用定义在静态类 WebHostDefaults 中如下所示的 4 个只读属性来得到它们的值。通过第 11 章的介绍可知，前三个配置项的名称同样以静态只读字段的形式定义在 HostDefaults 类型中。

```csharp
public static class WebHostDefaults
{
    public static readonly string EnvironmentKey   = "environment";
    public static readonly string ContentRootKey   = "contentRoot";
    public static readonly string ApplicationKey   = "applicationName";
    public static readonly string WebRootKey       = "webroot";;
}
```

```csharp
public static class HostDefaults
{
    public static readonly string EnvironmentKey = "environment";
    public static readonly string ContentRootKey = "contentRoot";
    public static readonly string ApplicationKey = "applicationName";
}
```

下面演示如何通过配置的方式来设置当前的承载环境。在如下这段实例程序中，我们调用 IWebHostBuilder 接口的 UseSetting 方法针对上述 4 个配置项做了相应的设置。由于针对 UseStartup<TStartup>方法的调用会设置应用的名称，所以通过调用 UseSetting 方法针对应用名称的设置需要放在后面才有意义。相对于当前目录（项目根目录）的两个子目录 "contents" 和 "contents/web" 是我们为 ContentRootPath 属性与 WebRootPath 属性设置的，由于系统会验证设置的目录是否存在，所以必须预先创建这两个目录。

```csharp
class Program
{
    static void Main()
    {
        Host.CreateDefaultBuilder().ConfigureWebHostDefaults(builder => builder
            .ConfigureLogging(options => options.ClearProviders())
            .UseStartup<Startup>()
            .UseSetting("environment", "Staging")
            .UseSetting("contentRoot",
                Path.Combine(Directory.GetCurrentDirectory(), "contents"))
            .UseSetting("webroot",
                Path.Combine(Directory.GetCurrentDirectory(), "contents/web"))
            .UseSetting("ApplicationName", "MyApp"))
        .Build()
        .Run();
    }

    public class Startup
    {
        public Startup(IWebHostEnvironment environment)
        {
            Console.WriteLine($"ApplicationName: {environment.ApplicationName}");
            Console.WriteLine($"EnvironmentName: {environment.EnvironmentName}");
            Console.WriteLine($"ContentRootPath: {environment.ContentRootPath}");
            Console.WriteLine($"WebRootPath: {environment.WebRootPath}");
        }
        public void Configure(IApplicationBuilder app) { }
    }
}
```

我们在注册的 Startup 类型的构造函数中注入了 IWebHostEnvironment 服务，并直接将这 4 个属性输出到控制台上。我们在目录 "C:\App" 下运行这个程序后，设置的 4 个与承载相关的属性会以图 11-14 所示的形式呈现在控制台上。（S1117）

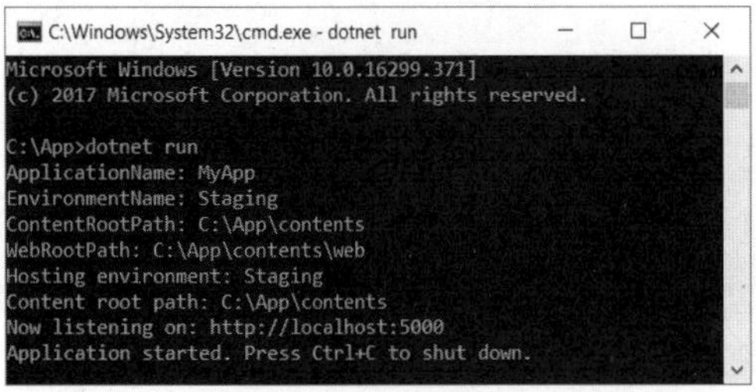

图 11-14　利用配置定义承载环境

由于 IWebHostEnvironment 服务提供的应用名称会被视为一个程序集名称，针对它的设置会影响类型的加载，所以我们基本上不会设置应用的名称。至于其他 3 个属性，除了采用最原始的方式设置相应的配置项，我们还可以直接调用 IWebHostBuilder 接口中如下 3 个对应的扩展方法来设置。通过第 10 章的介绍可知，IHostBuilder 接口也有类似的扩展方法。

```
public static class HostingAbstractionsWebHostBuilderExtensions
{
    public static IWebHostBuilder UseEnvironment(this IWebHostBuilder hostBuilder,
        string environment);
    public static IWebHostBuilder UseContentRoot(this IWebHostBuilder hostBuilder,
        string contentRoot);
    public static IWebHostBuilder UseWebRoot(this IWebHostBuilder hostBuilder,
        string webRoot);
}

public static class HostingHostBuilderExtensions
{
    public static IHostBuilder UseContentRoot(this IHostBuilder hostBuilder,
        string contentRoot);
    public static IHostBuilder UseEnvironment(this IHostBuilder hostBuilder,
        string environment);
}
```

11.4.3　针对环境的编程

对于同一个 ASP.NET Core 应用来说，我们添加的服务注册、提供的配置和注册的中间件可能会因部署环境的不同而有所差异。有了这个可以随意注入的 IWebHostEnvironment 服务，我们可以很方便地知道当前的部署环境并进行有针对性的差异化编程。

IHostEnvironment 接口提供了如下这个名为 IsEnvironment 的扩展方法，用于确定当前是否为指定的部署环境。除此之外，IHostEnvironment 接口还提供额外 3 个扩展方法来进行针对 3 种典型部署环境（开发、预发和产品）的判断，这 3 种环境采用的名称分别为 Development、Staging 和 Production，对应静态类型 EnvironmentName 的 3 个只读字段。

```
public static class HostEnvironmentEnvExtensions
{
    public static bool IsDevelopment(this IHostEnvironment hostEnvironment);
    public static bool IsProduction(this IHostEnvironment hostEnvironment);
```

```
    public static bool IsStaging(this IHostEnvironment hostEnvironment);
    public static bool IsEnvironment(this IHostEnvironment hostEnvironment,
        string environmentName);
}
public static class EnvironmentName
{
    public static readonly string Development      = "Development";
    public static readonly string Staging          = "Staging";
    public static readonly string Production       = "Production";
}
```

注册服务

下面先介绍针对环境的服务注册。ASP.NET Core 应用提供了两种服务注册方式：第一种是调用 IWebHostBuilder 接口的 ConfigureServices 方法；第二种是调用 UseStartup 方法或者 UseStartup<TStartup>方法注册一个 Startup 类型，并在其 ConfigureServices 方法中完成服务注册。对于第一种服务注册方式，用于注册服务的 ConfigureServices 方法具有一个参数类型为 Action<WebHostBuilderContext, IServiceCollection>的重载，所以我们可以利用提供的 WebHostBuilderContext 对象以如下所示的方式针对具体的环境注册相应的服务。

```
class Program
{
    public static void Main()
    {
        Host.CreateDefaultBuilder().ConfigureWebHostDefaults(builder => builder
            .ConfigureServices((context,svcs)=> {
                if (context.HostingEnvironment.IsDevelopment())
                {
                    svcs.AddSingleton<IFoobar, Foo>();
                }
                else
                {
                    svcs.AddSingleton<IFoobar, Bar>();
                }
            }))
            .Build()
            .Run();
    }
}
```

如果利用 Startup 类型来添加服务注册，我们就可以按照如下所示的方式通过构造函数注入的方式得到所需的 IWebHostEnvironment 服务，并在 ConfigureServices 方法中根据它提供的环境信息来注册对应的服务。另外，Startup 类型的 ConfigureServices 方法要么是无参的，要么具有一个类型为 IServiceCollection 的参数，所以我们无法直接在这个方法中注入 IWebHostEnvironment 服务。

```
public class Startup
{
```

```csharp
    private readonly IWebHostEnvironment _environment;

    public Startup(IWebHostEnvironment environment) => _environment = environment;
    public void ConfigureServices(IServiceCollection svcs)
    {
        if (_environment.IsDevelopment())
        {
            svcs.AddSingleton<IFoobar, Foo>();
        }
        else
        {
            svcs.AddSingleton<IFoobar, Bar>();
        }
    }
    public void Configure(IApplicationBuilder app) { }
}
```

除了在注册 Startup 类型中的 ConfigureServices 方法完成针对承载环境的服务注册，我们还可以将针对某种环境的服务注册实现在对应的 Configure{EnvironmentName}Services 方法中。上面定义的 Startup 类型完全可以改写成如下形式。

```csharp
public class Startup
{
    public void ConfigureDevelopmentServices(IServiceCollection svcs)
        => svcs.AddSingleton<IFoobar, Foo>();
    public void ConfigureServices(IServiceCollection svcs)
        => svcs.AddSingleton<IFoobar, Bar>()
    public void Configure(IApplicationBuilder app) {}
}
```

注册中间件

与服务注册类似，中间件的注册同样具有两种方式：一种是直接调用 IWebHostBuilder 接口的 Configure 方法；另一种则是调用注册的 Startup 类型的同名方法。不管采用何种方式，中间件都是借助 IApplicationBuilder 对象来注册的。由于针对应用程序的 IServiceProvider 对象可以通过其 ApplicationServices 属性获得，所以我们可以利用它提供承载环境信息的 IWebHostEnvironment 服务，进而按照如下所示的方式实现针对环境的中间件注册。

```csharp
class Program
{
    public static void Main()
    {
        Host.CreateDefaultBuilder().ConfigureWebHostDefaults(builder => builder
            .Configure(app=> {
                var environment = app.ApplicationServices
                    .GetRequiredService<IWebHostEnvironment>();
                if (environment.IsDevelopment())
                {
                    app.UseMiddleware<FooMiddleware>();
                }
                app
```

```
                .UseMiddleware<BarMiddleware>()
                .UseMiddleware<BazMiddleware>();
        }))
        .Build()
        .Run();
}
```

其实，用于注册中间件的 IApplicationBuilder 接口还有 UseWhen 的扩展方法。顾名思义，这个方法可以帮助我们根据指定的条件来注册对应的中间件。注册中间件的前提条件可以通过一个 Func<HttpContext, bool>对象来表示，对于某个具体的请求来说，只有对应的 HttpContext 对象满足该对象设置的断言，指定的中间件注册操作才会生效。

```
public static class UseWhenExtensions
{
    public static IApplicationBuilder UseWhen(this IApplicationBuilder app,
        Func<HttpContext, bool> predicate, Action<IApplicationBuilder> configuration);
}
```

如果调用 UseWhen 方法来实现针对具体环境注册对应的中间件，我们就可以按照如下所示的方式利用 HttpContext 来提供针对当前请求的 IServiceProvider 对象，进而得到承载环境信息的 IWebHostEnvironment 服务，最终根据提供的环境信息进行有针对性的中间件注册。

```
class Program
{
    public static void Main()
    {
        Host.CreateDefaultBuilder().ConfigureWebHostDefaults(builder => builder
            .Configure(app=> app
                .UseWhen(context=>context.RequestServices
                    .GetRequiredService<IWebHostEnvironment>().IsDevelopment(),
                    builder => builder.UseMiddleware<FooMiddleware>())
                .UseMiddleware<BarMiddleware>()
                .UseMiddleware<BazMiddleware>()))
            .Build()
            .Run();
    }
}
```

如果应用注册了 Startup 类型，那么针对环境的中间件注册就更加简单，因为用来注册中间件的 Configure 方法自身是可以注入任意依赖服务的，所以我们可以在该方法中按照如下所示的方式直接注入 IWebHostEnvironment 服务来提供环境信息。

```
public class Startup
{
    public void Configure(IApplicationBuilder app, IWebHostEnvironment environment)
    {
        if (environment.IsDevelopment())
        {
            app.UseMiddleware<FooMiddleware>();
        }
        app
```

```
            .UseMiddleware<BarMiddleware>()
            .UseMiddleware<BazMiddleware>();
    }
}
```

与服务注册类似，针对环境的中间件注册同样可以定义在对应的 Configure{Environment Name}方法中，上面这个 Startp 类型完全可以改写成如下形式。

```
public class Startup
{
    public void ConfigureDevelopment (IApplicationBuilder app)
    {
        app.UseMiddleware<FooMiddleware>();
    }

    public void Configure(IApplicationBuilder app)
    {
        app
            .UseMiddleware<BarMiddleware>()
            .UseMiddleware<BazMiddleware>();
    }
}
```

配置

上面介绍了针对环境的服务和中间件注册，下面介绍如何根据当前的环境来提供有针对性的配置。通过前面的介绍可知，IWebHostBuilder 接口提供了一个名为 ConfigureAppConfiguration 的方法，我们可以调用这个方法来注册相应的 IConfigureSource 对象。这个方法具有一个类型为 Action<WebHostBuilderContext, IConfigurationBuilder>的参数，所以可以通过提供的这个 WebHostBuilderContext 上下文得到提供环境信息的 IWebHostEnvironment 对象。

如果采用配置文件，我们可以将配置内容分配到多个文件中。例如，我们可以将与环境无关的配置定义在 Appsettings.json 文件中，然后针对具体环境提供对应的配置文件 Appsettings.{EnvironmentName}.json（如 Appsettings.Development.json、Appsettings.Staging.json 和 Appsettings.Production.json）。最终我们可以按照如下所示的方式将针对这两类配置文件的 IConfigureSource 注册到提供的 IConfigurationBuilder 对象上。

```
class Program
{
    public static void Main()
    {
        Host.CreateDefaultBuilder().ConfigureWebHostDefaults(builder => builder
            .ConfigureAppConfiguration((context, builder) => builder
                .AddJsonFile(path: "AppSettings.json", optional: false)
                .AddJsonFile(
                    path: $"AppSettings.{context.HostingEnvironment.EnvironmentName}.json",
                    optional: true)))
            .UseStartup<Startup>()
            .Build()
            .Run();
```

```
    }
}
```

11.5 初始化

一个 ASP.NET Core 应用的核心就是由一个服务器和一组有序中间件组成的请求处理管道，服务器只负责监听、接收和分发请求，以及最终完成对请求的响应，所以一个 ASP.NET Core 应用针对请求的处理能力和处理方式由注册的中间件来决定。一个 ASP.NET Core 在启动过程中的核心工作就是注册中间件，本节主要介绍应用启动过程中以中间件注册为核心的初始化工作。

11.5.1 Startup

由于 ASP.NET Core 应用承载于以 IHost/IHostBuilder 为核心的承载系统中，所以在启动过程中需要的所有操作都可以直接调用 IHostBuilder 接口相应的方法来完成，但是我们倾向于将这些代码单独定义在按照约定定义的 Startup 类型中。由于注册 Startup 的核心目的是注册中间件，所以 Configure 方法是必需的，用于注册服务的 ConfigureServices 方法和用来设置第三方依赖注入容器的 ConfigureContainer 方法是可选的。如下所示的代码片段体现了典型的 Startup 类型定义方式。

```
public class Startup
{
    public void Configure(IApplicationBuilder app);
    public void ConfigureServices(IServiceCollection services);
    public void ConfigureContainer(FoobarContainerBuilder container);
}
```

除了显式调用 IWebHostBuilder 接口的 UseStartup 方法或者 UseStartup<TStartup>方法注册一个 Startup 类型，如果另外一个程序集中定义了合法的 Startup 类型，我们可以通过配置将它作为启动程序集。作为启动程序集的配置项目的名称为 startupAssembly，对应静态类型 WebHostDefaults 的只读字段 StartupAssemblyKey。

```
public static class WebHostDefaults
{
    public static readonly string StartupAssemblyKey;
    ...
}
```

一旦启动程序集通过配置的形式确定下来，系统就会试着从该程序集中找到一个具有最优匹配度的 Startup 类型。下面列举了一系列 Startup 类型的有效名称，Startup 类型加载器正是按照这个顺序从启动程序集类型列表中进行筛选的，如果最终没有任何一个类型满足条件，那么系统会抛出一个 InvalidOperationException 异常。

- Startup{EnvironmentName}（全名匹配）。
- Startup（全名匹配）。
- {StartupAssembly}.Startup{EnvironmentName}（全名匹配）。
- {StartupAssembly}.Startup （全名匹配）。

- Startup{EnvironmentName}（任意命名空间）。
- Startup（任意命名空间）。

由此可以看出，当 ASP.NET Core 框架从启动程序集中定位 Startup 类型时会优先选择类型名称与当前环境名称相匹配的。为了使读者对这个选择策略有更加深刻的认识，下面做一个实例演示。我们利用 Visual Studio 创建一个名为 App 的控制台应用，并编写了如下这段简单的程序。在如下所示的代码片段中，我们将当前命令行参数作为配置源。我们既没有调用 IWebHostBuilder 接口的 Configure 方法注册任何中间件，也没有调用 UseStartup 方法或者 UseStartup<TStartup>方法注册 Startup 类型。

```
class Program
{
    static void Main(string[] args)
    {
        Host.CreateDefaultBuilder(args).ConfigureWebHostDefaults(builder => builder
            .ConfigureLogging(options => options.ClearProviders()))
        .Build()
        .Run();
    }
}
```

我们创建了另一个名为 AppStartup 的类库项目，并在其中定义了如下 3 个继承自抽象类 StartupBase 的类型。根据命名约定，StartupDevelopment 类型和 StartupStaging 类型分别针对 Development 环境与 Staging 环境，而 Startup 类型则不针对某个具体的环境（环境中性）。

```
namespace AppStartup
{
    public abstract class StartupBase
    {
        public StartupBase() => Console.WriteLine(this.GetType().FullName);
        public void Configure(IApplicationBuilder app) { }
    }

    public class StartupDevelopment          : StartupBase { }
    public class StartupStaging              : StartupBase { }
    public class Startup                     : StartupBase { }
}
```

由于基类 StartupBase 的构造函数会将自身类型的全名输出到控制台上，所以可以根据这个输出确定哪个 Startup 类型会被选用。我们采用命令行的形式多次启动 App 应用，并以命令行参数的形式指定启动程序集名称和当前环境名称，控制台上呈现的输出结果如图 11-15 所示。（S1118）

如图 11-15 所示，如果没有显式指定环境名称，当前应用就会采用默认的 Production 环境名称，所以不针对具体环境的 AppStartup.Startup 被选择作为 Startup 类型。当我们将环境名称分别显式设置为 Development 和 Staging 之后，被选择作为 Startup 类型的分别为 StartupDevelopment 和 StartupStaging。

```
C:\App>dotnet run /startupAssembly=AppStartup
AppStartup.Startup
C:\App>dotnet run /startupAssembly=AppStartup /environment=Development
AppStartup.StartupDevelopment
C:\App>dotnet run /startupAssembly=AppStartup /environment=Staging
AppStartup.StartupStaging
C:\App>dotnet run /startupAssembly=AppStartup /environment=Production
AppStartup.Startup
```

图 11-15　利用额外程序集来提供 Startup 类型

与具体承载环境进行关联除了可以体现在 Startup 类型的命名（Startup{EnvironmentName}）上，还可以体现在对方法的命名（Configure{EnvironmentName}、Configure{EnvironmentName}Services 和 Configure{EnvironmentName}Container）上。如下所示的这个 Startup 类型针对开发环境、预发环境和产品环境定义了对应的方法，如果还有其他的环境，不具有环境名称的 3 个方法将会被使用，在上面介绍服务注册和中间件注册时已经有明确的说明。

```csharp
public class Startup
{
    public void Configure(IApplicationBuilder app);
    public void ConfigureServices(IServiceCollection services);
    public void ConfigureContainer(FoobarContainerBuilder container);

    public void ConfigureDevelopment(IApplicationBuilder app);
    public void ConfigureDevelopmentServices(IServiceCollection services);
    public void ConfigureDevelopmentContainer(FoobarContainerBuilder container);

    public void ConfigureStaging(IApplicationBuilder app);
    public void ConfigureStagingServices(IServiceCollection services);
    public void ConfigureStagingContainer(FoobarContainerBuilder container);

    public void ConfigureProduction(IApplicationBuilder app);
    public void ConfigureProductionServices(IServiceCollection services);
    public void ConfigureProductionContainer(FoobarContainerBuilder container);
}
```

11.5.2　IHostingStartup

除了通过注册 Startup 类型来初始化应用程序，我们还可以通过注册一个或者多个 IHostingStartup 服务达到类似的目的。由于 IHostingStartup 服务可以通过第三方程序集来提供，如果第三方框架、类库或者工具需要在应用启动时做相应的初始化工作，就可以将这些工作实现在注册的 IHostingStart 服务中。如下所示的代码片段是服务接口 IHostingStartup 的定义，它只定义了一个唯一的 Configure 方法，该方法可以利用输入参数得到当前使用的 IWebHost Builder 对象。

```csharp
public interface IHostingStartup
{
    void Configure(IWebHostBuilder builder);
}
```

IHostingStartup 服务是通过如下所示的 HostingStartupAttribute 特性来注册的。从给出的定义可以看出这是一个针对程序集的特性，在构造函数中指定的就是注册的 IHostingStartup 类型。由于在同一个程序集中可以多次使用该特性（AllowMultiple=true），所以同一个程序集可以提供多个 IHostingStartup 服务类型。

```
[AttributeUsage((AttributeTargets) AttributeTargets.Assembly, Inherited=false,
    AllowMultiple=true)]
public sealed class HostingStartupAttribute : Attribute
{
    public Type HostingStartupType { get; }
    public HostingStartupAttribute(Type hostingStartupType);
}
```

如果希望某个程序集提供的 IHostingStartup 服务类型能够真正应用到当前程序中，我们需要采用配置的形式对程序集进行注册。注册 IHostingStartup 程序集的配置项名称为 hostingStartupAssemblies，对应静态类型 WebHostDefaults 的只读字段 HostingStartupAssembliesKey。通过配置形式注册的程序集名称以分号进行分隔。当前应用名称会作为默认的 IHostingStartup 程序集进行注册，如果针对 IHostingStartup 类型的注册定义在该程序集中，就不需要对该程序集进行显式配置。

```
public static class WebHostDefaults
{
    public static readonly string HostingStartupAssembliesKey;
    public static readonly string PreventHostingStartupKey;
    public static readonly string HostingStartupExcludeAssembliesKey;
}
```

这一特性还有一个全局开关。如果不希望第三方程序集对当前应用程序进行干预，我们可以通过配置项 preventHostingStartup 关闭这一特性，该配置项的名称对应 WebHostDefaults 的 PreventHostingStartupKey 属性。另外，对于布尔值类型的配置项，"true"（不区分大小写）和 "1" 都表示 True，其他值则表示 False。WebHostDefaults 还通过 HostingStartupExcludeAssembliesKey 属性定义了另一个配置项，其名称为 hostingStartupExcludeAssemblies，用于设置需要被排除的程序集列表。

下面通过对前面的程序略加修改来演示针对 IHostingStartup 服务的初始化。首先在 App 项目中定义了如下这个实现了 IHostingStartup 接口的类型 Foo，它实现的 Configure 方法会在控制台上打印出相应的文字以确定该方法是否被调用。这个自定义的 IHostingStartup 服务类型通过 HostingStartupAttribute 特性进行注册。IHostingStartup 相关的配置只有通过环境变量和调用 IWebHostBuilder 接口的 UseSetting 方法进行设置才有效，所以虽然我们采用命令行参数提供原始配置，但是必须调用 UseSetting 方法将它们应用到 IWebHostBuilder 对象上。

```
[assembly: HostingStartup(typeof(Foo))]

class Program
{
    static void Main(string[] args)
    {
        var config = new ConfigurationBuilder()
```

```
                .AddCommandLine(args)
                .Build();

        Host.CreateDefaultBuilder().ConfigureWebHostDefaults(builder => builder
            .ConfigureLogging(options => options.ClearProviders())
            .UseSetting("hostingStartupAssemblies", config["hostingStartupAssemblies"])
            .UseSetting("preventHostingStartup", config["preventHostingStartup"])
            .Configure(app => app.Run(context => Task.CompletedTask)))
        .Build()
        .Run();
    }
}

public class Foo : IHostingStartup
{
    public void Configure(IWebHostBuilder builder) => Console.WriteLine("Foo.Configure()");
}
```

另一个 AppStartup 项目包含如下两个自定义的 IHostingStartup 服务类型 Bar 和 Baz,我们采用如下方式利用 HostingStartupAttribute 特性对它们进行了注册。

```
[assembly: HostingStartup(typeof(Bar))]
[assembly: HostingStartup(typeof(Baz))]

public abstract class HostingStartupBarBase : IHostingStartup
{
    public void Configure(IWebHostBuilder builder)
        => Console.WriteLine($"{GetType().Name}.Configure()");
}
public class Bar : HostingStartupBarBase {}
public class Baz : HostingStartupBarBase {}
```

我们采用命令行以图 11-16 所示的形式两次启动 App 应用。对于第一次应用启动，由于对启动程序集 AppStartup 进行了显式设置，由它提供的两个 IHostingStartup 服务（Bar 和 Baz）都得以正常执行。而注册的 IHostingStartup 服务 Foo，由于被注册到当前应用程序对应的程序集，虽然我们没有将它显式地添加到启动程序集列表中，但它依然会执行，而且是在其他程序集注册的 IHostingStartup 服务之前执行。至于第二次应用启动，由于我们通过命令行参数关闭了针对 IHostingStartup 服务的初始化功能，所以 Foo、Bar 和 Baz 这 3 个自定义 IHostingStartup 服务都不会执行。（S1119）

图 11-16　利用注册的 IHostingStartup 服务对应用程序进行初始化

11.5.3　IStartupFilter

中间件的注册离不开 IApplicationBuilder 对象，注册的 IStartup 服务的 Configure 方法会利用该对象帮助我们完成中间件的构建与注册。调用 IWebHostBuilder 接口的 Configure 方法时，系统会注册一个类型为 DelegateStartup 的 IStartup 服务，DelegateStartup 会利用提供的 Action<IApplicationBuilder>对象完成中间件的构建与注册。

如果调用 IWebHostBuilder 接口的 UseStartup 方法或者 UseStartup<Startup>方法注册了一个 Startup 类型并且该类型没有实现 IStartup 接口，系统就会按照约定规则创建一个类型为 ConventionBasedStartup 的 IStartup 服务。如果注册的 Startup 类型实现了 IStartup 接口，意味着注册的就是 IStartup 服务。

除了采用上述两种方式利用系统提供的 IStartup 服务来注册中间件，我们还可以通过注册 IStartupFilter 服务来达到相同的目的。一个应用程序可以注册多个 IStartupFilter 服务，它们会按照注册的顺序组成一个链表。IStartupFilter 接口具有如下所示的唯一方法 Configure，中间件的注册体现在它返回的 Action<IApplicationBuilder>对象上，而作为唯一参数的 Action<IApplicationBuilder>对象则代表了针对后续中间件的注册。

```
public interface IStartupFilter
{
    Action<IApplicationBuilder> Configure(Action<IApplicationBuilder> next);
}
```

虽然注册中间件是 IStartup 对象和 IStartupFilter 对象的核心功能，但是两者之间还是不尽相同的，它们之间的差异在于：IStartupFilter 对象的 Configure 方法会在 IStartup 对象的 Configure 方法之前执行。正因为如此，如果需要将注册的中间件前置或者后置，就需要利用 IStartupFilter 对象来注册它们。

接下来我们同样会演示一个针对 IStartupFilter 的中间件注册的实例。首先定义如下两个中间件类型 FooMiddleware 和 BarMiddleware，它们派生于同一个基类 StringContentMiddleware。当 InvokeAsync 方法被执行时，中间件在将请求分发给后续中间件之前和之后会分别将一段预先指定的文字写入响应消息的主体内容中，它们代表了中间件针对请求的前置和后置处理。

```
public abstract class StringContentMiddleware
{
    private readonly RequestDelegate        _next;
    private readonly string                 _preContents;
    private readonly string                 _postContents;

    public StringContentMiddleware(RequestDelegate next, string preContents,
        string postContents)
    {
        _next = next;
        _preContents = preContents;
        _postContents = postContents;
    }
```

```csharp
    public async Task InvokeAsync(HttpContext context)
    {
        await context.Response.WriteAsync(_preContents);
        await _next(context);
        await context.Response.WriteAsync(_postContents);
    }
}

public class FooMiddleware : StringContentMiddleware
{
    public FooMiddleware (RequestDelegate next) : base(next, "Foo=>", "Foo") { }
}

public class BarMiddleware : StringContentMiddleware
{
    public BarMiddleware (RequestDelegate next) : base(next, "Bar=>", "Bar=>") { }
}
```

可以采用如下方式对 FooMiddleware 和 BarMiddleware 这两个中间件进行注册。具体来说，我们为中间件类型 FooMiddleware 创建了一个自定义的 IStartupFilter 类型 FooStartupFilter，FooStartupFilter 实现的 Configure 方法中注册了这个中间件。FooStartupFilter 最终通过 IWebHostBuilder 接口的 ConfigureServices 方法进行注册。至于中间件类型 BarMiddleware，我们调用 IWebHostBuilder 接口的 Configure 方法对它进行注册。

```csharp
class Program
{
    static void Main()
    {
        Host.CreateDefaultBuilder().ConfigureWebHostDefaults(builder => builder
            .ConfigureServices(svcs => svcs
                .AddSingleton<IStartupFilter, FooStartupFilter>())
            .Configure(app => app
                .UseMiddleware<BarMiddleware>()
                .Run(context => context.Response.WriteAsync("...=>"))))
            .Build()
            .Run();
    }
}

public class FooStartupFilter : IStartupFilter
{
    public Action<IApplicationBuilder> Configure(Action<IApplicationBuilder> next)
    {
        return app => {
            app.UseMiddleware<FooMiddleware>();
            next(app);
        };
    }
}
```

由于IStartupFilter的Configure方法会在IStartup的Configure方法之前执行，所以对于最终构建的请求处理管道来说，FooMiddleware中间件置于BarMiddleware中间件前面。换句话说，当管道在处理某个请求的过程中，FooMiddleware中间件的前置请求处理操作会在BarMiddleware中间件之前执行，而它的后置请求处理操作则在BarMiddleware中间件之后执行。在启动这个程序之后，如果利用浏览器对该应用发起请求，得到的输出结果如图11-17所示。（S1120）

图11-17　IStartupFilter和Startup注册的中间件

第 12 章

管道（中篇）

第 11 章利用一系列实例演示了 ASP.NET Core 应用的编程模式，并借此来体验 ASP.NET Core 管道对请求的处理流程。这个管道由一个服务器和多个有序排列的中间件构成。这看似简单，但 ASP.NET Core 真实管道的构建其实是一个很复杂的过程。由于这个管道对 ASP.NET Core 框架非常重要，为了使读者对此有深刻的认识，本章不会介绍真实的管道，而是按照类似的设计重建一个 Mini 版的 ASP.NET Core 框架。

12.1 中间件委托链

第 13 章会详细介绍 ASP.NET Core 请求处理管道的构建以及它对请求的处理流程，作为对这一部分内容的铺垫，笔者将管道最核心的部分提取出来构建一个 Mini 版的 ASP.NET Core 框架。较之真正的 ASP.NET Core 框架，虽然重建的模拟框架要简单很多，但是它们采用完全一致的设计。为了能够在真实框架中找到对应物，在定义接口或者类型时会采用真实的名称，但是在 API 的定义上会做最大限度的简化。

12.1.1 HttpContext

一个 HttpContext 对象表示针对当前请求的上下文。要理解 HttpContext 上下文的本质，需要从请求处理管道的层面来讲。对于由一个服务器和多个中间件构成的管道来说，面向传输层的服务器负责请求的监听、接收和最终的响应，当它接收到客户端发送的请求后，需要将请求分发给后续中间件进行处理。对于某个中间件来说，完成自身的请求处理任务之后，在大部分情况下需要将请求分发给后续的中间件。请求在服务器与中间件之间，以及在中间件之间的分发是通过共享上下文的方式实现的。

如图 12-1 所示，当服务器接收到请求之后，会创建一个通过 HttpContext 表示的上下文对象，所有中间件都在这个上下文中完成针对请求的处理工作。那么一个 HttpContext 对象究竟会携带什么样的上下文信息？一个 HTTP 事务（Transaction）具有非常清晰的界定，如果从服务器的角度来说就是始于请求的接收，而终于响应的回复，所以请求和响应是两个基本的要素，也是 HttpContext 承载的最核心的上下文信息。

图 12-1　中间件共享上下文

我们可以将请求和响应理解为一个 Web 应用的输入与输出，既然 HttpContext 上下文是针对请求和响应的封装，那么应用程序就可以利用这个上下文对象得到当前请求所有的输入信息，也可以利用它完成我们所需的所有输出工作。所以，我们为 ASP.NET Core 模拟框架定义了如下这个极简版本的 HttpContext 类型。

```
public class HttpContext
{
    public HttpRequest                  Request { get; }
    public HttpResponse                 Response { get; }
}

public class HttpRequest
{
    public Uri                          Url { get; }
    public NameValueCollection          Headers { get; }
    public Stream                       Body { get; }
}

public class HttpResponse
{
    public int                          StatusCode { get; set; }
    public NameValueCollection          Headers { get; }
    public Stream                       Body { get; }
}
```

如上面的代码片段所示，我们可以利用 HttpRequest 对象得到当前请求的地址、请求消息的报头集合和主体内容。利用 HttpResponse 对象，我们不仅可以设置响应的状态码，还可以添加任意的响应报头和写入任意的主体内容。

12.1.2　中间件

HttpContext 对象承载了所有与当前请求相关的上下文信息，应用程序针对请求的响应也利用它来完成，所以可以利用一个 Action<HttpContext>类型的委托对象来表示针对请求的处理，我们姑且将它称为请求处理器（Handler）。但 Action<HttpContext>仅仅是请求处理器针对"同步"编程模式的表现形式，对于面向 Task 的异步编程模式，这个处理器应该表示成类型为 Func<HttpContext,Task>的委托对象。

由于这个表示请求处理器的委托对象具有非常广泛的应用，所以我们为它专门定义了如下这个 RequestDelegate 委托类型，可以看出它就是对 Func<HttpContext,Task>委托的表达。一个 RequestDelegate 对象表示的是请求处理器，那么中间件在模型中应如何表达？

```
public delegate Task RequestDelegate(HttpContext context);
```

作为请求处理管道核心组成部分的中间件可以表示成类型为 Func<RequestDelegate, RequestDelegate>的委托对象。换句话说，中间件的输入与输出都是一个 RequestDelegate 对象。我们可以这样来理解：对于管道中的某个中间件（图 12-2 所示的第一个中间件）来说，后续中间件组成的管道体现为一个 RequestDelegate 对象，由于当前中间件在完成了自身的请求处理任务之后，往往需要将请求分发给后续中间件进行处理，所以它需要将后续中间件构成的 RequestDelegate 对象作为输入。

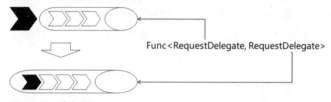

图 12-2　中间件

当代表当前中间件的委托对象执行之后，会将它自己"纳入"这个管道，那么代表新管道的 RequestDelegate 对象就成为该委托对象执行后的输出结果，所以中间件自然就表示成输入和输出类型均为 RequestDelegate 的 Func<RequestDelegate, RequestDelegate>对象。

12.1.3　中间件管道的构建

从事软件行业 10 多年来，笔者对架构设计越来越具有这样的认识：好的设计一定是"简单"的。所以在设计某个开发框架时笔者的目标是再简单点。上面介绍的请求处理管道的设计就具有"简单"的特质：Pipeline = Server + Middlewares。但是"再简单点"其实是可以的，我们可以将多个中间件组成一个单一的请求处理器。请求处理器可以通过 RequestDelegate 对象来表示，所以整个请求处理管道将具有更加简单的表达：Pipeline = Server + RequestDelegate（见图 12-3）。

图 12-3　Pipeline = Server + RequestDelegate

表示中间件的 Func<RequestDelegate, RequestDelegate>对象向表示请求处理器的 RequestDelegate 对象的转换是通过 IApplicationBuilder 对象来完成的。从接口命名可以看出，IApplicationBuilder 对象是用来构建"应用程序"（Application）的，实际上，由所有注册中间件构建的 RequestDelegate 对象就是对应用程序的表达，因为应用程序的意图完全是由注册的中间件达成的。

```
public interface IApplicationBuilder
```

```csharp
{
    RequestDelegate Build();
    IApplicationBuilder Use(Func<RequestDelegate, RequestDelegate> middleware);
}
```

如上所示的代码片段是模拟框架对 IApplicationBuilder 接口的简化定义。它的 Use 方法用来注册中间件，而 Build 方法则将所有的中间件按照注册的顺序组装成一个 RequestDelegate 对象。在如下所示的代码片段中，ApplicationBuilder 类型是对 IApplicationBuilder 接口的默认实现。我们给出的代码片段还体现了这样一个细节：当我们将注册的中间件转换成一个表示请求处理器的 RequestDelegate 对象时，会在管道的尾端添加一个处理器，用来响应一个状态码为 404 的响应。这个细节意味着如果没有注册任何的中间件或者注册的所有中间件都将请求分发给后续管道，那么应用程序会回复一个状态码为 404 的响应。

```csharp
public class ApplicationBuilder : IApplicationBuilder
{
    private readonly IList<Func<RequestDelegate, RequestDelegate>> _middlewares
        = new List<Func<RequestDelegate, RequestDelegate>>();

    public RequestDelegate Build()
    {
        RequestDelegate next = context =>
        {
            context.Response.StatusCode = 404;
            return Task.CompletedTask;
        };
        foreach (var middleware in _middlewares.Reverse())
        {
            next = middleware.Invoke(next);
        }
        return next;
    }

    public IApplicationBuilder Use(Func<RequestDelegate, RequestDelegate> middleware)
    {
        _middlewares.Add(middleware);
        return this;
    }
}
```

12.2 服务器

服务器在管道中的职责非常明确：负责 HTTP 请求的监听、接收和最终的响应。具体来说，启动后的服务器会绑定到指定的端口进行请求监听。一旦有请求抵达，服务器会根据该请求创建代表请求上下文的 HttpContext 对象，并将该上下文分发给注册的中间件进行处理。当中间件管道完成了针对请求的处理之后，服务器会将最终生成的响应回复给客户端。

12.2.1 IServer

在模拟的 ASP.NET Core 框架中,我们将服务器定义成一个极度简化的 IServer 接口。在如下所示的代码片段中,IServer 接口具有唯一的 StartAsync 方法,用来启动自身代表的服务器。服务器最终需要将接收的请求分发给注册的中间件,而注册的中间件最终会被 IApplicationBuilder 对象构建成一个代表请求处理器的 RequestDelegate 对象,StartAsync 方法的参数 handler 代表的就是这样一个对象。

```
public interface IServer
{
    Task StartAsync(RequestDelegate handler);
}
```

12.2.2 针对服务器的适配

面向应用层的 HttpContext 对象是对请求和响应的抽象与封装,但是请求最初是由面向传输层的服务器接收的,最终的响应也会由服务器回复给客户端。所有 ASP.NET Core 应用使用的都是同一个 HttpContext 类型,但是它们可以注册不同类型的服务器,应如何解决两者之间的适配问题?计算机领域有这样一句话:"任何问题都可以通过添加一个抽象层的方式来解决,如果解决不了,那就再加一层。"同一个 HttpContext 类型与不同服务器类型之间的适配问题自然也可以通过添加一个抽象层来解决。我们将定义在该抽象层的对象称为特性(Feature),特性可以视为对 HttpContext 某个方面的抽象化描述。

如图 12-4 所示,我们可以定义一系列特性接口来为 HttpContext 提供某个方面的上下文信息,具体的服务器只需要实现这些 Feature 接口即可。对于所有用来定义特性的接口,最重要的是提供请求信息的 IRequestFeature 接口和完成响应的 IResponseFeature 接口。

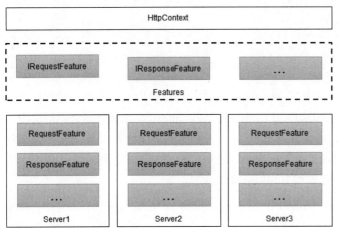

图 12-4　利用特性实现对不同服务器类型的适配

下面阐述用来适配不同服务器类型的特性在代码层面的定义。如下面的代码片段所示,我们定义了一个 IFeatureCollection 接口,用来表示存放特性的集合。可以看出,这是一个将 Type

和Object作为Key与Value的字典，Key代表注册Feature所采用的类型，而Value代表Feature对象本身，也就是说，我们提供的特性最终是以对应类型（一般为接口类型）进行注册的。为了便于编程，我们定义了Set<T>方法和Get<T>方法，用来设置与获取特性对象。

```
public interface IFeatureCollection : IDictionary<Type, object> { }

public class FeatureCollection : Dictionary<Type, object>, IFeatureCollection { }

public static partial class Extensions
{
    public static T Get<T>(this IFeatureCollection features)
        => features.TryGetValue(typeof(T), out var value) ? (T)value : default(T);

    public static IFeatureCollection Set<T>(this IFeatureCollection features, T feature)
    {
        features[typeof(T)] = feature;
        return features;
    }
}
```

最核心的两种特性类型就是分别用来表示请求和响应的特性，我们可以采用如下两个接口来表示。可以看出，IHttpRequestFeature接口和IHttpResponseFeature接口具有与类型HttpRequest和HttpResponse完全一致的成员定义。

```
public interface IHttpRequestFeature
{
    Uri                     Url { get; }
    NameValueCollection     Headers { get; }
    Stream                  Body { get; }
}
public interface IHttpResponseFeature
{
    int                     StatusCode { get; set; }
    NameValueCollection     Headers { get; }
    Stream                  Body { get; }
}
```

我们在前面给出了用于描述请求上下文的HttpContext类型的成员定义，下面介绍其具体实现。如下面的代码片段所示，表示请求和响应的HttpRequest与HttpResponse分别是由对应的特性（IHttpRequestFeature对象和IHttpResponseFeature对象）创建的。HttpContext对象本身则是通过一个表示特性集合的IFeatureCollection对象来创建的，它会在初始化过程中从这个集合中提取出对应的特性来创建HttpRequest对象和HttpResponse对象。

```
public class HttpContext
{
    public HttpRequest          Request { get; }
    public HttpResponse         Response { get; }

    public HttpContext(IFeatureCollection features)
    {
```

```
        Request = new HttpRequest(features);
        Response = new HttpResponse(features);
    }
}

public class HttpRequest
{
    private readonly IHttpRequestFeature _feature;

    public Uri Url
        => _feature.Url;
    public NameValueCollection Headers
        => _feature.Headers;
    public Stream Body
        => _feature.Body;

    public HttpRequest(IFeatureCollection features)
        => _feature = features.Get<IHttpRequestFeature>();
}

public class HttpResponse
{
    private readonly IHttpResponseFeature _feature;

    public NameValueCollection Headers
        => _feature.Headers;
    public Stream Body
        => _feature.Body;
    public int StatusCode
    {
        get => _feature.StatusCode;
        set => _feature.StatusCode = value;
    }

    public HttpResponse(IFeatureCollection features)
        => _feature = features.Get<IHttpResponseFeature>();
}
```

换句话说，我们利用 HttpContext 对象的 Request 属性提取的请求信息最初来源于 IHttpRequestFeature 对象，利用它的 Response 属性针对响应所做的任意操作最终都会作用到 IHttpResponseFeature 对象上。这两个对象最初是由注册的服务器提供的，这正是同一个 ASP.NET Core 应用可以自由地选择不同服务器类型的根源所在。

12.2.3 HttpListenerServer

在对服务器的职责和它与 HttpContext 的适配原理有了清晰的认识之后，我们可以尝试定义一个服务器。我们将接下来定义的服务器类型命名为 HttpListenerServer，因为它对请求的监听、接收和响应是由一个 HttpListener 对象来实现的。由于服务器接收到请求之后需要借助"特性"的适配来构建统一的请求上下文（即 HttpContext 对象），这也是中间件的执行上下文，所以提供针对性的特性实现是自定义服务类型的关键所在。

对 HttpListener 有所了解的读者都知道，当它在接收到请求之后同样会创建一个 HttpListenerContext 对象表示请求上下文。如果使用 HttpListener 对象作为 ASP.NET Core 应用的监听器，就意味着不仅所有的请求信息会来源于这个 HttpListenerContext 对象，我们针对请求的响应最终也需要利用这个上下文对象来完成。HttpListenerServer 对应特性所起的作用实际上就是在 HttpListenerContext 和 HttpContext 这两种上下文之间搭建起一座图 12-5 所示的桥梁。

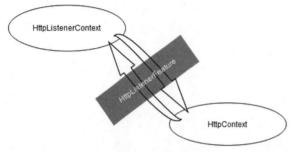

图 12-5　利用 HttpListenerFeature 适配 HttpListenerContext 和 HttpContext

图 12-5 中用来在 HttpListenerContext 和 HttpContext 这两个上下文类型之间完成适配的特性类型被命名为 HttpListenerFeature。如下面的代码片段所示，HttpListenerFeature 类型同时实现了针对请求和响应的特性接口 IHttpRequestFeature 与 IHttpResponseFeature。

```
public class HttpListenerFeature : IHttpRequestFeature, IHttpResponseFeature
{
    private readonly HttpListenerContext _context;
    public HttpListenerFeature(HttpListenerContext context)
        => _context = context;

    Uri IHttpRequestFeature.Url
        => _context.Request.Url;
    NameValueCollection IHttpRequestFeature.Headers
        => _context.Request.Headers;
    NameValueCollection IHttpResponseFeature.Headers
        => _context.Response.Headers;
    Stream IHttpRequestFeature.Body
        => _context.Request.InputStream;
    Stream IHttpResponseFeature.Body
        => _context.Response.OutputStream;
    int IHttpResponseFeature.StatusCode
    {
        get => _context.Response.StatusCode;
        set => _context.Response.StatusCode = value;
    }
}
```

创建 HttpListenerFeature 对象时需要提供一个 HttpListenerContext 对象，IHttpRequest Feature 接口的实现成员所提供的请求信息全部来源于这个 HttpListenerContext 上下文，IHttpResponseFeature 接口的实现成员针对响应的操作最终也转移到这个 HttpListenerContext 上下文上。如下所示的代码片段是针对 HttpListener 的服务器类型 HttpListenerServer 的完整定义。我们在创建 HttpListenerServer 对象的时候可以显式提供一组监听地址，如果没有提供，监听地址会默认设置"localhost:5000"。在实现的 StartAsync 方法中，我们启动了在构造函数中创建的 HttpListenerServer 对象，并且在一个无限循环中通过调用其 GetContextAsync 方法实现了针对请求的监听和接收。

```
public class HttpListenerServer : IServer
{
    private readonly HttpListener   _httpListener;
    private readonly string[]       _urls;

    public HttpListenerServer(params string[] urls)
    {
        _httpListener = new HttpListener();
        _urls = urls.Any() ? urls : new string[] { "http://localhost:5000/" };
    }

    public async Task StartAsync(RequestDelegate handler)
    {
        Array.ForEach(_urls, url => _httpListener.Prefixes.Add(url));
        _httpListener.Start();
        while (true)
        {
            var listenerContext = await _httpListener.GetContextAsync();
            var feature = new HttpListenerFeature(listenerContext);
            var features = new FeatureCollection()
                .Set<IHttpRequestFeature>(feature)
                .Set<IHttpResponseFeature>(feature);
            var httpContext = new HttpContext(features);
            await handler(httpContext);
            listenerContext.Response.Close();
        }
    }
}
```

当 HttpListener 监听到抵达的请求后，我们会得到一个 HttpListenerContext 对象，此时只需要利用它创建一个 HttpListenerFeature 对象，并且分别以 IHttpRequestFeature 接口和 IHttpResponseFeature 接口的形式注册到创建的 FeatureCollection 集合上即可。我们最终利用这个 FeatureCollection 集合创建出代表请求上下文的 HttpContext 对象，当将它作为参数调用由所有注册中间件共同构建的 RequestDelegate 对象时，中间件管道将接管并处理该请求。

12.3 承载服务

到目前为止，我们已经了解构成 ASP.NET Core 请求处理管道的两个核心要素（服务器和中间件），现在我们的目标是利用 .NET Core 承载服务来承载这一管道。毫无疑问，还需要通过实现 IHostedService 接口来定义对应的承载服务，为此我们定义了一个名为 WebHostedService 的承载服务。

12.3.1 WebHostedService

由于服务器是整个请求处理管道的"龙头"，所以从某种意义上来说，启动一个 ASP.NET Core 应用就是为了启动服务器，所以可以将服务的启动在 WebHostedService 承载服务中实现。如下面的代码片段所示，创建一个 WebHostedService 对象时，需要提供服务器对象和由所有注册中间件构建的 RequestDelegate 对象。在实现的 StartAsync 方法中，我们只需要调用服务器对象的 StartAsync 方法启动它即可。

```
public class WebHostedService : IHostedService
{
    private readonly IServer           _server;
    private readonly RequestDelegate   _handler;

    public WebHostedService(IServer server, RequestDelegate handler)
    {
        _server     = server;
        _handler    = handler;
    }

    public Task StartAsync(CancellationToken cancellationToken)
        => _server.StartAsync(_handler);
    public Task StopAsync(CancellationToken cancellationToken)
        => Task.CompletedTask;
}
```

到目前为止，我们基本上已经完成了所有的核心工作，如果能够将一个 WebHostedService 实例注册到 .NET Core 的承载系统中，它就能够帮助我们启动一个 ASP.NET Core 应用。为了使这个过程在编程上变得更加便利和"优雅"，我们定义了一个辅助的 WebHostBuilder 类型。

12.3.2 WebHostBuilder

要创建一个 WebHostedService 对象，必须显式地提供一个表示服务器的 IServer 对象，以及由所有注册中间件构建而成的 RequestDelegate 对象，WebHostBuilder 提供了更加便利和"优雅"的服务器与中间件注册方式。如下面的代码片段所示，WebHostBuilder 是对额外两个 Builder 对象的封装：一个是用来构建服务宿主的 IHostBuilder 对象，另一个是用来注册中间件并最终帮助我们创建 RequestDelegate 对象的 IApplicationBuilder 对象。

```
public class WebHostBuilder
```

```csharp
{
    public WebHostBuilder(IHostBuilder hostBuilder, IApplicationBuilder applicationBuilder)
    {
        HostBuilder         = hostBuilder;
        ApplicationBuilder  = applicationBuilder;
    }

    public IHostBuilder                HostBuilder { get; }
    public IApplicationBuilder         ApplicationBuilder { get; }
}
```

我们为 WebHostBuilder 定义了如下两个扩展方法：UseHttpListenerServer 方法完成了针对自定义的服务器类型 HttpListenerServer 的注册；Configure 方法提供了一个 Action<IApplicationBuilder>类型的参数，利用该参数来注册任意中间件。

```csharp
public static partial class Extensions
{
    public static WebHostBuilder UseHttpListenerServer(
        this WebHostBuilder builder, params string[] urls)
    {
        builder.HostBuilder.ConfigureServices(svcs => svcs
            .AddSingleton<IServer>(new HttpListenerServer(urls)));
        return builder;
    }

    public static WebHostBuilder Configure(this WebHostBuilder builder,
        Action<IApplicationBuilder> configure)
    {
        configure?.Invoke(builder.ApplicationBuilder);
        return builder;
    }
}
```

代表 ASP.NET Core 应用的请求处理管道最终是利用承载服务 WebHostedService 注册到 .NET Core 的承载系统中的，针对 WebHostedService 服务的创建和注册体现在为 IHostBuilder 接口定义的 ConfigureWebHost 扩展方法上。如下面的代码片段所示，ConfigureWebHost 方法定义了一个 Action<WebHostBuilder>类型的参数，利用该参数可以注册服务器、中间件及其他相关服务。

```csharp
public static partial class Extensions
{
    public static IHostBuilder ConfigureWebHost(this IHostBuilder builder,
        Action<WebHostBuilder> configure)
    {
        var webHostBuilder = new WebHostBuilder(builder, new ApplicationBuilder());
        configure?.Invoke(webHostBuilder);
        builder.ConfigureServices(svcs => svcs.AddSingleton<IHostedService>(provider => {
            var server = provider.GetRequiredService<IServer>();
            var handler = webHostBuilder.ApplicationBuilder.Build();
            return new WebHostedService(server, handler);
```

```
            }));
        return builder;
    }
}
```

在 ConfigureWebHost 方法中，我们创建了一个 ApplicationBuilder 对象，并利用它和当前的 IHostBuilder 对象创建了一个 WebHostBuilder 对象，然后将这个 WebHostBuilder 对象作为参数调用了指定的 Action<WebHostBuilder>委托对象。在此之后，我们调用 IHostBuilder 接口的 ConfigureServices 方法在依赖注入框架中注册了一个用于创建 WebHostedService 服务的工厂。对于由该工厂创建的 WebHostedService 对象来说，服务器来源于注册的服务，而作为请求处理器的 RequestDelegate 对象则由 ApplicationBuilder 对象根据注册的中间件构建而成。

12.3.3 应用构建

到目前为止，这个用来模拟 ASP.NET Core 请求处理管道的 Mini 版框架已经构建完成，下面尝试在它上面开发一个简单的应用。如下面的代码片段所示，我们调用静态类型 Host 的 CreateDefaultBuilder 方法创建了一个 IHostBuilder 对象，然后调用 ConfigureWebHost 方法并利用提供的 Action<WebHostBuilder>对象注册了 HttpListenerServer 服务器和 3 个中间件。在调用 Build 方法构建出作为服务宿主的 IHost 对象之后，我们调用其 Run 方法启动所有承载的 IHostedSerivce 服务。

```csharp
class Program
{
    static void Main()
    {
        Host.CreateDefaultBuilder()
            .ConfigureWebHost(builder => builder
                .UseHttpListenerServer()
                .Configure(app => app
                    .Use(FooMiddleware)
                    .Use(BarMiddleware)
                    .Use(BazMiddleware)))
            .Build()
            .Run();
    }

    public static RequestDelegate FooMiddleware(RequestDelegate next)
        => async context =>
        {
            await context.Response.WriteAsync("Foo=>");
            await next(context);
        };

    public static RequestDelegate BarMiddleware(RequestDelegate next)
        => async context =>
        {
            await context.Response.WriteAsync("Bar=>");
```

```
        await next(context);
    };

    public static RequestDelegate BazMiddleware(RequestDelegate next)
        => context => context.Response.WriteAsync("Baz");
}
```

由于中间件最终体现为一个类型为 Func<RequestDelegate, RequestDelegate>的委托对象，所以可以利用与之匹配的方法来定义中间件。演示实例中定义的 3 个中间件（FooMiddleware、BarMiddleware 和 BazMiddleware）对应的正是 3 个静态方法，它们调用 WriteAsync 扩展方法在响应中写了一段文字。

```
public static partial class Extensions
{
    public static Task WriteAsync(this HttpResponse response, string contents)
    {
        var buffer = Encoding.UTF8.GetBytes(contents);
        return response.Body.WriteAsync(buffer, 0, buffer.Length);
    }
}
```

应用启动之后，如果利用浏览器向应用程序采用的默认监听地址（"http://localhost:5000"）发送一个请求，得到的输出结果如图 12-6 所示。浏览器上呈现的文字正是注册的 3 个中间件写入的。

图 12-6　在模拟框架上构建的 ASP.NET Core 应用

第 13 章

管道（下篇）

有了第 11 章和第 12 章的铺垫，读者对 ASP.NET Core 框架的请求处理管道已经有了相对充分的了解。第 12 章使用少量的代码模拟了 ASP.NET Core 框架的实现，虽然两者在设计思想上完全一致，但是省略了太多的细节。本章会弥补这些细节，还原一个真实的 ASP.NET Core 框架。

13.1 请求上下文

ASP.NET Core 请求处理管道由一个服务器和一组有序排列的中间件构成，所有中间件针对请求的处理都在通过 HttpContext 对象表示的上下文中进行。由于应用程序总是利用服务器来完成对请求的接收和响应工作，所以原始请求上下文的描述由注册的服务器类型来决定。但是 ASP.NET Core 需要在上层提供具有一致性的编程模型，所以我们需要一个抽象的、不依赖具体服务器类型的请求上下文描述，这就是本节着重介绍的 HttpContext。

13.1.1 HttpContext

在第 12 章创建的模拟管道中，我们定义了一个简易版的 HttpContext 类，它只包含表示请求和响应的两个属性，实际上，真正的 HttpContext 具有更加丰富的成员定义。对于一个 HttpContext 对象来说，除了描述请求和响应的 Request 属性与 Response 属性，我们还可以通过它获取与当前请求相关的其他上下文信息，如用来表示当前请求用户的 ClaimsPrincipal 对象、描述当前 HTTP 连接的 ConnectionInfo 对象和用于控制 Web Socket 的 WebSocketManager 对象等。除此之外，我们还可以通过 Session 属性获取并控制当前会话，也可以通过 TraceIdentifier 属性获取或者设置调试追踪的 ID。

```
public abstract class HttpContext
{
    public abstract HttpRequest                 Request { get; }
    public abstract HttpResponse                Response { get; }

    public abstract ClaimsPrincipal             User { get; set; }
    public abstract ConnectionInfo              Connection { get; }
```

```
    public abstract WebSocketManager                    WebSockets { get; }
    public abstract ISession                            Session { get; set; }
    public abstract string                              TraceIdentifier { get; set; }

    public abstract IDictionary<object, object>         Items { get; set; }
    public abstract CancellationToken                   RequestAborted { get; set; }
    public abstract IServiceProvider                    RequestServices { get; set; }
    ...
}
```

当客户端中止请求（如请求超时）时，我们可以通过 RequestAborted 属性返回的 CancellationToken 对象接收到通知，进而及时中止正在进行的请求处理操作。如果需要针对整个管道共享一些与当前上下文相关的数据，我们可以将它保存在通过 Items 属性表示的字典中。HttpContext 的 RequestServices 返回的是针对当前请求的 IServiceProvider 对象，换句话说，该对象的生命周期与表示当前请求上下文的 HttpContext 对象绑定。对于一个 HttpContext 对象来说，表示请求和响应的 Request 属性与 Response 属性是它最重要的两个成员，请求通过如下这个抽象类 HttpRequest 表示。

```
public abstract class HttpRequest
{
    public abstract HttpContext                         HttpContext { get; }
    public abstract string                              Method { get; set; }
    public abstract string                              Scheme { get; set; }
    public abstract bool                                IsHttps { get; set; }
    public abstract HostString                          Host { get; set; }
    public abstract PathString                          PathBase { get; set; }
    public abstract PathString                          Path { get; set; }
    public abstract QueryString                         QueryString { get; set; }
    public abstract IQueryCollection                    Query { get; set; }
    public abstract string                              Protocol { get; set; }
    public abstract IHeaderDictionary                   Headers { get; }
    public abstract IRequestCookieCollection            Cookies { get; set; }
    public abstract string                              ContentType { get; set; }
    public abstract Stream                              Body { get; set; }
    public abstract bool                                HasFormContentType { get; }
    public abstract IFormCollection                     Form { get; set; }

    public abstract Task<IFormCollection>               ReadFormAsync(
        CancellationToken cancellationToken);
}
```

如上所示的抽象类 HttpRequest 是对 HTTP 请求的描述，它是 HttpContext 对象的只读属性 Request 的返回类型。我们可以利用 HttpRequest 对象获取与当前请求相关的各种信息，如请求的协议（HTTP 或者 HTTPS）、HTTP 方法、地址，也可以获取代表请求的 HTTP 消息的首部和主体。HttpRequest 中的属性/方法列表如表 13-1 所示。

表 13-1　HttpRequest 中的属性/方法列表

属性/方法	含　义
Body	读取请求主体内容的输入流对象
ContentLength	请求主体内容的字节数
ContentType	请求主体内容的媒体类型（如 text/xml、text/json 等）
Cookies	请求携带的 Cookie 列表，对应 HTTP 请求消息的 Cookie 首部。该属性的返回类型为 IRequestCookieCollection 接口，它具有与字典类似的数据结构，其 Key 和 Value 分别代表 Cookie 的名称与值
Form	请求提交的表单。该属性的返回类型为 IFormCollection，它具有一个与字典类似的数据结构，其 Key 和 Value 分别代表表单元素的名称与携带值。由于同一个表单中可以包含多个同名元素，所以 Value 是一个字符串列表
HasFormContentType	请求主体是否具有一个针对表单的媒体类型，一般来说，表单内容采用的媒体类型为 application/x-www-form-urlencoded 或者 multipart/form-data
Headers	请求首部列表。该属性的返回类型为 IHeaderDictionary，它具有一个与字典类似的数据结构，其 Key 和 Value 分别代表首部的名称与携带值。由于同一个请求中可以包含多个同名首部，所以 Value 是一个字符串列表
Host	请求目标地址的主机名（含端口号）。该属性返回的是一个 HostString 对象，它是对主机名称和端口号的封装
IsHttps	是否是一个采用 TLS/SSL 的 HTTPS 请求
Method	请求采用的 HTTP 方法
PathBase	请求的基础路径，一般体现为应用站点所在路径
Path	请求相对于 PathBase 的路径。如果当前请求的 URL 为 "http://www.artech.com/webapp/home/index"（PathBase 为 "/webapp"），那么 Path 属性返回 "/home/index"
Protocol	请求采用的协议及其版本，如 HTTP/1.1 代表针对 1.1 版本的 HTTP 协议。
Query	请求携带的查询字符串。该属性的返回类型为 IQueryCollection，它具有一个与字典类似的数据结构，其 Key 和 Value 分别代表以查询字符串形式定义的变量名称与值。由于查询字符串中可以定义多个同名变量（如 "?foobar=123&foobar=456"），所以 Value 是一个字符串列表
QueryString	请求携带的查询字符串。该属性返回一个 QueryString 对象，它的 Value 属性值代表整个查询字符串的原始表现形式，如 "{?foo=123&bar=456}"
Scheme	请求采用的协议前缀（"http" 或者 "https"）
ReadFormAsync	从请求的主体部分读取表单内容。该属性的返回类型为 IFormCollection，它具有一个与字典类似的数据结构，其 Key 和 Value 分别代表表单元素的名称与携带值。由于同一个表单可以包含多个同名元素，所以 Value 是一个字符串列表

在了解了表示请求的抽象类 HttpRequest 之后，下面介绍另一个与之相对的用于描述响应的 HttpResponse 类型。如下面的代码片段所示，HttpResponse 依然是一个抽象类，我们可以通过它定义的属性和方法来控制对请求的响应。从原则上讲，我们对请求所做的任意形式的响应都可以利用它来实现。

```
public abstract class HttpResponse
{
    public abstract HttpContext              HttpContext { get; }
    public abstract int                      StatusCode { get; set; }
    public abstract IHeaderDictionary        Headers { get; }
```

```csharp
public abstract Stream                    Body { get; set; }
public abstract long?                     ContentLength { get; set; }
public abstract IResponseCookies          Cookies { get; }
public abstract bool                      HasStarted { get; }

public abstract void OnStarting(Func<object, Task> callback, object state);
public virtual void OnStarting(Func<Task> callback);
public abstract void OnCompleted(Func<object, Task> callback, object state);
public virtual void RegisterForDispose(IDisposable disposable);
public virtual void OnCompleted(Func<Task> callback);
public virtual void Redirect(string location);
public abstract void Redirect(string location, bool permanent);
}
```

当通过表示当前上下文的 HttpContext 对象得到表示响应的 HttpResponse 对象之后，我们不仅可以将内容写入响应消息的主体部分，还可以设置响应状态码，并添加相应的报头。HttpResponse 中的属性/方法列表如表 13-2 所示。

表 13-2　HttpResponse 中的属性/方法列表

属性/方法	含 义
Body	将内容写入响应消息主体部分的输出流对象中
ContentLength	响应消息主体内容的长度（字节数）
ContentType	响应内容采用的媒体类型/MIME 类型
Cookies	返回一个用于设置（添加或者删除）响应 Cookie（对应响应消息的 Set-Cookie 首部）的 ResponseCookies 对象
HasStarted	表示响应是否已经开始发送。由于 HTTP 响应消息总是从首部开始发送，所以这个属性表示响应首部是否开始发送
Headers	响应消息的首部集合。该属性的返回类型为 IHeaderDictionary，它具有一个与字典类似的数据结构，其 Key 和 Value 分别代表首部的名称与携带值。由于同一个响应消息中可以包含多个同名首部，所以 Value 是一个字符串列表
StatusCode	响应状态码
OnCompleted	注册一个回调操作，以便在响应消息发送结束时自动执行
OnStarting	注册一个回调操作，以便在响应消息开始发送时自动执行
Redirect	发送一个针对指定目标的地址的重定向响应消息。参数 permanent 表示重定向类型，即状态码为 "302" 的暂时重定向或者状态码为 "301" 的永久重定向
RegisterForDispose	注册一个需要回收释放的对象，该对象对应的类型必须实现 IDisposable 接口，所谓的释放体现在对其 Dispose 方法的调用

13.1.2　服务器适配

由于应用程序总是利用这个抽象的 HttpContext 上下文来获取与当前请求有关的信息，需要完成的所有响应操作也总是作用在这个 HttpContext 对象上，所以不同的服务器与这个抽象的 HttpContext 需要进行 "适配"。通过第 12 章针对模拟框架的介绍可知，ASP.NET Core 框架会采用一种针对特性（Feature）的适配方式。

如图 13-1 所示，ASP.NET Core 框架为抽象的 HttpContext 定义了一系列标准的特性接口来对请求上下文的各个方面进行描述。在一系列标准的接口中,最核心的是用来描述请求的 IHttpRequestFeature 接口和描述响应的 IHttpResponseFeature 接口。我们在应用层使用的 HttpContext 上下文就是根据这样一组特性集合来创建的,对于某个具体的服务器来说,它需要提供这些特性接口的实现,并在接收到请求之后利用自行实现的特性来创建 HttpContext 上下文。

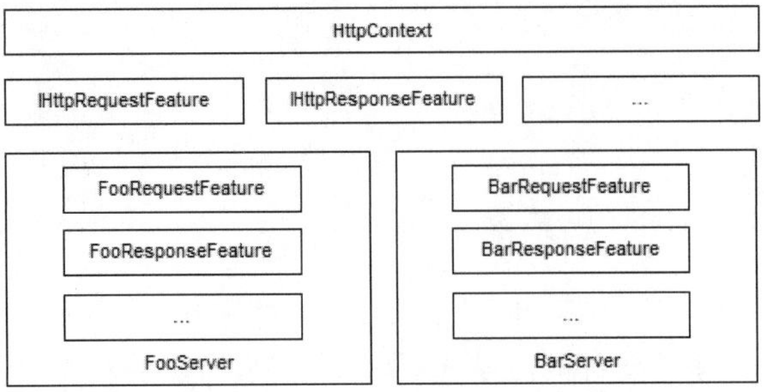

图 13-1 服务器与 HttpContext 之间针对 Feature 的适配

由于 HttpContext 上下文是利用服务器提供的特性集合创建的,所以可以统一使用抽象的 HttpContext 获取真实的请求信息,也能驱动服务器完成最终的响应工作。在 ASP.NET Core 框架中,由服务器提供的特性集合通过 IFeatureCollection 接口表示。第 12 章创建的模拟框架为 IFeatureCollection 接口提供了一个极简版的定义,实际上该接口具有更加丰富的成员定义。

```
public interface IFeatureCollection : IEnumerable<KeyValuePair<Type, object>>
{
    TFeature Get<TFeature>();
    void Set<TFeature>(TFeature instance);

    bool        IsReadOnly { get; }
    object      this[Type key] { get; set; }
    int         Revision { get; }
}
```

如上面的代码片段所示,一个 IFeatureCollection 对象本质上就是一个 Key 和 Value 类型分别为 Type 与 Object 的字典。通过调用 Set 方法可以将一个特性对象作为 Value,以指定的类型(一般为特性接口)作为 Key 添加到这个字典中,并通过 Get 方法根据该类型获取它。除此之外,特性的注册和获取也可以利用定义的索引来完成。如果 IsReadOnly 属性返回 True,就意味着不能注册新的特性或者修改已经注册的特性。整数类型的只读属性 Revision 可以视为 IFeatureCollection 对象的版本,不论是采用何种方式注册新的特性还是修改现有的特性,都将改变该属性的值。

具有如下定义的 FeatureCollection 类型是对 IFeatureCollection 接口的默认实现。它具有两个构造函数重载:默认无参构造函数帮助我们创建一个空的特性集合,另一个构造函数则需要指定一个 IFeatureCollection 对象来提供默认或者后备特性对象。对于采用第二个构造函数创建的 FeatureCollection 对象来说,当我们通过指定的类型试图获取对应的特性对象时,如果没有注册到当前 FeatureCollection 对象上,它会从这个后备的 IFeatureCollection 对象中查找目标特性。

```
public class FeatureCollection : IFeatureCollection
```

```
{
    //其他成员
    public FeatureCollection();
    public FeatureCollection(IFeatureCollection defaults);
}
```

对于一个 FeatureCollection 对象来说，它的 IsReadOnly 属性总是返回 False，所以它永远是可读可写的。对于调用默认无参构造函数创建的 FeatureCollection 对象来说，它的 Revision 属性默认返回零。如果我们通过指定另一个 IFeatureCollection 对象为参数调用第二个构造函数来创建一个 FeatureCollection 对象，前者的 Revision 属性值将成为后者同名属性的默认值。无论采用何种形式（调用 Set 方法或者索引）添加一个新的特性或者改变一个已经注册的特性，FeatureCollection 对象的 Revision 属性都将自动递增。上述这些特性都体现在如下所示的调试断言中。

```
var defaults = new FeatureCollection();
Debug.Assert(defaults.Revision == 0);

defaults.Set<IFoo>(new Foo());
Debug.Assert(defaults.Revision == 1);

defaults[typeof(IBar)] = new Bar();
Debug.Assert(defaults.Revision == 2);

FeatureCollection features = new FeatureCollection(defaults);
Debug.Assert(features.Revision == 2);
Debug.Assert(features.Get<IFoo>().GetType() == typeof(Foo));

features.Set<IBaz>(new Baz());
Debug.Assert(features.Revision == 3);
```

最初由服务器提供的 IFeatureCollection 对象体现在 HttpContext 类型的 Features 属性上。虽然特性最初是为了解决不同的服务器类型与统一的 HttpContext 上下文之间的适配设计的，但是它的作用不限于此。由于注册的特性是附加在代表当前请求的 HttpContext 上下文上，所以可以将任何基于当前请求的对象以特性的方式进行保存，它其实与 Items 属性的作用类似。

```
public abstract class HttpContext
{
    public abstract IFeatureCollection          Features { get; }
    ...
}
```

上述这种基于特性来实现不同类型的服务器与统一请求上下文之间的适配体现在 DefaultHttpContext 类型上，它是对 HttpContext 这个抽象类型的默认实现。DefaultHttpContext 具有一个如下所示的构造函数，作为参数的 IFeatureCollection 对象就是由服务器提供的特性集合。

```
public class DefaultHttpContext : HttpContext
{
    public DefaultHttpContext(IFeatureCollection features);
}
```

不论是组成管道的中间件还是建立在管道上的应用，在默认情况下都利用 DefaultHttp
Context 对象来获取当前请求的相关信息，并利用这个对象完成针对请求的响应。但是
DefaultHttpContext 对象在这个过程中只是一个"代理"，针对它的调用（属性或者方法）最终都
需要转发给由具体服务器创建的那个原始上下文，在构造函数中指定的 IFeatureCollection 对象
所代表的特性集合成为这两个上下文对象进行沟通的唯一渠道。对于定义在 DefaultHttpContext
中的所有属性，它们几乎都具有一个对应的特性，这些特性都对应一个接口。描述原始 HTTP
上下文的特性接口如表 13-3 所示。

表 13-3　描述原始 HTTP 上下文的特性接口

zFeature	属　　性	描　　述
IHttpRequestFeature	Request	获取描述请求的基本信息
IHttpResponseFeature	Response	控制对请求的响应
IHttpConnectionFeature	Connection	提供描述当前 HTTP 连接的基本信息
IItemsFeature	Items	提供用户存放针对当前请求的对象容器
IHttpRequestLifetimeFeature	RequestAborted	传递请求处理取消通知和中止当前请求处理
IServiceProvidersFeature	RequestServices	提供根据服务注册创建的 ServiceProvider
ISessionFeature	Session	提供描述当前会话的 Session 对象
IHttpRequestIdentifierFeature	TraceIdentifier	为追踪日志（Trace）提供针对当前请求的唯一标识
IHttpWebSocketFeature	WebSockets	管理 Web Socket

对于表 13-3 列举的众多特性接口，后续相关章节中都会涉及，目前只介绍表示请求和响应
的 IHttpRequestFeature 接口与 IHttpResponseFeature 接口。从下面给出的代码片段可以看出，这
两个接口具有与抽象类 HttpRequest 和 HttpResponse 一致的定义。对于 DefaultHttpContext 类型
来说，它的 Request 属性和 Response 属性返回的具体类型为 DefaultHttpRequest 与 Default
HttpResponse，它们分别利用这两个特性实现了定义在基类（HttpRequest 和 HttpResponse）的所
有抽象成员。

```
public interface IHttpRequestFeature
{
    Stream                  Body { get; set; }
    IHeaderDictionary       Headers { get; set; }
    string                  Method { get; set; }
    string                  Path { get; set; }
    string                  PathBase { get; set; }
    string                  Protocol { get; set; }
    string                  QueryString { get; set; }
    string                  Scheme { get; set; }
}

public interface IHttpResponseFeature
{
    Stream                  Body { get; set; }
    bool                    HasStarted { get; }
    IHeaderDictionary       Headers { get; set; }
    string                  ReasonPhrase { get; set; }
```

```
    int                     StatusCode { get; set; }
    void OnCompleted(Func<object, Task> callback, object state);
    void OnStarting(Func<object, Task> callback, object state);
}
```

13.1.3 获取上下文

如果第三方组件需要获取表示当前请求上下文的 HttpContext 对象，就可以通过注入 IHttpContextAccessor 服务来实现。IHttpContextAccessor 对象提供如下所示的 HttpContext 属性返回针对当前请求的 HttpContext 对象，由于该属性并不是只读的，所以当前的 HttpContext 也可以通过该属性进行设置。

```
public interface IHttpContextAccessor
{
    HttpContext HttpContext { get; set; }
}
```

ASP.NET Core 框架提供的 HttpContextAccessor 类型可以作为 IHttpContextAccessor 接口的默认实现。从如下所示的代码片段可以看出，HttpContextAccessor 将提供的 HttpContext 对象以一个 AsyncLocal<HttpContext>对象的方式存储起来，所以在整个请求处理的异步处理流程中都可以利用它得到同一个 HttpContext 对象。

```
public class HttpContextAccessor : IHttpContextAccessor
{
    private static AsyncLocal<HttpContext> _httpContextCurrent
        = new AsyncLocal<HttpContext>();
    public HttpContext HttpContext
    {
        get => _httpContextCurrent.Value;
        set => _httpContextCurrent.Value = value;
    }
}
```

针对 IHttpContextAccessor/HttpContextAccessor 的服务注册可以通过如下所示的 AddHttpContextAccessor 扩展方法来完成。由于它调用的是 IServiceCollection 接口的 TryAddSingleton<TService, TImplementation>扩展方法，所以不用担心多次调用该方法而出现服务的重复注册问题。

```
public static class HttpServiceCollectionExtensions
{
    public static IServiceCollection AddHttpContextAccessor(
        this IServiceCollection services)
    {
        services.TryAddSingleton<IHttpContextAccessor, HttpContextAccessor>();
        return services;
    }
}
```

13.1.4 上下文的创建与释放

利用注入的 IHttpContextAccessor 服务的 HttpContext 属性得到当前 HttpContext 上下文的前提是该属性在此之前已经被赋值，在默认情况下，该属性是通过默认注册的 IHttpContextFactory 服务赋值的。管道在开始处理请求前对 HttpContext 上下文的创建，以及请求处理完成后对它的回收释放都是通过 IHttpContextFactory 对象完成的。IHttpContextFactory 接口定义了如下两个方法：Create 方法会根据提供的特性集合来创建 HttpContext 对象，Dispose 方法则负责将提供的 HttpContext 对象释放。

```
public interface IHttpContextFactory
{
    HttpContext Create(IFeatureCollection featureCollection);
    void Dispose(HttpContext httpContext);
}
```

ASP.NET Core 框架提供如下所示的 DefaultHttpContextFactory 类型作为对 IHttpContextFactory 接口的默认实现，作为默认 HttpContext 上下文的 DefaultHttpContext 对象就是由它创建的。如下面的代码片段所示，在 IHttpContextAccessor 服务被注册的情况下，ASP.NET Core 框架将调用第二个构造函数来创建 HttpContextFactory 对象。在 Create 方法中，它根据提供的 IFeatureCollection 对象创建一个 DefaultHttpContext 对象，在返回该对象之前，它会将该对象赋值给 IHttpContextAccessor 对象的 HttpContext 属性。

```
public class DefaultHttpContextFactory : IHttpContextFactory
{
    private readonly IHttpContextAccessor    _httpContextAccessor;
    private readonly FormOptions             _formOptions;
    private readonly IServiceScopeFactory    _serviceScopeFactory;

    public DefaultHttpContextFactory(IServiceProvider serviceProvider)
    {
        _httpContextAccessor = serviceProvider.GetService<IHttpContextAccessor>();
        _formOptions = serviceProvider.GetRequiredService<IOptions<FormOptions>>().Value;
        _serviceScopeFactory = serviceProvider.GetRequiredService<IServiceScopeFactory>();
    }

    public HttpContext Create(IFeatureCollection featureCollection)
    {
        var httpContext = CreateHttpContext(featureCollection);
        if (_httpContextAccessor != null)
        {
            _httpContextAccessor.HttpContext = httpContext;
        }
        httpContext.FormOptions = _formOptions;
        httpContext.ServiceScopeFactory = _serviceScopeFactory;
        return httpContext;
    }

    private static DefaultHttpContext CreateHttpContext(
```

```
        IFeatureCollection featureCollection)
    {
        if (featureCollection is IDefaultHttpContextContainer container)
        {
            return container.HttpContext;
        }

        return new DefaultHttpContext(featureCollection);
    }
    public void Dispose(HttpContext httpContext)
    {
        if (_httpContextAccessor != null)
        {
            _httpContextAccessor.HttpContext = null;
        }
    }
}
```

如上面的代码片段所示，HttpContextFactory 在创建出 DefaultHttpContext 对象并将它设置到 IHttpContextAccessor 对象的 HttpContext 属性上之后，它还会设置 DefaultHttpContext 对象的 FormOptions 属性和 ServiceScopeFactory 属性，前者表示针对表单的配置选项，后者是用来创建服务范围的工厂。当 Dispose 方法执行的时候，DefaultHttpContextFactory 对象会将 IHttpContextAccessor 服务的 HttpContext 属性设置为 Null。

13.1.5　RequestServices

ASP.NET Core 框架中存在两个用于提供所需服务的依赖注入容器：一个针对应用程序，另一个针对当前请求。绑定到 HttpContext 上下文 RequestServices 属性上针对当前请求的 IServiceProvider 来源于通过 IServiceProvidersFeature 接口表示的特性。如下面的代码片段所示，IServiceProvidersFeature 接口定义了唯一的属性 RequestServices，可以利用它设置和获取与请求绑定的 IServiceProvider 对象。

```
public interface IServiceProvidersFeature
{
    IServiceProvider RequestServices { get; set; }
}
```

如下所示的 RequestServicesFeature 类型是对 IServiceProvidersFeature 接口的默认实现。如下面的代码片段所示，当我们创建一个 RequestServicesFeature 对象时，需要提供当前的 HttpContext 上下文和创建服务范围的 IServiceScopeFactory 工厂。RequestServicesFeature 对象的 RequestServices 属性提供的 IServiceProvider 对象来源于 IServiceScopeFactory 对象创建的服务范围，在请求处理过程中提供的 Scoped 服务实例的生命周期被限定在此范围之内。

```
public class RequestServicesFeature :
    IServiceProvidersFeature, IDisposable, IAsyncDisposable
{
    private readonly IServiceScopeFactory    _scopeFactory;
```

```csharp
private IServiceProvider                   _requestServices;
private IServiceScope                      _scope;
private bool                               _requestServicesSet;
private readonly HttpContext               _context;

public RequestServicesFeature(HttpContext context, IServiceScopeFactory scopeFactory)
{
    _context              = context;
    _scopeFactory         = scopeFactory;
}

public IServiceProvider RequestServices
{
    get
    {
        if (!_requestServicesSet && _scopeFactory != null)
        {
            _context.Response.RegisterForDisposeAsync(this);
            _scope                = _scopeFactory.CreateScope();
            _requestServices      = _scope.ServiceProvider;
            _requestServicesSet   = true;
        }
        return _requestServices;
    }

    set
    {
        _requestServices      = value;
        _requestServicesSet   = true;
    }
}

public ValueTask DisposeAsync()
{
    switch (_scope)
    {
        case IAsyncDisposable asyncDisposable:
            var vt = asyncDisposable.DisposeAsync();
            if (!vt.IsCompletedSuccessfully)
            {
                return Awaited(this, vt);
            }
            vt.GetAwaiter().GetResult();
            break;
        case IDisposable disposable:
            disposable.Dispose();
            break;
    }

    _scope                = null;
```

```
        _requestServices        = null;
        return default;

        static async ValueTask Awaited(RequestServicesFeature servicesFeature,
            ValueTask vt)
        {
            await vt;
            servicesFeature._scope = null;
            servicesFeature._requestServices = null;
        }
    }

    public void Dispose() => DisposeAsync().ConfigureAwait(false).GetAwaiter().GetResult();
}
```

为了在完成请求处理之后释放所有非 Singleton 服务实例，我们必须及时释放创建的服务范围。针对服务范围的释放实现在 DisposeAsync 方法中，该方法是针对 IAsyncDisposable 接口的实现。在服务范围被创建时，RequestServicesFeature 对象会调用表示当前响应的 HttpResponse 对象的 RegisterForDisposeAsync 方法将自身添加到需要释放的对象列表中，当响应完成之后，DisposeAsync 方法会自动被调用，进而将针对当前请求的服务范围联通该范围内的服务实例释放。

前面提及，除了创建返回的 DefaultHttpContext 对象，DefaultHttpContextFactory 对象还会设置用于创建服务范围的工厂（对应如下所示的 ServiceScopeFactory 属性）。用来提供基于当前请求依赖注入容器的 RequestServicesFeature 特性正是根据 IServiceScopeFactory 对象创建的。

```
public sealed class DefaultHttpContext : HttpContext
{
    public override IServiceProvider        RequestServices {get;set}
    public IServiceScopeFactory             ServiceScopeFactory { get; set; }
}
```

13.2　IServer + IHttpApplication

ASP.NET Core 的请求处理管道由一个服务器和一组中间件构成，但对于面向传输层的服务器来说，它其实没有中间件的概念。当服务器接收到请求之后，会将该请求分发给一个处理器进行处理，对服务器而言，这个处理器就是一个 HTTP 应用，此应用通过 IHttpApplication<TContext>接口来表示。由于服务器是通过 IServer 接口表示的，所以可以将 ASP.NET Core 框架的核心视为由 IServer 和 IHttpApplication<TContext>对象组成的管道（见图 13-2）。

图 13-2　由 IServer 和 IHttpApplication<TContext>对象组成的管道

13.2.1　IServer

由于服务器是整个请求处理管道的"龙头"，所以启动和关闭应用的最终目的是启动和关闭服务器。ASP.NET Core 框架中的服务器通过 IServer 接口来表示，该接口具有如下所示的 3 个成员，其中由服务器提供的特性就保存在其 Features 属性表示的 IFeatureCollection 集合中。
IServer 接口的 StartAsync<TContext>方法与 StopAsync 方法分别用来启动和关闭服务器。

```
public interface IServer : IDisposable
{
    IFeatureCollection Features { get; }

    Task StartAsync<TContext>(IHttpApplication<TContext> application,
        CancellationToken cancellationToken);
    Task StopAsync(CancellationToken cancellationToken);
}
```

服务器在开始监听请求之前总是绑定一个或者多个监听地址，这个地址是应用程序从外部指定的。具体来说，应用程序指定的监听地址会封装成一个特性，并且在服务器启动之前被添加到它的特性集合中。这个承载了监听地址列表的特性通过如下所示的 IServerAddressesFeature 接口来表示，该接口除了有一个表示地址列表的 Addresses 属性，还有一个布尔类型的 PreferHostingUrls 属性，该属性表示如果监听地址同时设置到承载系统配置和服务器上，是否优先考虑使用前者。

```
public interface IServerAddressesFeature
{
    ICollection<string>       Addresses { get; }
    bool                      PreferHostingUrls { get; set; }
}
```

正如前面所说，服务器将用来处理由它接收请求的处理器会被视为一个通过 IHttpApplication<TContext>接口表示的应用，所以可以将 ASP.NET Core 的请求处理管道视为 IServer 对象和 IHttpApplication<TContext>对象的组合。当调用 IServer 对象的 StartAsync<TContext>方法启动服务器时，我们需要提供这个用来处理请求的 IHttpApplication<TContext>对象。IHttpApplication<TContext>采用基于上下文的请求处理方式，泛型参数 TContext 代表的就是上下文的类型。在 IHttpApplication<TContext>处理请求之前，它需要先创建一个上下文对象，该上下文会在请求处理结束之后被释放。上下文的创建、释放和自身对请求的处理实现在该接口 3 个对应的方法（CreateContext、DisposeContext 和 ProcessRequestAsync）中。

```
public interface IHttpApplication<TContext>
{
    TContext CreateContext(IFeatureCollection contextFeatures);
    void DisposeContext(TContext context, Exception exception);
    Task ProcessRequestAsync(TContext context);
}
```

13.2.2　HostingApplication

ASP.NET Core 框架利用如下所示的 HostingApplication 类型作为 IHttpApplication<TContext>接口的默认实现，它使用一个内嵌的 Context 类型来表示处理请求的上下文。一个 Context 对象是对一个 HttpContext 对象的封装，同时承载了一些与诊断相关的信息。

```
public class HostingApplication : IHttpApplication<HostingApplication.Context>
{
```

```
...
public struct Context
{
    public HttpContext       HttpContext { get; set; }

    public IDisposable       Scope { get; set; }
    public long              StartTimestamp { get; set; }
    public bool              EventLogEnabled { get; set; }
    public Activity          Activity { get; set; }
}
}
```

HostingApplication 对象会在开始和完成请求处理，以及在请求过程中出现异常时发出一些诊断日志事件。具体来说，HostingApplication 对象会采用 3 种不同的诊断日志形式，包括基于 DiagnosticSource 和 EventSource 的诊断日志以及基于 .NET Core 日志系统的日志。Context 除 HttpContext 外的其他属性都与诊断日志有关。具体来说，Context 的 Scope 是为 ILogger 创建的针对当前请求的日志范围（第 9 章有对日志范围的详细介绍），此日志范围会携带唯一标识每个请求的 ID，如果注册 ILoggerProvider 提供的 ILogger 支持日志范围，它可以将这个请求 ID 记录下来，那么我们就可以利用这个 ID 将针对同一请求的多条日志消息组织起来做针对性分析。

HostingApplication 对象会在请求结束之后记录当前请求处理的耗时，所以它在开始处理请求时就会记录当前的时间戳，Context 的 StartTimestamp 属性表示开始处理请求的时间戳。它的 EventLogEnabled 属性表示针对 EventSource 的事件日志是否开启，而 Activity 属性则与针对 DiagnosticSource 的诊断日志有关，Activity 代表基于当前请求处理的活动。

虽然 ASP.NET Core 应用的请求处理完全由 HostingApplication 对象负责，但是该类型的实现逻辑其实是很简单的，因为它将具体的请求处理分发给一个 RequestDelegate 对象，该对象表示的正是所有注册中间件组成的委托链。在创建 HostingApplication 对象时除了需要提供 RequestDelegate 对象，还需要提供用于创建 HttpContext 上下文的 IHttpContextFactory 对象，以及与诊断日志有关的 ILogger 对象和 DiagnosticListener 对象，它们被用来创建上面提到过的 HostingApplicationDiagnostics 对象。

```
public class HostingApplication : IHttpApplication<HostingApplication.Context>
{
    private readonly RequestDelegate            _application;
    private HostingApplicationDiagnostics       _diagnostics;
    private readonly IHttpContextFactory        _httpContextFactory;

    public HostingApplication(RequestDelegate application, ILogger logger,
        DiagnosticListener diagnosticSource, IHttpContextFactory httpContextFactory)
    {
        _application          = application;
        _diagnostics          = new HostingApplicationDiagnostics(logger, diagnosticSource);
        _httpContextFactory   = httpContextFactory;
    }

    public Context CreateContext(IFeatureCollection contextFeatures)
```

```
{
    var context = new Context();
    var httpContext = _httpContextFactory.Create(contextFeatures);
    _diagnostics.BeginRequest(httpContext, ref context);
    context.HttpContext = httpContext;
    return context;
}

public Task ProcessRequestAsync(Context context)
    => _application(context.HttpContext);

public void DisposeContext(Context context, Exception exception)
{
    var httpContext = context.HttpContext;
    _diagnostics.RequestEnd(httpContext, exception, context);
    _httpContextFactory.Dispose(httpContext);
    _diagnostics.ContextDisposed(context);
}
}
```

如上面的代码片段所示，当 CreateContext 方法被调用时，HostingApplication 对象会利用 IHttpContextFactory 工厂创建出当前 HttpContext 上下文，并进一步将它封装成一个 Context 对象。在返回这个 Context 对象之前，它会调用 HostingApplicationDiagnostics 对象的 BeginRequest 方法记录相应的诊断日志。用来真正处理当前请求的 ProcessRequestAsync 方法比较简单，只需要调用代表中间件委托链的 RequestDelegate 对象即可。

对于用来释放上下文的 DisposeContext 方法来说，它会利用 IHttpContextFactory 对象的 Dispose 方法来释放创建的 HttpContext 对象。换句话说，HttpContext 上下文的生命周期是由 HostingApplication 对象控制的。完成针对 HttpContext 上下文的释放之后，HostingApplication 对象会利用 HostingApplicationDiagnostics 对象记录相应的诊断日志。Context 的 Scope 属性表示的日志范围就是在调用 HostingApplicationDiagnostics 对象的 ContextDisposed 方法时释放的。如果将 HostingApplication 对象引入 ASP.NET Core 的请求处理管道，那么完整的管道就体现为图 13-3 所示的结构。

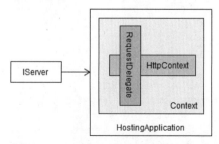

图 13-3　由 IServer 和 HostingApplication 组成的管道

13.2.3 诊断日志

很多人可能对 ASP.NET Core 框架自身记录的诊断日志并不关心，其实很多时候这些日志对纠错排错和性能监控提供了很有用的信息。例如，假设需要创建一个 APM（Application Performance Management）来监控 ASP.NET Core 处理请求的性能及出现的异常，那么我们完全可以将 HostingApplication 对象记录的日志作为收集的原始数据。实际上，目前很多 APM（如 Elastic APM 和 SkyWalking APM 等）针对 ASP.NET Core 应用的客户端都是利用这种方式收集请求调用链信息的。

日志系统

为了确定什么样的信息会被作为诊断日志记录下来，下面介绍一个简单的实例，将 HostingApplication 对象写入的诊断日志输出到控制台上。前面提及，HostingApplication 对象会将相同的诊断信息以 3 种不同的方式进行记录，其中包含第 9 章介绍的日志系统，所以我们可以通过注册对应 ILoggerProvider 对象的方式将日志内容写入对应的输出渠道。

整个演示实例如下面的代码片段所示：首先通过调用 IWebHostBuilder 接口的 ConfigureLogging 方法注册一个 ConsoleLoggerProvider 对象，并开启针对日志范围的支持。我们调用 IApplicationBuilder 接口的 Run 扩展方法注册了一个中间件，该中间件在处理请求时会利用表示当前请求上下文的 HttpContext 对象得到与之绑定的 IServiceProvider 对象，并进一步从中提取出用于发送日志事件的 ILogger<Program>对象，我们利用它写入一条 Information 等级的日志。如果请求路径为"/error"，那么该中间件会抛出一个 InvalidOperationException 类型的异常。

```
public class Program
{
    public static void Main()
    {
        Host.CreateDefaultBuilder()
            .ConfigureLogging(builder => builder.AddConsole(
                options => options.IncludeScopes = true))
            .ConfigureWebHostDefaults(builder => builder
                .Configure(app => app.Run(context =>
                    {
                        var logger = context.RequestServices
                            .GetRequiredService<ILogger<Program>>();
                        logger.LogInformation($"Log for event Foobar");
                        if (context.Request.Path == new PathString("/error"))
                        {
                            throw new InvalidOperationException(
                                "Manually throw exception.");
                        }
                        return Task.CompletedTask;
                    })))
            .Build()
            .Run();
}
```

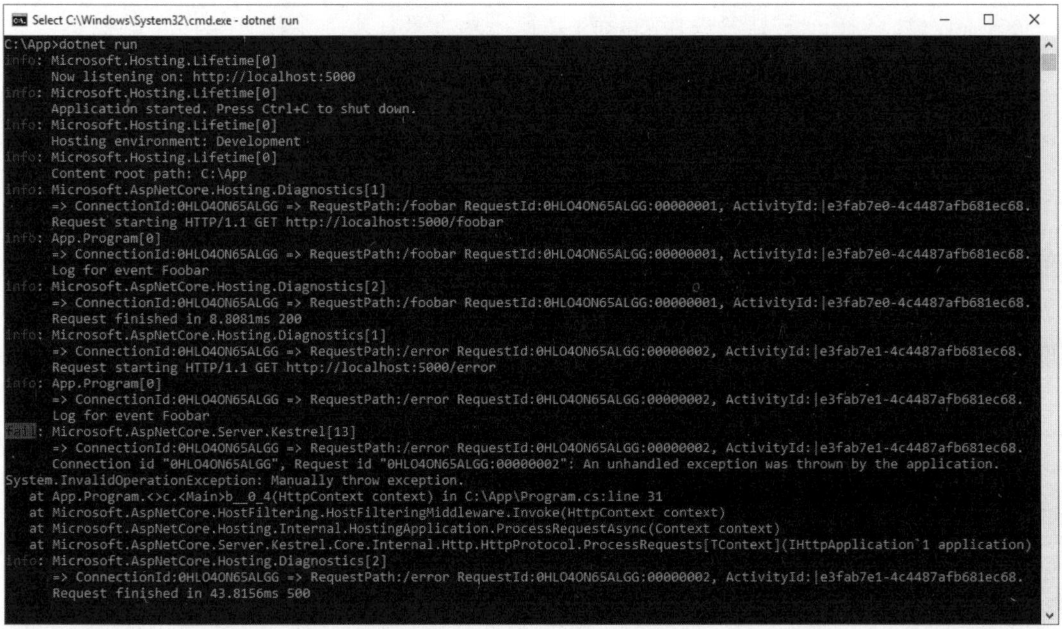

}
```

在启动程序之后,我们利用浏览器采用不同的路径("/foobar"和"/error")向应用发送了两次请求,演示程序的控制台上呈现的输出结果如图 13-4 所示。由于我们开启了日志范围的支持,所以被 ConsoleLogger 记录下来的日志都会携带日志范围的信息。日志范围的唯一标识被称为请求 ID(Request ID),它由当前的连接 ID 和一个序列号组成。从图 13-4 可以看出,两次请求的 ID 分别是"0HLO4ON65ALGG:00000001"和"0HLO4ON65ALGG:00000002"。由于采用的是长连接,并且两次请求共享同一个连接,所以它们具有相同的连接 ID("0HLO4ON65ALGG")。同一连接的多次请求将一个自增的序列号("00000001"和"00000002")作为唯一标识。(S1301)

图 13-4　捕捉 HostingApplication 记录的诊断日志

除了用于唯一表示每个请求的请求 ID,日志范围承载的信息还包括请求指向的路径,这也可以从图 13-4 所示的输出接口看出来。另外,上述请求 ID 实际上对应 HttpContext 类型的 TraceIdentifier 属性。如果需要进行跨应用的调用链跟踪,所有相关日志就可以通过共享 TraceIdentifier 属性构建整个调用链。

```
public abstract class HttpContext
{
    public abstract string  TraceIdentifier { get; set; }
    ...
}
```

对于两次采用不同路径的请求,控制台共捕获了 7 条日志,其中类别为 App.Program 的日志是应用程序自行写入的,HostingApplication 写入日志的类别为"Microsoft.AspNetCore.
```

Hosting.Diagnostics"。对于第一次请求的 3 条日志消息，第一条是在 HostingApplication 开始处理请求时写入的，我们利用这条日志获知请求的 HTTP 版本（HTTP/1.1）、HTTP 方法（GET）和请求 URL。对于包含主体内容的请求，请求主体内容的媒体类型（Content-Type）和大小（Content-Length）也会一并记录下来。当 HostingApplication 对象处理完请求后会写入第三条日志，日志承载的信息包括请求处理耗时（67.877 6 毫秒）和响应状态码（200）。如果响应具有主体内容，对应的媒体类型同样会被记录下来。（S1301）

对于第二次请求，由于我们人为抛出了一个异常，所以异常的信息被写入日志。但是如果足够仔细，就会发现这条等级为 Error 的日志并不是由 HostingApplication 对象写入的，而是作为服务器的 KestrelServer 写入的，因为该日志采用的类别为"Microsoft.AspNetCore.Server.Kestrel"。换句话说，HostingApplication 对象利用 ILogger 记录的日志中并不包含应用的异常信息。

DiagnosticSource 诊断日志

HostingApplication 采用的 3 种日志形式还包括基于 DiagnosticSource 对象的诊断日志，所以我们可以通过注册诊断监听器来收集诊断信息。如果通过这种方式获取诊断信息，就需要预先知道诊断日志事件的名称和内容荷载的数据结构。通过查看 HostingApplication 类型的源代码，我们会发现它针对"开始请求"、"结束请求"和"未处理异常"这 3 类诊断日志事件对应的名称，具体如下。

- 开始请求：Microsoft.AspNetCore.Hosting.BeginRequest。
- 结束请求：Microsoft.AspNetCore.Hosting.EndRequest。
- 未处理异常：Microsoft.AspNetCore.Hosting.UnhandledException。

至于针对诊断日志消息的内容荷载（Payload）的结构，上述 3 类诊断事件具有两个相同的成员，分别是表示当前请求上下文的 HttpContext 和通过一个 Int64 整数表示的当前时间戳，对应的数据成员的名称分别为 httpContext 和 timestamp。对于未处理异常诊断事件，它承载的内容荷载还包括一个额外的成员，那就是表示抛出异常的 Exception 对象，对应的成员名称为 exception。

既然我们已经知道事件的名称和诊断承载数据的成员，所以可以定义如下所示的 DiagnosticCollector 类型作为诊断监听器（需要针对 NuGet 包 "Microsoft.Extensions.DiagnosticAdapter" 的引用）。针对上述 3 类诊断事件，我们在 DiagnosticCollector 类型中定义了 3 个对应的方法，各个方法通过标注的 DiagnosticNameAttribute 特性设置了对应的诊断事件。我们根据诊断数据承载的结构定义了匹配的参数，所以 DiagnosticSource 对象写入诊断日志提供的诊断数据将自动绑定到对应的参数上。如果读者对如何监听并捕捉由 DiagnosticSource 对象发出的诊断日志事件不熟悉，可参阅第 8 章的相关内容。

```
public class DiagnosticCollector
{
    [DiagnosticName("Microsoft.AspNetCore.Hosting.BeginRequest")]
    public void OnRequestStart(HttpContext httpContext, long timestamp)
    {
```

```
        var request = httpContext.Request;
        Console.WriteLine($"\nRequest starting {request.Protocol} {request.Method}
            {request.Scheme}://{request.Host}{request.PathBase}{request.Path}");
        httpContext.Items["StartTimestamp"] = timestamp;
    }

    [DiagnosticName("Microsoft.AspNetCore.Hosting.EndRequest")]
    public void OnRequestEnd(HttpContext httpContext, long timestamp)
    {
        var startTimestamp = long.Parse(httpContext.Items["StartTimestamp"].ToString());
        var timestampToTicks = TimeSpan.TicksPerSecond / (double)Stopwatch.Frequency;
        var elapsed = new TimeSpan((long)(timestampToTicks *
            (timestamp - startTimestamp)));
        Console.WriteLine($"Request finished in {elapsed.TotalMilliseconds}ms
            {httpContext.Response.StatusCode}");
    }
    [DiagnosticName("Microsoft.AspNetCore.Hosting.UnhandledException")]
    public void OnException(HttpContext httpContext, long timestamp, Exception exception)
    {
        OnRequestEnd(httpContext, timestamp);
Console.WriteLine($"{exception.Message}\nType:{exception.GetType()}\nStacktrace:
        {exception.StackTrace}");
    }
}
```

可以在针对"开始请求"诊断事件的 OnRequestStart 方法中输出当前请求的 HTTP 版本、HTTP 方法和 URL。为了能够计算整个请求处理的耗时，我们将当前时间戳保存在 HttpContext 上下文的 Items 集合中。在针对"结束请求"诊断事件的 OnRequestEnd 方法中，我们将这个时间戳从 HttpContext 上下文中提取出来，结合当前时间戳计算出请求处理耗时，该耗时和响应的状态码最终会被写入控制台。针对"未处理异常"诊断事件的 OnException 方法则在调用 OnRequestEnd 方法之后将异常的消息、类型和跟踪堆栈输出到控制台上。

如下面的代码片段所示，在注册的 Startup 类型中，我们在 Configure 方法注入 DiagnosticListener 服务，并调用它的 SubscribeWithAdapter 扩展方法将上述 DiagnosticCollector 对象注册为诊断日志的订阅者。与此同时，我们调用 IApplicationBuilder 接口的 Run 扩展方法注册了一个中间件，该中间件会在请求路径为 "/error" 的情况下抛出一个异常。

```
public class Program
{
    public static void Main()
    {
        Host.CreateDefaultBuilder()
            .ConfigureLogging(builder => builder.ClearProviders())
            .ConfigureWebHostDefaults(builder => builder.UseStartup<Startup>())
            .Build()
            .Run();
    }
}
```

```csharp
public class Startup
{
    public void Configure(IApplicationBuilder app, DiagnosticListener listener)
    {
        listener.SubscribeWithAdapter(new DiagnosticCollector());
        app.Run(context =>
        {
            if (context.Request.Path == new PathString("/error"))
            {
                throw new InvalidOperationException("Manually throw exception.");
            }
            return Task.CompletedTask;
        });
    }
}
```

待演示实例正常启动后，可以采用不同的路径（"/foobar"和"/error"）对应用程序发送两个请求，服务端控制台会以图 13-5 所示的形式输出 DiagnosticCollector 对象收集的诊断信息。如果我们试图创建一个针对 ASP.NET Core 的 APM 框架来监控请求处理的性能和出现的异常，可以采用这样的方案来收集原始的诊断信息。（S1302）

图 13-5　利用注册的诊断监听器获取诊断日志

EventSource 事件日志

除了上述两种日志形式，HostingApplication 对象针对每个请求的处理过程中还会利用 EventSource 对象发出相应的日志事件。除此之外，在启动和关闭应用程序（实际上就是启动和关闭 IWebHost 对象）时，同一个 EventSource 对象还会被使用。这个 EventSource 类型采用的名称为 Microsoft.AspNetCore.Hosting，上述 5 个日志事件对应的名称如下。

- 启动应用程序：HostStart。
- 开始处理请求：RequestStart。
- 请求处理结束：RequestStop。
- 未处理异常：UnhandledException。

- 关闭应用程序：HostStop。

我们可以通过如下所示的实例来演示如何利用创建的 EventListener 对象来监听上述 5 个日志事件。如下面的代码片段所示，我们定义了派生于抽象类 EventListener 的 DiagnosticCollector。在启动应用前，我们创建了这个 DiagnosticCollector 对象，并通过注册其 EventSourceCreated 事件开启了针对目标名称为 Microsoft.AspNetCore.Hosting 的 EventSource 的监听。在注册的 EventWritten 事件中，我们将监听到的事件名称的负载内容输出到控制台上。

```csharp
public class Program
{
    private sealed class DiagnosticCollector : EventListener {}
    static void Main()
    {
        var listener = new DiagnosticCollector();
        listener.EventSourceCreated +=(sender, args) =>
        {
            if (args.EventSource.Name == "Microsoft.AspNetCore.Hosting")
            {
                listener.EnableEvents(args.EventSource, EventLevel.LogAlways);
            }
        };
        listener.EventWritten += (sender, args) =>
        {
            Console.WriteLine(args.EventName);
            for (int index = 0; index < args.PayloadNames.Count; index++)
            {
                Console.WriteLine($"\t{args.PayloadNames[index]} = {args.Payload[index]}");
            }
        };

        Host.CreateDefaultBuilder()
            .ConfigureLogging(builder => builder.ClearProviders())
            .ConfigureWebHostDefaults(builder => builder
                .Configure(app => app.Run(context =>
                {
                    if (context.Request.Path == new PathString("/error"))
                    {
                        throw new InvalidOperationException("Manually throw exception.");
                    }
                    return Task.CompletedTask;
                })))
            .Build()
            .Run();
    }
}
```

以命令行的形式启动这个演示程序后，从图 13-6 所示的输出结果可以看到名为 HostStart 的事件被发出。然后采用目标地址 "http://localhost:5000/foobar" 和 "http:// http://localhost:5000/error" 对应用程序发送两个请求，从输出结果可以看出，应用程序针对前者的处理过程会发出

RequestStart 事件和 RequestStop 事件，针对后者的处理则会因为抛出的异常发出额外的事件 UnhandledException。输入"Ctrl+C"关闭应用后，名称为 HostStop 的事件被发出。对于通过 EventSource 发出的 5 个事件，只有 RequestStart 事件会将请求的 HTTP 方法（GET）和路径（"/foobar"和"/error"）作为负载内容，其他事件都不会携带任何负载内容。（S1303）

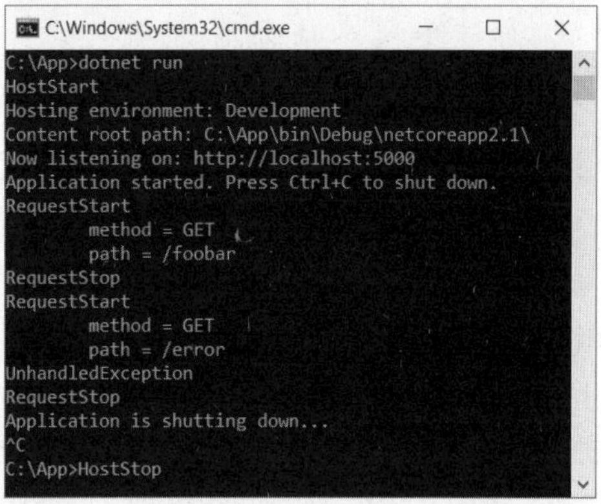

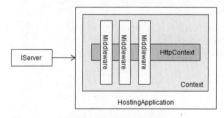

图 13-6　利用注册 EventListener 监听器获取诊断日志

13.3　中间件委托链

ASP.NET Core 应用默认的请求处理管道是由注册的 IServer 对象和 HostingApplication 对象组成的，后者利用一个在创建时提供的 RequestDelegate 对象来处理 IServer 对象分发给它的请求。而 RequestDelegate 对象实际上是由所有的中间件按照注册顺序创建的。换句话说，这个 RequestDelegate 对象是对中间件委托链的体现。如果将 RequestDelegate 替换成原始的中间件，那么 ASP.NET Core 应用的请求处理管道体现为图 13-7 所示的形式。

图 13-7　完整的请求处理管道

13.3.1　IApplicationBuilder

对于一个 ASP.NET Core 应用来说，它对请求的处理完全体现在注册的中间件上，所以"应用"从某种意义上来讲体现在通过所有注册中间件创建的 RequestDelegate 对象上。正因为如此，ASP.NET Core 框架才将构建这个 RequestDelegate 对象的接口命名为 IApplicationBuilder。IApplicationBuilder 是 ASP.NET Core 框架中的一个核心对象，我们将中间件注册在它上面，并且最终利用它来创建代表中间件委托链的 RequestDelegate 对象。

如下所示的代码片段是 IApplicationBuilder 接口的定义。该接口定义了 3 个属性：Application Services 属性代表针对当前应用程序的依赖注入容器，ServerFeatures 属性则返回服务器提供的特性集合，Properties 属性返回的字典则代表一个可以用来存放任意属性的容器。

```
public interface IApplicationBuilder
{
    IServiceProvider                    ApplicationServices { get; set; }
    IFeatureCollection                  ServerFeatures { get; }
    IDictionary<string, object>         Properties { get; }

    IApplicationBuilder Use(Func<RequestDelegate, RequestDelegate> middleware);
    RequestDelegate Build();
    IApplicationBuilder New();
}
```

通过第 12 章的介绍可知，ASP.NET Core 应用的中间件体现为一个 Func<RequestDelegate, RequestDelegate>对象，而针对中间件的注册则通过调用 IApplicationBuilder 接口的 Use 方法来完成。IApplicationBuilder 对象最终的目的就是根据注册的中间件创建作为代表中间件委托链的 RequestDelegate 对象，这个目标是通过调用 Build 方法来完成的。New 方法可以帮助我们创建一个新的 IApplicationBuilder 对象，除了已经注册的中间件，创建的 IApplicationBuilder 对象与当前对象具有相同的状态。

具有如下定义的 ApplicationBuilder 类型是对 IApplicationBuilder 接口的默认实现。ApplicationBuilder 类型利用一个 List<Func<RequestDelegate, RequestDelegate>>对象来保存注册的中间件，所以 Use 方法只需要将指定的中间件添加到这个列表中即可，而 Build 方法只需要逆序调用这些注册的中间件对应的 Func<RequestDelegate, RequestDelegate>对象就能得到我们需要的 RequestDelegate 对象。值得注意的是，Build 方法会在委托链的尾部添加一个额外的中间件，该中间件会将响应状态码设置为 404，所以应用在默认情况下会回复一个 404 响应。

```
public class ApplicationBuilder : IApplicationBuilder
{
    private readonly IList<Func<RequestDelegate, RequestDelegate>> middlewares
        = new List<Func<RequestDelegate, RequestDelegate>>();

    public IDictionary<string, object> Properties { get; }
    public IServiceProvider ApplicationServices
    {
        get { return GetProperty<IServiceProvider>("application.Services"); }
        set { SetProperty<IServiceProvider>("application.Services", value); }
    }

    public IFeatureCollection ServerFeatures
    {
        get { return GetProperty<IFeatureCollection>("server.Features"); }
    }

    public ApplicationBuilder(IServiceProvider serviceProvider)
    {
```

```csharp
        Properties = new Dictionary<string, object>();
        ApplicationServices = serviceProvider;
}

public ApplicationBuilder(IServiceProvider serviceProvider, object server)
    : this(serviceProvider)
    =>SetProperty("server.Features", server);

public IApplicationBuilder Use(Func<RequestDelegate, RequestDelegate> middleware)
{
    middlewares.Add(middleware);
    return this;
}

public IApplicationBuilder New()
    => new ApplicationBuilder(this);

public RequestDelegate Build()
{
    RequestDelegate app = context =>
    {
        context.Response.StatusCode = 404;
        return Task.FromResult(0);
    };
    foreach (var component in middlewares.Reverse())
    {
        app = component(app);
    }
    return app;
}

private ApplicationBuilder(ApplicationBuilder builder)
{
    this.Properties = new CopyOnWriteDictionary<string, object>(
        builder.Properties, StringComparer.Ordinal);
}

private T GetProperty<T>(string key)
{
    object value;
    return Properties.TryGetValue(key, out value) ? (T)value : default(T);
}

private void SetProperty<T>(string key, T value)
{
    Properties[key] = value;
}
}
```

由上面的代码片段可以看出，不论是通过 ApplicationServices 属性返回的 IService

Provider 对象，还是通过 ServerFeatures 属性返回的 IFeatureCollection 对象，它们实际上都保存在通过 Properties 属性返回的字典对象上。ApplicationBuilder 具有两个公共构造函数重载，其中一个构造函数具有一个类型为 Object 的 server 参数，但这个参数并不是表示服务器，而是表示服务器提供的 IFeatureCollection 对象。New 方法直接调用私有构造函数创建一个新的 ApplicationBuilder 对象，属性字典的所有元素会复制到新创建的 ApplicationBuilder 对象中。

ASP.NET Core 框架使用的 IApplicationBuilder 对象是通过注册的 IApplicationBuilderFactory 服务创建的。如下面的代码片段所示，IApplicationBuilderFactory 接口具有唯一的 CreateBuilder 方法，它会根据提供的特性集合创建相应的 IApplicationBuilder 对象。具有如下定义的 ApplicationBuilderFactory 类型是对该接口的默认实现，前面介绍的 ApplicationBuilder 对象正是由它创建的。

```
public interface IApplicationBuilderFactory
{
    IApplicationBuilder CreateBuilder(IFeatureCollection serverFeatures);
}

public class ApplicationBuilderFactory : IApplicationBuilderFactory
{
    private readonly IServiceProvider _serviceProvider;

    public ApplicationBuilderFactory(IServiceProvider serviceProvider)
        =>_serviceProvider = serviceProvider;

    public IApplicationBuilder CreateBuilder(IFeatureCollection serverFeatures)
        => new ApplicationBuilder(this._serviceProvider, serverFeatures);
}
```

13.3.2 弱类型中间件

虽然中间件最终体现为一个 Func<RequestDelegate, RequestDelegate>对象，但是在大部分情况下我们总是倾向于将中间件定义成一个 POCO 类型。通过第 11 章的介绍可知，中间件类型的定义具有两种形式：一种是按照预定义的约定规则来定义中间件类型，即弱类型中间件；另一种则是直接实现 IMiddleware 接口，即强类型中间件。下面介绍基于约定的中间件类型的定义方式，这种方式定义的中间件类型需要采用如下约定。

- 中间件类型需要有一个有效的公共实例构造函数，该构造函数必须包含一个 RequestDelegate 类型的参数，当前中间件通过执行这个委托对象将请求分发给后续中间件进行处理。这个构造函数不仅可以包含任意其他参数，对参数 RequestDelegate 出现的位置也不做任何约束。
- 针对请求的处理实现在返回类型为 Task 的 Invoke 方法或者 InvokeAsync 方法中，该方法的第一个参数表示当前请求对应的 HttpContext 上下文，对于后续的参数，虽然约定并未对此做限制，但是由于这些参数最终是由依赖注入框架提供的，所以相应的服务注册必须存在。

如下所示的代码片段就是一个典型的按照约定定义的中间件类型。我们在构造函数中注入了一个必需的 RequestDelegate 对象和一个 IFoo 服务。在用于请求处理的 InvokeAsync 方法中，除了包含表示当前 HttpContext 上下文的参数，我们还注入了一个 IBar 服务，该方法在完成自身请求处理操作之后，通过构造函数中注入的 RequestDelegate 对象可以将请求分发给后续的中间件。

```csharp
public class FoobarMiddleware
{
    private readonly RequestDelegate      _next;
    private readonly IFoo                 _foo;

    public FoobarMiddleware(RequestDelegate next, IFoo foo)
    {
        _next = next;
        _foo  = foo;
    }

    public async Task InvokeAsync(HttpContext context, IBar bar)
    {
        ...
        await _next(context);
    }
}
```

采用上述方式定义的中间件最终是通过调用 IApplicationBuilder 接口如下所示的两个扩展方法进行注册的。当我们调用这两个方法时，除了指定具体的中间件类型，还可以传入一些必要的参数，它们将作为调用构造函数的输入参数。对于定义在中间件类型构造函数中的参数，如果有对应的服务注册，ASP.NET Core 框架在创建中间件实例时可以利用依赖注入框架来提供对应的参数，所以在注册中间件时是不需要提供构造函数的所有参数的。

```csharp
public static class UseMiddlewareExtensions
{
    public static IApplicationBuilder UseMiddleware<TMiddleware>(
        this IApplicationBuilder app, params object[] args);
    public static IApplicationBuilder UseMiddleware(this IApplicationBuilder app,
        Type middleware, params object[] args);
}
```

由于 ASP.NET Core 应用的请求处理管道总是采用 Func<RequestDelegate, RequestDelegate> 对象来表示中间件，所以无论采用什么样的中间件定义方式，注册的中间件总是会转换成一个委托对象。那么上述两个扩展方法是如何实现这样的转换的？为了解决这个问题，我们采用极简的形式自行定义了第二个非泛型的 UseMiddleware 方法。

```csharp
public static class UseMiddlewareExtensions
{
    private static readonly MethodInfo GetServiceMethod = typeof(IServiceProvider)
        .GetMethod("GetService", BindingFlags.Public | BindingFlags.Instance);
    public static IApplicationBuilder UseMiddleware(this IApplicationBuilder app,
        Type middlewareType, params object[] args)
    {
```

```csharp
    ...
    var invokeMethod = middlewareType
        .GetMethods(BindingFlags.Instance | BindingFlags.Public)
        .Where(it => it.Name == "InvokeAsync" || it.Name == "Invoke")
        .Single();
    Func<RequestDelegate, RequestDelegate> middleware = next =>
    {
        var arguments = (object[])Array.CreateInstance(typeof(object),
            args.Length + 1);
        arguments[0] = next;
        if (args.Length > 0)
        {
            Array.Copy(args, 0, arguments, 1, args.Length);
        }
        var instance = ActivatorUtilities.CreateInstance(app.ApplicationServices,
            middlewareType, arguments);
        var factory = CreateFactory(invokeMethod);
        return context => factory(instance, context, app.ApplicationServices);
    };

    return app.Use(middleware);
}

private static Func<object, HttpContext, IServiceProvider, Task>
    CreateFactory(MethodInfo invokeMethod)
{
    var middleware = Expression.Parameter(typeof(object), "middleware");
    var httpContext = Expression.Parameter(typeof(HttpContext), "httpContext");
    var serviceProvider = Expression.Parameter(typeof(IServiceProvider),
        "serviceProvider");

    var parameters = invokeMethod.GetParameters();
    var arguments = new Expression[parameters.Length];
    arguments[0] = httpContext;
    for (int index = 1; index < parameters.Length; index++)
    {
        var parameterType = parameters[index].ParameterType;
        var type = Expression.Constant(parameterType, typeof(Type));
        var getService = Expression.Call(serviceProvider, GetServiceMethod, type);
        arguments[index] = Expression.Convert(getService, parameterType);
    }
    var converted = Expression.Convert(middleware, invokeMethod.DeclaringType);
    var body = Expression.Call(converted, invokeMethod, arguments);
    var lambda = Expression.Lambda<
        Func<object, HttpContext, IServiceProvider, Task>>(
        body, middleware, httpContext, serviceProvider);

    return lambda.Compile();
}
}
```

由于请求处理的具体实现定义在中间件类型的 Invoke 方法或者 InvokeAsync 方法上，所以注册这样一个中间件需要解决两个核心问题：其一，创建对应的中间件实例；其二，将针对中间件实例的 Invoke 方法或者 InvokeAsync 方法调用转换成 Func<RequestDelegate, RequestDelegate>对象。由于存在依赖注入框架，所以第一个问题很好解决，从上面给出的代码片段可以看出，我们最终调用静态类型 ActivatorUtilities 的 CreateInstance 方法创建出中间件实例。

由于 ASP.NET Core 框架对中间件类型的 Invoke 方法和 InvokeAsync 方法的声明并没有严格限制，该方法返回类型为 Task，它的第一个参数为 HttpContext 上下文，所以针对该方法的调用比较烦琐。要调用某个方法，需要先传入匹配的参数列表，有了 IServiceProvider 对象的帮助，针对输入参数的初始化就显得非常容易。我们只需要从表示方法的 MethodInfo 对象中解析出方法的参数类型，就能够根据类型从 IServiceProvider 对象中得到对应的参数实例。

如果有表示目标方法的 MethodInfo 对象和与之匹配的输入参数列表，就可以采用反射的方式来调用对应的方法，但是反射并不是一种高效的手段，所以 ASP.NET Core 框架采用表达式树的方式来实现针对 InvokeAsync 方法或者 Invoke 方法的调用。基于表达式树针对中间件实例的 InvokeAsync 方法或者 Invoke 方法的调用实现在前面提供的 CreateFactory 方法中，由于实现逻辑并不复杂，所以不需要再对提供的代码做详细说明。

13.3.3　强类型中间件

通过调用 IApplicationBuilder 接口的 UseMiddleware 扩展方法注册的是一个按照约定规则定义的中间件类型，由于中间件实例是在应用初始化时创建的，这样的中间件实际上是一个与当前应用程序具有相同生命周期的 Singleton 对象。但有时我们希望中间件对象采用 Scoped 模式的生命周期，即要求中间件对象在开始处理请求时被创建，在完成请求处理后被回收释放。

如果需要后面这种类型的中间件，就需要让定义的中间件类型实现 IMiddleware 接口。如下面的代码片段所示，IMiddleware 接口定义了唯一的 InvokeAsync 方法，用来实现对请求的处理。对于实现该方法的中间件类型来说，它可以利用输入参数得到针对当前请求的 HttpContext 上下文，还可以得到用来向后续中间件分发请求的 RequestDelegate 对象。

```
public interface IMiddleware
{
    Task InvokeAsync(HttpContext context, RequestDelegate next);
}
```

实现了 IMiddleware 接口的中间件是通过依赖注入的形式提供的，所以在调用 IAppplicationBuilder 接口的 UseMiddleware 扩展方法注册中间件类型之前需要做相应的服务注册。在一般情况下，我们只会在需要使用 Scoped 生命周期时才会采用这种方式来定义中间件，所以在进行服务注册时一般将生命周期模式设置为 Scoped，设置成 Singleton 模式也未尝不可，这就与按照约定规则定义的中间件没有本质区别。读者可能会有疑问，注册中间件服务时是否可以将生命周期模式设置为 Transient？实际上这与 Scoped 是没有区别的，因为中间件在同一个请求上下文中只会被创建一次。

对实现了 IMiddleware 接口的中间件的创建与释放是通过注册的 IMiddlewareFactory 服务来完成的。如下面的代码片段所示，IMiddlewareFactory 接口提供了如下两个方法：Create 方法会根据指定的中间件类型创建出对应的实例，Release 方法则负责释放指定的中间件对象。

```csharp
public interface IMiddlewareFactory
{
    IMiddleware Create(Type middlewareType);
    void Release(IMiddleware middleware);
}
```

ASP.NET Core 提供如下所示的 MiddlewareFactory 类型作为 IMiddlewareFactory 接口的默认实现，上面提及的中间件针对依赖注入的创建方式就体现在该类型中。如下面的代码片段所示，MiddlewareFactory 直接利用指定的 IServiceProvider 对象根据指定的中间件类型来提供对应的实例。由于依赖注入框架自身具有针对提供服务实例的生命周期管理策略，所以 MiddlewareFactory 的 Release 方法不需要对提供的中间件实例做具体的释放操作。

```csharp
public class MiddlewareFactory : IMiddlewareFactory
{
    private readonly IServiceProvider _serviceProvider;

    public MiddlewareFactory(IServiceProvider serviceProvider)
        => _serviceProvider = serviceProvider;
    public IMiddleware Create(Type middlewareType)
        => _serviceProvider.GetRequiredService(this._serviceProvider, middlewareType)
            as IMiddleware;
    public void Release(IMiddleware middleware) {}
}
```

了解了作为中间件工厂的 IMiddlewareFactory 接口之后，下面介绍 IApplicationBuilder 用于注册中间件的 UseMiddleware 扩展方法是如何利用它来创建并释放中间件的，为此我们编写了如下这段简写的代码来模拟相关的实现。如下面的代码片段所示，如果注册的中间件类型实现了 IMiddleware 接口，UseMiddleware 方法会直接创建一个 Func<RequestDelegate, RequestDelegate> 对象作为注册的中间件。

```csharp
public static class UseMiddlewareExtensions
{
    public static IApplicationBuilder UseMiddleware(this IApplicationBuilder app,
        Type middlewareType, params object[] args)
    {
        if (typeof(IMiddleware).IsAssignableFrom(middlewareType))
        {
            if (args.Length > 0)
            {
                throw new NotSupportedException(
                    "Types that implement IMiddleware do not support explicit arguments.");
            }
            app.Use(next =>
            {
                return async context =>
                {
```

```
                var middlewareFactory = context.RequestServices
                    .GetRequiredService<IMiddlewareFactory>();
                var middleware = middlewareFactory.Create(middlewareType);
                try
                {
                    await middleware.InvokeAsync(context, next);
                }
                finally
                {
                    middlewareFactory.Release(middleware);
                }
            };
        });
    }
    ...
}
```

当作为中间件的委托对象被执行时，它会从当前 HttpContext 上下文的 RequestServices 属性中获取针对当前请求的 IServiceProvider 对象，并由它来提供 IMiddlewareFactory 对象。在利用 IMiddlewareFactory 对象根据注册的中间件类型创建出对应的中间件对象之后，中间件的 InvokeAsync 方法被调用。在当前及后续中间件针对当前请求的处理完成之后，IMiddlewareFactory 对象的 Release 方法被调用来释放由它创建的中间件。

UseMiddleware 方法之所以从当前 HttpContext 上下文的 RequestServices 属性获取 IServiceProvider，而不是直接使用 IApplicationBuilder 的 ApplicationServices 属性返回的 IServiceProvider 来创建 IMiddlewareFactory 对象，是出于生命周期方面的考虑。由于后者采用针对当前应用程序的生命周期模式，所以不论注册中间件类型采用的生命周期模式是 Singleton 还是 Scoped，提供的中间件实例都是一个 Singleton 对象，所以无法满足我们针对请求创建和释放中间件对象的初衷。

上面的代码片段还反映了一个细节：如果注册了一个实现了 IMiddleware 接口的中间件类型，我们是不允许指定任何参数的，一旦调用 UseMiddleware 方法时指定了参数，就会抛出一个 NotSupportedException 类型的异常。

13.3.4 注册中间件

在 ASP.NET Core 应用请求处理管道构建过程中，IApplicationBuilder 对象的作用就是收集我们注册的中间件，并最终根据注册的先后顺序创建一个代表中间件委托链的 RequestDelegate 对象。在一个具体的 ASP.NET Core 应用中，利用 IApplicationBuilder 对象进行中间件的注册主要体现为如下 3 种方式。

- 调用 IWebHostBuilder 的 Configure 方法。
- 调用注册 Startup 类型的 Configure 方法。
- 利用注册的 IStartupFilter 对象。

如下所示的 IStartupFilter 接口定义了唯一的 Configure 方法，它返回的 Action<IApplication

Builder>对象将用来注册所需的中间件。作为该方法唯一输入参数的 Action<IApplicationBuilder>对象，则用来完成后续的中间件注册工作。IStartupFilter 接口的 Configure 方法比 IStartup 的 Configure 方法先执行，所以可以利用前者注册一些前置或者后置的中间件。

```
public interface IStartupFilter
{
    Action<IApplicationBuilder> Configure(Action<IApplicationBuilder> next);
}
```

13.4 应用的承载

到目前为止，我们知道 ASP.NET Core 应用的请求处理管道是由一个 IServer 对象和 IHttpApplication 对象构成的。我们可以根据需要注册不同类型的服务器，但在默认情况下，IHttpApplication 是一个 HostingApplication 对象。一个 HostingApplication 对象由指定的 RequestDelegate 对象来完成所有的请求处理工作，而后者代表所有中间件按照注册的顺序串联而成的委托链。所有的这一切都被 GenericWebHostService 整合在一起，在对这个承载 Web 应用的服务做进一步介绍之前，下面先介绍与它相关的配置选项。

13.4.1 GenericWebHostServiceOptions

GenericWebHostService 这个承载服务的配置选项类型为 GenericWebHostServiceOptions。如下面的代码片段所示，这个内部类型有 3 个属性，其核心配置选项由 WebHostOptions 属性承载。GenericWebHostServiceOptions 类型的 ConfigureApplication 属性返回的 Action<IApplicationBuilder> 对象用来注册中间件，启动过程中针对中间件的注册最终都会转移到这个属性上。

```
internal class GenericWebHostServiceOptions
{
    public WebHostOptions                        WebHostOptions { get; set; }
    public Action<IApplicationBuilder>           ConfigureApplication { get; set; }
    public AggregateException                    HostingStartupExceptions { get; set; }
}
```

第 11 章提出，可以利用一个外部程序集中定义的 IHostingStartup 实现类型来完成初始化任务，而 GenericWebHostServiceOptions 类型的 HostingStartupExceptions 属性返回的 AggregateException 对象就是对这些初始化任务执行过程中抛出异常的封装。一个 WebHostOptions 对象承载了与 IWebHost 相关的配置选项，虽然在基于 IHost/IHostBuilder 的承载系统中，IWebHost 接口作为宿主的作用已经不存在，但是 WebHostOptions 这个配置选项依然被保留下来。

```
public class WebHostOptions
{
    public string                    ApplicationName { get; set; }
    public string                    Environment { get; set; }
    public string                    ContentRootPath { get; set; }
    public string                    WebRoot { get; set; }
    public string                    StartupAssembly { get; set; }
    public bool                      PreventHostingStartup { get; set; }
    public IReadOnlyList<string>     HostingStartupAssemblies { get; set; }
```

```
    public IReadOnlyList<string>     HostingStartupExcludeAssemblies { get; set; }
    public bool                      CaptureStartupErrors { get; set; }
    public bool                      DetailedErrors { get; set; }
    public TimeSpan                  ShutdownTimeout { get; set; }

    public WebHostOptions() => ShutdownTimeout = TimeSpan.FromSeconds(5.0);
    public WebHostOptions(IConfiguration configuration);
    public WebHostOptions(IConfiguration configuration, string applicationNameFallback);
}
```

一个 WebHostOptions 对象可以根据一个 IConfiguration 对象来创建，当我们调用这个构造函数时，它会根据预定义的配置键从该 IConfiguration 对象中提取相应的值来初始化对应的属性。

```
public static class WebHostDefaults
{
    public static readonly string ApplicationKey              = "applicationName";
    public static readonly string StartupAssemblyKey          = "startupAssembly";
    public static readonly string DetailedErrorsKey           = "detailedErrors";
    public static readonly string EnvironmentKey              = "environment";
    public static readonly string WebRootKey                  = "webroot";
    public static readonly string CaptureStartupErrorsKey     = "captureStartupErrors";
    public static readonly string ServerUrlsKey               = "urls";
    public static readonly string ContentRootKey              = "contentRoot";
    public static readonly string PreferHostingUrlsKey        = "preferHostingUrls";
    public static readonly string PreventHostingStartupKey    = "preventHostingStartup";
    public static readonly string ShutdownTimeoutKey          = "shutdownTimeoutSeconds";

    public static readonly string HostingStartupAssembliesKey
        = "hostingStartupAssemblies";
    public static readonly string HostingStartupExcludeAssembliesKey
        = "hostingStartupExcludeAssemblies";
}
```

这些预定义的配置键作为静态只读字段被定义在静态类 WebHostDefaults 中，其中大部分在第 11 章已有相关介绍，本节只对此进行总结。WebHostOptions 属性列表如表 13-4 所示。值得注意的是，对于布尔类型的属性值（如 PreventHostingStartup 和 CaptureStartupErrors），配置项的值 "True"（不区分大小写）和 "1" 将转换为 True，其他的值将转换成 False。这个将配置项的值转换成布尔值的逻辑实现在 WebHostUtilities 的静态方法 ParseBool 中，如果我们有类似的需求可以直接调用这个方法。

表 13-4　WebHostOptions 属性列表

属　　性	配　置　键	说　　明
ApplicationName	applicationName	应用名称。如果调用 IWebHostBuilder 接口的 Configure 方法注册中间件，那么提供的 Action<IApplicationBuilder> 对象指向的目标方法所在的程序集名称将作为应用名称。如果调用 IWebHostBuilder 接口的 UseStartup 扩展方法，指定的 Startup 类型所在的程序集名称会作为应用名称
Environment	environment	应用当前的部署环境。如果没有显示指定，默认的环境名称为 Production

续表

属 性	配 置 键	说 明
ContentRootPath	contentRoot	存放静态内容文件的根目录。如果未做显式设置，默认为当前程序域的基础目录，对应 AppDomain 的 BaseDirectory 属性，静态类 AppContext 的 BaseDirectory 属性返回的也是这个目录
WebRoot	webroot	存放静态 Web 资源文件的根目录。如果未做显式设置，并且 ContentRootPath 目录下存在一个名为 wwwroot 的子目录，那么该目录将作为 Web 资源文件的根目录
StartupAssembly	startupAssembly	注册的 Startup 类型所在的程序集名称。如果调用 IWebHostBuilder 接口的 UseStartup 扩展方法，指定的 Startup 类型所在的程序集名称会作为该属性的值
PreventHostingStartup	preventHostingStartup	是否允许执行其他程序集中的初始化程序。如果这个开关并没有显式关闭，就可以在一个单独的程序集中利用 HostingStartupAttribute 特性注册一个实现了 IHostingStartup 接口的类型，它可以在应用启动时执行一些初始化操作
HostingStartupAssemblies	hostingStartupAssemblies	承载初始化程序的程序集列表，配置中的程序集名称之间采用分号分隔。ApplicationName 属性代表的程序集名称默认被添加到这个列表中
HostingStartupExcludeAssemblies	hostingStartupExcludeAssemblies	HostingStartupAssemblies 属性表示初始化程序的程序集列表中需要被排除的程序集
CaptureStartupErrors	captureStartupErrors	是否需要捕捉应用启动过程中出现的未处理异常。如果这个属性被显式设置为 True，出现的未处理异常并不会阻止应用的正常启动，但是这样的应用在接收到请求之后会返回一个状态码为 500 的响应
DetailedErrors	detailedErrors	如果 CaptureStartupErrors 属性被显式设置为 True，该属性表示是否需要在响应消息中输出详细的错误信息
ShutdownTimeout	shutdownTimeoutSeconds	应用关闭的超时时限，默认时限为 5 秒

13.4.2 GenericWebHostService

从如下所示的代码片段可以看出，GenericWebHostService 的构造函数中会注入一系列的依赖服务或者对象，其中包括用来提供配置选项的 IOptions<GenericWebHostServiceOptions>对象、作为管道"龙头"的服务器、用来创建 ILogger 对象的 ILoggerFactory 对象、用来发送相应诊断事件的 DiagnosticListener 对象、用来创建 HttpContext 上下文的 IHttpContextFactory 对象、用来创建 IApplicationBuilder 对象的 IApplicationBuilderFactory 对象、注册的所有 IStartupFilter 对象、承载当前应用配置的 IConfiguration 对象和代表当前承载环境的 IWebHostEnvironment 对象。在 GenericWebHostService 构造函数中注入的对象或者由它们创建的对象（如由 ILoggerFactory 对象创建的 ILogger 对象）最终会存储在对应的属性上。

```
internal class GenericWebHostService : IHostedService
{
    public GenericWebHostServiceOptions        Options { get; }
    public IServer                              Server { get; }
```

```csharp
    public ILogger                              Logger { get; }
    public ILogger                              LifetimeLogger { get; }
    public DiagnosticListener                   DiagnosticListener { get; }
    public IHttpContextFactory                  HttpContextFactory { get; }
    public IApplicationBuilderFactory           ApplicationBuilderFactory { get; }
    public IEnumerable<IStartupFilter>          StartupFilters { get; }
    public IConfiguration                       Configuration { get; }
    public IWebHostEnvironment                  HostingEnvironment { get; }

    public GenericWebHostService(IOptions<GenericWebHostServiceOptions> options,
        IServer server, ILoggerFactory loggerFactory,
        DiagnosticListener diagnosticListener, IHttpContextFactory httpContextFactory,
        IApplicationBuilderFactory applicationBuilderFactory,
        IEnumerable<IStartupFilter> startupFilters, IConfiguration configuration,
        IWebHostEnvironment hostingEnvironment);

    public Task StartAsync(CancellationToken cancellationToken);
    public Task StopAsync(CancellationToken cancellationToken);
}
```

由于 ASP.NET Core 应用是由 GenericWebHostService 服务承载的，所以启动应用程序本质上就是启动这个承载服务。承载 GenericWebHostService 在启动过程中的处理流程基本上体现在如下所示的 StartAsync 方法中，该方法中刻意省略了一些细枝末节的实现，如输入验证、异常处理、诊断日志事件的发送等。

```csharp
internal class GenericWebHostService : IHostedService
{
    public Task StartAsync(CancellationToken cancellationToken)
    {
        //1. 设置监听地址
        var serverAddressesFeature = Server.Features?.Get<IServerAddressesFeature>();
        var addresses = serverAddressesFeature?.Addresses;
        if (addresses != null && !addresses.IsReadOnly && addresses.Count == 0)
        {
            var urls = Configuration[WebHostDefaults.ServerUrlsKey];
            if (!string.IsNullOrEmpty(urls))
            {
                serverAddressesFeature.PreferHostingUrls = WebHostUtilities.ParseBool(
                    Configuration, WebHostDefaults.PreferHostingUrlsKey);

                foreach (var value in urls.Split(new[] { ';' },
                    StringSplitOptions.RemoveEmptyEntries))
                {
                    addresses.Add(value);
                }
            }
        }

        //2. 构建中间件管道
        var builder = ApplicationBuilderFactory.CreateBuilder(Server.Features);
```

```
        Action<IApplicationBuilder> configure = Options.ConfigureApplication;
        foreach (var filter in StartupFilters.Reverse())
        {
            configure = filter.Configure(configure);
        }
        configure(builder);
        var handler = builder.Build();

        //3. 创建HostingApplication对象
        var application = new HostingApplication(handler, Logger, DiagnosticListener,
            HttpContextFactory);

        //4. 启动服务器
        return Server.StartAsync(application, cancellationToken);
    }
}
```

我们将实现在 GenericWebHostService 类型的 StartAsync 方法中用来启动应用程序的流程划分为如下 4 个步骤。

- **设置监听地址**：服务器的监听地址是通过 IServerAddressesFeature 接口表示的特性来承载的，所以需要将配置提供的监听地址列表和相关的 PreferHostingUrls 选项（表示是否优先使用承载系统提供地址）转移到该特性中。
- **构建中间件管道**：通过调用 IWebHostBuilder 对象和注册的 Startup 类型的 Configure 方法针对中间件的注册会转换成一个 Action<IApplicationBuilder>对象，并复制给配置选项 GenericWebHostServiceOptions 的 ConfigureApplication 属性。GenericWebHostService 承载服务会利用注册的 IApplicationBuilderFactory 工厂创建出对应的 IApplicationBuilder 对象，并将该对象作为参数调用这个 Action<IApplicationBuilder>对象就能将注册的中间件转移到 IApplicationBuilder 对象上。但在此之前，注册 IStartupFilter 对象的 Configure 方法会优先被调用，IStartupFilter 对象针对前置中间件的注册就体现在这里。代表注册中间件管道的 RequestDelegate 对象最终通过调用 IApplicationBuilder 对象的 Build 方法返回。
- **创建 HostingApplication 对象**：在得到代表中间件管道的 RequestDelegate 之后，GenericWebHostService 对象进一步利用它创建出 HostingApplication 对象，该对象对于服务器来说就是用来处理由它接收请求的应用程序。
- **启动服务器**：将创建出的 HostingApplication 对象作为参数调用作为服务器的 IServer 对象的 StartAsync 方法后，服务器随之被启动。此后，服务器绑定到指定的地址监听抵达的请求，并为接收的请求创建出对应的 HttpContext 上下文，后续中间件将在这个上下文中完成各自对请求的处理任务。请求处理结束之后，生成的响应最终通过服务器回复给客户端。

关闭 GenericWebHostService 服务之后，只需要按照如下方式关闭服务器即可。除此之外，StopAsync 方法还会利用 EventSource 的形式发送相应的事件，我们在前面针对诊断日志的演示可以体验此功能。

```
internal class GenericWebHostService : IHostedService
{
    public async Task StopAsync(CancellationToken cancellationToken)
        => Server.StopAsync(cancellationToken);
}
```

13.4.3 GenericWebHostBuilder

要承载一个 ASP.NET Core 应用，只需要将 GenericWebHostService 服务注册到承载系统中即可。但 GenericWebHostService 服务具有针对其他一系列服务的依赖，所以在注册该承载服务之前需要先完成对这些依赖服务的注册。针对 GenericWebHostService 及其依赖服务的注册是借助 GenericWebHostBuilder 对象来完成的。

在传统的基于 IWebHost/IWebHostBuilder 的承载系统中，IWebHost 对象表示承载 Web 应用的宿主，它由对应的 IWebHostBuilder 对象构建而成，IWebHostBuilder 针对 IWebHost 对象的构建体现在它的 Build 方法上。由于通过该方法构建 IWebHost 对象是利用依赖注入框架提供的，所以 IWebHostBuilder 接口定义了两个 ConfigureServices 方法重载来注册这些依赖服务。如果注册的服务与通过 WebHostBuilderContext 对象表示的承载上下文（承载环境和配置）无关，我们一般调用第一个 ConfigureServices 方法重载，第二个方法可以帮助我们完成基于承载上下文的服务注册，我们也可以根据当前承载环境和提供的配置来动态地注册所需的服务。

```
public interface IWebHostBuilder
{
    IWebHost Build();

    string GetSetting(string key);
    IWebHostBuilder UseSetting(string key, string value);
    IWebHostBuilder ConfigureAppConfiguration(Action<WebHostBuilderContext,
        IConfigurationBuilder> configureDelegate);

    IWebHostBuilder ConfigureServices(Action<IServiceCollection> configureServices);
    IWebHostBuilder ConfigureServices(
        Action<WebHostBuilderContext, IServiceCollection> configureServices);
}
```

在基于 IWebHost/IWebHostBuilder 的承载系统中，WebHostBuilder 是对 IWebHostBuilder 接口的默认实现。如果采用基于 IHost/IHostBuilder 的承载系统，默认实现的 IWebHostBuilder 类型为 GenericWebHostBuilder。这个内部类型除了实现 IWebHostBuilder 接口，还实现了如下所示的两个内部接口（ISupportsStartup 和 ISupportsUseDefaultServiceProvider）。前面介绍 Startup 时提到过 ISupportsStartup 接口，它定义了一个用于注册中间件的 Configure 方法和一个用来注册 Startup 类型的 UseStartup 方法。ISupportsUseDefaultServiceProvider 接口则定义了唯一的 UseDefaultServiceProvider 方法，该方法用来对默认使用的依赖注入容器进行设置。

```
internal interface ISupportsStartup
{
    IWebHostBuilder Configure(
        Action<WebHostBuilderContext, IApplicationBuilder> configure);
```

```
        IWebHostBuilder UseStartup(Type startupType);
}

internal interface ISupportsUseDefaultServiceProvider
{
    IWebHostBuilder UseDefaultServiceProvider(
        Action<WebHostBuilderContext, ServiceProviderOptions> configure);
}
```

服务注册

下面通过简单的代码来模拟 GenericWebHostBuilder 针对 IWebHostBuilder 接口的实现。首先介绍用来注册依赖服务的 ConfigureServices 方法的实现。如下面的代码片段所示，GenericWebHostBuilder 实际上是对一个 IHostBuilder 对象的封装，针对依赖服务的注册是通过调用 IHostBuilder 接口的 ConfigureServices 方法实现的。

```
internal class GenericWebHostBuilder :
    IWebHostBuilder,
    ISupportsStartup,
    ISupportsUseDefaultServiceProvider
{
    private readonly IHostBuilder _builder;

    public GenericWebHostBuilder(IHostBuilder builder)
    {
        _builder = builder;
        ...
    }

    public IWebHostBuilder ConfigureServices(Action<IServiceCollection> configureServices)
        => ConfigureServices((_, services) => configureServices(services));

    public IWebHostBuilder ConfigureServices(
        Action<WebHostBuilderContext, IServiceCollection> configureServices)
    {
        _builder.ConfigureServices((context, services)
            => configureServices(GetWebHostBuilderContext(context), services));
        return this;
    }

    private WebHostBuilderContext GetWebHostBuilderContext(HostBuilderContext context)
    {
        if (!context.Properties.TryGetValue(typeof(WebHostBuilderContext), out var value))
        {
            var options = new WebHostOptions(context.Configuration,
                Assembly.GetEntryAssembly()?.GetName().Name);
            var webHostBuilderContext = new WebHostBuilderContext
            {
                Configuration       = context.Configuration,
                HostingEnvironment  = new HostingEnvironment(),
```

```
            };
            webHostBuilderContext.HostingEnvironment
                .Initialize(context.HostingEnvironment.ContentRootPath, options);
            context.Properties[typeof(WebHostBuilderContext)] = webHostBuilderContext;
            context.Properties[typeof(WebHostOptions)] = options;
            return webHostBuilderContext;
        }

        var webHostContext = (WebHostBuilderContext)value;
        webHostContext.Configuration = context.Configuration;
        return webHostContext;
    }
}
```

IHostBuilder 接口的 ConfigureServices 方法提供针对当前承载上下文的服务注册，通过 HostBuilderContext 对象表示的承载上下文包含两个元素，分别是表示配置的 IConfiguration 对象和表示承载环境的 IHostEnvironment 对象。而 ASP.NET Core 应用下的承载上下文是通过 WebHostBuilderContext 对象表示的，两个上下文之间的不同之处体现在针对承载环境的描述上，WebHostBuilderContext 上下文中的承载环境是通过 IWebHostEnvironment 对象表示的。GenericWebHostBuilder 在调用 IHostBuilder 对象的 ConfigureServices 方法注册依赖服务时，需要调用 GetWebHostBuilderContext 方法将提供的 WebHostBuilderContext 上下文转换成 HostBuilderContext 类型。

GenericWebHostBuilder 对象在构建时会以如下方式调用 ConfigureServices 方法注册一系列默认的依赖服务，其中包括表示承载环境的 IWebHostEnvironment 服务、用来发送诊断日志事件的 DiagnosticSource 服务和 DiagnosticListener 服务（它们返回同一个服务实例）、用来创建 HttpContext 上下文的 IHttpContextFactory 工厂、用来创建中间件的 IMiddlewareFactory 工厂、用来创建 IApplicationBuilder 对象的 IApplicationBuilderFactory 工厂等。除此之外，GenericWebHostBuilder 构造函数中还完成了针对 GenericWebHostServiceOptions 配置选项的设置，承载 ASP.NET Core 应用的 GenericWebHostService 服务也是在这里注册的。

```
internal class GenericWebHostBuilder :
    IWebHostBuilder,
    ISupportsStartup,
    ISupportsUseDefaultServiceProvider
{
    private readonly IHostBuilder        _builder;
    private AggregateException           _hostingStartupErrors;

    public GenericWebHostBuilder(IHostBuilder builder)
    {
        _builder = builder;
        _builder.ConfigureServices((context, services)=>
        {
            var webHostBuilderContext = GetWebHostBuilderContext(context);
            services.AddSingleton(webHostBuilderContext.HostingEnvironment);
            services.AddHostedService<GenericWebHostService>();
```

```
            DiagnosticListener instance = new DiagnosticListener("Microsoft.AspNetCore");
            services.TryAddSingleton(instance);
            services.TryAddSingleton<DiagnosticSource>(instance);
            services.TryAddSingleton<IHttpContextFactory, DefaultHttpContextFactory>();
            services.TryAddScoped<IMiddlewareFactory, MiddlewareFactory>();
            services.TryAddSingleton
                .<IApplicationBuilderFactory, ApplicationBuilderFactory>();

            var webHostOptions = (WebHostOptions)context
                .Properties[typeof(WebHostOptions)];
            services.Configure<GenericWebHostServiceOptions>(options=>
            {
                options.WebHostOptions = webHostOptions;
                options.HostingStartupExceptions = _hostingStartupErrors;
            });
        });
        ...
    }
}
```

配置的读写

除了两个 ConfigureServices 方法重载，IWebHostBuilder 接口的其他方法均与配置有关。基于 IHost/IHostBuilder 的承载系统涉及两种类型的配置：一种是在服务承载过程中供作为宿主的 IHost 对象使用的配置，另一种是供承载的服务或者应用消费的配置，前者是后者的子集。这两种类型的配置分别由 IHostBuilder 接口的 ConfigureHostConfiguration 方法和 ConfigureAppConfiguration 方法进行设置，GenericWebHostBuilder 针对配置的设置最终会利用这两个方法来完成。

GenericWebHostBuilder 提供的配置体现在它的字段_config 上，以键值对形式设置和读取配置的 UseSetting 方法与 GetSetting 方法操作的是_config 字段表示的 IConfiguration 对象。静态 Host 类型的 CreateDefaultBuilder 方法创建的 HostBuilder 对象会默认将前缀为"DOTNET_"的环境变量作为配置源，ASP.NET Core 应用则选择将前缀为"ASPNETCORE_"的环境变量作为配置源，这一点体现在如下所示的代码片段中。

```
internal class GenericWebHostBuilder :
    IWebHostBuilder,
    ISupportsStartup,
    ISupportsUseDefaultServiceProvider
{
    private readonly IHostBuilder            _builder;
    private readonly IConfiguration          _config;

    public GenericWebHostBuilder(IHostBuilder builder)
    {
        _builder = builder;
        _config = new ConfigurationBuilder()
            .AddEnvironmentVariables(prefix: "ASPNETCORE_")
```

```
        .Build();
    _builder.ConfigureHostConfiguration(config => config.AddConfiguration(_config));
    ...
}
public string GetSetting(string key) => _config[key];

public IWebHostBuilder UseSetting(string key, string value)
{
    _config[key] = value;
    return this;
}
```

如上面的代码片段所示，GenericWebHostBuilder 对象在构造过程中会创建一个 ConfigurationBuilder 对象，并将前缀为 "ASPNETCORE_" 的环境变量作为配置源。在利用 ConfigurationBuilder 对象创建 IConfiguration 对象之后，该对象体现的配置通过调用 IHostBuilder 对象的 ConfigureHostConfiguration 方法被合并到承载系统的配置中。由于 IHostBuilder 接口和 IWebHostBuilder 接口的 ConfigureAppConfiguration 方法具有相同的目的，所以 GenericWebHostBuilder 类型的 ConfigureAppConfiguration 方法直接调用 IHostBuilder 的同名方法。

```
internal class GenericWebHostBuilder :
    IWebHostBuilder,
    ISupportsStartup,
    ISupportsUseDefaultServiceProvider
{
    private readonly IHostBuilder _builder;

    public IWebHostBuilder ConfigureAppConfiguration(
        Action<WebHostBuilderContext, IConfigurationBuilder> configureDelegate)
    {
        _builder.ConfigureAppConfiguration((context, builder)
            => configureDelegate(GetWebHostBuilderContext(context), builder));
        return this;
    }
}
```

默认依赖注入框架配置

原生的依赖注入框架被直接整合到 ASP.NET Core 应用中，源于 GenericWebHostBuilder 类型 ISupportsUseDefaultServiceProvider 接口的实现。如下面的代码片段所示，在实现的 UseDefaultServiceProvider 方法中，GenericWebHostBuilder 会根据 ServiceProviderOptions 对象承载的配置选项完成对 DefaultServiceProviderFactory 工厂的注册。

```
internal class GenericWebHostBuilder :
    IWebHostBuilder,
    ISupportsStartup,
    ISupportsUseDefaultServiceProvider
{
```

```csharp
public IWebHostBuilder UseDefaultServiceProvider(
    Action<WebHostBuilderContext, ServiceProviderOptions> configure)
{
    _builder.UseServiceProviderFactory(context =>
    {
        var webHostBuilderContext = GetWebHostBuilderContext(context);
        var options = new ServiceProviderOptions();
        configure(webHostBuilderContext, options);
        return new DefaultServiceProviderFactory(options);
    });

    return this;
}
```

Startup

在大部分应用开发场景下，通常将应用启动时需要完成的初始化操作定义在注册的 Startup 中，按照约定定义的 Startup 类型旨在完成如下 3 个任务。

- 利用 Configure 方法或者 Configure{EnvironmentName}方法注册中间件。
- 利用 ConfigureServices 方法或者 Configure{EnvironmentName}Services 方法注册依赖服务。
- 利用 ConfigureContainer 方法或者 Configure{EnvironmentName}Container 方法对第三方依赖注入容器做相关设置。

上述 3 个针对 Startup 的设置最终都需要应用到基于 IHost/IHostBuilder 的承载系统上。由于 Startup 类型是注册到 GenericWebHostBuilder 对象上的，而 GenericWebHostBuilder 对象本质上是对 IHostBuilder 对象的封装，这些设置可以借助这个被封装的 IHostBuilder 对象被应用到承载系统上，具体的实现体现在如下几点。

- 将针对中间件的注册转移到 GenericWebHostServiceOptions 这个配置选项的 ConfigureApplication 属性上。
- 调用 IHostBuilder 对象的 ConfigureServices 方法来完成真正的服务注册。
- 调用 IHostBuilder 对象的 ConfigureContainer<TContainerBuilder>方法完成对依赖注入容器的设置。

上述 3 个在启动过程执行的初始化操作由 3 个对应的 Builder 对象（ConfigureBuilder、ConfigureServicesBuilder 和 ConfigureContainerBuilder）辅助完成，其中 Startup 类型的 Configure 方法或者 Configure{EnvironmentName}方法对应如下所示的 ConfigureBuilder 类型。ConfigureBuilder 对象由 Configure 方法或者 Configure{EnvironmentName}方法对应的 MethodInfo 对象创建而成，最终赋值给 GenericWebHostServiceOptions 配置选项 ConfigureApplication 属性的就是这个委托对象。如下所示的代码片段是 ConfigureBuilder 类型简化后的定义。

```csharp
public class ConfigureBuilder
{
    public MethodInfo MethodInfo { get; }
    public ConfigureBuilder(MethodInfo configure)
```

```
        => MethodInfo = configure;

    public Action<IApplicationBuilder> Build(object instance)
        => builder => Invoke(instance, builder);

    private void Invoke(object instance, IApplicationBuilder builder)
    {
        using (var scope = builder.ApplicationServices.CreateScope())
        {
            var serviceProvider = scope.ServiceProvider;
            var parameterInfos = MethodInfo.GetParameters();
            var parameters = new object[parameterInfos.Length];
            for (var index = 0; index < parameterInfos.Length; index++)
            {
                var parameterInfo = parameterInfos[index];
                parameters[index] =
                    parameterInfo.ParameterType == typeof(IApplicationBuilder)
                    ? builder
                    : serviceProvider.GetRequiredService(parameterInfo.ParameterType);
            }
            MethodInfo.InvokeWithoutWrappingExceptions(instance, parameters);
        }
    }
}
```

如下所示的 ConfigureServicesBuilder 和 ConfigureContainerBuilder 类型是简化后的版本，前者对应 Startup 类型的 ConfigureServices/Configure{EnvironmentName}Services 方法，后者对应 ConfigureContainer 方法或者 Configure{EnvironmentName}Container 方法。针对对应方法的调用会反映在 Build 方法返回的委托对象上。

```
public class ConfigureServicesBuilder
{
    public MethodInfo MethodInfo { get; }
    public ConfigureServicesBuilder2(MethodInfo configureServices)
        => MethodInfo = configureServices;
    public Func<IServiceCollection, IServiceProvider> Build(object instance)
        => services => Invoke(instance, services);
    private IServiceProvider Invoke(object instance, IServiceCollection services)
        => MethodInfo.InvokeWithoutWrappingExceptions(instance, new object[] { services })
        as IServiceProvider;
}

public class ConfigureContainerBuilder
{
    public MethodInfo MethodInfo { get; }
    public ConfigureContainerBuilder(MethodInfo configureContainerMethod)
        => MethodInfo = configureContainerMethod;
    public Action<object> Build(object instance) => container
        => Invoke(instance, container);
    private void Invoke(object instance, object container)
```

```
    => MethodInfo.InvokeWithoutWrappingExceptions(instance,
    new object[] { container });
}
```

通过第 11 章的介绍可知，Startup 类型的构造函数中是可以注入依赖服务的，但是可以在这里注入的依赖服务仅限于组成当前承载上下文的两个元素，即表示承载环境的 IHostEnvironment 对象或者 IWebHostEnvironment 对象和表示配置的 IConfiguration 对象。这一个特性是如下这个特殊的 IServiceProvider 实现类型决定的。

```
internal class GenericWebHostBuilder :
    IWebHostBuilder,
    ISupportsStartup,
    ISupportsUseDefaultServiceProvider
{
    private class HostServiceProvider : IServiceProvider
    {
        private readonly WebHostBuilderContext _context;
        public HostServiceProvider(WebHostBuilderContext context)
        {
            _context = context;
        }

        public object GetService(Type serviceType)
        {
            if (serviceType == typeof(IWebHostEnvironment)
                || serviceType == typeof(IHostEnvironment))
            {
                return _context.HostingEnvironment;
            }
            if (serviceType == typeof(IConfiguration))
            {
                return _context.Configuration;
            }
            return null;
        }
    }
}
```

如下所示的代码片段是 GenericWebHostBuilder 的 UseStartup 方法简化版本的定义。可以看出，Startup 类型是通过调用 IHostBuilder 对象的 ConfigureServices 方法来注册的。如下面的代码片段所示，GenericWebHostBuilder 对象会根据指定 Startup 类型创建出 3 个对应的 Builder 对象，然后利用上面的 HostServiceProvider 创建出 Startup 对象，并将该对象作为参数调用对应的 3 个 Builder 对象的 Build 方法构建的委托对象完成对应的初始化任务。

```
internal class GenericWebHostBuilder :
    IWebHostBuilder,
    ISupportsStartup,
    ISupportsUseDefaultServiceProvider
{
    private readonly IHostBuilder _builder;
```

```csharp
private readonly object _startupKey = new object();

public IWebHostBuilder UseStartup(Type startupType)
{
    _builder.Properties["UseStartup.StartupType"] = startupType;
    _builder.ConfigureServices((context, services) =>
    {
        if (_builder.Properties.TryGetValue("UseStartup.StartupType",
            out var cachedType) && (Type)cachedType == startupType)
        {
            UseStartup(startupType, context, services);
        }
    });

    return this;
}

private void UseStartup(Type startupType, HostBuilderContext context,
    IServiceCollection services)
{
    var webHostBuilderContext = GetWebHostBuilderContext(context);
    var webHostOptions = (WebHostOptions)context.Properties[typeof(WebHostOptions)];

    ExceptionDispatchInfo startupError = null;
    object instance = null;
    ConfigureBuilder configureBuilder = null;

    try
    {
        instance = ActivatorUtilities.CreateInstance(
            new HostServiceProvider(webHostBuilderContext), startupType);
        context.Properties[_startupKey] = instance;
        var environmentName = context.HostingEnvironment.EnvironmentName;
        BindingFlags bindingFlags = BindingFlags.Public | BindingFlags.Instance;

        //ConfigureServices
        var configureServicesMethod = startupType.GetMethod(
            $"Configure{environmentName}Services", bindingFlags)
            ?? startupType.GetMethod("ConfigureServices", bindingFlags);
        if (configureServicesMethod != null)
        {
            var configureServicesBuilder =
                new ConfigureServicesBuilder(configureServicesMethod);
            var configureServices = configureServicesBuilder.Build(instance);
            configureServices(services);
        }

        //ConfigureContainer
        var configureContainerMethod = startupType
            .GetMethod($"Configure{environmentName}Container", bindingFlags)
```

```csharp
                ?? startupType.GetMethod("ConfigureContainer", bindingFlags);
            if (configureContainerMethod != null)
            {
                var configureContainerBuilder =
                    new ConfigureBuilder(configureServicesMethod);
                _builder.Properties[typeof(ConfigureContainerBuilder)]
                    = configureContainerBuilder;
                var containerType = configureContainerBuilder.MethodInfo
                    .GetParameters()[0].ParameterType;
                var actionType = typeof(Action<,>)
                    .MakeGenericType(typeof(HostBuilderContext), containerType);
                var configure = GetType().GetMethod(nameof(ConfigureContainer),
                    BindingFlags.NonPublic | BindingFlags.Instance)
                    .MakeGenericMethod(containerType)
                    .CreateDelegate(actionType, this);

                //IHostBuilder.ConfigureContainer<TContainerBuilder>(
                //Action<HostBuilderContext, TContainerBuilder> configureDelegate)
                typeof(IHostBuilder).GetMethods().First(m => m.Name ==
                    nameof(IHostBuilder.ConfigureContainer))
                    .MakeGenericMethod(containerType)
                    .Invoke(_builder, BindingFlags.DoNotWrapExceptions, null,
                    new object[] { configure },null);
            }

            var configureMethod = startupType.GetMethod($"Configure{environmentName}",
                bindingFlags)
                ?? startupType.GetMethod("Configure", bindingFlags);
            configureBuilder = new ConfigureBuilder(configureMethod);
        }
        catch (Exception ex) when (webHostOptions.CaptureStartupErrors)
        {
            startupError = ExceptionDispatchInfo.Capture(ex);
        }

        //Configure
        services.Configure<GenericWebHostServiceOptions>(options =>
        {
            options.ConfigureApplication = app =>
            {
                startupError?.Throw();
                if (instance != null && configureBuilder != null)
                {
                    configureBuilder.Build(instance)(app);
                }
            };
        });
    }

    //用于创建IHostBuilder.ConfigureContainer<TContainerBuilder>(
```

```
//Action<HostBuilderContext, TContainerBuilder> configureDelegate)方法的参数
private void ConfigureContainer<TContainer>(HostBuilderContext context,
    TContainer container)
{
    var instance = context.Properties[_startupKey];
    var builder = (ConfigureContainerBuilder)context
        .Properties[typeof(ConfigureContainerBuilder)];
    builder.Build(instance)(container);
}
```

除了上面的 UseStartup 方法，GenericWebHostBuilder 还实现了 ISupportsStartup 接口的 Configure 方法。如下面的代码片段所示，该方法会将指定的用于注册中间件的 Action<WebHostBuilderContext, IApplicationBuilder>复制给作为配置选项的 GenericWebHostServiceOptions 对象的 ConfigureApplication 属性。由于注册 Startup 类型的目的也是通过设置 GenericWebHostServiceOptions 对象的 ConfigureApplication 属性来注册中间件，如果在调用了 Configure 方法时又注册了一个 Startup 类型，系统会采用"后来居上"的原则。

```
internal class GenericWebHostBuilder :
    IWebHostBuilder,
    ISupportsStartup,
    ISupportsUseDefaultServiceProvider
{
    public IWebHostBuilder Configure(
        Action<WebHostBuilderContext, IApplicationBuilder> configure)
    {
        _builder.ConfigureServices((context, services) =>
        {
            services.Configure<GenericWebHostServiceOptions>(options =>
            {
                var webhostBuilderContext = GetWebHostBuilderContext(context);
                options.ConfigureApplication =
                    app => configure(webhostBuilderContext, app);
            });
        });
        return this;
    }
}
```

除了直接调用 UseStartup 方法注册一个 Startup 类型，还可以利用配置注册 Startup 类型所在的程序集。GenericWebHostBuilder 对象在初始化过程中会按照约定的规则定位和加载 Startup 类型。通过第 11 章的介绍可知，GenericWebHostBuilder 对象会按照如下顺序从指定的程序集类型列表中筛选 Startup 类型。

- Startup{EnvironmentName}（全名匹配）。
- Startup（全名匹配）。
- {StartupAssembly}.Startup{EnvironmentName}（全名匹配）。
- {StartupAssembly}.Startup（全名匹配）。

- Startup{EnvironmentName}（任意命名空间）。
- Startup（任意命名空间）。

从指定启动程序集中加载 Startup 类型的逻辑体现在如下所示的 FindStartupType 方法中。在执行构造函数的最后阶段，如果 WebHostOptions 选项的 StartupAssembly 属性被设置了一个启动程序集，定义在该程序集中的 Startup 类型会被加载出来，并作为参数调用上面定义的 UseStartup 方法完成对它的注册。如果在此过程中抛出异常，并且将 WebHostOptions 选项的 CaptureStartupErrors 属性设置为 True，那么捕捉到的异常会通过设置 GenericWebHostService Options 对象的 ConfigureApplication 属性的方式重新抛出来。

```csharp
internal class GenericWebHostBuilder :
    IWebHostBuilder,
    ISupportsStartup,
    ISupportsUseDefaultServiceProvider
{
    public GenericWebHostBuilder(IHostBuilder builder)
    {
        ...
        if (!string.IsNullOrEmpty(webHostOptions.StartupAssembly))
        {
            try
            {
                var startupType = FindStartupType(webHostOptions.StartupAssembly,
                    webhostContext.HostingEnvironment.EnvironmentName);
                UseStartup(startupType, context, services);
            }
            catch (Exception ex) when (webHostOptions.CaptureStartupErrors)
            {
                var capture = ExceptionDispatchInfo.Capture(ex);
                services.Configure<GenericWebHostServiceOptions>(options =>
                {
                    options.ConfigureApplication = app => capture.Throw();
                });
            }
        }
    }

    private static Type FindStartupType(string startupAssemblyName, string environmentName)
    {
        var assembly = Assembly.Load(new AssemblyName(startupAssemblyName))
            ?? throw new InvalidOperationException(
                string.Format("The assembly '{0}' failed to load.", startupAssemblyName));
        var startupNameWithEnv = $"Startup{environmentName}";
        var startupNameWithoutEnv = "Startup";

        var type =
            assembly.GetType(startupNameWithEnv) ??
            assembly.GetType(startupAssemblyName + "." + startupNameWithEnv) ??
```

```
            assembly.GetType(startupNameWithoutEnv) ??
            assembly.GetType(startupAssemblyName + "." + startupNameWithoutEnv);
        if (null != type)
        {
            return type;
        }

        var types = assembly.DefinedTypes.ToList();
        type = types.Where(info => info.Name.Equals(startupNameWithEnv,
            StringComparison.OrdinalIgnoreCase)).FirstOrDefault()
            ?? types.Where(info => info.Name.Equals(startupNameWithoutEnv,
            StringComparison.OrdinalIgnoreCase)).FirstOrDefault();
        return type?? throw new InvalidOperationException(
            string.Format(
            "A type named '{0}' or '{1}' could not be found in assembly '{2}'.",
            startupNameWithEnv,
            startupNameWithoutEnv,
            startupAssemblyName));
    }
}
```

Hosting Startup

Startup 类型可定义在任意程序集中，并通过配置的方式注册到 ASP.NET Core 应用中。Hosting Startup 与之类似，我们可以将一些初始化操作定义在任意程序集中，在无须修改应用程序任何代码的情况下利用配置的方式实现对它们的注册。两者的不同之处在于：整个应用最终只会使用到一个 Startup 类型，但是采用 Hosting Startup 注册的初始化操作都是有效的。Hosting Startup 类型提供的方式将一些工具"附加"到一个 ASP.NET Core 应用中。

通过第 11 章的介绍可知，以 Hosting Startup 方法实现的初始化操作必须实现在 IHostingStartup 接口的实现类型中，该类型最终以 HostingStartupAttribute 特性的方式进行注册。Hosting Startup 相关的配置最终体现在 WebHostOptions 如下的 3 个属性上。

```
public class WebHostOptions
{
    public bool                      PreventHostingStartup { get; set; }
    public IReadOnlyList<string>     HostingStartupAssemblies { get; set; }
    public IReadOnlyList<string>     HostingStartupExcludeAssemblies { get; set; }
}
```

定义在 IHostingStartup 实现类型上的初始化操作会作用在一个 IWebHostBuilder 对象上，但最终对象并不是 GenericWebHostBuilder，而是如下所示的 HostingStartupWebHostBuilder 对象。从给出的代码片段可以看出，HostingStartupWebHostBuilder 对象实际上是对 GenericWebHostBuilder 对象的进一步封装，针对它的方法调用最终还是会转移到封装的 GenericWebHostBuilder 对象上。

```
internal class HostingStartupWebHostBuilder :
    IWebHostBuilder, ISupportsStartup, ISupportsUseDefaultServiceProvider
{
    private readonly GenericWebHostBuilder _builder;
```

```csharp
    private Action<WebHostBuilderContext, IConfigurationBuilder> _configureConfiguration;
    private Action<WebHostBuilderContext, IServiceCollection> _configureServices;

    public HostingStartupWebHostBuilder(GenericWebHostBuilder builder)
        =>_builder = builder;

    public IWebHost Build()
        => throw new NotSupportedException();

    public IWebHostBuilder ConfigureAppConfiguration(
        Action<WebHostBuilderContext, IConfigurationBuilder> configureDelegate)
    {
        _configureConfiguration += configureDelegate;
        return this;
    }

    public IWebHostBuilder ConfigureServices(
        Action<IServiceCollection> configureServices)
        => ConfigureServices((context, services) => configureServices(services));

    public IWebHostBuilder ConfigureServices(
        Action<WebHostBuilderContext, IServiceCollection> configureServices)
    {
        _configureServices += configureServices;
        return this;
    }
    public string GetSetting(string key) => _builder.GetSetting(key);

    public IWebHostBuilder UseSetting(string key, string value)
    {
        _builder.UseSetting(key, value);
        return this;
    }

    public void ConfigureServices(WebHostBuilderContext context,
        IServiceCollection services) => _configureServices?.Invoke(context, services);

    public void ConfigureAppConfiguration(WebHostBuilderContext context,
        IConfigurationBuilder builder)=> _configureConfiguration?.Invoke(context, builder);

    public IWebHostBuilder UseDefaultServiceProvider(
        Action<WebHostBuilderContext, ServiceProviderOptions> configure)
        => _builder.UseDefaultServiceProvider(configure);

    public IWebHostBuilder Configure(
        Action<WebHostBuilderContext, IApplicationBuilder> configure)
        => _builder.Configure(configure);

    public IWebHostBuilder UseStartup(Type startupType)=> _builder.UseStartup(startupType);
}
```

Hosting Startup 的实现体现在如下所示的 ExecuteHostingStartups 方法中，该方法会根据当前的配置和作为应用名称的入口程序集名称创建一个新的 WebHostOptions 对象，如果这个配置选项的 PreventHostingStartup 属性返回 True，就意味着人为关闭了 Hosting Startup 特性。在 Hosting Startup 特性没有被显式关闭的情况下，该方法会利用配置选项的 HostingStartupAssemblies 属性和 HostingStartupExcludeAssemblies 属性解析出启动程序集名称，并从中解析出注册的 IHostingStartup 类型。在通过反射的方式创建出对应的 IHostingStartup 对象之后，上面介绍的 HostingStartupWebHostBuilder 对象会被创建出来，并作为参数调用这些 IHostingStartup 对象的 Configure 方法。

```
internal class GenericWebHostBuilder :
    IWebHostBuilder,
    ISupportsStartup,
    ISupportsUseDefaultServiceProvider
{
    private readonly IHostBuilder           _builder;
    private readonly IConfiguration         _config;

    public GenericWebHostBuilder(IHostBuilder builder)
    {
        _builder    = builder;
        _config     = new ConfigurationBuilder()
            .AddEnvironmentVariables(prefix: "ASPNETCORE_")
            .Build();

        _builder.ConfigureHostConfiguration(config =>
        {
            config.AddConfiguration(_config);
            ExecuteHostingStartups();
        });
    }

    private void ExecuteHostingStartups()
    {
        var options = new WebHostOptions(
            _config, Assembly.GetEntryAssembly()?.GetName().Name);
        if (options.PreventHostingStartup)
        {
            return;
        }

        var exceptions = new List<Exception>();
        _hostingStartupWebHostBuilder = new HostingStartupWebHostBuilder(this);

        var assemblyNames = options.HostingStartupAssemblies
                .Except(options.HostingStartupExcludeAssemblies,
                    StringComparer.OrdinalIgnoreCase)
                .Distinct(StringComparer.OrdinalIgnoreCase);
```

```
            foreach (var assemblyName in assemblyNames)
            {
                try
                {
                    var assembly = Assembly.Load(new AssemblyName(assemblyName));
                    foreach (var attribute in
                        assembly.GetCustomAttributes<HostingStartupAttribute>())
                    {
                        var hostingStartup = (IHostingStartup)Activator
                            .CreateInstance(attribute.HostingStartupType);
                        hostingStartup.Configure(_hostingStartupWebHostBuilder);
                    }
                }
                catch (Exception ex)
                {
                    exceptions.Add(new InvalidOperationException(
                        $"Startup assembly {assemblyName} failed to execute. See the inner
                        exception for more details.", ex));
                }
            }
            if (exceptions.Count > 0)
            {
                _hostingStartupErrors = new AggregateException(exceptions);
            }
        }
}
```

由于调用 IHostingStartup 对象的 Configure 方法传入的 HostingStartupWebHostBuilder 对象是对当前 GenericWebHostBuilder 对象的封装,而这个 GenericWebHostBuilder 对象又是对 IHostBuilder 的封装,所以以 Hosting Startup 注册的初始化操作最终还是应用到了以 IHost/IHostBuilder 为核心的承载系统中。虽然 GenericWebHostBuilder 类型实现了 IWebHostBuilder 接口,但它仅仅是 IHostBuilder 对象的代理,其自身针对 IWebHost 对象的构建需求不复存在,所以它的 Build 方法会直接抛出异常。

```
internal class GenericWebHostBuilder :
    IWebHostBuilder,
    ISupportsStartup,
    ISupportsUseDefaultServiceProvider
{
    public IWebHost Build()=> throw new NotSupportedException(
        $"Building this implementation of {nameof(IWebHostBuilder)} is not supported.");
    ...
}
```

13.4.4 ConfigureWebHostDefaults

演示实例中针对 IWebHostBuilder 对象的应用都体现在 IHostBuilder 接口的 ConfigureWebHostDefaults 扩展方法上,它最终调用的其实是如下所示的 ConfigureWebHost 扩展方法。如下面的代码片段所示,ConfigureWebHost 方法会将当前 IHostBuilder 对象创建的

GenericWebHostBuilder 对象作为参数调用指定的 Action<IWebHostBuilder>委托对象。由于 GenericWebHostBuilder 对象相当于 IHostBuilder 对象的代理，所以这个委托中完成的所有操作最终都会转移到 IHostBuilder 对象上。

```
public static class GenericHostWebHostBuilderExtensions
{
    public static IHostBuilder ConfigureWebHost(
        this IHostBuilder builder, Action<IWebHostBuilder> configure)
    {
        var webhostBuilder = new GenericWebHostBuilder(builder);
        configure(webhostBuilder);
        return builder;
    }
}
```

顾名思义，ConfigureWebHostDefaults 方法会帮助我们做默认设置，这些设置实现在静态类型 WebHost 的 ConfigureWebDefaults 方法中。如下所示的 ConfigureWebDefaults 方法的实现，该方法提供的默认设置包括将定义在 Microsoft.AspNetCore.StaticWebAssets.xml 文件（物理文件或者内嵌文件）作为默认的 Web 资源、注册 KestrelServer、配置关于主机过滤（Host Filter）和 Http Overrides 相关选项、注册路由中间件，以及对用于集成 IIS 的 AspNetCoreModule 模块的配置。

```
public static class GenericHostBuilderExtensions
{
    public static IHostBuilder ConfigureWebHostDefaults(this IHostBuilder builder,
        Action<IWebHostBuilder> configure)
        => builder.ConfigureWebHost(webHostBuilder =>
        {
            WebHost.ConfigureWebDefaults(webHostBuilder);
            configure(webHostBuilder);
        });
}

public static class WebHost
{
    internal static void ConfigureWebDefaults(IWebHostBuilder builder)
    {
        builder.ConfigureAppConfiguration((ctx, cb) =>
        {
            if (ctx.HostingEnvironment.IsDevelopment())
            {
                StaticWebAssetsLoader.UseStaticWebAssets(ctx.HostingEnvironment);
            }
        });
        builder.UseKestrel((builderContext, options) =>
        {
            options.Configure(builderContext.Configuration.GetSection("Kestrel"));
        })
        .ConfigureServices((hostingContext, services) =>
```

```csharp
            {
                services.PostConfigure<HostFilteringOptions>(options =>
                {
                    if (options.AllowedHosts == null || options.AllowedHosts.Count == 0)
                    {
                        var hosts = hostingContext
                            .Configuration["AllowedHosts"]?.Split(new[] { ';' },
                            StringSplitOptions.RemoveEmptyEntries);
                        options.AllowedHosts = (hosts?.Length > 0 ? hosts : new[] { "*" });
                    }
                });
                services.AddSingleton<IOptionsChangeTokenSource<HostFilteringOptions>>(
                    new ConfigurationChangeTokenSource<HostFilteringOptions>(
                    hostingContext.Configuration));

                services.AddTransient<IStartupFilter, HostFilteringStartupFilter>();

                if (string.Equals("true",
                    hostingContext.Configuration["ForwardedHeaders_Enabled"],
                    StringComparison.OrdinalIgnoreCase))
                {
                    services.Configure<ForwardedHeadersOptions>(options =>
                    {
                        options.ForwardedHeaders =
                            ForwardedHeaders.XForwardedFor|ForwardedHeaders.XForwardedProto;

                        options.KnownNetworks.Clear();
                        options.KnownProxies.Clear();
                    });

                    services.AddTransient<IStartupFilter, ForwardedHeadersStartupFilter>();
                }

                services.AddRouting();
            })
            .UseIIS()
            .UseIISIntegration();
    }
}
```

附录A

实例演示 1

章 节	编 号	描 述
第 1 章	S101	.NET Core 控制台应用
	S102	ASP.NET Core 应用
	S103	指定服务端 URL
	S104	调用 ConfigureWebHostDefaults 方法设置 IHost 默认属性
	S105	ASP.NET Core MVC 应用
	S106	ASP.NET Core MVC 视图呈现
	S107	利用注册的 Startup 类型执行初始化操作
第 3 章	S301	[Cat]服务注册与消费
	S302	[Cat]泛型的服务注册与消费
	S303	[Cat]提取针对指定服务类型的所有服务实例
	S304	[Cat]服务实例的生命周期
第 4 章	S401	服务注册与消费
	S402	泛型的服务注册与消费
	S403	提取针对指定服务类型的所有服务实例
	S404	服务实例的生命周期
	S405	服务实例的释放
	S406	服务范围的验证
	S407	服务实例构建能力的验证
	S408	服务类型构造函数的选择[成功]
	S409	服务类型构造函数的选择[失败]
	S410	ServiceProviderEngine 与 ServiceProviderEngineScope
	S411	基于 IServiceProvider 服务的注入
	S412	与第三方 DI 框架整合
第 5 章	S501	呈现文件系统的目录结构[物理文件系统]
	S502	读取文件的内容[物理文件系统]
	S503	读取文件的内容[程序集内嵌文件系统]
	S504	监控文件的更新
	S505	呈现文件系统的目录结构[远程文件系统]
	S506	读取文件的内容[远程文件系统]
第 6 章	S601	以键值对的形式读取配置数据
	S602	读取结构化的配置数据

续表

章节	编号	描述
第6章	S603	将结构化的配置绑定为 POCO 对象
	S604	将 JSON 文件作为配置源
	S605	基于承载环境的配置提供方式
	S606	配置数据与配置源的及时同步
	S607	基于标量数据的配置绑定
	S608	利用自定义 TypeConverter 解决类型转换的问题
	S609	针对符合类型的配置绑定
	S610	针对集合和数组的配置绑定
	S611	针对集合和数组的配置绑定的差异
	S612	针对字典的配置绑定
	S613	针对环境变量的配置源
	S614	针对命令行参数的配置源
	S615	解决 XML 针对集合绑定的问题
	S616	自定义针对数据库的配置源
第7章	S701	将配置绑定为 Options 对象
	S702	提供具名的 Options 对象
	S703	利用依赖注入容器获取 Options 对象
	S704	Options 对象与配置源的同步
	S705	直接初始化默认 Options 对象的属性
	S706	直接初始化具名 Options 对象的属性
	S707	根据承载环境动态设置 Options 对象属性
	S708	验证 Options 数据的有效性
	S709	Options 对象的生命周期
	S710	通过自定义 IConfigureOptions 实现类型设置 Options
	S711	自定义 OptionsChangeTokenSource 实现 Options 数据的同步
第8章	S801	基于 TraceSource 的跟踪日志
	S802	通过设置 SourceLevels 过滤跟踪日志消息
	S803	自定义 TraceListener
	S804	基于 EventSource 的事件日志
	S805	自定义 EventListener
	S806	基于 DiagnosticListener 的诊断日志
	S807	强类型的诊断日志编程方式
	S808	利用 DefaultTraceListener 将日志输出到 TXT 文件
	S809	利用 DelimitedListTraceListener 将日志输出到 CSV 文件
	S810	利用 EventListener 收集事件日志的详细信息
	S811	针对事件日志消息的序列化
	S812	利用事件日志描述调用链
	S813	利用事件日志收集性能指标
	S814	利用 DiagnosticListener.AllListeners 收集诊断日志
	S815	诊断日志的强类型的事件订阅

续表

章节	编号	描述
第9章	S901	针对控制台和 Debugger 的日志输出
	S902	利用注入的 ILogger<T> 服务输出日志
	S903	针对 TraceSource 和 EventSource 的输出
	S904	基于等级的日志过滤
	S905	基于等级和类别的日志过滤
	S906	基于等级、类别和 LoggerProvider 的日志过滤
	S907	利用配置定义日志过滤规则
	S908	利用日志范围描述当前上下文
	S909	利用 LoggerMessage 收集日志
	S910	利用 LoggerMessage 设置日志范围
	S911	Logger<T> 采用的日志类别命名规则
	S912	利用日志范围描述逻辑调用链
	S913	利用自定义的 EventListener 接收由 EventSourceLogger 发出的日志
	S914	利用自定义的 EventListener 接收由 EventSourceLogger 发出的调用链信息
第10章	S1001	承载一个用于收集性能指标的服务
	S1002	依赖注入在服务承载中的应用
	S1003	配置选项在服务承载中的应用
	S1004	根据承载环境动态设置配置选项
	S1005	日志在服务承载中的应用
	S1006	利用配置定义日志的过滤规则
	S1007	利用 IHostApplicationLifetime 关闭当前应用
	S1008	与第三方 DI 框架整合
	S1009	利用 IHostEnvironment 服务得到承载环境信息
第11章	S1101	以 Func<RequestDelegate, RequestDelegate> 对象的形式注册中间件
	S1102	定义强类型的中间件
	S1103	按照约定定义中间件
	S1104	获取 ASP.NET Core 应用默认注册的服务
	S1105	将服务注册到 Startup 的构造函数中
	S1106	将服务注册到 Startup 的 Configure 方法中
	S1107	将服务注册到中间件的构造函数中
	S1108	将服务注册到中间件的 InvokeAsync 方法中
	S1109	ASP.NET Core MVC 应用的依赖注入
	S1110	服务的生命周期
	S1111	与第三方 DI 框架整合
	S1112	初始化配置
	S1113	以键值对的形式读取和设置配置数据
	S1114	以配置的方式指定服务器 URL
	S1115	配置合并
	S1116	注册 IConfigurationSource 对象
	S1117	通过配置定制承载环境
	S1118	根据当前承载环境动态选择 Startup 类型

续表

章节	编号	描述
第 11 章	S1119	通过注册 IHostingStartup 的方式执行初始化操作
	S1120	以 IStartupFilter 的形式注册中间件
第 12 章	S1201	Mini 版的 ASP.NET Core 框架
第 13 章	S1301	输出 ASP.NET Core 应用启动过程中输出的日志
	S1302	收集 ASP.NET Core 框架针对请求的诊断日志
	S1303	收集 ASP.NET Core 框架针对请求的事件日志

ASP.NET Core 3
框架揭秘（下册）

蒋金楠 ◎著

电子工业出版社
Publishing House of Electronics Industry
北京·BEIJING

内 容 简 介

本书主要阐述 ASP.NET Core 最核心的部分——请求处理管道。通过阅读本书，读者可以深刻、系统地了解 ASP.NET Core 应用在启动过程中管道的构建方式，以及请求在管道中的处理流程。本书还详细讲述了 .NET Core 跨平台的本质，以及多个常用的基础框架（如依赖注入、文件信息、配置选项和诊断日志等）。本书还对大部分原生的中间件提供了系统性介绍，采用"编程体验"、"总体设计"、"具体实现"和"灵活运用"的流程，使读者可以循序渐进地学习 ASP.NET Core 的每个功能模块。

本书可供所有 .NET 从业人员阅读与参考。

未经许可，不得以任何方式复制或抄袭本书之部分或全部内容。
版权所有，侵权必究。

图书在版编目（CIP）数据

ASP.NET Core 3 框架揭秘．下册 / 蒋金楠著．—北京：电子工业出版社，2020.5
ISBN 978-7-121-38462-2

Ⅰ．①A… Ⅱ．①蒋… Ⅲ．①网页制作工具－程序设计 Ⅳ．①TP393.092.2

中国版本图书馆 CIP 数据核字（2020）第 030760 号

责任编辑：张春雨　　　特约编辑：田学清
印　　刷：三河市良远印务有限公司
装　　订：三河市良远印务有限公司
出版发行：电子工业出版社
　　　　　北京市海淀区万寿路 173 信箱　　邮编：100036
开　　本：787×980　1/16　　印张：57.25　　字数：1390 千字
版　　次：2020 年 5 月第 1 版
印　　次：2020 年 5 月第 3 次印刷
定　　价：199.00 元（上下册）

凡所购买电子工业出版社图书有缺损问题，请向购买书店调换。若书店售缺，请与本社发行部联系，联系及邮购电话：（010）88254888，88258888。
质量投诉请发邮件至 zlts@phei.com.cn，盗版侵权举报请发邮件至 dbqq@phei.com.cn。
本书咨询联系方式：010-51260888-819，faq@phei.com.cn。

前言

写作源起

计算机图书市场存在一系列介绍 ASP.NET Web Forms、ASP.NET MVC、ASP.NET Web API 的图书，但是找不到一本专门介绍 ASP.NET 自身框架的图书，作为一名拥有 17 年工作经验的 .NET 开发者，笔者对此感到十分困惑。上述这些 Web 开发框架都是建立在 ASP.NET 底层框架之上的，底层 ASP.NET 框架才是根基所在。过去笔者接触过很多资深的 ASP.NET 开发人员，发现他们对 ASP.NET 框架大都没有进行深入了解。

2014 年，出版《ASP.NET MVC 5 框架揭秘》之后，笔者原本打算写 "ASP.NET 框架揭秘"。但在新书准备过程中，微软推出了 ASP.NET Core（当时被称为 ASP.NET 5，还没有 .NET Core 的概念）。所以，笔者将研究重点转移到 ASP.NET Core。

本书耗时 5 年左右，笔者投入了大量心血。2015 年年初，笔者开始了本书的写作，微软在 2016 年 6 月正式发布 .NET Core 1.0 时，本书的绝大部分内容就已经完成。随后，微软不断推出新的版本，本书的内容也在不断快速"迭代"中。本书正文部分共计 800 多页，但笔者在写作过程中删除的部分不少于这个数字。

有人认为自己每天只是做一些简单的编程工作，根本没有必要去了解底层原理和设计方面的内容。其实，不论我们从事何种层次的工作，最根本的目的只有一个——解决问题。解决方案分两种：一种是"扬汤止沸"，另一种是"釜底抽薪"。看到锅里不断沸腾的水，大多数人会选择不断地往锅里浇冷水，笔者希望这本书能够使读者看到锅底熊熊燃烧的薪火。

本书内容

ASP.NET Core 是一个全新的 Web 开发平台，为我们构建了一个可复用和可定制的请求处理管道，微软在它上面构建了 MVC、SignalR、GRPC、Orleans 这样广泛使用的 Web 框架，我们也可以利用它构建自己的 Web 框架（笔者曾经通过 ASP.NET Core 构建了一款 GraphQL 框架）。本书只关注最本质的东西，即 ASP.NET Core 请求处理管道，并不会涉及上述这些 Web 框架。本书的内容主要划分为如下 4 个部分。

跨平台的开发体验和实现原理

.NET Core 与传统 .NET Framework 最大的区别是跨平台，作为开篇入门材料，第 1 章通过几个简单的 Hello World 程序，让读者可以体验如何在 Windows、macOS、Linux 平台上开发 .NET Core 应用，以及通过 Docker 容器部署 ASP.NET Core 应用的乐趣。第 2 章将告诉读者 .NET Core 的跨平台究竟是如何实现的。

基础框架

ASP.NET Core 框架依赖于一些基础框架，其中最重要的是注入框架。由于依赖注入框架不但是构建 ASP.NET Core 请求处理管道的基石，而且依赖注入也是 ASP.NET Core 应用的基本编程模式，所以本书的第 3 章和第 4 章对依赖注入原理及依赖注入框架的设计与编程方式进行了详细介绍。

ASP.NET Core 应用具有很多读取文件内容的场景，所以它构建了一个抽象的文件系统，第 5 章会对这个文件系统的设计模型和两种实现方式（物理文件系统和程序集内嵌文件系统）进行详细介绍。

.NET Core 针对"配置"的支持是传统 .NET 开发人员所不能想象的，所以采用两章的篇幅对这一主题进行讲解：第 6 章旨在介绍支持多种数据源的配置系统；不论是开发 ASP.NET Core 应用还是组件，都可以采用 Options 模式来读取配置选项，第 7 章会着重讲述这种强类型的配置选项编程方式。

.NET Core 在错误诊断方面为我们提供了多种选择，第 8 章介绍了 5 种常用的记录诊断日志的方式。 .NET Core 还提供了一个支持多种输出渠道的日志系统，该日志系统在第 9 章进行了详细的介绍。

管道详解

.NET Core 的服务承载系统用来承载那些需要长时间运行的服务，ASP.NET Core 作为最重要的服务类型被承载于该系统中，第 10 章会对该服务承载系统进行系统介绍。由于请求处理管道是本书的核心所在，所以采用 3 章的篇幅进行介绍：第 11 章主要从编程模型的角度来认识管道；

第 12 章提供了一个极简版的模拟框架来展示 ASP.NET Core 框架的总体设计；第 13 章以这个模拟框架为基础，采用渐进的方式补充一些遗漏的细节，进而将 ASP.NET Core 框架真实的管道展现在读者眼前。

中间件

ASP.NET Core 框架的请求处理管道由服务器和中间件组成，管道利用服务器来监听和接收请求，并完成最终对请求的响应，应用针对请求的处理则体现在有序排列的中间件上。微软为我们提供了一系列原生的中间件，对这些中间件的介绍全部在下册。

这部分涉及用来处理文件请求（第 14 章）、路由（第 15 章）、异常（第 16 章）的中间件，也包括用来响应缓存（第 17 章）和会话（第 18 章）的中间件，还包括用来实现认证（第 19 章）、授权（第 20 章）、跨域资源共享（第 21 章）等与安全相关的中间件。

这部分还介绍了针对本地化（第 22 章）和健康检查（第 23 章）的中间件。除此之外，这部分还介绍了用来实现主机名过滤、HTTP 重写、设置基础路径等功能的中间件，这些零散的中间件全部在第 24 章进行介绍。

写作特点

本书是揭秘系列的第 6 本书。在过去的十来年里，笔者得到了很多热心读者的反馈，这些反馈对书中的内容基本上都持正面评价，但对写作技巧和表达方式的评价则不尽相同。每个作者都有属于自己的写作风格，每个读者的学习思维方式也不尽相同，两者很难出现百分之百的契合，但笔者还是决定在本书上做出改变。

本书内容采用了不一样的组织方式，笔者认为这样的方式更符合系统地学习一门全新技术的"流程"。对于每个模块，笔者采用"体验先行"的原则，提供一些简单的实例演示，使读者对当前模块的基本功能特性和编程模式具有大致的了解。同时，在编程体验中抽取一些核心对象，并利用它们构建当前模块的抽象模型，使读者只要读懂了这个模型也就了解了当前模块的总体设计。接下来我们从抽象转向具体，进一步深入介绍抽象模型的实现原理。为了使读者能够在真实项目中灵活自如地运用当前模块，笔者介绍了一些面向应用的扩展和最佳实践。总体来说，本书采用"编程体验"、"总体设计"、"具体实现"和"灵活运用"的流程，使读者能循序渐进地学习 ASP.NET Core 的每个功能模块。

本书综合运用 3 种不同的"语言"（文字语言、图表语言和编程语言）来讲述每个技术主题。一图胜千言，笔者在每章都精心设计了很多图表，这些具象的图表能够帮助读者理解技术模块的总体设计、执行流程和交互方式。除了利用编程语言描述应用编程接口（API），本书还提供了近 200 个实例，这些实例具有不同的作用，有的是为了演示某个实用的编程技巧或者最佳实践，有的是为了强调一些容易忽视但很重要的技术细节，有的是为了探测和证明所述的论点。

本书在很多地方会展示一些类型的代码，但是这些代码和真正的源代码是有差异的，两者的差异缘于以下几个原因：第一，源代码在版本更替中一直在发生改变；第二，由于篇幅的限制，笔者刻意删除了一些细枝末节的代码，如针对参数的验证、诊断日志的输出和异常处理等；第三，很多源代码其实都具有优化的空间。综上所述，本书提供的代码片段旨在揭示设计原理和实现逻辑，不是为了向读者展示源代码。

目标读者

虽然本书关注的是 ASP.NET Core 自身框架提供的请求处理管道，而不是具体某个应用编程框架（如 MVC、SignalR、GRPC 等），但是本书适合所有 .NET 技术从业人员阅读。

笔者认为任何好的设计都应该是简单的，唯有简单的设计才能应对后续版本更替中出现的复杂问题。从这个意义上讲，ASP.NET 框架就是好的设计。因为自正式推出的那一刻起，ASP.NET 框架的总体设计基本上没有发生改变。ASP.NET Core 的设计同样是好的设计，其简单的管道式设计在未来的版本更替中也不会发生太大的改变，既然是好的设计，它就应该是简单的。

正如上面所说，本书采用渐进式的写作方式，那些完全没有接触过 ASP.NET Core 的开发人员也可以通过本书深入、系统地掌握这门技术。由于本书提供的大部分内容都是独一无二的，即使是资深的 .NET 开发设计人员，也能在书中找到很多不甚了解的盲点。

关于作者

蒋金楠，同程艺龙技术专家。知名 IT 博主（多年来一直排名博客园第一位），拥有个人微信公众号"大内老 A"；2007—2018 年连续 12 次被评为微软 MVP（最有价值专家），也是少数跨多领域（Solutions Architect、Connected System、Microsoft Integration 和 ASP.NET/IIS 等）的 MVP 之一；畅销 IT 图书作者，先后出版了《WCF 全面解析》、《ASP.NET MVC 4 框架揭秘》、《ASP.NET MVC 5 框架揭秘》和《ASP.NET Web API 2 框架揭秘》等著作。

致谢

本书得以顺利出版离不开博文视点张春雨团队的辛勤努力，他们的专业水准和责任心为本书提供了质量保证。此外，徐妍妍在本书写作过程中做了大量的校对工作，在此表示衷心感谢。

本书支持

由于本书是随着 ASP.NET Core 一起成长起来的，并且随着 ASP.NET Core 的版本更替进行了多次"迭代"，所以书中某些内容最初是根据旧版本编写的，新版本对应的内容发生改变后相应内容可能没有及时更新。对于 ASP.NET Core 的每次版本升级，笔者基本上会尽可能将书中的内容做相应的更改，但其中难免有所疏漏。由于笔者的能力和时间有限，书中难免存在不足之处，恳请广大读者批评指正。

笔者博客：http://www.cnblogs.com/artech

笔者微博：http://www.weibo.com/artech

笔者电子邮箱：jinnan@outlook.com

笔者微信公众号：大内老 A

读者服务

- 获取本书配套素材、代码、视频、习题、模板、教程、课件资源
- 获取更多技术专家分享视频与学习资源
- 加入读者交流群，与更多读者互动、与本书作者互动

扫码回复：38462

目 录

第 14 章　静态文件 537
14.1　搭建文件服务器 537
14.1.1　发布物理文件 537
14.1.2　呈现目录结构 540
14.1.3　显示默认页面 541
14.1.4　映射媒体类型 544
14.2　处理文件请求 545
14.2.1　条件请求 545
14.2.2　区间请求 549
14.2.3　StaticFileMiddleware 552
14.3　处理目录请求 563
14.3.1　DirectoryBrowserMiddleware 563
14.3.2　DefaultFilesMiddleware 567

第 15 章　路由 571
15.1　路由映射 571
15.1.1　路由注册 571
15.1.2　设置内联约束 574
15.1.3　默认路由参数 576
15.1.4　特殊的路由参数 578
15.2　终结点的解析与执行 580
15.2.1　路由模式 580
15.2.2　终结点 586

ASP.NET Core 3 框架揭秘（下册）

15.2.3 中间件 593
15.3 路由约束 599
15.3.1 预定义的 IRouteConstraint 600
15.3.2 InlineConstraintResolver 602
15.3.3 自定义约束 603

第 16 章 异常处理 607
16.1 呈现错误信息 607
16.1.1 显示开发者异常页面 607
16.1.2 显示定制异常页面 610
16.1.3 针对响应状态码定制错误页面 612
16.2 开发者异常页面 615
16.2.1 IDeveloperPageExceptionFilter 616
16.2.2 显示编译异常信息 617
16.2.3 DeveloperExceptionPageMiddleware 622
16.3 异常处理器 624
16.3.1 ExceptionHandlerMiddleware 624
16.3.2 异常的传递与请求路径的恢复 626
16.3.3 清除缓存 629
16.4 响应状态码页面 631
16.4.1 StatusCodePagesMiddleware 632
16.4.2 阻止处理异常 632
16.4.3 注册 StatusCodePagesMiddleware 中间件 635

第 17 章 缓存 642
17.1 将数据缓存起来 642
17.1.1 将数据缓存在内存中 642
17.1.2 对数据进行分布式缓存 644
17.1.3 缓存整个 HTTP 响应 648
17.2 本地内存缓存 651
17.2.1 ICacheEntry 651
17.2.2 MemoryCacheEntryOptions 654
17.2.3 IMemoryCache 655
17.3 分布式缓存 663

		17.3.1	IDistributedCache	664
		17.3.2	基于 Redis 的分布式缓存	665
		17.3.3	基于 SQL Server 的分布式缓存	668
	17.4	响应缓存		670
		17.4.1	HTTP/1.1 Caching	670
		17.4.2	ResponseCachingMiddleware 中间件	673
		17.4.3	注册中间件	680

第 18 章 会话 681

18.1	利用会话保留"语境"		681
	18.1.1	设置和提取会话状态	681
	18.1.2	查看存储的会话状态	683
	18.1.3	查看 Cookie	685
18.2	会话状态的读写		686
	18.2.1	ISession	686
	18.2.2	DistributedSession	687
	18.2.3	ISessionStore	688
18.3	SessionMiddleware 中间件		689
	18.3.1	SessionOptions	689
	18.3.2	ISessionFeature	690
	18.3.3	SessionMiddleware	691

第 19 章 认证 694

19.1	认证、登录与注销		694
	19.1.1	认证票据	694
	19.1.2	基于 Cookie 的认证	695
	19.1.3	应用主页	696
	19.1.4	登录与注销	698
19.2	身份与用户		700
	19.2.1	IIdentity	700
	19.2.2	IPrincipal	707
19.3	认证模型		710
	19.3.1	认证票据	710
	19.3.2	认证处理器	713

XII | ASP.NET Core 3 框架揭秘（下册）

　　　19.3.3　认证服务 720
　　　19.3.4　服务注册 724
　　　19.3.5　AuthenticationMiddleware 727
　19.4　Cookie 认证方案 729
　　　19.4.1　AuthenticationHandler<TOptions> 729
　　　19.4.2　CookieAuthenticationHandler 735
　　　19.4.3　注册 CookieAuthenticationHandler 744

第 20 章　授权 747
　20.1　基于角色的权限控制 747
　　　20.1.1　用户与角色的映射 747
　　　20.1.2　根据角色授权 750
　　　20.1.3　预定义授权策略 754
　20.2　基于 "要求" 的授权 756
　　　20.2.1　IAuthorizationRequirement 756
　　　20.2.2　预定义的 IAuthorizationRequirement 实现类型 758
　　　20.2.3　授权检验 762
　20.3　基于 "策略" 的授权 767
　　　20.3.1　授权策略的构建 768
　　　20.3.2　授权策略的注册 769
　　　20.3.3　授权检验 770

第 21 章　跨域资源共享 773
　21.1　处理跨域资源 773
　　　21.1.1　跨域调用 API 773
　　　21.1.2　资源提供者显式授权 777
　　　21.1.3　基于策略的资源授权 779
　21.2　CORS 规范 780
　　　21.2.1　同源策略 781
　　　21.2.2　针对资源的授权 781
　　　21.2.3　获取授权的方式 782
　　　21.2.4　用户凭证 785
　21.3　CORS 模型 785
　　　21.3.1　CORS 策略 786

21.3.2 解析并应用授权结果 788
　　21.3.3 CorsMiddleware 中间件 790

第 22 章 本地化 793

22.1 提供本地化消息文本 793
　　22.1.1 提供对应语种的文本 793
　　22.1.2 自动设置语言文化 795
　　22.1.3 将本地化文本分而治之 798
　　22.1.4 直接注入 IStringLocalizer<T> 800

22.2 文本本地化 801
　　22.2.1 字符串本地化模型 801
　　22.2.2 基于 JSON 文件的本地化 804
　　22.2.3 基于资源文件的本地化 810

22.3 当前语言文化的设置 814
　　22.3.1 Culture 与 UICulture 814
　　22.3.2 IRequestCultureProvider 815
　　22.3.3 RequestLocalizationOptions 820
　　22.3.4 RequestLocalizationMiddleware 821

第 23 章 健康检查 822

23.1 检查应用的健康状况 822
　　23.1.1 确定当前应用是否可用 822
　　23.1.2 定制健康检查逻辑 823
　　23.1.3 改变响应状态码 825
　　23.1.4 细粒度的健康检查 826
　　23.1.5 定制响应内容 828
　　23.1.6 过滤 IHealthCheck 对象 830

23.2 设计与实现 832
　　23.2.1 IHealthCheck 832
　　23.2.2 HealthCheckService 838
　　23.2.3 HealthCheckMiddleware 842
　　23.2.4 针对 Entity Framework Core 的健康检查 846

23.3 发布健康报告 850
　　23.3.1 定期发布健康报告 850

23.3.2 IHealthCheckPublisher 851
23.3.3 HealthCheckPublisherHostedService 852

第 24 章　补遗 856

24.1　过滤主机名 856
24.1.1　实例演示 856
24.1.2　配置选项 857
24.1.3　HostFilteringMiddleware 中间件 858

24.2　HTTP 重写 859
24.2.1　实例演示 859
24.2.2　HttpMethodOverrideMiddleware 中间件 863
24.2.3　ForwardedHeadersMiddleware 中间件 864

24.3　基础路径 872
24.3.1　实例演示 873
24.3.2　UsePathBaseMiddleware 874

24.4　路由 876
24.4.1　实例演示 876
24.4.2　MapMiddleware 878
24.4.3　MapWhenMiddleware 880

附录 B　实例演示 2 882

第 14 章

静态文件

虽然 ASP.NET Core 是一款"动态"的 Web 服务端框架，但是由它接收并处理的大部分是针对静态文件的请求，最常见的是开发 Web 站点使用的 3 种静态文件（JavaScript 脚本、CSS 样式和图片）。ASP.NET Core 提供了 3 个中间件来处理针对静态文件的请求，利用它们不仅可以将物理文件发布为可以通过 HTTP 请求获取的 Web 资源，还可以将所在的物理目录的结构呈现出来。

14.1 搭建文件服务器

通过 HTTP 请求获取的 Web 资源大部分来源于存储在服务器磁盘上的静态文件。对于 ASP.NET Core 应用来说，如果将静态文件存储到约定的目录下，绝大部分文件类型都是可以通过 Web 的形式对外发布的。基于静态文件的请求由 3 个中间件负责处理，它们均定义在 NuGet 包"Microsoft.AspNetCore.StaticFiles"中，利用这 3 个中间件完全可以搭建一个基于 Web 的文件服务器，下面做相关的实例演示。

14.1.1 发布物理文件

我们创建的演示实例是一个简单的 ASP.NET Core 应用，它的项目结构如图 14-1 所示。在默认作为 WebRoot 的"wwwroot"目录下，可以将 JavaScript 脚本文件、CSS 样式文件和图片文件存放到对应的子目录（js、css 和 img）下。WebRoot 目录下的所有文件将自动发布为 Web 资源，客户端可以访问相应的 URL 来读取对应文件的内容。

针对具体某个静态文件的请求是通过一个名为 StaticFileMiddleware 的中间件来处理的。如下面的代码片段所示，承载 ASP.NET Core 应用的程序中调用 IApplicationBuilder 接口的 UseStaticFiles 扩展方法注册的就是这样一个中间件。

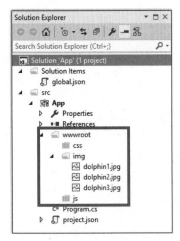

图 14-1　静态文件发布的项目结构

```
public class Program
{
    public static void Main()
    {
        Host.CreateDefaultBuilder()
            .ConfigureWebHostDefaults(builder=>builder.Configure(
                app => app.UseStaticFiles()))
            .Build()
            .Run();
    }
}
```

上述程序运行之后，就可以通过 GET 请求的方式来读取对应文件的内容。请求采用的 URL 由目标文件的路径决定。具体来说，目标文件相对于 WebRoot 目录的路径就是对应 URL 的路径，如 JPG 图片文件 "~/wwwroot/img/dolphin1.jpg" 对应的 URL 路径为 "/img/dolphin1.jpg"。如果直接利用浏览器访问这个 URL，目标图片就会直接以图 14-2 所示的形式显示出来。（S1401）

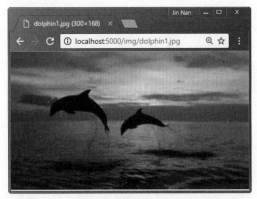

图 14-2　以 Web 形式请求发布的图片文件

上面通过一个简单的实例将 WebRoot 所在目录下的所有静态文件发布为 Web 资源，如果需要发布的静态文件存储在其他目录下呢？下面将上面演示的应用程序的一些文档存储在图 14-3 所示的"~/doc/"目录下，那么对应的程序又该如何编写？

图 14-3　发布 "~/doc/" 和 "~/wwwroot" 目录下的文件

ASP.NET Core 应用在大部分情况下都是利用一个 IFileProvider 对象来读取文件的（第 5 章对 IFileProvider 进行了系统介绍），针对静态文件的读取请求也不例外。对于 IApplicationBuilder 接口的 UseStaticFiles 扩展方法注册的 StaticFileMiddleware 中间件来说，它的内部维护着一个 IFileProvider 对象和请求路径的映射关系。如果调用 UseStaticFiles 方法没有指定任何参数，那么这个映射关系的请求路径就是应用的基地址（PathBase），对应的 IFileProvider 对象自然就是指向 WebRoot 目录的 PhysicalFileProvider 对象。

上述需求可以通过显式定制这个映射关系的方式来实现。如下面的代码片段所示，我们在现有程序的基础上额外添加了一次针对 UseStaticFiles 扩展方法的调用，在本次调用中指定一个对应的 Options 对象（一个类型为 StaticFileOptions 的对象）作为参数来定制请求路径（"/documents"）与对应 IFileProvider 对象（针对路径 "~/doc/" 的 PhysicalFileProvider 对象）之间的映射关系。

```
public class Program
{
    public static void Main()
    {
        var path = Path.Combine(Directory.GetCurrentDirectory(), "doc");
        var options = new StaticFileOptions
        {
            FileProvider = new PhysicalFileProvider(path),
            RequestPath = "/documents"
        };
        Host.CreateDefaultBuilder()
            .ConfigureWebHostDefaults(builder => builder.Configure(app => app
                .UseStaticFiles()
                .UseStaticFiles(options)))
            .Build()
            .Run();
    }
}
```

按照上面这段程序指定的映射关系,对于存储在"~/doc/"目录下的这个 PDF 文件（checklist.pdf）,对应 URL 的路径就应该是"/documents/checklist.pdf"。如果利用浏览器请求这个地址时,PDF 文件的内容就会按照图 14-4 所示的形式显示在浏览器上。（S1402）

图 14-4　以 Web 形式请求发布的 PDF 文件

14.1.2　呈现目录结构

上面的演示实例注册的 StaticFileMiddleware 中间件只会处理针对具体的某个静态文件的请求,如果利用浏览器发送一个针对目录的请求（如"http://localhost:5000/img/"）,得到的将是一个状态为"404 Not Found"的响应。如果希望浏览器呈现出目标目录的结构,就可以注册另一个名为 DirectoryBrowserMiddleware 的中间件。这个中间件会返回一个 HTML 页面,请求目录下的结构会以表格的形式显示在这个页面中。我们演示的程序可以按照如下方式调用 IApplicationBuilder 接口的 UseDirectoryBrowser 扩展方法来注册 DirectoryBrowserMiddleware 中间件。

```
public class Program
{
    public static void Main()
    {
        var path = Path.Combine(Directory.GetCurrentDirectory(), "doc");
        var fileProvider = new PhysicalFileProvider(path);

        var fileOptions = new StaticFileOptions
        {
            FileProvider       = fileProvider,
            RequestPath        = "/documents"
        };

        var diretoryOptions = new DirectoryBrowserOptions
        {
            FileProvider       = fileProvider,
            RequestPath        = "/documents"
        };

        Host.CreateDefaultBuilder()
```

```
            .ConfigureWebHostDefaults(builder => builder.Configure(app => app
                .UseStaticFiles()
                .UseStaticFiles(fileOptions)
                .UseDirectoryBrowser()
                .UseDirectoryBrowser(diretoryOptions)))
            .Build()
            .Run();
        }
}
```

当上面的应用启动之后，如果利用浏览器向针对某个目录的 URL（如 "http://localhost:5000/"或者 "http://localhost:5000/img/"）发起请求，目标目录的内容（包括子目录和文件）就会以图 14-5 所示的形式显示在一个表格中。可以看出，在呈现的表格中，当前目录的子目录和文件均会显示为链接。（S1403）

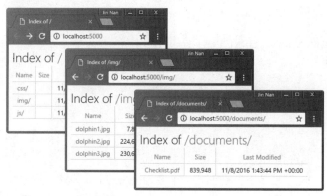

图 14-5　显示目录内容

14.1.3　显示默认页面

从安全的角度来讲，利用注册的 UseDirectoryBrowser 中间件会将整个目标目录的结构和所有文件全部暴露出来，所以这个中间件需要根据自身的安全策略谨慎使用。对于针对目录的请求，更加常用的处理策略就是显示一个保存在这个目录下的默认页面。默认页面文件一般采用如下 4 种命名约定：default.htm、default.html、index.htm 和 index.html。针对默认页面的呈现实现在一个名为 DefaultFilesMiddleware 的中间件中，我们演示的这个应用就可以按照如下方式调用 IApplicationBuilder 接口的 UseDefaultFiles 扩展方法来注册这个中间件。

```
public class Program
{
    public static void Main()
    {
        var path = Path.Combine(Directory.GetCurrentDirectory(), "doc");
        var fileProvider = new PhysicalFileProvider(path);

        var fileOptions = new StaticFileOptions
        {
```

```
            FileProvider         = fileProvider,
            RequestPath          = "/documents"
        };
        var diretoryOptions = new DirectoryBrowserOptions
        {
            FileProvider         = fileProvider,
            RequestPath          = "/documents"
        };
        var defaultOptions = new DefaultFilesOptions
        {
            RequestPath          = "/documents",
            FileProvider         = fileProvider,
        };

        Host.CreateDefaultBuilder()
            .ConfigureWebHostDefaults(builder => builder.Configure(app => app
                .UseDefaultFiles()
                .UseDefaultFiles(defaultOptions)
                .UseStaticFiles()
                .UseStaticFiles(fileOptions)
                .UseDirectoryBrowser()
                .UseDirectoryBrowser(diretoryOptions)))
            .Build()
            .Run();
    }
}
```

下面在"~/wwwroot/img/"目录和"~/doc"目录下分别创建一个名为 index.html 的默认页面，并且在该.html 文件的主体部分指定一段简短的文字（This is an index page!）。在应用启动之后，可以利用浏览器访问这两个目录对应的 URL（"http://localhost:5000/img/"和"http://localhost:5000/documents/"），图 14-6 显示的就是这个默认页面的内容。（S1404）

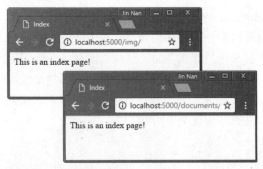

图 14-6　显示目录内容

必须在注册 StaticFileMiddleware 中间件和 DirectoryBrowserMiddleware 中间件之前注册 DefaultFilesMiddleware 中间件，否则它无法发挥作用。这是因为 DirectoryBrowserMiddleware 中间件和 DefaultFilesMiddleware 中间件处理的均是针对目录的请求，如果先注册 DirectoryBrowserMiddleware 中间件，那么显示的总是目录的结构；如果先注册用于显示默认页面的

DefaultFilesMiddleware 中间件，那么在默认页面不存在的情况下它会将请求分发给后续中间件，而 DirectoryBrowserMiddleware 中间件会接收请求的处理并将当前目录的结构呈现出来。

要先于 StaticFileMiddleware 中间件之前注册 DefaultFilesMiddleware 中间件是因为后者是通过采用 URL 重写的方式实现的，也就是说，这个中间件会将针对目录的请求改写成针对默认页面的请求，而最终针对默认页面的请求还需要依赖 StaticFileMiddleware 中间件来完成。

DefaultFilesMiddleware 中间件在默认情况下总是以约定的名称（default.htm、default.html、index.htm 和 index.html）在当前请求的目录下定位默认页面。如果作为默认页面的文件没有采用这样的约定命名（如我们将默认页面命名为 readme.html），就需要按照如下方式显式指定默认认页面的文件名。（S1405）

```
public class Program
{
    public static void Main()
    {
        var path = Path.Combine(Directory.GetCurrentDirectory(), "doc");
        var fileProvider = new PhysicalFileProvider(path);
        var fileOptions = new StaticFileOptions
        {
            FileProvider        = fileProvider,
            RequestPath         = "/documents"
        };
        var diretoryOptions = new DirectoryBrowserOptions
        {
            FileProvider        = fileProvider,
            RequestPath         = "/documents"
        };
        var defaultOptions1 = new DefaultFilesOptions();
        var defaultOptions2 = new DefaultFilesOptions
        {
            RequestPath         = "/documents",
            FileProvider        = fileProvider,
        };

        defaultOptions1.DefaultFileNames.Add("readme.html");
        defaultOptions2.DefaultFileNames.Add("readme.html");

        Host.CreateDefaultBuilder()
            .ConfigureWebHostDefaults(builder => builder.Configure(app => app
                .UseDefaultFiles(defaultOptions1)
                .UseDefaultFiles(defaultOptions2)
                .UseStaticFiles()
                .UseStaticFiles(fileOptions)
                .UseDirectoryBrowser()
                .UseDirectoryBrowser(diretoryOptions)))
            .Build()
            .Run();
    }
}
```

14.1.4 映射媒体类型

通过上面演示的实例可以看出，浏览器能够准确地将请求的目标文件的内容正常呈现出来。对 HTTP 协议具有基本了解的读者应该都知道：响应文件能够在浏览器上被正常显示的基本前提是响应报文通过 Content-Type 报头携带的媒体类型必须与内容一致。我们的实例演示了针对两种文件类型的请求，一种是 JPG 文件，另一种是 PDF 文件，对应的媒体类型分别是 image/jpg 和 application/pdf，那么用来处理静态文件请求的 StaticFileMiddleware 中间件是如何解析出对应的媒体类型的？

StaticFileMiddleware 中间件针对媒体类型的解析是通过一个 IContentTypeProvider 对象来完成的，默认采用的是该接口的实现类型 FileExtensionContentTypeProvider。顾名思义，FileExtensionContentTypeProvider 根据文件的扩展命名来解析媒体类型。FileExtensionContentTypeProvider 内部预定了数百种常用文件扩展名与对应媒体类型之间的映射关系，所以如果发布的静态文件具有标准的扩展名，那么 StaticFileMiddleware 中间件就能为对应的响应赋予正确的媒体类型。

如果某个文件的扩展名没有在预定义的映射之中，或者需要某个预定义的扩展名匹配不同的媒体类型，那么应该如何解决？同样是针对我们演示的这个实例，笔者将~/wwwroot/img/dolphin1.jpg 文件的扩展名改成.img，毫无疑问，StaticFileMiddleware 中间件将无法为针对该文件的请求解析出正确的媒体类型。这个问题具有若干不同的解决方案，第一种方案就是按照如下方式让 StaticFileMiddleware 中间件支持不能识别的文件类型，并为它们设置一个默认的媒体类型。（S1406）

```
public class Program
{
    public static void Main()
    {
        var options = new StaticFileOptions
        {
            ServeUnknownFileTypes    = true,
            DefaultContentType       = "image/jpg"
        };

        Host.CreateDefaultBuilder()
            .ConfigureWebHostDefaults(builder => builder.Configure(
                app => app.UseStaticFiles(options)))
            .Build()
            .Run();
    }
}
```

上述解决方案只能设置一种默认媒体类型，如果具有多种需要映射成不同媒体类型的文件类型，采用这种方案就达不到目的，所以最根本的解决方案还是需要将不能识别的文件类型和对应的媒体类型进行映射。由于 StaticFileMiddleware 中间件使用的 IContentTypeProvider 对象是可以定制的，所以可以按照如下方式显式地为该中间件指定一个 FileExtensionContentTypeProvider 对象，然后将缺失的映射添加到这个对象上。（S1407）

```csharp
public class Program
{
    public static void Main()
    {
        var contentTypeProvider = new FileExtensionContentTypeProvider();
        contentTypeProvider.Mappings.Add(".img", "image/jpg");
        var options = new StaticFileOptions
        {
            ContentTypeProvider = contentTypeProvider
        };
        Host.CreateDefaultBuilder()
            .ConfigureWebHostDefaults(builder => builder.Configure(
                app => app.UseStaticFiles(options)))
            .Build()
            .Run();
    }
}
```

14.2 处理文件请求

通过调用 IApplicationBuilder 接口的 UseStaticFiles 扩展方法注册的 StaticFileMiddleware 中间件旨在处理针对文件的请求。对于 StaticFileMiddleware 中间件处理请求的逻辑，大部分读者都应该想得到：根据请求的地址找到目标文件的路径，然后利用注册的 IContentTypeProvider 对象解析出与文件内容相匹配的媒体类型，后者将其作为响应报头 Content-Type 的值。StaticFileMiddleware 中间件最终利用 IFileProvider 对象读取文件的内容，并将其作为响应报文的主体。实际上，这个中间件在处理请求时所做的事情比前面的演示实例多，所以针对条件请求（Conditional Request）和区间请求（Range Request）的处理就没有体现在上面演示的实例中。

14.2.1 条件请求

条件请求就是客户端在发送 GET 请求获取某种资源时，会利用请求报头携带一些条件。服务端处理器在接收到这样的请求之后，会提取这些条件并验证目标资源当前的状态是否满足客户端指定的条件。只有在这些条件满足的情况下，目标资源的内容才会真正响应给客户端。

HTTP 条件请求

HTTP 条件请求作为一项标准记录在 HTTP 规范中。一般来说，一个 GET 请求在目标资源存在的情况下会返回一个状态码为"200 OK"的响应，目标资源的内容将直接存放在响应报文的主体部分。如果资源的内容不会轻易改变，那么我们希望客户端（如浏览器）在本地缓存获取的资源。对于针对同一资源的后续请求来说，如果资源内容不曾改变，那么资源内容就无须再次作为网络荷载予以响应。这就是条件请求需要解决的一个典型场景。

确定资源是否发生变化可以采用两种策略。第一种就是让资源的提供者记录最后一次更新资源的时间，资源的荷载内容（Payload）和这个时间戳将一并作为响应提供给作为请求发送者的客户端。客户端在缓存资源内容时也会保存这个时间戳。等到下次需要针对同一资源发送请求时,它会将这个时间戳一并发送出去,此时服务端就可以根据这个时间戳判断目标资源在上次响应之后是否被修改过,然后做出针对性的响应。第二种是针对资源的内容生成一个"标签",标签的一致性体现了资源内容的一致性,在 HTTP 规范中将这个标签称为 ETag(Entity Tag)。

下面从 HTTP 请求和响应报文的层面对条件请求进行详细介绍。对于 HTTP 请求来说,缓存资源携带的最后修改时间戳和 ETag 分别保存在名为 If-Modified-Since 与 If-None-Match 的报头中。报头名称体现的含义如下:只有目标资源在指定的时间之后被修改(If-Modified-Since)或者目前资源的状态与提供的 ETag 不匹配(If-None-Match)的情况下才会返回资源的荷载内容。

当服务端接收到针对某个资源的 GET 请求时,如果请求不具有上述两个报头或者根据这两个报头携带的信息判断资源已经发生改变,那么它返回一个状态码为"200 OK"的响应。除了将资源内容作为响应主体,如果能够获取到该资源最后一次修改的时间(一般精确到秒),那么格式化的时间戳还会通过一个名为 Last-Modified 的响应报头提供给客户端。针对资源自身内容生成的标签,则会以 ETag 响应报头的形式提供给客户端。反之,如果做出相反的判断,服务端就会返回一个状态码为"304 Not Modified"的响应,这个响应不包含主体内容。一般来说,这样的响应也会携带 Last-Modified 报头和 ETag 报头。

与条件请求相关的请求报头还有 If-Unmodified-Since 和 If-Match,它们具有与 If-Modified-Since 和 If-None-Match 完全相反的语义,分别表示如果目标资源在指定时间之后没有被修改(If-Unmodified-Since)或者目标资源目前的 ETag 与提供的 ETag 匹配的请求才会返回资源的内容荷载。针对这样的请求,如果根据携带的这两个报头判断出目标资源并不曾发生变化,服务端才会返回一个将资源荷载作为主体内容的"200 OK"响应,这样的响应也会携带 Last-Modified 报头和 ETag 报头。如果做出了相反的判断,服务端就会返回一个状态码为"412 Precondition Failed"的响应,表示资源目前的状态不满足请求设定的前置条件。表 14-1 列举了条件请求的响应状态码。

表 14-1 条件请求的响应状态码

请求报头	语 义	满足条件	不满足条件
If-Modified-Since	目标内容在指定时间戳之后是否有更新	200 OK	304 Not Modified
If-None-Match	目标内容的标签是否与指定的不一致	200 OK	304 Not Modified
If-Unmodified-Since	目标内容是否在指定时间戳之后没有更新	200 OK	412 Precondition Failed
If-Match	目标内容的标签是否与指定的一致	200 OK	412 Precondition Failed

针对静态文件的条件请求

下面通过实例演示的形式介绍 StaticFileMiddleware 中间件在针对条件请求方面做了什么。假设我们在 ASP.NET Core 应用中发布了一个文本文件(foobar.txt),内容为"abcdefghijklmnopqrstuvwxyz0123456789"(26 个字母+10 个数字),目标地址为"http://localhost:5000/foobar.txt"。然后直接针对这个地址发送一个普通的 GET 请求会得到什么样的响应?

```
HTTP/1.1 200 OK
Date: Wed, 18 Sep 2019 23:20:40 GMT
Content-Type: text/plain
Server: Kestrel
```

```
Content-Length: 39
Last-Modified: Wed, 18 Sep 2019 23:15:14 GMT
Accept-Ranges: bytes
ETag: "1d56e76ed13ed27"

abcdefghijklmnopqrstuvwxyz0123456789
```

从上面给出的请求与响应报文的内容可以看出，对于一个针对物理文件的 GET 请求，如果目标文件存在，服务器就会返回一个状态码为"200 OK"的响应。除了承载文件内容的主体，响应报文还有两个额外的报头，分别是表示目标文件最后修改时间的 Last-Modified 报头和作为文件内容标签的 ETag 报头。

现在客户端不但获得了目标文件的内容，还得到了该文件最后被修改的时间戳和标签，如果它只想确定这个文件是否被更新，并且在更新之后返回新的内容，那么它可以针对这个文件所在的地址再次发送一个 GET 请求，并将这个时间戳和标签通过相应的请求报头发送给服务端。我们知道这两个报头的名称分别是 If-Modified-Since 和 If-None-Match。由于我们没有修改文件的内容，所以服务器返回如下一个状态码为"304 Not Modified"的响应。这个不包括主体内容的响应报文同样具有相同的 Last-Modified 报头和 ETag 报头。

```
GET http://localhost:50000/foobar.txt HTTP/1.1
Host: localhost:50000
If-Modified-Since: Wed, 18 Sep 2019 23:15:14 GMT
If-None-Match: "1d56e76ed13ed27"

HTTP/1.1 304 Not Modified
Date: Wed, 18 Sep 2019 23:21:54 GMT
Content-Type: text/plain
Server: Kestrel
Last-Modified: Wed, 18 Sep 2019 23:15:14 GMT
Accept-Ranges: bytes
ETag: "1d56e76ed13ed27"
```

如果将 If-None-Match 报头修改成一个较早的时间戳，或者改变了 If-None-Match 报头的标签，服务端都将做出文件已经被修改的判断。在这种情况下，最初状态码为"200 OK"的响应会再次被返回，具体的请求和对应的响应体现在如下所示的代码片段中。

```
GET http://localhost:5000/foobar.txt HTTP/1.1
If-Modified-Since: Wed, 18 Sep 2019 01:01:01 GMT
Host: localhost:5000

HTTP/1.1 200 OK
Date: Wed, 18 Sep 2019 23:24:16 GMT
Content-Type: text/plain
Server: Kestrel
Content-Length: 39
Last-Modified: Wed, 18 Sep 2019 23:15:14 GMT
Accept-Ranges: bytes
ETag: "1d56e76ed13ed27"
```

```
abcdefghijklmnopqrstuvwxyz0123456789

GET http://localhost:50000/foobar.txt HTTP/1.1
Host: localhost:50000
If-None-Match: "abc123xyz456"

HTTP/1.1 200 OK
Date: Wed, 18 Sep 2019 23:26:03 GMT
Content-Type: text/plain
Server: Kestrel
Content-Length: 39
Last-Modified: Wed, 18 Sep 2019 23:15:14 GMT
Accept-Ranges: bytes
ETag: "1d56e76ed13ed27"

abcdefghijklmnopqrstuvwxyz0123456789
```

如果客户端想确定目标文件是否被修改，但是希望在未被修改的情况下才返回目标文件的内容，这样的请求就需要使用 If-Unmodified-Since 报头和 If-Match 报头来承载基准时间戳与标签。例如，对于如下两个请求携带的 If-Unmodified-Since 报头和 If-Match 报头，服务端都将做出文件尚未被修改的判断，所以文件的内容通过一个状态码为"200 OK"的响应返回。

```
GET http://localhost:5000/foobar.txt HTTP/1.1
If-Unmodified-Since: Wed, 18 Sep 2019 23:59:59 GMT
Host: localhost:5000

HTTP/1.1 200 OK
Date: Wed, 18 Sep 2019 23:27:57 GMT
Content-Type: text/plain
Server: Kestrel
Content-Length: 39
Last-Modified: Wed, 18 Sep 2019 23:15:14 GMT
Accept-Ranges: bytes
ETag: "1d56e76ed13ed27"

abcdefghijklmnopqrstuvwxyz0123456789

GET http://localhost:50000/foobar.txt HTTP/1.1
Host: localhost:50000
If-Match: "1d56e76ed13ed27"

HTTP/1.1 200 OK
Date: Wed, 18 Sep 2019 23:30:35 GMT
Content-Type: text/plain
Server: Kestrel
Content-Length: 39
Last-Modified: Wed, 18 Sep 2019 23:15:14 GMT
Accept-Ranges: bytes
ETag: "1d56e76ed13ed27"
```

abcdefghijklmnopqrstuvwxyz0123456789

如果目标文件当前的状态无法满足 If-Unmodified-Since 报头或者 If-Match 报头体现的条件，那么返回的将是一个状态码为"412 Precondition Failed"的响应，如下所示的代码片段就是这样的请求报文和对应的响应报文。

```
GET http://localhost:5000/foobar.txt HTTP/1.1
If-Unmodified-Since: Wed, 18 Sep 2019 01:01:01 GMT
Host: localhost:5000

HTTP/1.1 412 Precondition Failed
Date: Wed, 18 Sep 2019 23:31:53 GMT
Server: Kestrel
Content-Length: 0


GET http://localhost:50000/foobar.txt HTTP/1.1
Host: localhost:50000
If-Match: "abc123xyz456"

HTTP/1.1 412 Precondition Failed
Date: Wed, 18 Sep 2019 23:33:57 GMT
Server: Kestrel
Content-Length: 0
```

14.2.2　区间请求

大部分针对物理文件的请求都希望获取整个文件的内容，区间请求则与之相反，它希望获取某个文件部分区间的内容。区间请求可以通过多次请求来获取某个较大文件的全部内容，并实现断点续传。如果同一个文件同时存放到多台服务器，就可以利用区间请求同时下载不同部分的内容。与条件请求一样，区间请求也作为标准定义在 HTTP 规范之中。

HTTP 区间请求

如果希望通过一个 GET 请求获取目标资源的某个区间的内容，就需要将这个区间存放到一个名为 Range 的报头中。虽然 HTTP 规范允许指定多个区间，但是 StaticFileMiddleware 中间件只支持单一区间。分区所采用的计量单位，HTTP 规范并未做强制的规定，但是 StaticFileMiddleware 中间件支持的单位为 Byte，也就是说，它是以字节为单位对文件内容进行分区的。

Range 报头携带的分区信息采用的格式为 bytes={from}-{to}（{from}和{to}分别表示区间开始与结束的位置），如 bytes=1000-1999 表示获取目标资源从 1001 到 2000 共计 1000 字节（第 1 个字节的位置为 0）。如果{to}大于整个资源的长度，这样的区间依然被认为是有效的，它表示从{from}到资源的最后一个字节。如果区间被定义成 bytes={from}-这种形式，同样表示区间从{from}到资源的最后一个字节。采用 bytes=-{n}格式定义的区间则表示资源的最后 *n* 个字节。无论采用何种形式，如果{from}大于整个资源的总长度，这样的区间定义就被视为不合法。

如果请求的 Range 报头携带一个不合法的区间，服务端就会返回一个状态码为"416 Range

Not Satisfiable"的响应，否则返回一个状态码为"206 Partial Content"的响应，响应的主体将只包含指定区间的内容。返回的内容在整个资源的位置通过响应报头 Content-Range 来表示，采用的格式为{from}-{to}/{length}。除此之外，还有一个与区间请求相关的响应报头 Accept-Ranges，它表示服务端能够接受的区间类型。例如，前面针对条件请求的响应都具有一个 Accept-Ranges: bytes 报头，表示服务支持针对资源的区间划分。如果该报头的值被设置为 none，则意味着服务端不支持区间请求。

区间请求在某些时候也会验证资源内容是否发生改变。在这种情况下，请求会利用一个名为 If-Range 的报头携带一个时间戳或者整个资源（不是当前请求的区间）的标签。服务端在接收到请求之后会根据这个报头判断请求的整个资源是否发生变化，如果判断已经发生变化，它会返回一个状态码为"200 OK"的响应，响应主体将包含整个资源的内容。只有在判断资源并未发生变化的前提下，服务端才会返回指定区间的内容。

针对静态文件的区间请求

下面从 HTTP 请求和响应报文的角度来探讨 StaticFileMiddleware 中间件针对区间请求的支持。我们依然沿用前面演示条件请求的实例，该实例中作为目标文件的 foobar.txt 包含 26 个字母和 10 个数字，加上 UTF 文本文件初始的 3 个字符（EF、BB、BF），所以总长度为 39。我们发送如下两个请求分别获取前面 26 个字母（3-28）和后面 10 个数字（-10）。

```
GET http://localhost:50000/foobar.txt HTTP/1.1
Host: localhost:50000
Range: bytes=3-28

HTTP/1.1 206 Partial Content
Date: Wed, 18 Sep 2019 23:38:59 GMT
Content-Type: text/plain
Server: Kestrel
Content-Length: 26
Content-Range: bytes 3-28/39
Last-Modified: Wed, 18 Sep 2019 23:15:14 GMT
Accept-Ranges: bytes
ETag: "1d56e76ed13ed27"

abcdefghijklmnopqrstuvwxyz

GET http://localhost:50000/foobar.txt HTTP/1.1
Host: localhost:50000
Range: bytes=-10

HTTP/1.1 206 Partial Content
Date: Wed, 18 Sep 2019 23:39:51 GMT
Content-Type: text/plain
Server: Kestrel
Content-Length: 10
Content-Range: bytes 29-38/39
Last-Modified: Wed, 18 Sep 2019 23:15:14 GMT
```

```
Accept-Ranges: bytes
ETag: "1d56e76ed13ed27"

0123456789
```

  由于请求中指定了正确的区间,所以我们会得到两个状态码为"206 Partial Content"的响应,响应的主体仅包含目标区间的内容。除此之外,响应报头 Content-Range("bytes 3-28/39"和"bytes 29-38/39")指明了返回内容的区间范围和整个文件的总长度。目标文件最后修改的时间戳和标签同样会存在于响应报头 Last-Modified 与 ETag 之中。

  接下来我们发送如下所示的一个区间请求,并刻意指定一个不合法的区间(50-)。正如 HTTP 规范所描述的那样,在这种情况下可以得到一个状态码为"416 Range Not Satisfiable"的响应。

```
GET http://localhost:5000/foobar.txt HTTP/1.1
Host: localhost:5000
Range: bytes=50-

HTTP/1.1 416 Range Not Satisfiable
Date: Wed, 18 Sep 2019 23:43:21 GMT
Server: Kestrel
Content-Length: 0
Content-Range: bytes */39
```

  为了验证区间请求针对文件更新状态的检验,我们使用了请求报头 If-Range。在如下所示的这两个请求中,我们分别将一个基准时间戳和文件标签作为这个报头的值,显然服务端针对这两个报头的值都将做出"文件已经更新"的判断。根据 HTTP 规范的约定,这种请求会返回一个状态码为"200 OK"的响应,响应的主体将包含整个文件的内容。如下所示的响应报文就证实了这一点。

```
GET http://localhost:5000/foobar.txt HTTP/1.1
Range: bytes=-10
If-Range: Wed, 18 Sep 2019 01:01:01 GMT
Host: localhost:5000

HTTP/1.1 200 OK
Date: Wed, 18 Sep 2019 23:45:32 GMT
Content-Type: text/plain
Server: Kestrel
Content-Length: 39
Last-Modified: Wed, 18 Sep 2019 23:15:14 GMT
Accept-Ranges: bytes
ETag: "1d56e76ed13ed27"

abcdefghijklmnopqrstuvwxyz0123456789

GET http://localhost:50000/foobar.txt HTTP/1.1
User-Agent: Fiddler
Host: localhost:50000
Range: bytes=-10
```

```
If-Range: "123abc456"

HTTP/1.1 200 OK
Date: Wed, 18 Sep 2019 23:46:36 GMT
Content-Type: text/plain
Server: Kestrel
Content-Length: 39
Last-Modified: Wed, 18 Sep 2019 23:15:14 GMT
Accept-Ranges: bytes
ETag: "1d56e76ed13ed27"

abcdefghijklmnopqrstuvwxyz0123456789
```

14.2.3 StaticFileMiddleware

上面的实例演示、针对条件请求和区间请求的介绍，从提供的功能和特性的角度对 StaticFileMiddleware 中间件进行了全面的介绍，下面从实现原理的角度进一步介绍这个中间件。在此之前，先介绍 StaticFileMiddleware 类型的定义。

```
public class StaticFileMiddleware
{
    public StaticFileMiddleware(RequestDelegate next, IWebHostEnvironment hostingEnv,
        IOptions<StaticFileOptions> options, ILoggerFactory loggerFactory);
    public Task Invoke(HttpContext context);
}
```

如上面的代码片段所示，除了用来将当前请求分发给后续管道的参数 next，StaticFile Middleware 的构造函数还包含 3 个参数。其中，参数 hostingEnv 和参数 loggerFactory 分别表示当前承载环境与用来创建 ILogger 的 ILoggerFactory 对象，最重要的参数 options 表示为这个中间件指定的配置选项。至于具体可以提供什么样的配置选项，我们只需要了解 StaticFileOptions 类型提供了什么样的属性成员。StaticFileOptions 继承自如下所示的抽象类 SharedOptionsBase。基类 SharedOptionsBase 定义了请求路径与对应 IFileProvider 对象之间的映射关系（默认为 PhysicalFileProvider）。

```
public abstract class SharedOptionsBase
{
    protected SharedOptions SharedOptions { get; private set; }
    public PathString RequestPath
        { get => SharedOptions.RequestPath; set => SharedOptions.RequestPath = value; }
    public IFileProvider FileProvider
    {
        get => SharedOptions.FileProvider;
        set => SharedOptions.FileProvider = value;
    }
    protected SharedOptionsBase(SharedOptions sharedOptions)
        => SharedOptions = sharedOptions;
}

public class SharedOptions
```

```
{
    public PathString RequestPath { get; set; } = PathString.Empty;
    public IFileProvider FileProvider { get; set; }
}
```

定义在 StaticFileOptions 中的前三个属性都与媒体类型的解析有关，其中 ContentType Provider 属性返回一个根据请求相对地址解析出媒体类型的 IContentTypeProvider 对象。如果这个 IContentTypeProvider 对象无法正确解析出目标文件的媒体类型，就可以利用 Default ContentType 设置一个默认媒体类型。但只有将另一个名为 ServeUnknownFileTypes 的属性设置为 True，中间件才会采用这个默认设置的媒体类型。

```
public class StaticFileOptions : SharedOptionsBase
{
    public IContentTypeProvider                   ContentTypeProvider { get; set; }
    public string                                 DefaultContentType { get; set; }
    public bool                                   ServeUnknownFileTypes { get; set; }
    public HttpsCompressionMode                   HttpsCompression { get; set; }
    public Action<StaticFileResponseContext>      OnPrepareResponse { get; set; }

    public StaticFileOptions();
    public StaticFileOptions(SharedOptions sharedOptions);
}

public enum HttpsCompressionMode
{
    Default = 0,
    DoNotCompress,
    Compress
}

public class StaticFileResponseContext
{
    public HttpContext      Context { get; }
    public IFileInfo        File { get; }
}
```

StaticFileOptions 的 HttpsCompression 属性表示在压缩中间件存在的情况下，采用 HTTPS 方法请求的文件是否应该被压缩，该属性的默认值为 Compress（即默认情况下会对文件进行压缩）。StaticFileOptions 还有一个 OnPrepareResponse 属性，它返回一个 Action<StaticFileResponseContext>类型的委托对象，利用这个委托对象可以对最终的响应进行定制。作为输入的 StaticFileResponse Context 对象可以提供表示当前 HttpContext 上下文和描述目标文件的 IFileInfo 对象。

针对 StaticFileMiddleware 中间件的注册一般都是调用针对 IApplicationBuilder 对象的 UseStaticFiles 扩展方法来完成的。如下面的代码片段所示，我们共有 3 个 UseStaticFiles 方法重载可供选择。

```
public static class StaticFileExtensions
{
    public static IApplicationBuilder UseStaticFiles(this IApplicationBuilder app)
        => app.UseMiddleware<StaticFileMiddleware>();
```

```csharp
    public static IApplicationBuilder UseStaticFiles(this IApplicationBuilder app,
        StaticFileOptions options)
        => app.UseMiddleware<StaticFileMiddleware>(
            Options.Create<StaticFileOptions>(options));

    public static IApplicationBuilder UseStaticFiles(this IApplicationBuilder app,
        string requestPath)
    {
        var options = new StaticFileOptions
        {
            RequestPath = new PathString(requestPath)
        };
        return app.UseStaticFiles(options);
    }
}
```

IContentTypeProvider

StaticFileMiddleware 中间件针对静态文件请求的处理并不仅限于完成文件内容的响应，它还需要为目标文件提供正确的媒体类型。对于客户端来说，如果无法确定媒体类型，获取的文件就像是一部无法解码的天书，毫无价值。StaticFileMiddleware 中间件利用指定的 IContentTypeProvider 对象来解析媒体类型。如下面的代码片段所示，IContentTypeProvider 接口定义了唯一的方法 TryGetContentType，从而根据当前请求的相对路径来解析这个作为输出参数的媒体类型。

```csharp
public interface IContentTypeProvider
{
    bool TryGetContentType(string subpath, out string contentType);
}
```

StaticFileMiddleware 中间件默认使用的是一个具有如下定义的 FileExtensionContentTypeProvider 类型。顾名思义，FileExtensionContentTypeProvider 利用物理文件的扩展名来解析对应的媒体类型，并利用其 Mappings 属性表示的字典维护了扩展名与媒体类型之间的映射关系。常用的数百种标准的文件扩展名和对应的媒体类型之间的映射关系都会保存在这个字典中。如果发布的文件具有一些特殊的扩展名，或者需要将现有的某些扩展名映射为不同的媒体类型，都可以通过添加或者修改扩展名/媒体类型之间的映射关系来实现。

```csharp
public class FileExtensionContentTypeProvider : IContentTypeProvider
{
    public IDictionary<string, string> Mappings { get; }

    public FileExtensionContentTypeProvider();
    public FileExtensionContentTypeProvider(IDictionary<string, string> mapping);

    public bool TryGetContentType(string subpath, out string contentType);
}
```

实现原理

为了使读者对针对静态文件的请求在 StaticFileMiddleware 中间件的处理有更加深刻的认识，下面采用相对简单的代码来重新定义这个中间件。这个模拟中间件具有与 StaticFileMiddleware 相同的能力，它能够将目标文件的内容采用正确的媒体类型响应给客户端，同时能够处理条件请求和区间请求。（S1408）

StaticFileMiddleware 中间件处理针对静态文件请求的整个处理流程大体上可以划分为图 14-7 所示的 3 个步骤。

- **获取目标文件**：中间件根据请求的路径获取目标文件，并解析出正确的媒体类型。在此之前，中间件还会验证请求采用的 HTTP 方法是否有效，它只支持 GET 请求和 HEAD 请求。中间件还会获取文件最后被修改的时间，并根据这个时间戳和文件内容的长度生成一个标签，响应报文的 Last-Modified 报头和 ETag 报头的内容就来源于此。
- **条件请求解析**：获取与条件请求相关的 4 个报头（If-Match、If-None-Match、If-Modified-Since 和 If-Unmodified-Since）的值，根据 HTTP 规范计算出最终的条件状态。
- **响应请求**：如果是区间请求，中间件会提取相关的报头（Range 和 If-Range）并解析出正确的内容区间。中间件最终根据上面计算的条件状态和区间相关信息设置响应报头，并根据需要响应整个文件的内容或者指定区间的内容。

图 14-7 处理针对静态文件请求的流程

下面按照上述流程重新定义 StaticFileMiddleware 中间件，但在此之前需要先介绍预先定义的几个辅助性的扩展方法。如下面的代码片段所示，UseMethods 扩展方法用于确定请求是否采用指定的 HTTP 方法，而 TryGetSubpath 方法用于解析请求的目标文件的相对路径。TryGetContentType 方法会根据指定的 StaticFileOptions 携带的 IContentTypeProvider 对象解析出正确的媒体类型，而 TryGetFileInfo 方法则根据指定的路径获取描述目标文件的 IFileInfo 对象。IsRangeRequest 方法会根据是否携带 Rang 报头判断指定的请求是否是一个区间请求。

```
public static class Extensions
{
    public static bool UseMethods(this HttpContext context, params string[] methods)
        => methods.Contains(context.Request.Method, StringComparer.OrdinalIgnoreCase);

    public static bool TryGetSubpath(this HttpContext context, string requestPath,
        out PathString subpath)
        => new PathString(context.Request.Path).StartsWithSegments(requestPath,
            out subpath);

    public static bool TryGetContentType(this StaticFileOptions options,
        PathString subpath, out string contentType)
        => options.ContentTypeProvider.TryGetContentType(subpath.Value,
```

```csharp
            out contentType) ||
            (!string.IsNullOrEmpty(contentType = options.DefaultContentType) &&
            options.ServeUnknownFileTypes);

    public static bool TryGetFileInfo(this StaticFileOptions options, PathString subpath,
        out IFileInfo fileInfo)
        => (fileInfo = options.FileProvider.GetFileInfo(subpath.Value)).Exists;

    public static bool IsRangeRequest(this HttpContext context)
        => context.Request.GetTypedHeaders().Range != null;
}
```

模拟类型 StaticFileMiddleware 的定义如下。如果指定的 StaticFileOptions 没有提供 IFileProvider 对象,我们会创建一个针对 WebRoot 目录的 PhysicalFileProvider 对象。如果一个具体的 IContentTypeProvider 对象没有显式指定,我们使用的就是一个 FileExtensionContentTypeProvider 对象。这两个默认值分别解释了两个问题:为什么请求的静态文件将 WebRoot 作为默认的根目录,为什么目标文件的扩展名会决定响应的媒体类型。

```csharp
public class StaticFileMiddleware
{
    private readonly RequestDelegate       _next;
    private readonly StaticFileOptions     _options;

    public StaticFileMiddleware(RequestDelegate next, IWebHostEnvironment env,
        IOptions<StaticFileOptions> options)
    {
        _next                       = next;
        _options                    = options.Value;
        _options.FileProvider       = _options.FileProvider??env.WebRootFileProvider;
        _options.ContentTypeProvider = _options.ContentTypeProvider
            ?? new FileExtensionContentTypeProvider();
    }
    ...
}
```

上述 3 个步骤分别实现在对应的方法(TryGetFileInfo、GetPreconditionState 和 SendResponseAsync)中,所以 StaticFileMiddleware 中间件类型的 InvokeAsync 方法按照如下方式先后调用这 3 个方法完成对整个文件请求的处理。

```csharp
public class StaticFileMiddleware
{
    public async Task InvokeAsync(HttpContext context)
    {
        if (this.TryGetFileInfo(context, out var contentType, out var fileInfo,
            out var lastModified, out var etag))
        {
            var preconditionState = GetPreconditionState(
                context, lastModified.Value, etag);
            await SendResponseAsync(preconditionState, context, etag,
                lastModified.Value, contentType, fileInfo);
```

```
        return;
    }
    await _next(context);
}
...
```

下面重点介绍这 3 个方法的实现。首先介绍 TryGetFileInfo 方法是如何根据请求的路径获得描述目标文件的 IFileInfo 对象的。如下面的代码片段所示，如果目标文件存在，这个方法除了将目标文件的 IFileInfo 对象作为输出参数返回，与这个文件相关的数据（媒体类型、最后修改时间戳和封装标签的 ETag）也会一并返回。

```csharp
public class StaticFileMiddleware
{
    public bool TryGetFileInfo(HttpContext context, out string contentType,
        out IFileInfo fileInfo, out DateTimeOffset? lastModified,
        out EntityTagHeaderValue etag)
    {
        contentType = null;
        fileInfo = null;

        if (context.UseMethods("GET", "HEAD") &&
            context.TryGetSubpath(_options.RequestPath, out var subpath) &&
            _options.TryGetContentType(subpath, out contentType) &&
            _options.TryGetFileInfo(subpath, out fileInfo))
        {
            var last = fileInfo.LastModified;
            long etagHash = last.ToFileTime() ^ fileInfo.Length;
            etag = new EntityTagHeaderValue('\"' + Convert.ToString(etagHash, 16) + '\"');
            lastModified = new DateTimeOffset(last.Year, last.Month, last.Day, last.Hour,
                last.Minute, last.Second, last.Offset).ToUniversalTime();
            return true;
        }

        etag           = null;
        lastModified   = null;
        return false;
    }
}
```

GetPreconditionState 方法旨在获取与条件请求相关的 4 个报头（If-Match、If-None-Match、If-Modified-Since 和 If-Unmodified-Since）的值，并通过与目标文件当前的状态进行比较，进而得到一个最终的检验结果。针对这 4 个请求报头的检验最终会产生 4 种可能的结果，所以我们定义了如下所示的一个 PreconditionState 枚举来表示它们。

```csharp
private enum PreconditionState
{
    Unspecified          = 0,
    NotModified          = 1,
    ShouldProcess        = 2,
```

```
    PreconditionFailed            = 3,
}
```

对于定义在这个枚举类型中的 4 个选项，Unspecified 表示请求中不包含这 4 个报头。如果将请求报头 If-None-Match 的值与当前文件标签进行比较，或者将请求报头 If-Modified-Since 的值与文件最后修改时间进行比较确定目标文件不曾被更新，检验结果对应的枚举值为 NotModified，反之则对应的枚举值为 ShouldProcess。

如果目标文件当前的状态不满足 If-Match 报头或者 If-Unmodified-Since 报头表示的条件，那么检验结果对应的枚举值为 PreconditionFailed；反之，对应的枚举值为 ShouldProcess。如果请求携带多个报头，针对它们可能会得出不同的检验结果，那么值最大的那个将作为最终的结果。如下面的代码片段所示，GetPreconditionState 方法正是通过这样的逻辑得到表示最终条件检验结果的 PreconditionState 枚举的。

```csharp
public class StaticFileMiddleware
{
    private PreconditionState GetPreconditionState(HttpContext context,
        DateTimeOffset lastModified, EntityTagHeaderValue etag)
    {
        PreconditionState ifMatch,ifNonematch, ifModifiedSince, ifUnmodifiedSince;
        ifMatch = ifNonematch = ifModifiedSince = ifUnmodifiedSince =
            PreconditionState.Unspecified;

        var requestHeaders = context.Request.GetTypedHeaders();
        //If-Match:ShouldProcess or PreconditionFailed
        if (requestHeaders.IfMatch != null)
        {
            ifMatch = requestHeaders.IfMatch.Any(it
                => it.Equals(EntityTagHeaderValue.Any) || it.Compare(etag, true))
                ? PreconditionState.ShouldProcess
                : PreconditionState.PreconditionFailed;
        }

        //If-None-Match:NotModified or ShouldProcess
        if (requestHeaders.IfNoneMatch != null)
        {
            ifNonematch = requestHeaders.IfNoneMatch.Any(it
                => it.Equals(EntityTagHeaderValue.Any) || it.Compare(etag, true))
                ? PreconditionState.NotModified
                : PreconditionState.ShouldProcess;
        }

        //If-Modified-Since: ShouldProcess or NotModified
        if (requestHeaders.IfModifiedSince.HasValue)
        {
            ifModifiedSince = requestHeaders.IfModifiedSince < lastModified
                ? PreconditionState.ShouldProcess
                : PreconditionState.NotModified;
        }
```

```csharp
    //If-Unmodified-Since: ShouldProcess or PreconditionFailed
    if (requestHeaders.IfUnmodifiedSince.HasValue)
    {
        ifUnmodifiedSince = requestHeaders.IfUnmodifiedSince > lastModified
            ? PreconditionState.ShouldProcess
            : PreconditionState.PreconditionFailed;
    }

    //Return maximum.
    return new PreconditionState[] {
        ifMatch, ifNonematch, ifModifiedSince, ifUnmodifiedSince }.Max();
}
...
}
```

针对静态文件的处理最终在 **SendResponseAsync** 方法中实现，这个方法会设置相应的响应报头和状态码，如果需要，它还会将目标文件的内容写入响应报文的主体中。为响应选择什么样的状态码，设置哪些报头，以及响应主体内容的设置除了决定于 **GetPreconditionState** 方法返回的检验结果，与区间请求相关的两个报头（**Range** 和 **If-Range**）也是决定性因素之一。所以，我们定义了如下所示的 **TryGetRanges** 方法，用于解析这两个报头并计算出正确的区间。

```csharp
public class StaticFileMiddleware
{
    private bool TryGetRanges(HttpContext context, DateTimeOffset lastModified,
        EntityTagHeaderValue etag, long length,
        out IEnumerable<RangeItemHeaderValue> ranges)
    {
        ranges = null;
        var requestHeaders = context.Request.GetTypedHeaders();

        //Check If-Range
        var ifRange = requestHeaders.IfRange;
        if (ifRange != null)
        {
            bool ignore = (ifRange.EntityTag != null &&
                    !ifRange.EntityTag.Compare(etag, true)) ||
                (ifRange.LastModified.HasValue && ifRange.LastModified < lastModified);
            if (ignore)
            {
                return false;
            }
        }

        var list = new List<RangeItemHeaderValue>();
        foreach (var it in requestHeaders.Range.Ranges)
        {
            //Range:{from}-{to} Or {from}-
            if (it.From.HasValue)
            {
```

```csharp
                if (it.From.Value < length - 1)
                {
                    long to = it.To.HasValue
                        ? Math.Min(it.To.Value, length - 1)
                        : length - 1;
                    list.Add(new RangeItemHeaderValue(it.From.Value, to));
                }
            }
            //Range: -{size}
            else if (it.To.Value != 0)
            {
                long size = Math.Min(length, it.To.Value);
                list.Add(new RangeItemHeaderValue(length - size, length - 1));
            }
        }
        return ( ranges = list) != null;
    }
    ...
}
```

如上面的代码片段所示，TryGetRanges 方法先获取 If-Range 报头的值，并将它与目标文件当前的状态进行比较。如果当前状态不满足 If-Range 报头表示的条件，就意味着目标文件内容发生变化，那么请求 Range 报头携带的区间信息将自动被忽略。而 Range 报头携带的值具有不同的表现形式（如 bytes={from}-{to}、bytes={from}-和 bytes=-{size}），并且指定的端点有可能超出目标文件的长度，所以 TryGetRanges 方法定义了相应的逻辑来检验区间定义的合法性并计算出正确的区间范围。

对于区间请求，TryGetRanges 方法的返回值表示目标文件的当前状态是否与 If-Range 报头携带的条件相匹配。由于 HTTP 规范并未限制 Range 报头中设置的区间数量（原则上可以指定多个区间），所以 TryGetRanges 方法通过输出参数返回的区间信息是一个元素类型为 RangeItemHeaderValue 的集合。如果集合为空，就表示设置的区间不符合要求。

实现在 SendResponseAsync 方法中针对请求的处理基本上是指定响应状态码、设置响应报头和写入响应主体内容。我们将前两项工作实现在 HttpContext 如下所示的 SetResponseHeaders 扩展方法中。该方法不仅可以将指定的响应状态码应用到 HttpContext 上，还可以设置相应的响应报头。

```csharp
public static class Extensions
{
    public static void SetResponseHeaders(this HttpContext context, int statusCode,
        EntityTagHeaderValue etag, DateTimeOffset lastModified, string contentType,
        long contentLength, RangeItemHeaderValue range = null)
    {
        context.Response.StatusCode = statusCode;
        var responseHeaders = context.Response.GetTypedHeaders();
        if (statusCode < 400)
        {
            responseHeaders.ETag                                              = etag;
```

```
            responseHeaders.LastModified                              = lastModified;
            context.Response.ContentType                              = contentType;
            context.Response.Headers[HeaderNames.AcceptRanges]        = "bytes";
        }
        if (statusCode == 200)
        {
            context.Response.ContentLength = contentLength;
        }

        if (statusCode == 416)
        {
            responseHeaders.ContentRange = new ContentRangeHeaderValue(contentLength);
        }

        if (statusCode == 206 && range != null)
        {
            responseHeaders.ContentRange = new ContentRangeHeaderValue(
                range.From.Value, range.To.Value, contentLength);
        }
    }
}
```

如上面的代码片段所示，对于所有非错误类型的响应（主要是指状态码为"200 OK"、"206 Partial Content"和"304 Not Modified"的响应），除了表示媒体类型的 Content-Type 报头，还有 3 个额外的报头（Last-Modified、ETag 和 Accept-Range）。针对区间请求的两种响应（"206 Partial Content"和"416 Range Not Satisfiable"）都有一个 Content-Range 报头。

如下所示的代码片段是 SendResponseAsync 方法的完整定义。它会根据条件请求和区间请求的解析结果来决定最终采用的响应状态码。响应状态和相关响应报头的设置是通过调用上面的 SetResponseHeaders 方法来完成的。对于状态码为"200 OK"或者"206 Partial Content"的响应，SetResponseHeaders 方法会将整个文件的内容或者指定区间的内容写入响应报文的主体部分。而文件内容的读取则调用表示目标文件的 FileInfo 对象的 CreateReadStream 方法，并利用其返回的输出流来实现。

```
public class StaticFileMiddleware
{
    private async Task SendResponseAsync(PreconditionState state, HttpContext context,
        EntityTagHeaderValue etag, DateTimeOffset lastModified, string contentType,
        IFileInfo fileInfo)
    {
        switch (state)
        {
            //304 Not Modified
            case PreconditionState.NotModified:
                {
                    context.SetResponseHeaders(304, etag, lastModified, contentType,
                        fileInfo.Length);
                    break;
                }
```

```csharp
//416 Precondition Failded
case PreconditionState.PreconditionFailed:
    {
        context.SetResponseHeaders(412, etag, lastModified, contentType,
            fileInfo.Length);
        break;
    }
case PreconditionState.Unspecified:
case PreconditionState.ShouldProcess:
    {
        //200 OK
        if (context.UseMethods("HEAD"))
        {
            context.SetResponseHeaders(200, etag, lastModified, contentType,
                fileInfo.Length);
            return;
        }

        IEnumerable<RangeItemHeaderValue> ranges;
        if (context.IsRangeRequest() && this.TryGetRanges(context,
            lastModified, etag, fileInfo.Length, out ranges))
        {
            RangeItemHeaderValue range = ranges.FirstOrDefault();
            //416
            if (null == range)
            {
                context.SetResponseHeaders(416, etag, lastModified,
                    contentType, fileInfo.Length);
                return;
            }
            else
            {
                //206 Partial Content
                context.SetResponseHeaders(206, etag, lastModified,
                    contentType, fileInfo.Length, range);
                context.Response.GetTypedHeaders().ContentRange
                    = new ContentRangeHeaderValue(range.From.Value,
                        range.To.Value, fileInfo.Length);
                using (Stream stream = fileInfo.CreateReadStream())
                {
                    stream.Seek(range.From.Value, SeekOrigin.Begin);
                    await StreamCopyOperation.CopyToAsync(stream,
                        context.Response.Body, range.To - range.From + 1,
                        context.RequestAborted);
                }
                return;
            }
        }
        //200 OK
        context.SetResponseHeaders(200, etag, lastModified, contentType,
```

```
                fileInfo.Length);
            using (Stream stream = fileInfo.CreateReadStream())
            {
                await StreamCopyOperation.CopyToAsync(stream,
                    context.Response.Body, fileInfo.Length, context.RequestAborted);
            }
            break;
        }
    }
}
```

14.3 处理目录请求

对于 NuGet 包由"Microsoft.AspNetCore.StaticFiles"提供的 3 个中间件来说，StaticFile Middleware 中间件旨在处理针对具体静态文件的请求，其他两个中间件（DirectoryBrowser Middleware 和 DefaultFilesMiddleware）处理的均是针对某个目录的请求。

14.3.1 DirectoryBrowserMiddleware

与 StaticFileMiddleware 中间件一样，DirectoryBrowserMiddleware 中间件本质上还定义了一个请求基地址与某个物理目录之间的映射关系，而目标目录体现为一个 IFileProvider 对象。当这个中间件接收到匹配的请求后，会根据请求地址解析出对应目录的相对路径，并利用这个 IFileProvider 对象获取目录的结构。目录结构最终会以一个 HTML 文档的形式定义，而此 HTML 文档最终会被这个中间件作为响应的内容。

如下面的代码片段所示，DirectoryBrowserMiddleware 类型的第二个构造函数有 4 个参数。其中，第二个参数是代表当前执行环境的 IWebHostEnvironment 对象；第三个参数提供一个 HtmlEncoder 对象，当目标目录被呈现为一个 HTML 文档时，它被用于实现针对 HTML 的编码，如果没有显式指定（调用第一个构造函数），默认的 HtmlEncoder（HtmlEncoder.Default）会被使用；第四个类型为 IOptions<DirectoryBrowserOptions>的参数用于提供表示配置选项的 DirectoryBrowserMiddleware 的 DirectoryBrowserOptions 对象。与前面介绍的 StaticFileOptions 一样，DirectoryBrowserOptions 是 SharedOptionsBase 的子类。

```
public class DirectoryBrowserMiddleware
{
    public DirectoryBrowserMiddleware(RequestDelegate next, IWebHostEnvironment env,
        IOptions<DirectoryBrowserOptions> options)
    public DirectoryBrowserMiddleware(RequestDelegate next, IWebHostEnvironment hostingEnv,
        HtmlEncoder encoder, IOptions<DirectoryBrowserOptions> options);
    public Task Invoke(HttpContext context);
}

public class DirectoryBrowserOptions : SharedOptionsBase
{
    public IDirectoryFormatter Formatter { get; set; }
```

```csharp
    public DirectoryBrowserOptions();
    public DirectoryBrowserOptions(SharedOptions sharedOptions);
}
```

DirectoryBrowserMiddleware 中间件的注册可以通过 IApplicationBuilder 接口的 3 个 UseDirectoryBrowser 扩展方法来完成。在调用这些扩展方法时，如果没有指定任何参数，就意味着注册的中间件会采用默认配置。我们也可以显式地执行一个 DirectoryBrowserOptions 对象来对注册的中间件进行定制。如果我们只希望指定请求的路径，就可以直接调用第三个方法重载。

```csharp
public static class DirectoryBrowserExtensions
{
    public static IApplicationBuilder UseDirectoryBrowser(this IApplicationBuilder app)
        => app.UseMiddleware<DirectoryBrowserMiddleware>(Array.Empty<object>());

    public static IApplicationBuilder UseDirectoryBrowser(this IApplicationBuilder app,
        DirectoryBrowserOptions options)
    {
        var args = new object[] { Options.Create<DirectoryBrowserOptions>(options) };
        return app.UseMiddleware<DirectoryBrowserMiddleware>(args);
    }

    public static IApplicationBuilder UseDirectoryBrowser(this IApplicationBuilder app,
        string requestPath)
    {
        var options = new DirectoryBrowserOptions
        {
            RequestPath = new PathString(requestPath)
        };
        return app.UseDirectoryBrowser(options);
    }
}
```

DirectoryBrowserMiddleware 中间件的目的很明确，就是将目录下的内容（文件和子目录）格式化成一种可读的形式响应给客户端。针对目录内容的响应最终实现在一个 IDirectoryFormatter 对象上，DirectoryBrowserOptions 的 Formatter 属性设置和返回的就是这样的一个对象。如下面的代码片段所示，IDirectoryFormatter 接口仅包含一个 GenerateContentAsync 方法。当实现这个方法的时候，我们可以利用第一个参数获取当前 HttpContext 上下文。该方法的另一个参数返回一组 IFileInfo 的集合，每个 IFileInfo 代表目标目录下的某个文件或者子目录。

```csharp
public interface IDirectoryFormatter
{
    Task GenerateContentAsync(HttpContext context, IEnumerable<IFileInfo> contents);
}
```

在默认情况下，请求目录的内容在页面上是以一个表格的形式来呈现的，包含这个表格的 HTML 文档正是默认使用的 IDirectoryFormatter 对象生成的，该对象的类型为 HtmlDirectoryFormatter。如下面的代码片段所示，我们在构造一个 HtmlDirectoryFormatter 对象时需要指定一个 HtmlEncoder 对象，它就是在构造 DirectoryBrowserMiddleware 对象时提供的那个 Html

Encoder 对象。

```csharp
public class HtmlDirectoryFormatter : IDirectoryFormatter
{
    public HtmlDirectoryFormatter(HtmlEncoder encoder);
    public virtual Task GenerateContentAsync(HttpContext context,
        IEnumerable<IFileInfo> contents);
}
```

既然最复杂的工作（呈现目录内容）由 IDirectoryFormatter 完成，那么 DirectoryBrowserMiddleware 中间件自身的工作其实就会很少。为了更好地说明这个中间件在处理请求时具体做了些什么，可以采用一种比较容易理解的方式对 DirectoryBrowserMiddleware 类型重新定义。

```csharp
public class DirectoryBrowserMiddleware
{
    private readonly RequestDelegate           _next;
    private readonly DirectoryBrowserOptions   _options;

    public DirectoryBrowserMiddleware(RequestDelegate next, IWebHostEnvironment env,
        IOptions<DirectoryBrowserOptions> options) : this(next, env, HtmlEncoder.Default,
        options)
    {}

    public DirectoryBrowserMiddleware(RequestDelegate next, IWebHostEnvironment env,
        HtmlEncoder encoder, IOptions<DirectoryBrowserOptions> options)
    {
        _next                   = next;
        _options                = options.Value;
        _options.FileProvider   = _options.FileProvider ?? env.WebRootFileProvider;
        _options.Formatter      = _options.Formatter
                                  ?? new HtmlDirectoryFormatter(encoder);
    }

    public async Task InvokeAsync(HttpContext context)
    {
        //只处理 GET 请求和 HEAD 请求
        if (!new string[] { "GET", "HEAD" }.Contains(context.Request.Method,
            StringComparer.OrdinalIgnoreCase))
        {
            await _next(context);
            return;
        }

        //检验当前路径是否与注册的请求路径相匹配
        PathString path = new PathString(context.Request.Path.Value.TrimEnd('/') + "/");
        PathString subpath;
        if (!path.StartsWithSegments(_options.RequestPath, out subpath))
        {
            await _next(context);
            return;
        }
```

```csharp
        //检验目标目录是否存在
        IDirectoryContents directoryContents =
            _options.FileProvider.GetDirectoryContents(subpath);
        if (!directoryContents.Exists)
        {
            await _next(context);
            return;
        }

        //如果当前路径不以"/"作为后缀,会响应一个针对"标准"URL 的重定向
        if (!context.Request.Path.Value.EndsWith("/"))
        {
            context.Response.StatusCode = 302;
            context.Response.GetTypedHeaders().Location = new Uri(
                path.Value + context.Request.QueryString);
            return;
        }

        //利用 DirectoryFormatter 响应目录内容
        await _options.Formatter.GenerateContentAsync(context, directoryContents);
    }
}
```

如上面的代码片段所示,在最终利用注册的 IDirectoryFormatter 对象来响应目标目录的内容之前,DirectoryBrowserMiddleware 中间件会做一系列的前期工作:验证当前请求是否是 GET 请求或者 HEAD 请求;当前的 URL 是否与注册的请求路径相匹配,在匹配的情况下还需要验证目标目录是否存在。

这个中间件要求访问目录的请求路径必须以"/"作为后缀,否则会在目前的路径上添加这个后缀,并针对修正的路径发送一个 302 重定向。所以,利用浏览器发送针对某个目录的请求时,虽然 URL 没有指定"/"作为后缀,但浏览器会自动将这个后缀补上,这就是重定向导致的结果。

目录结构的呈现方式完全由 IDirectoryFormatter 对象完成,如果默认注册的 HtmlDirectoryFormatter 对象的呈现方式无法满足需求(如我们需要这个页面与现有网站保持相同的风格),就可以通过注册一个自定义的 DirectoryFormatter 来解决这个问题。下面通过一个简单的实例来演示如何定义一个 IDirectoryFormatter 实现类型。我们将自定义的 IDirectoryFormatter 实现类型命名为 ListDirectoryFormatter,因为它仅仅将所有文件或者子目录显示为一个简单的列表。

```csharp
public class ListDirectoryFormatter : IDirectoryFormatter
{
    public async Task GenerateContentAsync(HttpContext context,
        IEnumerable<IFileInfo> contents)
    {
        context.Response.ContentType = "text/html";
        await
context.Response.WriteAsync("<html><head><title>Index</title><body><ul>");
```

```
        foreach (var file in contents)
        {
            string href = $"{context.Request.Path.Value.TrimEnd('/')}/{file.Name}";
            await context.Response.WriteAsync(
                $"<li><a href='{href}'>{file.Name}</a></li>");
        }
        await context.Response.WriteAsync("</ul></body></html>");
    }
}

public class Program
{
    public static void Main()
    {
        var options = new DirectoryBrowserOptions
        {
            Formatter = new ListDirectoryFormatter()
        };
        Host.CreateDefaultBuilder()
            .ConfigureWebHostDefaults(builder => builder.Configure(
                app => app.UseDirectoryBrowser(options)))
            .Build()
            .Run();
    }
}
```

如上面的代码片段所示，ListDirectoryFormatter 最终响应的是一个完整的 HTML 文档，它的主体部分只包含一个通过<ul></ul>表示的无序列表，列表元素（<li>）是一个针对文件或者子目录的链接。在调用 UseDirectoryBrowser 扩展方法注册 DirectoryBrowserMiddleware 中间件时，需要将一个 ListDirectoryFormatter 对象设置为指定配置选项的 Formatter 属性。目录内容最终以图 14-8 所示的形式呈现在浏览器上。（S1409）

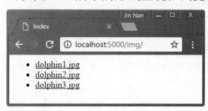

图 14-8　由自定义 ListDirectoryFormatter 呈现的目录内容

14.3.2　DefaultFilesMiddleware

DefaultFilesMiddleware 中间件的目的在于将目标目录下的默认文件作为响应内容。如果直接请求的就是这个默认文件，那么前面介绍的 StaticFileMiddleware 中间件就会将这个文件响应给客户端。如果能够将针对目录的请求重定向到这个默认文件上，一切问题就会迎刃而解。实际上，DefaultFilesMiddleware 中间件的实现逻辑很简单，它采用 URL 重写的形式修改了当前请求的地址，即将针对目录的 URL 修改成针对默认文件的 URL。

下面先介绍 DefaultFilesMiddleware 类型的定义。与其他两个中间件类似，DefaultFilesMiddleware 中间件的构造由一个 IOptions<DefaultFilesOptions>类型的参数来指定相关的配置选项。由于 DefaultFilesMiddleware 中间件本质上依然体现了请求路径与某个物理目录的映射，所以 DefaultFilesOptions 依然派生于 SharedOptionsBase。DefaultFilesOptions 的 DefaultFileNames 属性包含预定义的默认文件名，由此可以看到它默认包含 4 个名称（default.htm、default.html、index.htm 和 index.html）。

```
public class DefaultFilesMiddleware
{
    public DefaultFilesMiddleware(RequestDelegate next, IWebHostEnvironment hostingEnv,
        IOptions<DefaultFilesOptions> options);
    public Task Invoke(HttpContext context);
}

public class DefaultFilesOptions : SharedOptionsBase
{
    public IList<string> DefaultFileNames { get; set; }

    public DefaultFilesOptions() : this(new SharedOptions()){}
    public DefaultFilesOptions(SharedOptions sharedOptions) : base(sharedOptions)
    {
        this.DefaultFileNames = new List<string> {
            "default.htm", "default.html", "index.htm", "index.html" };
    }
}
```

DefaultFilesMiddleware 中间件的注册可以通过调用 IApplicationBuilder 接口的如下 3 个名为 UseDefaultFiles 的扩展方法来完成。从如下所示的代码片段可以看出，它们与用于注册 DirectoryBrowserMiddleware 中间件的 UseDirectoryBrowser 扩展方法具有一致的定义和实现方式。

```
public static class DefaultFilesExtensions
{
    public static IApplicationBuilder UseDefaultFiles(this IApplicationBuilder app)
        => app.UseMiddleware<DefaultFilesMiddleware>(Array.Empty<object>());

    public static IApplicationBuilder UseDefaultFiles(this IApplicationBuilder app,
        DefaultFilesOptions options)
    {
        var args = new object[] {Options.Create<DefaultFilesOptions>(options) };
        return app.UseMiddleware<DefaultFilesMiddleware>(args);
    }

    public static IApplicationBuilder UseDefaultFiles(this IApplicationBuilder app,
        string requestPath)
    {
        var options = new DefaultFilesOptions
        {
            RequestPath = new PathString(requestPath)
        };
```

```
        return app.UseDefaultFiles(options);
    }
}
```

下面采用一种易于理解的形式重新定义 DefaultFilesMiddleware 类型，以便于读者理解它的处理逻辑。如下面的代码片段所示，与前面介绍的 DirectoryBrowserMiddleware 中间件一样，DefaultFilesMiddleware 中间件会对请求做相应的验证。如果当前目录下存在某个默认文件，那么它会将当前请求的 URL 修改成指向这个默认文件的 URL。值得注意的是，DefaultFilesMiddleware 中间件同样要求访问目录的请求路径必须以 "/" 作为后缀，否则会在目前的路径上添加这个后缀并针对最终的路径发送一个重定向。

```
public class DefaultFilesMiddleware
{
    private RequestDelegate            _next;
    private DefaultFilesOptions        _options;

    public DefaultFilesMiddleware(RequestDelegate next, IWebHostEnvironment env,
        IOptions<DefaultFilesOptions> options)
    {
        _next                       = next;
        _options                    = options.Value;
        _options.FileProvider       = _options.FileProvider ?? env.WebRootFileProvider;
    }

    public async Task InvokeAsync(HttpContext context)
    {
        //只处理 GET 请求和 HEAD 请求
        if (!new string[] { "GET", "HEAD" }.Contains(context.Request.Method,
            StringComparer.OrdinalIgnoreCase))
        {
            await _next(context);
            return;
        }

        //检验当前路径是否与注册的请求路径相匹配
        PathString path = new PathString(context.Request.Path.Value.TrimEnd('/') + "/");
        PathString subpath;
        if (!path.StartsWithSegments(_options.RequestPath, out subpath))
        {
            await _next(context);
            return;
        }

        //检验目标目录是否存在
        if (!_options.FileProvider.GetDirectoryContents(subpath).Exists)
        {
            await _next(context);
            return;
        }
```

```csharp
    //检验当前目录是否包含默认文件
    foreach (var fileName in _options.DefaultFileNames)
    {
        if (_options.FileProvider.GetFileInfo($"{subpath}{fileName}").Exists)
        {
            //如果当前路径不以"/"作为后缀，会响应一个针对"标准"URL 的重定向
            if (!context.Request.Path.Value.EndsWith("/"))
            {
                context.Response.StatusCode = 302;
                context.Response.GetTypedHeaders().Location =
                    new Uri(path.Value + context.Request.QueryString);
                return;
            }
            //将针对目录的 URL 更新为针对默认文件的 URL
            context.Request.Path = new PathString($"{context.Request.Path}{fileName}");
        }
    }
    await _next(context);
}
```

由于 DefaultFilesMiddleware 中间件采用 URL 重写的方式来响应默认文件，默认文件的内容其实还是通过 StaticFileMiddleware 中间件予以响应的，所以针对后者的注册是必需的。也正是这个原因，DefaultFilesMiddleware 中间件需要优先注册，以确保 URL 重写发生在 StaticFileMiddleware 响应文件之前。

第 15 章

路　由

借助路由系统提供的请求 URL 模式与对应终结点（Endpoint）之间的映射关系，我们可以将具有相同 URL 模式的请求分发给应用的终结点进行处理。ASP.NET Core 的路由是通过 EndpointRoutingMiddleware 和 EndpointMiddleware 这两个中间件协作完成的，它们在 ASP.NET Core 平台上具有举足轻重的地位，因为 ASP.NET Core MVC 框架就建立在这个中间件之上。

15.1　路由映射

可以将一个 ASP.NET Core 应用视为一组终结点的组合，所谓的终结点可以理解为能够通过 HTTP 请求的形式访问的远程服务。每个终结点通过 RequestDelegate 对象来处理路由过来的请求。ASP.NET Core 的路由是通过 EndpointRoutingMiddleware 和 EndpointMiddleware 这两个中间件来实现的，这两个中间件类型都定义在 NuGet 包"Microsoft.AspNetCore.Routing"中。为了使读者对实现在 RouterMiddleware 的路由功能有一个大体的认识，下面先演示几个简单的实例。

15.1.1　路由注册

我们演示的这个 ASP.NET Core 应用是一个简易版的天气预报站点。如果用户希望获取某个城市在未来 N 天之内的天气信息，他可以直接利用浏览器发送一个 GET 请求并将对应城市（采用电话区号表示）和天数设置在 URL 中。如图 15-1 所示，为了得到成都未来两天的天气信息，我们将发送请求的路径设置为"weather/028/2"。对于采用路径"weather/0512/4"的请求，返回的自然就是苏州未来 4 天的天气信息。（S1501）

为了开发这个简单的应用，我们定义了如下所示的 WeatherReport 类型，表示某个城市在某段时间范围内的天气。如下面的代码片段所示，我们还定义了另一个 WeatherInfo 类型，表示具体某一天的天气。简单起见，我们让 WeatherInfo 对象只携带基本天气状况和气温区间的信息。创建一个 WeatherReport 对象时，我们会随机生成这些天气信息。

图 15-1　获取天气预报信息

```
public class WeatherReport
{
    private static string[]  _conditions = new string[] { "晴", "多云", "小雨" };
    private static Random    _random = new Random();

    public string                                City { get; }
    public IDictionary<DateTime, WeatherInfo>    WeatherInfos { get; }

    public WeatherReport(string city, int days)
    {
        City = city;
        WeatherInfos = new Dictionary<DateTime, WeatherInfo>();
        for (int i = 0; i < days; i++)
        {
            this.WeatherInfos[DateTime.Today.AddDays(i + 1)] = new WeatherInfo
            {
                Condition       = _conditions[_random.Next(0, 2)],
                HighTemperature = _random.Next(20, 30),
                LowTemperature  = _random.Next(10, 20)
            };
        }
    }

    public WeatherReport(string city, DateTime date)
    {
        City = city;
        WeatherInfos = new Dictionary<DateTime, WeatherInfo>
        {
            [date] = new WeatherInfo
            {
                Condition       = _conditions[_random.Next(0, 2)],
                HighTemperature = _random.Next(20, 30),
                LowTemperature  = _random.Next(10, 20)
            }
        };
    }
```

```
    public class WeatherInfo
    {
        public string Condition { get; set; }
        public double HighTemperature { get; set; }
        public double LowTemperature { get; set; }
    }
}
```

由于用于处理请求的处理器最终体现为一个 RequestDelegate 对象，所以我们定义了如下一个与这个委托类型具有一致声明的 WeatherForecast 方法来处理对应的请求。如下面的代码片段所示，我们在这个方法中直接调用 HttpContext 的 GetRouteData 扩展方法提取 Routing Middleware 中间件在路由解析过程中设置的路由参数。GetRouteData 扩展方法返回的是一个具有字典结构的对象，它的 Key 和 Value 分别代表路由参数的名称与值，通过预先定义的参数名（city 和 days）可以得到目标城市和预报天数。

```
public class Program
{
    private static Dictionary<string, string> _cities = new Dictionary<string, string>
    {
        ["010"]     = "北京",
        ["028"]     = "成都",
        ["0512"]    = "苏州"
    };

    public static async Task WeatherForecast(HttpContext context)
    {
        var city = (string)context.GetRouteData().Values["city"];
        city = _cities[city];
        int days = int.Parse(context.GetRouteData().Values["days"].ToString());
        var report = new WeatherReport(city, days);
        await RendWeatherAsync(context, report);
    }

    private static async Task RendWeatherAsync(HttpContext context, WeatherReport report)
    {
        context.Response.ContentType = "text/html;charset=utf-8";
        await context.Response.WriteAsync(
            "<html><head><title>Weather</title></head><body>");
        await context.Response.WriteAsync($"<h3>{report.city}</h3>");
        foreach (var it in report.WeatherInfos)
        {
            await context.Response.WriteAsync($"{it.Key.ToString("yyyy-MM-dd")}:");
            await context.Response.WriteAsync(
                $"{it.Value.Condition}({ it.Value.LowTemperature}℃ ~
                { it.Value.HighTemperature}℃)<br/><br/> ");
        }
        await context.Response.WriteAsync("</body></html>");
    }
```

```
    ...
}
```

有了这两个核心参数之后，我们可以据此生成一个 WeatherReport 对象，并将它携带的天气信息以一个 HTML 文档的形式响应给客户端，图 15-1 就是这个 HTML 文档在浏览器上的呈现效果。由于目标城市最初以电话区号的形式体现，所以在呈现天气信息的过程中我们还会根据区号获取具体城市的名称。简单起见，我们利用一个简单的字典来维护区号和城市之间的关系，并且只存储了 3 个城市而已。

下面完成所需的路由注册工作。如下面的代码片段所示，我们调用 IApplicationBuilder 的 UseRouting 方法和 UseEndpoints 方法分别完成针对 EndpointRoutingMiddleware 与 EndpointMiddleware 这两个终结点的注册。由于它们在进行路由解析过程中需要使用一些服务，所以可以调用 IServiceCollection 的 AddRouting 扩展方法来对它们进行注册。

```
public class Program
{
    public static void Main()
    {
        Host.CreateDefaultBuilder()
            .ConfigureWebHostDefaults(builder => builder
                .ConfigureServices(svcs => svcs.AddRouting())
                .Configure(app => app
                    .UseRouting()
                    .UseEndpoints(endpoints
                        => endpoints.MapGet("weather/{city}/{days}", WeatherForecast))))
            .Build()
            .Run();
    }
}
```

UseEndpoints 方法提供了一个 Action<IEndpointRouteBuilder>类型的参数，我们利用这个参数调用 IEndpointRouteBuilder 的 MapGet 方法提供了一个路由模板与对应处理器之间的映射。我们指定的路径模板为"weather/{city}/{days}"，其中携带两个路由参数（{city}和{days}），分别代表获取天气预报的目标城市和天数。由于针对天气请求的处理实现在 WeatherForecast 方法中，所以将指向这个方法的 RequestDelegate 对象作为第二个参数。MapGet 的后缀"Get"表示 HTTP 方法，这意味着与指定路由模板的模式相匹配的 GET 请求才会被路由到 WeatherForecast 方法对应的终结点。

15.1.2 设置内联约束

上面的演示实例注册的路由模板中定义了两个参数（{city}和{days}），分别表示获取天气预报的目标城市对应的区号和天数。区号应该具有一定的格式（以零开始的 3～4 位数字），而天数除了必须是一个整数，还应该具有一定的范围。由于我们在注册的时候并没有为这两个路由参数的值做任何约束，所以请求 URL 携带的任何字符都是有效的。而处理请求的 WeatherForecast 方法也并没有对提取的数据做任何验证，所以在执行过程中面对不合法的输入会直接抛出异常。如图 15-2 所示，由于请求 URL（"/weather/0512/iv"）指定的天数不合法，所以客户端接收到一个状态为"500 Internal Server Error"的响应。

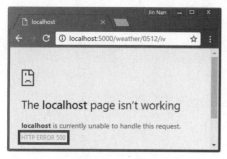

图 15-2　服务端发生异常而返回"500 Internal Server Error"的响应

为了确保路由参数值的有效性，在进行路由注册时可以采用内联（Inline）的方式直接将相应的约束规则定义在路由模板中。ASP.NET Core 为常用的验证规则定义了相应的约束表达式，我们可以根据需要为某个路由参数指定一个或者多个约束表达式。

如下面的代码片段所示，为了确保 URL 携带的是合法的区号，我们为路由参数{city}指定了一个针对正则表达式的约束（:regex(^0[1-9]{{2,3}}$)）。由于路由模板在被解析时会将{value}这样的字符理解为路由参数，如果约束表达式需要使用字符"{}"（如正则表达式^0[1-9]{2,3}$)），就需要采用"{{}}"进行转义。而路由参数{days}则应用了两个约束：第一个是针对数据类型的约束（:int），它要求参数值必须是一个整数；第二个是针对区间的约束（:range(1,4)），意味着我们的应用最多只提供未来 4 天的天气。

```
public class Program
{
    public static void Main()
    {
        var template = @"weather/{city:regex(^0\d{{2,3}}$)}/{days:int:range(1,4)}";
        Host.CreateDefaultBuilder()
            .ConfigureWebHostDefaults(builder => builder
                .ConfigureServices(svcs => svcs.AddRouting())
                .Configure(app => app
                    .UseRouting()
                    .UseEndpoints(routes => routes.MapGet(template, WeatherForecast))))
            .Build()
            .Run();
    }
    ...
}
```

如果在注册路由时应用了约束，那么 RoutingMiddleware 中间件在进行路由解析时除了要求请求路径必须与路由模板具有相同的模式，还要求携带的数据满足对应路由参数的约束条件。如果不能同时满足这两个条件，RoutingMiddleware 中间件将无法选择一个终结点来处理当前请求，在此情况下它会将请求直接递交给后续中间件进行处理。对于我们演示的这个实例来说，如果提供的是一个不合法的区号（1024）和预报天数（5），那么客户端都将得到图 15-3 所示的状态码为"404 Not Found"的响应。（S1502）

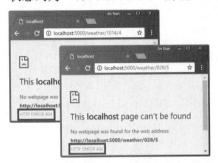

图 15-3　不满足路由约束而返回的"404 Not Found"响应

15.1.3　默认路由参数

路由注册时提供的路由模板（如"weather/{city}/{days}"）可以包含静态的字符（如weather），也可以包含动态的参数（如{city}和{days}），我们将后者称为路由参数。并非每个路由参数都是必需的，有的路由参数是默认的。还是以上面演示的实例来说，我们可以采用如下方式在路由参数名后面添加一个问号（?）将原本必需的路由参数变成可以默认的。默认的路由参数只能出现在路由模板尾部，这个应该不难理解。

```
public class Program
{
    public static void Main()
    {
        var template = "weather/{city?}/{days?}";
        Host.CreateDefaultBuilder()
            .ConfigureWebHostDefaults(builder => builder
                .ConfigureServices(svcs => svcs.AddRouting())
                .Configure(app => app
                    .UseRouting()
                    .UseEndpoints(routes => routes.MapGet(template, WeatherForecast))))
            .Build()
            .Run();
    }
    ...
}
```

既然路由变量占据的部分路径是可以默认的，那么即使请求的 URL 不具有对应的内容（如"weather"和"weather/010"），它与路由规则也是匹配的，但此时在路由参数字典中是找不到它们的。由于表示目标城市和预测天数的两个路由参数都是默认的，所以需要对处理请求的WeatherForecast 方法做相应的改动。下面的代码片段表明：如果请求 URL 为了显式提供对应参数的数据，那么它们的默认值分别为 010（北京）和 4（天），也就是说，应用默认提供北京未来 4 天的天气。

```
public class Program
```

```
{
    public static async Task WeatherForecast(HttpContext context)
    {
        var routeValues = context.GetRouteData().Values;
        var city = routeValues.TryGetValue("city", out var v1)
            ? (string)v1
            : "010";
        city = _cities[city];
        var days = routeValues.TryGetValue("days", out var v2)
            ? int.Parse(v2.ToString())
            : 4;
        var report = new WeatherReport(city, days);
        await RendWeatherAsync(context, report);
    }
    ...
}
```

针对上述改动，如果希望获取北京未来 4 天的天气状况，我们可以采用图 15-4 所示的 3 种 URL（"weather"、"weather/010" 和 "weather/010/4"），它们是完全等效的。（S1503）

图 15-4　不同 URL 针对默认路由参数的等效性

上面的程序相当于在进行请求处理时给予了默认路由参数一个默认值，实际上，路由参数默认值的设置还有一种更简单的方式，那就是按照如下所示的方式直接将默认值定义在路由模板中。如果采用这样的路由注册方式，针对 WeatherForecast 方法的改动就完全没有必要。（S1504）

```
public class Program
{
    public static void Main()
    {
        var template = "weather/{city=010}/{days=4}";
        Host.CreateDefaultBuilder()
            .ConfigureWebHostDefaults(builder => builder
                .ConfigureServices(svcs => svcs.AddRouting())
                .Configure(app => app
                    .UseRouting()
                    .UseEndpoints(routes => routes.MapGet(template, WeatherForecast))))
```

```
            .Build()
            .Run();
    }
    ...
}
```

15.1.4　特殊的路由参数

一个 URL 可以通过分隔符 "/" 划分为多个路径分段（Segment），路由模板中定义的路由参数一般来说会占据某个独立的分段（如 "weather/{city}/{days}"）。但也有例外情况，我们既可以在一个单独的路径分段中定义多个路由参数，也可以让一个路由参数跨越多个连续的路径分段。

下面先介绍在一个独立的路径分段中定义多个路由参数的情况。同样以前面演示的获取天气预报的路径为例，假设设计一种路径模式来获取某个城市某一天的天气信息，如 "/weather/010/2019.11.11" 这样一个 URL 可以获取北京在 2019 年 11 月 11 日的天气，那么路由模板为 "/weather/{city}/{year}.{month}.{day}"。

```
public class Program
{
    public static void Main()
    {
        var template = "weather/{city}/{year}.{month}.{day}";
        Host.CreateDefaultBuilder().ConfigureWebHostDefaults(builder => builder
            .ConfigureServices(svcs => svcs.AddRouting())
            .Configure(app => app.UseRouter(builder => builder
                .MapGet(template, WeatherForecast))))
            .Build()
            .Run();
    }

    public static async Task WeatherForecast(HttpContext context)
    {
        var values = context.GetRouteData().Values;
        var city = values["city"].ToString();
        city = _cities[city];
        int year = int.Parse(values["year"].ToString());
        int month = int.Parse(values["month"].ToString());
        int day = int.Parse(values["day"].ToString());
        var report = new WeatherReport(city, new DateTime(year, month, day));
        await RendWeatherAsync(context, report);
    }
    ...
}
```

由于 URL 采用了新的设计，所以我们按照如上形式对相关程序进行了相应的修改。现在我们采用 "/weather/{city}/{yyyy}.{mm}.{dd}" 这样的 URL，就可以获取某个城市指定日期的天气。如图 15-5 所示，我们采用请求路径 "/weather/010/2019.11.11" 可以获取北京在 2019 年 11 月 11

日的天气。（S1505）

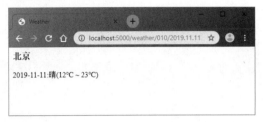

图 15-5　一个路径分段定义多个路由参数

对于上面设计的这个 URL 来说，我们采用"."作为日期分隔符，如果采用"/"作为日期分隔符（如 2019/11/11），这个路由默认应该如何定义？由于"/"也是路径分隔符，如果表示日期的路由变量也采用相同的分隔符，就意味着同一个路由参数跨越了多个路径分段，我们只能采用定义"通配符"的形式来达到这个目的。通配符路由参数采用{*variable}或者{**variable}的形式，星号（*）表示路径"余下的部分"，所以这样的路由参数只能出现在模板的尾端。对我们的实例来说，路由模板可以定义成"/weather/{city}/{*date}"。

```
public class Program
{
    public static void Main()
    {
        var template = "weather/{city}/{*date}";
        Host.CreateDefaultBuilder()
            .ConfigureWebHostDefaults(builder => builder
                .ConfigureServices(svcs => svcs.AddRouting())
                .Configure(app => app
                    .UseRouting()
                    .UseEndpoints(routes => routes.MapGet(template, WeatherForecast))))
            .Build()
            .Run();
    }

    public static async Task WeatherForecast(HttpContext context)
    {
        var values = context.GetRouteData().Values;
        var city = values["city"].ToString();
        city = _cities[city];
        var date = DateTime.ParseExact(values["date"].ToString(), "yyyy/MM/dd",
            CultureInfo.InvariantCulture);
        var report = new WeatherReport(city, date);
        await RendWeatherAsync(context, report);
    }
    ...
}
```

我们可以对程序做如上修改来使用新的 URL 模板（"/weather/{city}/{*date}"）。为了得到北京在 2019 年 11 月 11 日的天气，请求的 URL 可以替换成"/weather/010/2019/11/11"，返回的天

气信息如图 15-6 所示。（S1506）

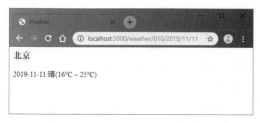

图 15-6　一个路由参数跨越多个路径分段

15.2　终结点的解析与执行

一个 Web 应用本质上体现为一组终结点的集合。终结点则体现为一个暴露在网络中可供外界采用 HTTP 协议调用的服务，路由的作用就是建立一个请求 URL 模式与对应终结点之间的映射关系。借助这个映射关系，客户端可以采用模式匹配的 URL 来调用对应的终结点。

除了利用图 15-7 所示的映射关系对请求进行路由解析，然后选择并执行与之匹配的终结点，路由系统还可以注册路由的 URL 模式和指定的路由参数值生成一个完整的 URL。我们将这两方面的工作称为两个路由方向（Routing Direction），前者为入栈路由（Inbound Routing），后者为出栈路由（Outbound Routing）。

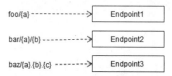

图 15-7　URL 模式与终结点的映射

15.2.1　路由模式

对于路由系统来说，作为路由目标的终结点总是关联一个具体的 URL 路径模式，我们将其称为路由模式（Route Pattern）。表示路由模式的 RoutePattern 是通过解析路由注册时提供的路由模板生成的，路由模式的基本组成元素通过抽象类型 RoutePatternPart 表示。

RoutePatternPart

RoutePatternPart 在路由模板中主要有两种类型：一种是静态文本，另一种是路由参数。例如，包含两段的路由模板"foo/{bar}"，第一段为静态文本，第二段为路由参数。由于花括号在路由模板中被用来定义路由参数，如果静态文本中包含"{"和"}"字符，就需要采用"{{"和"}}"进行转义。

其实除了上述这两种基本类型，RoutePatternPart 还有第三种类型。例如，如果采用字符串"files/{name}.{ext?}"来表示针对某个文件的路由模板，文件名（{name}）和扩展名（ext?）体现为路由参数，而它们之间的"."就是 RoutePattern 的第三种展现形式，被称为分隔符。路由系统对于分隔符具有特殊的匹配逻辑：如果分隔符后面跟的是一个可以默认的路由参数，请求地址在没有提供该参数值的情况下，分隔符是可以默认的。对于"files/{name}.{ext?}"这个路由模板来说，扩展名是可以默认的，如果请求地址没有提供扩展名，请求路径只需要提供文件名（如/files/foobar）即可。RoutePatternPart 的 3 种类型通过 RoutePatternPartKind 枚举表示。

```
public enum RoutePatternPartKind
{
    Literal,
    Parameter,
    Separator
}
```

如下所示的代码片段是 RoutePatternPart 的定义，可以看出这是一个抽象类。除了定义表示类型的 PartKind 只读属性，RoutePatternPart 还有 3 个布尔类型的属性（IsLiteral、IsParameter 和 IsSeparator），它们表示当前是否属于对应的类型。

```
public abstract class RoutePatternPart
{
    public RoutePatternPartKind PartKind { get; }

    public bool IsLiteral { get; }
    public bool IsParameter { get; }
    public bool IsSeparator { get; }
}
```

针对 RoutePatternPartKind 枚举体现的 3 种类型，路由系统提供 3 个针对 RoutePatternPart 的派生类，如下所示的代码片段是针对静态文本和分隔符的 RoutePatternLiteralPart 与 RoutePatternSeparatorPart 类型的定义，它们具有表示具体内容（静态文本内容和分隔符）的 Content 属性。

```
public sealed class RoutePatternLiteralPart : RoutePatternPart
{
    public string Content { get; }
}

public sealed class RoutePatternSeparatorPart : RoutePatternPart
{
    public string Content { get; }
}
```

由于路由参数在路由模板中有多种定义形式，所以对应的 RoutePatternParameterPart 类型的成员会多一些。RoutePatternParameterPart 的 Name 属性和 ParameterKind 属性表示路由参数的名称与类型。路由参数类型包括标准形式（如{foobar}）、默认形式（如{foobar?}或者{foobar?=123}）及通配符形式（如{*foobar}或者{**foobar}）。路由参数的这 3 种定义形式通过 RoutePatternParameterKind 枚举表示。

```
public sealed class RoutePatternParameterPart : RoutePatternPart
{
    public string                     Name           { get; }
    public RoutePatternParameterKind  ParameterKind  { get; }
    public bool                       IsOptional     { get; }
```

```
    public object                                   Default { get; }
    public bool                                     IsCatchAll { get; }
    public bool                                     EncodeSlashes { get; }

    public IReadOnlyList<RoutePatternParameterPolicyReference> ParameterPolicies { get; }
}

public enum RoutePatternParameterKind
{
    Standard,
    Optional,
    CatchAll
}
```

对于默认形式或者通配符形式对应的路由参数，对应 RoutePatternParameterPart 对象的 IsOptional 属性和 IsCatchAll 属性会返回 True。如果为参数定义了默认值，该值体现在 Default 属性上。对于两种通配符形式定义的路由参数，针对请求 URL 的解析来说并没有什么不同，它们之间的差异体现在路由系统根据它生成对应 URL 的时候。具体来说，对于提供的包含分隔符"/"的参数值（如 foo/bar），如果对应的路由参数采用{*variable}的方式，URL 格式化过程中会对分隔符进行编码（foo%2bar），倘若路由参数采用{**variable}的形式定义，提供的字符串将不做任何改变。RoutePatternParameterPart 的 EncodeSlashes 属性表示是否需要对路径分隔符"/"进行编码。

我们在定义路由参数时可以指定约束条件，路由系统将约束视为一种参数策略（Parameter Policy）。路由参数策略通过一个标记接口（不具有任何成员的接口）IParameterPolicy 表示路由参数策略，如下所示的 RoutePatternParameterPolicyReference 是对 IParameterPolicy 对象的进一步封装，它定义的 Content 属性表示策略的原始（字符串）表现形式。应用在路由参数上的策略定义体现在 RoutePatternParameterPart 的 ParameterPolicies 属性上。

```
public sealed class RoutePatternParameterPolicyReference
{
    public string               Content { get; }
    public IParameterPolicy     ParameterPolicy { get; }
}

public interface IParameterPolicy
{}
```

RoutePattern

在了解了作为路由模式的基本组成元素 RoutePatternPart 之后，下面介绍表示路由模式的 RoutePattern 如何定义。表示路由模式的 RoutePattern 对象是通过解析路由模板生成的，以字符串形式表示的路由模板体现为它的 RawText 属性。

```
public sealed class RoutePattern
{
    public string                                       RawText { get; }
    public IReadOnlyList<RoutePatternPathSegment>       PathSegments { get; }
```

```csharp
    public IReadOnlyList<RoutePatternParameterPart> Parameters { get; }
    public IReadOnlyDictionary<string, object>      Defaults { get; }
    public IReadOnlyDictionary<string, IReadOnlyList<RoutePatternParameterPolicyReference>>
                                                    ParameterPolicies { get; }

    public decimal                                  InboundPrecedence { get; }
    public decimal                                  OutboundPrecedence { get; }
    public IReadOnlyDictionary<string, object>      RequiredValues { get; }

    public RoutePatternParameterPart GetParameter(string name);
}
```

URL 的路径采用字符 "/" 作为分隔符，我们将分隔符内的内容称为段，路由模式下针对路径段的表示体现在如下所示的 RoutePatternPathSegment 类型上。RoutePatternPathSegment 类型的 Parts 属性返回一个 RoutePatternPart 对象的集合，表示构成该路径段的基本元素。如果 RoutePatternPathSegment 的 Parts 集合只包含一个元素（一般为静态文本或者路由参数），那么它被视为一个简短的路径段，其 IsSimple 属性会返回 True。

```csharp
public sealed class RoutePatternPathSegment
{
    public IReadOnlyList<RoutePatternPart>          Parts { get; }
    public bool                                     IsSimple { get; }
}
```

路由参数是路由模式的一个重要组成部分，RoutePattern 的 Parameters 属性返回的 RoutePatternParameterPart 列表是对所有路由参数的描述。路由参数的默认值会存放在 Defaults 属性表示的字典中，该字典对象的 Key 为路由参数的名称。RoutePattern 的 ParameterPolicies 属性同样返回一个字典对象，针对每个路由参数的参数策略被存放到该字典中。借助 RoutePattern 类型的 GetParameter 方法，我们可以通过指定路由参数的名称得到对应的 RoutePatternParameterPart 对象。

应用具有一个全局的路由表，其中包含若干注册的通过 RoutePattern 表示的路由模式，无论是入栈方向上针对请求 URL 的路由解析，还是出栈方向上生成完整的 URL，都需要从这个路由表中选择一个匹配的模式。如果注册的路由很多，就可能出现多个路由在模式上都与当前上下文匹配的情况，在这种状况下就需要为注册的路由模式指定不同的匹配的权重或者优先选择一个匹配度最高的路由模式，RoutePattern 类型的 InboundPrecedence 属性和 OutboundPrecedence 属性分别代表当前路由模式针对两个路由方向上的匹配优先级，数值越大表示匹配度越高。

RoutePattern 属性和 RequiredValues 属性与出栈 URL 的生成相关。"weather/{city=010}/{days=4}" 是本章开篇实例演示中定义的一个路由模板，如果根据指定的路由参数值（city=010，days=4）生成一个完整的 URL，由于提供的路由参数值为默认值，所以生成的如下所示的 3 个 URL 路径都是合法的。具体生成哪一种由 RequiredValues 属性来决定，该属性返回的字典中存放了生成 URL 时必须指定的路由参数默认值。

- weather。
- weather/010。
- weather/010/4。

RoutePatternFactory

静态类型 RoutePatternFactory 提供的一系列静态方法可以帮助我们根据路由模板字符串创建表示路由模式的 RoutePattern 对象。如下所示的 3 个静态 Parse 方法重载帮助我们根据指定的路由模板和其他相关数据，包括路由参数的默认值和参数策略，以及必需的路由参数值（对应 RoutePattern 的 RequiredValues 属性），生成了一个表示路由模式的 RoutePattern 对象。

```csharp
public static class RoutePatternFactory
{
    public static RoutePattern Parse(string pattern);
    public static RoutePattern Parse(string pattern, object defaults,
        object parameterPolicies);
    public static RoutePattern Parse(string pattern, object defaults,
        object parameterPolicies, object requiredValues);
    ...
}
```

下面通过一个简单的实例演示如何利用 RoutePatternFactory 对象解析指定的路由模板，并生成一个表示路由模式的 RoutePattern 对象。我们在一个 ASP.NET Core 应用程序中定义了如下所示的 Format 方法，该方法将指定的 RoutePattern 对象格式化成一个字符串。

```csharp
public class Program
{
    private static string Format(RoutePattern pattern)
    {
        var builder = new StringBuilder();
        builder.AppendLine($"RawText:{pattern.RawText}");
        builder.AppendLine($"InboundPrecedence:{pattern.InboundPrecedence}");
        builder.AppendLine($"OutboundPrecedence:{pattern.OutboundPrecedence}");
        var segments = pattern.PathSegments;
        builder.AppendLine("Segments");
        foreach (var segment in segments)
        {
            foreach (var part in segment.Parts)
            {
                builder.AppendLine($"\t{ToString(part)}");
            }
        }
        builder.AppendLine("Defaults");
        foreach (var @default in pattern.Defaults)
        {
            builder.AppendLine($"\t{@default.Key} = {@default.Value}");
        }

        builder.AppendLine("ParameterPolicies ");
        foreach (var policy in pattern.ParameterPolicies)
```

```csharp
        {
            builder.AppendLine($"\t{policy.Key} = {string.Join(',',
                policy.Value.Select(it => it.Content))}");
        }

        builder.AppendLine("RequiredValues");
        foreach (var required in pattern.RequiredValues)
        {
            builder.AppendLine($"\t{required.Key} = {required.Value}");
        }

        return builder.ToString();

        static string ToString(RoutePatternPart part)
        {
            if (part is RoutePatternLiteralPart literal)
            {
                return $"Literal: {literal.Content}";
            }
            if (part is RoutePatternSeparatorPart separator)
            {
                return $"Separator: {separator.Content}";
            }
            else
            {
                var parameter = (RoutePatternParameterPart)part;
                return $"Parameter: Name = {parameter.Name}; Default = {parameter.Default};
                    IsOptional = {parameter.IsOptional};
                    IsCatchAll = {parameter.IsCatchAll};
                    ParameterKind = {parameter.ParameterKind}";
            }
        }
    }
}
```

在如下所示的应用承载程序中，我们调用 RoutePatternFactory 类型的静态方法 Parse 解析指定的路由模板 "weather/{city:regex(^0\d{{2,3}}$)=010}/{days:int:range(1,4)=4}/{detailed?}"，并生成一个 RoutePattern 对象，该方法调用中还指定了 requiredValues 参数的值。我们调用 IApplicationBuilder 对象的 Run 方法注册了唯一的中间件，它会调用上面定义的 Format 方法将生成的 RoutePattern 对象格式化成字符串，并作为最终的响应内容。

```csharp
public class Program
{
    public static void Main()
    {
        var template =
@"weather/{city:regex(^0\d{{2,3}}$)=010}/{days:int:range(1,4)=4}/{detailed?}";
        var pattern = RoutePatternFactory.Parse(
            pattern: template,
```

```
            defaults: null,
            parameterPolicies: null,
            requiredValues: new { city = "010", days = 4 });

    Host.CreateDefaultBuilder()
        .ConfigureWebHostDefaults(builder => builder
            .Configure(app => app.Run(context =>
                context.Response.WriteAsync(Format(pattern)))))
        .Build()
        .Run();
    }
}
```

如果利用浏览器访问启动后的应用程序，得到的输出结果如图 15-8 所示，该结果结构化地展示了路由模式的原始文本、出入栈路由匹配权重、每个段的组成、路由参数的默认值和参数策略，以及生成 URL 必须提供的默认参数值。（S1507）

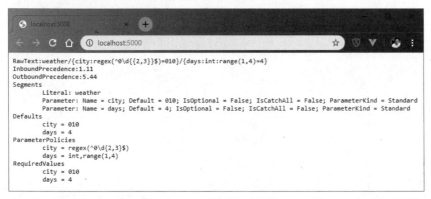

图 15-8　针对路由模式的解析

除了提供 Parse 方法解析指定的路由模板并生成表示路由模式的 RoutePattern 对象，RoutePatternFactory 还提供了用于解析其他与路由模式相关对象的静态方法，这些对象包括表示路径段的 RoutePatternPathSegment 对象、针对路由参数的 RoutePatternParameterPart 对象、针对参数策略的 RoutePatternParameterPolicyReference 对象等。由于篇幅有限，此处不再一一列举。

15.2.2　终结点

到目前为止，ASP.NET Core 提供了两种不同的路由解决方案。传统的路由系统以 IRouter 对象为核心，我们姑且将其称为 IRouter 路由。本章介绍的是最早发布于 ASP.NET Core 2.2 中的新路由系统，由于它采用基于终结点映射的策略，所以我们将其称为终结点路由。终结点路由自然以终结点为核心，所以先介绍终结点在路由系统中的表现形式。

之所以将应用划分为若干不同的终结点，是因为不同的终结点具有不同的请求处理方式。ASP.NET Core 应用可以利用 RequestDelegate 对象来表示 HTTP 请求处理器，每个终结点都封装了一个 RequestDelegate 对象并用它来处理路由给它的请求。如图 15-9 所示，除了请求处理器，终结点还提供了一个用来存放元数据的容器，路由过程中的很多行为都可以通过相应的元数据来控制。

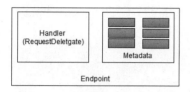

图 15-9　Endpoint = Handler + Metadata

Endpoint & EndpointBuilder

路由系统中的终结点通过如下所示的 Endpoint 类型表示。组成终结点的两个核心成员（请求处理器和元数据集合）分别体现为只读属性 RequestDelegate 和 Metadata。除此之外，终结点还有一个显示名称的只读属性 DisplayName。

```
public class Endpoint
{
    public string                               DisplayName { get; }
    public RequestDelegate                      RequestDelegate { get; }
    public EndpointMetadataCollection           Metadata { get; }

    public Endpoint(RequestDelegate requestDelegate, EndpointMetadataCollection metadata,
        string displayName);
}
```

终结点元数据集合体现为一个 EndpointMetadataCollection 对象。由于终结点并未对元数据的形式做任何限制，原则上任何对象都可以作为终结点的元数据，所以 EndpointMetadataCollection 对象本质上就是一个元素类型为 Object 的集合。如下面的代码片段所示，EndpointMetadataCollection 对象是一个只读列表，它包含的元数据需要在该集合被创建时被提供。

```
public sealed class EndpointMetadataCollection : IReadOnlyList<object>
{
    public object this[int index] { get; }
    public int Count { get; }

    public EndpointMetadataCollection(IEnumerable<object> items);
    public EndpointMetadataCollection(params object[] items);

    public Enumerator GetEnumerator();
    public T GetMetadata<T>() where T: class;
    public IReadOnlyList<T> GetOrderedMetadata<T>() where T: class;

    IEnumerator<object> IEnumerable<object>.GetEnumerator();
    IEnumerator IEnumerable.GetEnumerator();
}
```

我们可以调用泛型方法 GetMetadata<T>得到指定类型的元数据，由于多个具有相同类型的元数据可能会被添加到集合中，所以这个方法会采用"后来居上"的策略，返回最后被添加的元数据对象。如果没有指定类型的元数据，该方法会返回指定类型的默认值。如果希望按序返回指定类型的所有元数据，可以调用另一个泛型方法 GetOrderedMetadata<T>。

路由系统利用 EndpointBuilder 来构建表示终结点的 Endpoint 对象。如下面的代码片段所示，EndpointBuilder 是一个抽象类，针对终结点的构建体现在抽象的 Build 方法中。EndpointBuilder 定义了对应的属性来设置终结点的请求处理器、元数据和显示名称。

```
public abstract class EndpointBuilder
{
    public RequestDelegate      RequestDelegate { get; set; }
    public string               DisplayName { get; set; }
    public IList<object>        Metadata { get; }

    public abstract Endpoint Build();
}
```

RouteEndpoint & RouteEndpointBuilder

路由系统的终结点体现为一个 RouteEndpoint，它实际上是将映射的路由模式融入终结点中。如下面的代码片段所示，派生于 Endpoint 的 RouteEndpoint 类型有一个名为 RoutePattern 的只读属性，返回的正是表示路由模式的 RoutePattern 对象。除此之外，RouteEndpoint 类型还有另一个表示注册顺序的 Order 属性。

```
public sealed class RouteEndpoint : Endpoint
{
    public RoutePattern     RoutePattern { get; }
    public int              Order { get; }

    public RouteEndpoint(RequestDelegate requestDelegate, RoutePattern routePattern,
        int order, EndpointMetadataCollection metadata, string displayName);
}
```

RouteEndpoint 对象由 RouteEndpointBuilder 构建而成。如下面的代码片段所示，RouteEndpointBuilder 类型派生于抽象基类 EndpointBuilder。在重写的 Build 方法中，RouteEndpointBuilder 类型根据构造函数或者属性指定的信息创建出返回的 RouteEndpoint 对象。

```
public sealed class RouteEndpointBuilder : EndpointBuilder
{
    public RoutePattern     RoutePattern { get; set; }
    public int              Order { get; set; }

    public RouteEndpointBuilder(RequestDelegate requestDelegate, RoutePattern routePattern,
        int order)
    {
        base.RequestDelegate        = requestDelegate;
        RoutePattern                = routePattern;
        Order                       = order;
    }

    public override Endpoint Build()
        => new RouteEndpoint(base.RequestDelegate, RoutePattern, Order,
            new EndpointMetadataCollection((IEnumerable<object>) base.Metadata),
```

```
        base.DisplayName);
}
```

EndpointDataSource

路由系统中的终结点体现了针对某类请求的处理方式,它们的来源具有不同的表现形式,终结点的数据源通过 EndpointDataSource 表示。如图 15-10 所示,一个 EndpointDataSource 对象可以提供多个表示终结点的 Endpoint 对象,为应用提供相应的 EndpointDataSource 对象是路由注册的一项核心工作。

图 15-10　EndpointDataSource→Endpoints

如下面的代码片段所示,EndpointDataSource 是一个抽象类,除了表示提供终结点列表的只读属性 Endpoints,它还提供了一个 GetChangeToken 方法,我们可以利用这个方法返回的 IChangeToken 对象来感知数据源的变化。

```
public abstract class EndpointDataSource
{
    public abstract IReadOnlyList<Endpoint> Endpoints { get; }
    public abstract IChangeToken GetChangeToken();
}
```

路由系统提供了一个 DefaultEndpointDataSource 类型。如下面的代码片段所示,DefaultEndpointDataSource 通过重写 Endpoints 属性提供的终结点列表在构造函数中是显式指定的,其 GetChangeToken 方法返回的是一个不具有感知能力的 NullChangeToken 对象。

```
public sealed class DefaultEndpointDataSource : EndpointDataSource
{
    private readonly IReadOnlyList<Endpoint> _endpoints;
    public override IReadOnlyList<Endpoint> Endpoints => _endpoints;

    public DefaultEndpointDataSource(IEnumerable<Endpoint> endpoints)
        =>_endpoints = (IReadOnlyList<Endpoint>) new List<Endpoint>(endpoints);

    public DefaultEndpointDataSource(params Endpoint[] endpoints)
        =>_endpoints = (Endpoint[]) endpoints.Clone();

    public override IChangeToken GetChangeToken()
        => NullChangeToken.Singleton;
}
```

对于本章开篇演示的一系列路由实例来说,我们最终注册的实际上是一个类型为 ModelEndpointDataSource 的终结点数据源,它依然是一个未被公开的内部类型。要理解 ModelEndpointDataSource 针对终结点的提供机制,就必须了解另一个名为 IEndpointConventionBuilder 的接口。顾名思义,IEndpointConventionBuilder 体现了一种针对"约定"的终结点构建方式。

如下面的代码片段所示,该接口定义了一个唯一的 Add 方法,针对终结点构建的约定体现在该方法类型为Action<EndpointBuilder>的参数上。IEndpointConventionBuilder接口还有如下所示的3个扩展方法，用来为构建的终结点设置显示名称和元数据。

```
public interface IEndpointConventionBuilder
{
    void Add(Action<EndpointBuilder> convention);
}

public static class RoutingEndpointConventionBuilderExtensions
{
    public static TBuilder WithDisplayName<TBuilder>(this TBuilder builder,
        Func<EndpointBuilder, string> func) where TBuilder : IEndpointConventionBuilder
    {
        builder.Add(it=>it.DisplayName = func(it));
        return builder;
    }

    public static TBuilder WithDisplayName<TBuilder>(this TBuilder builder,
        string displayName) where TBuilder : IEndpointConventionBuilder
    {
        builder.Add(it => it.DisplayName = displayName);
        return builder;
    }
    public static TBuilder WithMetadata<TBuilder>(this TBuilder builder,
        params object[] items) where TBuilder : IEndpointConventionBuilder
    {
        builder.Add(it => Array.ForEach(items, item => it.Metadata.Add(item)));
        return builder;
    }
}
```

ModelEndpointDataSource 这个终结点数据源内部会使用一个名为 DefaultEndpointConventionBuilder 的类型，如下所示的代码片段给出了这两个类型的完整实现。从给出的代码片段可以看出，ModelEndpointDataSource 的 GetChangeToken 方法返回的依然是一个不具有感知能力的 NullChangeToken 对象。

```
internal class DefaultEndpointConventionBuilder : IEndpointConventionBuilder
{
    private readonly List<Action<EndpointBuilder>> _conventions;
    internal EndpointBuilder EndpointBuilder { get; }

    public DefaultEndpointConventionBuilder(EndpointBuilder endpointBuilder)
    {
        EndpointBuilder = endpointBuilder;
        _conventions = new List<Action<EndpointBuilder>>();
    }

    public void Add(Action<EndpointBuilder> convention)
        => _conventions.Add(convention);
```

```
    public Endpoint Build()
    {
        foreach (var convention in _conventions)
        {
            convention(EndpointBuilder);
        }
        return EndpointBuilder.Build();
    }
}

internal class ModelEndpointDataSource : EndpointDataSource
{
    private List<DefaultEndpointConventionBuilder> _endpointConventionBuilders;

    public ModelEndpointDataSource()
        => _endpointConventionBuilders = new List<DefaultEndpointConventionBuilder>();

    public IEndpointConventionBuilder AddEndpointBuilder(EndpointBuilder endpointBuilder)
    {
        var builder = new DefaultEndpointConventionBuilder(endpointBuilder);
        _endpointConventionBuilders.Add(builder);
        return builder;
    }

    public override IChangeToken GetChangeToken()=> NullChangeToken.Singleton;
    public override IReadOnlyList<Endpoint> Endpoints
        => _endpointConventionBuilders.Select(it => it.Build()).ToArray();
}
```

综上所述，ModelEndpointDataSource 最终采用图 15-11 所示的方式来提供终结点。当我们调用其 AddEndpointBuilder 方法为它添加一个 EndpointBuilder 对象时，它会利用这个 EndpointBuilder 对象创建一个 DefaultEndpointConventionBuilder 对象。DefaultEndpointConventionBuilder 针对终结点的构建最终还是落在 EndpointBuilder 对象上。

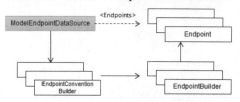

图 15-11　ModelEndpointDataSource→Endpoints

除了上述 ModelEndpointDataSource/DefaultEndpointConventionBuilder 类型，ASP.NET Core MVC 和 Razor Pages 框架分别根据自身的路由约定提供了针对 EndpointDataSource 和 IEndpointConventionBuilder 的实现。路由系统还提供了如下所示的 CompositeEndpointDataSource 类型。顾名思义，一个 CompositeEndpointDataSource 对象实际上是对一组 EndpointDataSource 对象的组合，它重写的 Endpoints 属性返回的终结点由作为组成成员的 EndpointDataSource 对象共

同提供。它的 GetChangeToken 方法返回的 IChangeToken 对象可以帮助我们感知其中任何一个 EndpointDataSource 对象的改变。

```
public sealed class CompositeEndpointDataSource : EndpointDataSource
{
    public IEnumerable<EndpointDataSource>   DataSources { get; }
    public override IReadOnlyList<Endpoint>  Endpoints { get; }

    public CompositeEndpointDataSource(
        IEnumerable<EndpointDataSource> endpointDataSources);
    public override IChangeToken GetChangeToken();
}
```

IEndpointRouteBuilder

表示终结点数据源的 EndpointDataSource 对象是借助 IEndpointRouteBuilder 对象注册的。我们可以在一个 IEndpointRouteBuilder 对象上注册多个 EndpointDataSource 对象，它们会被添加到 DataSources 属性表示的集合中。IEndpointRouteBuilder 接口还通过只读属性 ServiceProvider 提供了作为依赖注入容器的 IServiceProvider 对象。

```
public interface IEndpointRouteBuilder
{
    ICollection<EndpointDataSource>   DataSources { get; }
    IServiceProvider                  ServiceProvider { get; }

    IApplicationBuilder CreateApplicationBuilder();
}
```

IEndpointRouteBuilder 接口的 CreateApplicationBuilder 方法会帮助我们创建一个新的 IApplicationBuilder 对象。如果某个终结点针对请求处理的逻辑相对复杂，需要多个终结点协同完成，就可以将这些中间件注册到这个 IApplicationBuilder 对象上，然后利用它创建的 RequestDelegate 对象来处理路由的请求。如下所示的内部类型 DefaultEndpointRouteBuilder 是对 IEndpointRouteBuilder 接口的默认实现。

```
internal class DefaultEndpointRouteBuilder : IEndpointRouteBuilder
{
    public ICollection<EndpointDataSource>   DataSources { get; }
    public IServiceProvider                  ServiceProvider
        => ApplicationBuilder.ApplicationServices;
    public IApplicationBuilder               ApplicationBuilder { get; }

    public DefaultEndpointRouteBuilder(IApplicationBuilder applicationBuilder)
    {
        ApplicationBuilder = applicationBuilder;
        DataSources = new List<EndpointDataSource>();
    }

    public IApplicationBuilder CreateApplicationBuilder() => ApplicationBuilder.New();
}
```

本节的内容以终结点为核心，表示终结点的 Endpoint 对象来源于通过 EndpointDataSource

对象表示的数据源，EndpointDataSource 对象注册到 IEndpointRouteBuilder 对象上。以 IEndpointRouteBuilder、EndpointDataSource 和 Endpoint 为核心的终结点模型体现在图 15-12 中。

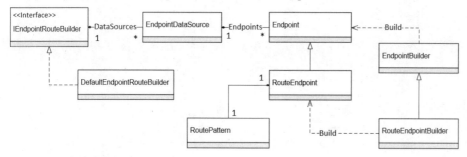

图 15-12　终结点数据流

15.2.3　中间件

针对终结点的路由是由 EndpointRoutingMiddleware 和 EndpointMiddleware 这两个中间件协同完成的。应用在启动之前会注册若干表示终结点的 Endpoint 对象（具体来说是包含路由模式的 RouteEndpoint 对象）。如图 15-13 所示，当应用接收到请求并创建 HttpContext 上下文之后，EndpointRoutingMiddleware 中间件会根据请求的 URL 及其他相关信息从注册的终结点中选择匹配度最高的那个。之后被选择的终结点会以一个特性（Feature）的形式附加到当前 HttpContext 上下文中，EndpointMiddleware 中间件最终提供这个终结点并用它来处理当前请求。

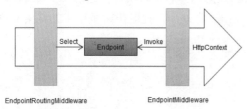

图 15-13　终结点的选择与执行

IEndpointFeature

EndpointRoutingMiddleware 中间件选择的终结点会以特性的形式存放在当前 HttpContext 上下文中，这个用来封装终结点的特性通过 IEndpointFeature 接口表示。如下面的代码片段所示，IEndpointFeature 接口通过唯一的属性 Endpoint 表示针对当前请求选择的终结点。我们可以针对 HttpContext 类型的 GetEndpoint 方法和 SetEndpoint 方法来获取与设置用来处理当前请求的终结点。

```
public interface IEndpointFeature
{
    Endpoint Endpoint { get; set; }
}

public static class EndpointHttpContextExtensions
{
    public static Endpoint GetEndpoint(this HttpContext context)
```

```
            =>context.Features.Get<IEndpointFeature>()?.Endpoint;
    public static void SetEndpoint(this HttpContext context, Endpoint endpoint)
    {
        var feature = context.Features.Get<IEndpointFeature>();
        if (feature != null)
        {
            feature.Endpoint = endpoint;
        }
        else
        {
            context.Features.Set<IEndpointFeature>(
                new EndpointFeature { Endpoint = endpoint });
        }
    }
    private class EndpointFeature : IEndpointFeature
    {
        public Endpoint Endpoint { get; set; }
    }
}
```

EndpointRoutingMiddleware

EndpointRoutingMiddleware 中间件利用一个 Matcher 对象选择出与当前 HttpContext 上下文相匹配的终结点，然后将选择的终结点以 IEndpointFeature 特性的形式附加到当前 HttpContext 上下文中。Matcher 只是一个内部抽象类型，针对终结点的选择和设置实现在它的 MatchAsync 方法中。如果匹配的终结点被成功选择出来，MatchAsync 方法还会提取出解析出来的路由参数，然后将它们逐个添加到表示当前请求的 HttpRequest 对象的 RouteValues 属性字典中。

```
internal abstract class Matcher
{
    public abstract Task MatchAsync(HttpContext httpContext);
}

public abstract class HttpRequest
{
    public virtual RouteValueDictionary RouteValues { get; set; }
}

public class RouteValueDictionary :
    IDictionary<string, object>, IReadOnlyDictionary<string, object>
{
    ...
}
```

EndpointRoutingMiddleware 中间件使用的 Matcher 由注册的 MatcherFactory 服务来提供。路由系统默认使用的 Matcher 类型为 DfaMatcher，它采用一种被称为确定有限状态自动机（Deterministic Finite Automaton，DFA）的形式从候选终结点中找到与当前请求匹配度最高的那个。由于篇幅有限，具体的细节此处不再展开介绍。DfaMatcher 最终会利用 DfaMatcherFactory

对象间接地创建出来，DfaMatcherFactory 类型派生于抽象类 MatcherFactory。

```
internal abstract class MatcherFactory
{
    public abstract Matcher CreateMatcher(EndpointDataSource dataSource);
}
```

对 Matcher 和 MatcherFactory 有了基本了解之后，我们将关注点转移到 EndpointRoutingMiddleware 中间件。如下所示的代码片段模拟了 EndpointRoutingMiddleware 中间件的实现逻辑。我们在构造函数中注入了用于提供注册终结点的 IEndpointRouteBuilder 对象和用来创建 Matcher 对象的 MatcherFactory 工厂。

```
internal class EndpointRoutingMiddleware
{
    private readonly RequestDelegate       _next;
    private readonly Task<Matcher>         _matcherAccessor;

    public EndpointRoutingMiddleware(RequestDelegate next,
        IEndpointRouteBuilder builder, MatcherFactory factory )
    {
        _next = next;
        _matcherAccessor = new Task<Matcher>(CreateMatcher);

        Matcher CreateMatcher()
        {
            var source = new CompositeEndpointDataSource(builder.DataSources);
            return factory.CreateMatcher(source);
        }
    }

    public async Task InvokeAsync(HttpContext httpContext)
    {
        var matcher = await _matcherAccessor;
        await matcher.MatchAsync(httpContext);
        await _next(httpContext);
    }
}
```

在实现的 InvokeAsync 方法中，我们只需要根据 IEndpointRouteBuilder 对象提供的终结点列表创建一个 CompositeEndpointDataSource 对象，并将其作为参数调用 MatcherFactory 工厂的 CreateMatcher 方法。该方法会返回一个 Matcher 对象，然后调用 Matcher 对象的 MatchAsync 方法选择出匹配的终结点，并以特性的方式附加到当前 HttpContext 上下文中。EndpointRoutingMiddleware 中间件一般通过如下所示的 UseRouting 扩展方法进行注册。

```
public static class EndpointRoutingApplicationBuilderExtensions
{
    public static IApplicationBuilder UseRouting(this IApplicationBuilder builder);
}
```

EndpointMiddleware

EndpointMiddleware 中间件的职责特别明确，就是执行由 EndpointRoutingMiddleware 中间件附加到当前 HttpContext 上下文中的终结点。EndpointRoutingMiddleware 中间件针对终结点的执行涉及如下所示的 RouteOptions 类型标识的配置选项。

```
public class RouteOptions
{
    public bool                              LowercaseUrls { get; set; }
    public bool                              LowercaseQueryStrings { get; set; }
    public bool                              AppendTrailingSlash { get; set; }
    public IDictionary<string, Type>         ConstraintMap { get; set; }

    public bool SuppressCheckForUnhandledSecurityMetadata { get; set; }
}
```

配置选项 RouteOptions 的前三个属性与路由系统针对 URL 的生成有关。具体来说，LowercaseUrls 属性和 LowercaseQueryStrings 属性决定是否会将生成的 URL 或者查询字符串转换成小写形式。AppendTrailingSlash 属性则决定是否会为生成的 URL 添加后缀 "/"。RouteOptions 的 ConstraintMap 属性表示的字典与路由参数的内联约束有关，它提供了在路由模板中实现的约束字符串（如 regex 表示正则表达式约束）与对应约束类型（正则表达式约束类型为 RegexRouteConstraint）之间的映射关系。

真正与 EndpointMiddleware 中间件相关的是 RouteOptions 的 SuppressCheckForUnhandledSecurityMetadata 属性，它表示目标终结点利用添加的元数据设置了一些关于安全方面的要求（主要是授权和跨域资源共享方面的要求），但是目前的请求并未经过相应的中间件处理（通过请求是否具有要求的报头判断），在这种情况下是否还有必要继续执行目标终结点。如果这个属性设置为 True，就意味着 EndpointMiddleware 中间件根本不会做这方面的检验。如下所示的代码片段模拟了 EndpointMiddleware 中间件对请求的处理逻辑。

```
internal class EndpointMiddleware
{
    private readonly RequestDelegate _next;
    private readonly RouteOptions    _options;

    public EndpointMiddleware(RequestDelegate next,IOptions<RouteOptions> optionsAccessor)
    {
        _next           = next;
        _options        = optionsAccessor.Value;
    }

    public Task InvokeAsync(HttpContext httpContext)
    {
        var endpoint = httpContext.GetEndpoint();
        if (null != endpoint)
        {
            if (!_options.SuppressCheckForUnhandledSecurityMetadata)
            {
```

```
            CheckSecurity();
        }
        return endpoint.RequestDelegate(httpContext);
    }
    return _next(httpContext);
}

private void CheckSecurity();
```

我们一般调用如下所示的 UseEndpoints 扩展方法来注册 EndpointMiddleware 中间件，该方法提供了一个类型为 Action<IEndpointRouteBuilder>的参数。通过前面的介绍可知，EndpointRoutingMiddleware 中间件会利用注入的 IEndpointRouteBuilder 对象来获取注册的表示终结点数据源的 EndpointDataSource，所以可以通过这个方法为 EndpointRoutingMiddleware 中间件注册终结点数据源。

```
public static class EndpointRoutingApplicationBuilderExtensions
{
    public static IApplicationBuilder UseEndpoints(this IApplicationBuilder builder,
        Action<IEndpointRouteBuilder> configure);
}
```

注册终结点

对于使用路由系统的应用程序来说，它的主要工作基本集中在针对 EndpointDataSource 的注册上。一般来说，当我们调用 IApplicationBuilder 接口的 UseEndpoints 扩展方法注册 Endpoint Middleware 中间件时，会利用提供的 Action<IEndpointRouteBuilder>委托对象注册所需的 EndpointDataSource 对象。IEndpointRouteBuilder 接口具有一系列的扩展方法，这些方法可以帮助我们注册所需的终结点。

如下所示的 Map 方法会根据提供的作为路由模式和处理器的 RoutePattern 对象与 RequestDelegate 对象创建一个终结点，并以 ModelEndpointDataSource 的形式予以注册。如下所示的代码片段还揭示了一个细节：对于作为请求处理器的 RequestDelegate 委托对象来说，其对应方法上标注的所有特性会以元数据的形式添加到创建的终结点上。

```
public static class EndpointRouteBuilderExtensions
{
    public static IEndpointConventionBuilder Map(this IEndpointRouteBuilder endpoints,
        RoutePattern pattern, RequestDelegate requestDelegate)
    {
        var builder = new RouteEndpointBuilder(requestDelegate, pattern, 0)
        {
            DisplayName = pattern.RawText
        };
        var attributes = requestDelegate.Method.GetCustomAttributes();
        if (attributes != null)
        {
            foreach (var attribute in attributes)
            {
```

```csharp
            builder.Metadata.Add(attribute);
        }
    }
    var dataSource = endpoints.DataSources
        .OfType<ModelEndpointDataSource>().FirstOrDefault()
        ?? new ModelEndpointDataSource();
    endpoints.DataSources.Add(dataSource);
    return dataSource.AddEndpointBuilder(builder);
}
```

HTTP 方法（Method）在 RESTful API 的设计中具有重要意义，几乎所有的终结点都会根据自身对资源的操作类型对请求采用 HTTP 方法做相应限制。如果需要为注册的终结点指定限定的 HTTP 方法，就可以调用如下所示的 MapMethods 方法。该方法会在 Map 方法的基础上为注册的终结点设置相应的显示名称，并针对指定的 HTTP 方法创建一个 HttpMethodMetadata 对象，然后作为元数据添加到注册的终结点上。

```csharp
public static class EndpointRouteBuilderExtensions
{
    public static IEndpointConventionBuilder MapMethods(
        this IEndpointRouteBuilder endpoints, string pattern,
        IEnumerable<string> httpMethods, RequestDelegate requestDelegate)
    {
        var builder = endpoints.Map(RoutePatternFactory.Parse(pattern), requestDelegate);
        builder.WithDisplayName($"{pattern} HTTP: {string.Join(", ", httpMethods)}");
        builder.WithMetadata(new HttpMethodMetadata(httpMethods));
        return builder;
    }
}
```

EndpointRoutingMiddleware 中间件在为当前请求筛选匹配的终结点时，针对 HTTP 方法的选择策略是通过 IHttpMethodMetadata 接口表示的元数据指定的，HttpMethodMetadata 类型正是对该接口的默认实现。如下面的代码片段所示，IHttpMethodMetadata 接口除了具有一个表示可接受 HTTP 方法列表的 HttpMethods 属性，还有一个布尔类型的只读属性 AcceptCorsPreflight，它表示是否接受针对跨域资源共享（Cross-Origin Resource Sharing，CORS）的预检（Preflight）请求。

```csharp
public interface IHttpMethodMetadata
{
    IReadOnlyList<string>     HttpMethods { get; }
    bool                      AcceptCorsPreflight { get; }
}

public sealed class HttpMethodMetadata : IHttpMethodMetadata
{
    public IReadOnlyList<string>     HttpMethods { get; }
    public bool                      AcceptCorsPreflight { get; }

    public HttpMethodMetadata(IEnumerable<string> httpMethods)
        : this(httpMethods, acceptCorsPreflight: false)
    {}
```

```csharp
public HttpMethodMetadata(IEnumerable<string> httpMethods, bool acceptCorsPreflight)
{
    HttpMethods = httpMethods.ToArray();
    AcceptCorsPreflight = acceptCorsPreflight;
}
}
```

路由系统还为 4 种常用的 HTTP 方法（GET、POST、PUT 和 DELETE）定义了相应的方法。从如下所示的代码片段可以看出，它们最终调用的都是 MapMethods 方法。我们在本章开篇演示的实例中正是调用其中的 MapGet 方法来注册终结点的。

```csharp
public static class EndpointRouteBuilderExtensions
{
    public static IEndpointConventionBuilder MapGet(this IEndpointRouteBuilder endpoints,
        string pattern, RequestDelegate requestDelegate)
        => MapMethods(endpoints, pattern, "GET", requestDelegate);
    public static IEndpointConventionBuilder MapPost(this IEndpointRouteBuilder endpoints,
        string pattern, RequestDelegate requestDelegate)
        => MapMethods(endpoints, pattern, "POST", requestDelegate);
    public static IEndpointConventionBuilder MapPut(this IEndpointRouteBuilder endpoints,
        string pattern, RequestDelegate requestDelegate)
        => MapMethods(endpoints, pattern, "PUT", requestDelegate);
    public static IEndpointConventionBuilder MapDelete(
        this IEndpointRouteBuilder endpoints, string pattern,
        RequestDelegate requestDelegate)
        => MapMethods(endpoints, pattern, "DELETE", requestDelegate);
}
```

调用 IApplicationBuilder 接口相应的扩展方法注册 EndpointRoutingMiddleware 中间件和 EndpointMiddleware 中间件时，必须确保它们依赖的服务已经被注册到依赖注入框架之中。针对路由服务的注册可以通过调用如下所示的 AddRouting 扩展方法重载来完成。

```csharp
public static class RoutingServiceCollectionExtensions
{
    public static IServiceCollection AddRouting(this IServiceCollection services);
    public static IServiceCollection AddRouting(this IServiceCollection services,
        Action<RouteOptions> configureOptions);
}
```

15.3 路由约束

表示路由终结点的 RouteEndpoint 对象包含以 RoutePattern 对象表示的路由模式，某个请求能够被成功路由的前提是它满足某个候选终结点的路由模式所体现的路由规则。具体来说，这不仅要求当前请求的 URL 路径必须满足路由模板指定的路径模式，还需要具体的字符内容满足对应路由参数上定义的约束。

路由系统采用 IRouteConstraint 接口来表示路由约束，该接口具有唯一的 Match 方法，该方法用来验证 URL 携带的参数值是否有效。路由约束在表示路由模式的 RoutePattern 对象中是以路由参数策略的形式存储在 ParameterPolicies 属性中的，所以 IRouteConstraint 接口派生于 IParameterPolicy 接口。通过 IRouteConstraint 接口表示的路由约束同时兼容传统 IRouter 路由系统和最新的终结点路由系统，所以 Match 方法具有一个表示 IRouter 对象的 route 参数。

```
public interface IRouteConstraint : IParameterPolicy
{
    bool Match(HttpContext httpContext, IRouter route, string routeKey,
        RouteValueDictionary values, RouteDirection routeDirection);
}

public enum RouteDirection
{
    IncomingRequest,
    UrlGeneration
}
```

针对路由参数约束的检验同时应用在两个路由方向上，即针对入栈请求的路由解析和针对 URL 的生成，当前应用的路由方向通过 Match 方法的 routeDirection 参数表示。Match 方法的第一个参数 httpContext 表示当前 HttpContext 上下文，routeKey 参数表示的其实是路由参数名称。如果当前的路由方向为 IncomingRequest，那么 Match 方法的 values 参数就代表解析出来的所有路由参数值；否则，该参数代表为生成 URL 提供的路由参数值。一般来说，我们只需要利用 routeKey 参数提供的参数名从 values 参数表示的字典中提取出当前参数值，并根据对应的规则加以验证即可。

15.3.1 预定义的 IRouteConstraint

路由系统定义了一系列原生的 IRouteConstraint 实现类型，我们可以使用它们解决很多常见的约束问题。即使现有的 IRouteConstraint 实现类型无法满足某些特殊的约束需求，我们也可以通过实现 IRouteConstraint 接口创建自定义的约束类型。对于路由约束的应用，除了直接创建对应的 IRouteConstraint 对象，还可以采用内联的方式直接在路由模板中为某个路由参数定义相应的约束表达式。这些以表达式定义的约束类型其实对应着一种具体的 IRouteConstraint 类型。表 15-1 列举了内联约束类型与 IRouteConstraint 类型。

表 15-1　内联约束类型与 IRouteConstraint 类型

内联约束类型	IRouteConstraint 类型	说　　明
int	IntRouteConstraint	要求路由参数值能够解析为一个 int 整数，如{variable:int}
bool	BoolRouteConstraint	要求参数值可以解析为一个 bool 值，如{ variable:bool}
datetime	DateTimeRouteConstraint	要求参数值可以解析为一个 DateTime 对象（采用 CultureInfo.InvariantCulture 进行解析），如{ variable:datetime}
decimal	DecimalRouteConstraint	要求参数值可以解析为一个 decimal 数字，如{ variable:decimal}
double	DoubleRouteConstraint	要求参数值可以解析为一个 double 数字，如{ variable:double}
float	FloatRouteConstraint	要求参数值可以解析为一个 float 数字，如{ variable:float}
guid	GuidRouteConstraint	要求参数值可以解析为一个 Guid，如{ variable:guid}
long	LongRouteConstraint	要求参数值可以解析为一个 long 整数，如{ variable:long}

续表

内联约束类型	IRouteConstraint 类型	说明
minlength	MinLengthRouteConstraint	要求参数值表示的字符串不小于指定的长度，如{ variable: minlength(5)}
maxlength	MaxLengthRouteConstraint	要求参数值表示的字符串不大于指定的长度，如{ variable: maxlength(10)}
length	LengthRouteConstraint	要求参数值表示的字符串长度限于指定的区间范围，如{ variable: length(5,10)}
min	MinRouteConstraint	最小值，如{ variable:min(5)}
max	MaxRouteConstraint	最大值，如{ variable:max(10)}
range	RangeRouteConstraint	要求参数值介于指定的区间范围，如{variable:range(5,10)}
alpha	AlphaRouteConstraint	要求参数的所有字符都是字母，如{variable:alpha}
regex	RegexInlineRouteConstraint	要求参数值表示的字符串与指定的正则表达式相匹配，如{variable: regex(^d{0[0-9]{{2,3}-d{2}-d{4}$)}}}$)
required	RequiredRouteConstraint	要求参数值不应该是一个空字符串，如{variable:required}
file	FileNameRouteConstraint	要求参数值可以作为一个包含扩展名的文件名，如{variable:file}
nonfile	NonFileNameRouteConstraint	与 FileNameRouteConstraint 刚好相反，这两个约束类型旨在区分针对静态文件的请求

为了使读者对这些 IRouteConstraint 实现类型有更加深刻的理解，我们选择一个用于限制变量值范围的 RangeRouteConstraint 类进行单独介绍。如下面的代码片段所示，RangeRouteConstraint 类型具有两个长整型的只读属性 Max 和 Min，它们分别表示约束范围的上限和下限。

```
public class RangeRouteConstraint : IRouteConstraint
{
    public long Max { get; }
    public long Min { get; }

    public RangeRouteConstraint(long min, long max)
    {
        Min = min;
        Max = max;
    }

    public bool Match(HttpContext httpContext, IRouter route, string routeKey,
        RouteValueDictionary values, RouteDirection routeDirection)
    {
        if (values.TryGetValue(routeKey, out var value) && value != null)
        {
            var valueString = Convert.ToString(value, CultureInfo.InvariantCulture);
            if (long.TryParse(valueString, NumberStyles.Integer,
                CultureInfo.InvariantCulture, out var longValue))
            {
                return longValue >= Min && longValue <= Max;
            }
        }
        return false;
    }
}
```

具体的约束检验实现在 Match 方法中。RangeRouteConstraint 在该方法中会根据被检验变量的名称（对应 routeKey 参数）从参数 values（表示路由检验生成的所有路由参数）中提取被验证的参数值，然后判断它是否在 Max 属性和 Min 属性表示的数值范围内。

15.3.2　InlineConstraintResolver

由于在进行路由注册时针对路由变量的约束直接以内联表达式的形式定义在路由模板中，所以路由系统需要解析约束表达式来创建对应类型的 IRouteConstraint 对象，这项任务由 IInlineConstraintResolver 对象来完成。如下面的代码片段所示，IInlineConstraintResolver 接口定义了唯一的 ResolveConstraint 方法，实现了路由约束从字符串表达式到 IRouteConstraint 对象之间的转换。

```
public interface IInlineConstraintResolver
{
    IRouteConstraint ResolveConstraint(string inlineConstraint);
}
```

DefaultInlineConstraintResolver 类型是对 IInlineConstraintResolver 接口的默认实现，如下面的代码片段所示，DefaultInlineConstraintResolver 具有一个字典类型的字段_inlineConstraintMap，表 15-1 列举的内联约束类型与 IRouteConstraint 类型之间的映射关系就保存在这个字典中。

```
public class DefaultInlineConstraintResolver : IInlineConstraintResolver
{
    private readonly IDictionary<string, Type> _inlineConstraintMap;
    public DefaultInlineConstraintResolver(IOptions<RouteOptions> routeOptions)
        =>_inlineConstraintMap = routeOptions.Value.ConstraintMap;
    public virtual IRouteConstraint ResolveConstraint(string inlineConstraint);
}

public class RouteOptions
{
    public IDictionary<string, Type> ConstraintMap { get; set; }
    ...
}
```

在根据提供的内联约束表达式创建对应的 IInlineConstraintResolver 对象时，DefaultInlineConstraintResolver 会根据指定表达式获得以字符串表示的约束类型和参数列表。通过解析出来的约束类型名称，它可以从 ConstraintMap 属性表示的映射关系中得到对应的 IRouteConstraint 类型。接下来它根据参数个数得到匹配的构造函数，然后将字符串表示的参数转换成对应的参数类型，并以反射的形式将它们传入构造函数，进而创建出相应的 IHttpRouteConstraint 对象。

对于一个通过指定的路由模板创建的 Route 对象来说，它在初始化时会利用 IServiceProvider 获取这个 IInlineConstraintResolver 对象，并用它来解析定义在路由模板中的所有内联约束表达式，最后将它们全部转换成具体的 IRouteConstraint 对象。针对 IInlineConstraintResolver 的服务注册就实现在 IServiceCollection 接口的 AddRouting 扩展方法中。

15.3.3 自定义约束

我们可以使用上述这些预定义的 IRouteConstraint 实现类型完成一些常用的约束，但是在一些对路由参数具有特定约束的应用场景中，我们不得不创建自定义的约束类型。例如，如果需要对资源提供针对多语言的支持，最好的方式是在请求的 URL 中提供目标资源所针对的 Culture。为了确保包含在 URL 中的是一个合法有效的 Culture，最好为此定义相应的约束。

下面将通过一个简单的实例来演示如何创建这样一个用于验证 Culture 的自定义路由约束。但在此之前需要先介绍使用这个约束最终实现的效果。在本例中我们创建了一个提供基于不同语言资源的 Web API，简单起见，我们仅仅提供针对相应 Culture 的文本数据。可以将资源文件作为文本资源进行存储，如图 15-14 所示，我们在一个 ASP.NET Core 应用中创建了两个资源文件，即 Resources.resx（语言文化中性）和 Resources.zh.resx（中文），并定义了一个名为 hello 的文本资源条目。（S1508）

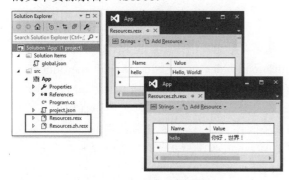

图 15-14　存储文本资源的两个资源文件

我们在演示程序中注册了一个模板为"resources/{lang:culture}/{resourceName:required}"的路由。路由参数{resourceName}表示获取的资源条目的名称（如 hello），这是一个必需的路由参数（路由参数应用了 RequiredRouteConstraint 约束）。另一个路由参数{lang}表示指定的语言，约束表达式名称 culture 对应的就是我们自定义的针对语言文化的约束类型 CultureConstraint。也正是因为这是一个自定义的路由约束，所以必须预先注册内联约束表达式名称 culture 和 CultureConstraint 类型之间的映射关系，在调用 AddRouting 方法时应将这样的映射添加到注册的 RouteOptions 之中。

```
public class Program
{
    public static void Main()
    {
        var template = "resources/{lang:culture}/{resourceName:required}";
        Host.CreateDefaultBuilder()
            .ConfigureWebHostDefaults(builder => builder
                .ConfigureServices(svcs => svcs
                    .AddRouting(options => options.ConstraintMap
                        .Add("culture",typeof(CultureConstraint))))
                .Configure(app => app
```

```
                .UseRouting()
                .UseEndpoints(routes => routes.MapGet(
                    template, BuildHandler(routes.CreateApplicationBuilder())))))
        .Build()
        .Run();
    static RequestDelegate BuildHandler(IApplicationBuilder app)
    {
        app.UseMiddleware<LocalizationMiddleware>("lang")
            .Run(async context =>
            {
                var values = context.GetRouteData().Values;
                var resourceName = values["resourceName"].ToString().ToLower();
                await context.Response.WriteAsync(
                Resources.ResourceManager.GetString(resourceName));
            });
        return app.Build();
    }
}
```

我们通过调用 UseEndpoints 扩展方法注册了路由终结点。该终结点的 RequestDelegate 对象是利用 IEndpointRouteBuilder 对象的 CreateApplicationBuilder 方法返回的 IApplicationBuilder 对象构建的。我们在这个 IApplicationBuilder 对象上注册了一个自定义的 LocalizationMiddleware 中间件，这个中间件可以实现针对多语言的本地化。至于资源内容的响应，我们将它实现在通过调用 IApplicationBuilder 对象的 Run 方法注册的中间件上。先从解析出来的路由参数中获取目标资源条目的名称，然后利用资源文件自动生成的 Resources 类型获取对应的资源内容并响应给客户端。

在揭秘自定义路由约束 CultureConstraint 及 LocalizationMiddleware 中间件的实现原理之前，需要先了解客户端采用什么样的形式获取某个资源条目针对某种语言的内容。如图 15-15 所示，直接利用浏览器采用与注册路由相匹配的 URL（"/resources/en/hello" 或者 "/resources/zh/hello"）不但可以获取目标资源的内容，而且显示的语言与我们指定的语言文化是一致的。如果指定一个不合法的语言（如 "xx"），将会违反我们自定义的约束，此时就会得到一个状态码为 "404 Not Found" 的响应。

图 15-15　采用相应的 URL 得到某个资源针对某种语言的内容

下面介绍针对语言文化的路由约束 CultureConstraint 究竟做了什么。如下面的代码片段所示，我们在 Match 方法中会试图获取作为语言文化内容的路由参数值，如果存在这样的路由参数，就可以利用它创建一个 CultureInfo 对象。如果这个 CultureInfo 对象的 EnglishName 属性名不以 "Unknown Language" 字符串作为前缀，我们就认为指定的是合法的语言文件。

```csharp
public class CultureConstraint : IRouteConstraint
{
    public bool Match(HttpContext httpContext, IRouter route, string routeKey,
        RouteValueDictionary values, RouteDirection routeDirection)
    {
        try
        {
            if (values.TryGetValue(routeKey, out object value))
            {
                return !new CultureInfo(value.ToString())
                    .EnglishName.StartsWith("Unknown Language");
            }
            return false;
        }
        catch
        {
            return false;
        }
    }
}
```

应用在运行的时候具有根据当前线程的语言文化属性选择对应匹配资源的能力。就我们演示实例提供的两个资源文件（Resources.resx 和 Resources.zh.resx）来说，如果当前线程的 UICulture 属性代表的是一个针对 "zh" 的语言文化，资源文件 Resources.zh.resx 就会被选择。对于其他语言文化，被选择的就是这个中性的 Resources.resx 文件。换句话说，如果要让应用程序选择某个我们希望的资源文件，就需要为当前线程设置相应的语言文化，实际上，LocalizationMiddleware 中间件就是这样做的。

```csharp
public class LocalizationMiddleware
{
    private readonly RequestDelegate        _next;
    private readonly string                 _routeKey;

    public LocalizationMiddleware(RequestDelegate next, string routeKey)
    {
        _next       = next;
        _routeKey   = routeKey;
    }

    public async Task InvokeAsync(HttpContext context)
    {
        var currentCulture = CultureInfo.CurrentCulture;
        var currentUICulture = CultureInfo.CurrentUICulture;
        try
```

```csharp
        {
            if (context.GetRouteData().Values.TryGetValue(_routeKey, out var culture))
            {
                CultureInfo.CurrentCulture = CultureInfo.CurrentUICulture =
                    new CultureInfo(culture.ToString());
            }
            await _next(context);
        }
        finally
        {
            CultureInfo.CurrentCulture = currentCulture;
            CultureInfo.CurrentUICulture = currentUICulture;
        }
    }
}
```

如上面的代码片段所示,LocalizationMiddleware 中间件的 InvokeAsync 方法被执行时,它会试图从路由参数中得到目标语言,代表路由参数名称的字段 _routeKey 是在构造函数中初始化的。如果存在这样的路由参数,它会据此创建一个 CultureInfo 对象并将其作为当前线程的 Culture 属性和 CultureInfo 属性。值得注意的是,在完成后续请求处理流程之后,我们需要将当前线程的语言文化恢复到之前的状态。

第 16 章

异常处理

由于 ASP.NET Core 是一个同时处理多个请求的 Web 应用框架，所以在处理某个请求过程中抛出的异常并不会导致整个应用的中止。出于安全方面的考量，为了避免敏感信息外泄，客户端在默认情况下并不会得到详细的出错信息，这无疑会在开发过程中增加查错和纠错的难度。对于生产环境来说，我们也希望最终用户能够根据具体的错误类型得到具有针对性并且友好的错误消息。ASP.NET Core 提供的相应的中间件可以帮助我们将定制化的错误信息呈现出来。

16.1 呈现错误信息

NuGet 包 "Microsoft.AspNetCore.Diagnostics" 中提供了几个与异常处理相关的中间件。当 ASP.NET Core 应用在处理请求过程中出现错误时，我们可以利用它们将原生的或者定制的错误信息作为响应内容发送给客户端。在着重介绍这些中间件之前，下面先演示几个简单的实例，从而使读者大致了解这些中间件的作用。

16.1.1 显示开发者异常页面

如果 ASP.NET Core 应用在处理某个请求时出现异常，它一般会返回一个状态码为 "500 Internal Server Error" 的响应。为了避免一些敏感信息的外泄，详细的错误信息并不会随着响应发送给客户端，所以客户端只会得到一个很泛化的错误消息。以如下所示的程序为例，它处理每个请求时都会抛出一个 InvalidOperationException 类型的异常。

```
public class Program
{
    public static void Main()
    {
        Host.CreateDefaultBuilder()
            .ConfigureWebHostDefaults(builder => builder.Configure(app => app.Run(
                context=> Task.FromException(
                new InvalidOperationException("Manually thrown exception...")))))
            .Build()
            .Run();
```

        }
}

利用浏览器访问这个应用总是会得到图 16-1 所示的错误页面。可以看出，这个页面仅仅告诉我们目标应用当前无法正常处理本次请求，除了提供的响应状态码（"HTTP ERROR 500"），它并没有提供任何有益于纠错的辅助信息。

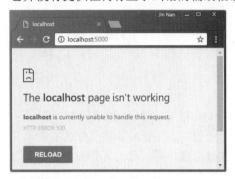

图 16-1　默认的错误页面

有人认为浏览器上虽然没有显示任何详细的错误信息，但这并不意味着 HTTP 响应报文中也没有携带任何详细的出错信息。实际上，针对通过浏览器发出的这个请求，服务端会返回如下这段 HTTP 响应报文。我们会发现响应报文根本没有主体部分，有限的几个报头也并没有承载任何与错误有关的信息。

```
HTTP/1.1 500 Internal Server Error
Date: Wed, 18 Sep 2019 23:38:59 GMT
Content-Length: 0
Server: Kestrel
```

由于应用并没有中断，浏览器上也并没有显示任何具有针对性的错误信息，开发人员在进行查错和纠错时如何准确定位到作为错误根源的那一行代码？这个问题有两种解决方案：一种是利用日志，因为 ASP.NET Core 应用在进行请求处理时出现的任何错误都会被写入日志，所以可以通过注册相应的 ILoggerProvider 对象来获取写入的错误日志，如可以注册一个 ConsoleLoggerProvider 对象将日志直接输出到宿主应用的控制台上。

另一种解决方案就是直接显示一个错误页面，由于这个页面只是在开发环境给开发人员看的，所以可以将这个页面称为开发者异常页面（Developer Exception Page）。开发者异常页面的呈现是利用一个名为 DeveloperExceptionPageMiddleware 的中间件完成的，我们可以采用如下所示的方式调用 IApplicationBuilder 接口的 UseDeveloperExceptionPage 扩展方法来注册这个中间件。

```
public class Program
{
    public static void Main()
    {
        Host.CreateDefaultBuilder()
            .ConfigureServices(svcs => svcs.AddRouting())
```

```
        .ConfigureWebHostDefaults(builder => builder.Configure(app => app
            .UseDeveloperExceptionPage()
            .UseRouting()
            .UseEndpoints(endpoints => endpoints.MapGet("{foo}/{bar}", HandleAsync))))
        .Build()
        .Run();

    static Task HandleAsync(HttpContext httpContext)
        => Task.FromException(new InvalidOperationException(
        "Manually thrown exception..."));
    }
}
```

一旦注册了 DeveloperExceptionPageMiddleware 中间件，ASP.NET Core 应用在处理请求过程中出现的异常信息就会以图 16-2 所示的形式直接出现在浏览器上，我们可以在这个页面中看到几乎所有的错误信息，包括异常的类型、消息和堆栈信息等。（S1601）

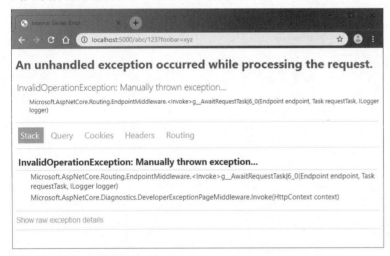

图 16-2　开发者异常页面（一）

开发者异常页面除了显示与抛出的异常相关的信息，还会以图 16-3 所示的形式显示与当前请求上下文相关的信息，其中包括当前请求 URL 携带的所有查询字符串、所有请求报头、Cookie 的内容和路由信息（终结点和路由参数）。如此详尽的信息无疑会极大地帮助开发人员尽快找出错误的根源。

通过 DeveloperExceptionPageMiddleware 中间件呈现的错误页面仅仅是供开发人员使用的，页面上往往会携带一些敏感的信息，所以只有在开发环境才能注册这个中间件，如下所示的代码片段体现了 Startup 类型中针对 DeveloperExceptionPageMiddleware 中间件正确的注册方式。

图16-3 开发者异常页面(二)

```
public class Startup
{
    public void Configure(IApplicationBuilder app, IWebHostEnvironment env)
    {
        if (env.IsDevelopment())
        {
            app.UseDeveloperExceptionPage();
        }
    }
}
```

16.1.2 显示定制异常页面

DeveloperExceptionPageMiddleware 中间件会将异常详细信息和基于当前请求的上下文直接呈现在错误页面中,这为开发人员的纠错诊断提供了极大的便利。但是在生产环境下,我们倾向于为最终的用户呈现一个定制的错误页面,这可以通过注册另一个名为 ExceptionHandlerMiddleware 的中间件来实现。顾名思义,这个中间件旨在提供一个异常处理器(Exception Handler)来处理抛出的异常。实际上,这个所谓的异常处理器就是一个 RequestDelegate 对象,ExceptionHandlerMiddleware 中间件捕捉到抛出的异常后利用它来处理当前的请求。

下面以上面创建的这个总是会抛出一个 InvalidOperationException 异常的应用为例进行介绍。

我们按照如下形式调用 IApplicationBuilder 接口的 UseExceptionHandler 扩展方法注册了 ExceptionHandlerMiddleware 中间件。这个扩展方法具有一个 ExceptionHandlerOptions 类型的参数，它的 ExceptionHandler 属性返回的就是这个作为异常处理器的 RequestDelegate 对象。

```csharp
public class Program
{
    public static void Main()
    {
        var options = new ExceptionHandlerOptions { ExceptionHandler = HandleAsync };
        Host.CreateDefaultBuilder()
            .ConfigureWebHostDefaults(builder => builder.Configure(app => app
                .UseExceptionHandler(options)
                .Run(context => Task.FromException(new InvalidOperationException(
                    "Manually thrown exception...")))))
            .Build()
            .Run();

        static Task HandleAsync(HttpContext context)
            => context.Response.WriteAsync("Unhandled exception occurred!");
    }
}
```

如上面的代码片段所示，这个作为异常处理器的 RequestDelegate 对象仅仅是将一个简单的错误消息（Unhandled exception occurred!）作为响应的内容。当我们利用浏览器访问该应用时，这个定制的错误消息会以图 16-4 所示的形式直接呈现在浏览器上。（S1602）

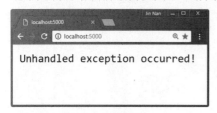

图 16-4　定制的错误页面

由于最终作为异常处理器的是一个 RequestDelegate 对象，而 IApplicationBuilder 对象具有根据注册的中间件来创建这个委托对象的能力，所以我们可以根据异常处理的需求将相应的中间件注册到某个 IApplicationBuilder 对象上，并最终利用它来创建作为异常处理器的 RequestDelegate 对象。如果异常处理需要通过一个或者多个中间件来完成，我们可以按照如下所示的形式调用另一个 UseExceptionHandler 方法重载。这个方法的参数类型为 Action<IApplicationBuilder>，我们调用它的 Run 方法注册了一个中间件来响应一个简单的错误消息。（S1603）

```csharp
public class Program
{
    public static void Main()
    {
        Host.CreateDefaultBuilder()
            .ConfigureWebHostDefaults(builder => builder.Configure(app => app
                .UseExceptionHandler(app2 => app2.Run(HandleAsync))
```

```
            .Run(context => Task.FromException(new InvalidOperationException(
                "Manually thrown exception...")))))
            .Build()
            .Run();

    static Task HandleAsync(HttpContext context)
        => context.Response.WriteAsync("Unhandled exception occurred!");
    }
}
```

上面这两种异常处理的形式都体现在提供一个 RequestDelegate 的委托对象来处理抛出的异常并完成最终的响应。如果应用已经设置了一个错误页面，并且这个错误页面有一个固定的路径，那么我们在进行异常处理的时候就没有必要提供这个 RequestDelegate 对象，只需要重定向到错误页面指向的路径即可。这种采用服务端重定向的异常处理方式可以采用如下所示的形式调用另一个 UseExceptionHandler 方法重载来完成，这个方法的参数表示的就是重定向的目标路径（"/error"），我们针对这个路径注册了一个路由来响应定制的错误消息。（S1604）

```
public class Program
{
    public static void Main()
    {
        Host.CreateDefaultBuilder()
            .ConfigureServices(svcs => svcs.AddRouting())
            .ConfigureWebHostDefaults(builder => builder.Configure(app => app
                .UseExceptionHandler("/error")
                .UseRouting()
                .UseEndpoints(endpoints => endpoints.MapGet("error", HandleAsync))
                .Run(context => Task.FromException(new InvalidOperationException(
                    "Manually thrown exception...")))))
            .Build()
            .Run();

    static Task HandleAsync(HttpContext context)
        => context.Response.WriteAsync("Unhandled exception occurred!");
    }
}
```

16.1.3　针对响应状态码定制错误页面

由于 Web 应用采用 HTTP 通信协议，所以我们应该尽可能迎合 HTTP 标准，并将定义在协议规范中的语义应用到程序中。异常或者错误的语义表达在 HTTP 协议层面主要体现在响应报文的状态码上，具体来说，HTTP 通信的错误大体分为如下两种类型。

- 客户端错误：表示因客户端提供不正确的请求信息而导致服务器不能正常处理请求，响应状态码的范围为 400～499。
- 服务端错误：表示服务器在处理请求过程中因自身的问题而发生错误，响应状态码的范围为 500～599。

正是因为响应状态码是对错误或者异常语义最重要的表达，所以在很多情况下我们需要针对不同的响应状态码来定制显示的错误信息。针对响应状态码对错误页面的定制可以借助一个 StatusCodePagesMiddleware 类型的中间件来实现，我们可以调用 IApplicationBuilder 接口相应的扩展方法来注册这个中间件。

DeveloperExceptionPageMiddleware 中间件和 ExceptionHandlerMiddleware 中间件都是在后续请求处理过程中抛出异常的情况下才会被调用的，而 StatusCodePagesMiddleware 中间件被调用的前提是后续请求处理过程中产生一个错误的响应状态码（范围为 400~599）。如果仅仅希望显示一个统一的错误页面，我们可以按照如下所示的形式调用 IApplicationBuilder 接口的 UseStatusCodePages 扩展方法注册这个中间件，传入该方法的两个参数分别表示响应采用的媒体类型和主体内容。

```
public class Program
{
    public static void Main()
    {
        Host.CreateDefaultBuilder()
            .ConfigureWebHostDefaults(webBuilder => webBuilder.Configure(app => app
                .UseStatusCodePages("text/plain", "Error occurred ({0})")
                .Run(context => Task.Run(() => context.Response.StatusCode = 500))))
            .Build()
            .Run();
    }
}
```

如上面的代码片段所示，应用程序在处理请求时总是将响应状态码设置为"500"，所以最终的响应内容将由注册的 StatusCodePagesMiddleware 中间件来提供。我们调用 UseStatusCodePages 方法时将响应的媒体类型设置为 text/plain，并将一段简单的错误消息作为响应的主体内容。值得注意的是，作为响应内容的字符串可以包含一个占位符（{0}），StatusCodePagesMiddleware 中间件最终会采用当前响应状态码来替换它。如果我们利用浏览器来访问这个应用，得到的错误页面如图 16-5 所示。（S1605）

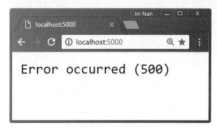

图 16-5　针对错误响应状态码定制的错误页面

如果我们希望针对不同的错误状态码显示不同的错误页面，那么就需要将具体的请求处理逻辑实现在一个状态码错误处理器中，并最终提供给 StatusCodePagesMiddleware 中间件。这个所谓的状态码错误处理器体现为一个 Func<StatusCodeContext, Task>类型的委托对象，作为输入的 StatusCodeContext 对象是对 HttpContext 上下文的封装，它同时承载着其他一些与错误处理相关的选项设置，我们将在本章后续部分对这个类型进行详细介绍。

对于如下所示的应用来说，它在处理任意一个请求时总是随机选择 400～599 的一个整数来作为响应的状态码，所以客户端返回的响应内容总是通过注册的 StatusCodePagesMiddleware 中间件来提供。在调用另一个 UseStatusCodePages 方法重载时，我们为注册的中间件指定一个 Func<StatusCodeContext, Task>对象作为状态码错误处理器。

```
public class Program
{
    private static readonly Random _random = new Random();
    public static void Main()
    {
        Host.CreateDefaultBuilder()
            .ConfigureWebHostDefaults(webBuilder => webBuilder.Configure(app => app
                .UseStatusCodePages(HandleAsync)
                .Run(context => Task.Run(
                    () => context.Response.StatusCode = _random.Next(400, 599)))))
            .Build()
            .Run();
    }

    static async Task HandleAsync(StatusCodeContext context)
    {
        var response = context.HttpContext.Response;
        if (response.StatusCode < 500)
        {
            await response.WriteAsync($"Client error ({response.StatusCode})");
        }
        else
        {
            await response.WriteAsync($"Server error ({response.StatusCode})");
        }
    }
}
```

我们指定的状态码错误处理器在处理请求时，根据响应状态码将错误分为客户端错误和服务端错误两种类型，并选择针对性的错误消息作为响应内容。当我们利用浏览器访问这个应用的时候，显示的错误消息将以图 16-6 所示的形式由响应状态码来决定。（S1606）

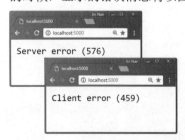

图 16-6　针对错误响应状态码定制的错误页面

在 ASP.NET Core 的世界里，针对请求的处理总是体现为一个 RequestDelegate 对象。如果请求的处理需要借助一个或者多个中间件来完成，就可以将它们注册到 IApplicationBuilder 对象上，并利用该对象将中间件管道转换成一个 RequestDelegate 对象。用于注册 StatusCodePagesMiddleware 中间件的 UseStatusCodePages 方法还有另一个重载，它允许我们采用这种方式来创建一个 RequestDelegate 对象来完成错误请求处理工作，所以上面演示的这个应用完全可以改写成如下形式。（S1607）

```csharp
public class Program
{
    private static readonly Random _random = new Random();
    public static void Main()
    {
        Host.CreateDefaultBuilder()
            .ConfigureWebHostDefaults(webBuilder => webBuilder.Configure(app => app
                .UseStatusCodePages(app2 => app2.Run(HandleAsync))
                .Run(context => Task.Run(
                    () => context.Response.StatusCode = _random.Next(400, 599)))))
            .Build()
            .Run();

        static async Task HandleAsync(HttpContext context)
        {
            var response = context.Response;
            if (response.StatusCode < 500)
            {
                await response.WriteAsync($"Client error ({response.StatusCode})");
            }
            else
            {
                await response.WriteAsync($"Server error ({response.StatusCode})");
            }
        }
    }
}
```

16.2 开发者异常页面

16.1 节通过几个简单的实例演示了如何呈现一个错误页面，该过程由 3 个对应的中间件来完成。下面先介绍用来呈现开发者异常页面的 DeveloperExceptionPageMiddleware 中间件，该中间件在捕捉到后续处理过程中抛出的异常之后会返回一个媒体类型为 text/html 的响应，后者在浏览器上会呈现一个错误页面。由于这是一个为开发者提供诊断信息的异常页面，所以可以将其称为开发者异常页面（Developer Exception Page）。该页面不仅会呈现异常的详细信息（类型、消息和跟踪堆栈等），还会出现与当前请求相关的上下文信息。如下所示的代码片段是 DeveloperExceptionPageMiddleware 中间件的定义。

```csharp
public class DeveloperExceptionPageMiddleware
```

```
{
    public DeveloperExceptionPageMiddleware(RequestDelegate next,
        IOptions<DeveloperExceptionPageOptions> options,
        ILoggerFactory loggerFactory, IWebHostEnvironment hostingEnvironment,
        DiagnosticSource diagnosticSource,
        IEnumerable<IDeveloperPageExceptionFilter> filters);

    public Task Invoke(HttpContext context);
}
```

如上面的代码片段所示，当我们创建一个 DeveloperExceptionPageMiddleware 对象的时候需要以参数的形式提供一个 IOptions<DeveloperExceptionPageOptions>对象，而 DeveloperExceptionPageOptions 对象携带着为这个中间件指定的配置选项，具体的配置选项体现在如下所示的两个属性（FileProvider 和 SourceCodeLineCount）上。

```
public class DeveloperExceptionPageOptions
{
    public IFileProvider         FileProvider { get; set; }
    public int                   SourceCodeLineCount { get; set; }
}
```

16.2.1　IDeveloperPageExceptionFilter

DeveloperExceptionPageMiddleware 中间件在默认情况下总是会呈现一个包含详细信息的错误页面，如果我们希望在呈现错误页面之前做一些额外的异常处理操作，或者希望完全按照自己的方式来处理异常，这个功能可以通过注册相应 IDeveloperPageExceptionFilter 对象的方式来实现。IDeveloperPageExceptionFilter 接口定义了如下所示的 HandleExceptionAsync 方法，用来实现自定义的异常处理操作。

```
public interface IDeveloperPageExceptionFilter
{
    Task HandleExceptionAsync(ErrorContext errorContext, Func<ErrorContext, Task> next);
}

public class ErrorContext
{
    public HttpContext        HttpContext { get; }
    public Exception          Exception { get; }

    public ErrorContext(HttpContext httpContext, Exception exception) ;
}
```

HandleExceptionAsync 方法提供的第一个参数是一个 ErrorContext 对象，它提供了当前的 HttpContext 上下文和抛出的异常。第二个参数表示的委托对象代表后续的异常操作，如果需要将抛出的异常分发给后续处理器做进一步处理，就需要显式地调用 Func<ErrorContext, Task>对象。在如下所示的演示实例中，我们通过实现 IDeveloperPageExceptionFilter 接口定义了一个 FakeExceptionFilter 类型，并将其注册到依赖注入框架中。

```
public class Program
{
```

```csharp
public static void Main()
{
    Host.CreateDefaultBuilder()
        .ConfigureWebHostDefaults(builder => builder
            .ConfigureServices(svcs=>svcs
                .AddSingleton<IDeveloperPageExceptionFilter, FakeExceptionFilter>())
            .Configure(app => app
                .UseDeveloperExceptionPage()
                .Run(context => Task.FromException(
                    new InvalidOperationException("Manually thrown exception...")))))
        .Build()
        .Run();
}

private class FakeExceptionFilter : IDeveloperPageExceptionFilter
{
    public Task HandleExceptionAsync(ErrorContext errorContext,
        Func<ErrorContext, Task> next)
        => errorContext.HttpContext.Response.WriteAsync(
            "Unhandled exception occurred!");
}
```

在 FakeExceptionFilter 类型实现的 HandleExceptionAsync 方法仅在响应的主体内容中写入了一条简单的错误消息（Unhandled exception occurred!），并没有显式调用该方法的参数 next 代表的"后续异常处理器"，所以 DeveloperExceptionPageMiddleware 中间件默认提供的错误页面并不会呈现出来，取而代之的就是图 16-7 所示的由注册 IDeveloperPageExceptionFilter 定制的错误页面。（S1608）

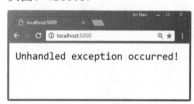

图 16-7　由注册 IDeveloperPageExceptionFilter 定制的错误页面

16.2.2　显示编译异常信息

我们编写的 ASP.NET Core 应用会先编译成程序集，然后部署并启动执行，为什么运行过程中还会出现"编译异常"？从 ASP.NET Core 应用层面来说，如果采用预编译模式，也就是说我们部署的不是源代码而是编译好的程序集，运行过程中根本就不存在编译异常的说法。但是在一个 ASP.NET Core MVC 应用中，视图文件（.cshtml）是支持动态运行时编译（Runtime Compilation）的。我们可以直接部署视图源文件，应用在执行过程中是可以动态地将它们编译成程序集的。换句话说，由于视图文件支持动态编译，所以可以在部署环境下直接修改视图文件的内容。

对于 DeveloperExceptionPageMiddleware 中间件来说，如果抛出的是普通的运行时异常，它会将异常自身的详细信息和当前请求上下文信息以 HTML 文档的形式呈现出来，前面演示的实例已经很好地说明了这一点。如果应用在动态编译视图文件时出现了编译异常，最终呈现出来的错误页面将具有不同的结构和内容，可以通过一个简单的实例演示 DeveloperExceptionPageMiddleware 中间件针对编译异常的处理。

为了支持运行时编译，我们需要为应用添加针对 NuGet 包 "Microsoft.AspNetCore.Mvc.Razor.RuntimeCompilation" 的依赖，并通过修改项目文件（.csproj）将 PreserveCompilationReferences 属性设置为 True，如下所示的代码片段是整个项目文件的定义。

```xml
<Project Sdk="Microsoft.NET.Sdk.Web">
  <PropertyGroup>
    <TargetFramework>netcoreapp3.0</TargetFramework>
    <PreserveCompilationReferences>true</PreserveCompilationReferences>
  </PropertyGroup>
  <ItemGroup>
    <PackageReference Include="Microsoft.AspNetCore.Mvc.Razor.RuntimeCompilation"
        Version="3.0.0" />
  </ItemGroup>
</Project>
```

我们通过如下所示的代码承载了一个 ASP.NET Core MVC 应用，并注册了 DeveloperExceptionPageMiddleware 中间件。为了支持针对 Razor 视图文件的运行时编译，在调用 IServiceCollection 接口的 AddControllersWithViews 扩展方法得到返回的 IMvcBuilder 对象之后，可以进一步调用该对象的 AddRazorRuntimeCompilation 扩展方法。

```csharp
public class Program
{
    public static void Main()
    {
        Host.CreateDefaultBuilder()
            .ConfigureWebHostDefaults(builder => builder
                .ConfigureServices(svcs => svcs
                    .AddRouting()
                    .AddControllersWithViews()
                    .AddRazorRuntimeCompilation())
                .Configure(app => app
                    .UseDeveloperExceptionPage()
                    .UseRouting()
                    .UseEndpoints(endpoints => endpoints.MapControllers())))
            .Build()
            .Run();
    }
}
```

我们定义了如下所示的 HomeController，它的 Action 方法 Index 会直接调用 View 方法将默认的视图呈现出来。根据约定，Action 方法 Index 呈现出来的视图文件对应的路径应该是 "~/views/home/index.cshtml"，我们为此在这个路径下创建了如下所示的视图文件。其中，

Foobar 是一个尚未被定义的类型。
```
public class HomeController : Controller
{
    [HttpGet("/")]
    public IActionResult Index() => View();
}

~/views/home/index.cshtml:
@{
    var value = new Foobar();
}
```

当我们利用浏览器访问 HomeController 的 Action 方法 Index 时，应用会动态编译目标视图。由于视图文件中使用了一个未定义的类型，动态编译会失败，响应的错误信息会以图 16-8 所示的形式出现在浏览器上。可以看出，错误页面显示的内容和结构与前面演示的实例是完全不一样的，我们不仅可以从这个错误页面中得到导致编译失败的视图文件的路径"Views/Home/Index.cshtml"，还可以直接看到导致编译失败的那一行代码。不仅如此，这个错误页面还直接将参与编译的源代码（不是定义在.cshtml 文件中的原始代码，而是经过转换处理生成的 C#代码）呈现出来。毫无疑问，如此详尽的错误页面对于开发人员的纠错是非常有价值的。（S1609）

图 16-8　显示在错误页面中的编译异常信息

一般来说，动态编译的过程如下：先将源代码（类似于.cshtml 这样的模板文件）转换成针对某种 .NET 语言（如 C#）的代码，然后进一步编译成 IL 代码。动态编译过程中抛出的异常类型一般会实现 ICompilationException 接口。如下面的代码片段所示，该接口具有一个唯一的属性 CompilationFailures，它返回一个元素类型为 CompilationFailure 的集合。编译失败的相关信息被封装在一个 CompilationFailure 对象之中，我们可以利用它得到源文件的路径（SourceFilePath）和内容（SourceFileContent），以及源代码转换后交付编译的内容。如果在内容转换过程已经发生错误，在这种情况下的 SourceFileContent 属性可能返回 Null。

```
public interface ICompilationException
{
    IEnumerable<CompilationFailure> CompilationFailures { get; }
}

public class CompilationFailure
{
    public string                          SourceFileContent { get; }
    public string                          SourceFilePath { get; }
    public string                          CompiledContent { get; }
    public IEnumerable<DiagnosticMessage>  Messages { get; }
    ...
}
```

CompilationFailure 类型还有一个名为 Messages 的只读属性，它返回一个元素类型为 DiagnosticMessage 的集合，一个 DiagnosticMessage 对象承载着一些描述编译错误的诊断信息。我们不仅可以借助 DiagnosticMessage 对象的相关属性得到描述编译错误的消息（Message 和 FormattedMessage），还可以得到发生编译错误所在源文件的路径（SourceFilePath）及范围，StartLine 属性和 StartColumn 属性分别表示导致编译错误的源代码在源文件中开始的行与列；EndLine 属性和 EndColumn 属性分别表示导致编译错误的源代码在源文件中结束的行与列（行数和列数分别从 1 与 0 开始计数）。

```
public class DiagnosticMessage
{
    public string    SourceFilePath { get; }
    public int       StartLine { get; }
    public int       StartColumn { get; }
    public int       EndLine { get; }
    public int       EndColumn { get; }

    public string    Message { get; }
    public string    FormattedMessage { get; }
    ...
}
```

从图 16-8 可以看出，错误页面会直接将导致编译失败的相关源代码显示出来。具体来说，它不仅将直接导致失败的源代码实现出来，还显示前后相邻的源代码。至于相邻源代码应该显示多少行，实际上是通过配置选项 DeveloperExceptionPageOptions 的 SourceCodeLineCount 属性控制的。

```
public class Program
{
    public static void Main()
    {
        var options = new DeveloperExceptionPageOptions { SourceCodeLineCount = 3 };
        Host.CreateDefaultBuilder()
            .ConfigureWebHostDefaults(builder => builder
                .ConfigureServices(svcs => svcs
                    .AddRouting()
                    .AddControllersWithViews()
                    .AddRazorRuntimeCompilation())
                .Configure(app => app
                    .UseDeveloperExceptionPage(options)
                    .UseRouting()
                    .UseEndpoints(endpoints => endpoints.MapControllers())))
            .Build()
            .Run();
    }
}
```

对于前面演示的这个实例来说，如果将前后相邻的 3 行代码显示在错误页面上，我们可以采用如上所示的方式为注册的 DeveloperExceptionPageMiddleware 中间件指定一个 DeveloperExceptionPageOptions 对象，并将它的 SourceCodeLineCount 属性设置为 3。与此同时，我们可以将视图文件（index.cshtml）改写成如下所示的形式，即在导致编译失败的那一行代码前后分别添加 4 行代码。

```
1:
2:
3:
4:
5:@{ var value = new Foobar();}
6:
7:
8:
9:
```

对于定义在视图文件中的 9 行代码，根据在注册 DeveloperExceptionPageMiddleware 中间件时指定的规则，最终显示在错误页面上的应该是第 2 行至第 8 行。如果利用浏览器访问相同的地址，这 7 行代码会以图 16-9 所示的形式出现在错误页面上。值得注意的是，如果我们没有对 SourceCodeLineCount 属性做显式设置，它的默认值为 6。（S1610）

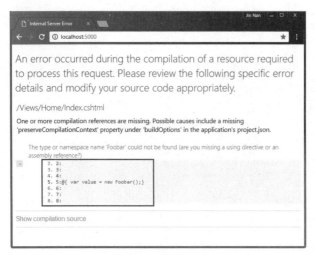

图 16-9　根据设置显示相邻源代码

16.2.3　DeveloperExceptionPageMiddleware

下面从 DeveloperExceptionPageMiddleware 类型的实现逻辑对该中间件针对异常页面的呈现做进一步讲解。如下所示的代码片段只保留了 DeveloperExceptionPageMiddleware 类型的核心代码，我们可以看到它的构造函数中注入了用来提供配置选项的 IOptions<DeveloperExceptionPageOptions>对象和一组 IDeveloperPageExceptionFilter 对象。

```csharp
public class DeveloperExceptionPageMiddleware
{
    private readonly RequestDelegate                    _next;
    private readonly DeveloperExceptionPageOptions      _options;
    private readonly Func<ErrorContext, Task>           _exceptionHandler;

    public DeveloperExceptionPageMiddleware(
        RequestDelegate next,
        IOptions<DeveloperExceptionPageOptions> options,
        ILoggerFactory loggerFactory,
        IWebHostEnvironment hostingEnvironment,
        DiagnosticSource diagnosticSource,
        IEnumerable<IDeveloperPageExceptionFilter> filters)
    {
        _next               = next;
        _options            = options.Value;
        _exceptionHandler   = context => context.Exception is ICompilationException
            ? DisplayCompilationException()
            : DisplayRuntimeException();
        ...

        foreach (var filter in filters.Reverse())
        {
```

```
            var nextFilter = _exceptionHandler;
            _exceptionHandler = errorContext =>
                filter.HandleExceptionAsync(errorContext, nextFilter);
        }
    }
    public async Task Invoke(HttpContext context)
    {
        try
        {
            await _next(context);
        }
        catch (Exception ex)
        {
            context.Response.Clear();
            context.Response.StatusCode = 500;
            await _exceptionHandler(new ErrorContext(context, ex));
            throw;
        }
    }
    private Task DisplayCompilationException();
    private Task DisplayRuntimeException();
}
```

被 DeveloperExceptionPageMiddleware 中间件用来作为异常处理器的是一个 Func<ErrorContext, Task>对象，通过字段_exceptionHandler 表示。当处理器在处理异常的时候，它会先调用注入的 IDeveloperPageExceptionFilter 对象，最后调用 DisplayRuntimeException 方法或者 DisplayCompilation Exception 方法来呈现"开发者异常页面"。如果某个注册的 IDeveloperPageExceptionFilter 阻止了后续的异常处理，整个处理过程将会就此中止。

在 Invoke 方法中，DeveloperExceptionPageMiddleware 中间件会直接将当前请求分发给后续的管道进行处理。如果抛出异常，它会根据该异常对象和当前 HttpContext 上下文创建一个 ErrorContext 对象，并将其作为参数调用作为异常处理器的 Func<ErrorContext, Task>委托对象。该中间件最终会回复一个状态码为"500 Internal Server Error"的响应。

我们一般调用 IApplicationBuilder 接口的如下所示的两个 UseDeveloperExceptionPage 扩展方法来注册 DeveloperExceptionPageMiddleware 中间件。我们可以利用作为配置选项的 DeveloperExceptionPageOptions 对象指定一个提供源文件的 IFileProvider 对象，也可以利用这个配置选项来控制导致异常源代码的前后行数。

```
public static class DeveloperExceptionPageExtensions
{
    public static IApplicationBuilder UseDeveloperExceptionPage(
        this IApplicationBuilder app)
        => app.UseMiddleware<DeveloperExceptionPageMiddleware>();

    public static IApplicationBuilder UseDeveloperExceptionPage(
        this IApplicationBuilder app,DeveloperExceptionPageOptions options)
```

```
        =>app.UseMiddleware<DeveloperExceptionPageMiddleware>(Options.Create(options));
}
```

16.3 异常处理器

DeveloperExceptionPageMiddleware 中间件错误页面可以呈现抛出的异常和当前请求上下文的详细信息，以辅助开发人员更好地进行纠错诊断工作。ExceptionHandlerMiddleware 中间件则主要面向最终用户，我们可以利用它来显示一个友好的定制化错误页面。

16.3.1 ExceptionHandlerMiddleware

由于 ExceptionHandlerMiddleware 中间件可以使用指定的 RequestDelegate 对象来作为异常处理器，所以我们可以将它视为一个"万能"的异常处理方案。按照惯例，下面先介绍 ExceptionHandlerMiddleware 类型的定义。

```
public class ExceptionHandlerMiddleware
{
    public ExceptionHandlerMiddleware(RequestDelegate next, ILoggerFactory loggerFactory,
        IOptions<ExceptionHandlerOptions> options, DiagnosticListener diagnosticListener);
    public Task Invoke(HttpContext context);
}

public class ExceptionHandlerOptions
{
    public RequestDelegate     ExceptionHandler { get; set; }
    public PathString          ExceptionHandlingPath { get; set; }
}
```

与 DeveloperExceptionPageMiddleware 类似，在创建一个 ExceptionHandlerMiddleware 对象时同样需要提供一个携带配置选项的对象，从上面的代码片段可以看出，配置选项由一个 ExceptionHandlerOptions 对象承载。一个 ExceptionHandlerOptions 对象通过其 ExceptionHandler 属性提供了一个作为异常处理器的 RequestDelegate 对象。如果希望应用在发生异常后自动重定向到某个指定的路径，该路径就可以利用 ExceptionHandlingPath 属性来指定。我们一般调用 IApplicationBuilder 接口的 UseExceptionHandler 扩展方法来注册 ExceptionHandlerMiddleware 中间件，这些重载的 UseExceptionHandler 扩展方法会采用如下方式完成中间件的注册工作。

```
public static class ExceptionHandlerExtensions
{
    public static IApplicationBuilder UseExceptionHandler(this IApplicationBuilder app)
        => app.UseMiddleware<ExceptionHandlerMiddleware>();

    public static IApplicationBuilder UseExceptionHandler(this IApplicationBuilder app,
        ExceptionHandlerOptions options)
        => app.UseMiddleware<ExceptionHandlerMiddleware>(Options.Create(options));

    public static IApplicationBuilder UseExceptionHandler(this IApplicationBuilder app,
        string errorHandlingPath)
        =>app.UseExceptionHandler(new ExceptionHandlerOptions
```

```
        {
            ExceptionHandlingPath = new PathString(errorHandlingPath)
        });
 
    public static IApplicationBuilder UseExceptionHandler(this IApplicationBuilder app,
        Action<IApplicationBuilder> configure)
    {
        IApplicationBuilder newBuilder = app.New();
        configure(newBuilder);

        return app.UseExceptionHandler(new ExceptionHandlerOptions
        {
            ExceptionHandler = newBuilder.Build()
        });
    }
}
```

ExceptionHandlerMiddleware 中间件处理请求的本质如下：在后续请求处理过程中出现异常的情况下，采用注册的异常处理器来处理当前请求，这个异常处理器就是 RequestDelegate 对象。该中间件采用的请求处理逻辑大体上可以通过如下所示的代码片段来体现。

```
public class ExceptionHandlerMiddleware
{
    private RequestDelegate            _next;
    private ExceptionHandlerOptions    _options;

    public ExceptionHandlerMiddleware(RequestDelegate next,
        IOptions<ExceptionHandlerOptions> options,...)
    {
        _next       = next;
        _options    = options.Value;
        ...
    }

    public async Task Invoke(HttpContext context)
    {
        try
        {
            await _next(context);
        }
        catch
        {
            context.Response.StatusCode = 500;
            context.Response.Clear();
            if (_options.ExceptionHandlingPath.HasValue)
            {
                context.Request.Path = _options.ExceptionHandlingPath;
            }
            var handler = _options.ExceptionHandler ?? _next;
            await handler(context);
        }
```

```
        }
}
```

如上面的代码片段所示，如果后续的请求处理过程中出现异常，ExceptionHandler Middleware 中间件会利用指定的作为异常处理器的 RequestDelegate 对象来完成最终的请求处理工作。如果创建 ExceptionHandlerMiddleware 对象时提供的 ExceptionHandlerOptions 对象携带了一个 RequestDelegate 对象，那么它将作为最终使用的异常处理器，否则作为异常处理器的实际上就是后续的中间件。换句话说，如果没有通过 ExceptionHandlerOptions 对象显式指定一个异常处理器，ExceptionHandlerMiddleware 中间件会在后续管道处理请求抛出异常的情况下将请求再次传递给后续管道。

在 ExceptionHandlerMiddleware 中间件利用异常处理器来处理请求之前，它会对请求做一些前置处理工作，其中包括将响应状态码设置为 500，并清空当前所有响应内容等。如果我们利用指定的 ExceptionHandlerOptions 对象的 ExceptionHandlingPath 属性设置了一个重定向路径，它会将该路径设置为当前请求的路径。除了包含前面代码片段的这些操作，Exception HandlerMiddleware 中间件实际上还执行了一些其他的操作。

16.3.2 异常的传递与请求路径的恢复

由于 ExceptionHandlerMiddleware 中间件总是利用一个作为异常处理器的 RequestDelegate 对象来完成最终的异常处理工作，为了使后者能够得到抛出的异常，该中间件应该采用某种方式将抛出的异常传递给它。除此之外，由于 ExceptionHandlerMiddleware 中间件会改变当前请求的路径，当整个请求处理完成之后，它必须将请求路径恢复成原始状态，否则前置的中间件就无法获取到正确的请求路径。

请求处理过程中抛出的异常和原始请求路径的恢复是通过相应的特性完成的。具体来说，传递这两者的特性分别通过 IExceptionHandlerFeature 接口和 IExceptionHandlerPathFeature 接口来表示。如下面的代码片段所示，后者继承前者，ExceptionHandlerFeature 类型同时实现了这两个接口。

```
public interface IExceptionHandlerFeature
{
    Exception Error { get; }
}

public interface IExceptionHandlerPathFeature : IExceptionHandlerFeature
{
    string Path { get; }
}

public class ExceptionHandlerFeature : IExceptionHandlerPathFeature,
{
    public Exception Error { get; set; }
    public string      Path { get; set; }
}
```

在 ExceptionHandlerMiddleware 中间件将代表当前请求的 HttpContext 上下文传递给处理器之前,它会按照如下所示的方式根据抛出的异常和原始请求路径创建一个 ExceptionHandlerFeature 对象,该对象最终被添加到 HttpContext 上下文的特性集合之中。当整个请求处理流程完全结束之后,ExceptionHandlerMiddleware 中间件会借助这个特性得到原始的请求路径,并将其重新应用到当前 HttpContext 上下文中。

```
public class ExceptionHandlerMiddleware
{
    ...
    public async Task Invoke(HttpContext context)
    {
        try
        {
            await _next(context);
        }
        catch(Exception ex)
        {
            context.Response.StatusCode = 500;

            var feature = new ExceptionHandlerFeature()
            {
                Error = ex,
                Path = context.Request.Path,
            };
            context.Features.Set<IExceptionHandlerFeature>(feature);
            context.Features.Set<IExceptionHandlerPathFeature>(feature);

            if (_options.ExceptionHandlingPath.HasValue)
            {
                context.Request.Path = _options.ExceptionHandlingPath;
            }
            RequestDelegate handler = _options.ExceptionHandler ?? _next;

            try
            {
                await handler(context);
            }
            finally
            {
                context.Request.Path = originalPath;
            }
        }
    }
}
```

在进行异常处理时,我们可以从当前 HttpContext 上下文中提取 ExceptionHandlerFeature 特性对象,进而获取抛出的异常和原始请求路径。如下面的代码片段所示,我们利用 HandleError 方法来呈现一个定制的错误页面。在这个方法中,我们正是借助 ExceptionHandlerFeature 特性得到抛出的异常的，并将其类型、消息及堆栈追踪信息显示出来。

```csharp
public class Program
{
    public static void Main()
    {
        Host.CreateDefaultBuilder()
            .ConfigureWebHostDefaults(builder => builder
                .ConfigureServices(svcs => svcs.AddRouting())
                .Configure(app => app
                    .UseExceptionHandler("/error")
                    .UseRouting()
                    .UseEndpoints(endpoints => endpoints.MapGet("error", HandleErrorAsync))
                    .Run(context => Task.FromException(new InvalidOperationException(
                        "Manually thrown exception")))))
            .Build()
            .Run();

        static async Task HandleErrorAsync(HttpContext context)
        {
            context.Response.ContentType = "text/html";
            var ex = context.Features.Get<IExceptionHandlerPathFeature>().Error;

            await context.Response.WriteAsync(
                "<html><head><title>Error</title></head><body>");
            await context.Response.WriteAsync($"<h3>{ex.Message}</h3>");
            await context.Response.WriteAsync($"<p>Type: {ex.GetType().FullName}");
            await context.Response.WriteAsync($"<p>StackTrace: {ex.StackTrace}");
            await context.Response.WriteAsync("</body></html>");
        }
    }
}
```

在上面这个应用中，我们注册了一个模板为"error"的路由指向 HandleError 方法。对于通过调用 UseExceptionHandler 扩展方法注册的 ExceptionHandlerMiddleware 中间件来说，我们将该路径设置为异常处理路径。对于任意从浏览器发出的请求，都会得到图 16-10 所示的错误页面。（S1611）

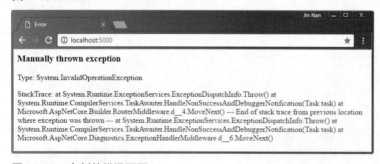

图 16-10　定制的错误页面

16.3.3 清除缓存

对于一个用于获取资源的 GET 请求来说，如果请求目标是一个相对稳定的资源，我们可以利用缓存避免相同资源的频繁获取和传输。对于作为资源提供者的 Web 应用来说，当它在处理请求的时候，除了将目标资源作为响应的主体内容，它还需要设置用于控制缓存的相关响应报头。由于缓存在大部分情况下只适用于成功状态的响应，如果服务端在处理请求过程中出现异常，之前设置的缓存报头是不应该出现在响应报文中的。对于 ExceptionHandlerMiddleware 中间件来说，清除缓存报头也是它负责的一项重要工作。

我们同样可以通过一个简单的实例来演示 ExceptionHandlerMiddleware 中间件针对缓存响应报头的清除。在如下所示的应用中，我们将针对请求的处理实现在 ProcessAsync 方法中，它有 50%的可能会抛出异常。不论是返回正常的响应内容还是抛出异常，这个方法都会先设置一个 Cache-Control 的响应报头，并将缓存时间设置为 1 小时（Cache-Control: max-age=3600）。

```
public class Program
{
    private static readonly Random _random = new Random();
    public static void Main()
    {
        Host.CreateDefaultBuilder()
            .ConfigureWebHostDefaults(builder => builder.Configure(app => app
                .UseExceptionHandler(app2 => app2.Run(HandleAsync))
                .Run(ProcessAsync)))
            .Build()
            .Run();

        static Task HandleAsync(HttpContext context)
            => context.Response.WriteAsync("Error occurred!");

        static async Task ProcessAsync(HttpContext context)
        {
            context.Response.GetTypedHeaders().CacheControl = new CacheControlHeaderValue
            {
                MaxAge = TimeSpan.FromHours(1)
            };

            if (_random.Next() % 2 == 0)
            {
                throw new InvalidOperationException("Manually thrown exception...");
            }
            await context.Response.WriteAsync("Succeed...");
        }
    }
}
```

通过调用 UseExceptionHandler 扩展方法注册的 ExceptionHandlerMiddleware 中间件在处理异常时会响应一个内容为"Error occurred!"的字符串。如下所示的两个响应报文分别对应正常响应和抛出异常的情况，我们会发现程序中设置的缓存报头 Cache-Control: max-age=3600 只会出现在状态码为 "200 OK" 的响应中。在状态码为 "500 Internal Server Error" 的响应中，则会出现 3 个与缓存相关的报头（Cache-Control、Pragma 和 Expires），它们的目的都是禁止缓存或者将缓存标识为过期。(S1612)

```
HTTP/1.1 200 OK
Date: Sat, 21 Sep 2019 11:25:27 GMT
Server: Kestrel
Cache-Control: max-age=3600
Content-Length: 10

Succeed...

HTTP/1.1 500 Internal Server Error
Date: Sat, 21 Sep 2019 11:26:11 GMT
Server: Kestrel
Cache-Control: no-cache
Pragma: no-cache
Expires: -1
Content-Length: 15

Error occurred!
```

ExceptionHandlerMiddleware 中间件针对缓存响应报头的清除体现在如下所示的代码片段中。可以看出，它通过调用 HttpResponse 对象的 OnStarting 方法注册了一个回调（ClearCacheHeaders），上述这 3 个缓存报头是在这个回调中设置的。除此之外，这个回调方法还会清除 ETag 报头。既然目标资源没有得到正常的响应，表示资源 "签名" 的 ETag 报头就不应该出现在响应报文中。

```
public class ExceptionHandlerMiddleware
{
    ...
    public async Task Invoke(HttpContext context)
    {
        try
        {
            await _next(context);
        }
        catch (Exception ex)
        {
            ...
            context.Response.OnStarting(ClearCacheHeaders, context.Response);
            var handler = _options.ExceptionHandler ?? _next;
            await handler(context);
        }
    }

    private Task ClearCacheHeaders(object state)
    {
        var response = (HttpResponse)state;
```

```
        response.Headers[HeaderNames.CacheControl]  = "no-cache";
        response.Headers[HeaderNames.Pragma]         = "no-cache";
        response.Headers[HeaderNames.Expires]        = "-1";
        response.Headers.Remove(HeaderNames.ETag);
        return Task.CompletedTask;
    }
}
```

16.4 响应状态码页面

StatusCodePagesMiddleware 中间件与 ExceptionHandlerMiddleware 中间件类似，它们都是在后续请求处理过程中"出错"的情况下利用一个错误处理器来接收针对当前请求的处理。它们之间的差异在于对"错误"的认定上：ExceptionHandlerMiddleware 中间件所谓的错误就是抛出异常；StatusCodePagesMiddleware 中间件则将 400~599 的响应状态码视为错误。如下面的代码片段所示，StatusCodePagesMiddleware 中间件也采用"标准"的定义方式，针对它的配置选项通过一个对应的对象以 Options 模式的形式提供给它。

```
public class StatusCodePagesMiddleware
{
    public StatusCodePagesMiddleware(RequestDelegate next,
        IOptions<StatusCodePagesOptions> options);
    public Task Invoke(HttpContext context);
}
```

除了对错误的认定方式，StatusCodePagesMiddleware 中间件和 ExceptionHandlerMiddleware 中间件对错误处理器的表达也不相同。ExceptionHandlerMiddleware 中间件的处理器是一个 RequestDelegate 委托对象，而 StatusCodePagesMiddleware 中间件的处理器则是一个 Func<StatusCodeContext, Task>委托对象。如下面的代码片段所示，配置选项 StatusCodePagesOptions 的唯一目的就是提供作为处理器的 Func<StatusCodeContext, Task>对象。

```
public class StatusCodePagesOptions
{
    public Func<StatusCodeContext, Task> HandleAsync { get; set; }
}
```

一个 RequestDelegate 对象相当于一个 Func<HttpContext, Task>类型的委托对象，而一个 StatusCodeContext 对象也是对一个 HttpContext 上下文的封装，这两个委托对象并没有本质上的不同。如下面的代码片段所示，除了从 StatusCodeContext 对象中获取当前 HttpContext 上下文，我们还可以通过其 Next 属性得到一个 RequestDelegate 对象，也可以利用它将请求再次分发给后续中间件进行处理。StatusCodeContext 对象的 Options 属性返回创建 StatusCodePagesMiddleware 中间件时指定的 StatusCodePagesOptions 对象。

```
public class StatusCodeContext
{
    public HttpContext              HttpContext     { get; }
    public RequestDelegate          Next            { get; }
    public StatusCodePagesOptions   Options         { get; }
```

```
    public StatusCodeContext(HttpContext context, StatusCodePagesOptions options,
        RequestDelegate next);
}
```

16.4.1　StatusCodePagesMiddleware

由于采用了针对响应状态码的错误处理策略，所以实现在 StatusCodePagesMiddleware 中间件的错误处理操作只会发生在当前响应状态码为 400~599 的情况下，如下所示的代码片段就体现了这一点。从下面给出的代码片段可以看出，StatusCodePagesMiddleware 中间件除了会查看当前响应状态码，还会查看响应内容及媒体类型。如果响应报文已经包含响应内容或者设置了媒体类型，StatusCodePagesMiddleware 中间件将不会执行任何操作，因为这正是后续中间件管道希望回复给客户端的响应，该中间件不应该再画蛇添足。

```
public class StatusCodePagesMiddleware
{
    private RequestDelegate        _next;
    private StatusCodePagesOptions _options;

    public StatusCodePagesMiddleware(RequestDelegate next,
        IOptions<StatusCodePagesOptions> options)
    {
        _next    = next;
        _options = options.Value;
    }

    public async Task Invoke(HttpContext context)
    {
        await _next(context);
        var response = context.Response;
        if ((response.StatusCode >= 400 && response.StatusCode <= 599) &&
            !response.ContentLength.HasValue &&
            string.IsNullOrEmpty(response.ContentType))
        {
            await _options.HandleAsync(new StatusCodeContext(context, _options, _next));
        }
    }
}
```

StatusCodePagesMiddleware 中间件对错误的处理非常简单，它只需要从 StatusCodePagesOptions 对象中提取出作为错误处理器的 Func<StatusCodeContext, Task>对象，然后创建一个 StatusCodeContext 对象作为输入参数调用这个委托对象即可。

16.4.2　阻止处理异常

通过 16.4.1 节的内容我们知道，如果某些内容已经被写入响应的主体部分，或者响应的媒体类型已经被预先设置，StatusCodePagesMiddleware 中间件就不会再执行任何错误处理操作。由于应用程序往往具有自身的异常处理策略，它们可能会显式地返回一个状态码为 400~599 的响应，在此情况下，StatusCodePagesMiddleware 中间件是不应该对当前响应做任何干预的。从这个意义上来讲，StatusCodePagesMiddleware 中间件仅仅是作为一种后备的错误处理机制而已。

更进一步来讲，如果后续的某个中间件返回了一个状态码为 400~599 的响应，并且这个响应只有报头集合没有主体（媒体类型自然也不会设置），那么按照我们在上面给出的错误处理逻辑来看，StatusCodePagesMiddleware 中间件还是会按照自己的策略来处理并响应请求。为了解决这种情况，我们必须赋予后续中间件能够阻止 StatusCodePagesMiddleware 中间件进行错误处理的功能。

阻止 StatusCodePagesMiddleware 中间件进行错误处理的功能是借助一个通过 IStatusCodePagesFeature 接口表示的特性来实现的。如下面的代码片段所示，IStatusCodePagesFeature 接口定义了唯一的 Enabled 属性，StatusCodePagesFeature 类型是对该接口的默认实现，它的 Enabled 属性默认返回 True。

```
public interface IStatusCodePagesFeature
{
    bool Enabled { get; set; }
}

public class StatusCodePagesFeature : IStatusCodePagesFeature
{
    public bool Enabled { get; set; } = true ;
}
```

StatusCodePagesMiddleware 中间件在将请求交付给后续管道之前，会创建一个 StatusCodePagesFeature 对象，并将其添加到当前 HttpContext 上下文的特性集合中。在最终决定是否执行错误处理操作的时候，它还会通过这个特性检验后续的某个中间件是否不希望其进行不必要的错误处理，如下所示的代码片段很好地体现了这一点。

```
public class StatusCodePagesMiddleware
{
    ...
    public async Task Invoke(HttpContext context)
    {
        var feature = new StatusCodePagesFeature();
        context.Features.Set<IStatusCodePagesFeature>(feature);

        await _next(context);
        var response = context.Response;
        if ((response.StatusCode >= 400 && response.StatusCode <= 599) &&
            !response.ContentLength.HasValue &&
            string.IsNullOrEmpty(response.ContentType) &&
            feature.Enabled)
        {
            await _options.HandleAsync(new StatusCodeContext(context, _options, _next));
        }
    }
}
```

下面通过一个简单的实例来演示如何利用 StatusCodePagesFeature 特性来屏蔽 StatusCodePagesMiddleware 中间件。在如下所示的代码片段中，我们将针对请求的处理定义在 ProcessAsync 方法中，该方法会返回一个状态码为"401 Unauthorized"的响应。我们通过随机数让这个方法在 50%的概率下利用 StatusCodePagesFeature 特性来阻止 StatusCodePagesMiddleware 中间件自身对错误的处理。我们通过调用 UseStatusCodePages 扩展方法注册的 StatusCodePagesMiddleware 中间件会直接响应一个内容为"Error occurred!"的字符串。

```csharp
public class Program
{
    private static readonly Random _random = new Random();
    public static void Main()
    {
        Host.CreateDefaultBuilder()
            .ConfigureWebHostDefaults(builder => builder.Configure(app => app
                .UseStatusCodePages(HandleAsync)
                .Run(ProcessAsync)))
            .Build()
            .Run();

        static Task HandleAsync(StatusCodeContext context)
            => context.HttpContext.Response.WriteAsync("Error occurred!");

        static Task ProcessAsync(HttpContext context)
        {
            context.Response.StatusCode = 401;
            if (_random.Next() % 2 == 0)
            {
                context.Features.Get<IStatusCodePagesFeature>().Enabled = false;
            }
            return Task.CompletedTask;
        }
    }
}
```

对于针对该应用的请求来说，我们会得到如下两种不同的响应。没有主体内容的响应是通过 ProcessAsync 方法产生的，这种情况发生在 StatusCodePagesMiddleware 中间件通过 StatusCodePagesFeature 特性被屏蔽的时候。有主体内容的响应则是 ProcessAsync 方法和 StatusCodePagesMiddleware 中间件共同作用的结果。（S1613）

```
HTTP/1.1 401 Unauthorized
Date: Sat, 21 Sep 2019 13:37:31 GMT
Server: Kestrel
Content-Length: 15

Error occurred!

HTTP/1.1 401 Unauthorized
Date: Sat, 21 Sep 2019 13:37:36 GMT
Server: Kestrel
Content-Length: 0
```

16.4.3 注册 StatusCodePagesMiddleware 中间件

我们在大部分情况下都会调用 IApplicationBuilder 接口相应的扩展方法来注册 StatusCodePagesMiddleware 中间件。对于 StatusCodePagesMiddleware 中间件的注册来说，除了 UseStatusCodePages 方法，还有其他方法可供选择。

UseStatusCodePages

我们可以调用如下所示的 3 个 UseStatusCodePages 扩展方法重载来注册 StatusCodePagesMiddleware 中间件。不论调用哪个重载，系统最终都会根据提供的 StatusCodePagesOptions 对象调用构造函数来创建这个中间件，而且 StatusCodePagesOptions 必须具有一个作为错误处理器的 Func<StatusCodeContext, Task>对象。

```csharp
public static class StatusCodePagesExtensions
{
    public static IApplicationBuilder UseStatusCodePages(this IApplicationBuilder app)
        => app.UseMiddleware<StatusCodePagesMiddleware>();

    public static IApplicationBuilder UseStatusCodePages(this IApplicationBuilder app,
        StatusCodePagesOptions options)
        => app.UseMiddleware<StatusCodePagesMiddleware>(Options.Create(options));

    public static IApplicationBuilder UseStatusCodePages(this IApplicationBuilder app,
        Func<StatusCodeContext, Task> handler)
        => app.UseStatusCodePages(new StatusCodePagesOptions
        {
            HandleAsync = handler
        });
}
```

由于 StatusCodePagesMiddleware 中间件最终的目的还是将定制的错误信息响应给客户端，所以可以在注册该中间件时直接指定响应的内容和媒体类型，这样的注册方式可以通过调用如下所示的 UseStatusCodePages 方法来完成。从如下所示的代码片段可以看出，通过参数 bodyFormat 指定的实际上是一个模板，它可以包含一个表示响应状态码的占位符（{0}）。

```csharp
public static class StatusCodePagesExtensions
{
    public static IApplicationBuilder UseStatusCodePages(this IApplicationBuilder app,
        string contentType, string bodyFormat)
    {
        return app.UseStatusCodePages(context =>
        {
            var body = string.Format(CultureInfo.InvariantCulture, bodyFormat,
                context.HttpContext.Response.StatusCode);
            context.HttpContext.Response.ContentType = contentType;
            return context.HttpContext.Response.WriteAsync(body);
        });
    }
}
```

UseStatusCodePagesWithRedirects

如果调用 UseStatusCodePagesWithRedirects 扩展方法,就可以使注册的 StatusCodePagesMiddleware 中间件向指定的路径发送一个客户端重定向。从如下所示的代码片段可以看出,参数 locationFormat 指定的重定向地址也是一个模板,它可以包含一个表示响应状态码的占位符 ({0})。我们可以指定一个完整的地址,也可以指定一个相对于 PathBase 的相对路径,后者需要包含表示基地址的前缀 "~/"。

```csharp
public static class StatusCodePagesExtensions
{
    public static IApplicationBuilder UseStatusCodePagesWithRedirects(
        this IApplicationBuilder app, string locationFormat)
    {
        if (locationFormat.StartsWith("~"))
        {
            locationFormat = locationFormat.Substring(1);
            return app.UseStatusCodePages(context =>
            {
                var location = string.Format(CultureInfo.InvariantCulture, locationFormat,
                    context.HttpContext.Response.StatusCode);
                context.HttpContext.Response.Redirect(
                    context.HttpContext.Request.PathBase + location);
                return Task.CompletedTask;
            });
        }
        else
        {
            return app.UseStatusCodePages(context =>
            {
                var location = string.Format(CultureInfo.InvariantCulture, locationFormat,
                    context.HttpContext.Response.StatusCode);
                context.HttpContext.Response.Redirect(location);
                return Task.CompletedTask;
            });
        }
    }
}
```

下面通过一个简单的应用来演示针对客户端重定向的错误页面呈现方式。我们在如下所示的应用中注册了一个路由模板为 "error/{statuscode}" 的路由,路由参数 statuscode 代表响应的状态码。在作为路由处理器的 HandleAsync 方法中,我们会直接响应一个包含状态码的字符串。我们调用 UseStatusCodePagesWithRedirects 方法注册 StatusCodePagesMiddleware 中间件时将重定义路径设置为 "error/{0}"。

```csharp
public class Program
{
    private static readonly Random _random = new Random();
    public static void Main()
    {
```

```
    Host.CreateDefaultBuilder()
        .ConfigureWebHostDefaults(builder => builder
            .ConfigureServices(svcs => svcs.AddRouting())
            .Configure(app => app
                .UseStatusCodePagesWithRedirects("~/error/{0}")
                .UseRouting()
                .UseEndpoints(endpoints => endpoints.MapGet(
                    "error/{statuscode}", HandleAsync))
                .Run(ProcessAsync)))
        .Build()
        .Run();

    static async Task HandleAsync(HttpContext context)
    {
        var statusCode = context.GetRouteData().Values["statuscode"];
        await context.Response.WriteAsync($"Error occurred ({statusCode})");
    }

    static Task ProcessAsync(HttpContext context)
    {
        context.Response.StatusCode = _random.Next(400, 599);
        return Task.CompletedTask;
    }
}
```

针对该应用的请求总是得到一个状态码为 400～599 的响应，StatusCodePagesMiddleware 中间件在此情况下会向指定的路径（"~/error/{statuscode}"）发送一个客户端重定向。由于重定向请求的路径与注册的路由相匹配，所以作为路由处理器的 HandleError 方法会响应图 16-11 所示的错误页面。（S1614）

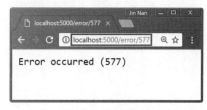

图 16-11　以客户端重定向的形式呈现错误页面

UseStatusCodePagesWithReExecute

除了可以采用客户端重定向的方式来呈现错误页面，还可以调用 UseStatusCodePagesWithReExecute 方法注册 StatusCodePagesMiddleware 中间件，并让它采用服务端重定向的方式来处理错误请求。如下面的代码片段所示，当我们调用这个方法的时候不仅可以指定重定向的路径，还可以指定查询字符串。这里作为重定向地址的参数 pathFormat 依旧是一个路径模板，它可以包含一个表示响应状态码的占位符（{0}）。

```
public static class StatusCodePagesExtensions
{
```

```csharp
public static IApplicationBuilder UseStatusCodePagesWithReExecute(
    this IApplicationBuilder app, string pathFormat, string queryFormat = null);
}
```

现在我们对前面演示的这个实例略做修改来演示采用服务端重定向呈现的错误页面。如下面的代码片段所示，我们将针对 UseStatusCodePagesWithRedirects 方法的调用替换成针对 UseStatusCodePagesWithReExecute 方法的调用。

```csharp
public class Program
{
    private static readonly Random _random = new Random();
    public static void Main()
    {
        Host.CreateDefaultBuilder()
            .ConfigureWebHostDefaults(builder => builder
                .ConfigureServices(svcs => svcs.AddRouting())
                .Configure(app => app
                    .UseStatusCodePagesWithReExecute("/error/{0}")
                    .UseRouting()
                    .UseEndpoints(endpoints => endpoints.MapGet(
                        "error/{statuscode}", HandleAsync))
                    .Run(ProcessAsync)))
            .Build()
            .Run();

        static async Task HandleAsync(HttpContext context)
        {
            var statusCode = context.GetRouteData().Values["statuscode"];
            await context.Response.WriteAsync($"Error occurred ({statusCode})");
        }

        static Task ProcessAsync(HttpContext context)
        {
            context.Response.StatusCode = _random.Next(400, 599);
            return Task.CompletedTask;
        }
    }
}
```

对于前面演示的实例，由于错误页面是通过客户端重定向的方式呈现的，所以浏览器地址栏显示的是重定向地址。我们在选择这个实例时采用了服务端重定向，虽然显示的页面内容并没有不同，但是地址栏上的地址是不会发生改变的，如图 16-12 所示。（S1615）

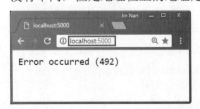

图 16-12　以服务端重定向的形式呈现错误页面

之所以命名为 UseStatusCodePagesWithReExecute，是因为通过这个方法注册的 StatusCodePagesMiddleware 中间件进行错误处理时，它仅仅将提供的重定向路径和查询字符串应用到当前 HttpContext 上下文，然后分发给后续管道重新执行。UseStatusCodePagesWithReExecute 方法中注册 StatusCodePagesMiddleware 中间件的实现总体上可以由如下所示的代码片段来体现。

```csharp
public static class StatusCodePagesExtensions
{
    public static IApplicationBuilder UseStatusCodePagesWithReExecute(
        this IApplicationBuilder app,
        string pathFormat,
        string queryFormat = null)
    {
        return app.UseStatusCodePages(async context =>
        {
            var newPath = new PathString(
                string.Format(CultureInfo.InvariantCulture, pathFormat,
                context.HttpContext.Response.StatusCode));
            var formatedQueryString = queryFormat == null ? null :
                 string.Format(CultureInfo.InvariantCulture, queryFormat,
                context.HttpContext.Response.StatusCode);

            context.HttpContext.Request.Path = newPath;
            context.HttpContext.Request.QueryString = newQueryString;
            await context.Next(context.HttpContext);
        });
    }
}
```

与 ExceptionHandlerMiddleware 中间件类似，StatusCodePagesMiddleware 中间件在处理请求的过程中会改变当前请求上下文的状态，具体体现在它会将指定的请求路径和查询字符串重新应用到当前请求上下文中。为了不影响前置中间件对请求的正常处理，StatusCodePagesMiddleware 中间件在完成自身处理流程之后必须将当前请求上下文恢复到原始状态。StatusCodePagesMiddleware 中间件依旧采用一个特性来保存原始路径和查询字符串。这个特性对应的接口是具有如下定义的 IStatusCodeReExecuteFeature，但是该接口仅仅包含两个针对路径的属性，并没有用于携带原始请求上下文的属性，但是默认实现类型 StatusCodeReExecuteFeature 包含了这个属性。

```csharp
public interface IStatusCodeReExecuteFeature
{
    string OriginalPath { get; set; }
    string OriginalPathBase { get; set; }
}

public class StatusCodeReExecuteFeature : IStatusCodeReExecuteFeature
{
    public string OriginalPath { get; set; }
```

```
    public string OriginalPathBase { get; set; }
    public string OriginalQueryString { get; set; }
}
```

在 StatusCodePagesMiddleware 中间件处理异常请求的过程中,在将指定的重定向路径和查询字符串应用到当前请求上下文之前,它会根据原始的上下文创建一个 StatusCodeReExecuteFeature 特性对象,并将其添加到当前 HttpContext 上下文的特性集合中。当整个请求处理过程结束之后,StatusCodePagesMiddleware 中间件还会将这个特性从当前 HttpContext 上下文中移除,并恢复原始的请求路径和查询字符串。如下所示的代码片段体现了 UseStatusCodePagesWithReExecute 方法的实现逻辑。

```
public static class StatusCodePagesExtensions
{
    public static IApplicationBuilder UseStatusCodePagesWithReExecute2(
        this IApplicationBuilder app,
        string pathFormat,
        string queryFormat = null)
    {
        return app.UseStatusCodePages(async context =>
        {
            var newPath = new PathString(
                string.Format(CultureInfo.InvariantCulture, pathFormat,
                context.HttpContext.Response.StatusCode));
            var formatedQueryString = queryFormat == null ? null :
                string.Format(CultureInfo.InvariantCulture, queryFormat,
                context.HttpContext.Response.StatusCode);
            var newQueryString = queryFormat == null
                ? QueryString.Empty : new QueryString(formatedQueryString);

            var originalPath = context.HttpContext.Request.Path;
            var originalQueryString = context.HttpContext.Request.QueryString;

            context.HttpContext.Features.Set<IStatusCodeReExecuteFeature>(
                new StatusCodeReExecuteFeature()
            {
                OriginalPathBase = context.HttpContext.Request.PathBase.Value,
                OriginalPath = originalPath.Value,
                OriginalQueryString = originalQueryString.HasValue
                    ? originalQueryString.Value : null,
            });

            context.HttpContext.Request.Path = newPath;
            context.HttpContext.Request.QueryString = newQueryString;
            try
            {
                await context.Next(context.HttpContext);
            }
```

```
            finally
            {
                context.HttpContext.Request.QueryString = originalQueryString;
                context.HttpContext.Request.Path = originalPath;
                context.HttpContext.Features.Set<IStatusCodeReExecuteFeature>(null);
            }
        });
    }
}
```

第 17 章

缓 存

缓存是提高应用程序性能最常用和最有效的"银弹"。借助 .NET Core 提供的缓存框架，我们不仅可以将数据缓存在应用进程的本地内存中，还可以采用分布式的形式将缓存数据存储在一个"中心数据库"中。ASP.NET Core 框架还借助一个中间件实现了所谓的"响应缓存"，即按照 HTTP 缓存规范对整个响应内容实施缓存。

17.1 将数据缓存起来

.NET Core 提供了两个独立的缓存框架：一个是针对本地内存的缓存，另一个是针对分布式存储的缓存。前者可以在不经过序列化的情况下直接将对象存储在当前应用程序进程的内存中，后者则需要将对象序列化成字节数组并存储到一个独立的"中心数据库"中。对于分布式缓存，.NET Core 提供了针对 Redis 和 SQL Server 的原生支持。除了这两个独立的缓存系统，ASP.NET Core 还借助一个中间件实现了响应缓存（Response Caching），即按照 HTTP 缓存规范对整个响应内容实施缓存。按照惯例，在对缓存进行系统介绍之前，需要先通过一些简单的实例介绍如何在一个 .NET Core 应用中使用缓存。

17.1.1 将数据缓存在内存中

相较于针对数据库和远程服务调用这种 IO 操作来说，针对内存的访问在性能上将获得不只一个数量级的提升，所以将数据对象直接缓存在应用进程的内容中具有最佳的性能优势。基于内存的缓存框架实现在 NuGet 包 "Microsoft.Extensions.Caching.Memory" 中，具体的缓存功能承载于通过 IMemoryCache 接口表示的服务对象。由于缓存的数据直接存放在内存中，并且不涉及持久化存储，所以无须考虑针对缓存对象的序列化问题，这种内存模式对缓存数据的类型也就没有任何限制。

缓存的操作主要是对缓存数据的读和写，这两个基本操作都是由上面介绍的 IMemoryCache 对象来完成的。如果在 ASP.NET Core 应用启动时对 IMemoryCache 服务做了注册，我们就可以以依赖注入的方式得到该服务对象并利用它进行缓存数据的读和写。在如下所示的演示程序中，我们调用 IServiceCollection 接口的 AddMemoryCache 扩展方法完成了针对

IMemoryCache 服务的注册。我们通过调用 IApplicationBuilder 接口的 Run 扩展方法注册了一个中间件，用来演示针对当前时间的缓存。

```
public class Program
{
    public static void Main()
    {
        Host.CreateDefaultBuilder()
            .ConfigureWebHostDefaults(builder => builder
                .ConfigureServices(svcs => svcs.AddMemoryCache())
                .Configure(app => app.Run(ProocessAsync)))
            .Build()
            .Run();
    }

    static async Task ProocessAsync(HttpContext httpContext)
    {
        var cache = httpContext.RequestServices.GetRequiredService<IMemoryCache>();
        if (!cache.TryGetValue<DateTime>("CurrentTime", out var currentTime))
        {
            cache.Set("CurrentTime", currentTime = DateTime.Now);
        }
        await httpContext.Response.WriteAsync($"{currentTime}({DateTime.Now})");
    }
}
```

在处理请求的过程中，我们利用 HttpContext 上下文得到针对当前请求的 IServiceProvider 对象，它提供了操作内存缓存的 IMemoryCache 服务。我们首先调用 IMemoryCache 接口的 TryGetValue<DateTime>扩展方法获取缓存的时间，使用的缓存项的 Key 为 "CurrentTime"。如果成功获取了缓存的时间，就直接将它和实时时间作为响应内容，否则调用 IMemoryCache 接口的 Set<DateTime>扩展方法将实时时间缓存起来。由于请求最终的响应内容是一个时间对，即缓存的时间和实时时间，所以通过浏览器访问该应用程序的时候，显示的时间在缓存过期之前是不变的（见图 17-1）。（S1701）

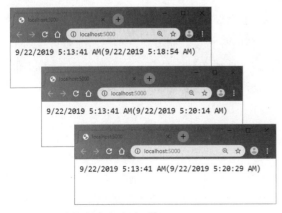

图 17-1　缓存在内存中的时间

17.1.2 对数据进行分布式缓存

虽然采用基于本地内存缓存可以获得最高的性能优势，但对于部署在集群的应用程序会出现缓存数据不一致的情况。对于这种部署场景，我们需要将数据缓存在某个独立的存储中心，以便让所有的 Web 服务器共享同一份缓存数据，我们将这种缓存形式称为分布式缓存。.NET Core 为分布式缓存提供了两种原生的存储形式：一种是基于 NoSQL 的 Redis 数据库，另一种是关系型数据库 SQL Server。

基于 Redis 的分布式缓存

Redis 是目前较为流行的 NoSQL 数据库，很多编程平台都将其作为分布式缓存的首选，下面演示在一个 ASP.NET Core 应用中如何采用基于 Redis 的分布式缓存。考虑到有的读者可能没有体验过 Redis，所以我们先简单介绍如何安装 Redis。Redis 最简单的安装方式就是采用 Chocolatey 命令行，Chocolatey 是 Windows 平台下一款优秀的软件包管理工具（类似于 NPM）。

```
PowerShell prompt :
iwr https://chocolatey.org/install.ps1 -UseBasicParsing | iex

CMD.exe:
@powershell -NoProfile -ExecutionPolicy Bypass -Command "iex ((New-Object
System.NET.WebClient).DownloadString('https://chocolatey.org/install.ps1'))" && SET
"PATH=%PATH%;%ALLUSERSPROFILE%\chocolatey\bin"
```

我们也可以采用 PowerShell（要求版本在 V3 以上）命令行或者普通 CMD.exe 命令行来安装 Chocolatey，具体的命令如上所示。在确保 Chocolatey 被正常安装的情况下，我们可以执行如下命令安装或者升级 64 位的 Redis。

```
C:\>choco install redis-64
C:\>choco upgrade redis-64
```

Redis 服务器的启动也很简单，我们只需要以命令行的形式执行 "redis-server" 命令即可。如果在执行该命令之后看到图 17-2 所示的输出，则表示本地的 Redis 服务器被正常启动，输出的结果会指定服务器采用的网络监听端口。

下面对上面演示的实例进行简单的修改，将基于内存的本地缓存切换到针对 Redis 数据库的分布式缓存。不论采用 Redis、SQL Server 还是其他的分布式存储方式，缓存的读和写都是通过 IDistributedCache 接口表示的服务对象来完成的。承载 Redis 分布式缓存框架的 NuGet 包 "Microsoft.Extensions.Caching.Redis"，我们需要手动添加针对该 NuGet 包的依赖。

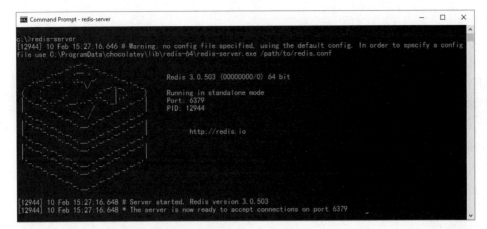

图 17-2 以命令行的形式启动 Redis 服务器

```
public class Program
{
    public static void Main()
    {
        Host.CreateDefaultBuilder()
            .ConfigureWebHostDefaults(builder => builder
                .ConfigureServices(svcs => svcs.AddDistributedRedisCache(options =>
                {
                    options.Configuration    = "localhost";
                    options.InstanceName     = "Demo";
                }))
                .Configure(app => app.Run(ProocessAsync)))
            .Build()
            .Run();
    }

    static async Task ProocessAsync(HttpContext httpContext)
    {
        var cache = httpContext.RequestServices
            .GetRequiredService<IDistributedCache>();
        var currentTime = await cache.GetStringAsync("CurrentTime");
        if (null == currentTime)
        {
            currentTime = DateTime.Now.ToString();
            await cache.SetAsync("CurrentTime", Encoding.UTF8.GetBytes(currentTime));
        }
        await httpContext.Response.WriteAsync($"{currentTime}({DateTime.Now})");
    }
}
```

从上面的代码片段可以看出，分布式缓存和内存缓存在总体编程模式上是一致的，我们需要先完成针对 IDistributedCache 服务的注册，然后利用依赖注入框架提供该服务对象来进行缓存数据的读和写。IDistributedCache 服务的注册是通过调用 IServiceCollection 接口的 AddDistributed

RedisCache 方法来完成的，我们在调用这个方法时提供了一个 RedisCacheOptions 对象，并利用它的 Configuration 属性和 InstanceName 属性设置 Redis 数据库的服务器与实例名称。

由于采用的是本地的 Redis 服务器，所以可以将 Configuration 属性设置为 localhost。其实，Redis 数据库并没有所谓的实例的概念，RedisCacheOptions 类型的 InstanceName 属性的目的在于当多个应用共享同一个 Redis 数据库时，缓存数据可以利用它进行区分。当缓存数据被保存到 Redis 数据库中的时候，Key 以 InstanceName 为前缀。应用程序启动后（确保 Redis 服务器被正常启动），如果我们利用浏览器来访问它，依然可以得到与图 17-1 类似的输出。（S1702）

对于基于内存的本地缓存来说，我们可以将任何类型的数据置于缓存之中，但是分布式缓存涉及网络传输和持久化存储，置于缓存中的数据类型只能是字节数组，所以我们需要自行负责对缓存对象的序列化和反序列化工作。如上面的代码片段所示，我们先将表示当前时间的 DateTime 对象转换成字符串，然后采用 UTF-8 编码进一步转换成字节数组。我们调用 IDistributedCache 接口的 SetAsync 方法缓存的数据是最终的字节数组。我们也可以直接调用 SetStringAsync 扩展方法将字符串编码为字节数组。在读取缓存数据时，我们调用的是 IDistributedCache 接口的 GetStringAsync 方法，它会将字节数组转换成字符串。

缓存数据在 Redis 数据库中是以散列（Hash）的形式存放的，对应的 Key 会将设置的 InstanceName 属性作为前缀。为了查看在 Redis 数据库中究竟存放了哪些数据，我们可以按照图 17-3 所示的形式执行 Redis 命令获取存储的数据。从图 17-3 呈现的输出结果可以看出，存入 Redis 数据库的不仅包括指定的缓存数据（Sub-Key 为 data），还包括其他两组针对该缓存条目的描述信息，对应的 Sub-Key 分别为 absexp 和 sldexp，表示缓存的绝对过期时间（Absolute Expiration Time）和滑动过期时间（Sliding Expiration Time）。

图 17-3　查看 Redis 数据库中存放的数据

基于 SQL Server 的分布式缓存

除了使用 Redis 这种主流的 NoSQL 数据库来支持分布式缓存，还可以使用关系型数据库 SQL Server。针对 SQL Server 的分布式缓存实现在 NuGet 包 "Microsoft.Extensions.Caching.SqlServer" 中，我们需要先确保该 NuGet 包被正常安装到演示的应用程序中。

所谓的针对 SQL Server 的分布式缓存，实际上就是将表示缓存数据的字节数组存放在 SQL Server 数据库的某个具有固定结构的数据表中，所以需要先创建这样一个缓存表，该表可以采用命令行的形式执行 "dotnet sql-cache create" 命令来创建。执行这个命令应该指定的参数可以按照如下形式通过执行 "dotnet sql-cache create --help" 命令来查看。从图 17-4 可以看出，该命令需要指定 3 个参数，它们分别表示缓存数据库的连接字符串、缓存表的 Schema 和名称。

图 17-4　dotnet sql-cache create 命令的帮助文档

接下来只需要以命令行的形式执行"dotnet sql-cache create"命令就可以在指定的数据库中创建缓存表。对于演示的实例来说，可以按照图 17-5 所示的方式执行"dotnet sql-cache create"命令，该命令会在本机一个名为 DemoDB 的数据库中（数据库需要预先创建好）创建一个名为 AspnetCache 的缓存表，该表采用 dbo 作为 Schema。

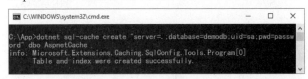

图 17-5　执行"dotnet sql-cache create"命令创建缓存表

在所有的准备工作完成之后，我们只需要对上面的程序做如下修改就可以将缓存存储方式从 Redis 数据库切换到针对 SQL Server 的数据库。由于采用的同样是分布式缓存，所以针对缓存数据的设置和提取的代码不用做任何改变，我们需要修改的地方仅仅是服务注册部分。如下面的代码片段所示，我们调用 IServiceCollection 接口的 AddDistributedSqlServerCache 扩展方法完成了对应的服务注册。在调用这个方法的时候，我们通过设置 SqlServerCacheOptions 对象 3 个属性的方式指定了缓存数据库的连接字符串、缓存表的 Schema 和名称。

```
public class Program
{
    public static void Main()
    {
        Host.CreateDefaultBuilder()
            .ConfigureWebHostDefaults(builder => builder
                .ConfigureServices(svcs => svcs.AddDistributedSqlServerCache(options =>
                {
                    options.ConnectionString =
                        "server=.;database=demodb;uid=sa;pwd=password";
                    options.SchemaName          = "dbo";
                    options.TableName           = "AspnetCache";
                }))
                .Configure(app => app.Run(async context =>
                {
                    var cache = context.RequestServices
                        .GetRequiredService<IDistributedCache>();
                    var currentTime = await cache.GetStringAsync("CurrentTime");
                    if (null == currentTime
```

```
                    {
                        currentTime = DateTime.Now.ToString();
                        await cache.SetAsync("CurrentTime",
                            Encoding.UTF8.GetBytes(currentTime));
                    }
                    await context.Response.WriteAsync($"{currentTime}({DateTime.Now})");
                })))
        .Build()
        .Run();
    }
}
```

若要查看最终存入 SQL Server 数据库中的缓存数据，我们只需要在数据库中查看对应的缓存表即可。对于演示实例缓存的时间戳，它会以图 17-6 所示的形式保存在我们创建的缓存表（AspnetCache）中。与基于 Redis 数据库的存储方式类似，与缓存数据的值一并存储的还包括缓存的过期信息。（S1703）

图 17-6　存储在缓存表中的数据

17.1.3　缓存整个 HTTP 响应

上面演示的两种缓存都要求利用注册的服务对象以手动方式存储和提取具体的缓存数据，而下面演示的缓存则不再基于某个具体的缓存数据，而是将服务端生成的 HTTP 响应的内容予以缓存，我们将这种缓存形式称为响应缓存（Response Caching）。标准的 HTTP 规范，不论是 HTTP 1.0+ 还是 HTTP 1.1，都对响应缓存做了详细规定，这是响应缓存的理论机制和指导思想。我们将在后续内容中详细介绍 HTTP 缓存规范，但在此之前需要先通过一个简单的实例来演示整个响应内容是如何借助一个名为 ResponseCachingMiddleware 的中间件被缓存起来的。

同样是采用基于时间的缓存场景，我们编写了如下所示的程序。如下面的代码片段所示，我们调用 IServiceCollection 接口的 AddResponseCaching 扩展方法注册了 ResponseCachingMiddleware 中间件依赖的服务，而这个中间件是通过调用 IApplicationBuilder 接口的 UseResponseCaching 扩展方法进行注册的。

```
public class Program
{
```

```
public static void Main()
{
    Host.CreateDefaultBuilder()
        .ConfigureWebHostDefaults(builder => builder
            .ConfigureServices(svcs => svcs.AddResponseCaching())
            .Configure(app => app
                .UseResponseCaching()
                .Run(ProcessAsync)))
        .Build()
        .Run();

    static async Task ProcessAsync(HttpContext httpContext)
    {
        var response = httpContext.Response;
        response.GetTypedHeaders().CacheControl = new CacheControlHeaderValue
        {
            Public = true,
            MaxAge = TimeSpan.FromSeconds(3600)
        };
        var isUtc = httpContext.Request.Query.ContainsKey("utc");
        await response.WriteAsync(
            isUtc ? DateTime.UtcNow.ToString() : DateTime.Now.ToString());
    }
}
```

对于最终实现的请求处理逻辑来说，我们仅仅是为响应添加了一个 Cache-Control 报头，并且将它的值设置为 public, max-age=3600（public 表示缓存的是可以被所有用户共享的公共数据，而 max-age 则表示过期时限，单位为秒）。真正写入响应的主体内容就是当前时间，但需要根据请求的查询字符串"utc"决定采用普通时间还是 UTC 时间。

要证明整个响应的内容是否被缓存，只需要验证在缓存过期之前具有相同路径的多个请求对应的响应是否具有相同的主体内容，所以可以采用 Fiddler 来发送请求并拦截响应的内容。如下所示的两组请求和响应是在不同时间发送的，可以看出，响应的内容是完全一致的。由于请求发送的时间不同，所以返回的缓存副本的"年龄"（对应响应报头 Age，由于第一个响应并不是从缓存中提取的，所以没有此报头）也是不同的。（S1704）

```
GET http://localhost:5000/ HTTP/1.1
Host: localhost:5000

HTTP/1.1 200 OK
Date: Sat, 21 Sep 2019 21:36:18 GMT
Server: Kestrel
Cache-Control: public, max-age=3600
Content-Length: 20

9/22/2019 5:36:18 AM
```

```
GET http://localhost:5000/ HTTP/1.1
Host: localhost:5000

HTTP/1.1 200 OK
Date: Sat, 21 Sep 2019 21:36:18 GMT
Server: Kestrel
Cache-Control: public, max-age=3600
Age: 38
Content-Length: 20

9/22/2019 5:36:18 AM


GET http://localhost:5000/ HTTP/1.1
Host: localhost:5000

HTTP/1.1 200 OK
Date: Sat, 21 Sep 2019 21:36:18 GMT
Server: Kestrel
Cache-Control: public, max-age=3600
Age: 68
Content-Length: 20

9/22/2019 5:36:18 AM
```

上面这两个请求的 URL 并没有携带 "utc" 查询字符串，所以返回的是一个非 UTC 时间。下面采用相同的方式生成一个试图返回 UTC 时间的请求。从下面给出的请求和响应的内容可以看出，虽然请求携带了查询字符串 "utc=true"，但是返回的依然是之前缓存的时间。由此可见，ResponseCachingMiddleware 中间件在默认情况下是针对请求的路径对响应实施缓存的，它会忽略请求 URL 携带的查询字符串，这显然不是我们希望看到的结果。

```
GET http://localhost:5000/?utc=true HTTP/1.1
Host: localhost:5000

HTTP/1.1 200 OK
Date: Sat, 21 Sep 2019 21:36:18 GMT
Server: Kestrel
Cache-Control: public, max-age=3600
Age: 330
Content-Length: 20

9/22/2019 5:36:18 AM
```

按照 REST 的原则，URL 是网络资源的标识，但是资源的表现形式（Representation）由一些参数来决定。这些参数可以体现为查询字符串，也可以体现为一些请求报头，如 Language 报头决定资源的描述语言，Content-Encoding 报头决定资源采用的编码方式。因此，针对响应的缓存不应该只考虑请求的路径，还应该综合考虑这些参数。

对于演示的这个实例，我们希望将查询字符串 "utc" 纳入缓存的范畴，这可以利用 IResponse

CachingFeature 接口表示的特性来实现。如下面的代码片段所示，在将当前时间写入响应之后，我们得到这个特性并设置了它的 VaryByQueryKeys 属性，该属性包含一组决定输出缓存的查询字符串名称，我们将查询字符串"utc"添加到这个列表中。（S1705）

```
public class Program
{
    public static void Main()
    {
        Host.CreateDefaultBuilder()
            .ConfigureWebHostDefaults(builder => builder
                .ConfigureServices(svcs => svcs.AddResponseCaching())
                .Configure(app => app
                    .UseResponseCaching()
                    .Run(ProcessAsync)))
            .Build()
            .Run();
    }

    static async Task ProcessAsync(HttpContext httpContext)
    {
        var response = httpContext.Response;
        response.GetTypedHeaders().CacheControl = new CacheControlHeaderValue
        {
            Public = true,
            MaxAge = TimeSpan.FromSeconds(3600)
        };
        var isUtc = httpContext.Request.Query.ContainsKey("utc");
        await response.WriteAsync(
            isUtc ? DateTime.UtcNow.ToString() : DateTime.Now.ToString());
        var feature = httpContext.Features.Get<IResponseCachingFeature>();
        feature.VaryByQueryKeys = new string[] { "utc" };
    }
}
```

17.2 本地内存缓存

在通过实例了解了缓存在 ASP.NET Core 应用中如何使用之后，下面对 .NET Core 的缓存框架进行系统介绍。首先介绍针对本地内存的缓存。如果采用基于内存的缓存，我们可以将任意类型的对象保存在缓存中，但是真正保存在内存中的其实不是指定的缓存数据对象，而是通过 ICacheEntry 接口表示的缓存条目（Cache Entry）。

17.2.1 ICacheEntry

我们提供的数据对象在真正被缓存之前需要被封装成一个 ICacheEntry 对象，该对象表示真正保存在内存中的一个缓存条目。ICacheEntry 接口的 Key 属性和 Value 属性分别表示缓存键与缓存数据，Size 属性表示缓存的容量，这个容量是应用程序在设置缓存时指定的，而不是自动计算出来的。

```csharp
public interface ICacheEntry : IDisposable
{
    object      Key { get; }
    object      Value { get; set; }
    long?       Size { get; set; }

    DateTimeOffset?         AbsoluteExpiration { get; set; }
    TimeSpan?               AbsoluteExpirationRelativeToNow { get; set; }
    TimeSpan?               SlidingExpiration { get; set; }
    IList<IChangeToken>     ExpirationTokens { get; }

    IList<PostEvictionCallbackRegistration>     PostEvictionCallbacks { get; }
    CacheItemPriority                           Priority { get; set; }
}
```

缓存数据仅仅是真实数据的一份副本而已，应用程序应该尽可能保证两者的一致性。缓存一致性可以通过过期策略来实现。当我们调用 IMemoryCache 接口的 TryGetValue 方法通过指定的 Key 试图获取对应的缓存数据时，该方法会进行过期检验，过期的内存条目会被直接从缓存字典中移除，此时该方法会返回 False。

具体来说，本地内存缓存会采用两种针对时间的过期策略，分别是针对绝对时间（Absolute Time）和滑动时间（Sliding Time）的过期策略。所谓绝对时间过期就是指缓存对象会在执行的某个时刻自动过期，ICacheEntry 接口的 AbsoluteExpiration 就是我们为缓存条目设置的绝对过期时间。除了直接设置一个过期时间点，还可以通过指定一个 TimeSpan 对象来设置 AbsoluteExpirationRelativeToNow 属性，在这种情况下，ICacheEntry 对象会基于当前时间来计算绝对过期时间。

如果采用绝对时间过期策略，就意味着不论缓存对象最近使用的频率如何，对应的 ICacheEntry 对象总是在指定的时间点之后过期。而滑动过期则与此相反，它会通过缓存条目最近是否被使用过来决定缓存是否应该过期。具体来说，我们可以为某个缓存条目设置一个 TimeSpan 对象表示的时间段，如果在最近这段时间内没有读取过该缓存条目，缓存将会过期。反之，针对缓存的每一次使用都会将过期时间延长，而指定的这个时间段就是延后的时长。针对滑动过期设置的过期时长通过 ICacheEntry 接口的 SlidingExpiration 属性来设置。

除了上述 3 个属性，ICacheEntry 接口还有一个与缓存过期有关的属性，即 ExpirationTokens，它返回一个 IChangeToken 的集合。一个 IChangeToken 对象一般与某个需要被监控的对象绑定，并在监控对象发生变化的情况下对外发送通知，在这里我们利用这些 IChangeToken 对象来判断当前缓存是否过期。也就是说，基于内存的缓存具有两种过期策略：一种是基于时间（绝对时间或者相对时间）的过期策略，另一种是利用 IChangeToken 对象来发送过期通知。假设内存的数据来自一个物理文件，那么最理想的方式就是让缓存在文件被修改之前永不过期，这种应用场景可以利用后一种缓存过期策略来实现。

虽然基于内存的缓存具有最好的性能，但是如果当前进程的内存资源被缓存数据大量占据，就没有足够的空间来放置程序运行过程中创建的对象。在这种情况下运行时的 GC 将会实施垃圾回收，但是如果缓存占据的内存不能被释放，垃圾回收的作用将大打折扣。为了解决这个问题，基于内存的缓存采用了一种被称为"内存压缩"的机制，该机制确保在运行时执行垃圾回收会按照相应的策略以一定的比率压缩缓存占据的内存空间。

所谓的压缩，实际上就是根据预定义的策略删除那些"重要性低"的 ICacheEntry 对象。通过 ICacheEntry 接口的 Priority 属性表示的优先级是判断缓存条目重要性的决定性因素之一。预定义的 4 种缓存优先级定义在如下这个类型为 CacheItemPriority 的枚举中。内存压缩只针对 Low、Normal 和 High 这 3 种优先级的 ICacheEntry 对象，从其命名可以看出优先级为 NeverRemove 的 ICacheEntry 对象在过期之前不会从内存中移除。

```
public interface ICacheEntry : IDisposable
{
    CacheItemPriority Priority { get; set; }
    ...
}

public enum CacheItemPriority
{
    Low,
    Normal,
    High,
    NeverRemove
}
```

综上所述，除了应用程序显式移除某个缓存条目，表示缓存条目的 ICacheEntry 对象还有可能因过期而被移除，或者因为优先等级低而在实施缓存压缩过程中被移除，但是我们不能保证被移除的 ICacheEntry 对象所携带的数据对于应用毫无用处。为了解决这个问题，我们可以注册一些在 ICacheEntry 对象被逐出（Eviction）时会被执行的回调，该回调通过如下所示的 PostEvictionDelegate 类型的委托表示。我们可以从该委托的输入参数中得到试图被移除 ICacheEntry 对象的 Key 和 Value，以及通过 EvictionReason 枚举表示的被移除的原因。而参数 state 是我们在注册这个回调时指定的。

```
public delegate void PostEvictionDelegate(object key, object value, EvictionReason reason,
    object state);
public enum EvictionReason
{
    None,
    Removed,
    Replaced,
    Expired,
    TokenExpired,
    Capacity
}
```

我们并不会直接将一个 PostEvictionDelegate 对象注册到某个 ICacheEntry 对象上，而是选择将该对象表示的回调及传入的状态对象（对应 state 参数）封装在如下所示的 PostEviction CallbackRegistration 对象中。表示缓存条目的 ICacheEntry 接口具有 PostEvictionCallbacks 属性，它返回一个 PostEvictionCallbackRegistration 对象的集合，所谓的回调注册正是向这个集合中添加相应的 PostEvictionCallbackRegistration 对象的过程。

```
public class PostEvictionCallbackRegistration
{
    public PostEvictionDelegate          EvictionCallback { get; set; }
    public object                        State { get; set; }
}

public interface ICacheEntry : IDisposable
{
    IList<PostEvictionCallbackRegistration> PostEvictionCallbacks { get; }
    ...
}
```

除了直接利用定义在 ICacheEntry 接口的成员设置某个缓存条目的相关属性，我们还可以调用该接口的一些扩展方法。由于直接针对属性的设置本身就非常简单、直接，这些扩展方法并不具有简化的作用，所以它们的意义其实并不大，唯一的好处是某个方法会为我们做一些基本的参数验证。

```
public static class CacheEntryExtensions
{
    public static ICacheEntry SetValue(this ICacheEntry entry, object value);
    public static ICacheEntry SetAbsoluteExpiration(this ICacheEntry entry,
        DateTimeOffset absolute);
    public static ICacheEntry SetAbsoluteExpiration(this ICacheEntry entry,
        TimeSpan relative);
    public static ICacheEntry SetSlidingExpiration(this ICacheEntry entry,
        TimeSpan offset);
    public static ICacheEntry AddExpirationToken(this ICacheEntry entry,
        IChangeToken expirationToken);

    public static ICacheEntry SetPriority(this ICacheEntry entry,
        CacheItemPriority priority);
    public static ICacheEntry SetSize(this ICacheEntry entry, long size);
    public static ICacheEntry SetValue(this ICacheEntry entry, object value)

    public static ICacheEntry RegisterPostEvictionCallback(this ICacheEntry entry,
        PostEvictionDelegate callback);
    public static ICacheEntry RegisterPostEvictionCallback(this ICacheEntry entry,
        PostEvictionDelegate callback, object state);
}
```

17.2.2　MemoryCacheEntryOptions

通过上面的介绍可知，缓存数据被封装成一个 ICacheEntry 对象，一个 ICacheEntry 对象除了封装指定的缓存数据，还承载着一些其他的控制信息，这些信息决定缓存何时失效，以及在内存压力较大时是否应该被移除。这些控制信息最终作为配置选项通过一个 MemoryCacheEntryOptions 对象表示，所以我们在该类型中可以看到与 ICacheEntry 接口类似的属性成员。

```csharp
public class MemoryCacheEntryOptions
{
    public DateTimeOffset?                      AbsoluteExpiration { get; set; }
    public TimeSpan?                            AbsoluteExpirationRelativeToNow { get; set; }
    public TimeSpan?                            SlidingExpiration { get; set; }
    public IList<IChangeToken>                  ExpirationTokens { get; }

    public CacheItemPriority                                       Priority { get; set; }
    public IList<PostEvictionCallbackRegistration>  PostEvictionCallbacks { get; }
}
```

与上面介绍的 ICacheEntry 接口类似，我们可以直接利用定义在 MemoryCacheEntryOptions 中的这些属性设置对应的缓存配置选项，缓存系统还帮助我们额外定义了如下所示的扩展方法。这种方式并不能带来任何编程方面的便利，开发人员可以根据自己的喜好选择缓存选项的设置方式。

```csharp
public static class MemoryCacheEntryExtensions
{
    public static MemoryCacheEntryOptions SetAbsoluteExpiration(
        this MemoryCacheEntryOptions options, DateTimeOffset absolute);
    public static MemoryCacheEntryOptions SetAbsoluteExpiration(
        this MemoryCacheEntryOptions options, TimeSpan relative);
    public static MemoryCacheEntryOptions SetSlidingExpiration(
        this MemoryCacheEntryOptions options, TimeSpan offset);
    public static MemoryCacheEntryOptions SetPriority(this MemoryCacheEntryOptions options,
        CacheItemPriority priority);
    public static MemoryCacheEntryOptions SetSize(this MemoryCacheEntryOptions options,
        long size)

    public static MemoryCacheEntryOptions AddExpirationToken(
        this MemoryCacheEntryOptions options, IChangeToken expirationToken);

    public static MemoryCacheEntryOptions RegisterPostEvictionCallback(
        this MemoryCacheEntryOptions options, PostEvictionDelegate callback);
    public static MemoryCacheEntryOptions RegisterPostEvictionCallback(
        this MemoryCacheEntryOptions options, PostEvictionDelegate callback, object state);
}
```

由于 IMemoryCache 对象最终存储的是 ICacheEntry 对象，所以一个 MemoryCacheEntryOptions 对象承载的缓存配置选项最终需要应用到对应的 ICacheEntry 对象上，这个过程可以通过调用 ICacheEntry 接口的 SetOptions 扩展方法来完成。

```csharp
public static class CacheEntryExtensions
{
    public static ICacheEntry SetOptions(this ICacheEntry entry,
        MemoryCacheEntryOptions options);
}
```

17.2.3 IMemoryCache

基于内存缓存的读和写最终落在通过 IMemoryCache 接口表示的服务对象上。

IMemoryCache 对象承载的操作非常明确，那就是针对缓存数据的设置（添加新的缓存或者替换现有缓存）、获取和移除，但是 IMemoryCache 接口只定义了分别用于获取和移除缓存的 TryGetValue 方法与 Remove 方法，并没有一个名为 Set 的方式帮助我们设置缓存。取而代之的是另一个 CreateEntry 方法，该方法仅仅根据我们指定的键创建一个 ICacheEntry 对象。

```csharp
public interface IMemoryCache : IDisposable
{
    ICacheEntry        CreateEntry(object key);
    bool               TryGetValue(object key, out object value);
    void               Remove(object key);
}
```

对于 IMemoryCache 接口的默认实现类型 MemoryCache 来说，它直接利用一个字典对象来保存添加的缓存条目。代表缓存条目的 ICacheEntry 对象通过它的 CreateEntry 方法创建，但是它是通过什么途径被添加到这个字典对象中的？其实，对于 MemoryCache 类型来说，它对缓存的设置发生在 ICacheEntry 对象的 Dispose 方法被调用的时候。当 MemoryCache 根据指定的 Key 创建一个 ICacheEntry 对象的时候，它会注册一个回调，当 ICacheEntry 对象的 Dispose 方法被执行时，注册的回调负责将 ICacheEntry 对象放到缓存字典之中。下面的这段程序很好地证实了这一点。

```csharp
var cache = new ServiceCollection()
    .AddMemoryCache()
    .BuildServiceProvider()
    .GetRequiredService<IMemoryCache>();
var entry = cache.CreateEntry("foobar");
entry.Value= "abc";
Debug.Assert(!cache.TryGetValue("foobar", out var value));
Debug.Assert(null == value);

entry.Dispose();
Debug.Assert(cache.TryGetValue("foobar", out value));
Debug.Assert(value.Equals("abc"));
```

在前面的演示实例中，我们调用 IMemoryCache 接口的 Set 方法采用"一步到位"的方式完成了针对缓存的设置，但 Set 方法只是针对 IMemoryCache 接口的一个扩展方法而已。除了实例演示中使用的 Set 方法，IMemoryCache 接口还有如下所示的 Set 扩展方法重载。对于这些方法来说，它们都会调用 CreateEntry 方法根据指定的 Key 创建一个新的 ICacheEntry 对象，在对该对象做相应设置之后，ICacheEntry 对象的 Dispose 方法会被调用，并将自身添加到缓存字典中。

```csharp
public static class CacheExtensions
{
    public static TItem Set<TItem>(this IMemoryCache cache, object key, TItem value);
    public static TItem Set<TItem>(this IMemoryCache cache, object key, TItem value,
        MemoryCacheEntryOptions options);
    public static TItem Set<TItem>(this IMemoryCache cache, object key, TItem value,
        IChangeToken expirationToken);
    public static TItem Set<TItem>(this IMemoryCache cache, object key, TItem value,
        DateTimeOffset absoluteExpiration);
```

```
public static TItem Set<TItem>(this IMemoryCache cache, object key, TItem value,
    TimeSpan absoluteExpirationRelativeToNow);
}
```

由于 IMemoryCache 接口的 TryGetValue 方法总是以 object 对象的形式返回缓存数据，为了便于编程，缓存系统为我们提供了如下这些泛型的扩展方法。TryGetValue 方法和 Get 方法仅仅实现了针对缓存数据的提取工作，而 GetOrCreate 扩展方法会在指定的缓存不存在的时候利用提供的委托对象重新设置缓存。

```
public static class CacheExtensions
{
    public static bool TryGetValue<TItem>(this IMemoryCache cache, object key,
        out TItem value);

    public static object Get(this IMemoryCache cache, object key);
    public static TItem Get<TItem>(this IMemoryCache cache, object key);

    public static TItem GetOrCreate<TItem>(this IMemoryCache cache, object key,
        Func<ICacheEntry, TItem> factory);
    public static Task<TItem> GetOrCreateAsync<TItem>(this IMemoryCache cache, object key,
        Func<ICacheEntry, Task<TItem>> factory);
}
```

MemoryCacheOptions

针对 IMemoryCache 接口的默认实现是定义在 NuGet 包 "Microsoft.Extensions.Caching.Memory" 的 MemoryCache 类型中。但是正式介绍该类型提供的默认缓存实现原理之前，需要先了解一个表示对应配置选项的 MemoryCacheOptions 类型（不要与配置选项类型 MemoryCacheEntryOptions 混淆）。如下面的代码片段所示，该类型实现了 IOptions<MemoryCacheOptions>接口[1]，所以我们可以采用 Options 模式将其注册为一个服务。

```
public class MemoryCacheOptions : IOptions<MemoryCacheOptions>
{
    public ISystemClock      Clock { get; set; }
    public TimeSpan          ExpirationScanFrequency { get; set; }
    public long?             SizeLimit { get; set; }
    public double            CompactionPercentage { get; set; }

    MemoryCacheOptions IOptions<MemoryCacheOptions>.Value { get; }
}
```

除了实现定义在 IOptions<MemoryCacheOptions>接口的 Value 属性（该属性会直接返回它自己），MemoryCacheOptions 类型还有其他 3 个配置选项属性。由于绝对过期的计算是基于某个具体的时间点，所以时间的同步（客户端与服务器的时间同步，以及集群中多台服务器之间时间同步）很重要，MemoryCacheOptions 类型的 Clock 属性可以帮助我们设置和返回这个同步时钟。

---

[1] 笔者认为将配置选项类型实现 IOptions<TOptions>接口或者 IOptionsMonitor<TOptions>接口并不是一种好的编程方式，配置选项类型只需要单纯地封装配置选项数据即可。实际上，这样的定义方式在微软的原生框架中也是极少见的。

ExpirationScanFrequency 属性表示的时长代表过期扫描的频率，也就是两次扫描所有缓存条目以确定它们是否过期的时间间隔，默认的间隔是 1 分。SizeLimit 属性和 CompactionPercentage 属性则与缓存压缩有关。SizeLimit 属性表示缓存的最大容量，如果没有显式设置就意味着对容量没有限制，如果对该属性进行了显式设置，提供的每个 ICacheEntry 的 Size 属性必须被赋值。CompactionPercentage 属性则表示每次实施缓存压缩时移除的缓存容量占当前总容量的百分比，默认值为 0.05（5%）。

缓存的设置

在了解了必要的储备知识之后，下面正式介绍 MemoryCache。如下面的代码片段所示，一个 MemoryCache 对象根据提供的 MemoryCacheOptions 对象创建。我们先介绍它承载的 3 个基本缓存操作是如何实现的。MemoryCache 对象直接利用一个 ConcurrentDictionary<object, CacheEntry>对象来保存添加的内存条目，所以添加、删除和获取都是基于这个字典对象的操作而已。也正是因为 MemoryCache 对象背后的存储结构是这样一个对象，所以它默认支持多线程并发，应用程序无须自行解决线程同步问题。

```
public class MemoryCache : IMemoryCache
{
    public MemoryCache(IOptions<MemoryCacheOptions> optionsAccessor)

    public ICacheEntry          CreateEntry(object key);
    public bool                 TryGetValue(object key, out object value);
    public void                 Remove(object key);
    ...
}
```

对于 MemoryCache 对象提供的几种基本的缓存操作来说，缓存设置相对来说比较复杂，因为它涉及针对过期缓存的检验。下面详细介绍其背后的流程。前面提及，MemoryCache 对象将提供的 CacheEntry 添加（包括替换现有的 CacheEntry）到缓存字典是利用注册到 CacheEntry 对象的一个回调来完成的，该回调会在 CacheEntry 的 Dispose 方法中被执行，下面介绍的实际上就是这个注册的回调所执行的操作。

MemoryCache 支持两种针对时间（绝对时间和滑动时间）的过期策略。绝对时间过期策略利用 ICacheEntry 对象的 AbsoluteExpiration 属性和 AbsoluteExpirationRelativeToNow 属性判断缓存是否过期，如果这两个属性都做了设置，究竟应该采用哪一个？其实，MemoryCache 在这种情况下会选择距离当前时间最近的那个时间，如当前时间为 1:00:00，AbsoluteExpiration 属性就被设置为 2:00:00，而 AbsoluteExpirationRelativeToNow 属性被设置为 30 分，缓存的绝对过期时间应该是 1:30:00。如果 AbsoluteExpirationRelativeToNow 属性被设置为 2 小时，那么缓存的绝对过期时间应该是 2:00:00。当 MemoryCache 在进行缓存设置时，它会根据这个原则计算出正确的绝对过期时间。

对于滑动时间过期来说，缓存是否过期取决于缓存最后一次被读取的时间。对于通过 MemoryCache 对象的 CreateEntry 方法创建的 CacheEntry 对象来说，它会携带这个最后被访问的时间戳，MemoryCache 在设置与读取缓存时都会将这个时间设置为当前系统时间，这里的"当前时间"是由 MemoryCacheOptions 的 Clock 属性返回的系统时钟来提供的。如果在创建 MemoryCache 时没有为 MemoryCacheOptions 指定这样一个系统时钟，它默认采用本地时间。

在为 CacheEntry 设置正确的绝对过期时间和最后访问时间戳之后，MemoryCache 就开始过期检验。具体采用的过期检验机制会在下面单独介绍。如果检验结果为过期，并且缓存字典中已经包含一个具有相同 Key 的 ICacheEntry 对象，后者将从缓存字典中移除。此时，注册在 ICacheEntry 对象（不是缓存字典中包含的那个 ICacheEntry 对象）的 PostEvictionCallbacks 属性上的所有回调会被执行。由于它被逐出的原因是过期，所以回调的参数 reason 将被设置为 EvictionReason.Expired。

如果提供的 CacheEntry 对象尚未过期，并且当前缓存字典中并没有一个对应的 CacheEntry 对象（它与提供的 CacheEntry 具有相同的 Key），那么提供的 CacheEntry 对象会直接添加到缓存字典中。如果这个 CacheEntry 对象的 ExpirationTokens 属性包含一个或者多个用于发送缓存过期通知的 IChangeToken 对象，并且其中任何一个的 HasChanged 属性为 True，那么该 CacheEntry 对象也会被视为过期。

如果缓存字典中已经存在一个与提供的 ICacheEntry 对象具有相同 Key 的缓存条目，那么该条目将直接被新的 ICacheEntry 对象替换。对于被替换的 ICacheEntry，注册在其 PostEvictionCallbacks 属性上的所有回调会被执行。由于它被逐出的原因是被新的 CacheEntry 替换，所以回调的参数 reason 将被设置为 EvictionReason.Replace。缓存设置流程如图 17-7 所示。

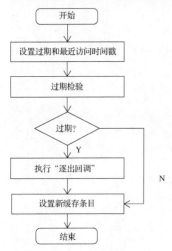

图 17-7　缓存设置流程

过期检验

对于针对时间（绝对时间或者滑动时间）的缓存过期策略，确定某个缓存条目是否过期的逻辑很明确，此处不再介绍。利用 IChangeToken 对象来发送过期通知的情况很少，下面利用一个简单实例来演示这种过期策略的应用。假设需要从一个物理文件中读取文件内容，为了最大限度地避免针对文件系统的 IO 操作，可以将文件内容进行缓存。缓存的内容将永久有效，直到物理文件的内容被修改，为此我们在一个控制台应用中编写了如下这段程序。

```csharp
public class Program
{
    public static async Task Main()
    {
        var fileProvider = new PhysicalFileProvider(Directory.GetCurrentDirectory());
        var fileName = "time.txt";
        var @lock = new object();

        var cache = new ServiceCollection()
            .AddMemoryCache()
            .BuildServiceProvider()
            .GetRequiredService<IMemoryCache>();

        var options = new MemoryCacheEntryOptions();
        options.AddExpirationToken(fileProvider.Watch(fileName));
        options.PostEvictionCallbacks.Add(
            new PostEvictionCallbackRegistration { EvictionCallback = OnEvicted });

        Write(DateTime.Now.ToString());
        cache.Set("CurrentTime", Read(), options);

        while (true)
        {
            Write(DateTime.Now.ToString());
            await Task.Delay(1000);
            if (cache.TryGetValue("CurrentTime", out string currentTime))
            {
                Console.WriteLine(currentTime);
            }
        }

        string Read()
        {
            lock (@lock)
            {
                return File.ReadAllText(fileName);
            }
        }

        void Write(string contents)
        {
            lock (@lock)
            {
                File.WriteAllText(fileName, contents);
            }
        }
    }
}
```

```
    void OnEvicted(object key, object value, EvictionReason reason, object state)
    {
        options.ExpirationTokens.Clear();
        options.AddExpirationToken(fileProvider.Watch(fileName));
        cache.Set("CurrentTime", Read(), options);
    }
}
```

如上面的代码片段所示,我们在当前目录下创建了一个名为 time.txt 的文件,并将当前时间作为内容写入该文件,然后利用依赖注入框架提供了一个 IMemoryCache 对象。我们创建了一个指向目标文件所在目录(实际上就是当前目录)的 PhysicalFileProvider 对象,并调用其 Watch 方法监控目标文件,返回的 IChangeToken 对象通过调用 AddExpirationToken 方法添加到创建的 MemoryCacheEntryOptions 对象的 ExpirationTokens 属性中。目标文件一旦发生改变,对应的缓存将被标识为过期,并在后续的扫描过程中从缓存字典中移除。为了使更新后的文件内容自动添加到缓存中,可以注册一个"缓存逐出"回调。

下面以 1 秒的间隔将表示当前时间的字符串作为内容覆盖 time.txt 文件,并随后读取和打印缓存的时间。值得注意的是,从修改文件到读取缓存之间具有 1 秒的间隔,其目的是确保缓存框架能够正常接收到目标文件的变化,并对缓存做出正确的过期处理。执行这个程序之后在控制台上得到的输出结果如图 17-8 所示,可以看出,输出的内容与目标文件的内容是同步的。(S1706)

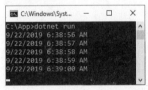

图 17-8　使物理文件的内容与缓存数据同步

关于 MemoryCache 对象针对所有 ICacheEntry 对象所做的过期检验,还有一点需要着重说明:虽然可以利用 MemoryCacheOptions 对象的 ExpirationScanFrequency 属性设置过期扫描的时间间隔,但是在后台并没有一个作业来帮助我们根据这个设定对所有的 ICacheEntry 对象进行扫描,真正的过期扫描发生在我们利用 MemoryCache 对象设置、提取、移除及实施内存压缩的时候。换句话说,如果我们在很长一段时间内没有执行过任何缓存操作,并且在这段时间内没有出现过垃圾回收,所有过期的 ICacheEntry 对象实际上还是保存在内存之中。MemoryCache 对象会记录下最近一次过期扫描的时间,对于下一次扫描请求,如果间隔时间小于设定,该请求就会被自动忽略。

缓存压缩

如果 MemoryCache 对象被设置为需要压缩缓存占用的内存空间(该选项通过 MemoryCacheOptions 类型的 CompactOnMemoryPressure 属性来设置,该属性默认返回 True,即需要在出现内存压力的情况下对缓存予以压缩),当运行时执行垃圾回收时,它总是以一个固定的比率将某些重要程度不高的 ICacheEntry 对象从缓存字典中移除。我们可以通过如下所示的实例来演示 MemoryCache 对象针对缓存的压缩。

```csharp
public class Program
{
    public static async Task Main()
    {
        var cache = new ServiceCollection()
            .AddMemoryCache(options=> {
                options.SizeLimit                       = 10;
                options.CompactionPercentage            = 0.2;
            })
            .BuildServiceProvider()
            .GetRequiredService<IMemoryCache>();

        for (int i = 1; i <= 5; i++)
        {
            cache.Set(i, i.ToString(), new MemoryCacheEntryOptions {
                Priority = CacheItemPriority.Low, Size = 1 });
        }
        for (int i = 6; i <= 10; i++)
        {
            cache.Set(i, i.ToString(), new MemoryCacheEntryOptions {
                Priority = CacheItemPriority.Normal, Size = 1 });
        }

        cache.Set(11, "11", new MemoryCacheEntryOptions
        {
            Priority = CacheItemPriority.Normal,
            Size = 1
        });
        await Task.Delay(1000);

        Console.WriteLine("Key\tValue");
        Console.WriteLine("--------------------");
        for (int i = 1; i <= 11; i++)
        {
            Console.WriteLine($"{i}\t{cache.Get<string>(i) ?? "N/A"}");
        }
    }
}
```

如上面的代码片段所示，我们利用依赖注入框架提供了一个 IMemoryCache 对象。在进行相关的服务注册时，我们将 MemoryCacheOptions 对象的 SizeLimit 属性和 CompactionPercentage 属性分别设置成 10 与 0.2。然后利用 MemoryCache 设置了 11 个 ICacheEntry 对象，这 11 个 ICacheEntry 对象具有相同的 Size（1），但是优先级不尽相同（前 5 个和后 6 个分别为 Low 与 Normal）。我们最终会尝试读取并打印这 11 个 ICacheEntry 的值，但是最终程序执行之后在控制台上输出的结果如图 17-9 所示。（S1707）

图 17-9 MemoryCache 以压缩缓存的方式清除 ICacheEntry 对象

由于我们将缓存的总容量设置为 10，所以当我们试图添加第 11 个 ICacheEntry 对象时已经超出了这个限制，此时会触发针对缓存的压缩。由于我们设置的压缩比为 20%，所以具有最低优先级的前两个 ICacheEntry 对象会从缓存字典中移除。这说明，当缓存总容量超出限制时针对缓存的设置会失败，如演示实例中的第 11 次缓存设置并没有成功。

服务注册

ASP.NET Core 应用中总是采用依赖注入的方式来使用 IMemoryCache 服务，所以在应用启动时需要进行相应的服务注册。通过前面演示的多个实例可知，针对 IMemoryCache 的服务注册是通过 IServiceCollection 接口的 AddMemoryCache 扩展方法来完成的。我们具有如下两个 AddMemoryCache 方法重载可供选择，前者采用 Singleton 模式在提供的 IServiceCollection 中添加了针对 MemoryCache 类型的注册，后者则在此基础上完成了针对 MemoryCacheOptions 的配置。

```
public static class MemoryCacheServiceCollectionExtensions
{
    public static IServiceCollection AddMemoryCache(this IServiceCollection services)
    {
        services.AddOptions();
        services.TryAdd(ServiceDescriptor.Singleton<IMemoryCache, MemoryCache>());
        return services;
    }

    public static IServiceCollection AddMemoryCache(this IServiceCollection services,
        Action<MemoryCacheOptions> setupAction)
    {
        services.AddMemoryCache();
        services.Configure(setupAction);
        return services;
    }
}
```

17.3　分布式缓存

如果将缓存数据直接保存在本地内存中，在集群的部署场景下会导致多台服务器缓存数据不一致，在这种情况下我们需要采用分布式缓存。分布式缓存不再将多个缓存副本分散于每台应用服务器内存中，而是采用独立的缓存存储。

17.3.1 IDistributedCache

分布式缓存的读和写实现在通过 IDistributedCache 接口表示的服务中。分布式缓存的操作除了设置、提取及移除，还包括刷新缓存。所谓的刷新是针对滑动时间过期策略而言的，每次刷新都会将对应缓存条目的最后访问时间设置为当前时间。IDistributedCache 接口为这 4 种基本的缓存操作提供了对应的方法，每组方法都包含同步版本和异步版本。由于采用集中式存储，必然涉及网络传输或者持久化存储，所以分布式缓存只支持字节数组这一种数据类型，应用程序需要自行解决针对缓存对象的序列化和反序列化问题。

```
public interface IDistributedCache
{
    void Set(string key, byte[] value, DistributedCacheEntryOptions options);
    Task SetAsync(string key, byte[] value, DistributedCacheEntryOptions options);

    byte[] Get(string key);
    Task<byte[]> GetAsync(string key);

    void Remove(string key);
    Task RemoveAsync(string key);

    void Refresh(string key);
    Task RefreshAsync(string key);
}
```

与基于内存的缓存一样，设置分布式缓存条目时，除了提供以字节数组表示的缓存数据，还需要提供一些其他的配置选项来决定设置的缓存条目何时过期，后者被封装成一个 DistributedCacheEntryOptions 对象。分布式缓存同样支持两种基于时间（绝对时间和滑动时间）的过期策略，绝对过期时间和滑动过期时间除了可以通过 DistributedCacheEntryOptions 相应的属性来指定，还可以调用对应的扩展方法。

```
public class DistributedCacheEntryOptions
{
    public DateTimeOffset? AbsoluteExpiration { get; set; }
    public TimeSpan? AbsoluteExpirationRelativeToNow { get; set; }
    public TimeSpan? SlidingExpiration { get; set; }
}

public static class DistributedCacheEntryExtensions
{
    public static DistributedCacheEntryOptions SetAbsoluteExpiration(
        this DistributedCacheEntryOptions options, DateTimeOffset absolute);
    public static DistributedCacheEntryOptions SetAbsoluteExpiration(
        this DistributedCacheEntryOptions options, TimeSpan relative);
    public static DistributedCacheEntryOptions SetSlidingExpiration(
        this DistributedCacheEntryOptions options, TimeSpan offset);
```

}
```

如果在设置分布式缓存时采用默认的过期配置，我们可以调用 IDistributedCache 接口的 Set 方法和 SetAsync 方法，这样只需要指定缓存条目的 Key 和 Value，而无须提供 Distributed CacheEntryOptions 对象。

```
public static class DistributedCacheExtensions
{
    public static void Set(this IDistributedCache cache, string key, byte[] value);
    public static Task SetAsync(this IDistributedCache cache, string key, byte[] value);
}
```

由于 Set 方法和 SetAsync 方法支持以字节数组表示的缓存数据，如果应用程序缓存的是一个具体的对象，那么它在被添加到缓存之前需要被序列化成字节数组。当应用程序从分布式缓存中提取出缓存数据之后，需要采用匹配的方式将其反序列成对应类型的对象。倘若我们缓存的是一个简单的字符串，针对缓存的设置和提取都可以通过调用如下这些扩展方法来实现，这些方法会采用 UTF-8 对字符串进行编码和解码。

```
public static class DistributedCacheExtensions
{
    public static void SetString(this IDistributedCache cache, string key, string value);
    public static void SetString(this IDistributedCache cache, string key, string value,
        DistributedCacheEntryOptions options);
    public static Task SetStringAsync(this IDistributedCache cache, string key,
        string value);
    public static Task SetStringAsync(this IDistributedCache cache, string key,
        string value, DistributedCacheEntryOptions options);

    public static string GetString(this IDistributedCache cache, string key);
    public static Task<string> GetStringAsync(this IDistributedCache cache, string key);
}
```

17.3.2　基于 Redis 的分布式缓存

Redis 是实现分布式缓存最好的选择，所以微软提供了针对 Redis 数据库的分布式缓存的原生支持。如图 17-10 所示，基于 Redis 数据库的分布式缓存实现在 NuGet 包 "Microsoft. Extensions.Caching.Redis" 中，而具体针对 Redis 数据库的访问则借助一个名为 StackExchange. Redis 的框架来完成。StackExchange.Redis 是 .NET 领域知名的 Redis 客户端框架，由于篇幅有限，所以本章不会涉及对 StackExchange.Redis 的介绍。

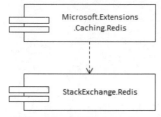

图 17-10　采用 StackExchange.Redis 访问 Redis 数据库
```

基于 Redis 数据库的分布式缓存具体实现在 RedisCache 类型上，它是对 IDistributedCache 接口的实现。如下面的代码片段所示，我们在创建一个 RedisCache 对象时需要提供一个 RedisCacheOptions 对象，后者承载了与目标数据库相关的配置选项。RedisCacheOptions 类型的 Configuration 属性相当于数据库的连接字符串，而 InstanceName 属性则是代表数据库实例的名称。但 InstanceName 仅仅是逻辑上的名称而已，在数据库服务器上并不存在一个对应的数据实例。分布式缓存在 Redis 数据库上是以散列（Hash）的形式存储的，这个 InstanceName 属性实际上会作为数据项对应 Key 的前缀。

```
public class RedisCache : IDistributedCache, IDisposable
{
    public RedisCache(IOptions<RedisCacheOptions> optionsAccessor);
    ...
}

public class RedisCacheOptions : IOptions<RedisCacheOptions>
{
    public string         Configuration { get; set; }
    public string         InstanceName { get; set; }
    RedisCacheOptions IOptions<RedisCacheOptions>.Value { get; }
}
```

当我们调用 RedisCache 对象的 Set 方法或者 SetAsync 方法写入缓存数据时，RedisCache 会将指定的缓存数据（字节数组）保存到 Redis 数据库中，与缓存数据一并保存的还包括缓存条目的绝对过期时间和滑动过期时间，具体的值是对应 DateTimeOffset 对象和 TimeSpan 对象 Ticks 属性的值（一个 Tick 代表 1 纳秒，即一千万分之一秒，1 毫秒等于 10 000 000 纳秒。DateTimeOffset 的 Ticks 数返回距离"0001 年 1 月 1 日午夜 12:00:00"这个基准时间点的纳秒数），这一点可以通过如下所示的实例进行演示。

```
public class Program
{
    public static void Main()
    {
        var cache = new ServiceCollection()
            .AddDistributedRedisCache(options =>
            {
                options.Configuration    = "localhost";
                options.InstanceName     = "Demo";
            })
            .BuildServiceProvider()
            .GetRequiredService<IDistributedCache>();

        var time = DateTimeOffset.UtcNow.AddHours(1);
        cache.SetString("Foobar", time.Ticks.ToString(), new DistributedCacheEntryOptions
        {
            AbsoluteExpiration       = time,
            SlidingExpiration        = TimeSpan.FromMinutes(1)
        });
```

        }
}
```

如上面的代码片段所示，我们针对本地 Redis 数据库创建了一个 RedisCache 对象，它采用的实例名称为 Demo。接下来我们利用此 RedisCache 对象写入了一个 Key 为 Foobar 的缓存，缓存的内容是 1 小时之后这个时间点的 Ticks 属性值，这个时间也是我们为缓存设置的绝对过期时间。除了绝对过期时间，我们还将滑动过期时间设置为 1 分（600 000 000 纳秒）。

在启动本地 Redis 服务的情况下运行这个程序后，缓存数据将会被保存起来。我们按照前面实例演示的方式运行 Redis 命令行来查看保存到 Redis 数据库中的缓存数据。由于缓存数据以散列的形式存放，并且 Key 会以指定的实例名称（Demo）为前缀，我们按照图 17-11 所示的方式执行命令"hgetall DemoFoobar"即可。从输出结果可以看出，缓存的值与绝对过期时间及滑动过期时间被保存起来，而具体的值就是对应的 Ticks 属性值。

图 17-11　查看 Redis 数据库中存放的数据

当我们利用 RedisCache 设置缓存或者根据提供的 Key 获取对应缓存数据时，对应的方法会帮助我们自动执行刷新操作，其目的在于为保存在 Redis 中的缓存项设置一个最新的过期时间。具体的实现逻辑其实很简单：RedisCache 会根据当前时间和保存在 Redis 数据库中的绝对过期时间与滑动过期时间计算一个新的过期时长，采用的算法为 Min (AbsoluteExpiration -Now, SlidingExpiration)。这个实时计算出的过期时长会立即应用到 Redis 数据库中对应的缓存数据项，最终借助 Redis 自身基于 TTL 的过期机制使该数据项在过期之后被自动清除。

最后介绍注册 RedisCache 服务的 AddDistributedRedisCache 扩展方法。如下面的代码片段所示，该方法除了采用 Singleton 模式完成了针对 RedisCache 的服务注册，还采用 Options 模式对相应的配置选项进行了设置。

```
public static class RedisCacheServiceCollectionExtensions
{
    public static IServiceCollection AddDistributedRedisCache(
        this IServiceCollection services, Action<RedisCacheOptions> setupAction)
    {
        services.AddOptions();
        services.Configure(setupAction);
        services.Add(ServiceDescriptor.Singleton<IDistributedCache, RedisCache>());
        return services;
    }
}
```
```

17.3.3 基于 SQL Server 的分布式缓存

除了适合做分布式缓存的 NoSQL 数据库，还有关系型数据库 SQL Server，基于后者的分布式缓存实现在 SqlServerCache 类型中，该类型由 NuGet 包 "Microsoft.Extensions.Caching.SqlServer" 提供。在正式介绍 SqlServerCache 之前，需要先了解承载配置选项的 SqlServerCacheOptions 类型。

```
public class SqlServerCache : IDistributedCache
{
    public SqlServerCache(IOptions<SqlServerCacheOptions> options);
    ...
}

public class SqlServerCacheOptions : IOptions<SqlServerCacheOptions>
{
    public string ConnectionString { get; set; }
    public string TableName { get; set; }
    public string SchemaName { get; set; }

    public TimeSpan           DefaultSlidingExpiration { get; set; }
    public ISystemClock       SystemClock { get; set; }
    public TimeSpan?          ExpiredItemsDeletionInterval { get; set; }

    SqlServerCacheOptions IOptions<SqlServerCacheOptions>.Value { get; }
}
```

由于需要将缓存数据及相关的控制信息保存到 SQL Server 数据库的某个具有预定义结构的表中，所以 SqlServerCacheOptions 需要提供目标数据库和数据表的信息。具体来说，它的 ConnectionString 属性表示目标数据库的连接字符串，而 TableName 属性和 SchemaName 属性分别表示缓存数据表的名称与 Schema。

我们在进行缓存设置时可以通过一个 DistributedCacheEntryOptions 对象设置绝对过期时间和滑动过期时间，如果这两者都没有设置，就意味着缓存永远不会过期，所以缓存记录将永远保存在数据库中，这无疑会导致缓存表的容量无限制扩大。为了解决这个问题，SqlServerCache 在这种情况下会采用基于滑动时间的缓存过期策略，默认的滑动过期时间由 SqlServerCacheOptions 的 DefaultSlidingExpiration 属性来表示，该属性的默认值为 20 分。至于定义在 SqlServerCacheOptions 中另一个与缓存过期有关的属性 SystemClock，返回的是一个用来提供同步时间的系统时钟。

与基于 Redis 数据库的分布式缓存相比，针对 SQL Server 数据库的分布式缓存与其差异最大的就是如何删除过期缓存条目。对于 Redis 数据库来说，它自身就具有为某个记录设置过期时间并在其过期之后将其清除的能力，而 SQL Server 数据库不具有这样的特性，所以 SqlServerCache 需要自行清除过期缓存记录。两次清除操作执行的时间间隔由 SqlServerCacheOptions 的 ExpiredItemsDeletionInterval 属性表示，如果创建 SqlServerCache 时提供的 SqlServerCacheOptions 没有显式指定这个属性，那么默认采用的时间间隔为 30 分。考虑到性能，针对过期缓存记录的清理不应该过于频繁，具体来说，这个间隔不应该小于 5 分，如果指定的间隔小于这个值将会抛出异常。

通过前面的实例演示可知，缓存表可以通过执行命令"dotnet sql-cache create"来创建。SQL Server 缓存数据表结构如图 17-12 所示，该表具有 5 行：Id 和 Value 分别表示缓存的 Key 与 Value；而 SlidingExpirationInSeconds 和 AbsoluteExpiration 则表示滑动过期时间（以秒为单位）与绝对过期时间；ExpiresAtTime 则是综合 SlidingExpirationInSeconds 和 AbsoluteExpiration 计算出的过期时间点，也就是说，SqlServerCache 判断某个缓存是否过期只需要将当前时间与这个时间进行比较即可。

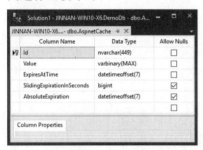

图 17-12　SQL Server 缓存数据表结构

调用 SqlServerCache 的 Set 方法或者 SetAsync 方法对缓存的设置，实际上就是在图 17-12 所示的这张表中添加或者修改一条缓存记录。对于缓存表中表示绝对过期时间的 AbsoluteExpiration 字段，缓存的绝对过期时间既可以通过 DistributedCacheEntryOptions 的 AbsoluteExpiration 属性来设置，也可以通过其 AbsoluteExpirationRelativeToNow 属性来设置。如果我们对这两个属性都做了设置，对于基于内存的缓存来说，它会选择距离当前时间最近的时间点作为缓存真正的绝对过期时间。但是基于 SqlServerCache 则会优先选择 AbsoluteExpirationRelativeToNow 属性，只有在该属性没有被显式赋值的情况下它才会选择 AbsoluteExpiration 属性。笔者认为采用一致规则也许是更好的选择。

当我们利用 SqlServerCache 的 Get 方法或者 GetAsync 方法提取缓存数据时，它会根据指定的 Key 从数据库中提取对应的缓存记录，只有在缓存记录尚未过期的情况下它才会返回缓存的数据，而判断某条缓存记录是否过期只需要让系统时钟提供的当前时间与其 ExpiresAtTime 字段表示的时间进行比较即可。如果调用 Refresh 方法或者 RefreshAsync 方法对指定的缓存进行刷新，SqlServerCache 会重新计算过期时间，并将该时间作为对应缓存记录的 ExpiresAtTime。

与基于内存的缓存一样，虽然通过 SqlServerCacheOptions 的 ExpiredItemsDeletionInterval 属性设置了清除过期缓存记录的时间间隔，但是我们在后台并没有一个长时间运行的作业定期做这样的清理工作。针对过期缓存记录的清理是在我们调用 Set（SetAsync）方法、Get（GetAsync）方法、Refresh（RefreshAsync）方法和 Remove（RemoveAsync）方法时触发的，所以它并不能保证过期的缓存记录在指定的时间间隔范围内被及时清除。

最后介绍注册 SqlServerCache 服务的 AddDistributedSqlServerCache 扩展方法。如下面的代

码片段所示，该方法除了采用 Singleton 模式完成了针对 SqlServerCache 的服务注册，还采用 Options 模式对相应的配置选项进行了设置。

```
public static class SqlServerCachingServicesExtensions
{
    public static IServiceCollection AddDistributedSqlServerCache(
        this IServiceCollection services, Action<SqlServerCacheOptions> setupAction)
    {
        services
            .AddOptions()
            .Add(ServiceDescriptor.Singleton<IDistributedCache, SqlServerCache>());
        services.Configure(setupAction);
        return services;
    }
}
```

17.4 响应缓存

上面介绍的缓存（内存缓存和分布式缓存）涉及的仅仅是单纯地对数据的存储，而接下来介绍的响应缓存则是从 HTTP 报文交换的角度针对服务器提供的 HTTP 响应报文进行缓存的。ASP.NET Core 的响应缓存是利用 ResponseCachingMiddleware 中间件实现的。ResponseCachingMiddleware 中间件按照标准的 HTTP 规范来操作缓存，所以在正式介绍这个中间件时，需要先介绍 HTTP 规范（HTTP/1.1）中针对缓存的描述。

17.4.1 HTTP/1.1 Caching

HTTP 规范下的缓存只针对方法为 GET 的请求或者 HEAD 的请求，这样的请求旨在获取 URL 所指向的资源或者描述资源的元数据。如果将资源的提供者和消费者称为原始服务器（Origin Server）与客户端（Client），那么所谓的缓存则是存在于这两者之间的一个 HTTP 处理部件，如图 17-13 所示。

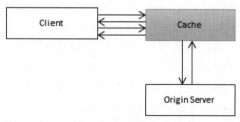

图 17-13　位于 Client 和 Origin Server 之间的 Cache

缓存会根据一定的规则在本地存储一份原始服务器提供的响应副本，并赋予它一个"保质期"，保质期内的副本可以直接用来作为后续匹配请求的响应，所以缓存能够避免客户端与原始服务器之间不必要的网络交互。即使过了保质期，缓存也不会直接从原始服务器中获取最新的响应副本，而是选择向其发送一个请求来检验目前的副本是否与最新的内容一致，如果原始服务器做出"一致"的答复,原本过期的响应副本又变得"新鲜"并且被继续使用。所以,缓存还能避免冗余资源在网络中的重复传输。

私有缓存和共享缓存

私有缓存为单一客户端存储响应副本,所以它不需要过多的存储空间,如浏览器利用私有缓存空间(本地物理磁盘或者内存)存储常用的响应文档,它的前进/后退、保存、查看源代码等操作访问的都是本地私有缓存的内容。有了私有缓存,我们还可以实现脱机浏览文档。

共享缓存又称为公共缓存,它存储的响应文档可以被所有的客户端共享,这种类型的缓存一般部署在一个私有网络的代理服务器上,我们将这样的服务器称为缓存代理服务器。缓存代理服务器可以从本地提取相应的响应副本对来自本网络的所有主机的请求予以响应,同时代表它们向原始服务器发送请求。响应报文以如下所示的形式采用 Cache-Control 报头区分私有缓存和共享缓存。

```
Cache-Control: public|private
```

响应的提取

缓存数据通常采用字典类型的存储结构,并通过提供的 Key 来定位目标缓存条目,那么对于 HTTP 缓存的响应报文,它采用的 Key 具有怎样的组成元素?一般来说,一个 GET 请求或者 HEAD 请求的 URL 会作为获取资源的标识,所以请求 URL 是组成缓存键最核心的元素。

当缓存接收到来自客户端的请求时,它会根据请求的 URL 选择与之匹配的响应副本。除了基本路径,请求 URL 可能还携带着一些查询字符串,至于查询字符串是否会作为选择的条件之一,HTTP/1.1 对此并没有明确的规定。通过前面演示的实例可知,ResponseCachingMiddleware 中间件在默认情况下并不会将携带的查询字符串作为缓存键的组成部分。

按照 REST 的原则,URL 实际上是网络资源的标识,但是响应的主体仅仅是资源的某种表现形式(Represetation)而已,相同的资源针对不同的格式、不同的语言将转换成完全不同的内容荷载,所以作为资源唯一标识的 URL 并不能唯一标识缓存的响应副本。由于相同资源的表现形式由某个或者多个请求报头来决定,所以缓存需要综合采用请求的 URL 以及这些请求报头来存储响应副本。为了提供用于存储响应内容荷载的请求报头名称,原始服务器在生成最初响应的时候会将它们存储在一个名为 Vary 的报头中。

下面列举一个关于报文压缩的典型例子。为了节约网络带宽,客户端希望原始服务器将响应的主体内容进行压缩,为此它会向服务器发送如下请求:请求报头集合中包含一个表示希望采用的压缩编码格式(gzip)的 Accept-Encoding 报头。原始服务器接收到该请求后将主体内容按照期望的格式进行压缩,并将压缩采用的编码保存到 Content-Encoding 响应报头中。除此之外,该响应还具有一个值为 Accept-Encoding 的 Vary 报头。

```
GET http://localhost/foobar HTTP/1.1
Host: localhost
Accept-Encoding: gzip
```

```
HTTP/1.1 200 OK
Date: Thu, 23 Feb 2017 15:24:05 GMT
Cache-Control: public, max-age=3600
Content-Encoding: gzip
Vary: Accept-Encoding

<<body>>
```

当缓存决定存储该响应副本的时候，会提取出响应的 Vary 报头提供的所有请求报头名称，并将对应的值作为存储该响应副本对应 Key 的组成部分。对于后续指向同一个 URL 的请求，只有在它们具有一致的报头值的情况下，对应的副本才会被选择。

新鲜度检验

缓存的数据仅仅是服务器提供响应的一份副本，两者之间应该尽可能保持一致。我们将确定缓存内容与真实内容一致性的检验过程称为再验证（Revalidation）。原则上，缓存可以在任何时候向服务端发出对缓存的响应内容实施再验证的请求，但是出于性能的考虑，缓存的再验证只会发生缓存在接收到客户端请求，并且认为本地存储的响应副本已经陈旧得需要再次确定一致性的时候。那么缓存如何确定自身存储的响应内容目前依旧是"新鲜"的？

一般来说，响应内容在某个时刻是否新鲜应该由作为提供者的原始服务器决定，服务器只需要为响应内容设置一个保质期即可。响应内容的保质期通过相应的报头来表示，HTTP/1.1 采用 Cache-Control：max-age = <seconds>报头，而 HTTP/10+使用的是 Expires 报头，前者携带的是采用以秒为单位的时长，后者采用的则是一个具体的过期时间点（一般采用 GMT 时间）。HTTP/1.1 之所以没有采用绝对过期时间点，主要是考虑到时间同步的问题。

```
Cache-Control: max-age=1800
Expires: Thu, 23 Feb 2017 02:48:10 GMT
```

当缓存接收到请求并按照上面的策略选择出匹配的响应副本之后，如果响应副本满足"新鲜度"要求，它会直接用来作为当前请求的响应。如果响应副本已经过期，缓存也不会直接将其丢弃，而是选择向原始服务器发送一个再验证请求以确定当前的响应副本是否与目前的数据一致。考虑到带宽及数据比较的代价，再验证请求并不会向原始服务器提供当前响应副本的内容供其比较，那么具体的一致性比较又是如何实现的？

确定资源是否发生变化可以采用两种策略。第一种就是让资源的提供者记录最后一次更新资源的时间，资源内容荷载和这个时间戳会作为响应的内容提供给客户端。客户端在缓存资源自身内容的同时会保存这个时间戳。等到下次需要针对同一资源发送请求的时候，它会将这个时间戳一并发送出去，原始服务器就可以根据这个时间戳判断目标资源在此时间之后是否被修改过。除了采用记录资源最后修改时间的方式，我们还可以针对资源的内容生成一个标签，标签的一致性体现了资源内容的一致性，在 HTTP 规范中将这个标签称为 ETag（Entity Tag）。

具体来说，原始服务器生成的响应会包含一个代表资源最后修改时间戳的 Last-Modified 报头，或者包含一个代表资源内容标签的 ETag 报头，也可以同时包含这两个报头。缓存向原始服务器发送的再验证请求会分别利用 If-Unmodified-Since 报头和 If-Match 报头携带这个时间戳与标签，原始服务器接收到请求之后利用它们判断资源是否发生改变。如果资源一直没有改变，它会返回一个状态码为 "304 Not Modified" 的响应，那么缓存会保留目前响应副本的主体内容，并更新相应的过期信息使其重新变得"新鲜"。如果资源已经改变，原始服务器返回一个状态码为 "200 OK" 的响应，该响应的主体会携带最新的资源。缓存在接收到响应之后，会使用新的响应副本覆盖现有过期的响应副本。

显式缓存控制

对于原始服务器生成的响应，如果它没有包含任何与缓存控制相关的信息，是否应该被存储，以及它在何时过期都由缓存自身采用的默认策略来决定。如果原始服务器希望缓存采用指定的策略对其生成的响应实施缓存，它就可以在响应中添加一些与缓存相关的报头。按照缓存约束程度（由紧到松）列举的报头有如下几个。

- Cache-Control:no-store：不允许缓存存储当前响应的副本，如果响应承载了一些敏感信息或者数据随时都会发生改变，应该使用这个报头阻止响应内容被缓存起来。
- Cache-Control:no-cache 或者 Pragma: no-cache：缓存可以在本地存储当前响应的副本，但是不论是否过期，该副本都需要经过再验证确定一致性之后才能提供给客户端。HTTP/1.1 保留 Pragma: no-cache 报头是为了兼容 HTTP/1.0+。
- Cache-Control:must-revalidate：缓存可以在本地存储当前响应的副本，但是在过期之后必须经过再验证确定内容一致性之后才能提供给客户端。
- Cache-Control:max-age：缓存可以在本地存储当前响应的副本，它在指定的时间内保持"新鲜"。
- Expires：缓存可以在本地存储当前响应的副本，并且在指定的时间点到来之前保持"新鲜"。

除了原始服务器，客户端有的时候对于响应副本的新鲜度同样具有自己的要求，这些要求可能高于或者低于缓存默认采用的新鲜度检验策略。客户端利用请求的 Cache-Control 报头来提供相应的缓存指令，具体可以使用的报头包括以下几个。

- Cache-Control:max-stale 或者 Cache-Control:max-state={seconds}：缓存可以提供过期的副本来响应当前请求，客户端可以设置一个以秒为单位的允许过期时间。
- Cache-Control:min-fresh={seconds}：缓存提供的响应副本必须在未来 *N* 秒内保持新鲜。
- Cache-Control:max-age={seconds}：缓存提供的响应副本在本地存储的时间不能超出指定的秒数。
- Cache-Control:no-cache 或者 Pragma: no-cache：客户端不接受未经过再验证的响应副本。
- Cache-Control:no-store：如果对应的响应副本存在，缓存就应该尽快将其删除。
- Cache-Control:only-if-cached：客户端只接受缓存的响应副本。

17.4.2　ResponseCachingMiddleware 中间件

ResponseCachingMiddleware 中间件就是对上述 HTTP/1.1 缓存的具体实现。当该中间件处理某个请求时,它会根据既定的策略判断该请求是否可以采用缓存的文档来对请求做出响应。如果不应该采用缓存的形式来处理该请求,那么它只需要直接将请求交给后续的管道进行处理即可;反之,该中间件会直接使用缓存的文档来响应请求。倘若本地并未存储对应的响应文档,ResponseCachingMiddleware 中间件会利用后续的管道生成此响应文档,该文档被用于响应请求之前会先被缓存起来。ResponseCachingMiddleware 中间件针对请求的处理由多个核心对象协作完成,下面会依次介绍这几个对象。虽然整个流程基本上是按照 HTTP/1.1 针对缓存的规范进行的,但还是显得相对复杂和烦琐,本节的内容可以作为选读内容。

ResponseCachingContext

ASP.NET Core 在很多地方都采用了基于上下文(Context)的设计,这个设计模式的一个典型特点就是为某项相对复杂的处理流程创建一个执行上下文。该上下文不仅为处理流程的各个环节的执行提供输入数据,处理后产生的输出也直接被放置到同一个上下文中。ResponseCachingMiddleware 中间件采用的基于响应缓存的请求处理同样采用这种模式,对应的上下文通过一个 ResponseCachingContext 对象来表示,该中间件在处理请求过程中的第一项操作就是根据表示当前请求 HttpContext 上下文创建一个 ResponseCachingContext 对象。

ResponseCachingContext 类型具有很多内部属性和方法成员,由于篇幅有限,此处仅列出了如下几个公共属性。如下面的代码片段所示,ResponseCachingContext 对象的 HttpContext 属性表示当前处理请求上下文,而 CachedEntryAge 属性表示响应文档被缓存的时间(以秒为单位),它将作为响应报头 Age 的值。ResponseTime 属性是 ResponseCachingMiddleware 中间件在试图利用缓存的文档来响应请求时由系统时钟提供的时间,它是后期对缓存进行过期检验的基础。

```
public class ResponseCachingContext
{
    public HttpContext              HttpContext { get; }
    public TimeSpan?                CachedEntryAge { get; }
    public DateTimeOffset?          ResponseTime { get; }
    public CachedVaryByRules        CachedVaryByRules { get; }
}
```

通过上面对 HTTP/1.1 缓存规范的介绍可知,缓存在对响应文档进行存储以及根据请求提取响应文档时不仅会考虑请求的路径,还会考虑一些指定的查询字符串和请求报头(包含在响应报头 Vary 携带的报头名称列表中),后者通过如下所示的 CachedVaryByRules 对象来表示。CachedVaryByRules 对象的 QueryKeys 属性和 Headers 属性分别表示查询字符串名称列表与 Vary 请求报头名称列表。除了这两个属性,CachedVaryByRules 对象还通过 VaryByKeyPrefix 属性携带一个前缀以保持唯一性。

```
public class CachedVaryByRules : IResponseCacheEntry
{
    public string                   VaryByKeyPrefix { get; set; }
    public StringValues             QueryKeys { get; set; }
    public StringValues             Headers { get; set; }
}
```

当 CachedVaryByRules 对象被创建时，一个 GUID 会被生成出来作为其 VaryByKeyPrefix 属性的值，而 Headers 属性的值则来源于当前响应的 Vary 报头。QueryKeys 属性携带的查询字符串名称列表则来源于一个通过 IResponseCachingFeature 接口表示的特性。在前面的实例演示中，我们正是使用这个特性来指定影响缓存的查询字符串名称列表的。

```
public interface IResponseCachingFeature
{
    string[] VaryByQueryKeys { get; set; }
}
```

IResponseCachingPolicyProvider

ResponseCachingMiddleware 中间件会根据既定的策略先判断当前请求是否能够采用缓存的文档来予以响应，这个所谓的策略体现在如下所示的 IResponseCachingPolicyProvider 接口的 5 个方法上。对于这 5 个方法来说，前三个是针对请求的缓存策略，AttemptResponseCaching 方法用于判断是否需要采用缓存的形式来处理当前请求。对于 ResponseCachingMiddleware 中间件来说，AttemptResponseCaching 方法相当于一个总开关，如果该方法返回 False，就意味着该请求将直接分发给后续中间件进行处理。AllowCacheLookup 方法则表示是否可以利用尚未过期的缓存文档在不需要进行再验证的情况下对当前请求予以响应。AllowCacheStorage 方法则表示请求对应的响应是否允许被缓存。

```
public interface IResponseCachingPolicyProvider
{
    bool AttemptResponseCaching(ResponseCachingContext context);
    bool AllowCacheLookup(ResponseCachingContext context);
    bool AllowCacheStorage(ResponseCachingContext context);

    bool IsResponseCacheable(ResponseCachingContext context);
    bool IsCachedEntryFresh(ResponseCachingContext context);
}
```

IResponseCachingPolicyProvider 接口的 IsResponseCacheable 方法与响应相关，它表示已经生成的响应是否需要被缓存起来。IsCachedEntryFresh 方法则用于缓存的新鲜度检验，它的返回值表示缓存的响应文档是否新鲜。默认实现 IResponseCachingPolicyProvider 接口的是一个名为 ResponseCachingPolicyProvider 的类型，它完全采用上述 HTTP/1.1 缓存规范来实现这 5 个方法。

默认的缓存策略如表 17-1 所示。

表 17-1 默认的缓存策略

方法	描述
AttemptResponseCaching	对于非 GET/HEAD 请求，或者请求携带一个 Authorization 报头，该方法返回 False，其他情况则返回 True
AllowCacheLookup	如果请求携带报头 Cache-Control:no-cache 或者 Pragma:no-cache，该方法返回 False，其他情况则返回 True
AllowCacheStorage	如果请求携带报头 Cache-Control:no-store，该方法返回 False，其他情况则返回 True

续表

方　　法	描　　述
IsResponseCacheable	对于如下这几种响应,该方法返回 False,其他情况则返回 True • 响应不具有一个 Cache-Control :public(只支持共享缓存而不支持私有缓存) • 响应携带报头Χαχηε–Χοντρολ:νο–χαχηε或者Χαχηε–Χοντρολ:νο–στορε • 响应携带报头Σετ–Χοοκιε(响应被设置了Χοοκιε) • 响应携带报头ςαρψ:*(表示每个请求应该单独对待) • 响应状态码不是200(成功的响应才能被缓存) • 响应已经过期
IsCachedEntryFresh	根据请求和响应携带的相关报头(Cache-Control、Expires、Date 和 Pragma),严格采用 HTTP /1.1 缓存规范描述的算法确定缓存是否新鲜

IResponseCachingKeyProvider

当 ResponseCachingMiddleware 中间件在存储响应文档或者根据请求提取缓存文档时总是先生成对应的 Key,这个 Key 由一个 IResponseCachingKeyProvider 对象生成。如果不需要考虑查询字符串或者 Vary 请求报头,ResponseCachingMiddleware 中间件只需要调用 CreateBaseKey 方法针对当前请求生成对应的 Key 即可。当它在存储响应文档的时候会调用 CreateStorageVaryByKey 方法生成对应的 Key,而在根据请求提取缓存响应文档的时候则会调用 CreateLookupVaryByKeys 方法生成一组候选的 Key。

```
public interface IResponseCachingKeyProvider
{
    string CreateBaseKey(ResponseCachingContext context);
    string CreateStorageVaryByKey(ResponseCachingContext context);
    IEnumerable<string> CreateLookupVaryByKeys(ResponseCachingContext context);
}
```

默认实现 IResponseCachingKeyProvider 接口的是一个名为 ResponseCachingKeyProvider 的类型,下面讨论该类型的 3 个方法会生成什么样的 Key。CreateBaseKey 方法会采用 {Method}{Delimiter}{Path}的格式生成对应的 Key,即生成的 Key 由请求的方法(GET 或者 HEAD)与路径组合而成。对于为存储的响应文档生成 Key 的 CreateStorageVaryByKey 方法来说,返回的 Key 由缓存上下文携带的 CachedVaryByRules 对象来生成。我们不需要关注具体的 Key 会采用什么样的格式,只需要知道这个 Key 会包含所有的 Vary 请求报头和 Vary 查询字符串的名称与请求携带的值。

用来存储响应文档的 Key 必然与用于提取缓存文档的 Key 保持一致,所以 ResponseCachingKeyProvider 类型的 CreateLookupVaryByKeys 方法会返回一个唯一的 Key,而这个 Key 就是直接调用 CreateStorageVaryByKey 方法生成的。但是对于缓存上下文携带的 CachedVaryByRules 对象来说,它的 Headers 属性值来源于响应的 Vary 报头,CreateLookupVaryByKeys 方法的目的在于根据当前请求生成用于提取对应缓存文档的 Key,而此时响应是不存在的,所以 CachedVaryByRules 对象的 Headers 属性应该为空。

为了解决这个问题,除了缓存具体的响应文档,ResponseCachingMiddleware 中间件还会将针对当前 HttpContext 上下文创建的 CachedVaryByRules 对象一并存储起来，后者采用的 Key 就是 CreateBaseKey 方法的返回值。所以，调用 CreateLookupVaryByKeys 方法时从缓存上下文中提取的 CachedVaryByRules 对象并不是针对当前请求的，而是之前缓存的。

IResponseCache

到目前为止，我们尚未提到响应文档是如何被缓存的，实际上，具体的缓存操作是由一个 IResponseCache 接口表示的服务完成的。该接口定义了如下两组方法，它们分别以同步和异步的方式实现了针对缓存的读写。在进行缓存设置时，我们可以为设置的缓存条目设置一个过期时间。从下面的代码片段可以看出，设置的缓存条目可以通过 IResponseCacheEntry 接口表示，但仅仅是一个不具有任何成员的标识接口而已，下面会讨论如此设计的目的。

```
public interface IResponseCache
{
    IResponseCacheEntry Get(string key);
    Task<IResponseCacheEntry> GetAsync(string key);

    void Set(string key, IResponseCacheEntry entry, TimeSpan validFor);
    Task SetAsync(string key, IResponseCacheEntry entry, TimeSpan validFor);
}

public interface IResponseCacheEntry{}
```

ASP.NET Core 默认采用的 IResponseCache 实现类型为 MemoryResponseCache，根据命名可知，它采用的是基于本地内存的缓存。如下面的代码片段所示，我们在创建一个 MemoryResponseCache 对象时需要提供一个 IMemoryCache 对象，真正的缓存设置和获取操作最终会落在这个对象上。

```
public class MemoryResponseCache : IResponseCache
{
    public MemoryResponseCache(IMemoryCache cache);
    public IResponseCacheEntry Get(string key);
    public Task<IResponseCacheEntry> GetAsync(string key);

    public void Set(string key, IResponseCacheEntry entry, TimeSpan validFor);
    public Task SetAsync(string key, IResponseCacheEntry entry, TimeSpan validFor);
}
```

当 ResponseCachingMiddleware 中间件利用 MemoryResponseCache 来缓存响应文档时，会创建一个 CachedResponse 对象来表示需要被缓存的响应文档。如下面的代码片段所示，我们可以利用这个对象得到响应创建的时间、状态码、报头集合和主体内容。

```
public class CachedResponse : IResponseCacheEntry
{
    public DateTimeOffset            Created { get; set; }
    public int                       StatusCode { get; set; }
    public IHeaderDictionary         Headers { get; set; }
    public Stream                    Body { get; set; }
}
```

虽然 ResponseCachingMiddleware 中间件提供的是一个 CachedResponse 对象，但是被 Memory

ResponseCache 真正存储在内存中的却是另一个类型为 MemoryCachedResponse（它仅仅是一个内部类型）的对象。由于 MemoryCachedResponse 类型并没有实现 IResponseCacheEntry 接口，所以调用 Get 方法或者 GetAsync 方法时，MemoryResponseCache 对象会将提取的 MemoryCachedResponse 对象还原成 CachedResponse 对象。

如果读者足够细心，应该可以意识到前面介绍的 CachedVaryByRules 实际上也是 IResponseCacheEntry 接口的实现类型之一。换句话说，CachedVaryByRules 对象实际上可以作为一个缓存条目被存储，这进一步佐证了前面提到的关于 ResponseCachingKeyProvider 对象利用缓存的 CachedVaryByRules 对象来计算用于提取缓存响应文档的 Key 的说法。这样一个对象在何时被存储和提取，下文会进行介绍。

响应缓存的实现

在认识了众多辅助对象之后，下面介绍 ResponseCachingMiddleware 中间件究竟是如何利用它们实现响应缓存的，但在此之前需要先了解 ResponseCachingMiddleware 类型的定义。如下面的代码片段所示，当我们在创建一个 ResponseCachingMiddleware 对象的时候，除了需要提供上述 3 个核心对象（它们分别是 IResponseCachingPolicyProvider、IResponseCachingKeyProvider 和 IResponseCache），还需要提供一个 LoggerFactory 对象用来生成记录日志的 Logger，以及承载相关配置选项的 ResponseCachingOptions 对象。

```
public class ResponseCachingMiddleware
{
    public ResponseCachingMiddleware(RequestDelegate next,
        IOptions<ResponseCachingOptions> options, ILoggerFactory loggerFactory,
        IResponseCachingPolicyProvider policyProvider, IResponseCache cache,
        IResponseCachingKeyProvider keyProvider);

    public Task Invoke(HttpContext httpContext);
}
```

下面先介绍 ResponseCachingOptions 对象包含的配置选项。如下面的代码片段所示，该类型只包含两个属性：MaximumBodySize 属性表示缓存的响应文档的主体内容允许的最大容量（以字节为单位），超过这个容量的响应文档将不会被缓存，该属性的默认值为 64MB；UseCaseSensitivePaths 属性表示在使用请求路径生成 Key 时是否需要考虑大小写的问题，由于路径在大部分情况下是不区分大小写的，所以这个属性默认返回 False。

```
public class ResponseCachingOptions
{
    public long MaximumBodySize { get; set; }
    public bool UseCaseSensitivePaths { get; set; }
}
```

当 ResponseCachingMiddleware 中间件开始处理分发给它的请求时，它会先创建一个作为缓存上下文的 ResponseCachingContext 对象，然后将这个上下文作为参数调用 IResponseCachingPolicyProvider 对象的 AttemptResponseCaching 方法，该方法会判断当前请求是否能够采用缓存机制来处理。如果不能采用缓存机制进行处理，那么它只需要将请求直接递交给后续管道处理即可。

如果可以利用缓存机制来处理当前请求，IResponseCachingPolicyProvider 对象的 AllowCacheLookup 方法就会被调用，该方法会判断是否允许在不经过再验证以确定一致性情况下，直接使用缓存的保持新鲜的文档来直接响应当前请求。如果允许，ResponseCaching Middleware 中间件会调用 IResponseCachingKeyProvider 对象的 CreateBaseKey 方法生成一个 Key，然后将这个 Key 作为参数调用 IResponseCache 对象的 GetAsync 方法得到一个表示缓存条目的 IResponseCacheEntry 对象。

如果这个 IResponseCacheEntry 是一个 CachedVaryByRules 对象，就意味着响应文档应该基于 Vary 请求报头或者 Vary 查询字符串（或者两者都包括）进行存储或者提取，ResponseCaching Middleware 中间件在此情况下会调用 IResponseCachingKeyProvider 的 CreateLookupVaryByKeys 方法生成用于提取响应文档的 Key。接下来它会再次利用这个 Key 调用 IResponseCache 对象的 GetAsync 方法获取缓存的响应文档。如果缓存的响应文档被成功提取出来，ResponseCaching Middleware 中间件就会调用 IResponseCachingPolicyProvider 对象的 IsCachedEntryFresh 方法判断它是否新鲜。对于新鲜的缓存文档，在对报头适当修正之后会直接用来响应当前请求。

如果这个 IResponseCacheEntry 并不是一个 CachedVaryByRules 对象，那么根据默认实现原理，它就是代表缓存响应文档的 CachedResponse 对象。在此情况下，ResponseCaching Middleware 中间件会采用与上面一致的方式来处理这个缓存响应文档。

如果调用 AllowCacheLookup 方法返回的结果是 False，或者针对当前请求无法获取对应的缓存文档，又或者获取的缓存文档已经不够新鲜并且需要进行再验证，ResponseCaching Middleware 中间件在这几种情况下都会调用 IResponseCachingPolicyProvider 对象的 AllowCache Storage 方法判断针对当前请求的响应是否可以被存储。在允许存储的情况下，Response CachingMiddleware 中间件会将请求递交给后续管道进行处理并生成响应。为了确定生成的响应能够被缓存，它会调用 IResponseCachingPolicyProvider 对象的 IsResponseCacheable 方法。如果该方法返回 True，ResponseCachingMiddleware 中间件就会对响应文档实施缓存。

具体来说，如果响应缓存需要考虑 Vary 请求报头或者 Vary 查询字符串，ResponseCaching Middleware 中间件就会针对当前 HttpContext 创建一个 CachedVaryByRules 对象并利用 IResponse Cache 对其进行缓存，对应的 Key 就是调用 IResponseCachingKeyProvider 对象的 CreateBaseKey 方法的返回值。在此之后，IResponseCachingKeyProvider 对象的 CreateStorageVaryByKey 方法才会被调用，并且生成用于存储响应文档的 Key。反之，如果无须考虑 Vary 请求报头或者 Vary 查询字符串，那么 ResponseCachingMiddleware 中间件只需要存储响应文档即可，对应的 Key 就是调用 IResponseCachingKeyProvider 对象的 CreateBaseKey 方法的返回结果。

上面就是 ResponseCachingMiddleware 中间件针对响应缓存的总体实现流程，但忽略了针对 IResponseCachingFeature 特性的注册和注销。当 ResponseCachingMiddleware 中间件将请求递交给后续管道进行处理之前，它会创建一个 IResponseCachingFeature 特性，并注册到代表当前请求上下文的 HttpContext 之上，所以我们的应用才可以利用它为当前响应设置 Vary 查询字符串名称列表。在整个请求处理完成之前，ResponseCachingMiddleware 中间件会负责将这个特性从当前 HttpContext 中删除。

17.4.3 注册中间件

针对 ResponseCachingMiddleware 中间件的注册是通过调用 IApplicationBuilder 接口的 UseResponseCaching 扩展方法来完成的。但仅仅调用这个扩展方法在 ASP.NET Core 管道上注册这个中间件是不够的，因为这个中间件针对响应缓存的实现是综合多个服务对象共同完成的，所以还需要预先注册这些服务。

```csharp
public static class ResponseCachingExtensions
{
    public static IApplicationBuilder UseResponseCaching(this IApplicationBuilder app)
    {
        return app.UseMiddleware<ResponseCachingMiddleware>();
    }
}
```

具体来说，ResponseCachingMiddleware 中间件主要依赖 3 个核心的服务对象，它们分别是根据 HTTP/1.1 缓存规范提供策略的 IResponseCachingPolicyProvider 对象、在设置和提取缓存时根据当前上下文生成 Key 的 IResponseCachingKeyProvider 对象、真正帮助我们完成响应缓存存取的 IResponseCache 对象。除此之外，由于默认的 ResponseCache 是一个 IMemoryResponseCache 对象，后者采用基于本地内存缓存的方式来存储响应文档，所以除了上述 3 个服务对象，还需要注册与内存缓存相关的服务。这些服务的注册可以通过 IServiceCollection 接口如下所示的两个 AddResponseCaching 扩展方法来完成。

```csharp
public static class ResponseCachingServicesExtensions
{
    public static IServiceCollection AddResponseCaching(this IServiceCollection services)
    {
        services.AddMemoryCache();
        services.TryAdd(ServiceDescriptor
            .Singleton<IResponseCachingPolicyProvider, ResponseCachingPolicyProvider>());
        services.TryAdd(ServiceDescriptor
            .Singleton<IResponseCachingKeyProvider, ResponseCachingKeyProvider>());
        services.TryAdd(ServiceDescriptor
            .Singleton<IResponseCache, MemoryResponseCache>());

        return services;
    }

    public static IServiceCollection AddResponseCaching(this IServiceCollection services,
        Action<ResponseCachingOptions> configureOptions)
    {
        services.Configure(configureOptions);
        services.AddResponseCaching();
        return services;
    }
}
```

第18章

会 话

HTTP 是一种采用请求/响应消息交换模式且无状态的传输协议。HTTP 协议旨在确保客户端将请求报文发送给目标服务器，并成功接收来自服务端的响应报文，这个基本的报文交换被称为一个 HTTP 事务（Transaction）。从协议角度来讲，即便在使用长连接的情况下，同一个客户端和服务器之间进行的多个 HTTP 事务也是完全独立的，所以需要在应用层为二者建立一个上下文来保存多次消息交换的状态，我们将其称为会话（Session）。

18.1 利用会话保留"语境"

客户端和服务器基于 HTTP 的消息交换就好比两个完全没有记忆能力的人在交流，每次单一的 HTTP 事务体现为一次"一问一答"的对话（Dialog）。每次对话对于交流双方来说都是全新的，他们不但不知道相同的问题是否在之前已经进行过讨论，甚至不知道彼此之间是否已经进行过交流。单一的对话毫无意义，有意义的是在同一语境下针对某个主题进行的多次对话，我们将其称为交谈（Conversation）。会话的目的就是在同一个客户端和服务器之间建立两者交谈的语境或者上下文（Context），ASP.NET Core 利用一个名为 SessionMiddleware 的中间件实现了会话。按照惯例，在介绍该中间件之前需要先利用几个简单的实例来演示如何在一个 ASP.NET Core 应用中利用会话来存储用户的状态。

18.1.1 设置和提取会话状态

每个会话都有一个被称为 Session Key 的标识（但不是唯一标识），会话状态以一个数据字典的形式将 Session Key 保存在服务端。当 SessionMiddleware 中间件在处理会话的第一个请求时，它会创建一个 Session Key，并基于它创建一个独立的数据字典来存储会话状态，应用程序设置的会话状态总是自动保存在当前会话对应的数据字典中。这个 Session Key 最终以 Cookie 的形式写入响应并返回客户端，客户端在每次发送请求时会自动附加这个 Cookie，从而使应用程序能够准确定位当前会话对应的数据字典。

下面利用一个简单的实例来演示会话状态的读写。ASP.NET Core 应用在默认情况下会利用缓存来存储会话状态，并且默认采用的是分布式缓存。由于演示实例采用基于 Redis 数据库的分布式缓存，所以需要添加针对 NuGet 包"Microsoft.Extensions.Caching.Redis"的依赖。如下面的代码片段所示，我们调用 IServiceCollection 接口的 AddDistributedRedisCache 扩展方法添加了基于 DistributedRedisCache 的服务注册。SessionMiddleware 中间件的注册则通过调用该 IApplicationBuilder 接口的 UseSession 扩展方法来完成。

```
public class Program
{
    public static void Main()
    {
        Host.CreateDefaultBuilder()
            .ConfigureWebHostDefaults(builder => builder
                .ConfigureServices(svcs => svcs
                    .AddDistributedRedisCache(
                        options => options.Configuration = "localhost")
                    .AddSession())
                .Configure(app => app
                    .UseSession()
                    .Run(ProcessAsync)))
            .Build()
            .Run();
    }

    static async Task ProcessAsync(HttpContext context)
    {
        var session = context.Session;
        await session.LoadAsync();
        string sessionStartTime;
        if (session.TryGetValue("SessionStartTime", out var value))
        {
            sessionStartTime = Encoding.UTF8.GetString(value);
        }
        else
        {
            sessionStartTime = DateTime.Now.ToString();
            session.SetString("SessionStartTime", sessionStartTime);
        }

        context.Response.ContentType = "text/html";
        await context.Response.WriteAsync(
            $"<html><body><ul><li>Session ID:{session.Id}</li>");
        await context.Response.WriteAsync(
            $"<li>Session Start Time:{sessionStartTime}</li>");
        await context.Response.WriteAsync(
            $"<li>Current Time:{DateTime.Now}</li></ul></table></body></html>");
    }
}
```

应用程序针对请求的处理体现在 ProcessAsync 方法中。在这个方法中，我们从当前

HttpContext 上下文中获取表示会话的 Session 对象，并调用其 TryGetValue 方法获取会话开始时间，这里使用的 Key 为 SessionStartTime。由于 TryGetValue 方法总是以字节数组的形式返回会话状态值，所以我们采用 UTF-8 编码转换成字符串形式。如果会话开始时间尚未设置，我们会调用 SetString 方法采用相同的 Key 进行设置。具体的响应体现为一个 HTML 文档，除了以会话状态保存的会话开始时间，该 HTML 文档的主体还会显示 Session ID 和实时时间。

在确保本地 Redis 服务器被启动的情况下，我们启动该应用程序并利用浏览器访问该站点，浏览器上呈现的输出结果如图 18-1 所示。如图 18-1 所示，我们利用 Chrome 先后两次访问目标站点，由于两次访问是在同一个会话中，所以 Session ID 和会话状态值都是一致的。但是对于最后一次利用 IE 的请求来说，则会开启一个新的会话，所以我们会看到不一样的值。（S1801）

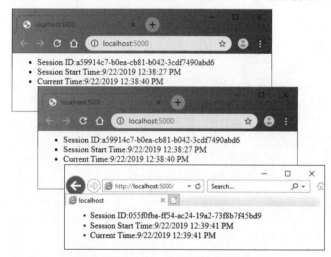

图 18-1　以会话状态保存的"会话开始时间"

18.1.2　查看存储的会话状态

会话状态在默认情况下采用分布式缓存的形式来存储，而我们的实例采用的是基于 Redis 数据库的分布式缓存，那么会话状态会以什么样的形式存储在 Redis 数据库中？由于缓存数据在 Redis 数据库中是以散列的形式存储的，所以我们只有知道具体的 Key 才能知道存储的值。缓存状态是基于作为会话标识的 Session Key 进行存储的，它与 Session ID 具有不同的值，到目前为止我们不能使用公布出来的 API 来获取它，但可以利用反射的方式来获取 Session Key。在默认情况下，表示 Session 的是一个 DistributedSession 对象，它通过如下所示的字段 _sessionKey 表示这个用来存储会话状态的 Session Key。

```
public class DistributedSession : ISession
{
    private readonly string _sessionKey;
}
```

接下来我们对上面演示的程序做简单的修改，从而使 Session Key 能够呈现出来。如下面的代码片段所示，我们可以采用反射的方式得到代表当前会话的 DistributedSession 对象的 _sessionKey 字段的值，并将它写入响应 HTML 文档的主体内容中。

```csharp
static async Task ProcessAsync(HttpContext context)
{
    var session = context.Session;
    await session.LoadAsync();
    string sessionStartTime;
    if (session.TryGetValue("SessionStartTime", out var value))
    {
        sessionStartTime = Encoding.UTF8.GetString(value);
    }
    else
    {
        sessionStartTime = DateTime.Now.ToString();
        session.SetString("SessionStartTime", sessionStartTime);
    }

    var field = typeof(DistributedSession).GetTypeInfo()
        .GetField("_sessionKey", BindingFlags.Instance | BindingFlags.NonPublic);
    var sessionKey = field.GetValue(session);

    context.Response.ContentType = "text/html";
    await context.Response.WriteAsync($"<html><body><ul><li>Session ID:{session.Id}</li>");
    await context.Response.WriteAsync($"<li>Session Key:{sessionKey}</li>");
    await context.Response.WriteAsync(
        $"<li>Session Start Time:{sessionStartTime}</li>");
    await context.Response.WriteAsync(
        $"<li>Current Time:{DateTime.Now}</li></ul></table></body></html>");
}
```

按照同样的方式启动应用后，我们使用浏览器访问目标站点得到的输出结果如图 18-2 所示，可以看到，Session Key 的值被正常呈现出来，它是一个不同于 Session ID 的 GUID。（S1802）

图 18-2　呈现当前会话的 Session Key

如果有这个保存当前会话状态的 Session Key，我们就可以按照图 18-3 所示的方式采用命令行的形式将存储在 Redis 数据库中的会话状态数据提取出来。当会话状态在采用默认的分布式缓存进行存储时，整个数据字典（包括 Key 和 Value）会采用预定义的格式序列化成字节数组，这基本上可以从图 18-3 体现出来。由图 18-3 还可以看出，基于会话状态的缓存默认采用的是基于滑动时间的过期策略，默认采用的滑动过期时间为 20 分（12 000 000 000 纳秒）。

```
Command Prompt - redis-cli
C:\Users\jinnan>redis-cli
127.0.0.1:6379> hgetall "c3177518-6b04-8ccf-42ad-0e4247c981dc"
1) "absexp"
2) "-1"
3) "sldexp"
4) "12000000000"
5) "data"
6) "\x02\x00\x00\x01\xc9\xd99\x82P\x98\x8bj5\xa3\x88\xcfp\x00\
xe0\x00\x10SessionStartTime\x00\x00\x00\x159/22/2019 12:46:43 P
M"
127.0.0.1:6379>
```

图 18-3　存储在 Redis 数据库中的会话状态

18.1.3　查看 Cookie

　　虽然整个会话状态数据存储在服务端，但是用来提取对应会话状态数据的 Session Key 需要以 Cookie 的形式由客户端来提供。如果请求没有以 Cookie 的形式携带 Session Key，SessionMiddleware 中间件就会将当前请求视为会话的第一次请求，在此情况下，它会生成一个 GUID 作为 Session Key，并最终以 Cookie 的形式返回客户端。

```
HTTP/1.1 200 OK
...
Set-Cookie:  .AspNetCore.Session=CfDJ8CYspSbYdOtFvhKqo9CYj2vdlf66AUAO2h2BDQ9%2FKoC2XILfJE2bk
IayyjXnXpNxMzMtWTceawO3eTWLV8KKQ5xZfsYNVlIf%2Fa175vwnCWFDeA5hKRyloWEpPPerphndTb8UJNv5R
68bGM8jP%2BjKVU7za2wgnEStgyV0ceN%2FryfW; path=/; httponly
```

　　如上所示的代码片段是响应报头中携带 Session Key 的 Set-Cookie 报头在默认情况下的表现形式。可以看出，Session Key 的值不仅是被加密的，更具有一个 httponly 标签，以防止 Cookie 值被跨站读取。在默认情况下，Cookie 采用的路径为 "/"。当我们使用同一个浏览器访问目标站点时，发送的请求将以如下形式附加上这个 Cookie。

```
GET http://localhost:5000/ HTTP/1.1
...
Cookie:  .AspNetCore.Session=CfDJ8CYspSbYdOtFvhKqo9CYj2vdlf66AUAO2h2BDQ9%2FKoC2XILfJE2b
kIayyjXnXpNxMzMtWTceawO3eTWLV8KKQ5xZfsYNVlIf%2Fa175vwnCWFDeA5hKRyloWEpPPerphndTb8UJNv5
R68bGM8jP%2BjKVU7za2wgnEStgyV0ceN%2FryfW
```

　　除了 Session Key，上面的内容还提到了 Session ID，读者可能不太了解两者具有怎样的区别。Session Key 和 Session ID 是两个不同的概念，上面演示的实例也证实了它们的值其实是不同的。Session ID 可以作为会话的唯一标识，但是 Session Key 不可以。换句话说，两个不同的 Session 肯定具有不同的 Session ID，但是它们可能共享相同的 Session Key。要正确理解两者之间的区别，就必须了解 SessionMiddleware 中间件究竟是如何处理会话的。当 SessionMiddleware 中间件接收到会话的第一个请求时，它会创建两个不同的 GUID 来分别表示 Session Key 和 Session ID。其中，Session ID 将作为会话状态的一部分被存储起来，而 Session Key 以 Cookie 的形式返回客户端。

　　会话是具有有效期的，会话的有效期基本决定了存储的会话状态数据的有效期，通过上面演示的实例可知，ASP.NET Core 应用的会话采用的默认过期时间为 20 分。在默认情况下，20 分之内的任意一次请求都会将会话的寿命延长至 20 分后。如果两次请求的时间间隔超过 20 分，就意味着会话过期，存储的会话状态数据（包括 Session ID）会被清除，但是请求携带可能还是原来的 Session Key。在这种情况下，SessionMiddleware 中间件会创建一个新的会话，该会话具有不同的 Session ID，但是整个会话状态依然沿用这个 Session Key，所以 Session Key 并不能唯一标识一个会话。

18.2 会话状态的读写

会话本质上就是在应用层面提供一个数据容器来保存客户端状态（会话状态），所以会话的核心功能就是针对会话状态的读写。会话状态在默认情况下采用分布式缓存的方式进行存储，那么具体的读写又是如何实现的？要回答这个问题，我们需要先了解在应用编程接口层面表示会话的 ISession 接口。

18.2.1 ISession

在应用编程接口层面，ASP.NET Core 应用的会话通过如下所示的 ISession 接口来表示。我们针对会话状态的所有操作（设置、提取、移除和清除）都是通过调用 ISession 接口相应的方法（Set、TryGetValue、Remove 和 Clear）来完成的。和分布式缓存一样，我们设置和提取的缓存状态的值体现为字节数组，所以应用程序需要自行完成序列化和反序列化的工作。除了这 4 个基本方法，我们还可以利用 Id 属性得到当前会话的 Session ID，通过 Keys 属性得到所有会话状态条目的 Key。除了上述这几个方法和属性，ISession 接口还包括 IsAvailable 属性以及 LoadAsync 方法和 CommitAsync 方法，它们具有怎样的作用？

```
public interface ISession
{
    string                  Id { get; }
    bool                    IsAvailable { get; }
    IEnumerable<string>     Keys { get; }

    void Set(string key, byte[] value);
    bool TryGetValue(string key, out byte[] value);
    void Remove(string key);
    void Clear();

    Task LoadAsync();
    Task CommitAsync();
}
```

对于针对基本操作的 4 个方法（Set、TryGetValue、Remove 和 Clear）来说，它们针对会话状态的设置、提取、移除和清除都是在内存中进行的。不仅如此，在调用这几个方法之前，ISession 对象需要确保后备存储（如 Redis 数据库）的会话状态被加载到内存之中。会话状态的异步加载可以直接调用 LoadAsync 方法来完成，而上述 4 个方法在会话状态未被加载的情况下会采用同步的方式加载它们。ISession 对象在会话状态尚未全部加载到内存之前处于不可用的状态，其 IsAvailable 属性返回 False。会话状态一旦被加载，IsAvailable 属性马上变成 True。在前面演示的实例中，我们会在操作缓存状态之前调用 ISession 对象的 LoadAsync 方法以异步的方式将所有的会话状态加载到内存中,这是一种推荐的做法。

由于作用于 ISession 对象上的 4 个基本会话状态操作都是针对内存的,这些操作最终需要通过 CommitAsync 方法做统一的提交。SessionMiddleware 中间件会在完成请求处理之前调用这个方法,该方法会将当前请求针对会话状态的改动保存到后备存储中。另外,只有在当前请求上下文中真正对会话状态做了相应改动的情况下,ISession 对象的 CommitAsync 方法才会真正执行提交操作。

除了调用 ISession 对象的 TryGetValue 方法判断指定的缓存状态项是否存在,并在存在的情况下通过输出参数以字节数组的形式返回缓存状态值,我们还可以调用如下所示的 Get 方法直接返回代表会话状态值的字节数组。如果指定的会话状态项不存在,Get 方法会直接返回 Null。

```
public static class SessionExtensions
{
    public static byte[] Get(this ISession session, string key);
}
```

由于 ISession 对象总是将会话状态的值表示为字节数组,所以应用程序总是需要自行解决序列化与反序列化的问题,但是如果会话状态的值类型是整数或者字符串这些简单的类型,针对它的设置和提取可以直接调用如下这几个扩展方法来完成。GetString 方法和 SetString 方法会采用 UTF-8 对字符串进行编码与解码。

```
public static class SessionExtensions
{
    public static int? GetInt32(this ISession session, string key);
    public static string GetString(this ISession session, string key);
    public static void SetInt32(this ISession session, string key, int value);
    public static void SetString(this ISession session, string key, string value);
}
```

18.2.2 DistributedSession

ASP.NET Core 应用在默认情况下会采用分布式缓存来存储会话状态,如下所示的 DistributedSession 类型就是针对 ISession 接口的默认实现。当我们在创建一个 DistributedSession 对象的时候,需要提供这个用来存储会话状态的 IDistributedCache 对象。除此之外,我们还需要提供 Session Key 和会话过期时间。DistributedSession 构造函数的参数 tryEstablishSession 是一个 Func<bool> 类型的委托对象,它的返回值表示当前会话是否存在,或者能否正常地建立会话。参数 isNewSessionKey 表示提供的 Session Key 并不是由请求的 Cookie 提供的,而是根据当前请求重新创建的。

```
public class DistributedSession : ISession
{
    public DistributedSession(IDistributedCache cache, string sessionKey,
        TimeSpan idleTimeout, Func<bool> tryEstablishSession, ILoggerFactory loggerFactory,
        bool isNewSessionKey);
    ...
}
```

DistributedSession 对象会使用一个字典对象来存储所有的会话状态,而针对缓存状态的设置、提取、移除和清除都是针对这个字典对象的操作。当这些操作被执行之前，DistributedSession 必须确保存储在分布式缓存中的会话状态已经被加载到这个数据字典之中，针对会话状态的加载通过调用 DistributedCache 对象的 Get 方法来完成，提供的 Session Key 将作为对应缓存项的 Key。以这种方式触发的会话状态的加载是以同步的方式进行的，如果希望采用异步的方式加载会话状态，我们可以在执行这些操作之前显式地调用其 LoadAsync 方法。

当 DistributedSession 对象的 CommitAsync 方法被执行时，它会将整个会话状态数据字典的内容和 Session ID 按照预定义的格式序列化成字节数组，然后调用 DistributedCache 对象的 SetAsync 方法将该字节数组作为值添加到分布式缓存之中。对应的 Key 自然就是最初提供的 Session Key，而提供的会话过期时间将作为缓存项的滑动过期时间。当加载会话状态时，IDistributedCache 对象的 Get 方法返回的就是这个被序列化的字节数组，该字节数组被反序列化后提取的 Session ID 将直接作为 DistributedSession 对象的 Id 属性，而其他数据将被添加在会话状态数据字典中。

值得注意的是，只有会话状态在当前请求上下文范围内已经被改动的前提下，DistributedSession 对象才会调用 IDistributedCache 对象的 SetAsync 方法重新设置会话状态对应的分布式缓存项。DistributedSession 判断当前会话状态是否被改动过的方法也很简单，它仅仅判断 Set 方法、Remove 方法和 Clear 方法是否被调用过。如果本次请求处理过程中并没有对当前会话状态做任何修改，DistributedSession 对象会调用 IDistributedCache 对象的 RefreshAsync 方法刷新对象的缓存性，进而达到延长（Renew）会话的目的。

在执行构造函数创建一个 DistributedSession 对象的时候，我们通过参数 tryEstablishSession 提供了一个 Func<bool>类型的委托来确定当前请求是否在一个现有的会话之中，或者当前能否创建一个新的会话，这个委托对象会在 Set 方法中被执行。具体来说，如果执行这个委托对象返回 False，Set 方法会抛出异常。一般来说，如果请求并没有携带一个有效的 Session Key，或者在响应已经开始发送的情况下调用 DistributedSession 对象的 Set 方法设置会话状态时，上述这个委托对象就会返回 False。

18.2.3　ISessionStore

ISessionStore 接口可以视为创建表示会话的 ISession 对象的工厂。如下面的代码片段所示，ISessionStore 接口具有唯一的 Create 方法，可以帮助我们创建一个 ISession 对象。该方法的 4 个参数与 DistributedSession 的构造函数定义的同名参数具有相同含义。

```
public interface ISessionStore
{
    ISession Create(string sessionKey, TimeSpan idleTimeout,
        Func<bool> tryEstablishSession, bool isNewSessionKey);
}
```

上面介绍的 DistributedSession 类型对应的 ISessionStore 实现类型是如下所示的 DistributedSessionStore。在实现的 Create 方法中，DistributedSessionStore 会利用提供的这些参数，连同创建 DistributedSessionStore 时提供的 IDistributedCache 对象和 ILoggerFactory 对象作为构造函数的

参数，用来创建返回的 DistributedSession 对象。

```
public class DistributedSessionStore : ISessionStore
{
    public DistributedSessionStore(IDistributedCache cache, ILoggerFactory loggerFactory) ;
    public ISession Create(string sessionKey, TimeSpan idleTimeout,
        Func<bool> tryEstablishSession, bool isNewSessionKey) ;
}
```

综上所述，应用程序利用 ISession 接口表示的服务来读写会话状态，ISessionStore 接口表示创建 ISession 对象的工厂。DistributedSessionStore 是对 ISessionStore 的默认实现，它创建的是一个 DistributedSession 对象，后者利用 IDistributedCache 对象表示的分布式缓存来存储会话状态。图 18-4 所示的 UML 体现了这些核心接口和类型之间的关系。

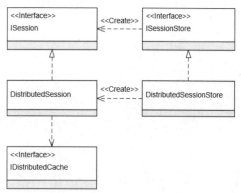

图 18-4　会话模型核心接口与类型之间的关系

18.3　SessionMiddleware 中间件

上面介绍了在应用编程接口层面表示会话的 ISession 接口和创建它的 ISessionStore 接口，以及在默认情况下采用分布式缓存存储会话状态的 DistributedSession 类型和对应的 DistributedSessionStore 类型。在请求处理过程中，利用注册的 ISessionStore 对象来创建表示会话的 ISession 对象是由 SessionMiddleware 中间件来完成的。但在介绍 SessionMiddleware 中间件之前，需要先介绍对应的配置选项类型 SessionOptions。

18.3.1　SessionOptions

由于保存会话状态的 Session Key 是通过 Cookie 进行传递的，所以 SessionOptions 承载的核心配置选项是 Cookie 属性表示的 CookieBuilder 对象。如下面的代码片段所示，SessionOptions 的 Cookie 属性返回的是一个 SessionCookieBuilder 的对象，它对 Cookie 的名称（.AspNetCore.Session）、路径（/）和安全策略（None）等做了一些默认设置。

```
public class SessionOptions
{
    public CookieBuilder       Cookie { get; set; }
    public TimeSpan            IdleTimeout { get; set; }     = TimeSpan.FromMinutes(20);
```

```csharp
    public TimeSpan                IOTimeout { get; set; }    = TimeSpan.FromMinutes(1);

    private class SessionCookieBuilder : CookieBuilder
    {
        public SessionCookieBuilder()
        {
            Name            = SessionDefaults.CookieName;
            Path            = SessionDefaults.CookiePath;
            SecurePolicy    = CookieSecurePolicy.None;
            SameSite        = SameSiteMode.Lax;
            HttpOnly        = true;
            IsEssential     = false;
        }

        public override TimeSpan? Expiration
        {
            get => null;
            set => throw new InvalidOperationException();
        }
    }
}

public static class SessionDefaults
{
    public static readonly string CookieName = ".AspNetCore.Session";
    public static readonly string CookiePath = "/";
}
```

  CookieBuilder 对象的 HttpOnly 属性表示响应的 Cookie 是否需要添加一个 httponly 标签，在默认情况下这个属性为 True。SameSite 属性表示是否会在生成的 Set-Cookie 中设置 SameSite 属性以阻止浏览器将它跨域发送，该属性的默认值为 Lax，表示 SameSite 属性会被设置。CookieBuilder 对象的 IsEssential 属性与 Cookie 的许可授权策略（Cookie Consent Policy）有关，该属性的默认值为 False，表示为了实现会话支持针对 Cookie 的设置不需要得到最终用户的显式授权。

  SessionOptions 的 IdleTimeout 属性表示会话过期时间，具体来说应该是客户端最后一次访问时间到会话过期之间的时长。如果这个属性未做显式设置，该属性会采用默认的会话过期时间 20 分，这个默认过期时间和其他属性的默认值都在前面演示的实例中做了证实。SessionOptions 的 IOTimeout 属性表示基于 ISessionStore 的会话状态的读取和提交所运行的最长时限，默认为 1 分。

18.3.2　ISessionFeature

  应用程序会使用 ISession 对象来读写会话状态，这个对象来源于当前 HttpContext 上下文的 Session 属性，那么这个属性又是如何得到这个 ISession 对象的呢？实际上，HttpContext 上下文的 Session 属性来源于一个通过 ISessionFeature 接口表示的特性。如下面的代码片段所示，

ISessionFeature 接口具有唯一的属性 Session，该属性用来设置和获取代表当前会话的 ISession 对象。SessionMiddleware 中间件使用的是实现该接口的 SessionFeature 类型。

```csharp
public interface ISessionFeature
{
    ISession Session { get; set; }
}

public class SessionFeature : ISessionFeature
{
    public ISession Session { get; set; }
}
```

18.3.3　SessionMiddleware

如下所示的代码片段是 SessionMiddleware 这个中间件的大致的实现。当我们创建一个 SessionMiddleware 对象时需要提供一个用于加密 Cookie 值的 IDataProtectionProvider 对象和用来创建 ISession 对象的 SessionStore。除此之外，我们还需要提供承载相关配置选项的 SessionOptions 对象。

```csharp
public class SessionMiddleware
{
    private readonly RandomNumberGenerator   _cryptoRandom;
    private const int                        SessionKeyLength = 36;
    private readonly RequestDelegate         _next;
    private readonly SessionOptions          _options;
    private readonly ILogger                 _logger;
    private readonly ISessionStore           _sessionStore;
    private readonly IDataProtector          _dataProtector;

    public SessionMiddleware(
        RequestDelegate next,
        ILoggerFactory loggerFactory,
        IDataProtectionProvider dataProtectionProvider,
        ISessionStore sessionStore,
        IOptions<SessionOptions> options)
    {
        _next              = next;
        _logger            = loggerFactory.CreateLogger<SessionMiddleware>();
        _dataProtector     = dataProtectionProvider.CreateProtector("SessionMiddleware");
        _options           = options.Value;
        _sessionStore      = sessionStore;
        _cryptoRandom      = RandomNumberGenerator.Create();
    }

    public async Task Invoke(HttpContext context)
    {
        string UnprotectSessionKey(string protectedKey)
        {
```

```csharp
        var padding = 3 - ((protectedKey.Length + 3) % 4);
        var padValue = padding == 0
            ? protectedKey
            : protectedKey + new string('=', padding);
        var rawData = Convert.FromBase64String(padValue);
        return Convert.ToBase64String(_dataProtector.Unprotect(rawData));
    }

    string ProtectSessionKey(string unProtectedKey)
    {
        var bytes = Encoding.UTF8.GetBytes(unProtectedKey);
        bytes = _dataProtector.Protect(bytes);
        return Convert.ToBase64String(bytes);
    }

    bool TryEstablishSession(string protectedKey)
    {
        var response = context.Response;
        response.OnStarting(_ => {
            if (!response.HasStarted)
            {
                var cookieOptions = _options.Cookie.Build(context);
                response.Cookies.Append(_options.Cookie.Name, protectedKey,
                    cookieOptions);
                response.Headers["Cache-Control"] = "no-cache";
                response.Headers["Pragma"] = "no-cache";
                response.Headers["Expires"] = "-1";
            }
            return Task.CompletedTask;
        }, null);
        return !response.HasStarted;
    }

    var isNewSessionKey = false;
    Func<bool> tryEstablishSession = () =>true;
    var cookieValue = context.Request.Cookies[_options.Cookie.Name];
    var sessionKey = UnprotectSessionKey(cookieValue);
    if (string.IsNullOrWhiteSpace(sessionKey) || sessionKey.Length != SessionKeyLength)
    {
        //Try establish a new session
        var guidBytes = new byte[16];
        _cryptoRandom.GetBytes(guidBytes);
        sessionKey = new Guid(guidBytes).ToString();
        cookieValue = ProtectSessionKey(sessionKey);
        tryEstablishSession = ()=>TryEstablishSession(cookieValue);
        isNewSessionKey = true;
    }

    var feature = new SessionFeature
    {
```

```
            Session = _sessionStore.Create(sessionKey, _options.IdleTimeout,
                _options.IOTimeout, tryEstablishSession, isNewSessionKey)
    };
    context.Features.Set<ISessionFeature>(feature);

    try
    {
        await _next(context);
    }
    finally
    {
        context.Features.Set<ISessionFeature>(null);
        await feature.Session?.CommitAsync(context.RequestAborted);
    }
}
```

实现在 SessionMiddleware 中间件的 Invoke 方法中针对请求的处理流程其实非常简单。SessionMiddleware 中间件会根据在 SessionOptions 中的设置从当前请求的 Cookie 中提取 Session Key。如果 Session Key 不存在，就意味着当前请求是开启新会话的第一个请求，在此情况下它会创建一个新的 GUID 作为 Session Key。该中间件接下来会调用 ISessionStore 对象的 Create 方法创建一个 ISession 对象。

对于针对 ISessionStore 对象的 Create 方法调用来说，SessionMiddleware 中间件除了会将 Session Key 和由 SessionOptions 提供的会话过期时间作为参数，它还需要提供一个 Func<bool> 类型的委托来帮助创建的 ISession 对象确定会话是否已经存在或者能够正常创建。对于提供的这个委托对象来说，它只有在会话不存在并且当前响应已经开始发送的情况下才会返回 False。

当 SessionMiddleware 中间件将请求递交给后续中间件进行处理之前，它会利用创建的 ISession 对象创建一个 SessionFeature 对象，并将该特性对象附加到当前 HttpContext 上下文，所以后续中间件以及应用程序才能从当前 HttpContext 上下文中得到这个 ISession 对象。SessionMiddleware 中间件最终会调用 ISession 对象的 CommitAsync 方法来对应用程序设置的会话状态进行提交。当响应开始发送时，SessionMiddleware 中间件会利用注册的回调将 Session Key 作为 Cookie 添加到响应报头集合中。

第 19 章

认 证

在安全领域，认证和授权是两个重要的主题。认证是安全体系的第一道屏障，是守护整个应用或者服务的第一道大门。当访问者请求进入的时候，认证体系通过验证对方的提供凭证确定其真实身份。认证体系只有在证实了访问者的真实身份的情况下才会允许其进入。ASP.NET Core 提供了多种认证方式，它们的实现都基于本章介绍的认证模型。

19.1 认证、登录与注销

认证是一个旨在确定请求访问者真实身份的过程，与认证相关的还有其他两个基本操作——登录和注销。要真正理解认证、登录和注销这 3 个核心操作的本质，就需要对 ASP.NET Core 采用的基于"票据"的认证机制有基本的了解。

19.1.1 认证票据

ASP.NET Core 应用的认证实现在一个名为 AuthenticationMiddleware 的中间件中，该中间件在处理分发给它的请求时会按照指定的认证方案（Authentication Scheme）从请求中提取能够验证用户真实身份的数据，我们一般将该数据称为安全令牌（Security Token）。ASP.NET Core 应用下的安全令牌被称为认证票据（Authentication Ticket），所以 ASP.NET Core 应用采用基于票据的认证方式。

AuthenticationMiddleware 中间件实现的整个认证流程涉及图 19-1 所示的 3 种针对认证票据的操作，即认证票据的颁发、检验和撤销。我们将这 3 个操作所涉及的 3 种角色称为票据颁发者（Ticket Issuer）、验证者（Authenticator）和撤销者（Ticket Revoker），在大部分场景下这 3 种角色由同一个主体来扮演。

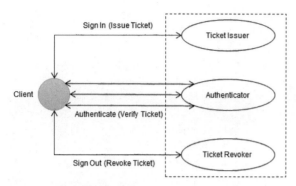

图 19-1　基于票据的认证

颁发认证票据的过程就是登录（Sign In）操作。一般来说，用户试图通过登录应用以获取认证票据的时候需要提供可用来证明自身身份的用户凭证（User Credential），最常见的用户凭证类型是"用户名 + 密码"。认证方在确定对方真实身份之后，会颁发一个认证票据，该票据携带着与该用户有关的身份、权限及其他相关的信息。

一旦拥有了由认证方颁发的认证票据，我们就可以按照双方协商的方式（如通过 Cookie 或者报头）在请求中携带该认证票据，并以此票据声明的身份执行目标操作或者访问目标资源。认证票据一般都具有时效性，一旦过期将变得无效。我们有的时候甚至希望在过期之前就让认证票据无效，以免别人使用它冒用自己的身份与应用进行交互，这就是注销（Sign Out）操作。

ASP.NET Core 的认证系统旨在构建一个标准的模型，用来完成针对请求的认证以及与之相关的登录和注销操作。按照惯例，在介绍认证模型的架构设计之前，需要通过一个简单的实例来演示如何在一个 ASP.NET Core 应用中实现认证、登录和注销的功能。

19.1.2　基于 Cookie 的认证

我们会采用 ASP.NET Core 提供的基于 Cookie 的认证方案。顾名思义，该认证方案采用 Cookie 来携带认证票据。为了使读者对基于认证的编程模式有深刻的理解，我们演示的这个应用将从一个空白的 ASP.NET Core 应用开始搭建。（S1901）

我们即将创建的这个 ASP.NET Core 应用主要处理 3 种类型的请求。应用的主页需要在登录之后才能访问，所以针对主页的匿名请求会被重定向到登录页面。在登录页面输入正确的用户名和密码之后，应用会自动重定向到应用主页，该页面会显示当前认证用户名并提供注销的链接。我们按照如下所示的方式利用路由来处理这 3 种类型的请求，其中登录和注销采用的是默认路径"Account/Login"与"Account/Logout"。

```
public class Program
{
    public static void Main()
    {
        Host.CreateDefaultBuilder()
            .ConfigureWebHostDefaults(builder => builder
                .ConfigureServices(svcs => svcs.AddRouting())
```

```csharp
            .Configure(app => app
                .UseRouting()
                .UseEndpoints(endpoints =>{
                    endpoints.Map(pattern: "/", RenderHomePageAsync);
                    endpoints.Map("Account/Login", SignInAsync);
                    endpoints.Map("Account/Logout", SignOutAsync);
                })))
        .Build()
        .Run();
}

public static async Task RenderHomePageAsync(HttpContext context)
{
    throw new NotImplementedException();
}

public static async Task SignInAsync(HttpContext context)
{
    throw new NotImplementedException();
}

public static async Task SignOutAsync(HttpContext context)
{
    throw new NotImplementedException();
}
}
```

19.1.3 应用主页

如下面的代码片段所示，我们调用 IApplicationBuilder 接口的 UseAuthentication 扩展方法就是为了注册用来实现认证的 AuthenticationMiddleware 中间件。该中间件的依赖服务是通过调用 IServiceCollection 接口的 AddAuthentication 扩展方法注册的。在注册这些基础服务时，我们还设置了默认采用的认证方案，静态类型 CookieAuthenticationDefaults 的 AuthenticationScheme 属性返回的就是 Cookie 认证方案的默认方案名称。

```csharp
public class Program
{
    public static void Main()
    {
        Host.CreateDefaultBuilder()
            .ConfigureWebHostDefaults(builder => builder
                .ConfigureServices(svcs => svcs
                    .AddRouting()
                    .AddAuthentication(options => options.DefaultScheme =
                        CookieAuthenticationDefaults.AuthenticationScheme)
                    .AddCookie())
                .Configure(app => app
                    .UseAuthentication()
                    .UseRouting()
```

```
            .UseEndpoints(endpoints =>{
                endpoints.Map(pattern: "/", RenderHomePageAsync);
                endpoints.Map("Account/Login", SignInAsync);
                endpoints.Map("Account/Logout", SignOutAsync);
            })))
        .Build()
        .Run();
    }
}
```

ASP.NET Core 提供了一个极具扩展性的认证模型，我们可以利用它支持多种认证方案，对认证方案的注册是通过 AddAuthentication 方法返回的一个 AuthenticationBuilder 对象来实现的。在上面提供的代码片段中，我们调用 AuthenticationBuilder 对象的 AddCookie 扩展方法完成了针对 Cookie 认证方案的注册。

演示实例的主页是通过如下所示的 RenderHomePageAsync 方法来呈现的。由于我们要求浏览主页必须是经过认证的用户，所以该方法会利用 HttpContext 上下文的 User 属性返回的 ClaimsPrincipal 对象判断当前请求是否经过认证。对于经过认证的请求，我们会响应一个简单的 HTML 文档，并在其中显示用户名和一个注销链接。

```
public class Program
{
    ...
    public static async Task RenderHomePageAsync(HttpContext context)
    {
        if (context?.User?.Identity?.IsAuthenticated == true)
        {
            await context.Response.WriteAsync(
                @"<html>
                    <head><title>Index</title></head>
                    <body>" +
                        $"<h3>Welcome {context.User.Identity.Name}</h3>" +
                        @"<a href='Account/Logout'>Sign Out</a>
                    </body>
                </html>");
        }
        else
        {
            await context.ChallengeAsync();
        }
    }
}
```

对于匿名请求，我们希望应用能够自动重定向到登录路径。从如上所示的代码片段可以看出，我们仅仅调用当前 HttpContext 上下文的 ChallengeAsync 扩展方法就完成了针对登录路径的重定向。前面提及，注册的登录和注销路径是基于 Cookie 的认证方案采用的默认路径，所以调用 ChallengeAsync 方法时根本不需要指定重定向路径。图 19-2 所示就是作为应用的主页在浏览器上呈现的效果。

图 19-2 应用主页

19.1.4 登录与注销

登录与注销分别实现在 SignInAsync 方法和 SignOutAsync 方法中,我们采用的是针对"用户名 + 密码"的登录方式,所以可以利用静态字段_accounts 来存储应用注册的账号。在静态构造函数中,我们添加密码均为"password"的 3 个账号(Foo、Bar 和 Baz)。

```
public class Program
{
    private static Dictionary<string, string> _accounts;

    static Program()
    {
        _accounts = new Dictionary<string, string>(StringComparer.OrdinalIgnoreCase);
        _accounts.Add("Foo", "password");
        _accounts.Add("Bar", "password");
        _accounts.Add("Baz", "password");
    }
}
```

如下所示的代码片段是用于处理登录请求的 SignInAsync 方法的定义,而 RenderLoginPageAsync 方法用来呈现登录页面。如下面的代码片段所示,对于 GET 请求,SignInAsync 方法会直接调用 RenderLoginPageAsync 方法来呈现登录界面。对于 POST 请求,我们会从提交的表单中提取用户名和密码,并对其实施验证。如果提供的用户名与密码一致,我们会根据用户名创建一个代表身份的 GenericIdentity 对象,并利用它创建一个代表登录用户的 ClaimsPrincipal 对象,RenderHomePageAsync 方法正是利用该对象来检验当前用户是否是经过认证的。有了 ClaimsPrincipal 对象,我们只需要将它作为参数调用 HttpContext 上下文的 SignInAsync 扩展方法即可完成登录,该方法最终会自动重定向到初始方法的路径,也就是我们的主页。

```
public class Program
{
    public static async Task SignInAsync(HttpContext context)
    {
        if (string.Compare(context.Request.Method, "GET") == 0)
        {
            await RenderLoginPageAsync(context, null, null, null);
        }
        else
        {
            var userName = context.Request.Form["username"];
            var password = context.Request.Form["password"];
```

```
            if (_accounts.TryGetValue(userName, out var pwd) && pwd == password)
            {
                var identity = new GenericIdentity(userName, "Passord");
                var principal = new ClaimsPrincipal(identity);
                await context.SignInAsync(principal);
            }
            else
            {
                await RenderLoginPageAsync(context, userName, password,
                    "Invalid user name or password!");
            }
        }
    }

    private static Task RenderLoginPageAsync(HttpContext context, string userName,
        string password, string errorMessage)
    {
        context.Response.ContentType = "text/html";
        return context.Response.WriteAsync(
            @"<html>
                <head><title>Login</title></head>
                <body>
                    <form method='post'>" +
                        $"<input type='text' name='username' placeholder='User name'
                            value ='{userName}'/>" +
                        $"<input type='password' name='password' placeholder='Password'
                            value ='{password}'/> " +
                        @"<input type='submit' value='Sign In' />
                    </form>" +
                    $"<p style='color:red'>{errorMessage}</p>" +
                @"</body>
            </html>");
    }
}
```

如果用户提供的用户名与密码不匹配，我们还是会调用 RenderLoginPageAsync 方法来呈现登录页面，该页面会以图 19-3 所示的形式保留用户的输入并显示错误消息。图 19-3 还反映了一个细节，调用 HttpContext 上下文的 ChallengeAsync 方法会将当前路径（主页路径"/"，经过编码后为"%2F"）存储在一个名为 ReturnUrl 的查询字符串中，SignInAsync 方法正是利用它实现对初始路径的重定向的。

图 19-3　登录页面

既然登录可以通过调用当前 HttpContext 上下文的 SignInAsync 扩展方法来完成，那么注销操作对应的自然就是 SignOutAsync 扩展方法。如下面的代码片段所示，我们定义在 Program 中的 SignOutAsync 扩展方法正是调用这个方法来注销当前登录状态的。我们在完成注销之后将应用重定向到主页。

```
public class Program
{
    ...
    public static async Task SignOutAsync(HttpContext context)
    {
        await context.SignOutAsync();
        context.Response.Redirect("/");
    }
}
```

19.2 身份与用户

认证是一个确定访问者真实身份的过程。ASP.NET Core 应用的认证系统通过 IPrincipal 接口表示接受认证的用户。一个用户可以具有一个或者多个身份，身份通过 IIdentity 接口来描述。ASP.NET Core 应用完全采用基于声明的认证与授权方式，声明对应一个 Claim 对象，我们可以利用它来描述用户的身份、权限和其他与用户相关的信息。

19.2.1 IIdentity

用户总是以某个声称的身份向目标应用发起请求，认证的目的在于确定请求者是否与其声称的这个身份相符。由于认证首先涉及的是对身份的鉴定，所以下面先介绍描述身份的 IIdentity 对象。用户身份总是具有一个确定的名称，该名称体现为 IIdentity 接口的 Name 属性。另一个布尔类型的 IsAuthenticated 属性表示身份是否经过认证，只有身份经过认证的用户才是值得信任的。AuthenticationType 属性则表示采用的认证类型。

```
public interface IIdentity
{
    string      Name { get; }
    bool        IsAuthenticated { get; }
    string      AuthenticationType { get; }
}
```

由于 ASP.NET Core 应用完全采用基于声明的认证与授权方式，这种方式对 IIdentity 对象的具体体现就是我们可以将任意与身份、权限及其他用户相关的信息以声明的形式附加到 IIdentity 对象之上，这样一个携带声明的身份对象可以通过 ClaimsIdentity 类型表示。但是在介绍 ClaimsIdentity 之前，需要先介绍表示声明的 Claim 类型。

Claim

声明是用户在某个方面的一种陈述（Statement）。一般来说，声明应该是身份得到确认之后由认证方赋予的，声明可以携带任何与认证用户相关的信息，它们可以描述用户的身份（如

E-mail 地址、电话号码或者指纹），也可以描述用户的权限（如拥有的角色或者所在的用户组）或者其他描述当前用户的基本信息（如性别、年龄和国籍等）。

声明通过如下所示的 Claim 类型来表示。由于声明最终会作为认证票据（Authentication Ticket）的一部分在网络中传递，所以 Claim 对象是可以被直接序列化的，因此其类型上标注了 SerializableAttribute 特性。它的 Subject 属性返回作为声明陈述主体的 ClaimsIdentity 对象。

```
[Serializable]
public class Claim
{
    public ClaimsIdentity                      Subject { get; }
    public string                              Type { get; }
    public string                              Value { get; }
    public string                              ValueType { get; }
    public IDictionary<string, string>         Properties { get; }
    public string                              Issuer { get; }
    public string                              OriginalIssuer { get; }
    ...
}
```

Claim 的 Type 属性和 Value 属性分别表示声明陈述的类型与对应的值。如果利用一个 Claim 对象来承载用户的 E-mail 地址，那么 Type 属性的值就是 EmailAddress，Value 属性的值就是具体的 E-mail 地址（如 foobar@outlook.com）。除了单纯采用键值对（Type 相当于 Key）陈述声明，如果需要附加一些额外信息，我们还可以将它们添加到由 Properties 属性表示的数据字典中。

由于声明可以用来陈述任意主题，所以声明的"值"针对不同的主题会具有不同的表现形式，或者具有不同的数据类型。例如，针对年龄的声明，它的值应该是一个整数；如果声明描述的是出生日期，那么对应的值应该是一个 DateTime 对象。虽然这些值最初具有不同的表现形式，但是它们最终都需要转化成字符串。为了能够在使用的时候将值还原，我们需要记录这个值原本的类型，Claim 对象的 ValueType 属性存在的目的就在于此。

由于声明承载了用户身份和权限信息，它们是之后进行授权的基础，所以声明信息必须是值得信任的。如果能够确保未被篡改，声明是否能够信任取决于它由谁颁发。Claim 对象的 Issuer 属性和 OriginalIssuer 属性代表声明的颁发者，前者代表当前颁发者，后者代表最初颁发者。在一般情况下，这两个属性会返回相同的值，两者的不同体现了这样的颁发场景：由某个颁发者生成的声明在最终被用来表示用户身份之前需要其他颁发者做进一步处理。

原则上，我们可以采用任何字符串来表示声明的类型，但对于一些常用的声明，微软定义了标准的类型，它们以常量的形式定义在静态类型 ClaimTypes 中。定义在这个类型中的标准声明类型有几十个，如下所示的代码片段仅列举了几个分别表示用户姓名（Surname 和 GivenName）、性别（Gender）及联系方式（Email、MobilePhone、PostalCode 和 StreetAddress）的声明类型，可以看出，它们都采用 URI 的形式来表示。如果要了解定义在 ClaimTypes 类型中所有的标准声明类型，可以参阅 MSDN 文档。

```
public static class ClaimTypes
```

```csharp
{
    public const string Surname =
        "http://schemas.xmlsoap.org/ws/2005/05/identity/claims/surname";
    public const string GivenName =
        "http://schemas.xmlsoap.org/ws/2005/05/identity/claims/givenname";
    public const string Gender =
        "http://schemas.xmlsoap.org/ws/2005/05/identity/claims/gender";
    public const string Email =
        "http://schemas.xmlsoap.org/ws/2005/05/identity/claims/emailaddress";
    public const string MobilePhone =
        "http://schemas.xmlsoap.org/ws/2005/05/identity/claims/mobilephone";
    public const string PostalCode =
        "http://schemas.xmlsoap.org/ws/2005/05/identity/claims/postalcode";
    public const string StreetAddress =
        "http://schemas.xmlsoap.org/ws/2005/05/identity/claims/streetaddress";
    ...
}
```

除了对声明类型进行标准化，微软采用同样的方式对常用的声明值类型（对应 ValueType 属性）进行了标准化，这些表示常用声明值类型的常量定义在如下所示的静态类型 ClaimValueTypes 中。与标准的声明类型一样，声明的这些标准值类型同样采用 URI 的形式来表示。不论是声明自身的类型还是它的值类型，我们应尽量使用这些标准的定义。

```csharp
public static class ClaimValueTypes
{
    public const string Base64Binary = "http://www.w3.org/2001/XMLSchema#base64Binary";
    public const string Base64Octet  = "http://www.w3.org/2001/XMLSchema#base64Octet";
    public const string Boolean      = "http://www.w3.org/2001/XMLSchema#boolean";
    public const string Date         = "http://www.w3.org/2001/XMLSchema#date";
    ...
}
```

ClaimsIdentity

在对表示声明的 Claim 类型有了充分了解之后，下面介绍 ClaimsIdentity 类型。顾名思义，一个 ClaimsIdentity 对象就是一个携带声明的 IIdentity 对象。如下面的代码片段所示，ClaimsIdentity 类型除了实现定义在 IIdentity 接口中的 3 个只读属性（Name、IsAuthenticated 和 AuthenticationType），还具有一个集合类型的 Claims 属性，用来存放携带的所有声明。

```csharp
public class ClaimsIdentity : IIdentity
{
    string    Name { get; }
    bool      IsAuthenticated { get; }
    string    AuthenticationType { get; }

    public virtual IEnumerable<Claim> Claims { get; }
    ...
}
```

一个 ClaimsIdentity 对象从本质上来说就是对一组 Claim 对象的封装，所以它提供了如下这些操作声明的方法。我们可以调用 AddClaim(s)方法和(Try)RemoveClaim 方法添加或者删除声明，也可以调用 FindAll 方法或者 FindFirst 方法从声明集合中查询所有或者第一个满足条件的声明，还可以调用 HasClaim 方法确定声明集合中是否包含某个满足指定过滤条件的声明。在调用 FindAll 方法、FindFirst 方法和 HasClaim 方法时，我们可以将指定的声明类型作为过滤条件，也可以直接将一个 Predicate<Claim>对象作为过滤条件。

```
public class ClaimsIdentity : IIdentity
{
    public virtual void AddClaim(Claim claim);
    public virtual void AddClaims(IEnumerable<Claim> claims);

    public virtual void RemoveClaim(Claim claim);
    public virtual bool TryRemoveClaim(Claim claim);

    public virtual IEnumerable<Claim> FindAll(Predicate<Claim> match);
    public virtual IEnumerable<Claim> FindAll(string type);

    public virtual Claim FindFirst(Predicate<Claim> match);
    public virtual Claim FindFirst(string type);

    public virtual bool HasClaim(Predicate<Claim> match);
    public virtual bool HasClaim(string type, string value);
    ...
}
```

除了表示认证类型的 AuthenticationType 属性需要在创建的时候指定，一个 ClaimsIdentity 对象提供的所有信息都是根据它携带的这些声明计算出来的，如它的 Name 属性其实就来源于一个表示用户名的声明。一个 ClaimsIdentity 对象往往携带与权限相关的声明，权限控制系统会利用这些声明确定是否允许当前用户访问目标资源或者执行目标操作。由于基于角色的授权方式是最常用的，为了方便获取当前用户的角色集合，ClaimsIdentity 对象会提供角色对应的声明类型。

```
public class ClaimsIdentity : IIdentity
{
    private string _authenticationType;
    private string _nameType;
    private string _roleType;

    public const string DefaultNameClaimType = ClaimTypes.Name;
    public const string DefaultRoleClaimType = ClaimTypes.Role;
    public const string DefaultIssuer = "LOCAL AUTHORITY";

    public string Name => FindFirst(this.NameClaimType)?.Value;

    public string NameClaimType => _nameType ?? DefaultNameClaimType;
    public string RoleClaimType => _roleType ?? DefaultRoleClaimType;
```

```csharp
public bool IsAuthenticated => !string.IsNullOrEmpty(_authenticationType);

public ClaimsIdentity(IEnumerable<Claim> claims, string authenticationType,
    string nameType, string roleType)
{
    ...
    _authenticationType     = authenticationType;
    _nameType               = nameType;
    _roleType               = roleType;
}
...
}
```

如上面的代码片段所示，ClaimsIdentity 类型中定义了两个针对声明类型的属性（NameClaimType 和 RoleClaimType），它们分别表示针对用户名和角色的声明所采用的类型名称。在默认情况下，这两种声明会采用标准的声明类型，对应 DefaultNameClaimType 和 DefaultRoleClaimType 这两个常量。如果不希望采用默认的声明类型，我们可以选择上面这个构造函数显式地指定一个不同的声明类型。对于表示用户名的 Name 属性来说，它实际上是返回对应声明的值。

上面给出的代码片段还反映出，表示身份是否经过认证的 IsAuthenticated 属性的值取决于 ClaimsIdentity 对象是否具有一个确定的认证类型。ClaimsIdentity 还定义了一个名为 DefaultIssuer 的常量来表示声明的默认颁发者，它的值为 "LOCAL AUTHORITY"，我们可以理解为 "本地认证中心"。除了上面介绍的这些属性，ClaimsIdentity 还有如下两个重要的属性：Actor 属性返回的是一个 ClaimsIdentity 对象；BootstrapContext 属性则直接返回一个 object 对象。这两个属性涉及一个重要的主题，我们姑且称其为身份委托（Identity Delegation）。

```csharp
public class ClaimsIdentity : IIdentity
{
    public ClaimsIdentity           Actor { get; set; }
    public object                   BootstrapContext { get; set; }
}
```

下面用传统的 ASP.NET 应用来说明究竟什么是身份委托。假设在一个基于 AD 的域环境中，我们在机器 MachineA 上的 IIS 服务器中部署了一个 ASP.NET 应用，并且采用 Windows 集成认证。虽然可以采用任何一个合法的域账号登录这个应用（假设这个账号为 Foo），但是应用自身的代码总是以启动 IIS 进程所使用的 Windows 账号来执行（我们假设这个账号为 Bar）。出于安全方面的考虑，Bar 通常是一个不具有太多权限的服务账号（如 Network Service 账号），如果在代码执行过程中涉及一些需要较高权限的操作（如访问某个受保护的文件），那么系统就会抛出异常，这就是图 19-4 所示的第一种场景。

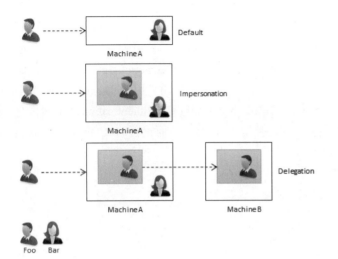

图 19-4　身份模拟与身份代理

为了解决这个问题，ASP.NET 提供了一种名为身份模拟（Identity Impersonation）的机制。所谓的身份模拟，就是让 ASP.NET 应用的部分代码可以直接在登录用户所使用的账号下执行。具体来说，我们可以根据登录用户的安全令牌（Security Token）创建一个模拟安全上下文，在这个上下文中的所有代码将以登录用户所使用的账号执行，这就是图 19-4 描述的第二种场景。

由于模拟安全上下文被严格限制在本机范围内，如果应用代码涉及远程调用，如它需要获取部署在另一台机器（MachineB）上的某个受保护的文件，那么身份模拟机制就无法发挥作用。在这种情况下，我们就需要使用身份委托（Identity Delegation）机制来解决这个问题，所谓的身份委托就是用户 Foo 让用户 Bar 作为代理并以自己的名义远程获取某项资源或者执行某项操作。实现身份代理的前提是代理人需要拥有被代理人的用户凭证（Credential），当代理人进行远程调用的时候需要提供被代理人的用户凭证，在完成认证之后，代理人才真正具有最初用户的代理资格，这就是图 19-4 描述的第三种场景。

委托是具有层级的，上面这个实例体现的是二级委托，既然有二级委托就会有三级委托、四级委托、N 级委托，所以委托其实是一个链式结构。现在我们将关注点再次转到 ClaimsIdentity 对象上，如果采用了身份委托，被委托人通过当前 ClaimsIdentity 的 Actor 属性来表示，该属性返回的依旧是一个 ClaimsIdentity 对象。被代理人可能是另一个用户的代理人，所以我们可以利用 Actor 属性表示图 19-5 所示的身份委托链（Identity Delegation Chain）。代理过程中被代理人提供的用户凭证会被封装到一个叫作安全令牌的可序列化对象中，具体采用何种类型的安全令牌取决于采用的认证类型，安全令牌被封装到 ClaimsIdentity 的 Bootstrap Context 属性中。

图 19-5　身份代理链

GenericIdentity

ClaimsIdentity 类型还具有一些子类，如在 Windows 认证下表示用户身份的 WindowsIdentity 就是它的派生类，但下面着重介绍的是另一个名为 GenericIdentity 的类型。顾名思义，GenericIdentity 表示一个泛化的身份，所以它是一个我们经常使用的 IIdentity 实现类型。GenericIdentity 重写了实现在基类中的 4 个属性，其中，Name 属性和 AuthenticationType 属性会直接返回构造函数中通过参数指定的用户名与认证类型（可默认），而重写的 Claims 属性其实有点多余，它实际上是直接返回基类的同名属性。

```
public class GenericIdentity : ClaimsIdentity
{
    public override string                      Name { get; }
    public override bool                        IsAuthenticated { get; }
    public override string                      AuthenticationType { get; }
    public override IEnumerable<Claim>          Claims { get; }

    public GenericIdentity(string name);
    public GenericIdentity(string name, string type)
}
```

对于一个 ClaimsIdentity 对象来说，表示是否经过认证的 IsAuthenticated 属性的值取决于它是否被设置了一个确定的认证类型。换句话说，如果 AuthenticationType 属性不是 Null 或者空字符串，那么它的 IsAuthenticated 属性就会返回 True。但是 GenericIdentity 重写的 IsAuthenticated 方法改变了这个默认逻辑，它的 IsAuthenticated 属性的值取决于它是否具有一个确定的用户名，如果表示用户名的 Name 属性是一个空字符串（由于构造函数做了验证，所以用户名不能为 Null），该属性就返回 False。

```
class Program
{
    static void Main()
    {
        var identity1 = new ClaimsIdentity(authenticationType: "Password");
        var identity2 = new ClaimsIdentity(new Claim[] {
            new Claim(ClaimTypes.Name, "Foobar") });
        var identity3 = new GenericIdentity(name: "", type: "Password");
        var identity4 = new GenericIdentity(name: "Foobar");

        Console.WriteLine("{0,-20}{1,-20}{2,-10}{3,-10}",
            "Type", "AuthenticationType", "Name", "IsAuthenticated");
        Console.WriteLine("----------------------------------------------------------------");

        foreach (var identity in new IIdentity[] {
            identity1, identity2, identity3, identity4 })
        {
            string name = string.IsNullOrEmpty(identity.Name) ? "-" : identity.Name;
            string authenticationType = string.IsNullOrEmpty(identity.AuthenticationType)
                ? "-" : identity.AuthenticationType;
            Console.WriteLine("{0,-20}{1,-20}{2,-10}{3,-10}",identity.GetType().Name,
                authenticationType, name, identity.IsAuthenticated);
```

```
        }
    }
}
```

在上面这段代码中,我们创建了 4 个 IIdentity 对象,包括两个 ClaimsIdentity 对象和两个 GenericIdentity 对象,它们具有一个共同的特征,那就是对于 Name 属性和 AuthenticationType 属性来说,只有其中一个有值。我们的目的是确定它们的 IsAuthenticated 属性的值与这两个属性具有何种关系。图 19-6 所示的输出结果证实了上述结论,即确定 ClaimsIdentity 和 GenericIdentity 是否被认证分别由它们的 AuthenticationType 属性与 Name 属性决定。(S1902)

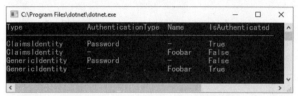

图 19-6 ClaimsIdentity 和 GenericIdentity 如何确定是否被认证

当我们创建一个 GenericIdentity 对象的时候,它除了直接将指定的用户名作为 Name 属性,还会在内部创建一个针对用户名的声明,该声明采用标准的类型(ClaimTypes.Name),指定的用户名就是该声明的值。如下所示的调试断言就体现了这一点。

```
var identity = new GenericIdentity("Foobar");
Debug.Assert(identity.HasClaim(ClaimTypes.Name, "Foobar"));
```

19.2.2 IPrincipal

对于 ASP.NET Core 应用的认证系统来说,接受认证的那个对象可能对应一个人,也可能对应一个应用、一个进程或者一个服务。不管这个对象是何种类型,我们统一采用一个具有如下定义的 IPrincipal 接口来表示。实际上,我们经常说的"用户"在大部分情况下指的就是一个 IPrincipal 对象。

```
public interface IPrincipal
{
    IIdentity Identity {  get; }
    bool IsInRole(string role);
}
```

一个表示认证用户的 IPrincipal 对象必须具有一个身份,该身份通过只读属性 Identity 来表示。IPrincipal 接口还有一个名为 IsInRole 的方法,用来确定当前用户是否被添加到指定的角色之中。也就是说,如果采用基于角色的授权方式,实际上我们可以直接调用这个方法来决定当前用户是否具有访问目标资源或者执行目标操作的权限。

ClaimsPrincipal

基于声明的认证与授权场景下的用户体现为一个 ClaimsPrincipal 对象,之所以被命名为 ClaimsPrincipal,是因为它使用 ClaimsIdentity 来表示其身份。一个 ClaimsPrincipal 对象代表一个用户,一个用户可以具有多个身份,所以一个 ClaimsPrincipal 对象是对多个 ClaimsIdentity 对象的封装。ClaimsPrincipal 的 Identities 属性用于返回这组 ClaimsIdentity 对象，我们可以调用 AddIdentity 方法或者 AddIdentities 方法为其添加任意的身份。对于实现的 IsInRole 方法来说，如果包含的任何一个 ClaimsPrincipal 具有基于角色的声明，并且该声明的值与指定的角色一致，该方法就会返回 True。

```csharp
public class ClaimsPrincipal : IPrincipal
{
    private static Func<IEnumerable<ClaimsIdentity>, ClaimsIdentity> _identitySelector;
    private List<ClaimsIdentity> _identities = new List<ClaimsIdentity>();

    static ClaimsPrincipal()
        => _identitySelector = ClaimsPrincipal.SelectPrimaryIdentity;

    public virtual IEnumerable<ClaimsIdentity> Identities
        => _identities.AsReadOnly();

    public IIdentity Identity
        => identitySelector ?? SelectPrimaryIdentity)(_identities);

    public static Func<IEnumerable<ClaimsIdentity>, ClaimsIdentity> PrimaryIdentitySelector
    {
        get { return _identitySelector; }
        set { _identitySelector = value; }
    }

    public virtual void AddIdentity(ClaimsIdentity identity) => _identities.Add(identity);
    public virtual void AddIdentities(IEnumerable<ClaimsIdentity> identities)
        => _identities.AddRange(identities);
    public bool IsInRole(string role)
        => _identities.Any(it => it.HasClaim(ClaimTypes.Role, role));

    private static ClaimsIdentity SelectPrimaryIdentity(
        IEnumerable<ClaimsIdentity> identities)
    {
        return identities.FirstOrDefault(it => it is WindowsIdentity)
            ?? identities.FirstOrDefault();
    }
    ...
}
```

虽然一个 ClaimsPrincipal 对象具有多个身份，但是它需要从中选择一个作为主身份（Primary Identity），它的 Identity 属性返回的就是作为主身份的 ClaimsIdentity 对象。如上所示的代码片段给出了主身份的选择策略，我们可以看到 ClaimsPrincipal 具有一个静态属性 PrimaryIdentitySelector，通过设置这个属性来提供一个类型为 Func<IEnumerable<ClaimsIdentity>, ClaimsIdentity>的委托对象来完成对主身份的选择。默认的选择策略体现在私有方法 SelectPrimaryIdentity 中，我们可以看到它在选择主身份的时候会优先选择 WindowsIdentity。

ClaimsPrincipal 具有如下一个 Claims 属性，用于返回 ClaimsIdentity 携带的所有声明，我们可以调用 FindAll 方法或者 FindFirst 方法获取满足指定条件的所有或者第一个声明，也可以调用 HasClaim 方法判断是否有一个或者多个 ClaimsIdentity 携带了某个指定条件的声明。除了这里提到的属性和方法，ClaimsPrincipal 还有一些额外的成员，由于篇幅有限，此处不再逐一介绍。

```
public class ClaimsPrincipal : IPrincipal
{
    public virtual IEnumerable<Claim> Claims { get; }

    public virtual IEnumerable<Claim> FindAll(Predicate<Claim> match);
    public virtual IEnumerable<Claim> FindAll(string type);
    public virtual Claim FindFirst(Predicate<Claim> match);
    public virtual Claim FindFirst(string type);
    public virtual bool HasClaim(Predicate<Claim> match);
    public virtual bool HasClaim(string type, string value);
    ...
}
```

GenericPrincipal

ClaimsPrincipal 同样具有一些预定义的子类，如针对 Windows 认证的 WindowsPrincipal 和针对 ASP.NET Roles 的 RolePrincipal，但这里着重介绍的是另一个名为 GenericPrincipal 的派生类。如下面的代码片段所示，当我们调用构造函数创建一个 GenericPrincipal 对象的时候，可以直接指定作为身份的 Identity 对象和角色列表。

```
public class GenericPrincipal : ClaimsPrincipal
{
    public override IIdentity Identity { get; }
    public GenericPrincipal(IIdentity identity, string[] roles);
    public override bool IsInRole(string role);
}
```

由于 GenericPrincipal 类型派生于 ClaimsPrincipal，所以它总是使用一个 ClaimsIdentity 对象来表示其身份，但是构造函数对应的参数类型却不是 ClaimsIdentity，而是 IIdentity 接口，这意味着我们在创建 GenericPrincipal 对象的时候可以指定一个任意类型的 IIdentity 对象。实际上，GenericPrincipal 的构造函数对此做了针对性处理，如果我们指定的不是一个 ClaimsIdentity 对象，它就会被转换成 ClaimsIdentity 类型。但是 GenericPrincipal 的 Identity 属性总是返回我们指定的那个 Identity，如下所示的调试断言就证实了这一点。

```
var principal = new GenericPrincipal(AnonymousIdentity.Instance, null);
Debug.Assert(ReferenceEquals(AnonymousIdentity.Instance, principal.Identity));
Debug.Assert(principal.Identities.Single() is ClaimsIdentity);

public class AnonymousIdentity : IIdentity
{
    public string AuthenticationType { get; }
    public bool IsAuthenticated { get; } = false;
    public string Name { get; }
    private AnonymousIdentity(){}

    public static readonly AnonymousIdentity Instance = new AnonymousIdentity();
```

当我们通过指定一个用户名和 N 个角色创建一个 GenericPrincipal 对象的时候，构造函数实际上会创建 N+1 个声明，其中，1 个是针对用户名的声明，N 个是针对角色的声明。这两种声明都采用标准的类型，这一点体现在如下所示的调试断言中。

```
var principal = new GenericPrincipal(new GenericIdentity("Foobar"),
    new string[] { "Role1", "Role2" });
Debug.Assert(principal.Claims.Count() == 3);
Debug.Assert(principal.HasClaim(ClaimTypes.Name, "Foobar"));
Debug.Assert(principal.HasClaim(ClaimTypes.Role, "Role1"));
Debug.Assert(principal.HasClaim(ClaimTypes.Role, "Role2"));
```

我们在前面介绍了用于表示用户身份的 IIdentity 接口和表示用户的 IPrincipal 接口，图 19-7 所示的 UML 体现了这两个接口及其实现类型的关系。

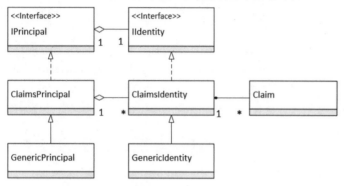

图 19-7　IIdentity 接口和 IPrincipal 接口及其实现类型的关系

19.3　认证模型

有了上面这个演示实例作为铺垫，同时了解了关于身份和用户的表达，读者理解 ASP.NET Core 的认证模型就会非常容易，实际上，前面演示的实例已经涉及认证模型的绝大部分成员。由于 ASP.NET Core 采用的是基于票据的认证，所以下面先介绍认证票据在认证模型中是如何表示的。

19.3.1　认证票据

认证票据通过一个 AuthenticationTicket 对象表示。如下面的代码片段所示，一个 AuthenticationTicket 对象实际上是对一个 ClaimsPrincipal 对象的封装。除了代表认证用户的 ClaimsPrincipal 对象，一个 ClaimsPrincipal 对象还有一个必需的 AuthenticationScheme 属性，该属性表示采用的认证方案名称。

```
public class AuthenticationTicket
{
    public string                           AuthenticationScheme { get; }
    public ClaimsPrincipal                  Principal { get; }
```

```csharp
    public AuthenticationProperties            Properties { get; }
    public AuthenticationTicket(ClaimsPrincipal principal, string authenticationScheme);
    public AuthenticationTicket(ClaimsPrincipal principal,
        AuthenticationProperties properties, string authenticationScheme);
}
```

AuthenticationProperties

AuthenticationTicket 的只读属性 Properties 返回一个 AuthenticationProperties 对象，它包含很多与当前认证上下文（Authentication Context）或者认证会话（Authentication Session）相关的信息，其中大部分属性是对认证票据的描述。一个 AuthenticationProperties 对象承载的所有数据都保存在其 Items 属性表示的数据字典中，如果需要为认证票据添加其他的描述信息，我们可以直接将它们添加到这个字典中。

```csharp
public class AuthenticationProperties
{
    public DateTimeOffset?                     IssuedUtc { get; set; }
    public DateTimeOffset?                     ExpiresUtc { get; set; }
    public bool?                               AllowRefresh { get; set; }
    public bool                                IsPersistent { get; set; }
    public string                              RedirectUri { get; set; }

    public IDictionary<string, string>         Items { get; }
}
```

出于安全性的考虑，我们不能让认证票据永久有效，而应该将其有效性限制在一个时间范围内。如果超出规定的限期，认证票据的持有人就必须利用其自身的凭证重新获取一张新的票据。认证票据的颁发时间和过期时间通过 AuthenticationProperties 对象的 IssuedUtc 属性与 ExpiresUtc 属性表示，根据这两个属性的命名可知，它们采用的均是 UTC 时间。

认证票据的过期策略可以采用绝对时间和滑动时间。假设我们设定认证票据的有效期为 30 分并在 1:00 获取一个认证票据，如果采用绝对时间，该票据总是在 1:30 过期。如果采用滑动时间，就意味着对认证票据的每一次使用都会将过期时间推迟到 30 分后。换句话说，如果每隔 29 分就使用一次认证票据，该票据将永远不会过期。如果我们采用滑动时间过期策略，实际上相当于使认证票据可以被"自动刷新"，AuthenticationProperties 的 AllowRefresh 属性决定认证票据是否可以被自动刷新。

AuthenticationProperties 的 IsPersistent 属性表示认证票据是否希望被客户端以持久化的形式保存起来。以浏览器作为客户端为例，如果认证票据被持久化存储，只要它尚未过期，即使多次重新启动浏览器也可以使用它，反之我们将不得不重新登录以获取新的认证票据。AuthenticationProperties 的 RedirectUri 属性携带着一个重定向地址，在不同情况下设置这个属性可以实现针对不同页面的重定向。例如，在登录成功后重定向到初始访问的页面，在注销之后重定向到登录页面，在访问受限的情况下重定向到我们定制的"访问拒绝"页面等。

TicketDataFormat

认证票据是一种私密性数据，请求携带的认证票据不仅是对 AuthenticationTicket 对象进行简单序列化之后的结果，中间还涉及对数据的加密，我们将这个过程称为对认证票据的格式化。认证票据的格式化通过一个 TicketDataFormat 对象表示的格式化器来完成。TicketDataFormat 实现了 ISecureDataFormat<TData>接口，如下面的代码片段所示，该接口定义了两组方法（Protect 和 Unprotect），分别实现针对数据对象的格式化和反格式化工作。Protect 方法和 Unprotect 方法都涉及一个名为 purpose 的参数，该参数表示当前格式化和反格式化的目的。

```
public interface ISecureDataFormat<TData>
{
    string Protect(TData data);
    string Protect(TData data, string purpose);
    TData Unprotect(string protectedText);
    TData Unprotect(string protectedText, string purpose);
}
```

TicketDataFormat 类型派生于基类 SecureDataFormat<TData>，后者实际上将数据的序列化/反序列化和加密/解密分开来实现。如下面的代码片段所示，SecureDataFormat<TData>将序列化/反序列化交给一个由 IDataSerializer<TData>接口表示的序列化器来完成，而加密/解密工作则由一个 IDataProtector 对象来负责。

```
public class SecureDataFormat<TData> : ISecureDataFormat<TData>
{
    public SecureDataFormat(IDataSerializer<TData> serializer, IDataProtector protector);
    public string Protect(TData data);
    public string Protect(TData data, string purpose);
    public TData Unprotect(string protectedText);
    public TData Unprotect(string protectedText, string purpose);
}

public interface IDataSerializer<TModel>
{
    byte[] Serialize(TModel model);
    TModel Deserialize(byte[] data);
}

public interface IDataProtector : IDataProtectionProvider
{
    byte[] Protect(byte[] plaintext);
    byte[] Unprotect(byte[] protectedData);
}
```

对于用来格式化认证票据的 TicketDataFormat 对象来说，它默认使用的序列化器通过 TicketSerializer 的静态属性 Default 返回一个 TicketSerializer 对象。而用来加密/解密认证票据的 IDataProtector 对象则需要手动指定，我们可以根据需要自定义相应的 IDataProtector 实现类型采用相应的算法来对认证票据实施加密。

```
public class TicketDataFormat : SecureDataFormat<AuthenticationTicket>
{
```

```csharp
    public TicketDataFormat(IDataProtector protector)
        : base(TicketSerializer.Default, protector){}
}
public class TicketSerializer : IDataSerializer<AuthenticationTicket>
{
    public virtual byte[] Serialize(AuthenticationTicket ticket);
    public virtual AuthenticationTicket Deserialize(byte[] data);

    public virtual AuthenticationTicket Read(BinaryReader reader);
    protected virtual Claim ReadClaim(BinaryReader reader, ClaimsIdentity identity);
    protected virtual ClaimsIdentity ReadIdentity(BinaryReader reader);

    public virtual void Write(BinaryWriter writer, AuthenticationTicket ticket);
    protected virtual void WriteClaim(BinaryWriter writer, Claim claim);
    protected virtual void WriteIdentity(BinaryWriter writer, ClaimsIdentity identity);

    public static TicketSerializer Default { get; }
}
```

上面主要介绍了表示认证票据的 AuthenticationTicket 以及对其进行格式化的 TicketDataFormat。AuthenticationTicket 和 TicketDataFormat 及其相关类型的关系如图 19-8 所示。

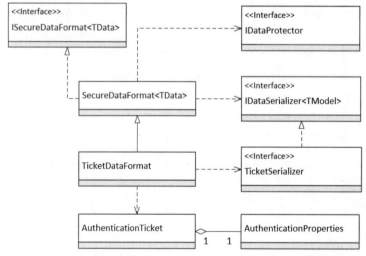

图 19-8　AuthenticationTicket 和 TicketDataFormat 及其相关类型的关系

19.3.2　认证处理器

得益于 ASP.NET Core 提供的这个极具扩展性的认证模型，我们可以为 ASP.NET Core 应用选择不同的认证方案。认证方案在认证模型中通过 AuthenticationScheme 类型标识，一个 AuthenticationScheme 对象的最终目的在于提供该方案对应的认证处理器类型。认证处理器在认证模型中通过 IAuthenticationHandler 接口表示，每种认证方案都对应针对该接口的实现类型，该类型承载了认证方案所需的所有操作。要对 ASP.NET Core 的认证具有充分的认识,我们需要先了解一个名为质询/响应(Challenge/Response)的认证模式。

质询/响应模式

质询/响应模式体现了这样一种消息交换模型:如果服务端(认证方)判断客户端(被认证方)没有提供有效的认证票据,它会向对方发送一个质询消息。客户端在接收到该消息后会重新提供一个合法的认证票据对质询予以响应。质询/响应式认证在 Web 应用中的实现比较有意思,因为质询体现为响应(Response),而响应体现为请求,但这两个响应代表完全不同的含义。前者代表一般意义上对认证方质询的响应,后者则表示认证方通过 HTTP 响应向对方发送质询。服务端通常会发送一个状态码为"401 Unauthorized"的响应作为质询消息。

服务端除了通过发送质询消息促使客户端提供一个有效的认证票据,如果通过认证的请求无权执行目标操作或者获取目标资源,它也会以质询消息的形式来通知客户端。一般来说,这样的质询消息体现为一个状态码为"403 Forbidden"的响应。虽然 IAuthenticationHandler 接口只是将前一种质询方法命名为 ChallengeAsync,后一种质询方法命名为 ForbidAsync,但是我们还是将两者统称为质询(Challenge)。

IAuthenticationHandler

如下面的代码片段所示,IAuthenticationHandler 接口定义了 4 个方法,其中 AuthenticateAsync 方法最为核心,因为认证中间件最终会调用它来对每个请求实施认证,而 ChallengeAsync 方法和 ForbidAsync 方法旨在实现前面介绍的两种类型的质询,这两个方法都利用一个类型为 AuthenticationProperties 的参数来传递当前上下文的信息。当某个 IAuthenticationHandler 对象被用来对请求实施认证之前,它的 InitializeAsync 方法会率先被调用以完成一些初始化的工作,该方法的两个参数分别是描述当前认证方案的 AuthenticationScheme 对象和当前 HttpContext 上下文。

```
public interface IAuthenticationHandler
{
    Task<AuthenticateResult> AuthenticateAsync();
    Task ChallengeAsync(AuthenticationProperties properties);
    Task ForbidAsync(AuthenticationProperties properties);
    Task InitializeAsync(AuthenticationScheme scheme, HttpContext context);
}
```

AuthenticateAsync 方法在完成对请求的认证之后,会将认证结果封装成一个 AuthenticateResult 对象。如下面的代码片段所示,认证结果具有成功、失败和 None 这 3 种状态。对于一个成功的认证结果,除了其 Succeeded 属性会返回 True,我们还可以从其 Principal 属性和 Ticket 属性得到表示认证用户的 ClaimsPrincipal 对象与表示认证票据的 AuthenticationTicket 对象。

```
public class AuthenticateResult
{
    public bool                             Succeeded { get; }
    public Exception                        Failure { get; protected set; }
    public bool                             None { get; protected set; }

    public ClaimsPrincipal                  Principal { get; }
```

```
    public AuthenticationTicket            Ticket { get; protected set; }
    public AuthenticationProperties        Properties { get; protected set; }
}
```

AuthenticateResult 为我们提供了如下 3 组静态方法，用来创建具有对应状态的 AuthenticateResult 对象。值得注意的是，如果调用 Fail 方法并以字符串的形式指定错误消息，该方法会根据错误消息创建一个 Exception 对象作为 AuthenticateResult 对象的 Failure 属性。

```
public class AuthenticateResult
{
    public static AuthenticateResult Success(AuthenticationTicket ticket);
    public static AuthenticateResult Fail(Exception failure);
    public static AuthenticateResult Fail(string failureMessage);
    public static AuthenticateResult Fail(Exception failure,
        AuthenticationProperties properties);
    public static AuthenticateResult Fail(string failureMessage,
        AuthenticationProperties properties);
    public static AuthenticateResult NoResult();
}
```

IAuthenticationHandler 对象表示的认证处理器承载了与认证相关的所有核心操作，但是我们只看到了用来认证请求的方法（AuthenticateAsync）和两种基于质询的方法（ChallengeAsync 和 ForbidAsync），并没有看到登录和注销操作对应的方法。这两个缺失的方法分别定义在派生于 IAuthenticationHandler 接口的其他两个接口中。如下面的代码片段所示，IAuthenticationSignOutHandler 接口定义的 SignOutAsync 方法用来完成注销操作，用于登录的 SignInAsync 方法则定义在 IAuthenticationSignInHandler 接口中。

```
public interface IAuthenticationSignOutHandler : IAuthenticationHandler
{
    Task SignOutAsync(AuthenticationProperties properties);
}

public interface IAuthenticationSignInHandler : IAuthenticationSignOutHandler,
{
    Task SignInAsync(ClaimsPrincipal user, AuthenticationProperties properties);
}
```

一般来说，一个完成的认证方案需要实现请求认证、登录和注销 3 个核心操作，所以对应的认证处理器类型一般会实现 IAuthenticationSignInHandler 接口。前面演示了针对 Cookie 的认证方案，该方案采用的认证处理器类型 CookieAuthenticationHandler 就实现了 IAuthenticationSignInHandler 接口。

除了上述两个接口，IAuthenticationHandler 接口还有如下这个特殊的派生接口 IAuthenticationRequestHandler。

对于一个普通的 IAuthenticationHandler 对象来说，认证中间件利用它来对当前请求实施认证之后总是将请求分发给下一个中间件做后续处理，而 IAuthenticationRequestHandler 对象则对请求处理具有更大的控制权，因为它可以决定针对当前请求的后续处理是否还有必要。如下面的代码片段所示，IAuthenticationRequestHandler 接口中定义了一个 HandleRequestAsync 方法，该方法用来处理当前请求。当认证中间件调用 HandleRequestAsync 方法对当前请求实施认证之后，如果该方法返回 True，那么整个请求处理流程将自此中止，认证中间件不再将请求分发给后续管道进行处理。

```
public interface IAuthenticationRequestHandler : IAuthenticationHandler
{
    Task<bool> HandleRequestAsync();
}
```

IAuthenticationHandlerProvider

被 AuthenticationMiddleware 中间件或者应用程序用来认证请求和完成登录注销操作的认证处理器对象是通过 IAuthenticationHandlerProvider 对象提供的。如下面的代码片段所示，IAuthenticationHandlerProvider 接口定义了唯一的 GetHandlerAsync 方法，从而根据当前 HttpContext 上下文和认证方案名称来提供对应的 IAuthenticationHandler 对象。

```
public interface IAuthenticationHandlerProvider
{
    Task<IAuthenticationHandler> GetHandlerAsync(HttpContext context,
        string authenticationScheme);
}
```

由于调用 GetHandlerAsync 方法会提供当前 HttpContext 上下文，这就意味着我们可以得到针对当前请求的 IServiceProvider 对象。假设需要提供的认证处理器的所有依赖服务都预先注册在依赖注入框架中，如果根据指定的认证方案名称可以得到对应的认证处理器类型，我们就可以利用这个作为依赖注入容器的 IServiceProvider 对象构建提供的 IAuthenticationHandler 对象。

我们在上面提到过，表示认证方案的 AuthenticationScheme 对象会为我们提供对应认证处理器的类型。如下所示的代码片段就是 AuthenticationScheme 类型的定义，我们所需的认证处理器类型就是通过它的 HandlerType 属性提供的。除此之外，AuthenticationScheme 还定义了分别表示认证方案名称和显示名称的 Name 属性与 DisplayName 属性。

```
public class AuthenticationScheme
{
    public string      Name { get; }
    public string      DisplayName { get; }
    public Type        HandlerType { get; }

    public AuthenticationScheme(string name, string displayName, Type handlerType);
}
```

AuthenticationScheme 能够为我们提供认证处理器的类型，那么现在的问题就变成如何根据认证方案名称得到对应的 AuthenticationScheme 对象，这个问题需要借助 IAuthenticationSchemeProvider 对象来解决。IAuthenticationSchemeProvider 对象不仅能够帮助我们提供所需的认证方案，应用采用的认证方案也是通过它来注册的。

如下面的代码片段所示，IAuthenticationSchemeProvider 接口除了定义了一个 GetSchemeAsync 方法根据指定的认证方案名称获取对应的 AuthenticationScheme 对象，还定义了相应的方法来为 5 种类型的操作（请求认证、登录、注销和两种质询）提供默认的认证方案。针对认证方案的注册体现在 AddScheme 方法上，注册的认证方案也可以通过 RemoveScheme 方法删除。我们可以调用 GetAllSchemesAsync 方法获取所有注册认证方案，而 GetRequestHandlerSchemesAsync 方法返回的认证方案是供 IAuthenticationRequestHandler 对象使用的。

```csharp
public interface IAuthenticationSchemeProvider
{
    Task<AuthenticationScheme> GetSchemeAsync(string name);

    Task<AuthenticationScheme> GetDefaultAuthenticateSchemeAsync();
    Task<AuthenticationScheme> GetDefaultChallengeSchemeAsync();
    Task<AuthenticationScheme> GetDefaultForbidSchemeAsync();
    Task<AuthenticationScheme> GetDefaultSignInSchemeAsync();
    Task<AuthenticationScheme> GetDefaultSignOutSchemeAsync();

    Task<IEnumerable<AuthenticationScheme>> GetAllSchemesAsync();
    Task<IEnumerable<AuthenticationScheme>> GetRequestHandlerSchemesAsync();

    void AddScheme(AuthenticationScheme scheme);
    void RemoveScheme(string name);
}
```

在了解了 IAuthenticationSchemeProvider 之后，再回到前面提出的关于如何提供认证处理器的问题。目前我们已经解决了根据指定的认证名称得到对应认证处理器类型的问题，所以能够根据 HttpContext 上下文提供的 IServiceProvider 对象创建 IAuthenticationHandler 对象，这个逻辑实现在如下所示的 AuthenticationHandlerProvider 类型上。

如下面的代码片段所示，AuthenticationHandlerProvider 的构造函数注入了一个用于提供注册认证方案的 IAuthenticationSchemeProvider 对象，在实现的 GetHandlerAsync 方法中，这个 IAuthenticationSchemeProvider 对象会根据指定的认证方案名称提供对应的 AuthenticationScheme 对象。一旦有了这个 AuthenticationScheme 对象，我们也就知道了对应 IAuthenticationHandler 的类型，进而可以利用 HttpContext 上下文提供的 IServiceProvider 对象得到对应的 IAuthenticationHandler 对象。

```csharp
public class AuthenticationHandlerProvider : IAuthenticationHandlerProvider
{
    private Dictionary<string, IAuthenticationHandler> _handlerMap
        = new Dictionary<string, IAuthenticationHandler>(StringComparer.Ordinal);

    public IAuthenticationSchemeProvider Schemes { get; }

    public AuthenticationHandlerProvider(IAuthenticationSchemeProvider schemes)
        => Schemes = schemes;

    public async Task<IAuthenticationHandler> GetHandlerAsync(HttpContext context,
        string authenticationScheme)
    {
        if (_handlerMap.TryGetValue(authenticationScheme, out var handler))
        {
```

```csharp
        return handler;
    }

    var scheme = await Schemes.GetSchemeAsync(authenticationScheme);
    if (scheme == null)
    {
        return null;
    }

    var serviceProvider = context.RequestServices;
    handler = (serviceProvider.GetService(scheme.HandlerType) ??
        ActivatorUtilities.CreateInstance(serviceProvider, scheme.HandlerType))
        as IAuthenticationHandler;
    if (handler != null)
    {
        await handler.InitializeAsync(scheme, context);
        _handlerMap[authenticationScheme] = handler;
    }
    return handler;
}
```

从上面提供的代码片段可以看出，AuthenticationHandlerProvider 为了避免对 IAuthenticationHandler 对象的重复创建，在内部对创建的实例做了缓存。除此之外，GetHandlerAsync 方法在返回提供的 IAuthenticationHandler 对象之前，会调用 InitializeAsync 方法对其进行初始化。

AuthenticationSchemeProvider

在了解了实现在 AuthenticationHandlerProvider 类型中针对认证处理器的默认提供机制之后，下面介绍认证方案的默认注册问题，这个问题的解决方案体现在作为 IAuthenticationSchemeProvider 接口的默认实现类型 AuthenticationSchemeProvider 上。AuthenticationSchemeProvider 对象利用一个字典对象维护注册认证方案的名称与对应 AuthenticationScheme 对象之间的映射关系，而这个映射字段最初的内容由 AuthenticationOptions 对象来提供。换句话说，认证方案最初其实是注册到配置选项 AuthenticationOptions 上的。

```csharp
public class AuthenticationSchemeProvider : IAuthenticationSchemeProvider
{
    public AuthenticationSchemeProvider(IOptions<AuthenticationOptions> options);

    public virtual Task<IEnumerable<AuthenticationScheme>> GetAllSchemesAsync();
    public virtual Task<AuthenticationScheme> GetDefaultAuthenticateSchemeAsync();
    public virtual Task<AuthenticationScheme> GetDefaultChallengeSchemeAsync();
    public virtual Task<AuthenticationScheme> GetDefaultForbidSchemeAsync();
    public virtual Task<AuthenticationScheme> GetDefaultSignInSchemeAsync();
    public virtual Task<AuthenticationScheme> GetDefaultSignOutSchemeAsync();
    public virtual Task<IEnumerable<AuthenticationScheme>> GetRequestHandlerSchemesAsync();
    public virtual Task<AuthenticationScheme> GetSchemeAsync(string name);

    public virtual void AddScheme(AuthenticationScheme scheme);
```

```
    public virtual void RemoveScheme(string name);
}
```

要了解认证方案是如何注册到配置选项 AuthenticationOptions 上的，就需要先了解用来构建认证方案的 AuthenticationSchemeBuilder 对象。如下面的代码片段所示，我们调用构造函数利用指定的认证方案名称创建了一个 AuthenticationSchemeBuilder 对象，并借助它的两个属性来设置认证方案的显示名称（可选）和认证处理器类型（必需），表示认证方案的 AuthenticationScheme 对象最终是由 Build 方法创建的。

```
public class AuthenticationSchemeBuilder
{
    public string       Name {get; }
    public string       DisplayName { get; set; }
    public Type         HandlerType { get; set; }

    public AuthenticationSchemeBuilder(string name)
        => Name = name;

    public AuthenticationScheme Build()
        => new AuthenticationScheme(Name, DisplayName, HandlerType);
}
```

如下所示的代码片段是 AuthenticationOptions 配置选项类型的完整定义，可以看出，真正注册到 AuthenticationOptions 对象上的其实是一个 AuthenticationSchemeBuilder 对象。AuthenticationScheme 通过其只读属性 SchemeMap 维护一组认证方案名称与对应 AuthenticationSchemeBuilder 对象之间的映射关系，当我们通过调用 AddScheme 注册一个认证方案时，该方法会创建一个 AuthenticationSchemeBuilder 对象并将其添加到映射字典中。

```
public class AuthenticationOptions
{
    private readonly IList<AuthenticationSchemeBuilder> _schemes
        = new List<AuthenticationSchemeBuilder>();
    public IEnumerable<AuthenticationSchemeBuilder> Schemes => _schemes;

    public string DefaultScheme { get; set; }
    public string DefaultAuthenticateScheme { get; set; }
    public string DefaultSignInScheme { get; set; }
    public string DefaultSignOutScheme { get; set; }
    public string DefaultChallengeScheme { get; set; }
    public string DefaultForbidScheme { get; set; }

    public IDictionary<string, AuthenticationSchemeBuilder> SchemeMap { get; }
        = new Dictionary<string, AuthenticationSchemeBuilder>(StringComparer.Ordinal);

    public void AddScheme(string name,
        Action<AuthenticationSchemeBuilder> configureBuilder)
    {
        if (SchemeMap.ContainsKey(name))
        {
```

```
            throw new InvalidOperationException("Scheme already exists: " + name);
        }
        var builder = new AuthenticationSchemeBuilder(name);
        configureBuilder(builder);
        _schemes.Add(builder);
        SchemeMap[name] = builder;
    }

    public void AddScheme<THandler>(string name, string displayName)
        where THandler : IAuthenticationHandler
        => AddScheme(name, b =>
        {
            b.DisplayName = displayName;
            b.HandlerType = typeof(THandler);
        });
}
```

AuthenticationOptions 还提供了一系列默认认证方案名称。如果在进行认证、登录、注销及发送质询（针对没有携带认证票据的匿名请求或者认证用户权限不够的请求）时没有显式提供采用的认证方案名称，这里设置的默认认证方案名称将会被采用。当利用 IOptions <AuthenticationOptions>对象创建 AuthenticationSchemeProvider 对象时，AuthenticationOptions 对象注册的这些认证方案信息将转移到 AuthenticationSchemeProvider 对象上。

19.3.3 认证服务

前面演示实例中的认证、登录和注销并没有直接调用作为认证处理器的 IAuthenticationHandler 对象的 AuthenticateAsync 方法、SignInAsync 方法和 SignOutAsync 方法，而是调用 HttpContext 上下文的同名扩展方法。如下面的代码片段所示，完整认证方案的 5 个核心操作都可以调用 HttpContext 上下文对应的扩展方法来完成。

```
public static class AuthenticationHttpContextExtensions
{
    public static Task<AuthenticateResult> AuthenticateAsync(this HttpContext context);
    public static Task<AuthenticateResult> AuthenticateAsync(this HttpContext context,
        string scheme);

    public static Task ChallengeAsync(this HttpContext context);
    public static Task ChallengeAsync(this HttpContext context,
        AuthenticationProperties properties);
    public static Task ChallengeAsync(this HttpContext context, string scheme);
    public static Task ChallengeAsync(this HttpContext context, string scheme,
        AuthenticationProperties properties);

    public static Task ForbidAsync(this HttpContext context);
    public static Task ForbidAsync(this HttpContext context,
        AuthenticationProperties properties);
    public static Task ForbidAsync(this HttpContext context, string scheme);
    public static Task ForbidAsync(this HttpContext context, string scheme,
```

```
    AuthenticationProperties properties);

public static Task SignInAsync(this HttpContext context, ClaimsPrincipal principal);
public static Task SignInAsync(this HttpContext context, ClaimsPrincipal principal,
    AuthenticationProperties properties);
public static Task SignInAsync(this HttpContext context, string scheme,
    ClaimsPrincipal principal);
public static Task SignInAsync(this HttpContext context, string scheme,
    ClaimsPrincipal principal, AuthenticationProperties properties);

public static Task SignOutAsync(this HttpContext context);
public static Task SignOutAsync(this HttpContext context,
    AuthenticationProperties properties);
public static Task SignOutAsync(this HttpContext context, string scheme);
public static Task SignOutAsync(this HttpContext context, string scheme,
    AuthenticationProperties properties);
}
```

HttpContext 的上述这些扩展方法与 IAuthenticationHandler 对象之间的适配是通过 IAuthenticationService 服务来实现的。如下面的代码片段所示，IAuthenticationService 接口同样定义了 5 个对应的方法。

```
public interface IAuthenticationService
{
    Task<AuthenticateResult> AuthenticateAsync(HttpContext context, string scheme);
    Task ChallengeAsync(HttpContext context, string scheme,
        AuthenticationProperties properties);
    Task ForbidAsync(HttpContext context, string scheme,
        AuthenticationProperties properties);
    Task SignInAsync(HttpContext context, string scheme, ClaimsPrincipal principal,
        AuthenticationProperties properties);
    Task SignOutAsync(HttpContext context, string scheme,
        AuthenticationProperties properties);
}
```

如下所示的 AuthenticationService 类型是对 IAuthenticationService 接口的默认实现。由于构造函数中注入了 IAuthenticationHandlerProvider 对象，所以 AuthenticationService 能够利用它得到对应的 IAuthenticationHandler 对象，并最后利用这个 IAuthenticationHandler 对象完成最终的操作。AuthenticationService 类型的构造函数还注入了 IAuthenticationSchemeProvider 对象，如果实现的方法指定的认证方案名称为 Null，AuthenticationService 对象就会利用这个 IAuthenticationSchemeProvider 对象提供默认的认证方案。

```
public class AuthenticationService : IAuthenticationService
{
    public IAuthenticationHandlerProvider      Handlers { get; }
    public IAuthenticationSchemeProvider       Schemes { get; }
    public IClaimsTransformation               Transform { get; }

    public AuthenticationService(IAuthenticationSchemeProvider schemes,
        IAuthenticationHandlerProvider handlers, IClaimsTransformation transform);
```

```csharp
    public virtual Task<AuthenticateResult> AuthenticateAsync(
        HttpContext context, string scheme);
    public virtual Task ChallengeAsync(HttpContext context, string scheme,
        AuthenticationProperties properties);
    public virtual Task ForbidAsync(HttpContext context, string scheme,
        AuthenticationProperties properties);
    public virtual Task SignInAsync(HttpContext context, string scheme,
        ClaimsPrincipal principal, AuthenticationProperties properties);
    public virtual Task SignOutAsync(HttpContext context, string scheme,
        AuthenticationProperties properties);
}
```

AuthenticationService 的构造函数中除了注入 IAuthenticationHandlerProvider 服务和 IAuthenticationSchemeProvider 服务，还注入了 IClaimsTransformation 服务。如下面的代码片段所示，IClaimsTransformation 接口提供的 TransformAsync 方法可以帮助我们实现针对 ClaimsPrincipal 对象的转换。认证模型默认提供的是如下这个没有实现任何转换操作的 NoopClaimsTransformation 类型，如果我们需要对表示认证用户的 ClaimsPrincipal 对象进行再加工，可以利用自定义的 IClaimsTransformation 服务来实现。

```csharp
public interface IClaimsTransformation
{
    Task<ClaimsPrincipal> TransformAsync(ClaimsPrincipal principal);
}

public class NoopClaimsTransformation : IClaimsTransformation
{
    public virtual Task<ClaimsPrincipal> TransformAsync(ClaimsPrincipal principal) =>
        Task.FromResult<ClaimsPrincipal>(principal);
}
```

如下所示的代码片段大体上展示了 AuthenticationService 类型的 AuthenticateAsync 方法的完整定义。如下面的代码片段所示，如果没有显式指定认证方案，AuthenticateAsync 方法会利用 IAuthenticationSchemeProvider 对象得到默认的认证方案名称，如果没有提供默认认证方案，该方法会直接抛出一个 InvalidOperationException 类型的异常。

```csharp
public class AuthenticationService : IAuthenticationService
{
    public virtual async Task<AuthenticateResult> AuthenticateAsync(
        HttpContext context, string scheme)
    {
        if (scheme == null)
        {
            var defaultScheme = await Schemes.GetDefaultAuthenticateSchemeAsync();
            scheme = defaultScheme?.Name;
            if (scheme == null)
            {
                throw new InvalidOperationException();
            }
```

```
        }
        var handler = await Handlers.GetHandlerAsync(context, scheme);
        if (handler == null)
        {
            throw await CreateMissingHandlerException(scheme);
        }

        var result = await handler.AuthenticateAsync();
        if (result != null && result.Succeeded)
        {
            var transformed = await Transform.TransformAsync(result.Principal);
            return AuthenticateResult.Success(new AuthenticationTicket(transformed,
                result.Properties, result.Ticket.AuthenticationScheme));
        }
        return result;
    }
    ...
}
```

接下来 AuthenticateAsync 方法会利用 IAuthenticationHandlerProvider 对象根据认证方案（显式指定或者默认注册）提供一个 IAuthenticationHandler 对象，并通过调用该对象的 AuthenticateAsync 方法来对请求实施认证。在认证成功的情况下，表示认证用户的 ClaimsPrincipal 对象会从认证结果中提取出来交给 IClaimsTransformation 对象进行加工或者转化。实现在 AuthenticationService 中的其他 4 个方法与 AuthenticateAsync 方法具有类似的实现逻辑。

上面的内容详细讲述了作为认证服务的 AuthenticationService 对象如何根据指定或者注册的认证方案获取作为认证处理器的 IAuthenticationHandler 对象，并利用它完成整个认证方案所需的 5 项核心操作，下面进行简单的概括。图 19-9 中的虚线表示对象之间的依赖关系，实线代表数据流向。对于 AuthenticationService 对象来说，它在默认情况下会利用 AuthenticationHandlerProvider 对象来提供所需的 IAuthenticationHandler。由于提供 IAuthenticationHandler 是在指定的 HttpContext 上下文中进行的，所以 AuthenticationHandlerProvider 对象可以得到针对当前请求的 IServiceProvider 对象。假设提供认证处理器类型的所有依赖服务都预先注册在依赖注入框架中，如果能够得到目标认证处理器的类型，AuthenticationHandlerProvider 就能利用这个 IServiceProvider 对象将所需的 IAuthenticationHandler 对象创建出来。

认证处理器类型是表示认证方案的 AuthenticationScheme 对象的核心组成部分，而认证方案的提供在默认情况下由 AuthenticationSchemeProvider 对象负责。由于应用程序会将认证方案注册到配置选项 AuthenticationOptions 上，而 AuthenticationSchemeProvider 对象正是根据这个配置选项创建的，所以它能获取所有注册的认证方案信息。

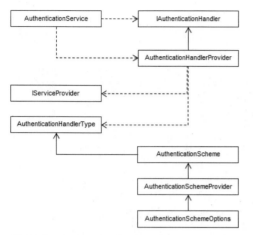

图 19-9　AuthenticationService→IAuthenticationHandler

19.3.4　服务注册

整个方案承载的 5 项核心操作（请求认证、登录、注销和两种类型的质询）在默认情况下都是先通过调用 AuthenticationService 服务相应的方法予以执行的（应用程序一般会直接调用 HttpContext 上下文相应的扩展方法，而这些方法调用还会转移到 AuthenticationService 对象上），而 AuthenticationService 对象最终会将方法调用转移到指定或者注册认证方案对应的认证处理器上。通过前面的介绍可知，AuthenticationService 能够利用表示认证处理器的 IAuthenticationHandler 对象建立在一个假设之上：认证处理器类型的所有依赖服务都预先注册在依赖注入框架中。下面介绍整个认证默认的基础服务是如何利用 IServiceCollection 接口的 AddAuthentication 方法进行注册的。

AddAuthentication

IServiceCollection 接口具有如下两个 AddAuthentication 方法重载。服务注册主要实现在第二个重载中，而认证模型的核心服务通过调用 AddAuthenticationCore 方法来完成。除此之外，AddAuthentication 方法调用其他的扩展方法完成了针对 IDataProtectorProvider、UrlEncoder 和 ISystemClock 这 3 个服务的注册。第一个重载在此基础上提供了针对认证配置选项 AuthenticationOptions 的设置，而认证方案最初就是注册在这个配置选项之上的。

```
public static class AuthenticationServiceCollectionExtensions
{
    public static AuthenticationBuilder AddAuthentication(this IServiceCollection services,
        Action<AuthenticationOptions> configureOptions)
    {
        services.Configure<AuthenticationOptions>(configureOptions);
        return services.AddAuthentication();
    }

    public static AuthenticationBuilder AddAuthentication(this IServiceCollection services)
    {
```

```
        services.AddAuthenticationCore();
        services.AddDataProtection();
        services.AddWebEncoders();
        services.TryAddSingleton<ISystemClock, SystemClock>();
        return new AuthenticationBuilder(services);
    }
}
```

IServiceCollection 接口用来注册认证模型核心服务的 AddAuthenticationCore 方法同样具有两个重载。从如下所示的代码片段可以看出，针对 IAuthenticationService、IClaimsTransformation、IAuthenticationHandlerProvider 和 IAuthenticationSchemeProvider 的服务注册就实现在这里。

```
public static class AuthenticationCoreServiceCollectionExtensions
{
    public static IServiceCollection AddAuthenticationCore(
        this IServiceCollection services)
    {
        services.TryAddScoped<IAuthenticationService, AuthenticationService>();
        services.TryAddSingleton<IClaimsTransformation, NoopClaimsTransformation>();
        services.TryAddScoped<IAuthenticationHandlerProvider,
            AuthenticationHandlerProvider>();
        services.TryAddSingleton<IAuthenticationSchemeProvider,
            AuthenticationSchemeProvider>();
        return services;
    }

    public static IServiceCollection AddAuthenticationCore(
        this IServiceCollection services, Action<AuthenticationOptions> configureOptions)
    {
        services.AddAuthenticationCore();
        services.Configure<AuthenticationOptions>(configureOptions);
        return services;
    }
}
```

AuthenticationBuilder

AddAuthentication 方法仅仅注册了定义在认证模型中的基础服务，与采用认证方案相关的服务则由该方法返回的 AuthenticationBuilder 对象做进一步注册。如下面的代码片段所示，一个 AuthenticationBuilder 对象实际上是对一个 IServiceCollection 对象的封装，我们可以调用它的两个 AddScheme<TOptions, THandler> 方法重载注册针对某种认证方案的认证处理器类型，并进一步设置对应的配置选项。

```
public class AuthenticationBuilder
{
    public virtual IServiceCollection Services { get; }
    public AuthenticationBuilder(IServiceCollection services)
        => Services = services;

    public virtual AuthenticationBuilder AddScheme<TOptions, THandler>(
        string authenticationScheme, string displayName, Action<TOptions> configureOptions
```

```
        where TOptions : AuthenticationSchemeOptions, new()
        where THandler : AuthenticationHandler<TOptions>
        => AddSchemeHelper<TOptions, THandler>(authenticationScheme, displayName,
            configureOptions);

    public virtual AuthenticationBuilder AddScheme<TOptions, THandler>(
        string authenticationScheme, Action<TOptions> configureOptions)
        where TOptions : AuthenticationSchemeOptions, new()
        where THandler : AuthenticationHandler<TOptions>
        => AddScheme<TOptions, THandler>(authenticationScheme, null,configureOptions);

    private AuthenticationBuilder AddSchemeHelper<TOptions, THandler>(
        string authenticationScheme, string displayName,
        Action<TOptions> configureOptions)
        where TOptions : class, new()
        where THandler : class, IAuthenticationHandler
    {
        Services.Configure<AuthenticationOptions>(o =>
        {
            o.AddScheme(authenticationScheme, scheme => {
                scheme.HandlerType = typeof(THandler);
                scheme.DisplayName = displayName;
            });
        });
        if (configureOptions != null)
        {
            Services.Configure(authenticationScheme, configureOptions);
        }
        Services.AddTransient<THandler>();
        return this;
    }
}
```

AuthenticationSchemeOptions

当我们调用 AuthenticationBuilder 的 AuthenticationBuilder 注册认证方案时，需要同时指定认证处理器和对应配置选项的类型。与认证方案相关的配置选项定义在如下所示的 AuthenticationSchemeOptions 的 Options 类型中，它为真正完成认证工作的认证处理器提供相应的设置。当我们为某种认证方案自定义一个认证处理器类型时，一般会将针对性的设置定义在派生于 AuthenticationSchemeOptions 的配置选项类型中。

```
public class AuthenticationSchemeOptions
{
    public string ClaimsIssuer { get; set; }

    public object    Events { get; set; }
    public Type      EventsType { get; set; }

    public string ForwardAuthenticate { get; set; }
    public string ForwardChallenge { get; set; }
```

```csharp
    public string  ForwardForbid  { get; set; }
    public string  ForwardSignIn  { get; set; }
    public string  ForwardSignOut { get; set; }
    public string  ForwardDefault { get; set; }
    public Func<HttpContext, string> ForwardDefaultSelector { get;  set; }

    public virtual void Validate();
    public virtual void Validate(string scheme);
}
```

AuthenticationSchemeOptions 定义了几组属性和方法成员。其中，ClaimsIssuer 属性表示在认证过程中创建的声明采用的颁发者名称，也就是前面介绍的 Claim 类型的 Issuer 属性。为了使某种认证方式具有更好的扩展性，我们往往希望应用程序可以对认证的流程进行干预，这个功能可以利用动态注册的事件（Events）对象或者类型来实现。具体来说，认证处理器可以在整个认证流程中触发若干事件，并将事件回调定义在一个类型中，那么应用程序就可以通过注册一个事件对象或者类型来达到对认证流程进行干预的目的。AuthenticationSchemeOptions 的 Events 属性与 EventsType 属性指的就是注册的事件对象和类型。

完整的认证流程会涉及一系列相关的操作，如认证票据的验证、登录、注销及发送认证质询等。对于两种不同的认证方案来说，它们针对某些操作可能具有相同的实现，或者某个认证操作完全可以借助另一种认证方案来完成。当我们在定义某个具体的认证处理器的时候，只需要实现必要的认证操作即可，如果某项认证操作可以采用另一种认证方案来完成，我们可以借助定义在 AuthenticationSchemeOptions 中一系列 ForwardXxx 属性和 ForwardDefaultSelector 方法转发给另一种替代的认证方案。

除了上述这些属性成员，AuthenticationSchemeOptions 还提供了两个 Validate 方法，用来验证设定的配置选项是否合法。在认证处理器初始化过程（InitializeAsync 方法被调用的时候）中，Validate 方法用来检验设置的配置选项针对指定的认证方案是否合法。

19.3.5 AuthenticationMiddleware

认证模型针对请求的认证最终是借助 AuthenticationMiddleware 中间件来完成的。由于具体认证的实现已经分散到前面介绍的若干服务类型上，所以实现在该中间件的认证逻辑就显得非常简单。下面的代码片段基本上体现了 AuthenticationMiddleware 中间件的完整实现。

```csharp
public class AuthenticationMiddleware
{
    private readonly RequestDelegate _next;
    public IAuthenticationSchemeProvider Schemes { get; set; }

    public AuthenticationMiddleware(RequestDelegate next,
        IAuthenticationSchemeProvider schemes)
    {
        _next       = next;
        Schemes     = schemes;
    }
```

```csharp
public async Task Invoke(HttpContext context)
{
    context.Features.Set<IAuthenticationFeature>(new AuthenticationFeature
    {
        OriginalPath         = context.Request.Path,
        OriginalPathBase     = context.Request.PathBase
    });

    // 先利用 IAuthenticationRequestHandler 来处理请求
    var handlers = context.RequestServices
        .GetRequiredService<IAuthenticationHandlerProvider>();
    foreach (var scheme in await Schemes.GetRequestHandlerSchemesAsync())
    {
        var handler = await handlers.GetHandlerAsync(context, scheme.Name)
            as IAuthenticationRequestHandler;
        if (handler != null && await handler.HandleRequestAsync())
        {
            return;
        }
    }

    var defaultAuthenticate = await Schemes.GetDefaultAuthenticateSchemeAsync();
    if (defaultAuthenticate != null)
    {
        var result = await context.AuthenticateAsync(defaultAuthenticate.Name);
        if (result?.Principal != null)
        {
            context.User = result.Principal;
        }
    }

    await _next(context);
}
```

如上面的代码片段所示，AuthenticationMiddleware 中间件类型的构造函数中注入了一个用于提供认证方案的 IAuthenticationSchemeProvider 服务。在用于处理请求的 Invoke 方法中，AuthenticationMiddleware 中间件先将当前请求的路径（Path）和基础路径（PathBase）利用一个通过 IAuthenticationFeature 接口表示的特性附加到当前的 HttpContext 上下文中。

在前面介绍认证处理器时，我们提到了一种通过 IAuthenticationRequestHandler 接口表示的认证处理器。这种处理器相当于一个中间件，因为它能够百分之百地掌控针对当前请求的处理，并决定是否还需要将请求分发给后续管道做进一步处理。从给出的代码片段可以看出，这样的请求处理器会率先被提取出来执行，而提供给它们的认证方案是调用 IAuthenticationSchemeProvider 对象的 GetRequestHandlerSchemesAsync 方法的返回结果。如果任何一个 IAuthenticationRequestHandler 对象的 HandleRequestAsync 方法返回 True，整个认证过程将就此中止。

如果当前应用并没有注册任何 IAuthenticationRequestHandler 处理器，或者它们并没有中止对当前请求的处理，那么 AuthenticationMiddleware 中间件就会利用 IAuthenticationSchemeProvider 对象提供一个默认的认证方案，并借助 IAuthenticationHandlerProvider 服务提供的 IAuthenticationHandler 对象来对当前请求实施认证。对于认证通过的请求，认证结果承载的 ClaimsPrincipal 对象将赋值给 HttpContext 上下文的 User 属性用来表示当前的认证用户。我们可以调用针对 IApplicationBuilder 接口的 UseAuthentication 方法来注册 AuthenticationMiddleware 中间件。

```
public static class AuthAppBuilderExtensions
{
    public static IApplicationBuilder UseAuthentication(this IApplicationBuilder app)
    => app.UseMiddleware<AuthenticationMiddleware>(Array.Empty<object>());
}
```

19.4 Cookie 认证方案

前面的实例演示了利用 Cookie 来携带认证票据的认证方案，下面详细介绍此认证方案的实现原理。通过前面的介绍可知，针对某个认证方案的核心功能体现在 IAuthenticationHandler 接口表示的认证处理器上，针对 Cookie 的认证方案实现在 CookieAuthenticationHandler 类型中。在具体介绍该类型之前，我们需要先了解其基类。

19.4.1 AuthenticationHandler<TOptions>

包括 CookieAuthenticationHandler 在内，认证模型提供的所有原生认证处理器类型都派生于如下所示的抽象类 AuthenticationHandler<TOptions>。如果我们需要实现额外的认证方案，对应的认证处理器最好也直接派生于这个基类。如下面的代码片段所示，AuthenticationHandler<TOptions>直接实现 IAuthenticationHandler 接口，其泛型参数类型 TOptions 表示承载对应认证方案的配置选项，它派生于基类 AuthenticationSchemeOptions。

```
public abstract class AuthenticationHandler<TOptions> : IAuthenticationHandler
    where TOptions : AuthenticationSchemeOptions, new()
{
    protected IOptionsMonitor<TOptions>     OptionsMonitor { get; }
    protected ILogger                       Logger { get; }
    protected UrlEncoder                    UrlEncoder { get; }
    protected ISystemClock                  Clock { get; }

    protected AuthenticationHandler1(IOptionsMonitor<TOptions> options,
        ILoggerFactory logger, UrlEncoder encoder, ISystemClock clock)
    {
        Logger          = logger.CreateLogger(this.GetType().FullName);
        UrlEncoder      = encoder;
        Clock           = clock;
        OptionsMonitor  = options;
    }
    ...
}
```

AuthenticationHandler<TOptions>类型在构造函数中注入了 4 个服务对象，其中包括用来提供实时配置数据的 IOptionsMonitor<TOptions>对象、创建 Logger 的 ILoggerFactory 对象、实现 URL 编码的 UrlEncoder 对象、用来提供同步系统时钟的 ISystemClock 对象。认证模型针对系统时钟提供了一个名为 SystemClock 的默认实现类型，它的 UtcNow 会返回本地当前的 UTC 时间。

```
public class SystemClock : ISystemClock
{
    public DateTimeOffset UtcNow { get; }
}
```

初始化（InitializeAsync）

通过前面对认证处理器的介绍可知，AuthenticationHandlerProvider 对象在根据表示当前 HttpContext 上下文和指定的认证方案名称来提供某个 IAuthenticationHandler 对象的时候，后者的 InitializeAsync 方法被调用以便完成一些初始化工作。下面介绍 AuthenticationHandler <TOptions>在这个方法中做了什么。

```
public abstract class AuthenticationHandler<TOptions>
    : IAuthenticationHandler
    where TOptions : AuthenticationSchemeOptions, new()
{
    public TOptions                             Options { get; private set; }
    protected IOptionsMonitor<TOptions>         OptionsMonitor { get; }
    protected HttpContext                       Context { get; private set; }
    public AuthenticationScheme                 Scheme { get; private set; }
    protected virtual object                    Events { get; set; }

    public async Task InitializeAsync(AuthenticationScheme scheme, HttpContext context)
    {
        Scheme      = scheme;
        Context     = context;

        Options = OptionsMonitor.Get(Scheme.Name) ?? new TOptions();
        Options.Validate(Scheme.Name);

        await InitializeEventsAsync();
        await InitializeHandlerAsync();
    }

    protected virtual async Task InitializeEventsAsync()
    {
        Events = Options.Events;
        if (Options.EventsType != null)
        {
            Events = Context.RequestServices.GetRequiredService(Options.EventsType);
        }
        Events = Events ?? await CreateEventsAsync();
    }
    protected virtual Task<object> CreateEventsAsync() => Task.FromResult(new object());
    protected virtual Task InitializeHandlerAsync() => Task.CompletedTask;
}
```

如上面的代码片段所示，IAuthenticationHandlerProvider 对象在 InitializeAsync 方法中会将提供的 HttpContext 上下文和认证方案类型通过对应的属性保存起来，然后从 IOptionsMonitor<TOptions>对象中获取当前的配置选项并将它赋值给 Options 属性。InitializeAsync 方法对每个请求都会调用一次，所以 Options 属性返回的实际上是针对当前请求的实时配置数据（类似于 IOptionsSnapshot<TOptions>）。

前面在介绍 AuthenticationSchemeOptions 类型的时候提到它具有两个属性：Events 和 EventsType，我们可以注册一个 Events 对象或者类型用来对整个认证流程实施干预，它们在 InitializeAsync 方法中被应用到对应的 IAuthenticationHandler 对象之上。如上面的代码片段所示，InitializeAsync 方法会初始化 Events 对象并将它赋值给对应的属性。这一切都实现在一个名为 InitializeEventsAsync 的虚方法上，如果我们具有针对 Events 的定制需要，可以重写这个方法。

InitializeAsync 方法调用了另一个名为 InitializeHandlerAsync 的虚方法，旨在执行一些额外的初始化操作，目前这个方法并没有执行任何具体的操作。如果自定义认证处理器类型的实现需要执行一些额外的初始化操作，就可以将它们实现在重写的 InitializeHandlerAsync 方法中。

认证（AuthenticateAsync）

下面介绍真正用来对请求实施认证的 AuthenticateAsync 方法是如何实现的。用于承载认证方案配置选项的 AuthenticationSchemeOptions 类型定义了一系列 ForwardXxx 方法，我们可以利用它们将认证过程中涉及的一些操作"转移"到其他兼容认证方案上，这一点直接体现在 AuthenticateAsync 方法的实现上。

```
public abstract class AuthenticationHandler<TOptions>
    : IAuthenticationHandler
    where TOptions : AuthenticationSchemeOptions, new()
{
    private Task<AuthenticateResult> _authenticateTask;

    public async Task<AuthenticateResult> AuthenticateAsync()
    {
        var target = ResolveTarget(Options.ForwardAuthenticate);
        if (target != null)
        {
            return await Context.AuthenticateAsync(target);
        }
        return await HandleAuthenticateOnceAsync();
    }

    protected virtual string ResolveTarget(string scheme)
    {
        var target = scheme
            ?? Options.ForwardDefaultSelector?.Invoke(Context)
            ?? Options.ForwardDefault;
        return string.Equals(target, Scheme.Name, StringComparison.Ordinal)
            ? null
```

```
            : target;
    }

    protected Task<AuthenticateResult> HandleAuthenticateOnceAsync()
        => _authenticateTask ??= HandleAuthenticateAsync();

    protected async Task<AuthenticateResult> HandleAuthenticateOnceSafeAsync()
    {
        try
        {
            return await HandleAuthenticateOnceAsync();
        }
        catch (Exception ex)
        {
            return AuthenticateResult.Fail(ex);
        }
    }

    protected abstract Task<AuthenticateResult> HandleAuthenticateAsync();
}
```

如上面的代码片段所示，AuthenticateAsync 方法会率先调用一个命名为 ResolveTarget 的虚方法，后者会根据提供的 AuthenticationSchemeOptions 配置选项来解析转移的目标认证方案。如果认证方案需要被转移，该方法只需要直接调用当前 HttpContext 上下文的 AuthenticateAsync 方法对采用新认证方案的请求实施认证即可。在确定了不需要认证转移的情况下，AuthenticateAsync 方法会调用 HandleAuthenticateOnceAsync 方法完成对请求的认证，真正的认证体现在抽象方法 HandleAuthenticateAsync 上。AuthenticationHandler<TOptions>还定义了一个名为 HandleAuthenticateOnceSafeAsync 的虚方法，用来提供更加"安全"的认证，因为它实现了针对异常的处理。

质询（ChallengeAsync 和 ForbidAsync）

我们将通过 IAuthenticationHandler 对象的 ChallengeAsync 方法和 ForbidAsync 方法完成的操作统一称为质询，前者旨在提供一个响应促使客户端提供一个合法的认证票据，后者则是告知认证用户无权执行当前操作或者获取当前资源。

```
public abstract class AuthenticationHandler<TOptions>
    : IAuthenticationHandler where TOptions : AuthenticationSchemeOptions, new()
{
    public async Task ChallengeAsync(AuthenticationProperties properties)
    {
        var target = ResolveTarget(Options.ForwardChallenge);
        if (target != null)
        {
            await Context.ChallengeAsync(target, properties);
            return;
        }
```

```
        properties = properties ?? new AuthenticationProperties();
        await HandleChallengeAsync(properties);
    }

    public async Task ForbidAsync(AuthenticationProperties properties)
    {
        var target = ResolveTarget(Options.ForwardForbid);
        if (target != null)
        {
            await Context.ForbidAsync(target, properties);
            return;
        }

        properties = properties ?? new AuthenticationProperties();
        await HandleForbiddenAsync(properties);
    }

    protected virtual Task HandleChallengeAsync(AuthenticationProperties properties)
    {
        Response.StatusCode = 401;
        return Task.CompletedTask;
    }

    protected virtual Task HandleForbiddenAsync(AuthenticationProperties properties)
    {
        Response.StatusCode = 403;
        return Task.CompletedTask;
    }
}
```

如上面的代码片段所示，跨方案认证转移机制同样应用在 ChallengeAsync 方法和 Handle ChallengeAsync 方法中。真正的质询体现在 HandleChallengeAsync 和 HandleForbiddenAsync 这两个虚方法上。从给出的代码片段可以看出，HandleChallengeAsync 方法和 HandleForbiddenAsync 方法分别回复一个状态码为"401 Unauthorized"与"403 Forbidden"的响应。如果需要提供不一样的质询响应，如重定向到登录和授权失败的页面，可以通过重写这两个方法来实现。实际上，AuthenticationHandler<TOptions>类型中还定义了一系列额外的属性和方法成员。由于篇幅的限制，此处不再赘述。

SignOutAuthenticationHandler<TOptions>

具有如下定义的抽象类 SignOutAuthenticationHandler<TOptions>派生于 AuthenticationHandler <TOptions>，同时作为针对 IAuthenticationSignOutHandler 接口的默认实现。如下面的代码片段所示，该类型将前面介绍的跨方案认证转移实现在 SignOutAsync 方法中，具体的注销操作则体现在抽象方法 HandleSignOutAsync 中。

```
public abstract class SignOutAuthenticationHandler<TOptions> :
    AuthenticationHandler<TOptions>,
    IAuthenticationSignOutHandler,
```

```csharp
    where TOptions: AuthenticationSchemeOptions, new()
{
    public SignOutAuthenticationHandler(IOptionsMonitor<TOptions> options,
        ILoggerFactory logger, UrlEncoder encoder, ISystemClock clock)
        : base(options, logger, encoder, clock)
    {}

    protected abstract Task HandleSignOutAsync(AuthenticationProperties properties);
    public virtual Task SignOutAsync(AuthenticationProperties properties)
    {
        string scheme = this.ResolveTarget(base.Options.ForwardSignOut);
        if (scheme != null)
        {
            return base.Context.SignOutAsync(scheme, properties);
        }
        return HandleSignOutAsync(properties ?? new AuthenticationProperties());
    }
}
```

SignInAuthenticationHandler<TOptions>

SignInAuthenticationHandler<TOptions>派生于 SignOutAuthenticationHandler<TOptions>，也是认证模型提供的针对 IAuthenticationSignInHandler 接口的默认实现。它与 SignOutAuthenticationHandler<TOptions>具有完全一致的定义模式，即在实现的 SignInAsync 方法中实现跨方案认证转移，并定义了 HandleSignInAsync 这个抽象方法，用来完成具体的登录操作。

```csharp
public abstract class SignInAuthenticationHandler<TOptions> :
    SignOutAuthenticationHandler<TOptions>,
    IAuthenticationSignInHandler,
    where TOptions: AuthenticationSchemeOptions, new()
{
    public SignInAuthenticationHandler(IOptionsMonitor<TOptions> options,
        ILoggerFactory logger, UrlEncoder encoder, ISystemClock clock)
        : base(options, logger, encoder, clock)
    {}

    protected abstract Task HandleSignInAsync(ClaimsPrincipal user,
        AuthenticationProperties properties);
    public virtual Task SignInAsync(ClaimsPrincipal user,
        AuthenticationProperties properties)
    {
        var scheme = this.ResolveTarget(base.Options.ForwardSignIn);
        if (scheme != null)
        {
            return base.Context.SignInAsync(scheme, user, properties);
        }
        return HandleSignInAsync(user, properties ?? new AuthenticationProperties());
    }
}
```

19.4.2 CookieAuthenticationHandler

针对 Cookie 的认证逻辑基本上都实现在 CookieAuthenticationHandler 类型中。在正式介绍这个认证处理器针对认证、登录和注销的实现原理之前，下面先介绍几个与其相关的类型。

CookieAuthenticationEvents

正如前面多次提及的，出于可扩展的目的，AuthenticationHandler<TOptions>采用一种特殊的事件（Event）机制使应用程序可以对整个认证流程实施干预。由于每种具体的认证方案具有各自不同的认证流程，所以作为基类的 AuthenticationHandler<TOptions>只能以弱类型的方式提供一个 Object 类型的认证事件，派生于这个抽象类的认证处理器定义了一个强类型的认证事件。CookieAuthenticationHandler 采用的认证事件类型就是具有如下定义的 CookieAuthenticationEvents。

```
public class CookieAuthenticationEvents
{
    public Func<CookieSigningInContext, Task>      OnSigningIn { get; set; }
    public Func<CookieSignedInContext, Task>       OnSignedIn { get; set; }
    public Func<CookieSigningOutContext, Task>     OnSigningOut { get; set; }
    public Func<CookieValidatePrincipalContext, Task> OnValidatePrincipal { get; set; }

    public Func<RedirectContext<CookieAuthenticationOptions>, Task>
        OnRedirectToAccessDenied { get; set; }
    public Func<RedirectContext<CookieAuthenticationOptions>, Task>
        OnRedirectToLogin { get; set; }
    public Func<RedirectContext<CookieAuthenticationOptions>, Task>
        OnRedirectToLogout { get; set; }
    public Func<RedirectContext<CookieAuthenticationOptions>, Task>
        OnRedirectToReturnUrl { get; set; }
}
```

如上面的代码片段所示，CookieAuthenticationEvents 定义了一系列委托类型的属性作为对应事件触发时的回调，从命名可以看出这些属性成员被调用的时机。我们将这些属性成员划分成两组：前一组会在登录、注销和验证代表用户的 ClaimsPrincipal 对象时被调用；后一组则与认证过程所需的重定向有关，我们可以利用它们控制对登录、注销、权限不足和初始访问页面的重定向。

从 CookieAuthenticationEvents 的定义可以看出，它从事的每项操作都是在一个上下文中进行的，这些上下文类型将如下所示的 BaseContext<TOptions>作为它们共同的基类。如下面的代码片段所示，我们可以利用这个 BaseContext<TOptions>对象得到当前的 HttpContext 上下文、配置选项和认证方案。BaseContext<TOptions>具有 PropertiesContext<TOptions>和 PrincipalContext<TOptions>两个派生类型，前者提供了承载当前认证会话信息的 AuthenticationProperties 对象，后者提供了代表当前认证用户的 ClaimsPrincipal 对象。

```
public abstract class BaseContext<TOptions> where TOptions: AuthenticationSchemeOptions
{
    public HttpContext      HttpContext { get; }
    public HttpRequest      Request { get; }
    public HttpResponse     Response { get; }
```

```csharp
    public TOptions                    Options { get; }
    public AuthenticationScheme        Scheme { get; }

    protected BaseContext(HttpContext context, AuthenticationScheme scheme,
        TOptions options);
}

public abstract class PropertiesContext<TOptions> : BaseContext<TOptions>
    where TOptions: AuthenticationSchemeOptions
{
    public virtual AuthenticationProperties Properties { get; protected set; }
    protected PropertiesContext(HttpContext context, AuthenticationScheme scheme,
        TOptions options, AuthenticationProperties properties);
}

public abstract class PrincipalContext<TOptions> : PropertiesContext<TOptions>
    where TOptions: AuthenticationSchemeOptions
{
    public virtual ClaimsPrincipal Principal { get; set; }
    protected PrincipalContext(HttpContext context, AuthenticationScheme scheme,
        TOptions options, AuthenticationProperties properties);
}
```

CookieAuthenticationEvents 针对登录和认证用户检验的 3 个上下文类型（CookieSigningInContext、CookieSignedInContext 和 CookieValidatePrincipalContext）派生于上面的 PrincipalContext<TOptions>类型，针对注销的 CookieSigningOutContext 则是 PropertiesContext<TOptions>的子类。如下面的代码片段所示，我们可以利用 CookieSigningInContext 和 CookieSigningOutContext 在完成登录与注销之前获取承载认证票据 Cookie 的配置选项（CookieOptions）。

```csharp
public class CookieSigningInContext : PrincipalContext<CookieAuthenticationOptions>
{
    public CookieOptions CookieOptions { get; set; }
    public CookieSigningInContext(HttpContext context, AuthenticationScheme scheme,
        CookieAuthenticationOptions options, ClaimsPrincipal principal,
        AuthenticationProperties properties, CookieOptions cookieOptions);
}

public class CookieSignedInContext : PrincipalContext<CookieAuthenticationOptions>
{
    public CookieSignedInContext(HttpContext context, AuthenticationScheme scheme,
        ClaimsPrincipal principal, AuthenticationProperties properties,
        CookieAuthenticationOptions options);
}

public class CookieValidatePrincipalContext : PrincipalContext<CookieAuthenticationOptions>
{
    public bool ShouldRenew { get; set; }

    public CookieValidatePrincipalContext(HttpContext context, AuthenticationScheme scheme,
        CookieAuthenticationOptions options, AuthenticationTicket ticket);
```

```
    public void RejectPrincipal();
    public void ReplacePrincipal(ClaimsPrincipal principal);
}

public class CookieSigningOutContext : PropertiesContext<CookieAuthenticationOptions>
{
    public CookieOptions CookieOptions { get; [CompilerGenerated] set; }
    public CookieSigningOutContext(HttpContext context, AuthenticationScheme scheme,
        CookieAuthenticationOptions options, AuthenticationProperties properties,
        CookieOptions cookieOptions);
}
```

CookieValidatePrincipalContext 通过相应的属性和方法来决定最终的检验结果。如果验证失败，我们可以直接调用其 RejectPrincipal 方法，它会将当前上下文的 Principal 属性设置为 Null，并以此拒绝了当前请求利用认证票据提供的 ClaimsPrincipal 对象。如果决定延长认证票据的过期时间，我们可以设置其 ShouldRenew 属性。而 ReplacePrincipal 方法使我们可以直接替换当前上下文中表示认证用户的 ClaimsPrincipal 对象。

CookieAuthenticationEvents 针对登录、注销、拒绝访问提示页面及初始请求路径的重定向是在如下所示的 RedirectContext<TOptions> 的上下文中进行的，它是 PropertiesContext<TOptions> 的派生类。我们可以通过设置 PropertiesContext<TOptions> 的 RedirectUri 属性来设置重定向的目标路径。

```
public class RedirectContext<TOptions> : PropertiesContext<TOptions> where TOptions:
    AuthenticationSchemeOptions
{
    public string RedirectUri { get; set; }
    public RedirectContext(HttpContext context, AuthenticationScheme scheme,
        TOptions options, AuthenticationProperties properties, string redirectUri);
}
```

CookieBuilder

CookieAuthenticationHandler 利用 Cookie 的形式来传递认证票据，与 Cookie 相关的属性由一个 CookieBuilder 对象来提供。如下面的代码片段所示，CookieBuilder 定义了一系列的属性，用来对这个承载认证票据的 Cookie 进行定制，我们可以利用它们设置 Cookie 的名称、路径、域名、过期时间和安全策略等属性。CookieBuilder 的 Build 方法返回的 CookieOptions 对象体现了 Cookie 最终应该具备的状态。

```
public class CookieBuilder
{
    public virtual string                   Name { get; set; }
    public virtual string                   Path { get; set; }
    public virtual string                   Domain { get; set; }
    public virtual bool                     HttpOnly { get; set; }
    public virtual bool                     IsEssential { get; set; }
    public virtual TimeSpan?                MaxAge { get; set; }
    public virtual TimeSpan?                Expiration { get; set; }
    public virtual SameSiteMode             SameSite { get; set; }
```

```csharp
    public virtual CookieSecurePolicy          SecurePolicy { get; set; }

    public CookieOptions Build(HttpContext context);
    public virtual CookieOptions Build(HttpContext context, DateTimeOffset expiresFrom);
}
```

ICookieManager

实现在 CookieAuthenticationHandler 中针对 Cookie 的操作是通过 ICookieManager 服务来完成的。在登录过程中，一旦成功验证了访问者的真实身份，CookieAuthenticationHandler 就会调用 CookieBuilder 的 Build 方法创建对应的 CookieOptions 对象，并据此创建承载认证票据的 Cookie。创建的 Cookie 最终通过调用 ICookieManager 对象的 AppendResponseCookie 方法被写入当前响应。

```csharp
public interface ICookieManager
{
    void AppendResponseCookie(HttpContext context, string key, string value,
        CookieOptions options);
    void DeleteCookie(HttpContext context, string key, CookieOptions options);
    string GetRequestCookie(HttpContext context, string key);
}
```

在 AuthenticationMiddleware 中间件针对请求实施认证的时候，ICookieManager 接口的 GetRequestCookie 方法会被用来从请求中提取承载认证票据的 Cookie。而 DeleteCookie 方法则用来删除这个承载认证票据的 Cookie 以达到注销的目的。如下所示的 ChunkingCookieManager 类型是对 ICookieManager 接口的默认实现。

```csharp
public class ChunkingCookieManager : ICookieManager
{
    public const int DefaultChunkSize = 0xfd2;
    public int? ChunkSize { get; set; }
    public bool ThrowForPartialCookies { get; set; }

    public void AppendResponseCookie(HttpContext context, string key, string value,
        CookieOptions options);
    public void DeleteCookie(HttpContext context, string key, CookieOptions options);
    public string GetRequestCookie(HttpContext context, string key);
}
```

CookieAuthenticationOptions

具有如下定义的 CookieAuthenticationOptions 承载了与 CookieAuthenticationHandler 相关的所有配置选项。我们可以利用它的 LoginPath 属性和 LogoutPath 属性设置登录与注销的路径，还可以利用 AccessDeniedPath 设置在访问权限不足的情况下的重定向路径。一般来说，如果我们以匿名的形式请求一个只允许认证用户才能访问的地址时，请求就会被重定向到登录路径，而当前的请求地址会以查询字符串的形式附加到重定向地址上，以便登录成功后还能回到原来的地方，ReturnUrlParameter 属性表示的就是这个查询字符串的名称。

```csharp
public class CookieAuthenticationOptions : AuthenticationSchemeOptions
```

```csharp
{
    public PathString              LoginPath { get; set; }
    public PathString              LogoutPath { get; set; }
    public PathString              AccessDeniedPath { get; set; }
    public string                  ReturnUrlParameter { get; set; }

    public TimeSpan                ExpireTimeSpan { get; set; }
    public bool                    SlidingExpiration { get; set; }
    public CookieBuilder           Cookie { get; set; }
    public ICookieManager          CookieManager { get; set; }

    public ITicketStore            SessionStore { get; set; }

    public ISecureDataFormat<AuthenticationTicket>   TicketDataFormat { get; set; }
    public IDataProtectionProvider                   DataProtectionProvider { get; set; }
    public CookieAuthenticationEvents                Events { get; set; }
}
```

认证票据一般都具有时效性，CookieAuthenticationHandler 类型利用 Cookie 的过期时间来控制认证票据的时效性。具体的过期时间通过 ExpireTimeSpan 属性来设置，而 SlidingExpiration 属性则表示采用的是针对 Cookie 被创建时间的绝对过期策略还是针对最近一次访问时间的滑动过期策略。除了设置过期时间，如果还需要对 Cookie 的其他属性进行设置，可以利用 Cookie 属性返回的 CookieBuilder 对象来完成。如果我们需要对 Cookie 的创建、提取和删除进行定制，可以自定义一个 ICookieManager 实现类型并通过 CookieManager 属性进行注册即可。

如果认证票据承载了用户身份、权限及其他个人信息，承载的数据可能会很大，大尺寸的认证票据附加到每个请求上必然影响应用的性能。为了解决这个问题，我们可以采用一种被称为认证会话（Authentication Session）的机制。具体来说，我们可以赋予每个认证票据一个唯一标识，并将票据的内容存储在服务端，那么请求的 Cookie 只需要存放这个唯一标识即可。CookieAuthenticationOptions 类型的 SessionStore 属性返回的这个 ITicketStore 对象就是为了帮助我们实现对认证票据的存储。具有如下定义的 ITicketStore 接口定义了 4 个方法，这 4 个方法实现了针对认证票据的存储、提取、移除和续期。

```csharp
public interface ITicketStore
{
    Task<string> StoreAsync(AuthenticationTicket ticket);
    Task<AuthenticationTicket> RetrieveAsync(string key);
    Task RemoveAsync(string key);
    Task RenewAsync(string key, AuthenticationTicket ticket);
}
```

CookieAuthenticationOptions 类型的 TicketDataFormat 帮助我们设置和获取用来格式化认证票据的格式化器，对认证票据承载的核心内容实施加密和解密的工作则由 DataProtectionProvider 属性提供的 IDataProtectionProvider 对象来完成。Events 属性就是前面介绍的用来定制或者干预认证流程的 CookieAuthenticationEvents 对象。

CookieAuthenticationHandler

在了解了上述这些辅助类型之后，下面正式介绍 CookieAuthenticationHandler 这个最核心的类型。为了使读者了解 CookieAuthenticationHandler 针对认证、登录、注销和质询这几个基本操作的实现原理，我们忽略了很多具体的细节，采用一种极简的方式重建了这个类型。

CookieAuthenticationHandler 派生自 SignInAuthenticationHandler<CookieAuthenticationOptions>，它定义的 CookieAuthenticationEvents 类型的 Events 属性覆盖了基类的同名属性（该属性返回类型为 Object），并且通过重写的 CreateEventsAsync 方法确保 CookieAuthenticationOptions 在没有提供 Events 对象或者类型的情况下该属性总是有一个默认值。

```csharp
public class CookieAuthenticationHandler
    : SignInAuthenticationHandler<CookieAuthenticationOptions>
{
    protected new CookieAuthenticationEvents Events
    {
        get { return (CookieAuthenticationEvents)base.Events; }
        set { base.Events = value; }
    }

    public CookieAuthenticationHandler(
        IOptionsMonitor<CookieAuthenticationOptions> options, ILoggerFactory logger,
        UrlEncoder encoder, ISystemClock clock) : base(options, logger, encoder, clock)
    {}

    protected override Task<object> CreateEventsAsync()
        => Task.FromResult<object>(new CookieAuthenticationEvents());
}
```

在重写的用于完成登录的 HandleSignInAsync 方法中，我们根据提供的配置选项构建了一个 CookieOptions 对象，代表用来传递认证票据的 Cookie 的相关设置。基于这个对象我们创建了一个 CookieSigningInContext 对象，并将其作为参数调用 CookieAuthenticationEvents 对象的 SigningIn 方法，这样利用 Events 注册的回调得以在登录前被执行。

```csharp
public class CookieAuthenticationHandler
    : SignInAuthenticationHandler<CookieAuthenticationOptions>
{
    protected override async Task HandleSignInAsync(ClaimsPrincipal user,
        AuthenticationProperties properties)
    {
        properties = properties ?? new AuthenticationProperties();
        var cookieOptions = Options.Cookie.Build(Context);

        var signInContext = new CookieSigningInContext(
            Context, Scheme, Options, user, properties, cookieOptions);
        await Events.SigningIn(signInContext);

        var ticket = new AuthenticationTicket(signInContext.Principal,
            signInContext.Properties, signInContext.Scheme.Name);
```

```
        var cookieValue = Options.TicketDataFormat.Protect(ticket, GetTlsTokenBinding());
        Options.CookieManager.AppendResponseCookie(
            Context, Options.Cookie.Name, cookieValue, signInContext.CookieOptions);

        var signedInContext = new CookieSignedInContext(
            Context, Scheme, signInContext.Principal, signInContext.Properties, Options);
        await Events.SignedIn(signedInContext);
    }

    private string GetTlsTokenBinding()
    {
        var binding = Context.Features.Get<ITlsTokenBindingFeature>()
            ?.GetProvidedTokenBindingId();
        return binding == null ? null : Convert.ToBase64String(binding);
    }
}
```

接下来创建代表认证票据的 AuthenticationTicket 对象，并利用 CookieAuthenticationOptions 配置选项提供的 ISecureDataFormat<AuthenticationTicket>对象对其进行加密。加密需要提供一个代表加密意图或者目的（Purpose）的字符串，该字符串是通过 GetTlsTokenBinding 方法提供的。GetTlsTokenBinding 方法涉及一种被称为 Token Binding 的协议，该协议旨在通过创建一个跨多个 TLS 会话和连接的绑定（Bindings）来防止安全令牌被黑客窃取或者冒用，该方法会试图提取该绑定的标识（ID）作为加密认证票据的目的。

加密的认证票据交由配置选项提供的 ICookieManager 对象写入当前响应的 Cookie 列表（Set-Cookie 报头）中。在此之后，一个 CookieSignedInContext 对象被创建出来并作为参数调用了 CookieAuthenticationEvents 对象的 SignedIn 方法，此时注册到 Events 对象上的 OnSignedIn 回调会被执行。

相较于登录操作，实现在重写的 HandleSignOutAsync 方法中的注销操作比较简单。如下面的代码片段所示，该方法会首先创建一个 CookieSigningOutContext 上下文，并将其作为参数调用 CookieAuthenticationEvents 对象的 SigningOut 来执行 Events 对象提供的注销回调。接下来配置选项提供的 ICookieManager 对象会用来删除承载认证票据的 Cookie，进而实现注销当前登录的目的。

```
public class CookieAuthenticationHandler
    : SignInAuthenticationHandler<CookieAuthenticationOptions>
{
    protected override async Task HandleSignOutAsync(AuthenticationProperties properties)
    {
        var cookieOptions = Options.Cookie.Build(Context);
        properties = properties ?? new AuthenticationProperties();
        var context = new CookieSigningOutContext(
            Context, Scheme, Options, properties, cookieOptions);

        await Events.SigningOut(context);
        Options.CookieManager.DeleteCookie(
            Context,
```

```
            Options.Cookie.Name,
            context.CookieOptions);
    }
    ...
}
```

在重写的用来对当前请求实施认证的 HandleAuthenticateAsync 方法中，配置选项提供的 ICookieManager 对象会被用来提取承载认证票据的 Cookie，提供的 ISecureDataFormat <AuthenticationTicket>对象会用来对 Cookie 值进行解密。如果解密之后能够得到一个有效的 AuthenticationTicket 对象，就意味着请求提供的是合法的认证票据，那么认证成功，反之则认证失败。

```
public class CookieAuthenticationHandler
    : SignInAuthenticationHandler<CookieAuthenticationOptions>
{
    protected override Task<AuthenticateResult> HandleAuthenticateAsync()
    {
        var cookie = Options.CookieManager.GetRequestCookie(Context, Options.Cookie.Name);
        var ticket = Options.TicketDataFormat.Unprotect(cookie, GetTlsTokenBinding());
        var result =  ticket == null
            ? AuthenticateResult.Fail("Unprotect ticket failed")
            : AuthenticateResult.Success(ticket);
        return Task.FromResult(result);
    }
}
```

对于定义在基类 AuthenticationHandler<TOptions>中的用于发送质询的 HandleChallengeAsync 方法和 HandleForbiddenAsync 方法来说，它们会分别回复一个状态码为"401 UnAuthorized"和"403 Forbidden"的响应，而一个 Web 应用更希望能够自动重定向到登录和拒绝访问提示页面，所以 CookieAuthenticationHandler 按照如下所示的方式重写了这两个方法，实现了这样的重定向功能。

```
public class CookieAuthenticationHandler
    : SignInAuthenticationHandler<CookieAuthenticationOptions>
{
    protected override async Task HandleChallengeAsync(AuthenticationProperties properties)
    {
        var redirectUri = properties.RedirectUri;
        if (string.IsNullOrEmpty(redirectUri))
        {
            redirectUri = OriginalPathBase + Request.Path + Request.QueryString;
        }

        var loginUri = Options.LoginPath +
            QueryString.Create(Options.ReturnUrlParameter, redirectUri);
        var redirectContext = new RedirectContext<CookieAuthenticationOptions>(
            Context, Scheme, Options, properties, BuildRedirectUri(loginUri));
        await Events.RedirectToLogin(redirectContext);
    }
```

```
protected override async Task HandleForbiddenAsync(AuthenticationProperties properties)
{
    var returnUrl = properties.RedirectUri;
    if (string.IsNullOrEmpty(returnUrl))
    {
        returnUrl = OriginalPathBase + Request.Path + Request.QueryString;
    }
    var accessDeniedUri = Options.AccessDeniedPath +
        QueryString.Create(Options.ReturnUrlParameter, returnUrl);
    var redirectContext = new RedirectContext<CookieAuthenticationOptions>(
        Context, Scheme, Options, properties, BuildRedirectUri(accessDeniedUri));
    await Events.RedirectToAccessDenied(redirectContext);
}
...
}
```

前面我们着重介绍了承载基于 Cookie 认证方案的 CookieAuthenticationHandler 的实现, 其中涉及它直接或者间接继承的一系列基类。由于表示认证处理器的 IAuthenticationHandler 接口是整个认证模型的核心, 所以笔者绘制了一张 UML, 如图 19-10 所示, 用于展示 ASP.NET Core 提供的几乎所有原生的 IAuthenticationHandler 实现类型。除了上述 CookieAuthenticationHandler, 还包括基于 OAuth 2.0 认证方案的 OAuthHandler<TOptions>、基于 Open ID 认证方案的 OpenIdConnectHandler, 以及基于 JWT(Json Web Token)的 JwtBeareHandler 等。

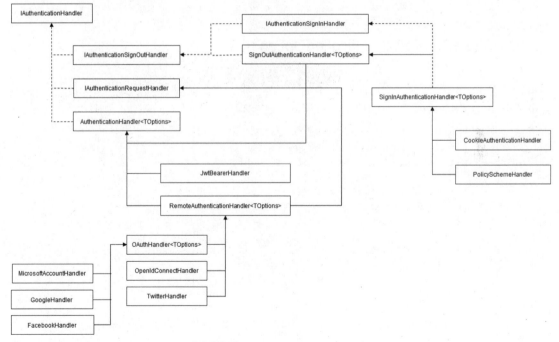

图 19-10　IAuthenticationHandler "全家桶"

19.4.3　注册 CookieAuthenticationHandler

在介绍了 CookieAuthenticationHandler 的实现原理之后，下面介绍如何在应用中注册这个认证处理器。CookieAuthenticationHandler 涉及很多配置选项，全部定义在 CookieAuthenticationOptions 这个配置选项类型中。如果应用程序没有对相应的配置选项进行设置，定义在 CookieAuthenticationDefaults 这个静态类型中的默认值会被使用。

CookieAuthenticationDefaults

如下面的代码片段所示，静态类型 CookieAuthenticationDefaults 以常量和静态只读属性的形式为基于 Cookie 的认证方案定义了一系列的默认配置选项，其中包括认证名称（Cookies）和 Cookie 前缀（".AspNetCore."），登录、注销和访问拒绝页面的路径（"/Account/Login"、"/Account/Logout" 和 "/Account/AccessDenied"），以及表示初始访问地址的查询字符串名称 "ReturnUrl"。

```
public static class CookieAuthenticationDefaults
{
    public const string                   AuthenticationScheme = "Cookies";
    public static readonly string         CookiePrefix = ".AspNetCore.";
    public static readonly string         ReturnUrlParameter = "ReturnUrl";
    public static readonly PathString     LoginPath = "/Account/Login";
    public static readonly PathString     LogoutPath = "/Account/Logout";
    public static readonly PathString     AccessDeniedPath = "/Account/AccessDenied";
}
```

PostConfigureCookieAuthenticationOptions

定义在静态类型 CookieAuthenticationDefaults 中的默认值最终通过一个 IPostConfigureOptions<TOptions>应用到配置选项上，该类型就是具有如下定义的 PostConfigureCookieAuthenticationOptions。如下面的代码片段所示，在实现的 PostConfigure 方法中，如果 CookieAuthenticationOptions 对象的某些属性尚未被赋值，它们将自动被赋予一个默认值。

```
public class PostConfigureCookieAuthenticationOptions :
    IPostConfigureOptions<CookieAuthenticationOptions>
{
    private readonly IDataProtectionProvider _dp;
    public PostConfigureCookieAuthenticationOptions(IDataProtectionProvider dataProtection)
        =>_dp = dataProtection;

    public void PostConfigure(string name, CookieAuthenticationOptions options)
    {
        options.DataProtectionProvider = options.DataProtectionProvider ?? _dp;
        if (string.IsNullOrEmpty(options.Cookie.Name))
        {
            options.Cookie.Name = CookieAuthenticationDefaults.CookiePrefix + name;
        }
        if (options.TicketDataFormat == null)
        {
            var dataProtector = options.DataProtectionProvider.CreateProtector(
```

```
            "Microsoft.AspNetCore.Authentication.Cookies.CookieAuthenticationMiddleware",
            name, "v2");
        options.TicketDataFormat = new TicketDataFormat(dataProtector);
    }
    if (options.CookieManager == null)
    {
        options.CookieManager = new ChunkingCookieManager();
    }
    if (!options.LoginPath.HasValue)
    {
        options.LoginPath = CookieAuthenticationDefaults.LoginPath;
    }
    if (!options.LogoutPath.HasValue)
    {
        options.LogoutPath = CookieAuthenticationDefaults.LogoutPath;
    }
    if (!options.AccessDeniedPath.HasValue)
    {
        options.AccessDeniedPath = CookieAuthenticationDefaults.AccessDeniedPath;
    }
}
```

AddCookie

前面提及，针对某个认证方案的认证处理器最终是通过 AuthenticationBuilder 对象来进行注册的，针对 CookieAuthenticationHandler 对象的注册就体现在它的 AddCookie 扩展方法重载上。如下面的代码片段所示，当我们在调用这些 AddCookie 扩展方法重载的时候，可以指定认证方案的名称，如果未显式指定，将采用默认的方案名称 Cookies。我们也可以利用指定的 Action<CookieAuthenticationOptions>对象设置相应的配置选项，如果某些选项没有被初始化，注册的 PostConfigureCookieAuthenticationOptions 对象确保它们总是有一个默认值。针对 Cookie AuthenticationHandler 对象的注册最终是通过调用 AuthenticationBuilder 类型的 AddScheme 方法来完成的。

```
public static class CookieExtensions
{
    public static AuthenticationBuilder AddCookie(this AuthenticationBuilder builder)
        => builder.AddCookie(CookieAuthenticationDefaults.AuthenticationScheme);

    public static AuthenticationBuilder AddCookie(this AuthenticationBuilder builder,
        string authenticationScheme)
        => builder.AddCookie(authenticationScheme, configureOptions: null);

    public static AuthenticationBuilder AddCookie(this AuthenticationBuilder builder,
        Action<CookieAuthenticationOptions> configureOptions)
        => builder.AddCookie(CookieAuthenticationDefaults.AuthenticationScheme,
            configureOptions);

    public static AuthenticationBuilder AddCookie(this AuthenticationBuilder builder,
```

```
        string authenticationScheme, Action<CookieAuthenticationOptions> configureOptions)
        => builder.AddCookie(authenticationScheme, null,configureOptions);

    public static AuthenticationBuilder AddCookie(this AuthenticationBuilder builder,
        string authenticationScheme, string displayName,
        Action<CookieAuthenticationOptions> configureOptions)
    {
        builder.Services.TryAddEnumerable(
ServiceDescriptor.Singleton<IPostConfigureOptions<CookieAuthenticationOptions>,
            PostConfigureCookieAuthenticationOptions>());
        return builder.AddScheme<CookieAuthenticationOptions,
            CookieAuthenticationHandler>(authenticationScheme, displayName,
            configureOptions);
    }
}
```

第 20 章

授 权

认证旨在确定用户的真实身份,而授权则是通过权限控制使用户只能做其允许做的事。授权的本质就是通过设置一个策略来决定究竟具有何种特性的用户会被授权访问某个资源或者执行某个操作。我们可以采用任何授权策略,如可以根据用户拥有的角色进行授权,也可以根据用户的等级和所在部门进行授权,有的授权甚至可以根据用户的年龄、性别和所在国家来进行。认证后的用户体现为一个 ClaimsPrincipal 对象,它携带的声明不仅仅用于描述用户的身份,还携带了上述这些构建授权策略的元素,所以授权实际上就是检查认证用户携带的声明是否与授权策略一致的过程。

20.1 基于角色的权限控制

ASP.NET Core 应用并没有对如何定义授权策略做硬性规定,所以我们完全根据用户具有的任意特性(如性别、年龄、学历、所在地区、宗教信仰、政治面貌等)来判断其是否具有获取目标资源或者执行目标操作的权限,但是针对角色的授权策略依然是最常用的。角色(或者用户组)实际上就是对一组权限集的描述,将一个用户添加到某个角色之中就是为了将对应的权限赋予该用户。按照惯例,在正式介绍 ASP.NET Core 应用的授权模型之前,需要先利用一个简单的实例演示来介绍基于角色的授权方式。

20.1.1 用户与角色的映射

对某个请求实施授权策略以确定该用户是否具有访问目标资源或者执行目标操作的权限只有在请求经过认证之后才有意义,所以我们需要先在应用中实现认证以及相关的登录与注销。这里我们采用第 19 章介绍的基于 Cookie 的认证方案。

要在一个 ASP.NET Core 应用中实现基于角色的授权,需要先解决如何存储用户、角色及其两者之间的映射关系。我们在本实例中直接使用 SQL Server 数据库来存储用户账号和权限信息,并将 Entity Framework Core 作为数据访问的手段,为此我们定义了如下 3 个实体类型:User 和 Role 分别表示用户与角色,UserRole 代表两者之间的映射关系。

```
public class User
```

```csharp
{
    public string                              UserName { get; set; }
    public string                              NormalizedUserName { get; set; }
    public string                              Password { get; set; }
    public virtual ICollection<UserRole>       Roles { get; } = new List<UserRole>();

    public User(){}
    public User(string userName, string password)
    {
        UserName            = userName;
        NormalizedUserName  = userName.ToUpper();
        Password            = password;
    }
}

public class Role
{
    public string                              RoleName { get; set; }
    public string                              NormalizedRoleName { get; set; }
    public virtual ICollection<UserRole>       Users { get; } = new List<UserRole>();

    public Role() {}
    public Role(string roleName)
    {
        RoleName            = roleName;
        NormalizedRoleName  = roleName.ToUpper();
    }
}

public class UserRole
{
    public string NormalizedUserName { get; set; }
    public string NormalizedRoleName { get; set; }
}
```

由于角色名称和用户名一般是不区分大小写的，所以可以利用 User 类型的 NormalizedUserName 属性和 Role 类型的 NormalizedRoleName 属性来表示标准化（实际上就是简单的大写形式）的用户与角色名称。在如下所示的 UserDbContext 的 DbContext 类型中，我们针对上述 3 个实体类型定义了对应的 DbSet<TEntity>（Users、Roles 和 UserRoles），它们可以帮助我们在数据库中生成 3 个对应的表。

```csharp
public class UserDbContext : DbContext
{
    public DbSet<User>      Users { get; set; }
    public DbSet<Role>      Roles { get; set; }
    public DbSet<UserRole>  UserRoles { get; set; }

    protected override void OnModelCreating(ModelBuilder modelBuilder)
    {
        modelBuilder.Entity<User>(builder =>
```

```
        {
            builder.HasKey(user => user.NormalizedUserName);
            builder.HasMany(user => user.Roles).WithOne().HasForeignKey(
                userRole => userRole.NormalizedUserName);
        });
        modelBuilder.Entity<Role>(builder =>
        {
            builder.HasKey(role => role.NormalizedRoleName);
            builder.HasMany(role => role.Users).WithOne().HasForeignKey(
                userRole => userRole.NormalizedRoleName);
        });
        modelBuilder.Entity<UserRole>(builder => builder.HasKey(
            userRole => new { userRole.NormalizedUserName, userRole.NormalizedRoleName }));
    }
    ...
}
```

在重写的 OnModelCreating 方法中，我们设置了 3 个实体类型对应的数据表（存储用户的 Users 表、存储角色的 Roles 表，以及两者的关联表 UserRoles）的主键和它们之间的关系（见图 20-1）。

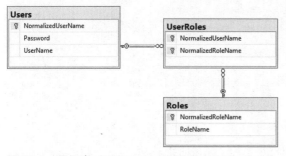

图 20-1　数据表（Users、Roles 和 UserRoles）

为了使演示的实例足够简单，我们直接在 UserDbContext 类型中生成了一些初始化数据。如下面的代码片段所示，UserDbContext 对象在实例化时会自动创建数据库，同时在 Users 表和 Roles 表中添加两个用户名（Foo 和 Bar）和一个角色（Admin），并将该角色赋予用户 Bar。

```
public class UserDbContext : DbContext
{
    public UserDbContext(DbContextOptions options) : base(options)
    {
        Database.EnsureCreated();
        if (Users.Find("FOO") == null)
        {
            Users.Add(new User("Foo", "password"));
            Users.Add(new User("Bar", "password"));
            Roles.Add(new Role("Admin"));
            UserRoles.Add(new UserRole
            {
                NormalizedUserName = "BAR",
                NormalizedRoleName = "ADMIN"
            });
```

```
            SaveChanges();
        }
    }
}
```

20.1.2　根据角色授权

数据库在被构建的过程中会自动创建两个用户（用户名分别为 Foo 和 Bar），其中后者（Bar）具有 Admin 角色。我们的演示实例体现的授权策略非常简单：只有拥有 Admin 角色的用户才能访问应用的主页。我们通过如下所示的代码定义了整个应用的雏形。整个应用旨在处理针对主页、登录、注销和访问拒绝这 4 种类型的请求，我们将它们实现在对应的方法中，并将它们作为对应的路由处理器。

```
public class Program
{
    public static void Main()
    {
        var connectionString = @"Server=(localdb)\\mssqllocaldb;Database=TestDb;
            Trusted_Connection=True;MultipleActiveResultSets=true";
        Host.CreateDefaultBuilder()
            .ConfigureWebHostDefaults(builder => builder
                .ConfigureServices(svcs => svcs
                    .AddDbContext<UserDbContext>(options =>
                        options.UseSqlServer(connectionString))
                    .AddRouting()
                    .AddAuthorization()
                    .AddAuthentication(options => options.DefaultScheme =
                        CookieAuthenticationDefaults.AuthenticationScheme).AddCookie())
                .Configure(app => app
                    .UseAuthentication()
                    .UseRouting()
                    .UseEndpoints(endpoints => {
                        endpoints.Map("", RenderHomePageAsync);
                        endpoints.Map("Account/Login", SignInAsync);
                        endpoints.Map("Account/Logout", SignOutAsync);
                        endpoints.Map("Account/AccessDenied", DenyAccessAysnc);
                    })))
            .Build()
            .Run();
    }

    public static async Task RenderHomePageAsync(HttpContext context)
    {
        throw new NotImplementedException();
    }

    public static async Task SignInAsync(HttpContext context)
    {
        throw new NotImplementedException();
```

```csharp
    }

    public static async Task SignOutAsync(HttpContext context)
    {
        throw new NotImplementedException();
    }

    public static async Task DenyAccessAysnc(HttpContext context)
    {
        throw new NotImplementedException();
    }
}
```

在应用启动时，我们需要注册一系列分别与路由、认证、授权及 Entity Framework Core 相关的服务。针对授权基础服务的注册是通过调用 IServiceCollection 接口的 AddAuthorization 扩展方法来完成的。由于我们使用上面定义的 UserDbContext 对象来存取用户和角色数据，所以需要添加针对这个 DbContext 的注册。

如下所示的代码片段是实现了登录和注销操作的 SignInAsync 方法与 SignOutAsync 方法的定义，由于第 19 章已经做过类似的实例演示，所以此外不再详述。对于登录来说，与之前演示的实例具有的唯一的不同之处在于，当我们成功验证了用户真实身份并创建了一个代表该用户的 ClaimsPrincipal 时，应将该用户拥有的角色提取出来。针对每个角色，我们创建了一个对应的 Claim 对象，并添加到 ClaimsPrincipal 对象的声明列表中。

```csharp
public class Program
{
    public static async Task SignInAsync(HttpContext context)
    {
        if (string.Compare(context.Request.Method, "GET") == 0)
        {
            await RenderLoginPageAsync(context, null, null, null);
        }
        else
        {
            string userName = context.Request.Form["username"];
            string password = context.Request.Form["password"];
            var dbContext = context.RequestServices.GetRequiredService<UserDbContext>();
            var user = await dbContext.Users.Include(it => it.Roles)
                .SingleOrDefaultAsync(it => it.UserName == userName.ToUpper());
            if (user?.Password == password)
            {
                var identity = new GenericIdentity(
                    userName, CookieAuthenticationDefaults.AuthenticationScheme);
                foreach (var role in user.Roles)
                {
                    identity.AddClaim(new Claim(ClaimTypes.Role, role.NormalizedRoleName));
                }
                var principal = new ClaimsPrincipal(identity);
                await context.SignInAsync(principal);
```

```csharp
            }
            else
            {
                await RenderLoginPageAsync(context, userName, password,
                    "Invalid user name or password!");
            }
        }
    }

    private static Task RenderLoginPageAsync(HttpContext context, string userName,
        string password, string errorMessage)
    {
        context.Response.ContentType = "text/html";
        return context.Response.WriteAsync(
            @"<html>
                <head><title>Login</title></head>
                <body>
                    <form method='post'>" +
                        $"<input type='text' name='username' placeholder='User name'
                            value ='{userName}'/>" +
                        $"<input type='password' name='password' placeholder='Password'
                            value ='{password}'/> " +
                        @"<input type='submit' value='Sign In' />
                    </form>" +
                        $"<p style='color:red'>{errorMessage}</p>" +
                @"</body>
            </html>");
    }

    public static async Task SignOutAsync(HttpContext context)
    {
        await context.SignOutAsync();
        await context.ChallengeAsync(new AuthenticationProperties { RedirectUri = "/" });
    }
}
```

在 SignOutAsync 方法中，除了调用当前 HttpContext 上下文的 SignOutAsync 扩展方法注销当前登录状态，我们还调用 ChallengeAsync 方法来发送一个质询。在调用 ChallengeAsync 方法时，我们利用作为输入参数的 AuthenticationProperties 对象的 RedirectUri 属性设置了针对主页的重定向地址。

由于我们希望具有 Admin 角色的用户才能浏览应用主页，所以作为路由处理器的 RenderHomePageAsync 方法包含对应的授权检验。如下面的代码片段所示，对于经过认证的请求，RenderHomePageAsync 方法会创建一个针对"ADMIN"角色的 RolesAuthorizationRequirement 对象，该对象代表了一种针对角色的认证策略，即要求授权用户至少拥有一个设定的角色。接下来，一个 IAuthorizationService 服务对象将从当前 HttpContext 上下文中提取出来，我们调用该对象的 AuthorizeAsync 方法针对 RolesAuthorizationRequirement 对象来检验当前登录用户是否具有足够的权限。

```csharp
public class Program
{
    public static async Task RenderHomePageAsync(HttpContext context)
    {
        if (context?.User?.Identity?.IsAuthenticated == true)
        {
            var requirement = new RolesAuthorizationRequirement(new string[] { "ADMIN" });
            var authorizationService = context.RequestServices
                .GetRequiredService<IAuthorizationService>();
            var result = await authorizationService.AuthorizeAsync(context.User, null,
                new IAuthorizationRequirement[] { requirement });
            if (result.Succeeded)
            {
                await context.Response.WriteAsync(
                @"<html>
                    <head><title>Index</title></head>
                    <body>" +
                        $"<h3>{context.User.Identity.Name}, you are authorized.</h3>" +
                        @"<a href='Account/Logout'>Sign Out</a>
                    </body>
                </html>");
            }
            else
            {
                await context.ForbidAsync();
            }
        }
        else
        {
            await context.ChallengeAsync();
        }
    }

    public static async Task DenyAccessAysnc(HttpContext context)
    {
        await context.Response.WriteAsync(
            @"<html>
                <head><title>Index</title></head>
                <body>" +
                    $"<h3>{context.User.Identity.Name}, your access is denied.</h3>" +
                    @"<a href='/Account/Logout'>Sign Out</a>
                </body>
            </html>");
    }
}
```

IAuthorizationService 接口的 AuthorizeAsync 方法会返回一个代表授权检验结果的 AuthorizationResult 对象。如果通过该对象确定用户具有对应的权限，我们会以图 20-2 所示的形式呈现主页的内容，否则调用当前 HttpContext 上下文的 ForbidAsync 扩展方法发送一个针对访问拒绝（Access Denied）的质询。由于我们提供了拒绝访问页面的路径"/Account/Access Denied"，所以由 DenyAccessAysnc 方法呈现的内容会以图 20-2 所示的形式出现在浏览器上。（S2001）

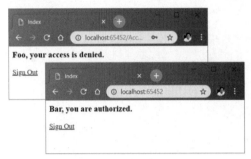

图 20-2　针对主页的授权

20.1.3　预定义授权策略

我们调用 IAuthorizationService 服务的 AuthorizeAsync 方法进行授权检验的时候，实际上是将授权要求定义在一个 RolesAuthorizationRequirement 对象中，这是一种比较烦琐的编程方式。一种推荐的做法是在应用启动的过程中创建一系列通过 AuthorizationPolicy 对象表示的授权规则，并通过指定一个唯一的名称对它们进行全局注册，那么后续就可以针对注册的策略名称进行授权检验。

为了演示这种针对预定义策略的授权方式，我们只需要对上面演示的实例略做修改即可。如下面的代码片段所示，在调用 IServiceCollection 接口的 AddAuthorization 扩展方法注册授权相关服务时，我们利用作为输入参数的 Action<AuthorizationOptions>对象对授权策略进行了全局注册。具体来说，我们针对角色"ADMIN"创建了一个 RolesAuthorizationRequirement 对象，然后以此创建了代表授权策略的 AuthorizationPolicy 对象，调用 AuthorizationOptions 对象的 AddPolicy 方法对此授权策略进行了注册。由于我们是针对主页的授权策略，所以可以将其命名为 HomePage。

```
public class Program
{
    public static void Main()
    {
        var connectionString = @"Server=(localdb)\\mssqllocaldb;Database=TestDb;
            Trusted_Connection=True;MultipleActiveResultSets=true";
        Host.CreateDefaultBuilder()
            .ConfigureWebHostDefaults(builder => builder
                .ConfigureServices(svcs => svcs
                    .AddDbContext<UserDbContext>(options => options
                        .UseSqlServer(connectionString))
                    .AddRouting()
                    .AddAuthorization(options => {
```

```csharp
                var requirement = new RolesAuthorizationRequirement
                    (new string[] { "ADMIN" });
                var policy = new AuthorizationPolicy(
                    new IAuthorizationRequirement[] { requirement }, new string[0]);
                options.AddPolicy("HomePage", policy);
            })
        .AddAuthentication(options => options.DefaultScheme =
            CookieAuthenticationDefaults.AuthenticationScheme)
        .AddCookie())
    .Configure(app => app
        .UseAuthentication()
        .UseRouter(builder => builder
            .MapRoute("", RenderHomePageAsync)
            .MapRoute("Account/Login", SignInAsync)
            .MapRoute("Account/Logout", SignOutAsync)
            .MapRoute("Account/AccessDenied", DenyAccessAysnc)))
    .Build()
    .Run();
}

public static async Task RenderHomePageAsync(HttpContext context)
{
    if (context?.User?.Identity?.IsAuthenticated == true)
    {
        var authorizationService = context.RequestServices
            .GetRequiredService<IAuthorizationService>();
        var result = await authorizationService
            .AuthorizeAsync(context.User,"HomePage");
        if (result.Succeeded)
        {
            await context.Response.WriteAsync(
            @"<html>
                <head><title>Index</title></head>
                <body>" +
                    $"<h3>{context.User.Identity.Name}, you are authorized.</h3>" +
                    @"<a href='Account/Logout'>Sign Out</a>
                </body>
            </html>");
        }
        else
        {
            await context.ForbidAsync();
        }
    }
    else
    {
        await context.ChallengeAsync();
    }
}
```

在呈现主页的 RenderHomePageAsync 方法中，我们依然调用 IAuthorizationService 服务的

AuthorizeAsync 方法来检验用户是否具有对应的权限，但这次采用的是另一个可以直接指定授权策略注册名称的 AuthorizeAsync 方法重载。（S2002）

20.2 基于"要求"的授权

通过前面演示的两个实例可知，ASP.NET Core 应用的授权是由通过 IAuthorizationService 接口表示的服务提供的。IAuthorizationService 服务提供了分别针对 IAuthorizationRequirement 和 AuthorizationPolicy 的授权方案，我们先来介绍前者。

20.2.1 IAuthorizationRequirement

IAuthorizationRequirement 接口表示授权访问目标资源或者操作在某个方面需要满足的要求（Requirement）。由于"授权要求"具有不同的表现形式，所以 IAuthorizationRequirement 仅仅是一个不具有任何成员的"标记接口"。另外，既然 IAuthorizationRequirement 接口体现了授权用户需要满足怎样的要求，也就是体现了如何检验某个用户是否满足对应的要求，所以大部分 IAuthorizationRequirement 接口的实现类型也实现了 IAuthorizationHandler 接口，后者提供的 HandleAsync 方法实现了对应的授权检验。我们将 IAuthorizationHandler 对象称为授权处理器，通过第 9 章的介绍可知，认证处理器是整个认证系统的核心，授权处理器也是整个授权系统的核心。

```
public interface IAuthorizationRequirement{}

public interface IAuthorizationHandler
{
    Task HandleAsync(AuthorizationHandlerContext context);
}
```

如上面的代码片段所示，IAuthorizationHandler 接口将针对授权要求的检验实现在唯一的方法 HandleAsync 中。该方法具有一个类型为 AuthorizationHandlerContext 的参数，代表授权检验的执行上下文。如下面的代码片段所示，我们可以从这个上下文对象中得到待检验的用户（User）、授权的目标资源（Resource），以及应用到授权目标上的所有 IAuthorizationRequirement 对象。

```
public class AuthorizationHandlerContext
{
    public virtual ClaimsPrincipal                          User { get; }
    public virtual object                                   Resource { get; }
    public virtual IEnumerable<IAuthorizationRequirement>   Requirements { get; }

    public AuthorizationHandlerContext(IEnumerable<IAuthorizationRequirement> requirements,
        ClaimsPrincipal user, object resource);
    ...
}
```

AuthorizationHandlerContext 上述 3 个属性体现了授权检验的输入，作为输出的授权结果则体现在如下 3 个属性中。授权成功和失败的标志通过 HasSucceeded 属性与 HasFailed 属性来表示，PendingRequirements 属性则返回尚未经过检验的 IAuthorizationRequirement 对象。对于一个刚刚被初始化的 AuthorizationHandlerContext 对象来说，Requirements 和 PendingRequirements 具有相同的元素。

```
public class AuthorizationHandlerContext
{
    public virtual bool                                         HasSucceeded { get; }
    public virtual bool                                         HasFailed { get; }
    public virtual IEnumerable<IAuthorizationRequirement>       PendingRequirements { get; }

    public virtual void Fail();
    public virtual void Succeed(IAuthorizationRequirement requirement);
    ...
}
```

在针对某个 IAuthorizationRequirement 对象实施授权检验的时候，如果不满足授权要求，我们可以直接调用 AuthorizationHandlerContext 上下文的 Fail 方法，该方法会将 HasFailed 属性设置为 True。反之，如果满足授权规则，我们可以将 IAuthorizationRequirement 对象作为参数调用 Succeed 方法，该对象会从 PendingRequirements 属性表示的列表中移除。只有在尚未调用 Fail 方法并且其 PendingRequirements 属性集合为空的情况下，AuthorizationHandlerContext 上下文的 HasSucceeded 属性才会返回 True。换句话说，授权成功的前提是必须满足所有 IAuthorizationRequirement 对象的授权要求。

由于描述目标资源所有授权要求的 IAuthorizationRequirement 对象都包含在 AuthorizationHandlerContext 上下文中，所以我们可以将针对不同授权要求的检验实现在同一个授权处理器类型中。但是这样的设计违反了"单一职责"的原则，所以大部分授权处理器只关注某种单一的授权要求，这样的 IAuthorizationHandler 实现类型一般会派生于如下所示的 AuthorizationHandler<TRequirement>的抽象类，泛型参数 TRequirement 表示对应的 IAuthorizationRequirement 实现类型。

```
public abstract class AuthorizationHandler<TRequirement> : IAuthorizationHandler
    where TRequirement : IAuthorizationRequirement
{
    public virtual async Task HandleAsync(AuthorizationHandlerContext context)
    {
        foreach (var requirement in context.Requirements.OfType<TRequirement>())
        {
            await HandleRequirementAsync(context, requirement);
        }
    }
    protected abstract Task HandleRequirementAsync(AuthorizationHandlerContext context,
        TRequirement requirement);
}
```

如上面的代码片段所示，当 AuthorizationHandler<TRequirement>的 HandleAsync 方法被执行的时候，它会从 AuthorizationHandlerContext 对象表示的授权上下文中提取出对应类型的 IAuthorizationRequirement 对象，并调用 HandleRequirementAsync 方法对它们逐一进行授权检验。

如果 AuthorizationHandler<TRequirement>是从授权要求的角度对职责进行单一化的，那么抽象类 AuthorizationHandler<TRequirement, TResource>则在此基础上将职责进一步细化到授权目标资源上。如下面的代码片段所示，当 AuthorizationHandler<TRequirement, TResource>的 HandleAsync 方法在进行授权检验的时候，它会进行资源类型的判断。

```csharp
public abstract class AuthorizationHandler<TRequirement, TResource> : IAuthorizationHandler
    where TRequirement : IAuthorizationRequirement
{
    public virtual async Task HandleAsync(AuthorizationHandlerContext context)
    {
        if (context.Resource is TResource)
        {
            foreach (var req in context.Requirements.OfType<TRequirement>())
            {
                await HandleRequirementAsync(context, req, (TResource)context.Resource);
            }
        }
    }
    protected abstract Task HandleRequirementAsync(AuthorizationHandlerContext context,
        TRequirement requirement, TResource resource);
}
```

20.2.2 预定义的 IAuthorizationRequirement 实现类型

在介绍了 IAuthorizationRequirement 接口和 IAuthorizationHandler 接口以及上述这些抽象基类之后，下面介绍几个原生的 IAuthorizationRequirement 实现类型，这些类型也实现了 IAuthorizationHandler 接口的授权处理器类型。

DenyAnonymousAuthorizationRequirement

DenyAnonymousAuthorizationRequirement 体现的授权要求非常简单，那就是拒绝未被验证的匿名用户访问目标资源，该类型派生于 AuthorizationHandler<DenyAnonymousAuthorizationRequirement>。如下面的代码片段所示，DenyAnonymousAuthorizationRequirement 通过表示用户的 ClaimsPrincipal 对象是否具有一个经过认证的身份来确定当前请求是否来源于匿名用户。

```csharp
public class DenyAnonymousAuthorizationRequirement :
    AuthorizationHandler<DenyAnonymousAuthorizationRequirement>, IAuthorizationRequirement
{
    protected override Task HandleRequirementAsync(AuthorizationHandlerContext context,
        DenyAnonymousAuthorizationRequirement requirement)
    {
        var user = context.User;
        var isAnonymous =
            user?.Identity == null ||
            !user.Identities.Any(i => i.IsAuthenticated);
        if (!isAnonymous)
        {
            context.Succeed(requirement);
        }
        return Task.CompletedTask;
```

```
        }
}
```

ClaimsAuthorizationRequirement

由于用户的权限大都以声明的形式保存在表示认证用户的 ClaimsPrincipal 对象上,所以授权检验实际上就是确定 ClaimsPrincipal 对象是否携带所需的授权声明,这样的授权检验是通过 ClaimsAuthorizationRequirement 对象来完成的。如下面的代码片段所示,该类型派生于 AuthorizationHandler<ClaimsAuthorizationRequirement>,定义中的两个属性(ClaimType 和 AllowedValues)表示声明的类型和候选值,它们都是在构造函数中被初始化的。

```
public class ClaimsAuthorizationRequirement :
    AuthorizationHandler<ClaimsAuthorizationRequirement>, IAuthorizationRequirement
{
    public string                       ClaimType { get; }
    public IEnumerable<string>          AllowedValues { get; }

    public ClaimsAuthorizationRequirement(string claimType,
        IEnumerable<string> allowedValues);
    protected override Task HandleRequirementAsync(AuthorizationHandlerContext context,
        ClaimsAuthorizationRequirement requirement);
}
```

如果我们创建 ClaimsAuthorizationRequirement 对象时只指定了声明类型,而没有指定声明的候选值,那么在进行授权检验的时候只要求表示当前用户的 ClaimsPrincipal 对象携带任意一个与指定类型一致的声明即可。反之,如果指定了声明的候选值,那么就需要进行声明值的比较。值得注意的是,针对声明类型的比较是不区分大小写的,但是针对声明值的比较则是区分大小写的,具体的授权检验体现在如下所示的代码片段中。

```
public class ClaimsAuthorizationRequirement :
    AuthorizationHandler<ClaimsAuthorizationRequirement>, IAuthorizationRequirement
{
    protected override Task HandleRequirementAsync(AuthorizationHandlerContext context,
        ClaimsAuthorizationRequirement requirement)
    {
        if (context.User != null)
        {
            var found = false;
            if (requirement.AllowedValues == null || !requirement.AllowedValues.Any())
            {
                found = context.User.Claims.Any(c => string.Equals(
                    c.Type, requirement.ClaimType, StringComparison.OrdinalIgnoreCase));
            }
            else
            {
                found = context.User.Claims.Any(c => string.Equals(c.Type,
                    requirement.ClaimType, StringComparison.OrdinalIgnoreCase) &&
                    requirement.AllowedValues.Contains(c.Value, StringComparer.Ordinal));
            }
            if (found)
```

```
            {
                context.Succeed(requirement);
            }
        }
        return Task.CompletedTask;
    }
}
```

NameAuthorizationRequirement

　　NameAuthorizationRequirement 类型旨在实现针对用户名的授权，也就是说，目标资源的访问授权给某个指定的用户。如下面的代码片段所示，授权用户的用户名体现为 NameAuthorizationRequirement 类型的 RequiredName 属性。当 HandleRequirementAsync 方法实施授权检验时，它要求表示当前用户的 ClaimsPrincipal 具有一个匹配的身份。值得注意的是，NameAuthorizationRequirement 类型采用的用户名比较是区分大小写的，笔者认为这一点是不合理的，因为用户名在大部分情况下是不区分大小写的。

```
public class NameAuthorizationRequirement :
    AuthorizationHandler<NameAuthorizationRequirement>, IAuthorizationRequirement
{
    public string RequiredName { get; }
    public NameAuthorizationRequirement(string requiredName) ;

    protected override Task HandleRequirementAsync(AuthorizationHandlerContext context,
        NameAuthorizationRequirement requirement)
    {
        if (context.User != null && context.User.Identities.Any(
            i => string.Equals(i.Name, requirement.RequiredName)))
        {
            context.Succeed(requirement);
        }
        return Task.CompletedTask;
    }
}
```

RolesAuthorizationRequirement

　　针对角色的授权是最常用的授权方式。在这种授权方式下，我们将目标资源与一组角色列表进行关联，如果用户拥有其中任意一个角色，则意味着该用户具有访问目标资源的权限，RolesAuthorizationRequirement 帮助我们完成针对角色的授权检验。如下面的代码片段所示，与目标资源关联的角色列表存储在 AllowedRoles 属性表示的集合中。在实现授权检验的 HandleRequirementAsync 方法中，它直接调用表示当前用户的 ClaimsPrincipal 的 IsInRole 方法判断该用户是否拥有指定的角色。

```
public class RolesAuthorizationRequirement :
    AuthorizationHandler<RolesAuthorizationRequirement>, IAuthorizationRequirement
{
    public IEnumerable<string> AllowedRoles { get; }
    public RolesAuthorizationRequirement(IEnumerable<string> allowedRoles);
```

```
protected override Task HandleRequirementAsync(AuthorizationHandlerContext context,
    RolesAuthorizationRequirement requirement)
{
    if (context.User != null && requirement.AllowedRoles.Any(
        role => context.User.IsInRole(role)))
    {
        context.Succeed(requirement);
    }
    return Task.CompletedTask;
}
```

AssertionRequirement

上面介绍的 4 种 AuthorizationRequirement 类型都派生于 AuthorizationHandler<TRequirement>，但 AssertionRequirement 类型则直接实现 IAuthorizationHandler 接口。一个 IAuthorizationHandler 对象针对授权规则的检验实际上体现为针对 AuthorizationHandlerContext 上下文的断言（Assert），该断言可以通过一个类型为 Func<AuthorizationHandlerContext, Task<bool>>的委托来表示，AssertionRequirement 对象会利用它来实施授权检验。如下面的代码片段所示，AssertionRequirement 对象使用的授权断言通过只读属性 Handler 表示，它提供的两个构造函数重载使我们可以以不同的形式初始化这个属性。HandleAsync 方法在实施授权检验的时候可以直接调用这个委托对象。

```
public class AssertionRequirement : IAuthorizationHandler, IAuthorizationRequirement
{
    public Func<AuthorizationHandlerContext, Task<bool>> Handler { get; }

    public AssertionRequirement(Func<AuthorizationHandlerContext, bool> handler)
        => Handler = context => Task.FromResult(handler(context));

    public AssertionRequirement(Func<AuthorizationHandlerContext, Task<bool>> handler)
        => Handler = handler;

    public async Task HandleAsync(AuthorizationHandlerContext context)
    {
        if (await this.Handler (context))
        {
            context.Succeed(this);
        }
    }
}
```

OperationAuthorizationRequirement

前面介绍的 5 个 IAuthorizationRequirement 实现类型同时实现了 IAuthorizationHandler 接口，但 OperationAuthorizationRequirement 类型并没有实现 IAuthorizationHandler 接口。授权旨在限制非法用户针对某个资源的访问或者对某项操作的执行，而 OperationAuthorizationRequirement

对象的目的在于将授权的目标对象映射到一个预定义的操作上，所以它只包含如下这个表示操作名称的 Name 属性。

```csharp
public class OperationAuthorizationRequirement : IAuthorizationRequirement
{
    public string Name { get; set; }
}
```

PassThroughAuthorizationHandler

PassThroughAuthorizationHandler 是一个特殊并且比较重要的授权处理器类型，其特殊之处在于它并没有实现针对某种具体规则的授权检验，但是 AuthorizationHandlerContext 上下文中所有的 IAuthorizationHandler 都是通过该对象驱动执行的。如下面的代码片段所示，当 PassThroughAuthorizationHandler 对象的 HandleAsync 方法被执行的时候，它会从 AuthorizationHandlerContext 的 Requirements 属性中提取所有 IAuthorizationHandler 对象，并逐个调用它们的 HandleAsync 方法来实施授权检验。

```csharp
public class PassThroughAuthorizationHandler : IAuthorizationHandler
{
    public async Task HandleAsync(AuthorizationHandlerContext context)
    {
        foreach (var handler in context.Requirements.OfType<IAuthorizationHandler>())
        {
            await handler.HandleAsync(context);
        }
    }
}
```

20.2.3 授权检验

应用程序最终针对授权的检验是通过 IAuthorizationService 服务来完成的。IAuthorizationService 接口定义了如下所示的 AuthorizeAsync 方法，该方法会根据提供的 IAuthorizationRequirement 对象列表实施授权检验，该方法用一个 ClaimsPrincipal 类型的参数（user）表示待检验的用户，而参数 resource 则表示授权的目标资源。

```csharp
public interface IAuthorizationService
{
    Task<AuthorizationResult> AuthorizeAsync(ClaimsPrincipal user, object resource,
        IEnumerable<IAuthorizationRequirement> requirements);
    ...
}
```

AuthorizationResult

授权检验的结果可以用如下所示的 AuthorizationResult 类型来表示。如果授权成功，它的 Succeeded 属性会返回 True；否则，授权失败的信息会保存在 Failure 属性返回的 AuthorizationFailure 对象中。由于 AuthorizationResult 类型只包含一个私有构造函数，所以针对 AuthorizationResult 对象的创建只能通过调用 Success 和 Failed 这两组静态工厂方法来完成。

```csharp
public class AuthorizationResult
```

```csharp
{
    public bool                              Succeeded { get; }
    public AuthorizationFailure              Failure { get; }

    private AuthorizationResult();

    public static AuthorizationResult Failed();
    public static AuthorizationResult Failed(AuthorizationFailure failure);
    public static AuthorizationResult Success();
}
public class AuthorizationFailure
{
    public bool                                          FailCalled { get; }
    public IEnumerable<IAuthorizationRequirement>        FailedRequirements { get; }

    private AuthorizationFailure();

    public static AuthorizationFailure ExplicitFail();
    public static AuthorizationFailure Failed(
        IEnumerable<IAuthorizationRequirement> failed);
}
```

承载授权失败信息的 AuthorizationFailure 同样只能通过静态方法 FailCalled 和 Failed Requirements 来创建，这两个方法分别用来初始化它的 FailCalled 属性和 FailedRequirements 属性。当我们调用 AuthorizationResult 类型的无参静态方法 Failed 来创建 AuthorizationResult 对象的时候，其内部会调用 ExplicitFail 方法来创建 AuthorizationFailure 对象。

IAuthorizationHandlerContextFactory

针对 IAuthorizationRequirement 对象的授权检验最终需要调用对应的授权处理器的 HandleAsync 方法，由于后者将一个 AuthorizationHandlerContext 对象作为授权检验的执行上下文，所以最终需要解决针对这个上下文的创建问题。授权模型中针对 AuthorizationHandler Context 上下文的创建是利用 IAuthorizationHandlerContextFactory 工厂完成的。

```csharp
public interface IAuthorizationHandlerContextFactory
{
    AuthorizationHandlerContext CreateContext(
        IEnumerable<IAuthorizationRequirement> requirements, ClaimsPrincipal user,
        object resource);
}
```

如上面的代码片段所示，IAuthorizationHandlerContextFactory 接口定义了唯一的方法 CreateContext，该方法用来创建作为授权上下文的 AuthorizationHandlerContext 对象。如下所示的 DefaultAuthorizationHandlerContextFactory 是对该接口的默认实现。

```csharp
public class DefaultAuthorizationHandlerContextFactory :
    IAuthorizationHandlerContextFactory
{
    public virtual AuthorizationHandlerContext CreateContext(
        IEnumerable<IAuthorizationRequirement> requirements,
```

```
        ClaimsPrincipal user, object resource)
        => new AuthorizationHandlerContext(requirements, user, resource);
}
```

IAuthorizationHandlerProvider

授权模型利用 IAuthorizationHandlerProvider 对象从授权上下文中提取作为授权处理器的 IAuthorizationHandler 对象。如下面的代码片段所示，IAuthorizationHandlerProvider 接口定义了唯一的 GetHandlersAsync 方法，该方法会从代表授权上下文的 AuthorizationHandlerContext 对象中提取一组 IAuthorizationHandler 对象。

```
public interface IAuthorizationHandlerProvider
{
    Task<IEnumerable<IAuthorizationHandler>> GetHandlersAsync(
        AuthorizationHandlerContext context);
}
```

如下所示的 DefaultAuthorizationHandlerProvider 类型是对 IAuthorizationHandlerProvider 接口的默认实现。从如下所示的代码片段可以看出，它的 GetHandlersAsync 方法返回的 IAuthorizationHandler 对象列表并不是从给定的 AuthorizationHandlerContext 上下文中提取的，而是直接在构造函数中注入的。

```
public class DefaultAuthorizationHandlerProvider : IAuthorizationHandlerProvider
{
    private readonly IEnumerable<IAuthorizationHandler> _handlers;

    public DefaultAuthorizationHandlerProvider(IEnumerable<IAuthorizationHandler> handlers)
        => _handlers = handlers;
    public Task<IEnumerable<IAuthorizationHandler>> GetHandlersAsync(
        AuthorizationHandlerContext context)
        => Task.FromResult<IEnumerable<IAuthorizationHandler>>(this._handlers);
}
```

IAuthorizationEvaluator

一次完整的授权检验会涉及对多个授权处理器的调用，它们都是在同一个 AuthorizationHandlerContext 上下文中进行的。由于每个授权处理器的授权结果都会写入 AuthorizationHandlerContext 上下文中，所以当所有授权处理器完成了各自的授权检验之后，我们需要根据授权上下文当前的状态对授权结果（成功或者失败）做出判断，并且最终会通过 IAuthorizationEvaluator 对象来完成。

```
public interface IAuthorizationEvaluator
{
    AuthorizationResult Evaluate(AuthorizationHandlerContext context);
}
```

如上面的代码片段所示，IAuthorizationEvaluator 接口具有唯一的方法 Evaluate，该方法会根据提供的授权上下文返回一个表示授权结果的 AuthorizationResult 对象。如下所示的 DefaultAuthorizationEvaluator 类型是对该接口的默认实现，可以看出，授权成功或者失败取决于 AuthorizationHandlerContext 对象的 HasSucceeded 属性。由于该属性只有在 Failed 属性返回

False,并且 PendingRequirements 集合为空的时候才会返回 True,所以只有满足 IAuthorizationRequirement 对象设置的所有的要求才会认为授权成功。

```
public class DefaultAuthorizationEvaluator : IAuthorizationEvaluator
{
    public AuthorizationResult Evaluate(AuthorizationHandlerContext context)
    {
        if (!context.HasSucceeded)
        {
            return AuthorizationResult.Failed(
                context.HasFailed
                ? AuthorizationFailure.ExplicitFail()
                : AuthorizationFailure.Failed(context.PendingRequirements));
        }
        return AuthorizationResult.Success();
    }
}
```

AuthorizationOptions

从命名可以看出,AuthorizationOptions 承载着与授权检验相关的配置选项,此处主要关注如下这个布尔类型的 InvokeHandlersAfterFailure 属性。该属性代表的含义如下:在调用一组授权处理器针对提供的授权上下文实施授权检验的过程中,如果没有通过某个授权处理器的授权检验,是否还有必要继续利用后续的授权处理器做后续检验?该属性默认返回 True,也就是说,不论最终的授权结果如何,所有的授权处理器都会参与授权检验。

```
public class AuthorizationOptions
{
    public bool InvokeHandlersAfterFailure { get; set; } = true;
    ...
}
```

DefaultAuthorizationService

DefaultAuthorizationService 类型是对 IAuthorizationService 接口的默认实现,下面介绍它最终如何利用上述这些服务对象来完成对 IAuthorizationRequiredment 对象的授权检验。DefaultAuthorizationService 以构造函数注入的方式提供了 IAuthorizationHandlerProvider 服务、IAuthorizationHandlerContextFactory 服务和 IAuthorizationEvaluator 服务,以及承载配置选项的 AuthorizationOptions 对象。

```
public class DefaultAuthorizationService : IAuthorizationService
{
    private readonly IAuthorizationHandlerContextFactory    _contextFactory;
    private readonly IAuthorizationEvaluator                _evaluator;
    private readonly IAuthorizationHandlerProvider          _handlers;
    private readonly ILogger                                _logger;
    private readonly AuthorizationOptions                   _options;
    private readonly IAuthorizationPolicyProvider           _policyProvider;

    public DefaultAuthorizationService(IAuthorizationPolicyProvider policyProvider,
```

```csharp
        IAuthorizationHandlerProvider handlers,
        ILogger<DefaultAuthorizationService> logger,
        IAuthorizationHandlerContextFactory contextFactory,
        IAuthorizationEvaluator evaluator,
        IOptions<AuthorizationOptions> options)
{
    _options            = options.Value;
    _handlers           = handlers;
    _policyProvider     = policyProvider;
    _logger             = logger;
    _evaluator          = evaluator;
    _contextFactory     = contextFactory;
}

public async Task<AuthorizationResult> AuthorizeAsync(ClaimsPrincipal user,
    object resource, IEnumerable<IAuthorizationRequirement> requirements)
{
    var authContext = _contextFactory.CreateContext(requirements, user, resource);
    var handlers = await _handlers.GetHandlersAsync(authContext);
    foreach (var handler in handlers)
    {
        await handler.HandleAsync(authContext);
        if (!_options.InvokeHandlersAfterFailure && authContext.HasFailed)
        {
            break;
        }
    }
    return _evaluator.Evaluate(authContext);
}
}
```

在实现的 AuthorizeAsync 方法中，IAuthorizationHandlerContextFactory 工厂率先被用来创建代表授权上下文的 AuthorizationHandlerContext 对象。然后 IAuthorizationHandlerProvider 服务会从该上下文中提取出所有代表授权处理器的 IAuthorizationHandler 对象。在将 AuthorizationHandlerContext 上下文作为参数依次调用这组 IAuthorizationHandler 对象的 HandleAsync 方法的过程中，如果当前授权结果为失败状态，并且 AuthorizationOptions 对象的 InvokeHandlersAfterFailure 属性返回 False，那么整个授权检验过程将立即中止。AuthorizeAsync 方法最终返回的是 IAuthorizationEvaluator 对象针对授权上下文评估的结果。

服务注册

DefaultAuthorizationService 及其依赖的服务是通过 IServiceCollection 接口的 AddAuthorization 扩展方法注册的。从如下所示的代码片段可以看出，注册这些服务采用的生命周期模式都是 Transient。对于注册的这些服务来说，除了包含注入 DefaultAuthorizationService 构造函数的服务，还有一个针对 IAuthorizationHandler 的服务注册，具体的实现类型为 PassThroughAuthorizationHandler。所以，在 DefaultAuthorizationHandlerProvider 的构造函数中注入的授权处理器集合其实只包含 PassThroughAuthorizationHandler 对象，该对象会从授权上下文中获取真正的 IAuthorizationHandler 对象来做最终的授权检验。

```csharp
public static class AuthorizationServiceCollectionExtensions
{
    public static IServiceCollection AddAuthorization(this IServiceCollection services)
    {
        services.TryAdd(ServiceDescriptor
          .Transient<IAuthorizationService, DefaultAuthorizationService>());
        services.TryAdd(ServiceDescriptor
          .Transient<IAuthorizationPolicyProvider,
          DefaultAuthorizationPolicyProvider>());
        services.TryAdd(ServiceDescriptor
           .Transient<IAuthorizationHandlerProvider,
          DefaultAuthorizationHandlerProvider>());
        services.TryAdd(ServiceDescriptor
          .Transient<IAuthorizationEvaluator, DefaultAuthorizationEvaluator>());
        services.TryAdd(ServiceDescriptor
          .Transient<IAuthorizationHandlerContextFactory,
          DefaultAuthorizationHandlerContextFactory>());
        services.TryAddEnumerable(ServiceDescriptor.Transient
            <IAuthorizationHandler, PassThroughAuthorizationHandler>());
        return services;
    }

    public static IServiceCollection AddAuthorization(this IServiceCollection services,
        Action<AuthorizationOptions> configure)
    {
        services.Configure<AuthorizationOptions>(configure);
        return services.AddAuthorization();
    }
}
```

20.3 基于"策略"的授权

如果在实施授权检验时总是针对授权的目标资源创建相应的 IAuthorizationRequirement 对象，这将是一项非常烦琐的工作，我们更加希望采用的编程模式如下：预先创建一组可复用的授权规则，在授权检验时提取对应的授权规则来确定用户是否具有访问目标资源的权限。基于策略的授权是通过 IAuthorizationService 服务如下所示的 AuthorizeAsync 方法重载来提供的，该方法的参数 policyName 代表的就是注册的授权策略名称。

```csharp
public interface IAuthorizationService
{
    ...
    Task<AuthorizationResult> AuthorizeAsync(ClaimsPrincipal user, object resource,
        string policyName);
}
```

20.3.1 授权策略的构建

授权策略在授权模型中体现为一个 AuthorizationPolicy 对象,该对象采用 Builder 模式利用对应的 AuthorizationPolicyBuilder 进行构建。

AuthorizationPolicy

如下所示的代码片段是表示授权策略的 AuthorizationPolicy 类型的定义。有的授权策略与采用的认证方式有关,所以 AuthorizationPolicy 类型利用其 AuthenticationSchemes 属性存储采用的认证方案名称。由于授权模型总是利用 IAuthorizationRequirement 对象来表达"授权要求",所以绝大部分授权策略还得利用它们做出最终的决策。AuthorizationPolicy 类型的 Requirements 属性存储了一组 IAuthorizationRequirement 对象。

```
public class AuthorizationPolicy
{
    public IReadOnlyList<string>                          AuthenticationSchemes { get; }
    public IReadOnlyList<IAuthorizationRequirement> Requirements { get; }

    public AuthorizationPolicy(IEnumerable<IAuthorizationRequirement> requirements,
        IEnumerable<string> authenticationSchemes);
}
```

AuthorizationPolicyBuilder

除了调用构造函数来创建 AuthorizationPolicy 对象,我们还可以利用一个 AuthorizationPolicyBuilder 对象以 Builder 模式来创建 AuthorizationPolicy 对象。如下面的代码片段所示,我们可以根据指定的一组认证方案名称或者现有的一个 AuthorizationPolicy 对象来创建 AuthorizationPolicyBuilder 对象,在这种情况下,构造函数会调用 AddAuthenticationSchemes 方法和 AddRequirements 方法来添加认证方案与 IAuthorizationRequirement 对象。我们需要的 AuthorizationPolicy 对象最终是由 Build 方法创建的。

```
public class AuthorizationPolicyBuilder
{
    public IList<string>                          AuthenticationSchemes { get; set; }
    public IList<IAuthorizationRequirement> Requirements { get; set; }

    public AuthorizationPolicyBuilder(params string[] authenticationSchemes);
    public AuthorizationPolicyBuilder(AuthorizationPolicy policy);

    public AuthorizationPolicyBuilder AddAuthenticationSchemes(params string[] schemes);
    public AuthorizationPolicyBuilder AddRequirements(
        params IAuthorizationRequirement[] requirements);

    public AuthorizationPolicy Build();
    ...
}
```

20.2.2 节介绍了一系列预定义的 IAuthorizationRequirement 实现类型,AuthorizationPolicyBuilder 类型为我们定义的如下这些方法用来创建它们并将其添加到 Requirements 集合中。

```csharp
public class AuthorizationPolicyBuilder
{
    public AuthorizationPolicyBuilder RequireAssertion(
        Func<AuthorizationHandlerContext, bool> handler);
    public AuthorizationPolicyBuilder RequireAssertion(
        Func<AuthorizationHandlerContext, Task<bool>> handler);

    public AuthorizationPolicyBuilder RequireAuthenticatedUser();

    public AuthorizationPolicyBuilder RequireClaim(string claimType);
    public AuthorizationPolicyBuilder RequireClaim(string claimType,
        IEnumerable<string> requiredValues);
    public AuthorizationPolicyBuilder RequireClaim(string claimType,
        params string[] requiredValues);

    public AuthorizationPolicyBuilder RequireRole(IEnumerable<string> roles);
    public AuthorizationPolicyBuilder RequireRole(params string[] roles);

    public AuthorizationPolicyBuilder RequireUserName(string userName);
}
```

一个 AuthorizationPolicy 对象的有效内容荷载就是一组认证方案列表和一组 IAuthorization Requirement 对象列表，有时我们需要将两个 AuthorizationPolicy 对象提供的这两组数据进行合并，所以 AuthorizationPolicyBuilder 类型提供了如下所示的 Combine 方法。除了这个实例方法，AuthorizationPolicy 类型还提供了两个静态的 Combine 方法，用来实现针对多个 AuthorizationPolicy 对象的合并。

```csharp
public class AuthorizationPolicyBuilder
{
    public AuthorizationPolicyBuilder Combine(AuthorizationPolicy policy);
}

public class AuthorizationPolicy
{
    public static AuthorizationPolicy Combine(IEnumerable<AuthorizationPolicy> policies);
    public static AuthorizationPolicy Combine(params AuthorizationPolicy[] policies);
}
```

20.3.2 授权策略的注册

针对授权策略的注册需要使用配置选项 AuthorizationOptions，而 DefaultAuthorization Service 对象会利用注入的 IAuthorizationPolicyProvider 对象来提供注册的授权策略。

AuthorizationOptions

前面已经介绍了定义在 AuthorizationOptions 类型中的 InvokeHandlersAfterFailure 属性，实际上，AuthorizationOptions 还有如下所示的其他成员。AuthorizationOptions 对象通过一个字典对象维护一组 AuthorizationPolicy 对象和对应名称的映射关系，我们可以调用两个 AddPolicy 方法来向这个字典中添加新的映射关系，也可以调用 GetPolicy 方法根据指定策略名称得到对应的 AuthorizationPolicy 对象。

```csharp
public class AuthorizationOptions
{
    public AuthorizationPolicy          DefaultPolicy { get; set; }
    public bool                         InvokeHandlersAfterFailure { get; set; }

    public void AddPolicy(string name, AuthorizationPolicy policy);
    public void AddPolicy(string name, Action<AuthorizationPolicyBuilder> configurePolicy);
    public AuthorizationPolicy GetPolicy(string name);
}
```

如果调用 GetPolicy 方法时指定的策略名称不存在，该方法就会返回 Null。在这种情况下，可以选择使用默认的授权策略，针对默认授权策略的设置可以通过 AuthorizationOptions 对象的 DefaultPolicy 属性来实现。

IAuthorizationPolicyProvider

作为对 IAuthorizationService 接口的默认实现，DefaultAuthorizationService 类型在实施授权检验时会利用注入的 IAuthorizationPolicyProvider 服务来提供注册的授权策略。如下面的代码片段所示，IAuthorizationPolicyProvider 接口定义了如下两个方法（GetDefaultPolicyAsync 和 GetPolicyAsync），用来提供默认和指定名称的授权策略。

```csharp
public interface IAuthorizationPolicyProvider
{
    Task<AuthorizationPolicy> GetDefaultPolicyAsync();
    Task<AuthorizationPolicy> GetPolicyAsync(string policyName);
}
```

DefaultAuthorizationPolicyProvider 类型是对 IAuthorizationPolicyProvider 接口的默认实现。如下面的代码片段所示，DefaultAuthorizationPolicyProvider 构造函数通过注入的 IOptions<AuthorizationOptions> 服务来提供承载配置选项的 AuthorizationOptions 对象，实现的两个方法正是利用此配置选项获得注册的授权策略的。前面介绍的用来注册授权基础服务的 AddAuthorization 扩展方法中提供了针对 DefaultAuthorizationPolicyProvider 的服务注册。

```csharp
public class DefaultAuthorizationPolicyProvider : IAuthorizationPolicyProvider
{
    private readonly AuthorizationOptions _options;
    public DefaultAuthorizationPolicyProvider(IOptions<AuthorizationOptions> options)
        => _options = options.Value;

    public Task<AuthorizationPolicy> GetDefaultPolicyAsync() =>
        Task.FromResult<AuthorizationPolicy>(this._options.DefaultPolicy);

    public virtual Task<AuthorizationPolicy> GetPolicyAsync(string policyName) =>
        Task.FromResult<AuthorizationPolicy>(this._options.GetPolicy(policyName));
}
```

20.3.3 授权检验

基于策略的授权在 DefaultAuthorizationService 类型中是通过如下所示的方式实现的。在实现的 AuthorizeAsync 方法中，DefaultAuthorizationService 对象会利用构造函数中注入的 IAuthorizationPolicyProvider 对象根据指定的策略名称得到对应的授权策略，并从表示授权策略的 AuthorizationPolicy 对象中得到所有的 IAuthorizationRequirement 对象。DefaultAuthorizationService 对象将这些 IAuthorizationRequirement 对象作为参数调用 AuthorizeAsync 方法重载来完成授权检验。从下面给出的代码片段可以看出，如果指定的策略名称尚未被注册，AuthorizeAsync 方法会直接抛出一个 InvalidOperationException 类型的异常，而不会选择默认的授权策略。

```csharp
public class DefaultAuthorizationService : IAuthorizationService
{
    ...
    private readonly IAuthorizationPolicyProvider _policyProvider;

    public DefaultAuthorizationService(IAuthorizationPolicyProvider policyProvider,
        IAuthorizationHandlerProvider handlers,
        ILogger<DefaultAuthorizationService> logger,
        IAuthorizationHandlerContextFactory contextFactory,
        IAuthorizationEvaluator evaluator,
        IOptions<AuthorizationOptions> options)
    {
        _policyProvider = policyProvider;
        ...
    }

    public Task<AuthorizationResult> AuthorizeAsync(ClaimsPrincipal user, object resource,
        IEnumerable<IAuthorizationRequirement> requirements);

    public async Task<AuthorizationResult> AuthorizeAsync(ClaimsPrincipal user,
        object resource, string policyName)
    {
        var policy = await _policyProvider.GetPolicyAsync(policyName);
        if (policy == null)
        {
            throw new InvalidOperationException($"No policy found: {policyName}.");
        }
        return service.AuthorizeAsync(user, resource, policy.Requirements);
    }
}
```

综上所述，应用程序最终利用 IAuthorizationService 服务针对目标操作或者资源实施授权检验，DefaultAuthorizationService 类型是对该服务接口的默认实现。IAuthorizationService 服务具体提供了两种授权检验模式，一种是针对提供的 IAuthorizationRequirement 对象列表实施授权，另一种则是针对注册的某个通过 AuthorizationPolicy 对象表示的授权策略，后者由注册的 IAuthorizationPolicyProvider 服务提供。

最终的授权检验都会落实到通过 IAuthorizationHandler 接口表示的授权处理器上，它们由注册的 IAuthorizationHandlerProvider 服务提供。所有 IAuthorizationHandler 对象均在同一个通过

AuthorizationHandlerContext 对象表示的授权上下文中实施授权检验，该上下文由注册的 IAuthorizationHandlerContextFactory 服务创建。当所有 IAuthorizationHandler 对象完成了授权检验之后，注册的 IAuthorizationEvaluator 服务会根据授权上下文的状态得到最终的授权结果，该结果体现为一个 AuthorizationResult 对象。组成授权模式的这些核心接口和服务体现在图 20-3 所示的 UML 中。

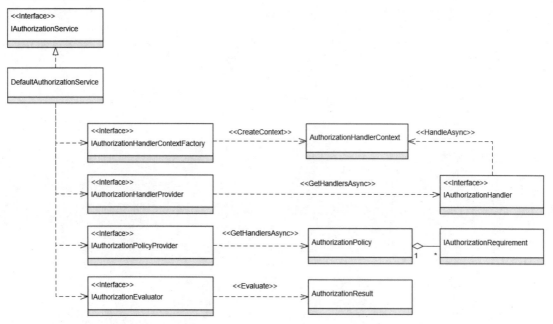

图 20-3　授权模型的核心接口与类型

第 21 章

跨域资源共享

同源策略是所有浏览器都必须遵循的一项安全原则，它的存在决定了浏览器在默认情况下无法对跨域请求的资源做进一步处理。为了实现跨域资源的共享，W3C 制定了 CORS 规范，从而使授权的客户端可以处理跨域调用返回的资源。ASP.NET Core 利用 CorsMiddleware 中间件提供了针对 CORS 规范的实现。

21.1 处理跨域资源

ASP.NET Core 应用利用 CorsMiddleware 中间件按照标准的 CORS 规范实现了资源的跨域共享。按照惯例，在正式介绍 CorsMiddleware 中间件的实现原理之前，我们先通过几个简单的实例介绍如何利用该中间件来实现浏览器环境中的 Web API 跨域调用。

21.1.1 跨域调用 API

为了方便在本机环境下模拟跨域 Web API 调用，我们通过修改 Host 文件的方式将本地地址（"127.0.0.1"）映射为多个不同的域名。所以，我们以管理员（Administrator）身份打开文件"%windir%\System32\drivers\etc\hosts"，并以如下所示的方式添加了针对 4 个域名的映射。

```
127.0.0.1        www.foo.com
127.0.0.1        www.bar.com
127.0.0.1        www.baz.com
127.0.0.1        www.qux.com
```

我们的演示程序由图 21-1 所示的两个 ASP.NET Core 应用程序构成。我们将 Web API 定义在 Web API 项目中，WebApp 是一个 JavaScript 应用程序，它会在浏览器环境下以跨域请求的方式调用承载于 WebApi 应用中的 Web API。

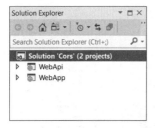

图 21-1　演示实例解决方案结构

如下所示的代码片段是 WebApi 应用的完整定义。我们在该应用程序中定义了表示联系人的 Contact 类型，API 响应内容是以 JSON 格式表示的 Contact 对象数组。我们还为 WebApi 应用设置了监听端口号（8080）和基础路径（"/contacts"）。

```
public class Program
{
    public static void Main()
    {
        Host.CreateDefaultBuilder()
            .ConfigureWebHostDefaults(builder => builder
                .UseUrls("http://0.0.0.0:8080")
                .Configure(app => app
                    .UsePathBase("/contacts")
                    .Run(ProcessAsync)))
            .Build()
            .Run();

        static Task ProcessAsync(HttpContext httpContext)
        {
            var response = httpContext.Response;
            response.ContentType = "application/json";
            var contacts = new Contact[]
            {
                new Contact("张三", "123", "zhangsan@gmail.com"),
                new Contact("李四","456", "lisi@gmail.com"),
                new Contact("王五", "789", "wangwu@gmail.com")
            };
            return response.WriteAsync(JsonConvert.SerializeObject(contacts));
        }
    }
}

public class Contact
{
    public string Name { get; }
    public string PhoneNo { get; }
    public string EmailAddress { get; }
    public Contact(string name, string phoneNo, string emailAddress)
    {
        Name        = name;
        PhoneNo     = phoneNo;
```

```
            EmailAddress    = emailAddress;
    }
}
```

下面的代码片段展示了 WebApp 应用程序的完整定义。该应用会呈现一个包含联系人列表的 Web 页面，我们在该页面中采用 jQuery 以 AJAX 的方式调用上面这个 Web API 获取呈现的联系人列表。我们将 AJAX 请求的目标地址设置为 "http://www.gux.com:8080/contacts"。在 AJAX 请求的回调操作中，可以将返回的联系人以无序列表的形式呈现出来。

```
public class Program
{
    public static void Main()
    {
        Host.CreateDefaultBuilder()
            .ConfigureWebHostDefaults(builder => builder
                .UseUrls("http://0.0.0.0:3721")
                .Configure(app => app.Run(ProcessAsync)))
            .Build()
            .Run();

        static async Task ProcessAsync(HttpContext httpContext)
        {
            httpContext.Response.ContentType = "text/html";
            var html =
            @"<html>
            <body>
                <ul id='contacts'></ul>
                <script src='http://code.jquery.com/jquery-3.3.1.min.js'></script>
                <script>
                $(function()
                {
                    var url = 'http://www.gux.com:8080/contacts';
                    $.getJSON(url, null, function(contacts) {
                        $.each(contacts, function(index, contact)
                        {
                            var html = '<li><ul>';
                            html += '<li>Name: ' + contact.Name + '</li>';
                            html += '<li>Phone No:' + contact.PhoneNo + '</li>';
                         html += '<li>Email Address: ' + contact.EmailAddress + '</li>';
                            html += '</ul>';
                            $('#contacts').append($(html));
                        });
                    });
                });
                </script>
            </body>
            </html>";
            await httpContext.Response.WriteAsync(html);
        }
    }
}
```

然后先后启动应用程序 WebApi 和 WebApp。如果利用浏览器采用映射的域名（www.foo.com）访问 WebApp 应用，就会发现我们期待的联系人列表并没有呈现出来。如果按 F12 键查看开发工具，就会发现图 21-2 所示的关于 CORS 的错误，具体的错误消息为 "Access to XMLHttpRequest at 'http://www.gux.com:8080/contacts' from origin 'http://www.foo.com:3721' has been blocked by CORS policy: No 'Access-Control-Allow-Origin' header is present on the requested resource."。（S2101）

图 21-2 跨域访问导致联系人无法呈现

有的读者认为是因为 AJAX 调用发生错误，所以联系人列表就没有出现在响应内容中。如果我们利用抓包工具捕捉 AJAX 请求和响应的内容，就会捕获到如下所示的 HTTP 报文。因此，AJAX 调用其实是成功的，但浏览器阻止了针对跨域请求返回数据的进一步处理。另外，如下请求具有一个名为 Origin 的报头，表示的正是 AJAX 请求的 "源"，也就是跨域（Cross-Orgin）中的 "域"。

```
GET http://www.gux.com:8080/contacts HTTP/1.1
Host: www.gux.com:8080
Connection: keep-alive
Accept: application/json, text/javascript, */*; q=0.01
Origin: http://www.foo.com:3721
User-Agent: Mozilla/5.0 (Windows NT 10.0; Win64; x64) AppleWebKit/537.36 (KHTML, like Gecko) Chrome/70.0.3538.67 Safari/537.36
Referer: http://www.foo.com:3721/
Accept-Encoding: gzip, deflate
Accept-Language: en-US,en;q=0.9,zh-CN;q=0.8,zh;q=0.7


HTTP/1.1 200 OK
Date: Tue, 24 Sep 2019 13:23:45 GMT
Server: Kestrel
Content-Length: 205

[{"Name":"张三","PhoneNo":"123","EmailAddress":"zhangsan@gmail.com"},{"Name":"李四",
"PhoneNo":"456","EmailAddress":"lisi@gmail.com"},{"Name":"王五","PhoneNo":"789",
"EmailAddress":"wangwu@gmail.com"}]
```

21.1.2　资源提供者显式授权

下面通过注册 CorsMiddleware 中间件来解决上面遇到的 API 跨域调用的问题。CorsMiddleware 中间件完全是基于 W3C CORS 规范实现的，该规范采用"由资源提供者显式授权"的策略来确定资源消费者是否具有进一步操作返回资源的权限。

对于我们演示的实例来说，作为资源提供者的 WebApi 应用如果希望将提供的资源授权给某个应用程序，可以将作为资源消费者应用程序的"域"添加到授权域列表中，具体的做法体现在如下所示的代码片段中。我们调用 IApplicationBuilder 接口的 UseCors 扩展方法完成了针对 CorsMiddleware 中间件的注册，并在注册该中间件时指定了两个授权的"域"（"http://www.foo.com:3721"和"http://www.bar.com:3721"）。CorsMiddleware 中间件涉及的服务则通过调用 IServiceCollection 接口的 AddCors 扩展方法进行注册。

```
public class Program
{
    public static void Main()
    {
        Host.CreateDefaultBuilder()
            .ConfigureWebHostDefaults(builder => builder
                .UseUrls("http://0.0.0.0::8080")
                .ConfigureServices(svcs=>svcs.AddCors())
                .Configure(app => app
                    .UsePathBase("/contacts")
                    .UseCors(cors => cors.WithOrigins(
                        "http://www.foo.com:3721",
                        "http://www.bar.com:3721"))
                    .Run(ProcessAsync)))
            .Build()
            .Run();
        ...
    }
}
```

由于 WebApi 应用对"http://www.foo.com:3721"和"http://www.bar.com:3721"这两个域进行了显式授权，如果采用它们来访问 WebApp 应用程序，浏览器上就会呈现出图 21-3 所示的联系人列表。倘若将浏览器地址栏的 URL 设置成未被授权的"http://www.baz.com:3721"，我们依然得不到想要的显示结果。（S2102）

下面从 HTTP 消息交换的角度来介绍这次由 WebApi 应用响应的报文有何不同。如下所示的是 WebApi 针对地址为"http://www.foo.com:3721"的响应报文，可以看出它多了两个名称分别为 Vary 和 Access-Control-Allow-Origin 的报头。前者与缓存有关，它要求在对响应报文实施缓存的时候，选用的 Key 应该包含请求的 Origin 报头值，它提供给浏览器授权访问当前资源的域。

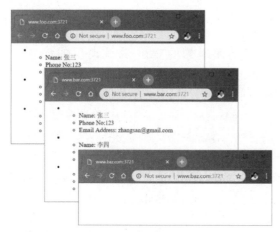

图 21-3　针对域的显式授权

```
HTTP/1.1 200 OK
Date: Tue, 24 Sep 2019 13:32:02 GMT
Server: Kestrel
Vary: Origin
Access-Control-Allow-Origin: http://www.foo.com:3721
Content-Length: 205

[{"Name":"张三","PhoneNo":"123","EmailAddress":"zhangsan@gmail.com"},{"Name":"李四",
"PhoneNo":"456","EmailAddress":"lisi@gmail.com"},{"Name":"王五","PhoneNo":"789",
"EmailAddress":"wangwu@gmail.com"}]
```

对于我们演示的实例来说，当 AJAX 调用成功并返回联系人列表之后，浏览器正是利用 Access-Control-Allow-Origin 报头确定当前请求采用的域是否有权对获取的资源做进一步处理的。只有在授权明确之后，浏览器才会执行回调操作将返回的数据呈现出来。从演示的实例可以看出，"跨域资源共享"所谓的"域"是由协议前缀（如"http://"或者"https://"）、主机名（或者域名）和端口号组成的，但在很多情况下，资源提供在授权的时候往往只需要考虑域名，这样的授权策略可以采用如下所示的方式来解决。（S2103）

```
public class Program
{
    public static void Main()
    {
        var allowedOrigins = new HashSet<string>(StringComparer.OrdinalIgnoreCase)
        {
            "www.foo.com",
            "www.bar.com"
        };

        Host.CreateDefaultBuilder()
            .ConfigureWebHost(builder => builder
                .UseKestrel()
                .UseUrls("http://0.0.0.0:8080")
                .ConfigureServices(svcs => svcs.AddCors())
```

```
                .Configure(app => app
                    .UsePathBase("/contacts")
                    .UseCors(cors => cors.SetIsOriginAllowed(
                        origin => allowedOrigins.Contains(new Uri(origin).Host)))
                    .Run(ResponseAsync)))
            .Build()
            .Run();
        ...
    }
}
```

如上面的代码片段所示，我们在调用 IApplicationBuilder 对象的 UseCors 扩展方法注册 CorsMiddleware 中间件时调用 CorsPolicyBuilder 对象的 SetIsOriginAllowed 方法来设置授权策略。对于这个方法来说，设置的资源授权策略通过作为参数的 Func<string, bool>对象来表示，该委托对象的输入参数表示请求的域，所以可以利用提供的委托对象实现任何我们想要的资源授权策略。在上面演示的代码中，我们指定了一组授权的域名，并利用提供的委托对象完成了针对域名的资源授权。

21.1.3 基于策略的资源授权

从提供的演示实例可以看出，CORS 本质上还是属于授权的问题，所以我们采用类似于第 20 章介绍的方式将资源授权的规则定义成相应的策略，CorsMiddleware 中间件就可以针对某个预定义的策略来实施跨域资源授权。在调用 IServiceCollection 接口的 AddCors 扩展方法时可以采用如下所示的方式注册一个默认的 CORS 策略，该策略达到的效果与上面演示的实例完全一致。（S2104）

```
public class Program
{
    public static void Main()
    {
        var allowedOrigins = new HashSet<string>(StringComparer.OrdinalIgnoreCase)
        {
            "www.foo.com",
            "www.bar.com"
        };

        Host.CreateDefaultBuilder()
            .ConfigureWebHost(builder => builder
                .UseKestrel()
                .UseUrls("http://0.0.0.0::8080")
                .ConfigureServices(svcs => svcs.AddCors(
                    options => options.AddDefaultPolicy(builder => builder
                        .SetIsOriginAllowed(origin => allowedOrigins
                            .Contains(new Uri(origin).Host)))))
                .Configure(app => app
                    .UsePathBase("/contacts")
                    .UseCors()
```

```
                    .Run(ResponseAsync)))
                .Build()
                .Run();
        ...
    }
}
```

除了注册一个默认的匿名 CORS 策略，我们还可以注册一个非默认的具名策略。如下面的代码片段所示，我们在调用 AddCors 扩展方法时注册了一个名为 foobar 的 CORS 策略，在调用 IApplicationBuilder 接口的 UseCors 扩展方法注册 CorsMiddleware 中间件时显式地使用该策略来实施 CORS 授权。（S2105）

```
public class Program
{
    public static void Main()
    {
        var allowedOrigins = new HashSet<string>(StringComparer.OrdinalIgnoreCase)
        {
            "www.foo.com",
            "www.bar.com"
        };

        Host.CreateDefaultBuilder()
            .ConfigureWebHost(builder => builder
                .UseKestrel()
                .UseUrls("http://0.0.0.0::8080")
                .ConfigureServices(svcs => svcs.AddCors(
                    options => options.AddPolicy("foobar", builder => builder
                        .SetIsOriginAllowed(origin => allowedOrigins
                            .Contains(new Uri(origin).Host)))))
                .Configure(app => app
                    .UsePathBase("/contacts")
                    .UseCors("foobar")
                    .Run(ResponseAsync)))
            .Build()
            .Run();
        ...
    }
}
```

21.2 CORS 规范

　　Internet 目前已经无所不在。虽然目前访问 Internet 的客户端越来越多，但是浏览器依旧是一个常用的入口。随着 Web 开放的程度越来越高，通过浏览器跨域获取资源的需求已经变得非常普遍。提及浏览器的核心竞争力，安全性必然是其重要组成部分，而提及浏览器的安全就不得不提到同源策略。

21.2.1 同源策略

同源策略是浏览器的一项最基本的安全策略。毫不夸张地说，浏览器的整个安全体系均建立在此基础之上。同源策略限制了"源"自 A 站点的脚本只能操作"同源"页面的 DOM，"跨源"操作来源于 B 站点的页面将会被拒绝。所谓的"同源站点"，必须要求它们的 URI 在如下 3 个方面保持一致。

- 主机名称（域名/子域名或者 IP 地址）。
- 端口号。
- 网络协议（Scheme，分别采用"http"和"https"协议的两个 URI 被视为不同源）。

值得注意的是，对于一段 JavaScript 脚本来说，其"源"与存储的地址无关，而是取决于脚本被加载的页面。如果在同一个页面中通过如下所示的<script>标签引用了来源于不同地方（"http://www.artech.top/"和"http://www.jinnan.me/"）的两个 JavaScript 脚本，那么它们均与当前页面同源。基于 JSONP 跨域资源共享就是利用了这个特性。

```
<script src="http://www.artech.top/scripts/common.js"></script>
<script src="http://www.jinnan.me/scripts/utility.js"></script>
```

除了<script>标签，HTML 还提供了其他一些具有 src 属性的标签（如<img>、<iframe>和<link>等），它们均具有跨域加载资源的能力，同源策略对它们不做限制。对于这些具有 src 属性的 HTML 标签来说，标签的每次加载都伴随着针对目标地址的一次 GET 请求。同源策略及跨域资源共享在大部分情况下针对的是 AJAX 请求，如果请求指向一个异源地址，浏览器在默认情况下不允许读取返回的内容。

21.2.2 针对资源的授权

基于 Web 的资源共享涉及两个基本角色，即资源的提供者和消费者。CORS 旨在定义一种规范，从而使浏览器在接收到从提供者获取的资源时能够决定是否应该将此资源分发给消费者做进一步处理。CORS 根据资源提供者的显式授权来决定目标资源是否应该与消费者分享。换句话说，浏览器需要得到提供者的授权之后才会将其提供的资源分发给消费者。那么资源的提供者应如何进行资源的授权，并将授权的结果告之浏览器？

具体的实现其实很简单。如果浏览器自身提供对 CORS 的支持，由它发送的请求会携带一个名为 Origin 的报头表明请求页面所在的站点。对于前面演示实例中调用 Web API 获取联系人列表的请求来说，它就具有如下一个 Origin 报头。

```
Origin: http://www.foo.com:3721
```

资源获取请求被提供者接收之后，可以根据该报头确定提供的资源需要与谁共享。资源提供者的授权结果通过一个名为 Access-Control-Allow-Origin 的响应报头来承载，它表示得到授权的站点列表。一般来说，如果资源的提供者认可当前请求的 Origin 报头携带的站点，那么它会将该站点作为 Access-Control-Allow-Origin 报头的值。

除了指定具体的"源"并对其做针对性授权，资源提供者还可以将 Access-Control-Allow-Origin 报头的值设置为"*"，从而对所有消费者进行授权。换言之，如果做了这样的设置，就意味着由其提供的是一种公共资源，所以在做此设置之前需要慎重。如果资源请求被拒绝，资源提供者可以将此响应报头值设置为 null，或者让响应不具有此报头。

当浏览器接收到包含资源的响应之后，会提取 Access-Control-Allow-Origin 报头的值。如果此值为"*"或者提供的站点列表包含此前请求的站点（即请求的 Origin 报头的值），就意味着资源的消费者获得了提供者授予的权限，在此情况下浏览器会允许 JavaScript 程序操作获取的资源。如果此响应报头不存在或者其值为 null，客户端 JavaScript 程序针对资源的操作就会被拒绝。

资源提供者除了通过设置 Access-Control-Allow-Origin 报头对提供的资源进行授权，还可以通过设置另一个名为 Access-Control-Expose-Headers 的报头对响应报头进行授权。具体来说，此 Access-Control-Expose-Headers 报头用于设置一组直接暴露给客户端 JavaScript 程序的响应报头，没有在此列表的响应报头对客户端 JavaScript 程序是不可见的。采用这种方式对响应报头的授权对简单响应报头来说是无效的，对于 CORS 规范来说，这里所谓的简单响应报头（Simple Response Header）包含如下 6 种，也就是说，它们是不需要授权访问的公共响应报头。

- Cache-Control。
- Content-Language。
- Content-Type。
- Expires。
- Last-Modified。
- Pragma。

用于实现 AJAX 请求的 XMLHttpRequest 具有一个 getResponseHeader 方法，调用该方法会返回一组响应报头的列表。按照这里介绍的针对响应报头的授权原则，只有在 Access-Control-Expose-Headers 报头中指定的报头和简单响应报头才会包含在该方法返回的列表中。

21.2.3 获取授权的方式

W3C 的 CORS 规范将跨域资源请求划分为两种类型，即简单请求（Simple Request）和预检请求（Preflight Request）。要弄清楚 CORS 规范将哪些类型的跨域资源请求划分为简单请求的范畴，需要额外了解几个定义在 CORS 规范中的概念，其中包括简单（HTTP）方法（Simple Method）、简单（请求）报头（Simple Header）和自定义请求报头（Author Request Header/ Custom Request Header）。CORS 规范将 GET、HEAD 和 POST 这 3 个 HTTP 方法视为简单 HTTP 方法，而将请求报头 Accept、Accept-Language、Content-Language 以及采用如下 3 种媒体类型的 Content-Type 报头称为简单请求报头。

- application/x-www-form-urlencoded。
- multipart/form-data。
- text/plain。

请求报头包含两种类型：一种是通过浏览器自动生成的报头，另一种则是由 JavaScript 程序自行添加的报头（如调用 XMLHttpRequest 对象的 setRequestHeader 方法可以为生成的 AJAX 请求添加任意报头），后者被称为自定义报头。

在了解了什么是简单 HTTP 方法、简单请求报头和自定义请求报头之后，下面介绍 CORS 规范定义的简单请求和预检请求。可以将跨域获取 Web 资源人为地划分为两个环节，即获取授权信息和获取资源。如果采用简单请求模式，就相当于将这两个环节合并到一个 HTTP 事务中进行，即针对资源请求的响应报文中同时包含请求的资源和授权信息。在请求满足如下两个条件的情况下，浏览器会采用简单请求模式来完成跨域资源请求。

- 请求采用简单 HTTP 方法。
- 请求携带的均为简单报头。

在其他情况下，浏览器应该采用一种被称为预检请求模式的机制来完成跨域资源请求。所谓的预检机制，就是浏览器在发送真正的跨域资源请求前，它会先发送一个采用 OPTIONS 方法的预检请求。预检请求报文不包含主体内容，用户凭证相关的报头也会被剔除。预检请求的报头列表中会携带一些反映真实资源请求的信息。除了代表请求页面所在站点的 Origin 报头，如下所示的是两个典型的 CORS 预检请求报头。

- Access-Control-Request-Method：跨域资源请求采用的 HTTP 方法。
- Access-Control-Request-Headers：跨域资源请求携带的自定义报头列表。

以前面演示的实例来说，假设将 WebApp 中用来呈现 HTML 页面的 RenderPage 方法做如下所示的改写，即让它在页面加载的时候发送一个采用 PUT 方法的 AJAX 请求来修改联系人信息。除此之外，该请求还携带着一个名为 x-foo-bar 的自定义报头。

```
public class Program
{
    ...
    private static async Task RenderPage(HttpContext httpContext)
    {
        httpContext.Response.ContentType = "text/html";
        var html =
            @"<html>
            <body>
                <script src='http://code.jquery.com/jquery-3.3.1.min.js'></script>
                <script>
                $(function()
                {
                    $.ajax({
                        url: 'http://www.gux.com:8080/contacts/foobar',
                        headers: {
                            'x-foo-bar': 'foobar'
                        },
                        type: 'PUT',
                        data: {
                            name            : 'foobar',
                            phoneNo         : '123456',
                            emailAddress    : 'foobar@outlook.com'
```

```
                    });
                });
            </script>
        </body>
    </html>";
    await httpContext.Response.WriteAsync(html);
}
```

由于 PUT 方法并非一个简单的 HTTP 方法，所以浏览器在试图分发这个 AJAX 请求之前，会先发送如下这样一个预检请求来获得授权信息。可以看出，这是一个不包含主体内容的 OPTIONS 请求，除了具有一个表示请求域的 Origin 报头，它还具有一个表示 HTTP 方法的 Access-Control-Request-Method 报头。除此之外，自定义的报头 x-foo-bar 也会包含在 Access-Control-Request-Headers 报头之中。

```
OPTIONS http://www.gux.com:8080/contacts HTTP/1.1
Host: www.gux.com:8080
Connection: keep-alive
Access-Control-Request-Method: PUT
Origin: http://www.foo.com:3721
User-Agent: Mozilla/5.0 (Windows NT 10.0; Win64; x64) AppleWebKit/537.36 (KHTML, like Gecko) Chrome/70.0.3538.67 Safari/537.36
Access-Control-Request-Headers: x-foo-bar
Accept: */*
Accept-Encoding: gzip, deflate
Accept-Language: en-US,en;q=0.9,zh-CN;q=0.8,zh;q=0.7
```

资源的提供者在接收到预检请求之后会根据其提供的信息实施授权检验，具体的检验包括确定请求站点是否值得信任，以及请求采用 HTTP 方法和自定义报头是否被允许。如果预检请求没有通过授权检验，资源提供者一般会返回一个状态码为 "400，Bad Request" 的响应，反之则返回一个状态码为 "200 OK" 或者 "204 No Content" 的响应，授权相关信息会包含在响应报头中。除了上面介绍的 Access-Control-Allow-Origin 报头和 Access-Control-Expose-Headers 报头，预检请求的响应还具有如下 3 个典型的报头。

- Access-Control-Allow-Methods：跨域资源请求允许采用的 HTTP 方法列表。
- Access-Control-Allow-Headers：跨域资源请求允许携带的自定义报头列表。
- Access-Control-Max-Age：浏览器可以将响应结果进行缓存的时间（单位为秒），针对响应的缓存是为了使浏览器避免频繁地发送预检请求。

浏览器在接收到预检响应之后，会根据响应报头确定真正的跨域资源请求能否被接受。浏览器只有在确定服务端一定会授权的情况下才会发送真正的跨域资源请求。如果预检响应满足如下条件，浏览器认为真正的跨域资源请求会被授权。

- 通过请求的 Origin 报头表示的源站点必须存在于 Access-Control-Allow-Origin 报头标识的站点列表中，或者 Access-Control-Allow-Origin 报头的值为 "*"。
- 响应报头 Access-Control-Allow-Methods 不存在，或者预检请求的 Access-Control-

Request-Method 报头表示的请求方法在其列表之内。
- 预检请求的 Access-Control-Request-Headers 报头存储的报头名称均在响应报头 Access-Control-Allow-Headers 表示的报头列表之内。

预检响应结果会被浏览器缓存，在 Access-Control-Max-Age 报头设定的时间内，缓存的结果将被浏览器用于授权检验，所以在此期间不会再发送预检请求。对于上面发送的跨域 PUT 请求，服务端在授权检验通过的情况下会返回与如下类似的响应。对于如下状态码为"204 No Content"的响应，它的 Access-Control-Allow-Headers 报头和 Access-Control-Allow-Methods 报头会携带请求提供的 HTTP 方法与自定义报头名称，另一个值为"*"的 Access-Control-Allow-Origin 报头表示对请求域不做任何限制。

```
HTTP/1.1 204 No Content
Date: Sun, 21 Oct 2018 13:53:20 GMT
Server: Kestrel
Access-Control-Allow-Headers: x-foo-bar
Access-Control-Allow-Methods: PUT
Access-Control-Allow-Origin: *
```

21.2.4 用户凭证

在默认情况下，利用 XMLHttpRequest 发送的 AJAX 请求不会携带与用户凭证相关的敏感信息。携带了 Cookie、HTTP-Authentication 报头及客户端 X.509 证书（采用支持客户端证书的 TLS/SSL）的请求会被视为携带了用户凭证。如果要将用户凭证附加到 AJAX 请求上，就需要将 XMLHttpRequest 对象的 withCredentials 属性设置为 True。

对于 CORS 规范来说，是否支持用户凭证也是授权检验的一个重要环节。换句话说，只有在服务端显式允许请求提供用户凭证的前提下，携带了用户凭证的请求才会被认为是有效的。对于 W3C 的 CORS 规范来说，服务端利用响应报头 Access-Control-Allow-Credentials 来表明它是否允许请求携带用户凭证。

如果客户端 JavaScript 程序利用一个将 withCredentials 属性设置为 True 的 XMLHttpRequest 对象发送了一个跨域资源请求，但是得到的响应却不包含一个值为 True 的响应报头 Access-Control-Allow-Credentials，那么针对获取资源的操作将会被浏览器拒绝。

上面对 W3C 的 CORS 规范做了概括性介绍，由于篇幅所限，很多细节并没有涉及。如果读者对此有兴趣，可直接阅读 W3C 的官方文档。

21.3 CORS 模型

ASP.NET Core 应用针对跨域资源的授权实现在 CorsMiddleware 中间件中，对应的 NuGet 包为 "Microsoft.AspNetCore.Cors"。CorsMiddleware 中间件实际上是对 CORS 规范的实现，下面介绍 CORS 制定的这些规范是如何落实到这个中间件上的。CORS 最终体现为对资源的授权，具体的授权规则被定义在 CORS 策略上。

21.3.1 CORS 策略

CorsMiddleware 中间件在接收到针对跨域资源的请求（包括简单请求和预检请求）时，总是会根据预先指定的 CORS 策略来确定最终针对资源请求的授权结果，授权结果最终以 CORS 规范中定义的一系列响应报头的形式返回客户端。授权策略在整个 CORS 模型中通过 CorsPolicy 类型表示。

CorsPolicy

CORS 策略利用相应的规则来帮助资源提供者确定当前请求资源能否授权给由 Origin 报头表示的消费者，所以授权规则可以体现为一个 Func<string, bool>类型的委托对象，CorsPolicy 类型的 IsOriginAllowed 属性返回的就是这样一个委托对象。如果资源提供者具有一组候选的授权域列表，就可以将它们添加到 Origins 属性表示的集合中。对于提供的公共资源，我们可以将 AllowAnyOrigin 属性设置为 True。

```
public class CorsPolicy
{
    public Func<string, bool>        IsOriginAllowed { get; set; }
    public IList<string>             Origins { get; }
    public bool                      AllowAnyOrigin { get; }
    ...
}
```

CORS 规则除了与跨域资源请求的 Origin 报头表示的请求域有关，有时还与请求采用的 HTTP 方法和自定义报头有关。如果对请求采用的 HTTP 方法有要求，就可以将许可的 HTTP 方法添加到 Methods 属性表示的列表中，否则可以将 AllowAnyMethod 属性设置为 True 来支持任意 HTTP 方法。针对自定义请求报头的要求与之类似，CorsPolicy 类型的 Headers 属性表示许可的自定义报头名称，如果将 AllowAnyHeader 属性设置为 True，就意味着允许请求携带任何自定义报头。ExposedHeaders 属性表示能够暴露给客户端的响应报头列表。

```
public class CorsPolicy
{
    public IList<string>             Methods { get; }
    public bool                      AllowAnyMethod { get; }

    public IList<string>             Headers { get; }
    public bool                      AllowAnyHeader { get; }
    public IList<string>             ExposedHeaders { get; }
    ...
}
```

通过上面对 CORS 规范的介绍可知，CORS 规则还与请求是否携带用户凭证有关。如果允许请求携带与用户凭证相关的请求报头，就可以将 SupportsCredentials 设置为 True。而 PreflightMaxAge 属性对应 CORS 规范定义的 Access-Control-Max-Age 报头，表示缓存的预检响应的有效时长。

```
public class CorsPolicy
{
```

```
    public bool         SupportsCredentials { get; set; }
    public TimeSpan?    PreflightMaxAge { get; set; }
    ...
}
```

CorsPolicyBuilder

表示 CORS 策略的 CorsPolicy 对象可以采用 Builder 模式借助 CorsPolicyBuilder 对象来创建。如下面的代码片段所示，我们可以根据一组授权的域列表或者一组 CorsPolicy 对象来创建 CorsPolicyBuilder 对象，它的 Build 方法会帮助我们创建出最终的 CorsPolicy 对象。对于上面介绍的定义在 CorsPolicy 类型中的规则属性，我们可以调用 CorsPolicyBuilder 对应的方法来指定。

```
public class CorsPolicyBuilder
{
    public CorsPolicyBuilder(params string[] origins);
    public CorsPolicyBuilder(CorsPolicy policy);

    public CorsPolicyBuilder AllowAnyHeader();
    public CorsPolicyBuilder AllowAnyMethod();
    public CorsPolicyBuilder AllowAnyOrigin();
    public CorsPolicyBuilder AllowCredentials();
    public CorsPolicyBuilder DisallowCredentials();
    public CorsPolicyBuilder SetIsOriginAllowed(Func<string, bool> isOriginAllowed);
    public CorsPolicyBuilder SetIsOriginAllowedToAllowWildcardSubdomains();
    public CorsPolicyBuilder SetPreflightMaxAge(TimeSpan preflightMaxAge);
    public CorsPolicyBuilder WithExposedHeaders(params string[] exposedHeaders);
    public CorsPolicyBuilder WithHeaders(params string[] headers);
    public CorsPolicyBuilder WithMethods(params string[] methods);
    public CorsPolicyBuilder WithOrigins(params string[] origins);

    public CorsPolicy Build();
}
```

CorsOptions

CorsMiddleware 中间件使用的 CORS 策略最终会注册到 CorsOptions 对象表示的配置选项上。如下面的代码片段所示，一个 CorsOptions 对象利用一个字典类型的内部属性（PolicyMap）维护一组 CorsPolicy 对象与注册名称之间的映射关系。我们可以调用两个 AddPolicy 方法重载添加新的 CorsPolicy 对象，也可以调用 GetPolicy 方法根据注册的策略名称提取对应的 CorsPolicy 对象。

```
public class CorsOptions
{
    internal IDictionary<string, CorsPolicy> PolicyMap { get; }

    public void AddDefaultPolicy(CorsPolicy policy);
    public void AddDefaultPolicy(Action<CorsPolicyBuilder> configurePolicy);

    public void AddPolicy(string name, CorsPolicy policy);
    public void AddPolicy(string name, Action<CorsPolicyBuilder> configurePolicy);
    public CorsPolicy GetPolicy(string name);
}
```

```
    public string DefaultPolicyName { get; set; }
}
```

一个 ASP.NET Core 应用在全局范围内存在一个默认的 CORS 策略，默认 CORS 策略的名称通过 CorsOptions 对象的 DefaultPolicyName 属性进行设置，该属性的默认值为 __DefaultCorsPolicy。针对默认 CORS 策略的注册可以通过 CorsOptions 类型的两个 AddDefaultPolicy 方法重载来完成。

ICorsPolicyProvider

虽然 ASP.NET Core 应用将 CORS 策略注册到 CorsOptions 对象上，但是 CorsMiddleware 中间件并不会直接从该 Options 对象上提取 CORS 策略，CORS 模型针对策略的提供会交给一个 ICorsPolicyProvider 对象来负责。如下面的代码片段所示，ICorsPolicyProvider 接口提供了唯一的 GetPolicyAsync 方法，该方法会根据当前 HttpContext 上下文和指定的策略名称来提供对应的 CorsPolicy 对象。

```
public interface ICorsPolicyProvider
{
    Task<CorsPolicy> GetPolicyAsync(HttpContext context, string policyName);
}
```

作为对 ICorsPolicyProvider 接口的默认实现，具有如下定义的 DefaultCorsPolicyProvider 会利用作为配置选项的 CorsOptions 对象来提供针对当前请求的 CORS 策略。如下所示的代码片段还体现了一个细节：如果指定的策略名称为 Null，GetPolicyAsync 方法返回的将是注册的默认策略。

```
public class DefaultCorsPolicyProvider : ICorsPolicyProvider
{
    private readonly CorsOptions _options;

    public DefaultCorsPolicyProvider(IOptions<CorsOptions> options)
        =>_options = options.Value;
    public Task<CorsPolicy> GetPolicyAsync(HttpContext context, string policyName)
        =>Task.FromResult<CorsPolicy>(_options.GetPolicy(policyName
        ?? t_options.DefaultPolicyName));
}
```

21.3.2 解析并应用授权结果

在了解了 CORS 策略及其注册与提供机制之后，下面介绍 CorsMiddleware 中间件是如何根据 CORS 策略对接收到的请求进行解析并得出最终授权结果的，但在此之前需要先介绍表示资源授权结果的 CorsResult 类型。

CorsResult

针对跨域资源的授权结果最终会应用到针对请求的响应报文中。具体来说，组成授权结果的每个元素大都体现在针对某个 CORS 响应报头的设置中。例如，CorsResult 类型的 Allowed Origin 属性对应表示授权请求域的 Access-Control-Allow-Origin 报头，AllowedMethods 属性和 AllowedHeaders 属性则分别表示服务端允许的 HTTP 方法列表与自定义请求报头列表的 Access-

Control-Allow-Methods 报头和 Access-Control-Request-Headers 报头。

```
public class CorsResult
{
    public string              AllowedOrigin { get; set; }
    public IList<string>       AllowedMethods { get; }
    public IList<string>       AllowedHeaders { get; }
    public IList<string>       AllowedExposedHeaders { get; }
    public TimeSpan?           PreflightMaxAge { get; set; }
    public bool                SupportsCredentials { get; set; }
    public bool                VaryByOrigin { get; set; }
}
```

CorsResult 类型的 AllowedExposedHeaders 属性对应的响应报头为 Access-Control-Expose-Headers，表示允许暴露给客户端的自定义响应报头列表。PreflightMaxAge 属性对应的 Access-Control-Max-Age 报头表示预检响应缓存的时间。如果允许跨域请求携带与用户凭证相关的报头，SupportsCredentials 属性就会返回 True，在此情况下的响应具有一个值为 True 的 Access-Control-Allow-Credentials 报头。如果 VaryByOrigin 属性返回 True，响应将具有一个值为 Origin 的 Vary 报头，指示针对请求的"域"对响应报文实施缓存。

ICorsService

ICorsService 接口表示的服务旨在完成两方面的任务：其一，根据指定的 CORS 策略对跨域资源请求实施授权检验，并最终得到表示授权结果的 CorsResult 对象，这个任务实现在它的 EvaluatePolicy 方法中；其二，将授权结果以报头的形式应用到当前的响应报文中，这项任务是通过调用 ApplyResult 方法来完成的。

```
public interface ICorsService
{
    CorsResult EvaluatePolicy(HttpContext context, CorsPolicy policy);
    void ApplyResult(CorsResult result, HttpResponse response);
}
```

CORS 模型定义如下所示的 CorsService 类型作为对 ICorsService 接口的默认实现，它将具体的授权检验分别实现在 EvaluateRequest 方法和 EvaluatePreflightRequest 方法中，前者针对简单请求，后者针对预检请求。由于前面已经对 CORS 检验规则做了详细介绍，针对实现在 CorsService 中具体的授权检验和响应逻辑在这里就不再赘述。

```
public class CorsService : ICorsService
{
    public CorsService(IOptions<CorsOptions> options);
    public CorsService(IOptions<CorsOptions> options, ILoggerFactory loggerFactory);

    public CorsResult EvaluatePolicy(HttpContext context, CorsPolicy policy);
    public CorsResult EvaluatePolicy(HttpContext context, string policyName);

    public virtual void EvaluatePreflightRequest(HttpContext context, CorsPolicy policy,
        CorsResult result);
    public virtual void EvaluateRequest(HttpContext context, CorsPolicy policy,
        CorsResult result);
```

```
    public virtual void ApplyResult(CorsResult result, HttpResponse response);
}
```

服务注册

CorsMiddleware 中间件在处理跨域资源请求时依赖的服务通过 IServiceCollection 接口的两个 AddCors 扩展方法来注册。如下面的代码片段所示，这两个扩展方法重载完成了针对 ICorsService 接口和 ICorsPolicyProvider 接口的服务注册。我们还可以利用参数 setupAction 提供的 Action<CorsOptions> 对象对配置选项做进一步设置，如利用它来注册 CORS 策略。

```
public static class CorsServiceCollectionExtensions
{
    public static IServiceCollection AddCors(this IServiceCollection services)
    {
        services.AddOptions();
        services.TryAdd(ServiceDescriptor.Transient<ICorsService, CorsService>());
        services.TryAdd(ServiceDescriptor
            .Transient<ICorsPolicyProvider, DefaultCorsPolicyProvider>());
        return services;
    }

    public static IServiceCollection AddCors(this IServiceCollection services,
        Action<CorsOptions> setupAction)
    {
        services.AddCors();
        services.Configure<CorsOptions>(setupAction);
        return services;
    }
}
```

21.3.3　CorsMiddleware 中间件

由于核心操作都实现在 ICorsService 服务中，所以 CorsMiddleware 中间件最终会利用它来处理跨域请求，也就是说，CorsMiddleware 类型的 3 个构造函数都需要注入 ICorsService 服务。由于 CorsMiddleware 中间件总是采用一个具体的 CORS 策略来对跨域资源请求实施授权，所以在初始化该中间件时还需要提供一个表示 CORS 策略的 CorsPolicy 对象。

```
public class CorsMiddleware
{
    public CorsMiddleware(RequestDelegate next, ICorsService corsService,
        CorsPolicy policy);
    public CorsMiddleware(RequestDelegate next, ICorsService corsService,
        ICorsPolicyProvider policyProvider);
    public CorsMiddleware(RequestDelegate next, ICorsService corsService,
        ICorsPolicyProvider policyProvider, string policyName);

    public Task Invoke(HttpContext context);
}
```

CorsMiddleware 类型的 3 个构造函数分别体现了 3 种 CORS 策略的提供方式。调用第一个构造函数可以显式提供一个创建好的 CorsPolicy 对象，调用第二个构造函数可以由指定的 ICorsPolicyProvider 对象来提供一个默认的 CORS 策略。如果希望指定一个非默认的 CORS 策略，可以调用第三个构造函数显式指定 CORS 策略的注册名称。CorsMiddleware 中间件针对跨域资源请求的处理逻辑体现在如下所示的代码片段中。

```csharp
public class CorsMiddleware
{
    private readonly RequestDelegate        _next;
    private readonly ICorsService           _corsService;
    private readonly ICorsPolicyProvider    _corsPolicyProvider;
    private readonly CorsPolicy             _policy;
    private readonly string                 _corsPolicyName;

    public async Task Invoke(HttpContext context)
    {
        if (context.Request.Headers.ContainsKey("Origin"))
        {
            var corsPolicy = _policy
                ?? await _corsPolicyProvider?.GetPolicyAsync(context, _corsPolicyName);
            if (corsPolicy != null)
            {
                var corsResult = _corsService.EvaluatePolicy(context, corsPolicy);
                _corsService.ApplyResult(corsResult, context.Response);

                var accessControlRequestMethod =
                    context.Request.Headers["Access-Control-Request-Method"];
                if (string.Equals(
                        context.Request.Method,
                        "OPTIONS",
                        StringComparison.OrdinalIgnoreCase) &&
                    !StringValues.IsNullOrEmpty(accessControlRequestMethod))
                {
                    context.Response.StatusCode = 204;
                    return;
                }
            }
        }
        await _next(context);
    }
    ...
}
```

针对 CorsMiddleware 中间件的注册可以通过调用 IApplicationBuilder 接口的如下所示的 3 个 UseCors 扩展方法来完成，它们分别调用上述 3 个构造函数来创建注册的 CorsMiddleware 对象。第一个方法重载注册的中间件会采用注入的 ICorsPolicyProvider 对象提供的默认 CORS 策略；第二个方法重载可以利用提供的 CorsPolicyBuilder 对象来构建 CORS 策略；第三个方法重载注册的 CorsMiddleware 中间件会采用指定名称的 CORS 策略来处理跨域资源请求。

```
public static class CorsMiddlewareExtensions
{
    public static IApplicationBuilder UseCors(this IApplicationBuilder app);
    public static IApplicationBuilder UseCors(this IApplicationBuilder app,
        Action<CorsPolicyBuilder> configurePolicy);
    public static IApplicationBuilder UseCors(this IApplicationBuilder app,
        string policyName);
}
```

综上所述，CorsMiddleware 中间件最终会利用注册的 ICorsService 服务来处理跨域资源请求，ICorsService 服务旨在完成两方面任务：根据提供的通过 CorsPolicy 对象表示的 CORS 策略检验是否应该对请求的资源予以授权，并将最终通过 CorsResult 对象表示的授权结果落实到当前的响应报文中。CorsService 是对 ICorsService 接口的默认实现。

CorsMiddleware 中间件可以利用注册的 ICorsPolicyProvider 对象来提供表示 CORS 策略的 CorsPolicy 对象，对于作为默认实现的 DefaultCorsPolicyProvider 类型来说，它提供的 CORS 策略来源于承载配置选项的 CorsOptions 对象，所以可以利用 CorsOptions 对象来注册我们所需的 CORS 策略。当我们在创建表示 CORS 策略的 CorsPolicy 对象的时候，可以利用 CorsPolicyBuilder 对象以 Builder 模式进行构建。图 21-4 所示的 UML 展示了 CORS 模型的核心接口与类型。

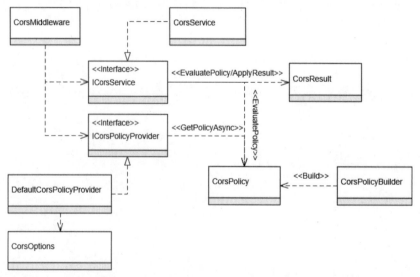

图 21-4　CORS 模型的核心接口与类型

第 22 章

本地化

如果要开发面向不同语种用户的网站,就不得不考虑本地化(Localization)和全球化(Globalization)的问题。本地化或者全球化涉及的范围很大,本章只关注语言选择的问题,即如何根据请求携带的语言文化信息来提供对应语言的字符串文本。这里主要涉及两个方面的功能实现:一是如何利用注册的中间件识别请求携带的语言文化信息,并利用它对当前执行上下文的语言文化属性进行定制;二是如何根据当前执行环境的语言文化属性来提供对应的字符串文本。

22.1 提供本地化消息文本

按照本书秉承的"体验先行"原则,在对 ASP.NET Core 应用针对本地化模块的设计和实现原理展开介绍之前,我们先介绍针对本地化的编程模式。下面通过一个简单的 ASP.NET Core MVC 应用来演示如何根据请求携带的语言文化信息来提供对应语种的消息文本。

22.1.1 提供对应语种的文本

.NET 应用通常采用资源文件来存储针对多语言的资源,资源文件在 .NET Core 中得到了很好的传承,所以下面演示的实例将针对多语种的字符串文本存储在相应的资源文件中。在演示的实例程序中,我们将字符串资源定义在图 22-1 所示的 SharedResource.resx 文件中。

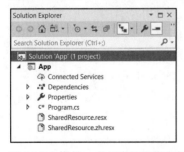

图 22-1 提供一个共享的资源文件

我们提供了两个同名的资源文件，SharedResource.zh.resx 文件存储的是针对中文的字符串文本，而另一个不带任何语言文化扩展名的 SharedResource.resx 文件则用来存储"语言文化中性"的资源。如图 22-2 所示，我们在这两个资源文件中定义了一个名为"Greeting"的字符串资源条目，具体的文本内容分别被定义为中文（"你好，世界！"）和英文（"Hello World!"）。

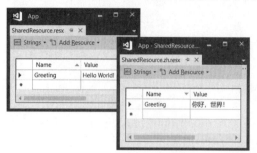

图 22-2　将针对不同语种的文本定义在对应的资源文件中

完成了针对资源文件的定义之后，下面需要完成相应的编程工作。从如下所示的代码片段可以看出，我们演示的实际上是一个 ASP.NET Core MVC 应用。在针对 IWebHostBuilder 接口的 ConfigureServices 扩展方法调用中，我们调用 IServiceCollection 接口的 AddLocalization 扩展方法注册了与本地化相关的服务。

```
public class Program
{
    public static void Main()
    {
        Host.CreateDefaultBuilder()
            .ConfigureWebHostDefaults(builder => builder
                .ConfigureServices(svcs => svcs
                    .AddLocalization()
                    .AddRouting()
                    .AddControllers())
                .Configure(app => app
                    .UseRouting()
                    .UseEndpoints(endpoints => endpoints.MapControllers())))
            .Build()
            .Run();
    }
}
```

我们为 ASP.NET Core MVC 应用提供了一个唯一的 Controller（HomeController）。如下面的代码片段所示，我们在 HomeController 的构造函数中注入了一个 IStringLocalizerFactory 服务，并调用它的 Create 方法创建了一个用于提供本地化字符串的 IStringLocalizer 对象。对于 Create 方法的两个参数，我们可以暂时将它们理解为资源文件名和所在程序集名称。

```
public class HomeController
{
    private readonly IStringLocalizer _localizer;
    public HomeController(IStringLocalizerFactory localizerFactory)
```

```csharp
    => _localizer = localizerFactory.Create("SharedResource", "App");

    [HttpGet("/")]
    public string Index(string culture)
    {
        if (!string.IsNullOrEmpty(culture))
        {
            CultureInfo.CurrentCulture =
            CultureInfo.CurrentUICulture =
            new CultureInfo(culture);
        }
        return _localizer.GetString("Greeting");
    }
}
```

在映射为根路径（"/"）的 Action 方法 Index 中，我们利用参数 culture 获得请求携带的语种信息，并利用它创建出相应的 CultureInfo 对象对当前线程的 CurrentCulture 属性和 CurrentUICulture 属性进行了设置。最后将文本资源条目名称"Greeting"作为参数调用 IStringLocalizer 对象的 GetString 扩展方法，并将返回的字符串作为当前 Action 方法的返回值。

根据 ASP.NET Core MVC 应用的参数绑定约定，我们可以利用请求的查询字符串来提供 Action 方法 Index 的参数 culture。如图 22-3 所示，如果请求携带的是针对中文的语言文化（"zh"或者"zh-CN"），那么响应的内容就是中文。如果请求没有对语言文化进行显式设置，或者设置的是一个不支持的语种（"fr-FR"），那么响应的内容会来源于 SharedResource.resx 这个"语言文化中性"的后备资源文件。（S2201）

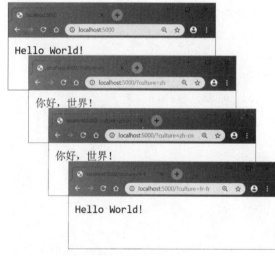

图 22-3　针对不同语言文化的响应

22.1.2　自动设置语言文化

在上面的实例演示中，由于我们在 Action 方法 Index 中通过手动设置当前线程的 Culture 属性和UICulture属性,所以本地化模块能够根据该设置选择出对应的资源文件。针对当前线程语言文化属性的设置其实可以利用 RequestLocalizationMiddleware 中间件来完成。为了演示 RequestLocalizationMiddleware 中间件针对当前语言文化上下文的自动化设置,我们对上面的程序做了如下改动。(S2202)

```
public class Program
{
    public static void Main()
    {
        Host.CreateDefaultBuilder()
            .ConfigureWebHostDefaults(builder => builder
                .ConfigureServices(svcs=>svcs
                    .AddLocalization()
                    .AddRouting()
                    .AddControllers())
                .Configure(app=>app
                    .UseRequestLocalization(options => options
                        .AddSupportedCultures("en", "zh")
                        .AddSupportedUICultures("en", "zh"))
                    .UseRouting()
                    .UseEndpoints(endpoints=>endpoints.MapControllers())))
            .Build()
            .Run();
    }
}
```

我们调用 IApplicationBuilder 对象的 UseRequestLocalization 扩展方法完成了对 RequestLocalizationMiddleware 中间件的注册。在注册该中间件的同时,我们将"en"和"zh"这两种语言文化添加到配置选项的 Culture 与 UICulture 列表中。由于 RequestLocalizationMiddleware 中间件帮助我们实现了针对当前线程语言文化的自动化设置,所以 HomeController 的 Action 方法 Index 方法只需要直接返回 IStringLocalizer 对象提供的本地化文本。再次运行修改后的程序,并以图 22-3 所示的方式利用查询字符串来指定语言文化,依然可以得到我们所需的本地化文本。

```
public class HomeController
{
    private readonly IStringLocalizer _localizer;
    public HomeController(IStringLocalizerFactory localizerFactory)
        => _localizer = localizerFactory.Create("SharedResource", "App");

    [HttpGet("/")]
    public string Index() => _localizer.GetString("Greeting");
}
```

针对本地化资源的请求除了可以采用查询字符串来指定语言文化,我们还有其他选择。HTTP 请求用一个名为 Accept-Language 的报头来表示客户端可以接受的语言,该报头是请求针对语言文化最贴切的表达。RequestLocalizationMiddleware 中间件在解析请求携带的语言文化信息时会将 Accept-Language 报头作为首选的来源。如下所示的代码片段是分别携带不同 Accept-Language 报头的 4 个请求访问我们的演示实例得到的响应。

```
GET http://localhost:5000/ HTTP/1.1
Host: localhost:5000

HTTP/1.1 200 OK
Date: Tue, 24 Sep 2019 13:35:02 GMT
Content-Type: text/plain; charset=utf-8
Server: Kestrel
Content-Length: 12

Hello World!


GET http://localhost:5000/ HTTP/1.1
Accept-Language: zh
Host: localhost:5000

HTTP/1.1 200 OK
Date: Tue, 24 Sep 2019 13:42:07 GMT
Content-Type: text/plain; charset=utf-8
Server: Kestrel
Content-Length: 18

你好,世界!


GET http://localhost:5000/ HTTP/1.1
Accept-Language: zh-CN
Host: localhost:5000

HTTP/1.1 200 OK
Date: Tue, 24 Sep 2019 13:50:03 GMT
Content-Type: text/plain; charset=utf-8
Server: Kestrel
Content-Length: 18

你好,世界!


GET http://localhost:5000/ HTTP/1.1
Accept-Language: fr-FR
Host: localhost:5000

HTTP/1.1 200 OK
Date: Tue, 24 Sep 2019 13:52:09 GMT
Content-Type: text/plain; charset=utf-8
Server: Kestrel
Content-Length: 12

Hello World!
```

HTTP 请求希望语言文化还可以通过 Cookie 的形式进行传递。具体的 Cookie 名称为.AspNetCore.Culture，我们可以将 Cookie 的值以 c={Culture}|uic={UICulture}的形式来指定 Culture 和 UICulture。针对 Cookie 对请求语言文化的设置体现在如下所示的 3 个 HTTP 请求和响应报文中。

```
GET http://localhost:5000/ HTTP/1.1
Cookie: .AspNetCore.Culture=c=zh|uic=zh
Host: localhost:5000

HTTP/1.1 200 OK
Date: Tue, 24 Sep 2019 13:56:55 GMT
Content-Type: text/plain; charset=utf-8
Server: Kestrel
Content-Length: 18

你好，世界!


GET http://localhost:5000/ HTTP/1.1
Cookie: .AspNetCore.Culture=c=zh-CN|uic=zh-CN
Host: localhost:5000

HTTP/1.1 200 OK
Date: Tue, 24 Sep 2019 14:30:04 GMT
Content-Type: text/plain; charset=utf-8
Server: Kestrel
Content-Length: 18

你好，世界!


GET http://localhost:5000/ HTTP/1.1
Cookie: .AspNetCore.Culture=c=fr-FR|uic=fr-FR
Host: localhost:5000

HTTP/1.1 200 OK
Date: Tue, 24 Sep 2019 14:33:07 GMT
Content-Type: text/plain; charset=utf-8
Server: Kestrel
Content-Length: 12

Hello World!
```

22.1.3　将本地化文本分而治之

上面演示的这个实例试图将整个应用涉及的本地化文本统一存储在一个资源文件中，对于一个简单的应用程序来说，这没有什么问题，但一旦涉及过多的本地化文本，这种单一存储方式就很难维护。为了解决这个问题，我们需要对涉及的本地化文本进行分组，并将它们分散到不同的资源文件中。如果将针对资源文件的拆分做到极致，我们就可以将每个类型中使用的本地化文本定义在独立的资源文件中。

上面演示的这个 ASP.NET Core MVC 应用仅仅提供了一个唯一的 Controller（HomeController），接下来我们额外定义两个 Controller 类型，并且将它们分别命名为 FooController 和 BarController。如下面的代码片段所示，我们为 FooController 类型和 BarController 类型定义了一个抽象基类 BaseController，它有一个 IStringLocalizer 类型只读属性 Localizer，并在构造函数中通过注册的 IStringLocalizerFactory 服务进行创建。

```
public abstract class BaseController
{
    public IStringLocalizer Localizer { get; }
    public BaseController(IStringLocalizerFactory localizerFactory)
        => Localizer = localizerFactory.Create(GetType());
}

public class FooController : BaseController
{
    public FooController(IStringLocalizerFactory localizerFactory) : base(localizerFactory)
    {}

    [HttpGet("/foo")]
    public string Index() => Localizer.GetString("Greeting");
}

public class BarController : BaseController
{
    public BarController(IStringLocalizerFactory localizerFactory) : base(localizerFactory)
    {}

    [HttpGet("/bar")]
    public string Index() => Localizer.GetString("Greeting");
}
```

在 BaseController 的构造函数中，我们调用 IStringLocalizerFactory 的 Create 方法重载来创建用于提供本地化文本的 IStringLocalizer 对象，当前 Controller 的类型会作为调用 Create 方法的参数。FooController 和 BarController 定义了唯一的 Action 方法 Index，它们直接返回 IStringLocalizer 对象的 GetString 扩展方法得到的本地化文本。

如果需要将 FooController 和 BarController 中使用的本地化字符串分别存储于各自的资源文件中，就可以按照图 22-4 所示的方式在当前的项目中创建两组以它们类型命名的资源文件。为了确定最终返回的本地化文本究竟来源于哪个资源文件，可以将字符串资源条目的文本添加上前缀"[Foo]"和"[Bar]"。

图 22-4　针对本地化文本的消费类型创建的资源文件

在启动演示程序之后，我们利用浏览器采用路由匹配的 URL（"/foo"和"/bar"）访问定义在 FooController 和 BarController 的两个 Action 方法，并采用查询字符串的方式指定客户端希望的语言文化。如图 22-5 所示，服务端能够根据请求的语言文化选择对应语种的响应文本，从具体文本内容可以看出，FooController 和 BarController 提供的本地化字符串来源于它们各自对应的资源文件。（S2203）

图 22-5　根据类型定位资源文件

22.1.4　直接注入 IStringLocalizer<T>

上面演示的 3 个实例都是在 Controller 类型的构造函数中注入 IStringLocalizerFactory 工厂来创建用于提供本地化字符串文本的 IStringLocalizer 对象的，但还有更简单的方式，就是直接注入泛型的 IStringLocalizer<T>服务。下面对上面演示程序中的两个 Controller 进行改造。如下面的代码片段所示，FooController 和 BarController 继承的基类变成了泛型的抽象类 BaseController<T>，我们在它们的构造函数中直接注入对应的 IStringLocalizer<T>服务，改造后的程序在功能上没有本质的区别。（S2204）

```
public abstract class BaseController<T>
{
    public IStringLocalizer Localizer { get; }
```

```csharp
    public BaseController(IStringLocalizer<T> localizer) => Localizer = localizer;
}

public class FooController : BaseController<FooController>
{
    public FooController(IStringLocalizer<FooController> localizer) : base(localizer)
    { }

    [HttpGet("/foo")]
    public string Index() => Localizer.GetString("Greeting");
}

public class BarController : BaseController<BarController>
{
    public BarController(IStringLocalizer<BarController> localizer) : base(localizer)
    { }

    [HttpGet("/bar")]
    public string Index() => Localizer.GetString("Greeting");
}
```

22.2 文本本地化

从上面演示的 3 个实例可以看出,应用程序返回的本地化文本是通过 IStringLocalizer 对象提供的,通过 IStringLocalizerFactory 接口表示的服务对象是提供 IStringLocalizer 对象的工厂。这两个核心对象连同表示本地化字符串的 LocalizedString 共同构成字符串本地化模型。

22.2.1 字符串本地化模型

构成字符串本地化模型的 3 个接口/类型(LocalizedString、IStringLocalizer 和 IStringLocalizerFactory)定义在 NuGet 包 "Microsoft.Extensions.Localization.Abstractions"中,下面先介绍由 IStringLocalizer 对象提供的表示本地化字符串的 LocalizedString 类型。

LocalizedString

表示本地化字符串的 LocalizedString 类型具有 4 个只读属性:表示名称和值(文本)的 Name 属性与 Value 属性必须在构造函数中显式指定;SearchedLocation 属性表示所在的位置,对于来源于资源文件的 LocalizedString 来说,该属性一般会返回内嵌在程序集中的资源文件的路径;如果根据名称在资源存储中找不到对应的字符串资源条目,返回的 LocalizedString 对象的 ResourceNotFound 属性为 True。

```csharp
public class LocalizedString
{
    public string      Name { get; }
    public string      Value { get; }
    public string      SearchedLocation { get; }
    public bool        ResourceNotFound { get; }
```

```
    ...
}
```

LocalizedString 类型用如下 3 个构造函数重载来初始化其必需的 Name 属性和 Value 属性，以及可选的 SearchedLocation 属性（默认为 Null）和 SearchedLocation 属性（默认为 False）。值得注意的是，即使调用第三个构造函数并将 ResourceNotFound 属性设置为 True，也不能将 Value 属性设置为 Null。

```
public class LocalizedString
{
    public LocalizedString(string name, string value);
    public LocalizedString(string name, string value, bool resourceNotFound);
    public LocalizedString(string name, string value, bool resourceNotFound,
        string searchedLocation);
    ...
}
```

LocalizedString 类型还重写了针对字符串类型的隐式类型转化器，当一个 LocalizedString 对象转换成一个字符串时，会直接返回其 Value 属性。除此之外，LocalizedString 类型还重写了 ToString 方法，该方法返回的依然是其 Value 属性。

```
public class LocalizedString
{
    public static implicit operator string(LocalizedString localizedString);
    public override string ToString();
    ...
}
```

IStringLocalizer

IStringLocalizer 对象以索引的形式提供 LocalizedString 对象，LocalizedString 对象的名称会作为索引键。在很多情况下，LocalizedString 对象的 Value 属性表示的文本可能是一个包含占位符（如{0}，{1}，…，{N}）的模板，所以 IStringLocalizer 接口定义的另一个索引使我们可以额外提供一组替换占位符的参数列表，索引最终会返回一个格式化的字符串。

```
public interface IStringLocalizer
{
    LocalizedString this[string name] { get; }
    LocalizedString this[string name, object[] arguments] { get; }

    IEnumerable<LocalizedString> GetAllStrings(bool includeParentCultures);
}
```

除了根据指定的名称来提供对应的本地化字符串，还可以调用 IStringLocalizer 对象的 GetAllStrings 方法得到由它提供的所有本地化字符串。该方法的 includeParentCultures 参数表示返回的集合是否包含针对"Parent Culture"的本地化字符串。例如，如果当前语言文化为"zh-CN"，那么调用 GetAllStrings 方法并将参数设置为 True，返回的结果可能会包括针对语言文化"zh"的 LocalizedString 对象。

如果根据消费类型来组织和管理本地化字符串，如演示实例中为每个 Controller 类型指定了单独的资源文件，就可以利用泛型的 IStringLocalizer<T>对象来提供与指定的泛型参数 T 关联的本地化字符串。如下面的代码片段所示，IStringLocalizer<T>直接派生于 IStringLocalizer 接口，自身并没有定义额外的成员。

```
public interface IStringLocalizer<T> : IStringLocalizer
{}
```

IStringLocalizer 接口还有如下 3 个扩展方法，其中就包括上面演示实例中多次使用的 GetString 方法，这 3 个方法本质上还是调用 IStringLocalizer 对象的索引和 GetAllStrings 方法而已。在调用 GetAllStrings 方法时会将 includeParentCultures 参数指定为 True。

```
public static class StringLocalizerExtensions
{
    public static LocalizedString GetString( this IStringLocalizer localizer,
        string name)
        => localizer [name];

    public static LocalizedString GetString( this IStringLocalizer localizer,
        string name, params object[] arguments)
        => localizer [name, arguments];

    public static IEnumerable<LocalizedString> GetAllStrings(
        this IStringLocalizer localizer)
        => stringLocalizer.GetAllStrings(includeParentCultures: true);
}
```

IStringLocalizerFactory

作为 IStringLocalizer 对象的工厂，IStringLocalizerFactory 对象支持两种创建 IStringLocalizer 对象的方式。如下面的代码片段所示，IStringLocalizerFactory 接口用来提供 IStringLocalizer 对象的两个 Create 方法是按照资源文件的语义来定义的。第一个 Create 方法重载根据资源类型来提供 IStringLocalizer 对象，第二个 Create 方法根据资源文件的路径和基础名称来提供 IStringLocalizer 对象。

```
public interface IStringLocalizerFactory
{
    IStringLocalizer Create(Type resourceSource);
    IStringLocalizer Create(string baseName, string location);
}
```

StringLocalizer<TResourceSource>

泛型 StringLocalizer<TResourceSource>是对 IStringLocalizer<TResourceSource>的默认实现。如下面的代码片段所示，一个 StringLocalizer<TResourceSource>对象实际上是对另一个 IStringLocalizer 对象的封装。在构造函数中，StringLocalizer<TResourceSource>利用注入的 IStringLocalizerFactory 工厂根据泛型参数类型创建封装的 IStringLocalizer 对象。

```
public class StringLocalizer<TResourceSource> : IStringLocalizer<TResourceSource>
{
    private IStringLocalizer _localizer;
    public StringLocalizer(IStringLocalizerFactory factory)
        => _localizer = factory.Create(typeof(TResourceSource));
```

```csharp
    public virtual IStringLocalizer WithCulture(CultureInfo culture)
        => _localizer.WithCulture(culture);
    public virtual LocalizedString this[string name]
    {
        get => _localizer[name];
    }

    public virtual LocalizedString this[string name, params object[] arguments]
    {
        get => _localizer[name, arguments];
    }

    public IEnumerable<LocalizedString> GetAllStrings(bool includeParentCultures)
        => _localizer.GetAllStrings(includeParentCultures);
}
```

综上所述,文本本地化模型通过 LocalizedString 类型来表示本地化字符串文本。LocalizedString 对象由通过 IStringLocalizer 接口表示的字符串本地化器提供,IStringLocalizerFactory 则表示创建或者提供 IStringLocalizer 对象的工厂。泛型的 IStringLocalizer<TResourceSource>接口表示与具体数据源关联的字符串本地化器,具体的数据源由作为泛型参数的 TResourceSource 类型来定位。作为对该接口默认实现的 StringLocalizer<TResourceSource>类型实际上是对另一个 IStringLocalizer 对象的封装,后者由指定的 IStringLocalizerFactory 对象根据作为泛型参数的 TResourceSource 类型创建。文本本地化模型的核心接口和类型以及它们之间的关系体现在图 22-6 所示的 UML 中。

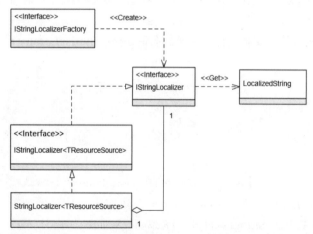

图 22-6　文本本地化模型的核心接口和类型以及它们之间的关系

22.2.2　基于 JSON 文件的本地化

到目前为止,读者对字符串本地化模型已经有了充分的了解,下面在此基础上介绍一个简单的针对 JSON 文件存储的实现,但需要先介绍这个扩展最终达到的效果。我们依然使用上面演示的实例，上面将 FooController 和 BarController 所需的本地化文本定义在各自对应的资源文件中，现在我们将它们定义在对应的 JSON 文件中。如图 22-7 所示，我们依然采用与资源文件类似的命名约定，可以将对应的 JSON 文件命名为 FooController.json 和 BarController.json。

图 22-7　将针对不同语种的文本定义在对应的 JSON 文件中

如下所示的代码片段就是分别定义在 JSON 文件 FooController.json 和 BarController.json 中的本地化文本。我们为每条本地化的文本条目定义一个 Value 属性表示"语言文化中性"的文本，针对具体语言文化的本地化文本则定义在 Translations 节点下面。

```
FooController.json
{
    "Greeting": {
        "Value": "[Foo]Hello World!",
        "Translations": {
            "zh": "[Foo]你好，世界",
            "fr": "[Foo]Bonjour le monde"
        }
    }
}

BarController.json
{
    "Greeting": {
        "Value": "[Bar]Hello World!",
        "Translations": {
            "zh": "[Bar]你好，世界",
            "fr": "[Bar]Bonjour le monde"
        }
    }
}
```

由于采用了基于 JSON 文件的本地化文本的定义方式，所以上面演示的程序就可以改写成如下形式。具体的差别体现在：我们调用 IServiceCollection 接口的 AddJsonLocalizer 扩展方法注册了自定义的针对 JSON 文件的 IStringLocalizerFactory 实现类型。由于 JSON 文件中定义了基于法语（"fr"）的本地化文本，所以我们在调用 IApplicationBuilder 接口的 UseRequestLocalization 扩展方法注册 RequestLocalizationMiddleware 中间件时添加了针对该语言的支持。启动改写的程序后，我们依然可以按照上面的方式呈现出与请求语种对应的文本内容。（S2205）

```
public class Program
```

```csharp
{
    public static void Main()
    {
        Host.CreateDefaultBuilder()
            .ConfigureWebHostDefaults(builder => builder
                .ConfigureServices((context, svcs)=>svcs
                    .AddLocalization()
                    .AddJsonLocalizer(context.HostingEnvironment.ContentRootFileProvider)
                    .AddRouting()
                    .AddControllers())
                .Configure(app=>app
                    .UseRequestLocalization(options => options
                        .AddSupportedCultures("en", "zh", "fr")
                        .AddSupportedUICultures("en", "zh", "fr"))
                    .UseRouting()
                    .UseEndpoints(endpoints=>endpoints.MapControllers())))
            .Build()
            .Run();
    }
}

public class FooController
{
    private readonly IStringLocalizer _localizer;
    public FooController(IStringLocalizer<FooController> localizer)
        => _localizer = localizer;

    [HttpGet("/foo")]
    public string Index() => _localizer.GetString("Greeting");
}

public class BarController
{
    private readonly IStringLocalizer _localizer;
    public BarController(IStringLocalizer<BarController> localizer)
        => _localizer = localizer;

    [HttpGet("/bar")]
    public string Index() => _localizer.GetString("Greeting");
}
```

下面介绍上面演示的基于 JSON 文件的本地化文本存储是如何实现的。首先为 JSON 文件中的本地化文本条目定义一个对应的 LocalizedStringEntry 类型。它的 Translations 属性通过一个字典表示对具体的语言文化的"翻译",该字典对象的 Key 属性和 Value 属性类型分别为 CultureInfo 与字符串。它的 Value 属性提供一条与具体语种无关的文本作为"后备"。

```csharp
public class LocalizedStringEntry
{
    public string                              Value { get; set; }
    public IDictionary<CultureInfo, string>    Translations { get; set; }
```

}

我们通过实现 IStringLocalizer 接口定义了如下所示的 DictionaryStringLocalizer 类型。之所以采用这样的命名是因为它提供的本地化文本是通过一个 Dictionary<string, LocalizedStringEntry>对象来承载的。

```
public class DictionaryStringLocalizer : IStringLocalizer
{
    private readonly Dictionary<string, LocalizedStringEntry>       _entries;

    public DictionaryStringLocalizer(
        Dictionary<string, LocalizedStringEntry> entries)
    {
        _entries = new Dictionary<string, LocalizedStringEntry>(entries);
        _culture = culture;
    }

    public LocalizedString this[string name]
        => GetString(name, _culture ?? CultureInfo.CurrentUICulture);

    public LocalizedString this[string name, params object[] arguments]
    {
        get
        {
            var raw = this[name];
            return raw.ResourceNotFound
                ? raw
                : new LocalizedString(name, string.Format(raw.Value, arguments));
        }
    }

    public IEnumerable<LocalizedString> GetAllStrings(bool includeParentCultures)
    {
        var culture = _culture ?? CultureInfo.CurrentUICulture;
        foreach (var item in _entries)
        {
            if (includeParentCultures)
            {
                yield return GetString(item.Key, culture);
            }
            else
            {
                yield return item.Value.Translations.TryGetValue(culture, out var text)
                    ? new LocalizedString(item.Key, text)
                    : new LocalizedString(item.Key, item.Key, true);
            }
        }
    }

    public IStringLocalizer WithCulture(CultureInfo culture)
        => new NotImplementedException (_entries, culture);
```

```csharp
    private LocalizedString GetString(string name, CultureInfo culture)
    {
        if (!_entries.TryGetValue(name, out var entry))
        {
            return new LocalizedString(name, name, true);
        }

        if (entry.Translations.TryGetValue(culture, out var message))
        {
            return new LocalizedString(name, message);
        }

        if (culture == CultureInfo.InvariantCulture)
        {
            return new LocalizedString(name, entry.Value, true);
        }

        return GetString(name, culture.Parent);
    }
}
```

上面的 DictionaryStringLocalizer 对象最终是通过对应的 JsonStringLocalizerFactory 工厂创建的。由于需要读取 JSON 文件来提供本地化文本,所以我们需要在创建 JsonStringLocalizerFactory 对象时提供一个 IFileProvider 对象。在实现的两个 Create 方法中,我们按照约定解析 JSON 文件的路径并利用 IFileProvider 对象读取文件的内容。读取的内容最后被反序列化成一个 Dictionary<string, LocalizedStringEntry>对象,利用这个字典创建出返回的 DictionaryStringLocalizer 对象。为了避免重复读取物理文件,我们将创建的 DictionaryStringLocalizer 对象进行了缓存。

```csharp
public class JsonStringLocalizerFactory : IStringLocalizerFactory
{
    private readonly ConcurrentDictionary<string, IStringLocalizer> _localizers;
    private readonly IFileProvider                                   _fileProvider;

    public JsonStringLocalizerFactory(IFileProvider fileProvider)
    {
        _localizers = new ConcurrentDictionary<string, IStringLocalizer>();
        _fileProvider = fileProvider;
    }

    public IStringLocalizer Create(Type resourceSource)
    {
        var path = ParseFilePath(resourceSource);
        return _localizers.GetOrAdd(path, _ =>
        {
            return CreateStringLocalizer(_);
        });
    }
```

```csharp
public IStringLocalizer Create(string baseName, string location)
{
    var path = ParseFilePath(location, baseName);
    return _localizers.GetOrAdd(path, _ =>
    {
        return CreateStringLocalizer(_);
    });
}

private IStringLocalizer CreateStringLocalizer(string path)
{
    var file = _fileProvider.GetFileInfo(path);
    if (!file.Exists)
    {
        return new DictionaryStringLocalizer(
            new Dictionary<string, LocalizedStringEntry>());
    }
    using (var stream = file.CreateReadStream())
    {
        var buffer = new byte[stream.Length];
        stream.Read(buffer, 0, buffer.Length);
        var dictionary = (Dictionary<string, LocalizedStringEntry>)
            JsonConvert.DeserializeObject(Encoding.UTF8.GetString(buffer),
            typeof(Dictionary<string, LocalizedStringEntry>));
        return new DictionaryStringLocalizer(dictionary);
    }
}

private string ParseFilePath(string location, string baseName)
{
    var path = location + "." + baseName;
    return path
        .Replace("..", ".")
        .Replace('.', Path.DirectorySeparatorChar) + ".json";
}

private string ParseFilePath(Type resourceSource)
{
    var rootNS = resourceSource.Assembly.GetCustomAttribute<RootNamespaceAttribute>()
        ?.RootNamespace ?? new AssemblyName(resourceSource.Assembly.FullName).Name;
    return resourceSource.FullName.StartsWith(rootNS)
        ? resourceSource.FullName.Substring(rootNS.Length + 1) + ".json"
        : resourceSource.FullName + ".json";
}
}
```

在演示实例中对 JsonStringLocalizerFactory 的注册定义在如下所示的针对 IServiceCollection 接口的 AddJsonLocalizer 扩展方法中。我们在该方法中利用指定的 IFileProvider 对象创建出 JsonStringLocalizerFactory 对象，并将其注册成 Singleton 服务实例。

```csharp
public static class ServiceCollectionExtensions
{
```

```
public static IServiceCollection AddJsonLocalizer(this IServiceCollection services,
    IFileProvider    fileProvider)
{
    services.Replace(ServiceDescriptor.Singleton<IStringLocalizerFactory>(
        new JsonStringLocalizerFactory(fileProvider)));
    return services;
}
```

22.2.3 基于资源文件的本地化

上面介绍了自定义的针对 JSON 文件的本地化实现，下面介绍默认提供的针对资源文件的本地化实现，后者的核心体现为通过 ResourceManagerStringLocalizer 类型表示的字符串本地化器。顾名思义，这个自定义的 ResourceManagerStringLocalizer 类型会利用一个 ResourceManager 对象来读取资源文件的内容。

ResourceManager

一般来说，在项目中定义的.resx 文件最终都会内嵌到编译后的程序集中。第 5 章提及，内嵌于程序集中的文件系统并没有目录的概念，所以.resx 文件针对项目根目录的路径最终都体现在内嵌的文件名上。如果需要利用 ResourceManager 加载某个内嵌的资源文件，就必须知道该文件的名称，所以我们有必要再次了解程序集在编译过程中是如何确定内嵌资源文件的文件名的。

假设我们创建了如图 22-8 所示的一个 .NET Core 项目，并在不同的目录下（"/" "/Foo" "/Foo/Bar"）添加了两个同名的资源文件（Resources.resx 和 Resources.zh.resx）。在项目文件 App.csproj 中，我们将程序集名称和根命名空间分别设置成 App 与 Artech.App。

```
<Project Sdk="Microsoft.NET.Sdk.Web">
  <PropertyGroup>
    ...
    <AssemblyName>App</AssemblyName>
    <RootNamespace>Artech.App</RootNamespace>
  </PropertyGroup>
  ...
</Project>
```

图 22-8　定义在不同目录下的资源文件

当这样一个项目被编译之后，生成程序集的清单（Manifest）会采用如下 3 个.mresource 条目来描述内嵌的 3 个"语言文化中性"资源文件（Resources.resx）。可以看出，以当前项目的根命名空间（不是程序集名称）作为前缀的内嵌资源文件的名称反映了原文件所在的路径（路径分隔符"/"替换成"."），具体的格式可以表示成"{RootNamespace}.{Path}.{FileName}"。

```
.mresource public Artech.App.Foo.Bar.Resources.resources
{
  // Offset: 0x00000000 Length: 0x000000B4
}
.mresource public Artech.App.Foo.Resources.resources
{
  // Offset: 0x000000B8 Length: 0x000000B4
}
.mresource public Artech.App.Resources.resources
{
  // Offset: 0x00000170 Length: 0x000000B4
}
```

针对具体语言文化的 3 个资源文件（Resources.zh.resx）会被编译到 zh\App.resources.dll 程序集中，这种专门用来承载与具有某种语言文化相关的程序集被称为卫星程序集（Satellite Assembly）。经过编译的卫星程序集名称会以.resources.dll 作为扩展名，并保存在对应语种命名的子目录下。可以通过部署卫星程序集的方式来实现针对某种语言文化的支持。我们编译的这个卫星程序集用类似的.mresource 条目来描述 3 个对应的内嵌资源文件。

```
.mresource public Artech.App.Foo.Bar.Resources.zh.resources
{
  //Offset: 0x00000000 Length: 0x000000B4
}
.mresource public Artech.App.Foo.Resources.zh.resources
{
  //Offset: 0x000000B8 Length: 0x000000B4
}
.mresource public Artech.App.Resources.zh.resources
{
  //Offset: 0x00000170 Length: 0x000000B4
}
```

在了解了编译器针对内嵌资源文件的命名规则之后，下面介绍用来读取资源文件的 ResourceManager 类型。如下面的代码片段所示，如果需要创建一个 ResourceManager 对象来读取内嵌在某个程序集中的资源文件，就需要指定所在程序集和内嵌资源文件的 BaseName，后者表示不包含扩展名（如.resources 或者 en-US.resources）的内嵌资源文件名称。如果目标资源文件为/Foo/Resources.resx，就需要将 BaseName 设置为"Artech.App.Foo.Resources"。

```
public class ResourceManager
{
    public ResourceManager(string baseName, Assembly assembly);
    public virtual string GetString(string name);
    public virtual string GetString(string name, CultureInfo culture);
    ...
```

对于定义在资源文件中的字符串文本，我们可以调用 ResourceManager 对象的两个 GetString 方法重载来读取它们。GetString 方法会根据指定的 CultureInfo 对象来定位对应的资源文件，如果在调用该方法时没有显式指定 CultureInfo 对象，那么当前线程的 UICulture（不是 Culture）会默认被使用。

ResourceManagerStringLocalizer

如下所示的代码片段是 ResourceManagerStringLocalizer 的定义。如给出的代码片段所示，我们在创建该对象时需要指定一个用来读取目标资源文件的 ResourceManager 对象，对于两个索引返回的 LocalizedString 对象来说，它们的本地化文本是通过调用 ResourceManager 对象的 GetString 方法返回的。

```
public class ResourceManagerStringLocalizer : IStringLocalizer
{
    public ResourceManagerStringLocalizer(ResourceManager resourceManager,
        Assembly resourceAssembly, string baseName, IResourceNamesCache resourceNamesCache,
        ILogger logger);

    public virtual IEnumerable<LocalizedString> GetAllStrings(bool includeParentCultures);

    public virtual LocalizedString this[string name] { get; }
    public virtual LocalizedString this[string name, object[] arguments] { get; }
    ...
}
```

ResourceManagerStringLocalizer 类型的构造函数还接受一个 IResourceNamesCache 接口的参数，该接口定义了如下所示的 GetOrAdd 方法，该方法对提取的本地化文本实施缓存。如下所示的 ResourceNamesCache 类型是对该接口的默认实现。在实现的 GetAllStrings 方法中，ResourceManagerStringLocalizer 会利用它对提取出来的本地化文本针对具体的语种实施缓存以提高性能。

```
public interface IResourceNamesCache
{
    IList<string> GetOrAdd(string name, Func<string, IList<string>> valueFactory);
}

public class ResourceNamesCache : IResourceNamesCache
{
    private readonly ConcurrentDictionary<string, IList<string>> _cache
        = new ConcurrentDictionary<string, IList<string>>();
    public IList<string> GetOrAdd(string name, Func<string, IList<string>> valueFactory)
        => _cache.GetOrAdd(name, valueFactory);
}
```

ResourceManagerStringLocalizerFactory

ResourceManagerStringLocalizer 由如下所示的 ResourceManagerStringLocalizerFactory 工厂来创建，该工厂类型涉及一个名为 LocalizationOptions 的配置选项，它的 ResourcesPath 属性表

示存放原始资源文件的根目录。如果需要将所有的资源文件统一存放到某个固定的目录下，就需要对该属性做相应设置。这个路径还可以利用如下所示的ResourceLocationAttribute特性进行设置。

```csharp
public class ResourceManagerStringLocalizerFactory : IStringLocalizerFactory
{
    public ResourceManagerStringLocalizerFactory(
        IOptions<LocalizationOptions> localizationOptions, ILoggerFactory loggerFactory);

    public IStringLocalizer Create(Type resourceSource);
    public IStringLocalizer Create(string baseName, string location);
}

public class LocalizationOptions
{
    public string ResourcesPath { get;  set; }
}

[AttributeUsage(AttributeTargets.Assembly, AllowMultiple=false, Inherited=false)]
public class ResourceLocationAttribute : Attribute
{
    public string ResourceLocation { get; }
    public ResourceLocationAttribute(string resourceLocation);
}
```

如果要创建一个 ResourceManagerStringLocalizer 对象，就必须提供一个具体的 ResourceManager 对象。如果要创建一个 ResourceManager 对象，就需要先明确资源文件所在的程序集和 BaseName。如果调用第二个 Create 方法重载来创建 ResourceManagerStringLocalizer 对象，ResourceManagerStringLocalizerFactory 工厂会将第二个参数（location）视为程序集名称。

如果 ResourceManagerStringLocalizer 对象是通过指定的类型创建的，ResourceManagerStringLocalizerFactory 就会将指定类型所在的程序集作为资源文件所在的程序集，所以需要确保两者的一致性。对资源文件 BaseName 的解析比较复杂，具体的逻辑可以通过如下这个公式来表示。如果找不到目标资源，就可以根据这个规则确定指定的类型与资源文件的路径是否匹配。

```
BaseName = {RootNameSpace} + {ResourcePath} + {TrimedTypeName}
RootNameSpace = 程序集根命名空间（如果没有，则用程序集名称代替）
ResourcePath = 通过 LocalizationOptions 或者 ResourceLocationAttribute
设置（优先选择）的路径（路径分隔符需要替换成"."）
TrimedTypeName = 指定类型的全名剔除程序集名（不是根命名空间）前缀
```

服务注册

针对 ResourceManagerStringLocalizerFactory 工厂的服务注册体现在如下这两个针对 IServiceCollection 接口的 AddLocalization 扩展方法上。从给出的代码片段可以看出，这两个方法还完成了针对 IStringLocalizer<>服务的注册。如果需要对资源文件的根目录进行设置，就可以调用第二个 AddLocalization 方法重载对作为配置选项的 LocalizationOptions 做相应的设置。

```csharp
public static class LocalizationServiceCollectionExtensions
{
    public static IServiceCollection AddLocalization(this IServiceCollection services)
    {
        services.AddOptions();
        AddLocalizationServices(services);
        return services;
    }

    public static IServiceCollection AddLocalization(
        this IServiceCollection services,
        Action<LocalizationOptions> setupAction)
    {
        AddLocalizationServices(services, setupAction);
        return services;
    }

    internal static void AddLocalizationServices(IServiceCollection services)
    {
        services.TryAddSingleton<IStringLocalizerFactory,
            ResourceManagerStringLocalizerFactory>();
        services.TryAddTransient(typeof(IStringLocalizer<>),
typeof(StringLocalizer<>));
    }

    internal static void AddLocalizationServices(
        IServiceCollection services,
        Action<LocalizationOptions> setupAction)
    {
        AddLocalizationServices(services);
        services.Configure(setupAction);
    }
}
```

22.3 当前语言文化的设置

IStringLocalizer 对象总是根据当前线程的语言文化来提供对应的本地化文本，所以在此之前请求携带的语言文化需要应用到当前线程上，此操作是由 RequestLocalizationMiddleware 中间件来完成的。该中间件及其相关类型由 NuGet 包 "Microsoft.AspNetCore.Localization" 来提供。在具体介绍 RequestLocalizationMiddleware 中间件之前，需要先了解与线程相关联的 Culture 属性与 UICulture 属性。

22.3.1 Culture 与 UICulture

通过 CultureInfo 对象表示的语言文化是描述线程执行上下文的一项重要信息，表示线程的 Thread 类型具有如下两个分别表示当前的 Culture 和 UICulture 的属性。一般来说，UICulture 属性决定采用的语种，而数据类型（如数字、日期时间和货币等）的格式、类型转换、排序等行为规则由 Culture 属性决定。

```
public sealed class Thread : CriticalFinalizerObject, _Thread
{
    public CultureInfo CurrentCulture { get; set; }
    public CultureInfo CurrentUICulture { get; set; }
    ...
}

public class CultureInfo : ICloneable, IFormatProvider
{
    public static CultureInfo CurrentCulture { get; set; }
    public static CultureInfo CurrentUICulture { get; set; }
    ...
}
```

如上面的代码片段所示，CultureInfo 类型具有两个同名的静态属性，它们返回的其实就是当前线程的 Culture 和 UICulture，两者之间的同一性可以通过如下所示的调试断言来证明。RequestLocalizationMiddleware 中间件需要做的就是从当前请求中提取出语言文化信息，并对这两个属性进行设置。

```
Debug.Assert(ReferenceEquals(CultureInfo.CurrentCulture,
    Thread.CurrentThread.CurrentCulture));
Debug.Assert(ReferenceEquals(CultureInfo.CurrentUICulture,
    Thread.CurrentThread.CurrentUICulture));
```

22.3.2　IRequestCultureProvider

RequestLocalizationMiddleware 中间件在对当前线程的 Culture 属性和 UICulture 属性进行设置之前，需要先确定当前请求携带了怎样的语言文化信息，这项工作由 IRequestCultureProvider 服务来完成。如下面的代码片段所示，IRequestCultureProvider 接口定义了唯一的 DetermineProviderCultureResult 方法，该方法会根据指定的 HttpContext 上下文来提取请求所携带的语言文化。

```
public interface IRequestCultureProvider
{
    Task<ProviderCultureResult> DetermineProviderCultureResult(HttpContext httpContext);
}
```

IRequestCultureProvider 对象针对语言文化的解析结果通过如下所示的 ProviderCultureResult 类型来表示，它的两个只读属性 Cultures 和 UICultures 分别表示解析出来的 Culture 与 UICulture 列表。由于请求携带的语言文化并不一定是唯一的，如请求报头 Accept-Language 可以携带多个具有不同权重的语言（如 "fr-CH, fr;q=0.9, en;q=0.8, de;q=0.7, *;q=0.5"），所以解析出来的 Culture 和 UICulture 通过元素为 StringSegment 的列表来表示。

```
public class ProviderCultureResult
{
    public IList<StringSegment> Cultures { get; }
    public IList<StringSegment> UICultures { get; }

    public ProviderCultureResult(StringSegment culture);
    public ProviderCultureResult(IList<StringSegment> cultures);
    public ProviderCultureResult(StringSegment culture, StringSegment uiCulture);
```

```csharp
    public ProviderCultureResult(IList<StringSegment> cultures,
        IList<StringSegment> uiCultures);
}
```

ProviderCultureResult 类型定义了 4 个构造函数来初始化它的 Cultures 属性和 UICultures 属性，如果调用构造函数只提供了 Culture（列表）或者 UICulture（列表），它们会同时赋值给这两个属性。如果 IRequestCultureProvider 对象只能从请求中解析出 Culture 或者 UICulture，就意味着当前线程 Culture 和 UICulture 将具有相同的设置。

如下所示的抽象类型 RequestCultureProvider 实现了 IRequestCultureProvider 接口，我们可以利用其 Options 属性来提取和设置相关的配置选项，后面会着重介绍对应的配置选项类型 RequestLocalizationOptions。这个抽象类提供了一个静态的 NullProviderCultureResult 属性，如果派生类实现的 DetermineProviderCultureResult 方法无法解析出明确的语言文化，可以直接返回这个属性表示的 Task<ProviderCultureResult>对象。NuGet 包 "Microsoft.AspNetCore.Localization" 提供了一系列针对 IRequestCultureProvider 接口的实现类型，它们都派生于这个抽象基类。

```csharp
public abstract class RequestCultureProvider : IRequestCultureProvider
{
    protected static readonly Task<ProviderCultureResult>   NullProviderCultureResult;
    public RequestLocalizationOptions                        Options { get; set; }

    public abstract Task<ProviderCultureResult> DetermineProviderCultureResult(
        HttpContext httpContext);
}
```

QueryStringRequestCultureProvider

在本章开篇的演示实例中，我们大都采用请求 URL 的查询字符串来指定希望的响应语言，之所以可以采用这种方式来指定语言文化是因为 QueryStringRequestCultureProvider 是默认注册的 IRequestCultureProvider 成员。如下面的代码片段所示，QueryStringRequestCultureProvider 派生于上面介绍的抽象基类 RequestCultureProvider，它的 QueryStringKey 属性和 UIQueryStringKey 属性分别表示 Culture 与 UICulture 对应的查询字符串的名称。

```csharp
public class QueryStringRequestCultureProvider : RequestCultureProvider
{
    public string QueryStringKey { get; set; }
    public string UIQueryStringKey { get; set; }

    public override Task<ProviderCultureResult> DetermineProviderCultureResult(
        HttpContext httpContext);
}
```

如果没有对 QueryStringKey 属性和 UIQueryStringKey 属性进行显式设置，那么它们的值为 "culture" 和 "ui-culture"。前面演示的实例在请求 URL 中以查询字符串 "?culture={culture}" 的形式来指定希望的响应语言，其实我们也可以替换成 "?ui-culture={culture}" 的形式。

CookieRequestCultureProvider

CookieRequestCultureProvider 对象会从请求提供的 Cookie 中解析出希望的语言文化。如下面的代码片段所示，CookieRequestCultureProvider 类型提供了一个 CookieName 属性，用来设置携带语言文化的 Cookie 名称，该属性的默认值为.AspNetCore.Culture。
```
public class CookieRequestCultureProvider : RequestCultureProvider
{
    public string CookieName { get; set; }
    public override Task<ProviderCultureResult> DetermineProviderCultureResult(
        HttpContext httpContext);
    public static string MakeCookieValue(RequestCulture requestCulture);
    public static ProviderCultureResult ParseCookieValue(string value);
    ...
}
public class RequestCulture
{
    public CultureInfo Culture { get; }
    public CultureInfo UICulture { get; }
    ...
}
```

Cookie 携带的语言文化信息必须采用"c={culture}|uic={ui-culture}"这样的格式来指定，所以前面演示的实例将 Cookie 设置成"Cookie: .AspNetCore.Culture=c=zh|uic=zh"的形式。Cookie RequestCultureProvider 针对 Cookie 值的生成和解析实现在它的两个静态方法 MakeCookieValue 和 ParseCookieValue 中，如下代码片段给出的调试断言体现了这两个方法的作用。

```
var culture = new RequestCulture(new CultureInfo("en-US"), new CultureInfo("zh-CN"));
var cookieValue = "c=en-US|uic=zh-CN";
var result = CookieRequestCultureProvider.ParseCookieValue(cookieValue);

Debug.Assert(cookieValue == CookieRequestCultureProvider.MakeCookieValue(culture));
Debug.Assert(result.Cultures.Single() == culture.Culture.ToString());
Debug.Assert(result.UICultures.Single() == culture.UICulture.ToString());
```

AcceptLanguageHeaderRequestCultureProvider

请求报头 Accept-Language 携带的语言列表是通过 AcceptLanguageHeaderRequestCultureProvider 对象提取出来的。如下面的代码片段所示，该类型具有一个名为 MaximumAcceptLanguageHeaderValuesToTry 的属性，用于设置最终接受的语种数量，该属性的默认值为 3。例如，请求的 Accept-Language 报头以 "en,zh,fr,ja" 的形式提供了 4 种语言，在默认情况下，AcceptLanguageHeaderRequestCultureProvider 只会选择前面 3 种作为候选语言。

```
public class AcceptLanguageHeaderRequestCultureProvider : RequestCultureProvider
{
    public int MaximumAcceptLanguageHeaderValuesToTry { get; set; }
    public override Task<ProviderCultureResult> DetermineProviderCultureResult(
        HttpContext httpContext);
}
```

由于请求报头 Accept-Language 携带的语言是可以包含权重的（如 en;q=0.2, zh;q=0.6, fr;q=0.4），所以 AcceptLanguageHeaderRequestCultureProvider 对象在根据 MaximumAcceptLanguageHeaderValuesToTry 属性设定的数量得到候选语言列表之后，它还会对它们针对权重进行排序（降序）。

RouteDataRequestCultureProvider

上面介绍的 3 个针对 IRequestCultureProvider 接口的实现类型都是由 NuGet 包"Microsoft.AspNetCore.Localization"提供的,并且是默认注册的。RouteDataRequestCultureProvider 类型则不同,它所在的 NuGet 包的名称为"Microsoft.AspNetCore.Localization.Routing"。顾名思义,RouteDataRequestCultureProvider 帮助我们提取通过路由参数表示的语言文化信息。由于 RouteDataRequestCultureProvider 对象并没有被默认注册,如果我们希望采用基于路由参数的语言文化解析方式,就只能对它进行显式注册。

与基于查询字符串的 QueryStringRequestCultureProvider 类似,RouteDataRequestCultureProvider 类型同样定义了两个属性(RouteDataStringKey 和 UIRouteDataStringKey)来设置 Culture 和 UICulture 对应的路由参数名,它们的默认值分别为 culture 和 ui-culture,这与针对查询字符串的表示是一致的。

```
public class RouteDataRequestCultureProvider : RequestCultureProvider
{
    public string RouteDataStringKey { get; set; }
    public string UIRouteDataStringKey { get; set; }

    public override Task<ProviderCultureResult> DetermineProviderCultureResult(
        HttpContext httpContext);
}
```

下面通过一个简单的实例来演示如何借助 RouteDataRequestCultureProvider 对象使我们可以利用路由参数来确定请求希望的语言。首先定义了一个如下所示的简单的中间件类型 FoobarMiddleware,并在构造函数中注入了一个 IStringLocalizer<FoobarMiddleware>对象。在用于处理请求的 InvokeAsync 方法中,我们利用此对象得到一个名为"Greeting"的本地化文本并写入响应。

```
public class FoobarMiddleware
{
    public FoobarMiddleware(RequestDelegate _) {}

    public Task InvokeAsync(HttpContext context,
        IStringLocalizer<FoobarMiddleware> localizer)
    {
        context.Response.ContentType = "text/plain;charset=utf-8";
        return context.Response.WriteAsync(localizer.GetString("Greeting"));
    }
}
```

在作为程序入口的 Main 方法中,我们注册了一个针对模板"/{ui-culture}"的路由,路由模板中定义的参数{ui-culture}表示的正是 RouteDataRequestCultureProvider 对象采用的表示 UICulture 属性的路由参数名称。我们针对这个路由分支注册了两个中间件:一个是用于设置当前语言文化属性的 RequestLocalizationMiddleware 中间件,另一个是上面定义的 FoobarMiddleware 中间件。

```
public class Program
{
    public static void Main()
```

```
{
    Host.CreateDefaultBuilder()
        .ConfigureWebHostDefaults(builder => builder
            .ConfigureServices(svcs=>svcs
                .AddLocalization()
                .AddRouting())
            .Configure(app=>app
                .UseRouting()
                .UseEndpoints(endpoints=>endpoints.MapGet("{ui-culture}", app.New()
                    .UseRequestLocalization(options => options
                        .AddSupportedCultures("zh", "en")
                        .AddSupportedUICultures("zh", "en")
                        .RequestCultureProviders.Insert(1,
                            new RouteDataRequestCultureProvider()))
                    .UseMiddleware<FoobarMiddleware>()
                    .Build())))
        .Build()
        .Run();
}
```

当我们调用 UseRequestLocalization 扩展方法注册 RequestLocalizationMiddleware 中间件时，除了指定支持的 Culture 和 UICulture（"en" 和 "zh"），还创建了一个 RouteDataRequestCultureProvider 对象，并将其插入 RequestLocalizationOptions 对象的 RequestCultureProviders 属性表示的 IRequestCultureProvider 列表中。值得注意的是，RouteDataRequestCultureProvider 对象是被作为第二个元素插入列表中的，具体的原因会在后面介绍 RequestLocalizationMiddleware 中间件时进行说明。

由于 FoobarMiddleware 中间件利用注入的 IStringLocalizer<FoobarMiddleware>对象来提供本地化文本，所以我们在当前项目中添加了图 22-9 所示的两个资源文件（FoobarMiddleware.resx 和 FoobarMiddleware.zh.resx），并在其中添加了名为 "Greeting" 的字符串资源条目。应用启动后，如果采用浏览器直接访问该应用，我们可以以图 22-10 所示的方式直接在请求 URL 的路径中指定我们希望的语言。（S2206）

图 22-9　为 FoobarMiddleware 中间件定义的资源文件

图 22-10 在请求 URL 的路径中指定语言

22.3.3　RequestLocalizationOptions

RequestLocalizationMiddleware 中间件涉及的相关配置选项定义在 RequestLocalizationOptions 类型中。如下面的代码片段所示，用来解析当前请求语言文化的 IRequestCultureProvider 对象就注册在由 RequestCultureProviders 属性表示的列表中。

```
public class RequestLocalizationOptions
{
    public IList<IRequestCultureProvider>      RequestCultureProviders { get; set; }
    public IList<CultureInfo>                  SupportedCultures { get; set; }
    public IList<CultureInfo>                  SupportedUICultures { get; set; }
    public bool                                FallBackToParentCultures { get; set; }
    public bool                                FallBackToParentUICultures { get; set; }
    public RequestCulture                      DefaultRequestCulture { get; set; }

    public RequestLocalizationOptions AddSupportedCultures(params string[] cultures);
    public RequestLocalizationOptions AddSupportedUICultures(params string[] uiCultures);
    public RequestLocalizationOptions SetDefaultCulture(string defaultCulture);
}
```

RequestLocalizationOptions 类型的 SupportedCultures 属性和 SupportedUICultures 属性表示当前应用支持的语言文化列表。当 RequestLocalizationMiddleware 中间件利用注册的 IRequestCultureProvider 对象解析出 Culture 列表和 UICulture 列表之后，需要与这个列表做一个交集，该交集会作为当前线程的候选语言文化。我们可以调用两个对应的方法 AddSupportedCultures 和 AddSupportedUICultures 将指定的语言文化添加到这两个列表之中。另外，需要支持多语言的应用应显式地将支持的语言文化添加到这两个列表中。

RequestLocalizationOptions 类型的两个布尔类型的属性 FallBackToParentCultures 和 FallBackToParentUICultures 会影响上述这个交集的计算逻辑。如果 IRequestCultureProvider 对象提供的语言文化（如 zh-CN）并不存在于当前支持的列表中，但是它的"Parent"（"zh"）是支持的语言之一，RequestLocalizationMiddleware 中间件就会选择其 Parent 作为当前的语言文化。如果应用程序并不希望采用这种基于 Parent 的语言文化后备（Fallback）策略，就可以将这两个属性（FallBackToParentCultures 和 FallBackToParentUICultures）显式设置为 False。

另外，表示语言文化的 CultureInfo 用一个 Parent 属性来表示自己的"父亲"。严格来讲，语言"zh-CN"的 Parent 是"zh-Hans",后者的 Parent 才是"zh",表示中性语言文化的 CultureInfo 是它们的"鼻祖"。上述这些语言文化的继承关系体现在如下所示的调试断言中。但是 RequestLocalizationMiddleware 中间件并不会利用 CultureInfo 的 Parent 来得到其"父亲",如果当前的语言文化为"foo-BAR",它总是将"foo"作为其 Parent。

```
Debug.Assert(new CultureInfo("zh-CN").Parent.Name == "zh-Hans");
Debug.Assert(new CultureInfo("zh-Hans").Parent.Name == "zh");
Debug.Assert(new CultureInfo("zh").Parent == CultureInfo.InvariantCulture);
```

如果 RequestLocalizationMiddleware 中间件并没有利用注册的 IRequestCultureProvider 对象提取出有效的 Culture 和 UICulture,或者提取出来的 Culture 和 UICulture 都没有在应用支持的语言文化范围内,此时它会使用通过 DefaultRequestCulture 属性表示的默认语言文化。默认的语言文化也可以通过调用 SetDefaultCulture 方法来指定。DefaultRequestCulture 属性会在 RequestLocalizationOptions 对象构造过程中被初始化,并且采用的是当前线程的 CurrentCulture 属性和 CurrentUICulture 属性。

22.3.4 RequestLocalizationMiddleware

如下所示的代码片段就是最终设置当前线程 Culture 和 UICulture 的 RequestLocalization Middleware 中间件的定义,可以看出,用来承载配置选项的 RequestLocalizationOptions 对象以 IOptions<RequestLocalizationOptions>服务的形式被注入构造函数中。

```
public class RequestLocalizationMiddleware
{
    public RequestLocalizationMiddleware(RequestDelegate next,
        IOptions<RequestLocalizationOptions> options);
    public Task Invoke(HttpContext context);
}
```

RequestLocalizationMiddleware 中间件实现在其 Invoke 方法中针对请求的处理逻辑很简单:首先,该中间件会从 RequestLocalizationOptions 对象中提取出所有注册的 IRequestCulture Provider 对象,并按照列表的顺序利用它们从当前请求中提取出对应的语言文化信息。如果提取出来的语言文化(Culture 或者 UICulture)刚好是当前应用支持的,那么 RequestLocalization Middleware 中间件会将它们分别设置为当前线程的 CurrentCulture 属性和 CurrentUICulture 属性。如果无法从请求中解析出有效的语言文化信息,那么 RequestLocalizationOptions 对象提供的默认设置将会被使用。

需要注意的是,RequestLocalizationMiddleware 中间件会严格按照顺序从 RequestLocalization Options 对象的 RequestCultureProviders 属性列表中提取 IRequestCultureProvider 对象,所以前面的 IRequestCultureProvider 对象具有优先选择权。默认注册的 3 个 IRequestCultureProvider 类型的顺序为 QueryStringRequestCultureProvider、CookieRequestCultureProvider 和 AcceptLanguageHeaderRequest CultureProvider,所以应用程序总是优先选择请求查询字符串携带的语言文化。这也是上面演示实例中将注册的 RouteDataRequestCultureProvider 对象插入 QueryStringRequestCultureProvider 对象和 CookieRequestCultureProvider 对象之间的原因。

第23章

健康检查

现代化的应用及服务的部署场景主要体现在集群化、微服务和容器化，这一切都建立在针对部署应用或者服务的健康检查上。提到健康检查，读者想到的可能就是通过发送"心跳"请求以确定目标应用或者服务的可用性。其实采用 ASP.NET Core 来开发 Web 应用或者服务，可以直接利用框架提供的原生健康检查功能。

23.1 检查应用的健康状况

ASP.NET Core 框架的健康检查功能是通过 HealthCheckMiddleware 中间件完成的。我们不仅可以利用该中间件确定当前应用的可用性，还可以注册相应的 IHealthCheck 对象来完成针对不同方面的健康检查。例如，当前应用部署或者依赖一些组件或者服务，我们可以注册相应的 IHealthCheck 对象来完成针对它们的健康检查。下面通过实例演示一些典型的健康检查应用场景。

23.1.1 确定当前应用是否可用

对于部署于集群或者容器的应用或者服务来说，它需要对外暴露一个终结点，以便负载均衡器或者容器编排框架可以利用该终结点确定是否可用。ASP.NET Core 应用可以通过如下所示的编程方式来提供这个健康检查终结点。

```
public class Program
{
    public static void Main()
    {
        Host.CreateDefaultBuilder()
            .ConfigureWebHostDefaults(builder => builder
                .ConfigureServices(svcs => svcs.AddHealthChecks())
                .Configure(app => app.UseHealthChecks("/healthcheck")))
            .Build()
            .Run();
    }
}
```

如上面的代码片段所示，我们调用 IApplicationBuilder 接口的 UseHealthChecks 扩展方法注册了 HealthCheckMiddleware 中间件，该方法提供的参数"/healthcheck"是为健康检查终结点指定的路径。HealthCheckMiddleware 中间件依赖的服务通过调用 IServiceCollection 接口的 AddHealthChecks 扩展方法进行注册。在程序正常运行的情况下，如果利用浏览器向注册的健康检查路径"/healthcheck"发送一个简单的 GET 请求，就可以得到图 23-1 所示的"健康报告"。（S2301）

图 23-1　健康检查结果

如下所示的代码片段是健康检查响应报文的内容。可以看出，这是一个状态码为"200OK"且媒体类型为 text/plain 的响应，其主体内容就是健康状态的字符串描述。在大部分情况下，发送健康检查请求希望得到的是目标应用或者服务当前实时的健康状况，所以响应报文是不应该被缓存的，如下所示的响应报文的 Cache-Control 报头和 Pragma 报头也体现了这一点。

```
HTTP/1.1 200 OK
Date: Sat, 29 Jun 2019 02:24:30 GMT
Content-Type: text/plain
Server: Kestrel
Cache-Control: no-store, no-cache
Pragma: no-cache
Expires: Thu, 01 Jan 1970 00:00:00 GMT
Content-Length: 7

Healthy
```

23.1.2　定制健康检查逻辑

对于前面演示的实例来说，只要应用正常启动，它就被视为"健康"（完全可用），这种情况有时候可能并不是我们希望的。有的时候应用在启动之后需要做一些初始化的工作，如需要将一些被频繁请求的资源加载到内存中，我们希望在这些初始化操作完成之前当前应用处于不可用的状态，那么正常的请求就不会被导流进来。这样的需求就需要我们自行实现具体的健康检查逻辑。

在如下所示的演示实例中，我们将健康检查的逻辑实现在内嵌的 Check 方法中，该方法会随机返回 3 种健康状态（Healthy、Unhealthy 和 Degraded）。在调用 IServiceCollection 接口的 AddHealthChecks 扩展方法注册所需依赖服务之后，我们进一步调用作为返回结果的 IHealthChecksBuilder 对象的 AddCheck 方法注册了一个 IHealthCheck 对象，该对象会调用 Check 方法来决定最终的健康状态。

```csharp
public class Program
{
    public static void Main()
    {
        var random = new Random();

        Host.CreateDefaultBuilder()
            .ConfigureWebHostDefaults(builder => builder
                .ConfigureServices(svcs => svcs.AddHealthChecks()
                    .AddCheck("default", Check))
                .Configure(app => app.UseHealthChecks("/healthcheck")))
            .Build()
            .Run();

        HealthCheckResult Check()
        {
            return (random.Next(1, 4)) switch
            {
                1 => HealthCheckResult.Unhealthy(),
                2 => HealthCheckResult.Degraded(),
                _ => HealthCheckResult.Healthy(),
            };
        }
    }
}
```

如下所示的代码片段是针对 3 种健康状态的响应报文，可以看出它们的状态码是不同的。具体来说，针对健康状态 Healthy 和 Degraded，响应码都是"200 OK"，因为此时的应用或者服务均会被视为可用（Available）状态，两者之间只是完全可用和部分可用的区别。状态为 Unhealthy 的服务被视为不可用（Unavailable），所以响应状态码为"Service Unavailable"。（S2302）

```
HTTP/1.1 200 OK
Date: Sat, 29 Jun 2019 02:46:29 GMT
Content-Type: text/plain
Server: Kestrel
Cache-Control: no-store, no-cache
Pragma: no-cache
Expires: Thu, 01 Jan 1970 00:00:00 GMT
Content-Length: 7

Healthy

HTTP/1.1 503 Service Unavailable
Date: Sat, 29 Jun 2019 02:47:00 GMT
Content-Type: text/plain
Server: Kestrel
Cache-Control: no-store, no-cache
```

```
Pragma: no-cache
Expires: Thu, 01 Jan 1970 00:00:00 GMT
Content-Length: 9

Unhealthy

HTTP/1.1 200 OK
Date: Sat, 29 Jun 2019 02:47:19 GMT
Content-Type: text/plain
Server: Kestrel
Cache-Control: no-store, no-cache
Pragma: no-cache
Expires: Thu, 01 Jan 1970 00:00:00 GMT
Content-Length: 8

Degraded
```

23.1.3 改变响应状态码

前面我们已经简单解释了 3 种健康状态与对应的响应状态码，我们可以选择不同的状态码。虽然健康检查默认响应状态码的设置是合理的，但是不能通过状态码来区分 Healthy 和 Unhealthy 这两种可用状态，可以通过如下所示的方式来改变默认的响应状态码设置。

```csharp
public class Program
{
    public static void Main()
    {
        var random = new Random();
        var options = new HealthCheckOptions
        {
            ResultStatusCodes = new Dictionary<HealthStatus, int>
            {
                [HealthStatus.Healthy] = 299,
                [HealthStatus.Degraded] = 298,
                [HealthStatus.Unhealthy] = 503
            }
        };

        Host.CreateDefaultBuilder()
            .ConfigureWebHostDefaults(builder => builder
                .ConfigureServices(svcs => svcs.AddHealthChecks()
                    .AddCheck("default", Check))
                .Configure(app => app.UseHealthChecks("/healthcheck", options)))
            .Build()
            .Run();

        HealthCheckResult Check()
        {
```

```
            return (random.Next(1, 4)) switch
            {
                1 => HealthCheckResult.Unhealthy(),
                2 => HealthCheckResult.Degraded(),
                _ => HealthCheckResult.Healthy(),
            };
        }
    }
}
```

如上面的代码片段所示，我们在调用 IApplicationBuilder 接口的 UseHealthChecks 扩展方法注册 HealthCheckMiddleware 中间件时提供了一个 HealthCheckOptions 对象作为对应的配置选项。HealthCheckOptions 对象通过 ResultStatusCodes 属性返回的字典维护了这 3 种健康状态与对应响应状态码之间的映射关系。在上面的演示实例中，我们将针对 Healthy 和 Unhealthy 这两种健康状态对应的响应状态码分别设置为"299"与"298"。新的修改体现在如下所示的响应报文中。（S2303）

```
HTTP/1.1 299
Date: Sat, 29 Jun 2019 03:06:16 GMT
Content-Type: text/plain
Server: Kestrel
Cache-Control: no-store, no-cache
Pragma: no-cache
Expires: Thu, 01 Jan 1970 00:00:00 GMT
Content-Length: 7

Healthy

HTTP/1.1 298
Date: Sat, 29 Jun 2019 03:06:15 GMT
Content-Type: text/plain
Server: Kestrel
Cache-Control: no-store, no-cache
Pragma: no-cache
Expires: Thu, 01 Jan 1970 00:00:00 GMT
Content-Length: 8

Degraded
```

23.1.4 细粒度的健康检查

如果当前应用承载或者依赖了若干组件或者服务，就可以针对它们做细粒度的健康检查。前面的演示实例通过注册 IHealthCheck 对象对应用级别的健康检查进行了定制，我们可以采用同样的形式为某个组件或者服务注册相应的 IHealthCheck 对象来确定它们的健康状况。

```
public class Program
{
    public static void Main()
```

```
{
    var random = new Random();
    Host.CreateDefaultBuilder()
        .ConfigureWebHostDefaults(builder => builder
            .ConfigureServices(svcs => svcs.AddHealthChecks()
                .AddCheck("foo", Check)
                .AddCheck("bar", Check)
                .AddCheck("baz", Check))
            .Configure(app => app.UseHealthChecks("/healthcheck")))
        .Build()
        .Run();

    HealthCheckResult Check()
    {
        return (random.Next(1, 4)) switch
        {
            1 => HealthCheckResult.Unhealthy(),
            2 => HealthCheckResult.Degraded(),
            _ => HealthCheckResult.Healthy(),
        };
    }
}
```

假设当前应用承载了 3 个服务，分别命名为 foo、bar 和 baz，我们可以采用如下所示的方式为它们注册 3 个 IHealthCheck 对象来完成针对它们的健康检查。由于注册的 3 个 IHealthCheck 对象采用同一个 Check 方法决定最后的健康状态，所以最终具有 27 种不同的组合。针对 3 个服务的 27 种健康状态组合最终会产生如下 3 种不同的响应报文。（S2304）

```
HTTP/1.1 200 OK
Date: Sat, 29 Jun 2019 13:42:44 GMT
Content-Type: text/plain
Server: Kestrel
Cache-Control: no-store, no-cache
Pragma: no-cache
Expires: Thu, 01 Jan 1970 00:00:00 GMT
Content-Length: 7

Healthy


HTTP/1.1 200 OK
Date: Sat, 29 Jun 2019 13:42:44 GMT
Content-Type: text/plain
Server: Kestrel
Cache-Control: no-store, no-cache
Pragma: no-cache
Expires: Thu, 01 Jan 1970 00:00:00 GMT
```

```
Content-Length: 8

Degraded

HTTP/1.1 503 Service Unavailable
Date: Sat, 29 Jun 2019 13:42:44 GMT
Content-Type: text/plain
Server: Kestrel
Cache-Control: no-store, no-cache
Pragma: no-cache
Expires: Thu, 01 Jan 1970 00:00:00 GMT
Content-Length: 9

Unhealthy
```

可以看出，健康检查响应并没有返回针对具体 3 个服务的健康状态，而是返回针对整个应用的整体健康状态，这个状态是根据 3 个服务当前的健康状态组合计算出来的。具体的计算逻辑很简单，按照严重程度，3 种健康状态的顺序应该是 Unhealthy > Degraded > Healthy，组合中最严重的健康状态就是应用整体的健康状态。

按照这个逻辑，如果应用的整体健康状态为 Healthy，就意味着 3 个服务的健康状态都是 Healthy；如果应用的整体健康状态为 Degraded，就意味着至少有一个服务的健康状态为 Degraded，并且没有 Unhealthy；如果其中某个服务的健康状态为 Unhealthy，应用的整体健康状态就是 Unhealthy。

23.1.5 定制响应内容

上面演示的实例虽然注册了相应的 IHealthCheck 对象来检验独立服务的健康状况，但是最终得到的依然是应用的整体健康状态，我们有可能得到一份详细的针对所有服务的"健康诊断书"。所以，我们将演示程序做了如下所示的改写。

```
public class Program
{
    public static void Main()
    {
        var random = new Random();

        var options = new HealthCheckOptions
        {
            ResponseWriter = ReportAsync
        };

        Host.CreateDefaultBuilder()
            .ConfigureWebHostDefaults(builder => builder
                .ConfigureServices(svcs => svcs.AddHealthChecks()
                    .AddCheck("foo", Check, new string[] { "foo1", "foo2" })
                    .AddCheck("bar", Check, new string[] { "bar1", "bar2" })
```

```csharp
            .AddCheck("baz", Check, new string[] { "baz1", "baz2" }))
        .Configure(app => app.UseHealthChecks("/healthcheck", options)))
    .Build()
    .Run();

static Task ReportAsync(HttpContext context, HealthReport report)
{
    context.Response.ContentType = "application/json";
    var settings = new JsonSerializerSettings();
    settings.Formatting = Formatting.Indented;
    settings.Converters.Add(new StringEnumConverter());
    return context.Response.WriteAsync(
        JsonConvert.SerializeObject(report, settings));
}

HealthCheckResult Check()
{
    return (random.Next(1, 4)) switch
    {
        1 => HealthCheckResult.Unhealthy("Unavailable"),
        2 => HealthCheckResult.Degraded("Degraded"),
        _ => HealthCheckResult.Healthy("Normal"),
    };
}
```

对演示实例的改写体现在以下几点：首先，为 Check 方法返回的表示健康检查结果的 HealthCheckResult 对象设置了对应的描述性文字（Normal、Degraded 和 Unavailable）；其次，在调用 IHealthChecksBuilder 接口的 AddCheck 方法注册相应的 IHealthCheck 对象时，指定了两个的标签（Tag），如针对服务 foo 的 IHealthCheck 对象的标签设置为 foo1 和 foo2；最后，在调用 IApplicationBuilder 接口的 UseHealthChecks 扩展方法注册 HealthCheckMiddleware 中间件时，指定了作为配置选项的 HealthCheckOptions 对象，并通过设置其 ResponseWriter 属性的方式完成了对健康报告的呈现。

HealthCheckOptions 的 ResponseWriter 属性返回的是一个 Func<HttpContext, HealthReport, Task>对象，其中 HealthReport 对象是 HealthCheckMiddleware 中间件针对健康报告的表达。我们设置的委托对象对应的方法为 ReportAsync，该方法会直接将指定的 HealthReport 对象序列化成 JSON 格式并作为响应的主体内容。我们并没有设置相应的状态码，所以可以直接在浏览器中看到图 23-2 所示的这份完整的健康报告。（S2305）

图 23-2 完整的健康报告

23.1.6 过滤 IHealthCheck 对象

HealthCheckMiddleware 中间件提取注册的 IHealthCheck 对象在完成具体的健康检查工作之前，我们可以对它们做进一步过滤。针对 IHealthCheck 对象的过滤逻辑依然是利用 HealthCheckOptions 配置选项来实现的，我们需要设置的是它的 Predicate 属性，该属性返回一个 Func<HealthCheckRegistration, bool>类型的委托对象，其中 HealthCheckRegistration 对象代表针对 IHealthCheck 对象的注册。

在前面演示的实例中，我们为注册的 IHealthCheck 对象指定了相应的标签，该标签不仅会出现在图 23-2 所示的健康报告中，还可以作为过滤条件。在如下所示的代码片段中，我们通过设置配置选项 HealthCheckOptions 对象的 Predicate 属性使之选择 Tag 前缀不为"baz"的 IHealthCheck 对象。

```
public class Program
{
```

```csharp
public static void Main()
{
    var random = new Random();
    var options = new HealthCheckOptions
    {
        ResponseWriter    = ReportAsync,
        Predicate = reg => reg.Tags.Any(
            tag => !tag.StartsWith("baz", StringComparison.OrdinalIgnoreCase))
    };

    Host.CreateDefaultBuilder()
        .ConfigureWebHostDefaults(builder => builder
            .ConfigureServices(svcs => svcs.AddHealthChecks()
                .AddCheck("foo", Check, new string[] { "foo1", "foo2" })
                .AddCheck("bar", Check, new string[] { "bar1", "bar2" })
                .AddCheck("baz", Check, new string[] { "baz1", "baz2" }))
            .Configure(app => app.UseHealthChecks("/healthcheck", options)))
        .Build()
        .Run();

    static Task ReportAsync(HttpContext context, HealthReport report)
    {
        context.Response.ContentType = "application/json";
        var settings = new JsonSerializerSettings
        {
            Formatting = Formatting.Indented
        };
        settings.Converters.Add(new StringEnumConverter());
        return context.Response.WriteAsync(
            JsonConvert.SerializeObject(report, settings));
    }

    HealthCheckResult Check()
    {
        return (random.Next(1, 4)) switch
        {
            1 => HealthCheckResult.Unhealthy("Unavailable"),
            2 => HealthCheckResult.Degraded("Degraded"),
            _ => HealthCheckResult.Healthy("Normal"),
        };
    }
}
```

由于我们设置的过滤规则相当于忽略了针对服务 baz 的健康检查，所以利用浏览器查看健康报告时就看不到对应的健康状态，如图 23-3 所示。（S2306）

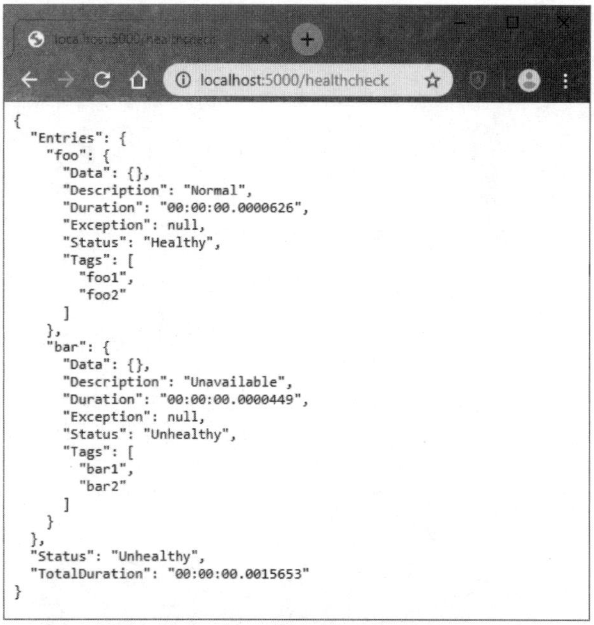

```
{
    "Entries": {
        "foo": {
            "Data": {},
            "Description": "Normal",
            "Duration": "00:00:00.0000626",
            "Exception": null,
            "Status": "Healthy",
            "Tags": [
                "foo1",
                "foo2"
            ]
        },
        "bar": {
            "Data": {},
            "Description": "Unavailable",
            "Duration": "00:00:00.0000449",
            "Exception": null,
            "Status": "Unhealthy",
            "Tags": [
                "bar1",
                "bar2"
            ]
        }
    },
    "Status": "Unhealthy",
    "TotalDuration": "00:00:00.0015653"
}
```

图 23-3　部分 IHealthCheck 过滤后的健康报告

23.2　设计与实现

前面的 6 个实例演示了健康检查的典型用法，下面深入介绍 HealthCheckMiddleware 中间件针对健康检查请求的处理逻辑。由于核心的健康检查由注册的 IHealthCheck 对象完成，所以下面先介绍这个核心对象。

23.2.1　IHealthCheck

如下所示的代码片段是 IHealthCheck 接口的定义，健康检查的逻辑就实现在 CheckHealthAsync 方法中。该方法将一个类型为 HealthCheckContext 的上下文对象作为健康检查的输入，而最终的诊断结果则通过一个 HealthCheckResult 对象来表示。一般来说，健康检查不是一个耗时的操作，或者如果健康检查本身花费了太多的时间，就意味着对应的应用或者服务是不健康的，所以 CheckHealthAsync 方法还提供了一个 CancellationToken 类型的参数来及时中止进行中的健康检查。

```
public interface IHealthCheck
{
    Task<HealthCheckResult> CheckHealthAsync(HealthCheckContext context,
        CancellationToken cancellationToken = new CancellationToken());
}
```

下面介绍表示健康检查结果的 HealthCheckResult 类型的定义。如下面的代码片段所示，这是一个结构，而不是一个类，设计者可能考虑到健康检查是一个高频的调用，所以将 Health

CheckResult 设计成一个结构以获得更好的性能。健康检查结果的核心是通过 Status 属性表示的健康状态,这是一个枚举类型,3 个枚举项体现了 3 种对应的健康状态。

```
public struct HealthCheckResult
{
    public HealthStatus                         Status { get; }
    public string                               Description { get; }
    public Exception                            Exception { get; }
    public IReadOnlyDictionary<string, object>  Data { get; }

    public HealthCheckResult(HealthStatus status, string description = null,
        Exception exception = null,
        IReadOnlyDictionary<string, object> data = null);
}

public enum HealthStatus
{
    Unhealthy,
    Degraded,
    Healthy
}
```

具体的 IHealthCheck 对象在创建 HealthCheckResult 对象的时候,除了可以指定必要的健康状态,还可以利用 Description 属性提供一些针对状态的描述,甚至可以在其 Data 属性表示的字典中添加任何的辅助数据。对于非健康状态(Degraded 和 Unhealthy)的两种结果,我们还可以将健康检查过程中捕获的异常赋值给 Exception 属性。我们虽然可以调用构造函数来创建 HealthCheckResult 对象,但在更多情况下我们倾向于调用如下 3 个静态方法来创建针对具体状态的 HealthCheckResult 对象。

```
public struct HealthCheckResult
{
    public static HealthCheckResult Healthy(string description = null,
        IReadOnlyDictionary<string, object> data = null);
    public static HealthCheckResult Degraded(string description = null,
        Exception exception = null,
        IReadOnlyDictionary<string, object> data = null);
    public static HealthCheckResult Unhealthy(string description = null,
        Exception exception = null,
        IReadOnlyDictionary<string, object> data = null);
}
```

作为健康检查输入的 HealthCheckContext 对象并不是对表示当前 HttpContext 上下文的封装,而是对当前 IHealthCheck 对象注册信息的封装,具体体现为 Registration 属性返回的 HealthCheckRegistration 对象。一个 HealthCheckRegistration 对象的核心是 Factory 属性返回 Func<IServiceProvider, IHealthCheck>委托,它根据指定的 IServiceProvider 对象为我们提供注册的 IHealthCheck 对象。我们在注册 IHealthCheck 对象的时候必须指定一个名称,该注册名称体现在对应 HealthCheckRegistration 对象的 Name 属性上。HealthCheckRegistration 类型的 FailureStatus 属性表示在健康检查操作失败的情况下应该采用的健康状态,在默认情况下该属性的值为Unhealthy，但是就微软目前提供的源代码来看，该属性似乎没有使用到。

```csharp
public sealed class HealthCheckContext
{
    public HealthCheckRegistration Registration { get; set; }
}

public sealed class HealthCheckRegistration
{
    public Func<IServiceProvider, IHealthCheck>     Factory { get; set; }
    public string                                   Name { get; set; }
    public HealthStatus                             FailureStatus { get; set; }
    public TimeSpan                                 Timeout { get; set; }
    public ISet<string>                             Tags { get; }

    public HealthCheckRegistration(string name, IHealthCheck instance,
        HealthStatus? failureStatus, IEnumerable<string> tags);
    public HealthCheckRegistration(string name,
        Func<IServiceProvider, IHealthCheck> factory,
        HealthStatus? failureStatus, IEnumerable<string> tags);
    public HealthCheckRegistration(string name, IHealthCheck instance,
        HealthStatus? failureStatus, IEnumerable<string> tags, TimeSpan? timeout);
    public HealthCheckRegistration(string name,
        Func<IServiceProvider, IHealthCheck> factory, HealthStatus? failureStatus,
        IEnumerable<string> tags, TimeSpan? timeout);
}
```

我们在注册 IHealthCheck 对象的时候还可以指定健康检查的超时时限，这个设置体现在对应 HealthCheckRegistration 对象的 Timeout 属性上。在 HealthCheckMiddleware 中间件调用对应 IHealthCheck 对象的 CheckHealthAsync 方法的时候，此超时时限将用来创建作为参数的 CancellationToken 对象。HealthCheckRegistration 类型的 Tags 属性返回的集合用来存放我们设置的标签，通过前面演示的实例可知，设置的标签不仅体现在最终的健康报告中，还可以参与定义过滤条件。

IHealthChecksBuilder

我们一般利用 IHealthChecksBuilder 对象来完成针对 IHealthCheck 对象的注册。如下面的代码片段所示，IHealthChecksBuilder 接口提供的 Add 方法用来添加一个 HealthCheckRegistration 对象。如果注册的 IHealthCheck 对象具有对其他服务的依赖，可以将依赖服务的注册添加到 Services 属性表示的 IServiceCollection 对象中。

```csharp
public interface IHealthChecksBuilder
{
    IServiceCollection Services { get; }
    IHealthChecksBuilder Add(HealthCheckRegistration registration);
}
```

如下所示的内部类型 HealthChecksBuilder 是对 IHealthChecksBuilder 接口的默认实现。从 Add 方法的定义可以看出，HealthChecksBuilder 对象自身并没有保存添加的 HealthCheckRegistration 对

象，而是将其存放到作为配置选项的 HealthCheckServiceOptions 对象中。

```csharp
internal class HealthChecksBuilder : IHealthChecksBuilder
{
    public IServiceCollection Services { get; }

    public HealthChecksBuilder(IServiceCollection services)
        => Services = services;

    public IHealthChecksBuilder Add(HealthCheckRegistration registration)
    {
        Services.Configure<HealthCheckServiceOptions>(options =>
        {
            options.Registrations.Add(registration);
        });

        return this;
    }
}

public sealed class HealthCheckServiceOptions
{
    public ICollection<HealthCheckRegistration> Registrations { get; }
}
```

IHealthChecksBuilder 接口如下所示的方法可以帮助我们完成对 IHealthCheck 对象的注册。我们既可以调用 AddCheck 扩展方法重载注册一个提供的 IHealthCheck 对象，也可以调用泛型的 AddCheck<T>方法或者 AddTypeActivatedCheck<T>方法注册 IHealthCheck 接口的实现类型。

```csharp
public static class HealthChecksBuilderAddCheckExtensions
{
    public static IHealthChecksBuilder AddCheck(this IHealthChecksBuilder builder,
        string name, IHealthCheck instance, HealthStatus? failureStatus = null,
        IEnumerable<string> tags = null, TimeSpan? timeout = null)
        => builder.Add(new HealthCheckRegistration(name, instance, failureStatus,
            tags, timeout));

    public static IHealthChecksBuilder AddCheck(this IHealthChecksBuilder builder,
        string name, IHealthCheck instance, HealthStatus? failureStatus,
        IEnumerable<string> tags)
        => AddCheck(builder, name, instance, failureStatus, tags, default);

    public static IHealthChecksBuilder AddCheck<T>(
        this IHealthChecksBuilder builder, string name,
        HealthStatus? failureStatus = null, IEnumerable<string> tags = null,
        TimeSpan? timeout = null) where T : class, IHealthCheck
        => builder.Add(new HealthCheckRegistration(name, s => ActivatorUtilities
            .GetServiceOrCreateInstance<T>(s), failureStatus, tags, timeout));

    public static IHealthChecksBuilder AddCheck<T>(
        this IHealthChecksBuilder builder, string name, HealthStatus? failureStatus,
```

```csharp
        IEnumerable<string> tags) where T : class, IHealthCheck
        => AddCheck<T>(builder, name, failureStatus, tags, default);

    public static IHealthChecksBuilder AddTypeActivatedCheck<T>(
        this IHealthChecksBuilder builder, string name,
        HealthStatus? failureStatus, IEnumerable<string> tags, TimeSpan timeout,
        params object[] args) where T : class, IHealthCheck
        => builder.Add(new HealthCheckRegistration(name, s => ActivatorUtilities
            .CreateInstance<T>(s, args), failureStatus, tags, timeout));

    public static IHealthChecksBuilder AddTypeActivatedCheck<T>(
        this IHealthChecksBuilder builder, string name, params object[] args)
        where T : class, IHealthCheck
        => AddTypeActivatedCheck<T>(builder, name, failureStatus: null,
            tags: null, args);

    public static IHealthChecksBuilder AddTypeActivatedCheck<T>(
        this IHealthChecksBuilder builder, string name,
        HealthStatus? failureStatus, params object[] args)
        where T : class, IHealthCheck
        => AddTypeActivatedCheck<T>(builder, name, failureStatus, tags: null, args);

    public static IHealthChecksBuilder AddTypeActivatedCheck<T>(
        this IHealthChecksBuilder builder, string name,
        HealthStatus? failureStatus, IEnumerable<string> tags, params object[] args)
        where T : class, IHealthCheck
        => builder.Add(new HealthCheckRegistration(name, s => ActivatorUtilities
            .CreateInstance<T>(s, args), failureStatus, tags));
}
```

AddCheck<T>方法和 GetServiceOrCreateInstance<T>方法的差异体现在它们利用依赖注入框架提供对应 IHealthCheck 对象的方式上。如果直接将 IHealthCheck 接口的实现类型或者实例注册到依赖注入框架中，AddCheck<T>方法调用的是 ActivatorUtilities 类型的 GetServiceOrCreateInstance<T>方法，意味着它会复用现有的 Scoped 服务条例或者 Singleton 服务实例。GetServiceOrCreateInstance<T>方法调用的是 ActivatorUtilities 类型的 CreateInstance<T>方法，意味着它总是创建一个新的 IHealthCheck 对象。

DelegateHealthCheck

DelegateHealthCheck 是对 IHealthCheck 接口的实现。顾名思义，一个 DelegateHealthCheck 对象利用提供的委托对象来完成健康检查工作。如下面的代码片段所示，这是一个内部类型，最终用来完成健康检查的是一个类型为 Func<CancellationToken, Task<HealthCheckResult>>的委托对象。

```csharp
internal sealed class DelegateHealthCheck : IHealthCheck
{
    private readonly Func<CancellationToken, Task<HealthCheckResult>> _check;
    public DelegateHealthCheck(
        Func<CancellationToken, Task<HealthCheckResult>> check)
```

```csharp
        =>_check = check;

    public Task<HealthCheckResult> CheckHealthAsync(HealthCheckContext context,
        CancellationToken cancellationToken = default)
        => _check(cancellationToken);
}
```

IHealthChecksBuilder 接口的如下这些扩展方法最终注册的都是一个 DelegateHealthCheck 对象。这些扩展方法都要求提供一个用来实施健康检查的委托对象，具体的委托对象可以是直接提供给 DelegateHealthCheck 对象的 Func<CancellationToken, Task<HealthCheckResult>>对象，还可以是更简单的 Func<Task<HealthCheckResult>>对象，甚至是 Func<HealthCheckResult>对象和 Func<CancellationToken, HealthCheckResult>对象。

```csharp
public static class HealthChecksBuilderDelegateExtensions
{
    public static IHealthChecksBuilder AddCheck(this IHealthChecksBuilder builder,
        string name, Func<HealthCheckResult> check, IEnumerable<string> tags = null,
        TimeSpan? timeout = default)
    {
        var instance = new DelegateHealthCheck((ct) => Task.FromResult(check()));
        return builder.Add(new HealthCheckRegistration(name, instance,
            failureStatus: null, tags, timeout));
    }

    public static IHealthChecksBuilder AddCheck(this IHealthChecksBuilder builder,
        string name, Func<HealthCheckResult> check, IEnumerable<string> tags)
        => AddCheck(builder, name, check, tags, default);

    public static IHealthChecksBuilder AddCheck(this IHealthChecksBuilder builder,
        string name, Func<CancellationToken, HealthCheckResult> check,
        IEnumerable<string> tags = null, TimeSpan? timeout = default)
    {
        var instance = new DelegateHealthCheck((ct) => Task.FromResult(check(ct)));
        return builder.Add(new HealthCheckRegistration(name, instance,
            failureStatus: null, tags, timeout));
    }

    public static IHealthChecksBuilder AddCheck(this IHealthChecksBuilder builder,
        string name, Func<CancellationToken, HealthCheckResult> check,
        IEnumerable<string> tags)
        => AddCheck(builder, name, check, tags, default);

    public static IHealthChecksBuilder AddAsyncCheck(
        this IHealthChecksBuilder builder, string name,
        Func<Task<HealthCheckResult>> check,
        IEnumerable<string> tags = null, TimeSpan? timeout = default)
    {
        var instance = new DelegateHealthCheck((ct) => check());
        return builder.Add(new HealthCheckRegistration(name, instance,
            failureStatus: null, tags, timeout));
```

```csharp
}

public static IHealthChecksBuilder AddAsyncCheck(
    this IHealthChecksBuilder builder, string name,
    Func<Task<HealthCheckResult>> check, IEnumerable<string> tags)
    => AddAsyncCheck(builder, name, check, tags, default);

public static IHealthChecksBuilder AddAsyncCheck(
    this IHealthChecksBuilder builder, string name,
    Func<CancellationToken, Task<HealthCheckResult>> check,
    IEnumerable<string> tags = null, TimeSpan? timeout = default)
{
    var instance = new DelegateHealthCheck((ct) => check(ct));
    return builder.Add(new HealthCheckRegistration(name, instance,
        failureStatus: null, tags, timeout));
}

public static IHealthChecksBuilder AddAsyncCheck(
    this IHealthChecksBuilder builder, string name,
    Func<CancellationToken, Task<HealthCheckResult>> check,
    IEnumerable<string> tags)
    => AddAsyncCheck(builder, name, check, tags, default);
}
```

23.2.2 HealthCheckService

HealthCheckMiddleware 中间件其实并没有直接采用注册的 IHealthCheck 对象来实施健康检查，而是间接地利用 HealthCheckService 服务来驱动注册的 IHealthCheck 对象进行健康检查。如下面的代码片段所示，抽象类 HealthCheckService 定义了两个用来完成健康检查的 CheckHealthAsync 方法重载。

```csharp
public abstract class HealthCheckService
{
    public Task<HealthReport> CheckHealthAsync(CancellationToken cancellationToken
        = new CancellationToken())
        => CheckHealthAsync(null, cancellationToken);

    public abstract Task<HealthReport> CheckHealthAsync(
        Func<HealthCheckRegistration, bool> predicate,
        CancellationToken cancellationToken = new CancellationToken());
}
```

抽象 CheckHealthAsync 方法的第一个参数类型为 Func<HealthCheckRegistration, bool>，该委托对象用来对注册的 IHealthCheck 对象进行过滤。该方法执行之后会得到一份完整的健康报告，该报告用如下所示的 HealthReport 类型表示。

```csharp
public sealed class HealthReport
{
    public IReadOnlyDictionary<string, HealthReportEntry>   Entries { get; }
    public HealthStatus                                      Status { get; }
    public TimeSpan                                          TotalDuration { get; }
```

```
    public HealthReport(IReadOnlyDictionary<string, HealthReportEntry> entries,
        TimeSpan totalDuration);
}
```

HealthCheckService 服务总是驱动注册的 IHealthCheck 对象完成最终的健康检查。通过前面的介绍可知,IHealthCheck 对象在完成健康检查之后会将结果封装成一个 HealthCheckResult 对象。HealthCheckService 服务在生成健康报告的时候,会将 HealthCheckResult 对象转换成如下所示的代表健康报告条目的 HealthReportEntry 对象。生成的 HealthReportEntry 对象会存放到 Entries 属性表示的字典中,该字典的 Key 就是注册 IHealthCheck 对象时指定的名称,即对应 HealthCheckRegistration 对象的 Name 属性。

```
public struct HealthReportEntry
{
    public HealthStatus                                 Status { get; }
    public string                                       Description { get; }
    public TimeSpan                                     Duration { get; }
    public Exception                                    Exception { get; }
    public IReadOnlyDictionary<string, object>          Data { get; }
    public IEnumerable<string>                          Tags { get; }

    public HealthReportEntry(HealthStatus status, string description,
        TimeSpan duration, Exception exception,
        IReadOnlyDictionary<string, object> data);
    public HealthReportEntry(HealthStatus status, string description,
        TimeSpan duration, Exception exception,
        IReadOnlyDictionary<string, object> data, IEnumerable<string> tags = null);
}
```

HealthReportEntry 对象的 Status 属性、Description 属性、Duration 属性和 Data 属性均来源于 HealthCheckResult 对象的同名属性,表示标签的 Tags 来源于 HealthCheckRegistration 对象的同名属性。HealthReport 类型的 Status 属性表示应用整体的健康状态,正如前面实例演示的那样,该状态与所有 HealthReportEntry 条目的最严重健康状态一致。HealthReport 类型还定义了一个 TotalDuration 属性,表示执行整个健康检查所花费的时间。

如下所示的内部 DefaultHealthCheckService 行继承了抽象类 HealthCheckService。从给出的代码片段可以看出,表示配置选项的 HealthCheckServiceOptions 对象采用 Options 模式注入构造函数中。通过前面对 HealthChecksBuilder 的介绍可知,为注册的 IHealthCheck 对象创建的 HealthCheckRegistration 对象正是存放在这个 HealthCheckServiceOptions 对象中。

```
internal class DefaultHealthCheckService : HealthCheckService
{
    public DefaultHealthCheckService(IServiceScopeFactory scopeFactory,
        IOptions<HealthCheckServiceOptions> options,
        ILogger<DefaultHealthCheckService> logger);

    public override Task<HealthReport> CheckHealthAsync(
        Func<HealthCheckRegistration, bool> predicate,
        CancellationToken cancellationToken = new CancellationToken());
}
```

在如下所示的代码片段中，我们采用简单的代码模拟了 DefaultHealthCheckService 服务针对健康检查的实现。实现在 CheckHealthAsync 方法中的健康检查逻辑非常简单：DefaultHealthCheckService 对象利用构造函数中注入的 HealthCheckServiceOptions 配置选项得到针对所有 IHealthCheck 对象的注册（体现为一系列的 HealthCheckRegistration 对象），然后根据方法传入的 Func<HealthCheckRegistration, bool>对它们做进一步过滤。DefaultHealthCheckService 对象接下来会创建出 HealthCheckContext 对象执行上下文，并调用注册的每个 IHealthCheck 对象的 CheckHealthAsync 方法在此上下文中完成各自的检查工作。返回的这些 HealthReportEntry 对象会用来创建表示健康报告的 HealthReport 对象。

```csharp
internal class DefaultHealthCheckService : HealthCheckService
{
    private readonly IServiceScopeFactory                        _scopeFactory;
    private readonly ICollection<HealthCheckRegistration>        _registrations;

    public DefaultHealthCheckService(IServiceScopeFactory scopeFactory,
        IOptions<HealthCheckServiceOptions> options)
    {
        _scopeFactory       = scopeFactory;
        _registrations      = options.Value.Registrations;
    }

    public override async Task<HealthReport> CheckHealthAsync(
        Func<HealthCheckRegistration, bool> predicate,
        CancellationToken cancellationToken = default)
    {
        var registrations = predicate == null
            ? _registrations
            : _registrations.Where(predicate);

        var stopwatch = Stopwatch.StartNew();
        using (var scope = _scopeFactory.CreateScope())
        {
            var tasks = registrations.Select(registration => RunCheckAsync(
                scope.ServiceProvider, registration, cancellationToken));
            var result = await Task.WhenAll(tasks);
            return new HealthReport(result.ToDictionary(it => it.Name,
                it => it.Entry), stopwatch.Elapsed);
        }
    }

    private async Task<(string Name, HealthReportEntry Entry)> RunCheckAsync(
        IServiceProvider serviceProvider, HealthCheckRegistration registration,
        CancellationToken cancellationToken)
    {
        cancellationToken.ThrowIfCancellationRequested();
        var check = registration.Factory(serviceProvider);
```

```csharp
    var stopwatch = Stopwatch.StartNew();
    var context = new HealthCheckContext
    {
        Registration = registration
    };
    HealthReportEntry entry;
    CancellationTokenSource tokenSource = null;
    try
    {
        var token = cancellationToken;
        if (registration.Timeout > TimeSpan.Zero)
        {
            tokenSource = CancellationTokenSource
                .CreateLinkedTokenSource(cancellationToken);
            tokenSource.CancelAfter(registration.Timeout);
            token = tokenSource.Token;
        }
        var result = await check.CheckHealthAsync(context, token);
        entry = new HealthReportEntry(result.Status, result.Description,
            stopwatch.Elapsed, result.Exception, result.Data, registration.Tags);
        return (registration.Name, entry);
    }
    catch (OperationCanceledException ex)
        when (!cancellationToken.IsCancellationRequested)
    {
        entry = new HealthReportEntry(HealthStatus.Unhealthy,
            "A timeout occured while running check.", stopwatch.Elapsed, ex,
            null);
    }
    catch (Exception ex) when (ex as OperationCanceledException == null)
    {
        entry = new HealthReportEntry(HealthStatus.Unhealthy, ex.Message,
            stopwatch.Elapsed, ex, null);
    }
    finally
    {
        tokenSource?.Dispose();
    }
    return (registration.Name, entry);
}
```

DefaultHealthCheckService 构造函数中注入了 IServiceScopeFactory 工厂，当 CheckHealthAsync 方法在利用 HealthCheckRegistration 对象来提供对应注册的 IHealthCheck 对象的时候，它会利用这个 IServiceScopeFactory 工厂创建一个代表服务范围的 IServiceScope 对象，所需的 IServiceProvider 对象正是由这个 IServiceScope 对象提供的。这样做是为了确保依赖的服务能够根据其注册的生命周期模式得到释放。

如果某个 IHealthCheck 对象在进行健康检查的过程中抛出异常，它就会返回一个健康状态为 Unhealthy 的对象，笔者认为这里的实现有待商榷。由于 HealthCheckRegistration 类型中定义的 FailureStatus 属性用来表示在健康检查失败的情况下采用的健康状态，所以该属性似乎应该应用在这里。

23.2.3　HealthCheckMiddleware

在正式介绍 HealthCheckMiddleware 中间件之前，需要先了解该中间件采用的配置选项类型 HealthCheckOptions。如下面的代码片段所示，HealthCheckOptions 类型的 Predicate 属性返回的是一个类型为 Func<HealthCheckRegistration, bool>的委托对象，用来对注册的 IHealthCheck 对象实施过滤，该属性的默认值为 Null，意味着在默认情况下所有注册的 IHealthCheck 对象都将被使用。ResultStatusCodes 属性返回的字典提供了健康状态与最终响应状态码的映射关系，在默认情况下，Healthy 和 Degraded 这两种可用的状态都会产生一个状态码为 "200 OK" 的响应，而 Unhealthy 状态代表应用或者服务不可用，对应的响应状态码为 "503 Service Unavailable"。

```
public class HealthCheckOptions
{
    public Func<HealthCheckRegistration, bool>      Predicate { get; set; }
    public IDictionary<HealthStatus, int>           ResultStatusCodes { get; set; }
    public Func<HttpContext, HealthReport, Task>    ResponseWriter { get; set; }
    public bool                                     AllowCachingResponses { get; set; }

    public HealthCheckOptions()
    {
        ResultStatusCodes = new Dictionary<HealthStatus, int>
        {
            {HealthStatus.Healthy, 200},
            {HealthStatus.Degraded, 200},
            {HealthStatus.Unhealthy, 503},
        };
        ResponseWriter = (context, report) =>
        {
            context.Response.ContentType = "text/plain";
            return context.Response.WriteAsync(report.Status.ToString());
        };
    }
}
```

HealthCheckOptions 类型的 ResponseWriter 属性返回的 Func<HttpContext, HealthReport, Task>对象会被用来设置响应的主体内容。在默认情况下，健康请求的响应报文采用的媒体类型为 text/plain，具体的内容就是应用整体健康状态的文字描述，这是由 ResponseWriter 属性的默认值决定的。HealthCheckOptions 类型的 AllowCachingResponses 属性决定了是否希望健康检查响应被客户端缓存，在一般情况下我们都希望在进行健康检查的时候能得到实时的健康状况，所以该属性的默认值为 False。

HealthCheckMiddleware 中间件在利用 HealthCheckService 服务完成绝大部分的健康检查工作并得到作为健康报告的 HealthReport 对象之后，需要对请求做出最终的响应。我们采用如下所示的代码片段模拟了HealthCheckMiddleware中间件针对健康检查请求的处理流程。

```csharp
public class HealthCheckMiddleware
{
    private readonly RequestDelegate        _next;
    private readonly HealthCheckOptions     _healthCheckOptions;
    private readonly HealthCheckService     _healthCheckService;

    public HealthCheckMiddleware(RequestDelegate next,
        HealthCheckOptions healthCheckOptions,
        HealthCheckService healthCheckService)
    {
        _next                = next;
        _healthCheckOptions  = healthCheckOptions;
        _healthCheckService  = healthCheckService;
    }

    public async Task InvokeAsync(HttpContext httpContext)
    {
        var report = await _healthCheckService.CheckHealthAsync(
            _healthCheckOptions.Predicate, httpContext.RequestAborted);
        httpContext.Response.StatusCode =
            _healthCheckOptions.ResultStatusCodes[report.Status];

        if (!_healthCheckOptions.AllowCachingResponses)
        {
            var headers = httpContext.Response.Headers;
            headers["Cache-Control"] = "no-store, no-cache";
            headers["Pragma"] = "no-cache";
            headers["Expires"] = "Thu, 01 Jan 1970 00:00:00 GMT";
        }
        await _healthCheckOptions.ResponseWriter(httpContext, report);
    }
}
```

由于HealthCheckMiddleware中间件依赖HealthCheckService服务，该服务是在IServiceCollection接口的AddHealthChecks扩展方法中被注册的。如下面的代码片段所示，该方法返回一个封装了当前IServiceCollection集合的HealthChecksBuilder对象，我们可以利用它完成进一步的服务注册。

```csharp
public static class HealthCheckServiceCollectionExtensions
{
    public static IHealthChecksBuilder AddHealthChecks(
        this IServiceCollection services)
    {
        services.TryAddSingleton<HealthCheckService, DefaultHealthCheckService>();
        ...
        return new HealthChecksBuilder(services);
    }
}
```

}

我们可以调用如下一系列 UseHealthChecks 扩展方法来注册 HealthCheckMiddleware 中间件。在调用这些方法的时候需要指定健康检查终结点的路径和端口号（可以默认，整数和字符串类型均可），还可以提供一个承载配置选项的 HealthCheckOptions 对象。注册的 HealthCheckMiddleware 中间件只有在当前请求的路径和端口与设置相匹配的情况下才会用来处理当前请求。

```
public static class HealthCheckApplicationBuilderExtensions
{
    public static IApplicationBuilder UseHealthChecks(this IApplicationBuilder app,
        PathString path)
        => UseHealthChecksCore(app, path, null);

    public static IApplicationBuilder UseHealthChecks(this IApplicationBuilder app,
        PathString path, HealthCheckOptions options)
        => UseHealthChecksCore(app, path, null, Options.Create(options));

    public static IApplicationBuilder UseHealthChecks(this IApplicationBuilder app,
        PathString path, int port)
        =>UseHealthChecksCore(app, path, new int?(port), Array.Empty<object>());

    public static IApplicationBuilder UseHealthChecks(this IApplicationBuilder app,
        PathString path, string port)
    {
        int? parsedPort = string.IsNullOrEmpty(port)
            ? (int?)null
            : int.Parse(port);
        UseHealthChecksCore(app, path, parsedPort);
        return app;
    }

    public static IApplicationBuilder UseHealthChecks(this IApplicationBuilder app,
        PathString path, int port, HealthCheckOptions options)
        =>UseHealthChecksCore(app, path, port, Options.Create(options));

    public static IApplicationBuilder UseHealthChecks(this IApplicationBuilder app,
        PathString path, string port, HealthCheckOptions options)
    {
        int? parsedPort = string.IsNullOrEmpty(port)
            ? (int?)null
            : int.Parse(port);
        return UseHealthChecksCore(app, path, parsedPort, Options.Create(options));
    }

    private static IApplicationBuilder UseHealthChecksCore(IApplicationBuilder app,
        PathString path, int? port, params object[] arguments)
    {
        bool Match(HttpContext context)
```

```
        {
            if (port.HasValue && context.Connection.LocalPort != port.Value)
            {
                return false;
            }
            return context.Request.Path.StartsWithSegments(path, out var remaining)
                && string.IsNullOrEmpty(remaining);
        }
        return app.MapWhen(Match, builder
            => builder.UseMiddleware<HealthCheckMiddleware>(arguments));
    }
}
```

用来构建路由终结点的 IEndpointRouteBuilder 接口具有如下所示的两个 MapHealthChecks 扩展方法,意味着 HealthCheckMiddleware 中间件还可以采用终结点路由的形式进行注册,这种注册方式的好处是可以将健康检查终结点的路径设置为一个包含路由参数占位符的路径模板。

```
public static class HealthCheckEndpointRouteBuilderExtensions
{
    public static IEndpointConventionBuilder MapHealthChecks(
        this IEndpointRouteBuilder endpoints, string pattern)
        => MapHealthChecksCore(endpoints, pattern, null);

    public static IEndpointConventionBuilder MapHealthChecks(
        this IEndpointRouteBuilder endpoints, string pattern,
        HealthCheckOptions options)
        => MapHealthChecksCore(endpoints, pattern, options);

    private static IEndpointConventionBuilder MapHealthChecksCore(
        IEndpointRouteBuilder endpoints, string pattern, HealthCheckOptions options)
    {
        var handler = endpoints.CreateApplicationBuilder()
            .UseMiddleware<HealthCheckMiddleware>(
                options == null ? null : Options.Create(options))
            .Build();
        return endpoints.Map(pattern, handler).WithDisplayName("Health checks");
    }
}
```

到目前为止,我们已经从设计和实现的层面介绍了 HealthCheckMiddleware 中间件进行健康检查的整个流程,下面以图 23-4 所示的 UML 对该中间件的整体设计进行总结:该中间件利用注册的 HealthCheckService 服务来完成核心的健康检查操作,并得到一份通过 HealthReport 对象表示的健康报告,它自身则根据 HealthCheckOptions 对象提供的配置选项将健康报告承载的内容响应给客户端。

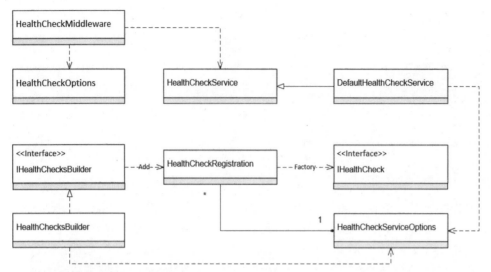

图 23-4　健康检查的核心接口和类型

健康检查系统默认注册的 HealthCheckService 服务的类型为 DefaultHealthCheckService，它利用作为配置选项的 HealthCheckServiceOptions 对象得到所有 HealthCheckRegistration 对象，然后利用它们提供对应的 IHealthCheck 对象。在对这些 IHealthCheck 做相应的过滤之后，DefaultHealthCheckService 对象会利用它们各自完成对应的健康检查工作并得到一份通过 HealthCheckResult 对象标识的健康检查结果。这些表示健康检查结果的 HealthCheckResult 对象会转换成 HealthReportEntry 对象，DefaultHealthCheckService 对象最终将所有的 HealthReportEntry 对象收集起来生成一份通过 HealthReport 对象标识的健康报告。

DefaultHealthCheckService 对象之所以能够利用 HealthCheckServiceOptions 配置选项得到承载了所有 IHealthCheck 注册信息的 HealthCheckRegistration 对象，源于健康检查系统采用的注册方式。具体来说，针对 IHealthCheck 对象的注册以及针对其他依赖服务的注册都通过 IHealthCheckBuilder 对象完成，作为对该接口的默认实现，HealthCheckBuilder 对象将封装了 IHealthCheck 注册信息的 HealthCheckRegistration 对象存放到了 HealthCheckServiceOptions 配置选项中。

23.2.4　针对 Entity Framework Core 的健康检查

如果应用涉及针对数据库的访问，就需要通过健康检查的方式来确定数据库的可用性。如果采用 Entity Framework Core 作为数据库存取框架，就可以利用健康检查来确定某个具体的 DbContext 是否可用。针对 DbContext 的健康检查是通过 DbContextHealthCheck<TContext>这个内部类型完成的，该类型由 NuGet 包 "Microsoft.Extensions.Diagnostics.HealthChecks.EntityFrameworkCore" 来提供。

基于 DbContext 的健康检查

下面先通过一个简单的实例来介绍如何利用该注册 DbContextHealthCheck<TContext>对象来检验对应 DbContext 的可用性。该类型所在的 NuGet 包并不在"Microsoft.NET.Sdk.Web" SDK 的默认依赖范围内,所以需要添加对 NuGet 包"Microsoft.Extensions.Diagnostics.HealthChecks.EntityFrameworkCore"的依赖。除此之外,我们的演示实例采用的数据库类型为 SQL Server,所以还需要为演示程序添加针对 NuGet 包"Microsoft.EntityFrameworkCore.SqlServer"的依赖。在添加了上述两个 NuGet 包的依赖之后,我们定义了如下两个空的 DbContext 类型(FooContext 和 BarContext)。

```
public class FooContext : DbContext
{
    public FooContext(DbContextOptions<FooContext> options) : base(options)
    {}
}

public class BarContext : DbContext
{
    public BarContext(DbContextOptions<BarContext> options) : base(options)
    {}
}
```

在如下所示的演示程序中,我们调用 IServiceCollection 接口的 AddDbContext<TDbContext>扩展方法注册了针对 Entity Framework Core 的相关服务,并完成了针对上述两个自定义 DbContext 类型的注册。我们为 FooContext 和 BarContext 分别指定了两个针对 SQL Server 的连接字符串,其中前者是有效的,后者是无效的。

```
public class Program
{
    public static void Main()
    {
        var options = new HealthCheckOptions
        {
            ResponseWriter = ReportAsync
        };

        Host.CreateDefaultBuilder()
            .ConfigureWebHostDefaults(builder => builder
                .ConfigureServices(svcs => svcs
                    .AddDbContext<FooContext>(options => options
                        .UseSqlServer("{valid connection string}"))
                    .AddDbContext<BarContext>(options => options
                        .UseSqlServer("invalid connection string"))
                    .AddHealthChecks()
                        .AddDbContextCheck<FooContext>()
                        .AddDbContextCheck<BarContext>())
                .Configure(app => app.UseHealthChecks("/healthcheck", options)))
            .Build()
            .Run();
```

```
static Task ReportAsync (HttpContext context, HealthReport report)
{
    var builder = new StringBuilder();
    builder.AppendLine($"Status: {report.Status}");
    foreach (var name in report.Entries.Keys)
    {
        builder.AppendLine($"    {name}: {report.Entries[name].Status}");
    }
    return context.Response.WriteAsync(builder.ToString());
}
```

在调用 IServiceCollection 接口的 AddHealthChecks 扩展方法得到 IHealthChecksBuilder 对象之后，我们针对两个 DbContext 类型（FooContext 和 BarContext）调用 AddDbContextCheck<TContext>扩展方法，它们会创建并注册对应的 DbContextHealthCheck<TContext>对象。为了得到针对两个 DbContext 类型的健康报告，我们依然在注册 HealthCheckMiddleware 中间件时指定了对应的配置选项，并通过指定其 ResponseWriter 属性将每个健康检查条目的状态呈现在响应内容中。如果直接利用浏览器发送健康检查请求，就会得到图 23-5 所示的健康检查报告。由于 FooContext 连接的数据库是可用的，所以针对它的状态为 Healthy。BarContext 采用的连接字符串是无效的，所以它对应的状态为 Unhealthy。（S2307）

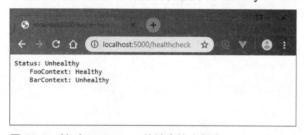

图 23-5　针对 DbContext 的健康检查报告

DbContextHealthCheck<TContext>

针对 Entity Framework Core 的健康检查由 DbContextHealthCheck<TContext>这个内部类型完成，是对 IHealthCheck 接口的实现，DbContextHealthCheckOptions<TContext>类型代表它采用的配置选项。如下面的代码片段所示，该配置选项类型具有唯一的属性成员 CustomTestQuery，并且提供了一个委托对象来完成真正的健康检查工作。

```
internal sealed class DbContextHealthCheckOptions<TContext>
    where TContext : DbContext
{
    public Func<TContext, CancellationToken, Task<bool>>
        CustomTestQuery { get; set; }
}
```

如下所示的代码片段体现了 DbContextHealthCheck<TContext>针对健康检查的实现逻辑。

由于上述配置选项是由注入的 IOptionsMonitor<DbContextHealthCheckOptions<TContext>>对象提供的，所以 DbContextHealthCheck<TContext>可以获得最新的配置选项。在实现的 CheckHealthAsync 方法中，如果配置选项的 CustomTestQuery 属性进行了设置，该属性返回的委托对象将完成健康检查工作，否则该方法会直接调用 DbContext 对象的 CanConnectAsync 方法判断目标数据库的连接是否正常，并以此来决定最终的健康状态。

```
internal sealed class DbContextHealthCheck<TContext> : IHealthCheck
    where TContext : DbContext
{
    private static readonly Func<TContext, CancellationToken,
        Task<bool>> DefaultTestQuery = (dbContext, cancellationToken) =>
    dbContext.Database.CanConnectAsync(cancellationToken);

    private readonly TContext _dbContext;
    private readonly IOptionsMonitor<DbContextHealthCheckOptions<TContext>>
        _options;

    public DbContextHealthCheck(TContext dbContext,
        IOptionsMonitor<DbContextHealthCheckOptions<TContext>> options)
    {
        _dbContext = dbContext;
        _options = options;
    }

    public async Task<HealthCheckResult> CheckHealthAsync(
        HealthCheckContext context, CancellationToken cancellationToken = default)
    {
        var options = _options.Get(context.Registration.Name);
        var testQuery = options.CustomTestQuery ?? DefaultTestQuery;

        if (await testQuery(_dbContext, cancellationToken))
        {
            return HealthCheckResult.Healthy();
        }

        return HealthCheckResult.Unhealthy();
    }
}
```

演示实例中用来注册 DbContextHealthCheck<TContext>的 AddDbContextCheck<TContext>扩展方法的定义如下。该方法所有的参数都是默认的，如果没有显式指定注册名称，指定 DbContext 类型的名称将作为注册名称。我们还可以指定健康检查失败情况下的健康状态（默认为 Unhealthy）、标签列表和用来完成具体健康检查工作的 Func<TContext, CancellationToken, Task<bool>>委托对象。

```
public static class EntityFrameworkCoreHealthChecksBuilderExtensions
{
    public static IHealthChecksBuilder AddDbContextCheck<TContext>(
        this IHealthChecksBuilder builder,
```

```
            string name                                                  = null,
            HealthStatus? failureStatus                                  = default,
            IEnumerable<string> tags                                     = default,
            Func<TContext, CancellationToken, Task<bool>> customTestQuery = default)
            where TContext : DbContext
{
    name = name ?? typeof(TContext).Name;
    if (customTestQuery != null)
    {
        builder.Services.Configure<DbContextHealthCheckOptions<TContext>>(name,
            options => options.CustomTestQuery = customTestQuery);
    }
    return builder.AddCheck<DbContextHealthCheck<TContext>>(name, failureStatus,
        tags);
}
```

23.3 发布健康报告

除了针对具体的请求返回当前的健康报告，我们还能以设定的间隔定期收集和发布健康报告。这个功能很有用，如可以利用这个功能将收集的健康报告发送给 APM（Application Performance Management）系统。

23.3.1 定期发布健康报告

我们通过如下实例来介绍健康报告的定期发布。健康报告的发布是通过 IHealthCheck Publisher 服务来完成的。我们实现该接口的实现类型 ConsolePublisher 将健康报告输出到控制台上，后续内容会给出 ConsolePublisher 类型的定义。如下所示的代码片段展示了如何利用 ConsolePublisher 对象将收集的健康报告以设定的时间间隔输出到控制台上。

```
public class Program
{
    public static void Main()
    {
        var random = new Random();
        Host.CreateDefaultBuilder()
            .ConfigureWebHostDefaults(builder => builder
                .ConfigureLogging(logging=>logging.ClearProviders())
                .ConfigureServices(svcs => svcs.AddHealthChecks()
                    .AddCheck("foo", Check)
                    .AddCheck("bar", Check)
                    .AddCheck("baz", Check)
                    .AddConsolePublisher()
                    .ConfigurePublisher(options=>
                        options.Period = TimeSpan.FromSeconds(5)))
                .Configure(app => app.UseHealthChecks("/healthcheck")))
            .Build()
            .Run();
```

```
    HealthCheckResult Check()
    {
        switch (random.Next(1, 4))
        {
            case 1:  return HealthCheckResult.Unhealthy();
            case 2:  return HealthCheckResult.Degraded();
            default: return HealthCheckResult.Healthy();
        }
    }
}
```

如上面的代码片段所示，我们注册了 3 个 DelegateHealthCheck 对象，它们会随机返回针对 3 种状态的健康状态。ConsolePublisher 对象的注册体现在针对 AddConsolePublisher 方法的调用上，这是专门为 IHealthChecksBuilder 接口定义的扩展方法。紧随其后调用的 ConfigurePublisher 方法也是自定义的扩展方法，我们利用它将健康报告发布间隔设置为 5 秒。程序运行之后，当前应用的健康报告会以图 23-6 所示的形式输出到控制台上。（S2308）

图 23-6　健康报告的定期发布

23.3.2　IHealthCheckPublisher

健康报告的发布实现在通过 IHealthCheckPublisher 接口表示的服务中。我们可以在同一个应用中注册多个 IHealthCheckPublisher 服务，如可以注册多个这样的服务将健康报告分别输出到控制台、日志文件或者直接发送给另一个健康报告处理服务。如下面的代码片段所示，IHealthCheckPublisher 接口采用唯一的 PublishAsync 方法来发布通过参数 report 表示的健康报告。

```
public interface IHealthCheckPublisher
{
    Task PublishAsync(HealthReport report, CancellationToken cancellationToken);
}
```

如下所示的是演示实例使用了 ConsolePublisher 类型的定义。在实现的 PublishAsync 方法中，我们将表示健康报告的 HealthReport 对象格式化成字符串并输出到控制台。该过程涉及针对 StringBuilder 对象的使用，由于我们采用对象池的方式来使用这个对象，所以在构造函数中注入了 ObjectPoolProvider 对象。

```
public class ConsolePublisher : IHealthCheckPublisher
```

```csharp
{
    private readonly ObjectPool<StringBuilder> _stringBuilderPool;

    public ConsolePublisher(ObjectPoolProvider provider)
    {
        _stringBuilderPool = provider.CreateStringBuilderPool();
    }

    public Task PublishAsync(HealthReport report,
        CancellationToken cancellationToken)
    {
        cancellationToken.ThrowIfCancellationRequested();
        var builder = _stringBuilderPool.Get();
        try
        {
            builder.AppendLine($"Status: {report.Status}
                [{DateTimeOffset.Now.ToString("yy-MM-dd hh:mm:ss")}]");
            foreach (var name in report.Entries.Keys)
            {
                builder.AppendLine($"    {name}: {report.Entries[name].Status}");
            }
            Console.WriteLine(builder);
            return Task.CompletedTask;
        }
        finally
        {
            _stringBuilderPool.Return(builder);
        }
    }
}
```

用来发布健康报告的 IHealthCheckPublisher 服务需要注册到依赖注入框架中，前面演示实例中将针对 ConsolePublisher 对象的注册实现在 IHealthChecksBuilder 接口的 AddConsolePublisher 扩展方法中，如下所示的代码片段就是这个扩展方法的定义。

```csharp
public static class Extensions
{
    public static IHealthChecksBuilder AddConsolePublisher(
        this IHealthChecksBuilder builder)
    {
        builder.Services.AddSingleton<IHealthCheckPublisher, ConsolePublisher>();
        return builder;
    }
}
```

23.3.3　HealthCheckPublisherHostedService

IHealthCheckPublisher 服务针对健康报告的收集和发布是利用 HealthCheckPublisherHostedService 服务来驱动的，这是一个实现了 IHostedService 的承载服务。在给出该类型的定义之前，下面先介绍对应的配置选项类型 HealthCheckPublisherOptions。

```csharp
public sealed class HealthCheckPublisherOptions
{
    public TimeSpan                                 Delay { get; set; }
    public TimeSpan                                 Period { get; set; }
    public TimeSpan                                 Timeout { get; set; }

    public Func<HealthCheckRegistration, bool>      Predicate { get; set; }
}
```

除了前面演示实例中用来控制健康报告发布时间间隔的 Period 属性，HealthCheckPublisherOptions 类型，还有两个额外的 TimeSpan 类型的属性，其中 Delay 属性表示健康发布服务启动之后到开始收集发布工作之间的时延，这个设置可以确保在各项初始化工作尽可能正常完成之后才开始收集健康报告；Timeout 属性表示 IHealthCheckPublisher 对象发布健康报告的超时时间。Period 属性、Delay 属性和 Timeout 属性的默认设置分别为 30 秒、5 秒和 30 秒。HealthCheckPublisherOptions 类型还有另一个名为 Predicate 的属性，该对象用来对注册的 IHealthCheck 对象实施过滤。

HealthCheckPublisherHostedService 对象针对健康报告的收集和发布逻辑基本体现在如下所示的代码片段中。它的构造函数中注入了 3 个核心对象，分别是用来生成健康报告的 HealthCheckService 对象、用来发布健康报告的一组 IHealthCheckPublisher 对象和配置选项。

```csharp
internal sealed class HealthCheckPublisherHostedService : IHostedService
{
    private readonly HealthCheckService                     _healthCheckService;
    private readonly HealthCheckPublisherOptions            _options;
    private readonly IEnumerable<IHealthCheckPublisher>     _publishers;
    private readonly CancellationTokenSource                _stopSource;

    private Timer                                           _timer;

    public HealthCheckPublisherHostedService(
        HealthCheckService healthCheckService,
        IOptions<HealthCheckPublisherOptions> optionsAccessor,
        IEnumerable<IHealthCheckPublisher> publishers)
    {
        _healthCheckService     = healthCheckService;
        _options                = optionsAccessor.Value;
        _publishers             = publishers;
        _stopSource             = new CancellationTokenSource();
    }

    public Task StartAsync(CancellationToken cancellationToken)
    {
        var restoreFlow = false;
        try
        {
            if (!ExecutionContext.IsFlowSuppressed())
            {
                restoreFlow = true;
```

```csharp
                ExecutionContext.SuppressFlow();
            }
            _timer = new Timer(Tick, null, _options.Delay, _options.Period);
            return Task.CompletedTask;
        }
        finally
        {
            if (restoreFlow)
            {
                ExecutionContext.RestoreFlow();
            }
        }

        async void Tick(object state) => await RunAsync();
    }

    private async Task RunAsync()
    {
        var stopwatch = Stopwatch.StartNew();

        CancellationTokenSource source = null;
        try
        {
            var timeout = _options.Timeout;
            source = CancellationTokenSource
                .CreateLinkedTokenSource(_stopSource.Token);
            source.CancelAfter(timeout);

            await Task.Yield();
            var token = source.Token;
            var report = await _healthCheckService
                .CheckHealthAsync(_options.Predicate, token);
            var tasks = _publishers.Select(it => it.PublishAsync(report, token));
            await Task.WhenAll();
        }
        catch {}
        finally
        {
            source?.Dispose();
        }
    }

    public Task StopAsync(CancellationToken cancellationToken)
    {
        _stopSource.Cancel();
        _timer?.Dispose();
        _timer = null;
        return Task.CompletedTask;
    }
}
```

在实现的 StartAsync 方法中，HealthCheckPublisherHostedService 服务利用配置选项提供的延迟时间和间隔时间创建了一个 Timer 对象，该对象用来实现对健康报告的定期发布。在为这个 Timer 对象提供的回调中，HealthCheckService 服务会用来检验当前应用的健康状况并生成通过 HealthReport 对象表示的健康报告，此健康报告会同时分发给所有 IHealthCheckPublisher 对象进行发布。以 HealthCheckPublisherHostedService 类型为核心的健康报告发布模型如图 23-7 所示。

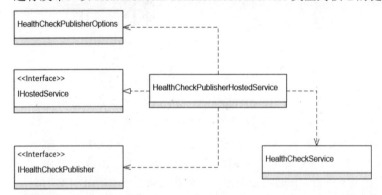

图 23-7　以 HealthCheckPublisherHostedService 类型为核心的健康报告发布模型

针对 HealthCheckPublisherHostedService 对象的注册实现在 IServiceCollection 接口的 AddHealthChecks 扩展方法中（该方法同时注册了 HealthCheckService 服务）。由于 HealthCheckPublisherOptions 对象承载的配置选项并没有一个专门的扩展方法来设置，所以前面演示的实例特意定义了如下所示的 ConfigurePublisher 扩展方法。

```
public static class HealthCheckServiceCollectionExtensions
{
    public static IHealthChecksBuilder AddHealthChecks(
        this IServiceCollection services)
    {
        services.TryAddSingleton<HealthCheckService, DefaultHealthCheckService>();
        services.TryAddEnumerable(ServiceDescriptor
            .Singleton<IHostedService, HealthCheckPublisherHostedService>());
        return new HealthChecksBuilder(services);
    }
}

public static class Extensions
{
    public static IHealthChecksBuilder ConfigurePublisher(
        this IHealthChecksBuilder builder,
        Action<HealthCheckPublisherOptions> configure)
    {
        builder.Services.Configure(configure);
        return builder;
    }
}
```

第24章

补　遗

本章会补充几个简单且常用的中间件。HostFilteringMiddleware 中间件通过对主机名的过滤强制请求提供有效的主机名。如果因为通信网络或者软件和硬件的原因无法发送指定方法（如 PUT、DELETE 和 OPTIONS 等）的请求，我们可以借助 HttpMethodOverrideMiddleware 中间件完成针对 HTTP 方法的改写功能。如果请求在到达最终服务器之前经过多方转发导致 HttpContext 上下文无法反映出请求的最初发起者，就可以利用 ForwardedHeadersMiddleware 中间件来解决。UsePathBaseMiddleware 中间件可以帮助我们设置当前应用的基础路径，而 MapMiddleware 中间件和 MapWhenMiddleware 中间件则提供了最简单与最直接的路由实现。

24.1　过滤主机名

如果应用程序对请求采用的主机名（Host Name）有要求，则可以利用 HostFilteringMiddleware 中间件对请求采用主机名进行验证。在使用 HostFilteringMiddleware 中间件时，我们可以指定一组有效的主机名，该中间件在处理请求时会验证当前请求采用的主机名是否在此范围之内，并拒绝采用不合法主机名的请求。在具体介绍 HostFilteringMiddleware 中间件之前，我们先通过简单的实例来了解如何利用该中间件完成针对主机名的过滤。

24.1.1　实例演示

为了方便在本机环境下模拟采用不同的域名（主机名称）访问我们的应用程序，可以通过修改 hosts 文件（"%windir%\system32\drivers\etc\hosts"）将本地的 IP 地址（"127.0.0.1"）映射为不同的域名。如下所示是我们映射的 3 个域名。

```
127.0.0.1 www.foo.com
127.0.0.1 www.bar.com
127.0.0.1 www.baz.com
```

在如下这段简单的演示程序中，我们通过调用 IApplicationBuilder 接口的 UseHostFiltering 扩展方法注册了 HostFilteringMiddleware 中间件，还调用 IServiceCollection 接口的 AddHostFiltering 扩展方法注册了该中间件依赖的服务。

```
public class Program
{
    public static void Main()
    {
        Host.CreateDefaultBuilder()
            .ConfigureWebHostDefaults(builder => builder
                .ConfigureServices(svcs => svcs.AddHostFiltering(options =>
                {
                    options.AllowedHosts.Add("www.foo.com");
                    options.AllowedHosts.Add("www.bar.com");
                }))
                .Configure(app => app
                    .UseHostFiltering()
                    .Run(contenxt => contenxt.Response
                        .WriteAsync($"{contenxt.Request.Host} is valid!"))))
            .Build()
            .Run();
    }
}
```

在针对 AddHostFiltering 扩展方法的调用中，我们利用作为参数的 Action<HostFilteringOptions>委托对象对配置选项做了相应设置。具体来说，可以将两个许可的域名（www.foo.com 和 www.bar.com）添加到 HostFilteringOptions 对象的 AllowedHosts 属性集合中。如图 24-1 所示，当我们采用域名访问该应用的时候，浏览器会呈现正常的响应内容。如果将域名换成 www.baz.com 之后，得到的就是一个状态码为 "400 Bad Request" 的响应，并提示采用的域名不合法。（S2401）

图 24-1　HostFilteringMiddleware 中间件针对主机名的过滤

24.1.2　配置选项

HostFilteringMiddleware 中间件定义在 NuGet 包 "Microsoft.AspNetCore.HostFiltering" 中。在具体介绍该中间件针对主机过滤的处理逻辑之前，我们需要先了解通过 HostFilteringOptions 类型标识的配置选项。

```csharp
public class HostFilteringOptions
{
    public IList<string>        AllowedHosts { get; set; }
    public bool                 AllowEmptyHosts { get; set; }
    public bool                 IncludeFailureMessage { get; set; }
}
```

如上面的代码片段所示，HostFilteringOptions 类型具有一个字符串列表类型的 AllowedHosts 属性，表示允许的主机名称，上面演示实例设置的合法域名正是被添加到该列表中的。除了指定一个确定的主机名称（如 www.foobar.com），还可以将添加的主机名设定为如下 3 种特殊的形式，它们都表示匹配任意的主机名称。如果 AllowedHosts 属性表示的列表中包含这样的主机名称，就意味着关闭了针对主机名称的过滤。

- *。
- 0.0.0.0。
- ::（针对 IP V6）。

HostFilteringMiddleware 中间件提取的主机名称来源于请求的 Host 报头。按照 HTTP/1.1 的规定，请求报文的 Host 报头是允许为空的，HTTP 1.0 甚至不要求请求报文提供 Host 报头。如果无法从请求的 Host 报头中提取出有效的主机名，HostFilteringMiddleware 中间件又会对当前请求做何处理？这就涉及 HostFilteringOptions 类型的 AllowEmptyHosts 属性。

顾名思义，HostFilteringOptions 类型的 AllowEmptyHosts 属性表示不具有 Host 报头或者 Host 报头值为空的请求是否是合法的。该属性的默认值为 True，意味着在默认情况下这样的请求是合法的；如果将该属性显式设置为 False，HostFilteringMiddleware 中间件在处理这类请求时会返回一个状态码为 "400 Bad Request" 的响应。

HostFilteringOptions 类型还有一个名为 IncludeFailureMessage 的属性，表示对于采用无效主机名的请求，HostFilteringMiddleware 中间件除了返回一个状态码为 "400 Bad Request" 的响应，响应的主体内容是否还需要包含预定义的出错信息。该属性的默认值为 True，图 24-1 所示的错误消息会以一个 HTML 文档的形式包含在响应的主体内容中。

24.1.3 HostFilteringMiddleware 中间件

如下所示的代码片段体现了 HostFilteringMiddleware 中间件类型的定义。它的构造函数通过注入的 IOptionsMonitor<HostFilteringOptions> 服务来提供承载配置选项的 HostFilteringOptions 对象，所以该中间件会自动感知配置选项内容的改变，并及时将更新的配置选项应用到后续请求处理过程中。Invoke 方法中针对主机名称过滤的具体实现前面已有介绍，这里就不再提供 Invoke 方法的实现代码。

```csharp
public class HostFilteringMiddleware
{
    public HostFilteringMiddleware(RequestDelegate next,
        ILogger<HostFilteringMiddleware> logger,
        IOptionsMonitor<HostFilteringOptions> optionsMonitor);
    public Task Invoke(HttpContext context);
}
```

HostFilteringMiddleware 中间件的注册实现在 IApplicationBuilder 接口如下所示的 UseHostFiltering 扩展方法中，IServiceCollection 接口的 AddHostFiltering 扩展方法则提供针对 HostFilteringOptions 配置选项的设置。

```csharp
public static class HostFilteringBuilderExtensions
{
    public static IApplicationBuilder UseHostFiltering(this IApplicationBuilder app)
     => app.UseMiddleware<HostFilteringMiddleware>();
}

public static class HostFilteringServicesExtensions
{
    public static IServiceCollection AddHostFiltering(
        this IServiceCollection services,
        Action<HostFilteringOptions> configureOptions)
     => services.Configure<HostFilteringOptions>(configureOptions);
}
```

24.2 HTTP 重写

本节旨在介绍两个可以改写 HTTP 请求消息的中间件，它们分别是用来改写请求 HTTP 方法的 HttpMethodOverrideMiddleware 中间件以及用来改写客户端 IP 地址、主机名称和协议类型（HTTP 或者 HTTPS）的 ForwardedHeadersMiddleware 中间件。按照惯例，我们先通过几个简单的实例来了解这两个中间件针对请求 HTTP 消息的改写功能。

24.2.1 实例演示

HTTP 方法（HTTP Method）对于 REST 架构风格的 Web API 来说是一个非常重要的元素，它基本上体现了请求针对目标资源的操作类型。由于目标操作基本上会关联一个固定的 HTTP 方法，所以请求能否被正常路由不仅取决于 URL 是否正确，还取决于它的 HTTP 方法是否与目标操作相匹配。由于一些网络设置、客户端软件或者选用服务器的限制，在一些场景下只允许发送或者接收 GET 请求和 POST 请求，这就要求服务端在进行路由之前改写当前请求的 HTTP 方法。

改写 HTTP 方法

X-Http-Method-Override 虽然不属于 HTTP 标准规定的请求报头，但是是事实上的标准，各个厂商基本都接受利用它来表示请求希望被改写的 HTTP 方法，携带此报头的一般要求是一个 POST 请求。下面演示针对该请求报头的 HTTP 方法的改写。

```csharp
public class Program
{
    public static void Main()
    {
        Host.CreateDefaultBuilder()
            .ConfigureWebHostDefaults(builder => builder
```

```
            .Configure(app => app
                .UseHttpMethodOverride()
                .Run(contenxt => contenxt.Response.WriteAsync(
                    $"HTTP Method: {contenxt.Request.Method}"))))
        .Build()
        .Run();
    }
}
```

如上面的代码片段所示,我们调用 IApplicationBuilder 接口的 UseHttpMethodOverride 扩展方法注册了 HttpMethodOverrideMiddleware 中间件。在后续通过调用 Run 扩展方法注册的中间件中,我们将当前(针对该中间件)的 HTTP 方法写入响应的主体内容。应用启动后,我们发送如下这个 POST 请求,该请求具有一个 X-Http-Method-Override 报头,它将希望被改写的目标 HTTP 方法设置为 PUT。从得到的响应来看,HttpMethodOverrideMiddleware 中间件确实将 POST 请求改写成 PUT 请求。(S2402)

```
POST http://localhost:5000/ HTTP/1.1
X-Http-Method-Override: PUT
Host: localhost:5000
Content-Length: 0

HTTP/1.1 200 OK
Date: Thu, 26 Sep 2019 22:40:34 GMT
Server: Kestrel
Content-Length: 16

HTTP Method: PUT
```

除了利用 X-Http-Method-Override 报头携带希望改写的目标 HTTP 方法,对于一个媒体类型为 application/x-www-form-urlencoded 的请求,还可以利用一个预定义的字段来存储改写的目标 HTTP 方法。如果采用这种方法来演示针对 HTTP 方法的改写,就需要对上面的程序做如下修改。

```
public class Program
{
    public static void Main()
    {
        Host.CreateDefaultBuilder()
            .ConfigureWebHostDefaults(builder => builder
                .Configure(app => app
                    .UseHttpMethodOverride(new HttpMethodOverrideOptions
                    {
                        FormFieldName = "Http-Method-Override"
                    })
                    .Run(contenxt => contenxt.Response.WriteAsync(
                        $"HTTP Method: {contenxt.Request.Method}"))))
            .Build()
            .Run();
    }
}
```

如上面的代码片段所示,在调用 IApplicationBuilder 接口的 UseHttpMethodOverride 扩展方

法注册 HttpMethodOverrideMiddleware 中间件时，我们提供了一个 HttpMethodOverrideOptions 配置选项。HttpMethodOverrideOptions 类型的 FormFieldName 属性表示携带 HTTP 方法（改写后）的表单字段的名称，我们将该字段名设置为 Http-Method-Override。

改写后的应用启动后，我们发送了如下这个请求。该请求的 Content-Type 被设置为 application/x-www-form-urlencoded，作为主体内容的表单包含一个唯一的字段 Http-Method-Override，可以利用它将改写的 HTTP 方法设置为 PUT。发送请求后，我们得到了与上面一致的响应。（S2403）

```
POST http://localhost:5000/ HTTP/1.1
Content-Type: application/x-www-form-urlencoded
Host: localhost:5000
Content-Length: 24

Http-Method-Override=PUT

HTTP/1.1 200 OK
Date: Thu, 26 Sep 2019 22:49:06 GMT
Server: Kestrel
Content-Length: 16

HTTP Method: PUT
```

改写传输信息

从传输层面来讲，Web 服务器只会将 TCP 连接的另一端视为客户端，但是应用程序视角的客户端一般指的是最初发送请求的终端。在大部分部署场景下，两者之间都会存在代理或者负载均衡器这样的中间节点（以下统称为代理），所以双方理解的客户端就存在不一致的情况。不仅如此，客户端与代理之间以及代理与服务器采用的协议也可能不一致，客户端与代理之间可能采用 HTTPS，代理与服务器之间则可能采用 HTTP。为了解决这个问题，厂商都会遵循这样一个事实标准（de-facto Standard）：代理会在转发的请求上添加如下 3 个报头来表示原始客户端的主机名、IP 地址和传输协议。

- X-Forwarded-Host。
- X-Forwarded-For。
- X-Forwarded-Proto。

ForwardedHeadersMiddleware 中间件的目的在于从请求中提取上述这 3 个报头，当前的 HttpContext 上下文中承载的对应信息会根据它们进行修正，而修正之前的内容会转存到另外 3 个请求报头中，它们对应的名称如下。

- X-Original-Host。
- X-Original-For。
- X-Original-Proto。

下面通过如下这段简单的程序来介绍 ForwardedHeadersMiddleware 中间件针对与传输相关的 3 个属性（客户端 IP 地址、主机名称和传输类型）的改写。如下面的代码片段所示，我们调用

IApplicationBuilder 接口的 UseForwardedHeaders 扩展方法注册了 ForwardedHeadersMiddleware 中间件。在通过调用 Run 方法注册的中间件中，我们从当前 HttpContext 上下文中提取客户端的 IP 地址、主机名称和协议类型，以及上述 3 个 X-Original-报头值，并将它们写入响应报文的主体内容中。

```csharp
public class Program
{
    public static void Main()
    {
        var options = new ForwardedHeadersOptions
        {
            ForwardedHeaders = ForwardedHeaders.All
        };

        Host.CreateDefaultBuilder()
            .ConfigureWebHostDefaults(builder => builder
                .Configure(app => app
                    .UseForwardedHeaders(options)
                    .Run(ProcessAsync)))
            .Build()
            .Run();

        static async Task ProcessAsync(HttpContext context)
        {
            var request         = context.Request;
            var response        = context.Response;
            var headers         = request.Headers;

            await response.WriteAsync(
                $"Remote IP:{context.Connection.RemoteIpAddress}\n");
            await response.WriteAsync($"Host:{request.Host}\n");
            await response.WriteAsync($"Scheme:{request.Scheme}\n");

            await response.WriteAsync(
                $"X-Original-For:{headers["X-Original-For"]}\n");
            await response.WriteAsync(
                $"X-Original-Host:{headers["X-Original-Host"]}\n");
            await response.WriteAsync(
                $"X-Original-Proto:{headers["X-Original-Proto"]}");
        }
    }
}
```

应用启动之后，我们向它发送了如下所示的测试请求。如下面的代码片段所示，我们在请求中通过添加的 3 个 X-Forwarded-报头设置了原始客户端的 IP 地址、主机名称和协议类型。在得到的响应中，我们发现服务端从当前 HttpContext 上下文得到的信息与这 3 个 X-Forwarded-报头携带的数据是一致的。至于 ForwardedHeadersMiddleware 中间件修正之前的原始值，则可以从请求的 3 个 X-Original-报头中提取出来。（S2404）

```
GET http://localhost:5000/ HTTP/1.1
```

```
X-Forwarded-Host: www.foobar.com
X-Forwarded-For: 192.168.0.1
X-Forwarded-Proto: https
Host: localhost:5000


HTTP/1.1 200 OK
Date: Thu, 26 Sep 2019 22:59:05 GMTServer: Kestrel
Content-Length: 143


Remote IP:192.168.0.1
Host:www.foobar.com
Scheme:https
X-Original-For: [::1]:49417
X-Original-Host: localhost:5000
X-Original-Proto: http
```

24.2.2　HttpMethodOverrideMiddleware 中间件

　　HttpMethodOverrideMiddleware 中间件解决了由于客户端或者服务器的局限而无法发送具有正确 HTTP 方法的请求问题。具体来说，客户端可以在请求中携带与目标终结点相匹配的 HTTP 方法，该 HTTP 方法会被 HttpMethodOverrideMiddleware 中间件提取出来并应用到当前 HttpContext 上下文中，那么后续的中间件和应用程序就能按照客户端希望的 HTTP 方法语义来处理请求。在对该中间件进行详细讲述之前，需要先了解其对应的配置选项。

　　通过前面演示的实例可知，希望被改写的目标 HTTP 方法可以通过两种方式来传递：一种是通过 X-Http-Method-Override 报头，这是一个几乎被所有厂商接受的事实标准；另一种是通过一个预定义的表单元素来携带。对于一个采用媒体类型为 application/x-www-form-urlencoded 的请求来说，被改写的 HTTP 方法还可以通过一个预定义的表单元素来携带。HttpMethodOverrideMiddleware 中间件对应的配置选项通过 HttpMethodOverrideOptions 类型表示，它具有如下所示的唯一的 FormFieldName 属性，该属性表示的就是携带 HTTP 方法的表单字段名。

```
public class HttpMethodOverrideOptions
{
    public string FormFieldName { get; set; }
}
```

　　如下所示的代码片段是 HttpMethodOverrideMiddleware 类型的完整定义，该中间件通过在构造函数中注入的 IOptions<HttpMethodOverrideOptions>对象来获取当前的配置选项。在用于处理请求的 Invoke 方法中，它会先确定当前处理的是否是一个 POST 请求，因为针对 HTTP 方法的改写仅仅针对 POST 请求。

```
public class HttpMethodOverrideMiddleware
{
    private readonly RequestDelegate                _next;
    private readonly HttpMethodOverrideOptions      _options;

    public HttpMethodOverrideMiddleware(RequestDelegate next,
```

```csharp
        IOptions<HttpMethodOverrideOptions> options)
{
    _next      = next;
    _options   = options.Value;
}

public async Task Invoke(HttpContext context)
{
    if (string.Equals(context.Request.Method, "POST",
        StringComparison.OrdinalIgnoreCase))
    {
        if (_options.FormFieldName != null)
        {
            if (context.Request.HasFormContentType)
            {
                var form = await context.Request.ReadFormAsync();
                var methodType = form[_options.FormFieldName];
                if (!string.IsNullOrEmpty(methodType))
                {
                    context.Request.Method = methodType;
                }
            }
        }
        else
        {
            var xHttpMethodOverrideValue =
                context.Request.Headers["X-Http-Method-Override"];
            if (!string.IsNullOrEmpty(xHttpMethodOverrideValue))
            {
                context.Request.Method = xHttpMethodOverrideValue;
            }
        }
    }
    await _next(context);
}
}
```

请求可以采用两种方式携带目标 HTTP 方法，HttpMethodOverrideMiddleware 中间件会优先采用表单元素携带的 HTTP 方法，然后选择 X-Http-Method-Override 报头携带的 HTTP 方法。提取出的 HTTP 方法最终被赋值给表示当前请求的 HttpRequest 对象的 Method 属性。

24.2.3 ForwardedHeadersMiddleware 中间件

ForwardedHeadersMiddleware 中间件旨在处理与请求转发场景相关的 3 个请求报头。如果 HttpMethodOverrideMiddleware 中间件的作用是让最终的 HttpContext 上下文能够体现客户端希望采用的 HTTP 方法，那么 ForwardedHeadersMiddleware 中间件就是为了让服务端感觉到与它连接的就是最初的客户端。前者篡改了当前请求真实的 HTTP 方法，后者则改变了传输信息。

具体来说，被 ForwardedHeadersMiddleware 中间件篡改的传输信息包括网络连接的远程 IP

地址、请求的主机名和传输协议类型。在表示当前请求上下文的 HttpContext 对象中，上述这组信息分别通过表示连接信息的 ConnectionInfo 对象的 RemoteIpAddress 属性以及表示请求的 HttpRequest 对象的 Host 属性和 Scheme 属性来表示，ForwardedHeadersMiddleware 中间件最终需要改写的就是这 3 个属性。在对上述 3 个属性进行改写之前，ForwardedHeadersMiddleware 中间件会将原始信息保存到 3 个预定义的请求报头中。

```
public abstract class HttpContext
{
    public abstract ConnectionInfo    Connection { get; }
    public abstract HttpRequest       Request { get; }
    ...
}

public abstract class ConnectionInfo
{
    public abstract IPAddress RemoteIpAddress { get; set; }
    ...
}

public abstract class HttpRequest
{
    public abstract HostString        Host { get; set; }
    public abstract string            Scheme { get; set; }
    ...
}
```

在具体介绍 ForwardedHeadersMiddleware 中间件针对传输信息的改写逻辑之前，我们需要先了解承载配置选项的 ForwardedHeadersOptions 类型。如下面的代码片段所示，ForwardedHeadersOptions 具有两组属性：前面一组表示待处理请求中携带初始客户端 IP 地址、主机名和协议类型的 3 个请求报头名称；后面一组则表示最终被中间件转存的 3 个请求报头名称。这 6 个属性的默认值来源于定义在静态类型 ForwardedHeadersDefaults 中的 6 个只读属性。

```
public class ForwardedHeadersOptions
{
    public string ForwardedForHeaderName { get; set; }
    public string ForwardedHostHeaderName { get; set; }
    public string ForwardedProtoHeaderName { get; set; }

    public string OriginalForHeaderName { get; set; }
    public string OriginalHostHeaderName { get; set; }
    public string OriginalProtoHeaderName { get; set; }
    ...
}

public static class ForwardedHeadersDefaults
{
    public static string XForwardedForHeaderName { get; } = "X-Forwarded-For";
    public static string XForwardedHostHeaderName { get; } = "X-Forwarded-Host";
    public static string XForwardedProtoHeaderName { get; } = "X-Forwarded-Proto";
```

```csharp
    public static string XOriginalForHeaderName { get; } = "X-Original-For";
    public static string XOriginalHostHeaderName { get; } = "X-Original-Host";
    public static string XOriginalProtoHeaderName { get; } = "X-Original-Proto";
}
```

X-Forwarded-报头

除了上述 6 个属性，ForwardedHeadersOptions 类型还有其他一系列属性，为了加强读者对它们的理解，下面采用实例演示的方式对它们进行逐一讲述。ForwardedHeadersMiddleware 中间件涉及 3 个与 HTTP 报文转发相关的报头，它可以有选择地处理其中 1 个或者 2 个，也可以全部忽略它们。ForwardedHeadersOptions 类型的 ForwardedHeaders 属性的作用就是指定该中间件需要处理的 X-Forwarded-报头。

```csharp
public class ForwardedHeadersOptions
{
    public ForwardedHeaders ForwardedHeaders { get; set; }
    ...
}

[Flags]
public enum ForwardedHeaders
{
    None            = 0,
    XForwardedFor   = 1,
    XForwardedHost  = 2,
    XForwardedProto = 4,
    All             = 7,
}
```

下面演示通过设置 ForwardedHeadersOptions 配置选项的 ForwardedHeaders 属性来有选择地处理 X-Forwarded-报头。如下面的代码片段所示，我们在调用 UseForwardedHeaders 扩展方法注册 ForwardedHeadersMiddleware 中间件时提供了一个 ForwardedHeadersOptions 对象，可以将该配置选项的 ForwardedHeaders 属性设置成 XForwardedFor | XForwardedProto。这个选项设置决定了该中间件只会处理 X-Forwarded-For 报头和 X-Forwarded-Proto 报头。

```csharp
public class Program
{
    public static void Main()
    {
        var options = new ForwardedHeadersOptions
        {
            ForwardedHeaders = ForwardedHeaders.XForwardedFor |
                ForwardedHeaders.XForwardedProto
        };

        Host.CreateDefaultBuilder()
            .ConfigureWebHostDefaults(builder => builder
                .Configure(app => app
                    .UseForwardedHeaders(options)
                    .Run(ProcessAsync)))
```

```
            .Build()
            .Run();

    static async Task ProcessAsync(HttpContext context)
    {
        var request         = context.Request;
        var response        = context.Response;
        var headers         = request.Headers;

        await response.WriteAsync(
            $"Remote IP:{context.Connection.RemoteIpAddress}\n");
        await response.WriteAsync($"Host:{request.Host}\n");
        await response.WriteAsync($"Scheme:{request.Scheme}\n");

        await response.WriteAsync(
            $"X-Original-For:{headers["X-Original-For"]}\n");
        await response.WriteAsync(
            $"X-Original-Host:{headers["X-Original-Host"]}\n");
        await response.WriteAsync(
            $"X-Original-Proto:{headers["X-Original-Proto"]}");
    }
}
```

如下所示的代码片段是针对上述这个应用发送的请求和得到的响应。可以看出，虽然请求中包含了 3 个 X-Forwarded-For 报头，但是对于表示当前请求上下文的 HttpContext 来说，它的主机名称依然没有改变，从 X-Original-Host 报头中也得不到任何值。（S2405）

```
GET http://localhost:5000/ HTTP/1.1
X-Forwarded-Host: www.foobar.com
X-Forwarded-For: 192.168.0.1
X-Forwarded-Proto: https
Host: localhost:5000


HTTP/1.1 200 OK
Date: Thu, 26 Sep 2019 23:22:02 GMT
Server: Kestrel
Content-Length: 120

Remote IP:192.168.0.1
Host:localhost:5000
Scheme:https
X-Original-For:[::1]:54329
X-Original-Host:
X-Original-Proto:http
```

请求的多次转发

原始的客户端到目标服务器之间可能存在多个中间节点，所以 HTTP 报文在抵达服务器之前可能经过了多次转发。一般来说，某个代理在对请求进行转发之前会将针对它的客户端（可能是原始的客户端，也可能是上游代理）的 IP 地址、主机名称和协议名称追加到上述 3 个 X-Forwarded-报头上，所以 ForwardedHeadersMiddleware 中间件处理的这 3 个 X-Forwarded-报头可能包含多个值，定义在配置选项类型 ForwardedHeadersOptions 中的如下这些属性就与这种场景有关。

```
public class ForwardedHeadersOptions
{
    public IList<IPAddress>     KnownProxies{ get; set; }
    public IList<string>        AllowedHosts { get; set; }
    public IList<IPNetwork>     KnownNetworks{ get; set; }
    public int?                 ForwardLimit { get; set; }
    public bool                 RequireHeaderSymmetry { get; set; }
}
```

虽然客户端可以利用 X-Forwarded-For 报头和 X-Forwarded-Host 报头指定任意的 IP 地址与主机名，但是它们能否被接受则由 ForwardedHeadersMiddleware 中间件来决定。配置选项 ForwardedHeadersOptions 的 AllowedHosts 属性和 KnownProxies 属性表示的就是一组有效的主机名称与 IP 地址。

除了采用穷举的方式指定有效 IP 地址范围的方式来检验代理有效性，我们还可以根据 IP 地址所在的网段来决定代理是否有效。ForwardedHeadersOptions 类型的 KnownNetworks 属性表示的就是代理的有效网段列表，每个网段通过一个 IPNetwork 对象表示。IPNetwork 类型的定义如下，它的 Prefix 属性表示网络的 IP 地址前缀，而 PrefixLength 属性则表示对子网掩码（Net Mask）的体现。在默认情况下，本机 IP 地址（"127.0.0.1"或者"::"）会默认作为可接受的 IP 地址，而网段"127.0.0.0/8"也默认被添加到可被接受的网段范围。

```
public class IPNetwork
{
    public IPNetwork(IPAddress prefix, int prefixLength);

    public bool         Contains(IPAddress address);
    public IPAddress    Prefix { get; }
    public int          PrefixLength { get; }
}

public class ForwardedHeadersOptions
{
    public ForwardedHeadersOptions()
    {
        KnownProxies =  new List<IPAddress> {
            IPAddress.IPv6Loopback
        };
        KnownNetworks = new List<IPNetwork> {
            new IPNetwork(IPAddress.Loopback, 8)
        };
        AllowedHosts = new List<string>();
    }
}
```

如果请求因多次转发而导致 X-Forwarded-报头携带了多个值，则可以设置 ForwardedHeadersOptions 对象的 ForwardLimit 属性来决定只处理最后 N 个值。由于 ForwardLimit 属性在默认情况下的值为 1，所以 ForwardedHeadersMiddleware 中间件总是选择针对最后一次转发的值。如果需要忽略这个限制，就需要显式地将该属性设置为 Null。需要着重强调的是，针对 ForwardLimit 属性的设置只有在对 KnownProxies 属性和 KnownNetworks 属性做了相应设置的前提下才有效。

为了演示 ForwardLimit 属性针对路由节点数量的选择，我们按照如下所示的方式将前面演示实例中使用的配置选项 ForwardedHeadersOptions 的 ForwardedHeaders 属性设置为 All，并将网段"192.168.0.0/28"添加到 KnownNetworks 属性表示的已知网络区间中。（S2406）

```
public class Program
{
    public static void Main()
    {
        var options = new ForwardedHeadersOptions
        {
            ForwardedHeaders = ForwardedHeaders.All
        };
        options.KnownNetworks.Add(new IPNetwork(
            IPAddress.Parse("192.168.0.0"), 28));

        Host.CreateDefaultBuilder()
            .ConfigureWebHostDefaults(builder => builder
                .Configure(app => app
                    .UseForwardedHeaders(options)
                    .Run(ProcessAsync)))
            .Build()
            .Run();

        static async Task ProcessAsync(HttpContext context)
        {
            var request        = context.Request;
            var response       = context.Response;
            var headers        = request.Headers;

            await response.WriteAsync(
                $"Remote IP:{context.Connection.RemoteIpAddress}\n");
            await response.WriteAsync($"Host:{request.Host}\n");
            await response.WriteAsync(
                $"Scheme:{request.Scheme}\n");

            await response.WriteAsync(
                $"X-Original-For:{headers["X-Original-For"]}\n");
            await response.WriteAsync(
                $"X-Original-Host:{headers["X-Original-Host"]}\n");
            await response.WriteAsync(
                $"X-Original-Proto:{headers["X-Original-Proto"]}");
```

```
            }
        }
}
```

我们针对改写后的演示实例发送了如下所示的请求,后面给出的是得到的响应。如下面的代码片段所示,我们利用 3 个 X-Forwarded-报头指定了 4 个 IP 地址("192.168.0.1"、"192.168.0.2"、"192.168.0.3"和"192.168.0.4")、主机名称(www.foo.com、www.bar.com、www.baz.com 和 www.gux.com)及对应的协议类型(http、http、https 和 http)。由于 ForwardLimit 属性的默认值为 1,所以 ForwardedHeadersMiddleware 中间件最终会选择最后一组报头值"192.168.0.4/www.gux.com/http"。

```
GET http://localhost:5000/ HTTP/1.1
X-Forwarded-Host: www.foo.com, www.bar.com, www.baz.com, www.gux.com
X-Forwarded-For: 192.168.0.1, 192.168.0.2, 192.168.0.3, 192.168.0.4
X-Forwarded-Proto: http, http, https,http
Host: localhost:5000


HTTP/1.1 200 OK
Date: Thu, 26 Sep 2019 23:34:10 GMT
Server: Kestrel
Content-Length: 130

Remote IP:192.168.0.4
Host:www.gux.com
Scheme:http
X-Original-For:[::1]:54441
X-Original-Host:localhost:5000
X-Original-Proto:http
```

然后将 ForwardedHeadersOptions 对象的 ForwardLimit 属性设置为 2,并采用与上面完全一致的请求,我们会得到如下所示的响应。在这种情况下,X-Forwarded-报头携带的最后两组值才会被视为候选项,所以最终选择的是倒数第 2 组报头值"192.168.0.3/www.baz.com/https"。(S2407)

```
GET http://localhost:5000/ HTTP/1.1
X-Forwarded-Host: www.foo.com, www.bar.com, www.baz.com, www.gux.com
X-Forwarded-For: 192.168.0.1, 192.168.0.2, 192.168.0.3, 192.168.0.4
X-Forwarded-Proto: http, http, https,http
Host: localhost:5000


HTTP/1.1 200 OK
Date: Sun, 11 Aug 2019 00:21:51 GMT
Server: Kestrel
Content-Length: 131


HTTP/1.1 200 OK
Date: Thu, 26 Sep 2019 23:37:17 GMT
Server: Kestrel
```

```
Content-Length: 131

Remote IP:192.168.0.3
Host:www.baz.com
Scheme:https
X-Original-For:[::1]:54507
X-Original-Host:localhost:5000
X-Original-Proto:http
```

如果不希望对 X-Forwarded-报头携带的报头值在数量上做任何限制，我们就可以将 ForwardLimit 属性显式设置为 Null。针对同样的请求，此时被 ForwardedHeadersMiddleware 中间件最终选择的是最初被转发的路由信息"192.168.0.1/www.foo.com/http"。（S2408）

```
GET http://localhost:5000/ HTTP/1.1
X-Forwarded-Host: www.foo.com, www.bar.com, www.baz.com, www.gux.com
X-Forwarded-For: 192.168.0.1, 192.168.0.2, 192.168.0.3, 192.168.0.4
X-Forwarded-Proto: http, http, https,http
Host: localhost:5000


HTTP/1.1 200 OK
Date: Sun, 11 Aug 2019 00:27:26 GMT
Server: Kestrel
Content-Length: 130

Remote IP:192.168.0.1
Host:www.foo.com
Scheme:http
X-Original-For:[::1]:58580
X-Original-Host:localhost:5000
X-Original-Proto:http
```

对于前面演示的实例，由于我们将"192.168.0.0/28"设置成已知网络区间，所以 X-Forwarded-For 报头携带的 4 个 IP 地址都会被视为合法。现在我们将 X-Forwarded-For 报头携带的第三个值设置为不在此区间范围内的 IP 地址"192.168.0.254"，ForwardedHeadersMiddleware 中间件在这种情况下会选择这个未知 IP 地址对应的路由信息。

```
GET http://localhost:5000/ HTTP/1.1
X-Forwarded-Host: www.foo.com, www.bar.com, www.baz.com, www.gux.com
X-Forwarded-For: 192.168.0.1, 192.168.0.2, 192.168.0.254, 192.168.0.4
X-Forwarded-Proto: http, http, https,http
Host: localhost:5000


HTTP/1.1 200 OK
Date: Thu, 26 Sep 2019 23:40:50 GMT
Server: Kestrel
Content-Length: 130

Remote IP:192.168.0.1
Host:www.foo.com
```

```
Scheme:http
X-Original-For:[::1]:54578
X-Original-Host:localhost:5000
X-Original-Proto:http
```

一般来说，3 个 X-Forwarded-报头携带的地址/主机名/协议类型在数量上应该是相同的。但是在默认情况下，ForwardedHeadersMiddleware 中间件并不会对此做相应的验证。如果我们希望该中间件在处理请求时预先验证 3 个 X-Forwarded-报头是否携带了同等数量的值，就可以将 ForwardedHeadersOptions 对象的 RequireHeaderSymmetry 属性显式设置为 True。

注册中间件

上面介绍了 ForwardedHeadersOptions 配置选项，下面介绍 ForwardedHeadersMiddleware 中间件类型的定义。如下面的代码片段所示，ForwardedHeadersMiddleware 类型构造函数中利用注入的 IOptions<ForwardedHeadersOptions>对象来提供表示配置选项的 ForwardedHeadersOptions 对象，并在实现的 Invoke 方法中根据对应的设置来处理分发给它的请求。至于具体的请求处理逻辑，我们在前面介绍 ForwardedHeadersOptions 配置选项时已经有所阐述，此处不再赘述。

```
public class ForwardedHeadersMiddleware
{
    public ForwardedHeadersMiddleware(RequestDelegate next,
        ILoggerFactory loggerFactory, IOptions<ForwardedHeadersOptions> options);
    public Task Invoke(HttpContext context);
}
```

我们在前面的演示实例中总是调用 IApplicationBuilder 接口的 UseForwardedHeaders 扩展方法来注册 ForwardedHeadersMiddleware 中间件，如下所示的代码片段体现了该方法针对中间件的注册。

```
public static class ForwardedHeadersExtensions
{
    public static IApplicationBuilder UseForwardedHeaders(
        this IApplicationBuilder builder)
        => builder.UseMiddleware<ForwardedHeadersMiddleware>();

    public static IApplicationBuilder UseForwardedHeaders(
        this IApplicationBuilder builder, ForwardedHeadersOptions options)
        => builder.UseMiddleware<ForwardedHeadersMiddleware>(
            Options.Create(options));
}
```

24.3 基础路径

标准的 URL 采用的格式为 "protocol ://hostname[:port]/path/[;parameters][?query]#fragment"，主机名称后边的就是路径（Path）。ASP.NET Core 管道在创建 HttpContext 上下文的时会根据 URL 来解析请求的路径，具体的解析过程由设置的基础路径（PathBase）来决定。换句话说，HttpContext 上下文体现的请求路径（对应 HttpRequest 对象的 Path 属性）与请求 URL 的路径可能是不一致的，它们之间的映射关系取决于我们为应用设置了怎样的基础路径（对应 HttpRequest 对象的 PathBase 属性），HttpRequest 对象的路径实际上是针对基础路径的相对路径。

24.3.1 实例演示

下面利用一个简单的实例演示如何利用注册的 UsePathBaseMiddleware 中间件来设置基础路径。我们创建了 3 个 ASP.NET Core 应用（App1、App2 和 App3），如下面的代码片段所示，我们在应用 App2 和 App3 中调用 IApplicationBuilder 接口的 UsePathBase 扩展方法注册了 UsePathBaseMiddleware 中间件，并将基础路径设置为 "/foo" 和 "/foo/bar"。至于应用 App1，我们并没有做任何相关设置。

```
App1
public class Program
{
    public static void Main()
    {
        Host.CreateDefaultBuilder()
            .ConfigureWebHostDefaults(builder => builder
                .UseUrls("http://localhost:1000")
                .Configure(app => app
                    .Run(HandleAsync)))
            .Build()
            .Run();
    }
}

App2
public class Program
{
    public static void Main()
    {
        Host.CreateDefaultBuilder()
            .ConfigureWebHostDefaults(builder => builder
                .UseUrls("http://localhost:2000")
                .Configure(app => app
                    .UsePathBase("/foo")
                    .Run(HandleAsync)))
            .Build()
            .Run();
    }
}

App3
public class Program
{
    public static void Main()
    {
        Host.CreateDefaultBuilder()
```

```
            .ConfigureWebHostDefaults(builder => builder
                .UseUrls("http://localhost:2000")
                .Configure(app => app
                    .UsePathBase("/foo/bar")
                    .Run(HandleAsync)))
            .Build()
            .Run();
    }
}
```

3个应用程序中通过调用 IApplicationBuilder 接口的 Run 扩展方法提供的 RequestDelegate 委托对象指向一个 ProcessAsync 方法。如下面的代码片段所示，ProcessAsync 方法会将请求的 URL（通过请求的 Scheme、Host、PathBase 和 Path 属性组合而成）、设置的基础路径（请求的 PathBase 属性）和路径（请求的 Path 属性）呈现出来。

```
static Task ProcessAsync(HttpContext httpContext)
{
    var request       = httpContext.Request;
    var response      = httpContext.Response;
    var url = $"{request.Scheme}://{request.Host}{request.PathBase}{request.Path}";
    return response.WriteAsync(
        $"Url: {url}\nPathBase: {request.PathBase}\nPath: {request.Path}");
}
```

当上述 3 个应用程序启动之后，我们利用浏览器来分别访问它们。如图 24-2 所示，虽然采用的请求 URL 具有相同的路径（"/foo/bar/baz"），但是对于 3 个应用程序来说，由于它们具有不同的基础路径（默认为"/"），所以解析出来的路径是不同的。从输出结果可以看出，请求的（相对）路径和设置的基础路径合并在一起组成的绝对路径就是请求 URL 的路径。（S2409）

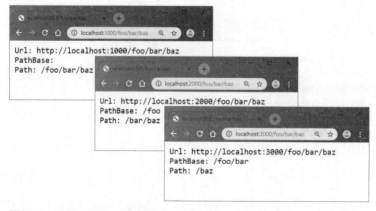

图 24-2　URL Path = PathBase + Path

24.3.2　UsePathBaseMiddleware

基础路径虽然是作为 ASP.NET Core 应用的一个全局设置，但是它最终需要落实到表示每个请求的 HttpRequest 对象的 PathBase 属性上，所以 UsePathBaseMiddleware 中间件的目的就是拦截匹配的请求并将对应 HttpRequest 的 PathBase 属性赋值为设置的基础路径。UsePathBase

Middleware 中间件针对基础路径的设置体现在如下所示的代码片段中。

```csharp
public class UsePathBaseMiddleware
{
    private readonly RequestDelegate _next;
    private readonly PathString      _pathBase;

    public UsePathBaseMiddleware(RequestDelegate next, PathString pathBase)
    {
        _next     = next;
        _pathBase = pathBase;
    }

    public async Task Invoke(HttpContext context)
    {
        PathString matchedPath;
        PathString remainingPath;

        if (context.Request.Path.StartsWithSegments(_pathBase, out matchedPath,
            out remainingPath))
        {
            var originalPath                 = context.Request.Path;
            var originalPathBase             = context.Request.PathBase;
            context.Request.Path             = remainingPath;
            context.Request.PathBase         = originalPathBase.Add(matchedPath);

            try
            {
                await _next(context);
            }
            finally
            {
                context.Request.Path         = originalPath;
                context.Request.PathBase     = originalPathBase;
            }
        }
        else
        {
            await _next(context);
        }
    }
}
```

　　如上面的代码片段所示，UsePathBaseMiddleware 中间件只会选择当前路径（默认为请求 URL 的路径）以指定基础路径为前缀的请求。在设置了当前请求的基础路径之后，Path 属性表示的路径也会做相应的调整。当后续中间件完成了针对当前请求处理之后，UsePathBase Middleware 中间件还会将请求的基础路径和路径恢复到之前的状态，所以它针对请求基础路径和路径的修改不会对前置中间件造成任何影响。如下所示的代码片段是用于注册 UsePathBase Middleware 中间件的 UsePathBase 扩展方法的定义。

```csharp
public static class UsePathBaseExtensions
{
    public static IApplicationBuilder UsePathBase(this IApplicationBuilder app,
        PathString pathBase)
    {
        pathBase = pathBase.Value?.TrimEnd('/');
        if (!pathBase.HasValue)
        {
            return app;
        }
        return app.UseMiddleware<UsePathBaseMiddleware>(pathBase);
    }
}
```

24.4 路由

第 15 章详细介绍了 ASP.NET Core 提供的基于终结点的路由是一个功能强大、特性丰富的路由系统，并且几乎能够解决我们遇到的所有路由需求。但是针对一些简单的路由场景，我们还可以利用下面介绍的 MapMiddleware 中间件和 MapWhenMiddleware 中间件来完成。

24.4.1 实例演示

在正式介绍 MapMiddleware 中间件和 MapWhenMiddleware 中间件的实现原理之前，我们先通过两个简单的实例演示如何利用它们实现一些简单的路由功能。Web 应用中的路由本质上就是为某个请求动态映射目标终结点的过程，ASP.NET Core 应用中用来处理请求的终结点通过 RequestDelegate 委托对象来表示，MapMiddleware 中间件实现的路由功能建立在请求路径与对应的终结点分支（体现为一个 RequestDelegate 对象）的映射关系上。

```csharp
public class Program
{
    public static void Main()
    {
        Host.CreateDefaultBuilder()
            .ConfigureWebHostDefaults(builder => builder.Configure(app=>app
                .Map("/foo", Foo)
                .Map("/bar", Bar)
                .Map("/baz", Baz)))
            .Build()
            .Run();

        static void Foo(IApplicationBuilder app) => app.Run(
            context => context.Response.WriteAsync("Endpoint foo is selected."));
        static void Bar(IApplicationBuilder app) => app.Run(
            context => context.Response.WriteAsync("Endpoint bar is selected."));
        static void Baz(IApplicationBuilder app) => app.Run(
            context => context.Response.WriteAsync("Endpoint baz is selected."));
    }
}
```

}
```

如上面的代码片段所示,我们调用 IApplicationBuilder 接口的 Map 扩展方法注册了分别映射到不同的路径上("/foo"、"/bar"和"/baz")上的 3 个 MapMiddleware 中间件,表示映射终结点的 3 个对应 RequestDelegate 对象是通过指定的 Action<IApplicationBuilder>对象构建的。我们可以利用指定的 IApplicationBuilder 对象注册任意的中间件,并利用它最终构建的表示中间件管道的 RequestDelegate 对象来处理映射的请求。在演示的实例中,我们通过调用 IApplicationBuilder 接口的 Run 扩展方法注册了一个简单的中间件来处理映射的请求,采用 3 个指定路径的请求得到的响应内容如图 24-3 所示。(S2410)

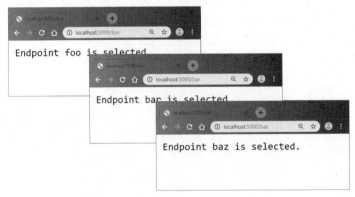

**图 24-3　基于 MapMiddleware 中间件的路由**

下面演示基于 MapWhenMiddleware 中间件的路由。MapMiddleware 中间件采用基于路径的终结点选择策略,而 MapWhenMiddleware 中间件则允许我们自行控制针对目标终结点的选择,因为它提供的中间件选择策略体现为一个 Func<HttpContext, bool>对象。借助 MapWhenMiddleware 中间件,上面演示的程序可以改写成如下形式。(S2411)

```
public class Program
{
    public static void Main()
    {
        Host.CreateDefaultBuilder()
            .ConfigureWebHostDefaults(builder => builder.Configure(app => app
                .MapWhen(context => context.Request.Path == "/foo", Foo)
                .MapWhen(context => context.Request.Path == "/bar", Bar)
                .MapWhen(context => context.Request.Path == "/baz", Baz)))
            .Build()
            .Run();

        static void Foo(IApplicationBuilder app) => app.Run(
            context => context.Response.WriteAsync("Endpoint foo is selected."));
        static void Bar(IApplicationBuilder app) => app.Run(
            context => context.Response.WriteAsync("Endpoint bar is selected."));
        static void Baz(IApplicationBuilder app) => app.Run(
            context => context.Response.WriteAsync("Endpoint baz is selected."));
```
```

        }
    }

如上面的代码片段所示，MapWhenMiddleware 中间件通过调用 IApplicationBuilder 接口的 MapWhen 扩展方法进行注册，该方法依然利用提供的 Action<IApplicationBuilder>对象来构建作为中间件的 RequestDelegate 对象。我们利用指定的 Func<HttpContext, bool>对象来完成请求路径与目标终结点的映射。

24.4.2　MapMiddleware

MapMiddleware 中间件通过指定的请求路径与目标终结点的映射关系来完成针对请求的路由，该映射关系通过对应的配置选项类型 MapOptions 标识。如下面的代码片段所示，MapOptions 通过 PathMatch 属性表示映射的路径，目标终结点则体现为一个 RequestDelegate 委托对象。

```
public class MapOptions
{
    public PathString         PathMatch { get; set; }
    public RequestDelegate    Branch { get; set; }
}
```

如下所示的代码片段是 MapMiddleware 中间件的定义。从给出的代码片段可以看出，该中间件进行路径匹配时采用的不是完全匹配，而是前缀匹配，也就是说，它只要求请求的路径以指定的路径为前缀。对于路径不匹配的请求，该中间件会直接分发给后续中间件进行处理。

```
public class MapMiddleware
{
    private readonly RequestDelegate        _next;
    private readonly MapOptions             _options;

    public MapMiddleware(RequestDelegate next, MapOptions options)
    {
        _next       = next;
        _options    = options;
    }

    public async Task Invoke(HttpContext context)
    {
        PathString matchedPath;
        PathString remainingPath;

        if (context.Request.Path.StartsWithSegments(_options.PathMatch,
            out matchedPath, out remainingPath))
        {
            var path        = context.Request.Path;
            var pathBase    = context.Request.PathBase;
            context.Request.PathBase = pathBase.Add(matchedPath);
            context.Request.Path = remainingPath;

            try
            {
                await _options.Branch(context);
```

```
            }
            finally
            {
                context.Request.PathBase        = pathBase;
                context.Request.Path            = path;
            }
        }
        else
        {
            await _next(context);
        }
    }
}
```

从上面给出的代码片段可以看出，MapMiddleware 中间件在确认了处理的请求具有匹配路径之后，它还会修正当前请求的基础路径和相对路径。具体来说，MapMiddleware 中间件会在现有基础路径的基础上附加指定的映射路径作为当前请求新的基础路径，并根据新的基础路径重新计算相对路径。在完成了针对当前请求的处理之后，MapMiddleware 中间件会将请求的基础路径和相对路径恢复到最初的状态。我们利用如下所示的演示实例来加深读者对这个特性的理解。

```
public class Program
{
    public static void Main()
    {
        Host.CreateDefaultBuilder()
            .ConfigureWebHostDefaults(builder => builder.Configure(app => app
                .UsePathBase("/foo")
                .Map("/bar", appBuilder => appBuilder.Run(ProcessAsync))))
            .Build()
            .Run();
    }

    private static Task ProcessAsync(HttpContext httpContext)
    {
        var request     = httpContext.Request;
        var response    = httpContext.Response;
        var url =
            $"{request.Scheme}://{request.Host}{request.PathBase}{request.Path}";
        return response.WriteAsync(
            $"Url: {url}\nPathBase: {request.PathBase}\nPath: {request.Path}");
    }
}
```

如上面的代码片段所示，我们在调用 IApplicationBuilder 的 Map 扩展方法注册 MapMiddleware 中间件之前，调用 UsePathBase 扩展方法利用注册的 UsePathBaseMiddleware 中间件将基础路径设置为"/foo"。注册的 MapMiddleware 中间件映射的路径为"/bar"，对应的终结点会将当前请求 URL、基础路径和相对路径作为响应内容。针对这个演示实例，如果请求 URL 的路径被设置为"/foo/bar/baz"，浏览器上将会呈现出图 24-4 所示的结果。（S2412）

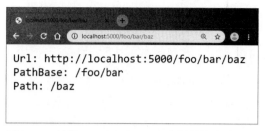

图 24-4 基于 MapMiddleware 中间件的路由

虽然 MapMiddleware 中间件映射的终结点处理器是利用作为配置选项的 MapOptions 对象提供的，但是考虑到在大部分情况下我们会利用注册的中间件来处理被路由的请求，所以用于注册该中间件的 Map 扩展方法会提供一个 Action<IApplicationBuilder>类型的参数。Map 扩展方法针对 MapMiddleware 中间件的注册体现在如下所示的代码片段中。

```
public static class MapExtensions
{
    public static IApplicationBuilder Map(this IApplicationBuilder app,
        PathString pathMatch, Action<IApplicationBuilder> configuration)
    {
        if (pathMatch.HasValue && pathMatch.Value.EndsWith(
            "/", StringComparison.Ordinal))
        {
            throw new ArgumentException(
                "The path must not end with a '/'", nameof(pathMatch));
        }
        var branchBuilder = app.New();
        configuration(branchBuilder);
        var branch = branchBuilder.Build();

        var options = new MapOptions
        {
            Branch = branch,
            PathMatch = pathMatch,
        };
        return app.Use(next => new MapMiddleware(next, options).Invoke);
    }
}
```

24.4.3 MapWhenMiddleware

与 MapMiddleware 中间件的不同之处在于：MapWhenMiddleware 中间件不再利用指定的请求路径来选择映射的路由分支，而是利用一个 Func<HttpContext, bool>对象来动态决定目标终结点，所以它为我们提供了更加灵活的"有条件"的路由。如下所示的代码片段是 MapWhenMiddleware 中间件对应配置选项类型 MapWhenOptions 的定义，该中间件映射的路由分支体现在 Branch 属性提供的 RequestDelegate 委托对象上，它的 Predicate 属性返回的 Func<HttpContext, bool>对象用来判断当前请求是否与期望路由规则相匹配。

```
public class MapWhenOptions
```

```
{
    public Func<HttpContext, bool>      Predicate { get;set; }
    public RequestDelegate              Branch { get; set; }
}
```

实现在 MapWhenMiddleware 中间件的路由逻辑非常简单。如下面的代码片段所示，在实现的 Invoke 方法中，它直接将 HttpContext 上下文作为参数调用通过 MapWhenOptions 对象提供的 Func<HttpContext, bool>委托对象，如果返回结果为 True，意味着满足设置的路由规则，则直接执行作为映射终结点处理器的 RequestDelegate 即可。如果返回结果为 False，则直接将请求分发给后续中间件。

```
public class MapWhenMiddleware
{
    private readonly RequestDelegate _next;
    private readonly MapWhenOptions  _options;

    public MapWhenMiddleware(RequestDelegate next, MapWhenOptions options)
    {
        _next       = next;
        _options    = options;
    }

    public async Task Invoke(HttpContext context)
        => options.Predicate(context)
            ? _options.Branch(context)
            : _next(context);
}
```

注册 MapWhenMiddleware 中间件的 MapWhen 扩展方法同样提供了一个 Action<IApplicationBuilder>类型的参数，这使我们可以利用它注册一组中间件来处理匹配的路由请求。IApplicationBuilder 接口的 MapWhen 扩展方法针对 MapWhenMiddleware 中间件的注册体现在如下所示的代码片段中。

```
public static class MapWhenExtensions
{
    public static IApplicationBuilder MapWhen(this IApplicationBuilder app,
        Func<HttpContext, bool> predicate,
        Action<IApplicationBuilder> configuration)
    {
        var branchBuilder = app.New();
        configuration(branchBuilder);
        var branch = branchBuilder.Build();

        var options = new MapWhenOptions
        {
            Predicate       = predicate,
            Branch          = branch,
        };
        return app.Use(next => new MapWhenMiddleware(next, options).Invoke);
    }
}
```

附录B

实例演示 2

章 节	编 号	描 述
第 14 章	S1401	将物理文件发布为 Web 资源
	S1402	映射存放 Web 资源的目录
	S1403	呈现目录结构
	S1404	为目录定制默认页面（采用约定文件名）
	S1405	为目录定制默认页面（自定义文件名）
	S1406	设置默认媒体类型
	S1407	基于文件扩展名的媒体类型映射
	S1408	模拟 StaticFileMiddleware 中间件的实现原理
	S1409	自定义 IDirectoryFormatter
第 15 章	S1501	基本路由映射
	S1502	设置内联约束
	S1503	定义默认路由参数
	S1504	为默认的路由参数设置默认值
	S1505	一个路径段由多个路由参数组成
	S1506	跨越多路径段的路由参数
	S1507	解析路由模式
	S1508	自定义路由约束
第 16 章	S1601	显示开发者异常页面
	S1602	显示定制异常页面
	S1603	为定制异常页面设置异常处理器
	S1604	为定制异常页面设置错误页面路径
	S1605	针对响应状态码定制错误页面
	S1606	为状态码异常页面设置处理器
	S1607	将注册的中间件作为状态码异常页面的处理器
	S1608	自定义 IDeveloperPageExceptionFilter
	S1609	显示编译异常信息
	S1610	设置显示错误源代码行数
	S1611	获取 ExceptionHandlerFeature 特性并输出其承载信息
	S1612	模拟异常处理中间件对缓存的清除
	S1613	阻止 StatusCodePagesMiddleware 中间件的异常处理
	S1614	以客户端的形式呈现状态码异常页面

续表

章　节	编　号	描　述
第 16 章	S1615	以服务端的形式呈现状态码异常页面
第 17 章	S1701	将数据缓存在内存中
	S1702	基于 Redis 的分布式缓存
	S1703	基于 SQL Server 的分布式缓存
	S1704	缓存整个 HTTP 响应
	S1705	根据指定的查询字符串对响应实施缓存
	S1706	物理文件与缓存的同步
	S1707	缓存的压缩策略
第 18 章	S1801	设置和提取会话状态
	S1802	查看存储的会话状态
第 19 章	S1901	用户认证的最简实现
	S1902	ClaimsIdentity 和 GenericIdentity 的差别
第 20 章	S2001	根据角色授权
	S2002	预定义授权策略
第 21 章	S2101	跨域调用 API
	S2102	资源提供者显式授权（指定域名）
	S2103	资源提供者显式授权（指定授权规则）
	S2104	基于策略的资源授权（默认策略）
	S2105	基于策略的资源授权（具名策略）
第 22 章	S2201	提供对应语种的文本
	S2202	自动设置语言文化
	S2203	将本地化文本分而治之
	S2204	利用注入的 IStringLocalizer<T>服务提供本地化文本
	S2205	自定义 IStringLocalizer 实现基于 JSON 的本地换文本存储
	S2206	自定义 IRequestCultureProvider 提供请求的语言文化
第 23 章	S2301	确定当前应用是否可用
	S2302	定制健康检查逻辑
	S2303	改变健康检查响应状态码
	S2304	细粒度的健康检查
	S2305	定制健康检查响应内容
	S2306	过滤 IHealthCheck 对象
	S2307	针对 Entity Framework Core 的健康检查
	S2308	定期发布健康报告
第 24 章	S2401	过滤主机名
	S2402	改写 HTTP 方法（利用 X-Http-Method-Override 报头）
	S2403	改写 HTTP 方法（利用表单元素）
	S2404	改写传输信息
	S2405	基于 ForwardedHeadersMiddleware 的报头转发
	S2406	ForwardedHeadersMiddleware 中间件针对请求的多次转发（转发数量＝1）
	S2407	ForwardedHeadersMiddleware 中间件针对请求的多次转发（转发数量＝2）
	S2408	ForwardedHeadersMiddleware 中间件针对请求的多次转发（不限转发数量）

续表

章节	编号	描述
第 24 章	S2409	利用 UsePathBaseMiddleware 中间件设置基础路径
	S2410	基于 MapMiddleware 中间件的路由
	S2411	基于 MapWhenMiddleware 中间件的路由
	S2412	MapMiddleware 中间件针对基础路径的处理